2015 6th International Conference on Power Electronics Systems and Applications (PESA 2015)

AA002401

Hong Kong
15 – 17 December 2015

IEEE Catalog Number: CFP15591-POD
ISBN: 978-1-5090-0064-7

Copyright © 2015 by the Institute of Electrical and Electronic Engineers, Inc
All Rights Reserved

Copyright and Reprint Permissions: Abstracting is permitted with credit to the source. Libraries are permitted to photocopy beyond the limit of U.S. copyright law for private use of patrons those articles in this volume that carry a code at the bottom of the first page, provided the per-copy fee indicated in the code is paid through Copyright Clearance Center, 222 Rosewood Drive, Danvers, MA 01923.

For other copying, reprint or republication permission, write to IEEE Copyrights Manager, IEEE Service Center, 445 Hoes Lane, Piscataway, NJ 08854. All rights reserved.

***This publication is a representation of what appears in the IEEE Digital Libraries. Some format issues inherent in the e-media version may also appear in this print version.**

IEEE Catalog Number: CFP15591-POD
ISBN (Print-On-Demand):   978-1-5090-0064-7
ISBN (Online):   978-1-5090-0063-0

Additional Copies of This Publication Are Available From:

Curran Associates, Inc
57 Morehouse Lane
Red Hook, NY 12571 USA
Phone: (845) 758-0400
Fax: (845) 758-2633
E-mail: curran@proceedings.com
Web: www.proceedings.com

TABLE OF CONTENTS

SIMULATION OF THERMAL PROPERTIES OF THE LIQUID METAL BATTERIES..1
W. Wang ; K. Wang

COMPARISON OF DETECTION METHODS FOR POWER QUALITY IN MICRO-GRID................................11
Wang Huihui ; Wang Ping

DESIGN AND DEVELOPMENT OF CLOSED LOOP MODEL OF AN ADJUSTABLE SPEED
PERMANENT MAGNET SYNCHRONOUS MOTOR DRIVE USING PI CONTROLLER 19
P. K. Biswas ; A. Banerjee ; C. Sain

ESTIMATION OF VEHICLE STATES AND ROAD FRICTION BASED ON DEKF...................................24
Ying Xu ; Banglin Deng ; Gang Xu

OPERATION STRATEGY OF STAND-ALONE MICROGRID...31
W. Wei ; J. Lei ; X. Guo ; P. Yang ; Z. Xu ; Y. Zhang ; S. Zhou

DESIGN OF MGCC FOR USER-SIDE MICROGRID...39
B. Tian ; J. Lei ; X. Guo ; P. Yang ; H. Yuan ; Z. Xu ; S. Zhou ; Ting He

STUDY ON BLACK START STRATEGY OF MULTI-MICROGRIDS WITH PV AND ENERGY
STORAGE SYSTEMS CONSIDERING GENERAL SITUATIONS 45
L. Yu ; J. Lei ; X. Guo ; P. Yang ; Z. Zeng ; Z. Xu ; Q. Zheng

DESIGN OPTIMIZATION OF LINEAR INDUCTION MOTOR 51
A. Kumar ; M. A. Hasan ; M. J. Akhtar ; S. K. Parida ; R. K. Behera

DESIGN TECHNIQUE OF COUPLED INDUCTOR FILTER FOR SUPPRESSING SWITCHING
RIPPLES IN PWM CONVERTERS 56
Hyeon-gyu Choi ; Jung-Ik Ha

ANALYSIS OF THE INFLUENCE OF PHOTOVOLTAIC SOURCES INTEGRATED INTO
DISTRIBUTION NETWORK IN DIFFERENT WAYS..60
X. Ma ; J. Lei ; X. Guo ; P. Yang ; Jiajun Peng ; Z. Xu ; S. Zhou

DESIGN CONSIDERATIONS FOR HIGH FREQUENCY DCM FLYBACK CONVERTER........................66
Jun Lee ; Kyung-Hwan Lee ; Jung-Ik Ha

CONTROL METHOD CONSIDERING CURRENT AND VOLTAGE LIMITS IN MAGNETIC
MANIPULATION SYSTEMS...72
Jun Lee ; Jung-Ik Ha

THE RESEARCH OF BATTERY MODEL AND INTELLIGENT CHARGING IN THE RAIL
TRANSPORT SYSTEM 77
Qin Zhuqian ; Huang Shenghua ; Zou Xunhao ; Kan Guangqiang

A HIERARCHICAL GROUP CONTROL METHOD OF ELECTRICAL LOADS IN SMART
HOME 81
Zhaojing Yin ; Yanbo Che ; Wei He

A RESEARCH OF THE STRATEGY OF ELECTRIC VEHICLE ORDERED CHARGING BASED
ON THE DEMAND SIDE RESPONSE...87
Jidong Wang ; Yuhao Yang ; Wei He

A RESEARCH FOR THE INTERACTIVE BEHAVIOR BETWEEN ELECTRIC VEHICLE AND
RESIDENTIAL ENERGY MANAGEMENT SYSTEM BASED ON PROBABILITY
DISTRIBUTION 92
Jidong Wang ; Kaijie Fang ; Wei He

A NEW METHOD TO SUPPRESS THE COMMUTATION TORQUE RIPPLE FOR BLDC
MOTOR BASED ON ZETA CONVERTER 98
Zhen Chen ; Hancheng Zhang ; Xiangdong Liu ; Hengzai Hu ; Jing Zhao ; Congzhe Gao

MODELING AND CONTROL OF DAB CONVERTER FOR SOLAR MICRO-GRID
APPLICATION 104
R. K. Behera ; O. Ojo

PROPULSION SYSTEM DESIGN OF ELECTRIC VEHICLE 109
M. J. Akhtar ; R. K. Behera ; S. K. Parida

COMPARATIVE ANALYSIS OF 2-LEVEL AND MULTI-LEVEL INVERTER FED COUPLED IM
DRIVES BASED ON V/F AND DTC TECHNIQUES 114
S. Misal ; T. Swetha ; M. Wandalkar ; Chang Yan Tai

THE COMPARATIVE STUDY BETWEEN DIFFERENT PERFORMANCE INDICES OF A
PERMANENT MAGNET SYNCHRONOUS MOTOR DRIVE ON VARIABLE SENSOR ANGLE 122
A. Banerjee ; P. K. Biswas ; C. Sain

HIGH POWER HIGH VOLTAGE GAIN INTERLEAVED DC-DC BOOST CONVERTER APPLICATION 128
Jiexun Liu ; Dawei Gao ; Yue Wang

A BIDIRECTIONAL FLYBACK CELL EQUALIZER FOR SERIES-CONNECTED LITHIUM IRON PHOSPHATE BATTERIES 134
D. Yang ; S. Li ; G. Qi

A CONTROL SCHEME FOR A HIGH SPEED RAILWAY TRACTION SYSTEM BASED ON HIGH POWER PMSM 139
S. F. Zhao ; X. Y. Huang ; Y. T. Fang ; J. Li

STATE-OF-CHARGE ESTIMATION OF LITHIUM IRON PHOSPHATE BATTERY USING EXTREME LEARNING MACHINE 147
Zhihao Wang ; Daiming Yang

A SENSORLESS VECTOR STRATEGY FOR THE PMSM USING IMPROVED SLIDING MODE OBSERVER AND FUZZY PI SPEED CONTROLLER 152
Hui Deng ; Guang-Zhong Cao ; Su-Dan Huang ; Lai-Juan Shi ; Zhi-Ming He

DEVELOPMENT OF THE THREE-DIMENSIONAL SCANNING SYSTEM BASED ON MONOCULAR VISION 158
Yu-Xin Liang ; Guang-Zhong Cao ; Hong Qiu ; Su-Dan Huang ; Shou-Qin Zhou

DEVELOPMENT OF A 4-DOF SCARA ROBOT WITH 3R1P FOR PICK-AND-PLACE TASKS 163
Wen-Bo Li ; Guang-Zhong Cao ; Xiao-Qin Guo ; Su-Dan Huang

YAW ANGLE CONTROL OF A BOXFISH-LIKE ROBOT BASED ON CASCADE PID CONTROL ALGORITHM 168
Hanbo Deng ; Wei Wang ; Wenguang Luo ; Guangming Xie

GENETIC ALGORITHM BASED BACK-PROPAGATION NEURAL NETWORK APPROACH FOR FAULT DIAGNOSIS IN LITHIUM-ION BATTERY SYSTEM 173
Zuchang Gao ; C. S. Chin ; Wai Lok Woo ; Junbo Jia ; Wei Da Toh

STEP-LESS VOLTAGE REGULATION ON RADIAL FEEDER WITH OLTC TRANSFORMER-DVR HYBRID 179
V. Verma ; R. Gour

SOME CONSIDERATIONS OF ARC PROTECTION AND BREAKER DESIGN FOR HIGH FREQUENCY AC POWER DISTRIBUTION SYSTEMS 185
Junfeng Liu ; C. D. Xu ; Jun Zeng ; K. W. E. Cheng

A NOVEL INITIAL ROTOR POSITION DETECTION METHOD FOR PM SYNCHRONOUS MOTORS 193
Zou Xunhao ; Shenghua Huang ; Qin Zhuqian ; Xiaodong Hu ; Kan Guangqiang

CASCADED HIGH GAIN MICRO-CONVERTER FOR STORAGE-LESS PV FED RURAL TELECOM SYSTEMS 197
V. Verma ; V. Arora

STATIC MODEL OF A 2X25KV AC TRACTION SYSTEM 203
M. Plakhova ; B. Mohamed ; P. Arboleya

GRAPH THEORY APPROACH FOR A 2X25KV AC BIVOLTAGE TRACTION SYSTEM 208
M. Plakhova ; B. Mohamed ; P. Arboleya

A HYBRID OPEN MP/MPI PARALLEL COMPUTING MODEL DESIGN ON THE SM CLUSTER 213
Ying Xu ; Tie Zhang

COMPARISON AND ANALYSIS OF DBD-OZONIZERS POWERED BY CURRENT-AND VOLTAGE-MODE POWER SUPPLIES 218
Le Wang ; Xiongmin Tang ; Shuaijie Luan ; Sizhe Chen

RESEARCH OF PHASE-SHIFT SERIES-LOAD RESONANT POWER SUPPLY FOR DBD-OZONE GENERATOR 223
Shuaijie Luan ; Xiongmin Tang ; Le Wang ; Sizhe Chen

A CONTROL METHOD FOR PERMANENT-MAGNET SYNCHRONOUS MOTOR WITH UNBALANCED CABLE RESISTOR 230
Zou Xunhao ; Shenghua Huang ; Qin Zhuqian ; Xiaodong Hu ; Kan Guangqiang

DESIGN OF A WIRELESS CHARGING SYSTEM COMPOSITED OF ANTENNA AND CIRCUITS IN HF BAND 233
C. F. Huang ; Yu-Wei Weng

ENERGY STORAGE VARIATION RATIO AND POWER EFFICIENCY OF CUK CONVERTER IN CONTINUOUS MODE 236
Wenzheng Xu ; K. W. E. Cheng

STUDY OF AC-DC CONVERTERS WITH ISOLATED TRANSFORMERS APPLIED TO HIGH FREQUENCY AC-DC POWER CONVERSION 242
X. D. Xue ; J. Mei ; R. S. Raghu ; J. F. Liu ; K. W. E. Cheng

INTEGRATION DEVELOPMENT FOR SUPERCAPACITOR CONTROLLED DISTRIBUTED GENERATION SYSTEM 250
Xiaolin Wang ; K. W. E. Cheng ; Yongquan Nie

HIGH FREQUENCY AC AUXILIARY POWER SOURCE FOR FUTURE VEHICLES 255
R. S. Raghu ; Junfeng Liu ; X. D. Xue ; K. W. E. Cheng

MEASUREMENT AND ANALYSIS OF EMI RADIATED BY A HIGH-FREQUENCY AC DISTRIBUTION SYSTEM IN VEHICLES 261
Z. Y. Jiang ; X. D. Xue ; J. Mei ; E. K. W. Cheng

CHARACTERIZATION AND MODELING OF COPPER FOIL CONDUCTOR FOR HIGH FREQUENCY POWER DISTRIBUTION 266
C. D. Xu ; K. W. E. Cheng ; R. S. Raghu ; J. F. Liu

ENERGY FACTOR ANALYSIS FOR SWITCHED RELUCTANCE MOTORS UNDER VARIOUS CONDITIONS 270
Jingwei Zhu ; K. W. E. Cheng

LUMINOUS FLUX METER FAILURES AND POTENTIAL SOLUTION 276
W. F. Chen ; K. W. E. Cheng

AN ADAPTIVE MODULATION SCHEME FOR FUNDAMENTAL FREQUENCY SWITCHED MULTILEVEL INVERTER WITH UNBALANCED AND VARYING VOLTAGE SOURCES 280
Y. C. Fong ; K. W. E. Cheng

LARGE SIGNAL STABILITY ANALYSIS OF AIRCRAFT ELECTRIC POWER SYSTEM BASED ON AVERAGED-VALUE MODEL 286
Yanbo Che ; Xiaokun Liu ; Zhangang Yang

SCENARIO METHOD IN ECONOMICAL DISPATCH OF MICROGRID CONNECTED TO DISTRIBUTION SYSTEM 291
C. Qin ; P. Dong ; Y. Lin ; Y. Feng

SHORTING-TIME TRANSMISSION LINE MAINTENANCE SCHEDULING METHOD BASED ON CREDIBILITY THEORY 295
Y. Lin ; P. Dong ; C. Qin ; Y. Feng

MODEL OF INVERTER IN MORE ELECTRIC AIRCRAFT BASED ON GENERALIZED STATE SPACE AVERAGING APPROACH 301
Yanbo Che ; Guojian Liu ; Zhangang Yang ; Xiaokun Liu

HIGH POWER CAPACITIVE POWER TRANSFER FOR ELECTRIC VEHICLE CHARGING APPLICATIONS 306
C. Mi

THE DESIGN OF AN ELECTRIC WING-IN-GROUND-EFFECT (WIG) VEHICLE AS PART OF AN URBAN AIR TRANSIT SYSTEM 310
E. Bodlak

AN ESS CHARGE BALANCING METHOD BASED ON CURRENT ALLOCATION WITH MULTI-SOURCE POWER CONVERTERS FOR ELECTRIC MICROCARS 320
Y. C. Fong ; K. W. E. Cheng

BOOTSTRAP GATE DRIVER AND OUTPUT FILTER OF AN SC-BASED MULTILEVEL INVERTER FOR AIRCRAFT APU 326
Yuanmao Ye ; K. W. E. Cheng

NEW LED LIGHTING DESIGN FOR ROAD VEHICLES 331
W. F. Chen ; K. W. E. Cheng

SOFT-SWITCHING TOPOLOGIES FOR SWITCHED RELUCTANCE MOTORS IN THE APPLICATION OF ELECTRIC VEHICLES 335
Jingwei Zhu ; K. W. E. Cheng

IDEAS FOR FUTURE ELECTRIC AIRCRAFT SYSTEM 342
S. R. Raman ; K. W. E. Cheng

APPLICATION OF CUK CONVERTER TOGETHER WITH BATTERY TECHNOLOGIES ON THE LOW VOLTAGE DC SUPPLY FOR ELECTRIC VEHICLES 346
Wenzheng Xu ; K. W. E. Cheng ; K. W. Chan

A NEW TWO-DEGREE OF FREEDOM SWITCHED RELUCTANCE MOTOR FOR ELECTRIC VESSEL 351
S. Y. Li ; K. W. Cheng

DESIGN OF ELECTRIC VESSEL BASED ON CONCENTRATED PHOTOVOLTAIC FOR DENSITY ENERGY SOURCE 357
S. Y. Li ; K. W. Cheng

FUTURE BODY DESIGN FOR ELECTRIC VESSEL AND AIRCRAFT 362
Xiaolin Wang ; G. Sin ; K. W. E. Cheng

HYBRIDIZATION OF ENERGY STORAGE SYSTEMS FOR ELECTRIC TRANSPORTATION BY MEANS OF BIDIRECTIONAL POWER ELECTRONIC CONVERTERS 367
R. Georgious ; J. Garcia

LOAD ANALYSIS FOR THE ASYMMETRIC BILATERAL LINEAR SWITCHED RELUCTANCE GENERATOR 373
Qianlong Li ; Jun Xia ; J. F. Pan ; Bo Zhang ; N. Cheung

COORDINATION POSITION CONTROL OF LINEAR SWITCHED RELUCTANCE MACHINES 377
Weiyu Wang ; Jun Xia ; J. F. Pan ; Bo Zhang ; N. Cheung

CLUSTER FLIGHT ALGORITHMS FOR DISTRIBUTED SATELLITE BASED ON CYCLIC PURSUIT 380
Bo Zhang ; Yanhui Yun ; Jianjun Luo ; N. C. Cheung ; J. F. Pan

DEVELOPMENT OF THE EQUIVALENT MAGNETIC CIRCUIT MODEL FOR A SURFACE-INTERIOR PERMANENT MAGNET SYNCHRONOUS MOTOR 385
M. Si ; Xiangyu Yang ; Shiwei Zhao ; J. Si

IMPLEMENTATION OF THE DERIVATIVE-BASED TECHNIQUE FOR SOLVING A 2X25KV AC BIVOLTAGE TRACTION SYSTEM 389
M. Plakhova

DESIGN AND CONSTRUCTION OF A DAB CONVERTER FOR INTEGRATION OF ENERGY STORAGE SYSTEMS IN POWER ELECTRONIC APPLICATIONS 396
S. Saeed

HALF-BRIDGE CURRENT SOURCE BIDIRECTIONAL RESONANT CONVERTER FOR SUPERCAPACITOR STORAGE IN TRACTION SYSTEMS 403
M. Karadeniz

DESIGN OPTIMIZATIONS OF OUTER-ROTOR PERMANENT MAGNET SYNCHRONOUS MACHINES WITH FRACTIONAL-SLOT AND CONCENTRATED-WINDING CONFIGURATIONS IN LIGHTWEIGHT ELECTRIC VEHICLES 411
D. Wu ; W. Fei ; P. C. K. Luk

ANALYSIS AND DESIGN OF V-SPOKE FERRITE INTERIOR PERMANENT MAGNET MACHINE FOR TRACTION APPLICATIONS 419
Bing Xia ; Weizhong Fei ; P. Luk

THE DESIGN AND APPLICATION OF AN UNMANNED SURFACE VEHICLE POWERED BY SOLAR AND WIND ENERGY 425
X. Q. Zhou ; L. L. Ling ; J. M. Ma ; H. L. Tian ; Q. S. Yan ; G. F. Bai ; S. Y. Liu ; L. Dong

INTEGRATED PIEZOELECTRIC ENERGY HARVESTING AND STRUCTURAL HEALTH MONITORING FOR TRANSPORTATION INFRASTRUCTURE 435
N. Kaur ; S. Balguvhar

ADVANCES IN THE APPLICATION OF POWER ELECTRONICS TO RAILWAY TRACTION 439
K.-K. Lee

ACHIEVING MOBILITY ON DEMAND USING AUTONOMOUS VEHICLES 443
M. H. Ang

RECENT PROGRESS IN DEVELOPMENTS OF ON-LINE ELECTRIC VEHICLES 447
S. Y. Choi ; C. T. Rim

ZERO EMISSION ELECTRIC VESSEL DEVELOPMENT 455
K. W. E. Cheng ; X. D. Xue ; K. H. Chan

Author Index

2015 6th International Conference on
Power Electronics Systems and Applications

15th - 17th December 2015

6th PESA

Electric Transportation - Automotive, Vessel and Aircraft

CONFERENCE

Email: eepesa15@polyu.edu.hk
http://www.pesa-polyu-2015.com/

THE HONG KONG
POLYTECHNIC UNIVERSITY
香港理工大學

POWER ELECTRONICS RESEARCH CENTRE

All rights reserved. No part of this publication may be reproduced or transmitted in any form or by any means, electronic or mechanical, including photocopying recording or any information storage or retrieval system, without permission in writing from the publisher.

First edition December 2015

Printed in Hong Kong by Reprographic Unit
The Hong Kong Polytechnic University

Published by
Power Electronics Research Centre
The Hong Kong Polytechnic University
Hung Hom, Kowloon, Hong Kong

Edited by
K.W. Eric Cheng

Production Team
Ms. Wang Xiaolin
Mr. Raghu Raman S
Mr. William F. Chen
Ms. Xu Cuidong
Mr. George Sin

Disclaimer
Any options, findings, conclusions or recommendations expressed in this material / even do not reflect the view of The Hong Kong Polytechnic University

Simulation of Thermal Properties of the Liquid Metal Batteries

WANG Wei   WANG Kangli

School of Electrical & Electronic Engineering, Huazhong University of Science & Technology, Wuhan, P.R. of China

Abstract—With the development of intermittent renewable energy technologies (such as wind and solar) large-scale access into the electric grid, a low cost high efficiency energy storage medium is indeed to greatly improve the ability of the power grid adopting the renewable energy., among all kinds of energy storage technologies, batteries have long been considered a very promising choice owing to their seldom pollution, high efficiency, low maintenance, and flexibility in power and energy characteristics, besides its long service life. Here we describe the liquid metal battery which is a newly developed and can be self-heated. This cheap and efficient energy storage medium will meets the specifications for power grid applications. This paper focuses to analysis the self-heated mechanism in this new battery, which is the vital factor to ensure the startup and reliability of the battery. The analysis is based on the finite element numerical methods with a three-dimensional thermal simulation model of liquid metal battery, and compares the temperature distribution under various current densities, different battery electrolyte sizes respectively. Simulation results show that both the current densities and the thickness of the electrolyte have a critical value above which the liquid metal battery can achieve a self-heating state. We also verify the simulation results theoretically from the view of mathematical formula, and reveal the ways to get self-heated for this battery.

Keywords—Liquid metal battery; Self-heating; Temperature distribution; Thermal model

I. INTRODUCTION

With the large-scale deployment of renewable energy sources, massive and cheap electricity storage becomes indispensable since the major part of renewable electricity generation (solar, wind) is inherently fluctuating. Storage is thus essential to balance supply and demand, and is the key to stabilize the power grid. Given that the potential for pumped storage hydropower is largely exhausted. electrolytically generated hydrogen, partly processed to synthesized hydrocarbons, seems to be the only viable option for long-term large-scale storage on the TWh scale. However, the total efficiency of the conversion chain is relatively low, due to the multitude of process steps involved. Electrochemical energy storage (EES) shows generally higher efficiencies, but needs improvements towards larger capacities at affordable lower costs. If these demands can be met, EES will be an attractive candidate for short-term and mid-term stationary electricity storage [1].

Liquid metal batteries (LMBs) are currently discussed as a means to provide economic grid-scale energy storage. The original model of this battery can be traced back to the thermal regeneration metal battery and the double metal molten salt battery introduced in the General Motors Corporation and the Argonne National Laboratory Research in the 1960s[3]-[7].LMBs work in high temperature environment. It comprises a liquid lithium negative electrode, a molten salt electrolyte and a liquid positive electrode, which self-segregate by density into three distinct layers owing to the immiscibility of the contiguous salt and metal phases [2]. On discharge, the cathode metal migrates to the anode through the molten salt electrolyte after ionization. At the anode/electrolyte interface, the cathode metal alloys with the anode metal, while the charging performs in the opposite process.

Liquidity endows liquid metal batteries with superior kinetics and transport properties, thereby allowing for fast charging and discharging and high current densities. But the LMBs have the disadvantages of relatively low cell voltage (typically below 1V) and the high operation temperature. The characteristics of the LMBs are determined by the temperature in the chemical reaction pool. The battery temperature is impacted by the heat generation during the course of charge and discharge accompanied with chemical electrochemical reaction process and mass transport process, the inter environment for battery operating conditions, and the convection, and the external heat dissipation of the battery. Low temperature may decrease the chemical reaction rate and high temperature may decrease the ionic conductivity. The non-uniform temperature distribution and rapid change temperature of the LMBs will lead to the early damage and the thermal runaway, even lead to security incident. Therefore the thermal management is crucial for the normal operation of LMBs. Thermal management is the project to handle the heat transfer and heat dissipation of the battery. Based on the thermal model, we can analyses the distribution of the temperature of the battery in time and space domain, and explain the possible thermal runaway phenomenon in the working condition, thereby optimizing the battery structure and thermal management strategies to improve the safety and utility. Thermal management is critical to the successful operation of any battery technology and now gets increasing attention [9]-[12]. At present, the research on the thermal model of the battery is concentrated on the lithium battery. The original thermal model of the lithium battery usually analyses its average temperature by assuming that the battery heats uniformly [13]-[14], this model is modified by considering the electric current field caused by ear layout. When the theory of the porous electrode model is put forward, the research on the thermal model of lithium integrates with the internal material transportation and electrochemical reaction of the battery [15]-[16].

Considering the complexity of the practical structure of the battery, it usually needs lots of manpower and material resources to study the electricity and heat performance in an experimental. Instead, the numerical simulation technology can help study battery thermal behavior and effectively shorten the design period, thus save the time and cost. Besides, the numerical simulation can provide some

978-1-5090-0064-7/15 $31.00 © 2015 IEEE

insightful observations of the internal distribution of some vital parameters.

The paper is based on the finite element numerical methods, with a three-dimensional thermal simulation model of liquid metal battery. And we analyses the temperature distribution under various current densities, different battery electrolyte sizes respectively, revealing the critical condition of the battery's self-heating and the matching relationship between the battery and the load in high efficiency. The simulation provides theoretical guidance for the design and optimization of the battery structure.

II. THE SIMULATION AND CALCULATION

1. The Influence of Electrolyte Thickness on the Thermal Effect of the Battery

We use the liquid metal battery as the study objection, and choose the lead antimony alloy as the positive electrode active material, and lithium metal as the anode material. The battery's open circuit voltage is 0.9V. The geometric model of the simulation cell is shown in fig 1.

2. Simulation Model

The geometric model depicts the geometric structure of the battery, while we use the electrothermal coupling model to calculate the temperature and depict the temperature distribution of the battery. To simplify the model calculation, improve the practicability and the convergence of the model, we usually do some reasonable assumption as follows:

1) The materials in the battery are isotropic and have homogeneous physical properties;
2) The fluidity of the battery's internal electrolyte is very poor; the effect of the internal convection is negligible.

The essence of the thermal simulation of the liquid metal battery is the battery internal energy of partial differential conservation equations listed as followed:

$$\rho c_p \frac{\partial T}{\partial t} = \lambda_x \frac{\partial^2 T}{\partial x^2} + \lambda_y \frac{\partial^2 T}{\partial y^2} + \lambda_z \frac{\partial^2 T}{\partial z^2} + \dot{q} \qquad (1\text{-}1)$$

where ρ is the average density of the battery, c_p is the average specific heat capacity of the battery, T is the temperature; t is the time, q is the heat rate of the battery.

Considering that the single battery volume is small, we regard the battery as a uniform heating element in the calculation. The main thermal source comes from the battery core heating. Because it is very difficult to measure the heat generation rate accurately in real-time, we usually use the Bernardi heat generation rate equation to calculate the battery core heat:

$$\dot{q} = \frac{I}{V}(E_{oc} - U - T \frac{dE_{oc}}{dT}) \qquad (1\text{-}2)$$

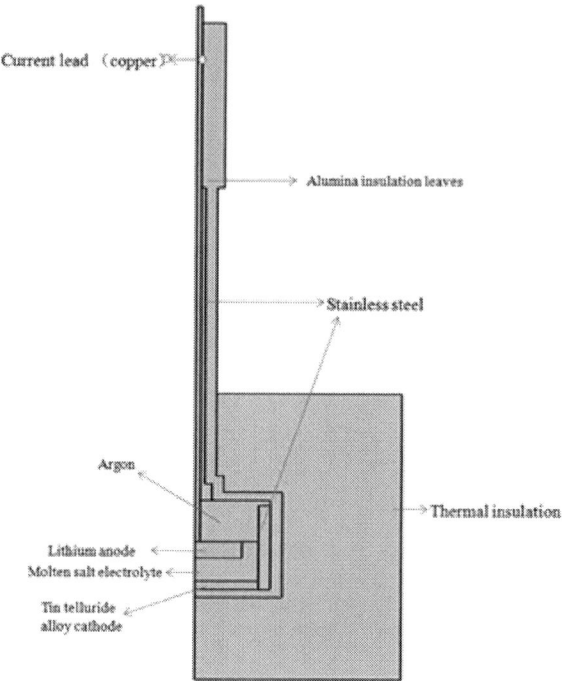

Fig.1: Geometric model of the liquid metal battery

where I is the current, V is the battery volume, Eoc is the battery electromotive force, U is the battery voltage, T is the initial temperature of the battery, dEOC/dT is the temperature coefficient of the battery voltage varying with temperature, in the calculation, Eoc-U equals IR(R is the battery resistance).

Strictly speaking, due to the current densities vary with the different part of the battery in charging and discharging process, it will lead to slightly different in the thermal physical parameters. But in the electric-current thermal model, we take the same thermal physical parameters; the differences are mainly reflected in the model of heat generation rate

According to Newton's law of cooling, the thermal boundary condition of the liquid metal battery model can be described as followed:

$$-\lambda(\frac{\partial T}{\partial n}) = h(T_{sur} - T_\infty) \qquad (2\text{-}3)$$

where h is the convection heat transfer coefficient between the cell surface and the surrounding fluid, T_∞ is the temperature of the surrounding fluid, Tsuris the surface temperature of the battery, λ is the heat conduction coefficient of battery surface material, and n is the vector direction of vertical battery surface.

3. Thermal physical parameters

Thermal physical parameters are important for the battery simulation process, reliable parameters provide guarantee for the accuracy of the simulation results under different working conditions, and the thermal and physical parameters of the materials are listed in the table 1. The material parameters of the positive electrode, the negative electrode and the electrolyte refers to [2] [8] [17], and the other material parameters refer to the manual [18].

978-1-5090-0064-7/15 $31.00 © 2015 IEEE

Table 1: Thermal-physics properties parameters of liquid metal battery

Material	Density/ (kg·m⁻³)	Specific heat /(J·kg⁻¹·K⁻¹)	Thermal conductivity /(J·m⁻¹·K⁻¹)
Positive electrode	6500	700	10
Negative electrode	534	3600	30
Alumina	3900	900	0.2
Electrolyte	2000	1000	5
Stainless steel	7850	475	44.5
Copper	8700	385	100
Insulation layer	1800	2394	0.01

III. THE SIMULATION AND CALCULATION

1. Heat effect and temperature distribution of the single battery

In discharge process of the liquid metal battery, as the internal heat generation rate is high, external convection is relatively low, thus battery heating rate is greater than the rate of heat loss, and then heat accumulation leads to the increase of core temperature for the single battery. The simulation results show that the high temperature distributes in the center area of the single battery. Due to convective heat transfer, the temperature of the cell surface is low. And for the axial thermal conductivity of cell is less than the radial thermal conductivity, which makes heat in the diameter direction conduct less than the axial direction.in an equal time.

Fig.2 shows the temperature distribution of the liquid metal battery in the current densities of 0.7A/cm2, the overall

temperature distribution of the battery is shown in fig.2 (a), to make clear of the internal temperature distribution of the battery, we get half battery in the YOZ direction and the temperature distribution is shown in the fig.2 (b). The heat transfer coefficient of the cell surface convection is 3W/ (m2.K).

The heat mainly comes from the chemical reactions in the electrolyte of the LMBs; therefore the thickness of the electrolyte has a great impact on the heat generation and temperature distribution of the battery. To study the thermal effect of the liquid metal battery under different thickness of the electrolyte, we choose 8mm、10mm、12mm、14mm thickness-electrolyte respectively in the current densities of 0.7A/cm2 to simulate. The

simulation results are shown in fig. 3. And the temperature distribution field of the half section in YOZ direction from the thickness of 8mm to 12mm is shown in fig.3 (a)、fig.3 (b)、fig.3(c)、fig.3 (d) respectively.

LMBs work in high temperature, they won't work when the temperature is below a certain value, and therefore it's very important for us to clarify the effecting factors on the critical temperature. And for optimal battery operation and high energy efficiency, it is also very crucial to make clear of the relationship between the battery and the load.

2. The influence of electrolyte thickness on the thermal effect of the battery

The heat mainly comes from the chemical reactions in the electrolyte of the LMBs; therefore the thickness of the electrolyte has a great impact on the heat generation and temperature distribution of the battery. To study the thermal effect of the liquid metal battery under different thickness of the electrolyte, we choose 8mm、10mm、12mm、14mm thickness-electrolyte respectively in the current densities of 0.7A/cm2 to simulate.

The simulation results are shown in fig. 3. And the temperature distribution field of the half section in YOZ direction from the thickness of 8mm to 12mm is shown in fig.3 (a)、fig.3 (b)、fig.3(c)、fig.3 (d) respectively.

From fig.3, we can see that the temperature distribution in the center area of the battery is almost the same, namely high temperature is concentrated in battery electrolyte region, and the temperature gradually decreases near the edge of the battery. And we can also see that with the increase of the thickness of the electrolyte, the battery

978-1-5090-0064-7/15 $31.00 © 2015 IEEE

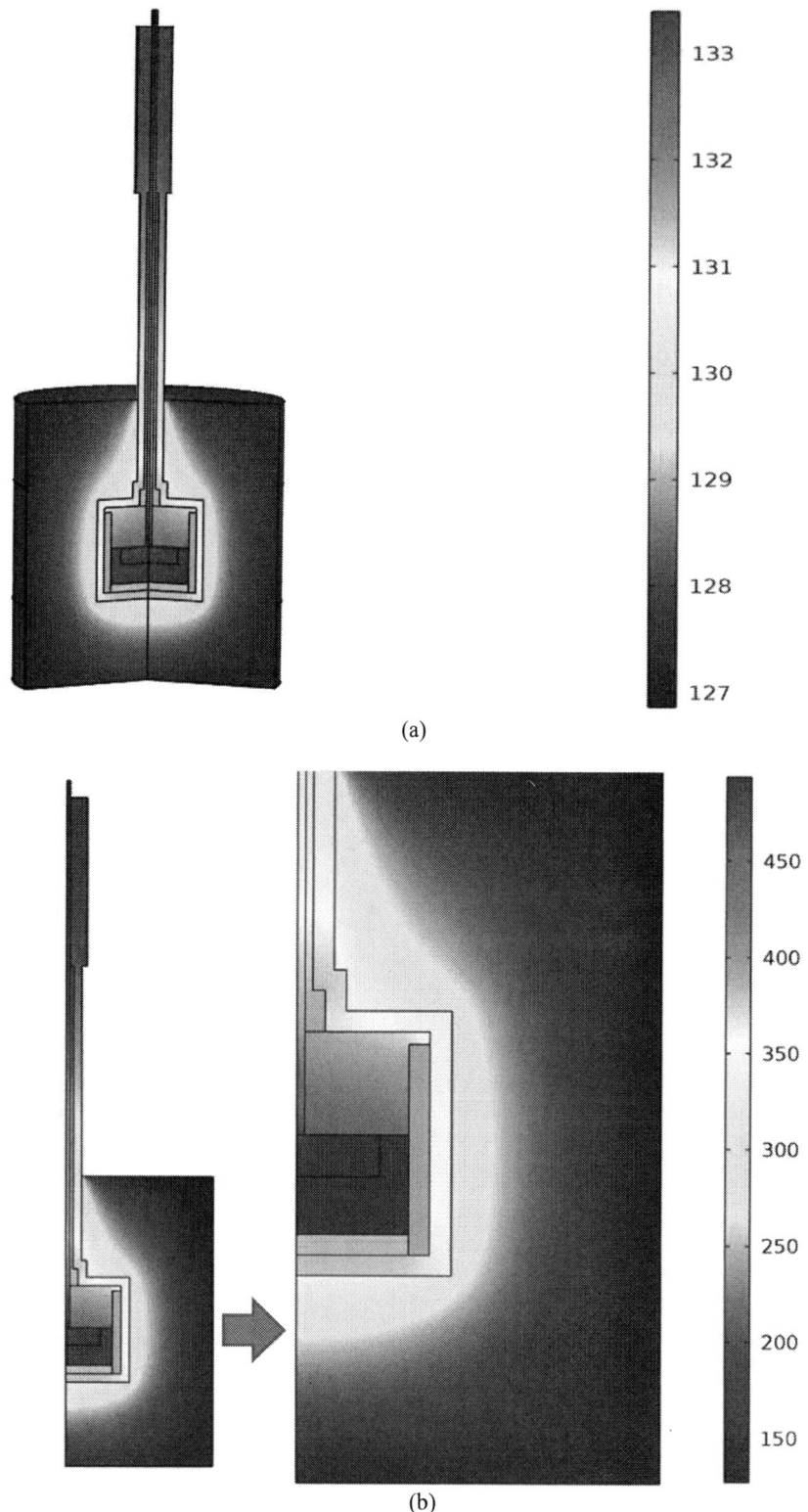

(a)

(b)

Fig.2: The whole temperature field distribution of the liquid metal battery (a) the whole temperature field distribution, (b) temperature distribution of the YOZ half section.

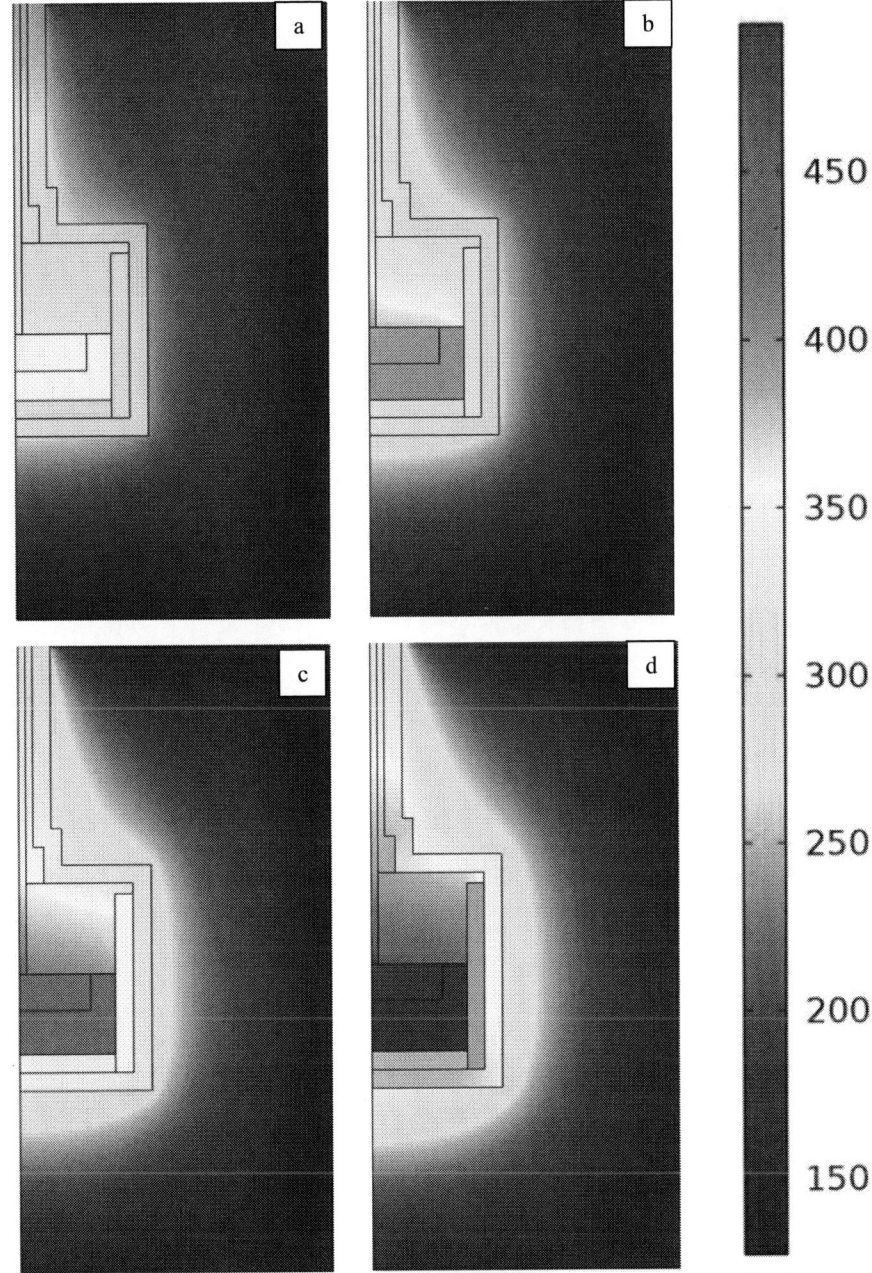

Fig.3: The temperature distribution field of different electrolyte thickness under 0.7A/cm2 current density

temperature of the center gradually increases, and the highest temperature is linear to the thickness of the electrolyte. As the liquid metal battery is a high temperature battery, the electrolyte, the anode and the cathode part usually work in more than a certain degree. To ensure battery's normal working, we need to understand the relationship between the temperature and the electrolyte thickness. Therefore we calculate the average temperature under different thickness of the electrolyte as fig.4. Fig.4 shows the relationship between the average temperature of the battery and the thickness of the battery in 0.7 A/cm2 current density.

From the fig.4, we can see that the average temperature of the battery is linear to the electrolyte thickness. Using the linear interpolation method, we can calculate that the critical electrolyte size under the critical average temperature of the battery (is about 450 degree) is about 13mm. In other words, the electrolyte thickness of the battery should be less than a certain size to guarantee the battery's self heating, but it may lead to low efficiency in too large electrolyte thickness. Therefore, to ensure normal operation of the battery, we need to design a reasonable electrolyte size according to the current density when we design the battery structure.

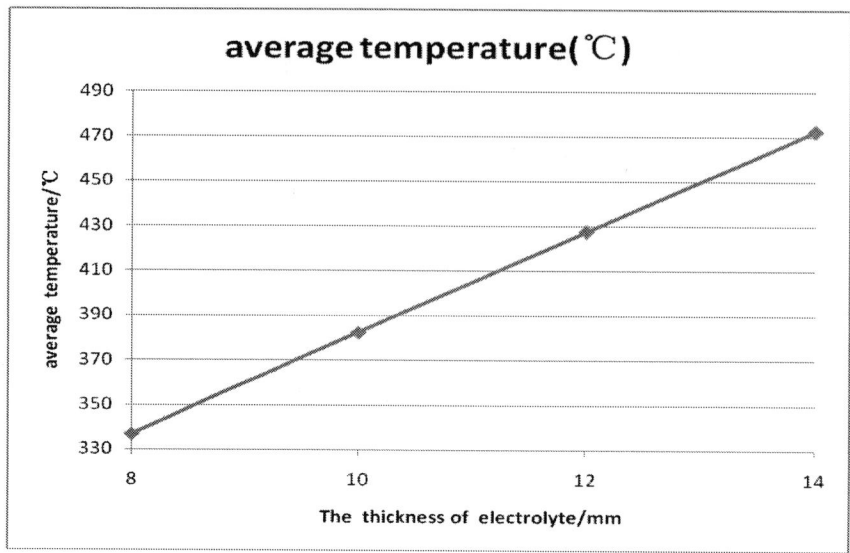

Fig.4: The relationship between the average temperature of the battery and the electrolyte thickness under 0.7A/cm2 current density

3. The influence of current density on the thermal effect of the battery

To study the thermal effect of the liquid metal battery under different current density, we choose 0.1A/cm2、0.3A/cm2、0.5A/cm2、0.7A/cm2、0.9A/cm2 current density respectively in the electrolyte thickness of 10mm to simulate. The simulation results are shown in fig.5. And the temperature distribution field of the half section in YOZ direction from the current density of 0.1A/cm2 to 0.9A/cm2 is shown in fig.5 (a)、fig.5 (b)、fig.5(c)、fig.5 (d)、fig.5 (e) respectively.

From fig. 5, we can see that the battery temperature still has the same distribution, namely high temperature is concentrated in battery electrolyte region, and the temperature gradually decreases near the edge of the battery. And we can also see that with the increase of the current density, the battery temperature increases rapidly, and the temperature increases obviously faster than that of the electrolyte thickness. Because the mathematical relationship between the core heating power of the battery, the current and the battery resistance is described as followed:

$$W = I^2 R \qquad (2\text{-}1)$$

where W is the core heating power of the battery, I is the current, R is the battery resistance.

From the equation (3-1), we know that the heating power is proportional to the square of current, whereas the electrolyte thickness is proportional to the resistance, namely the heating power is proportional to the thickness, which explains why the effect of current density on the temperature distribution of the battery is more obvious than that of the electrolyte thickness.
To reveal the relationship between the current density and the heat generation of the battery, we make the relation curve as shown in fig.6.

The simulation results also show that there is a critical current density for realizing the self-heating of the battery, in other words, the LMBs may not work at a lower current density, and it may lead to low energy efficiency at a higher current density, therefore we need to choose a reasonable current density according to the load and the size of the LMBs.

4. The analysis of the factors effecting on the energy efficiency of the liquid metal battery

The thermal efficiency of the battery is directly related to the energy efficiency, we can go on the following derivation to understand the factors affecting on the thermal efficiency, and the mathematical relationship between current density the electrolyte thickness and the thermal efficiency.

Assuming that liquid metal battery supplies power to load and the load is equivalent to a resistance R, the internal resistance of the cell is r, the terminal voltage of the battery is U, then the battery current can be described as followed:

$$I = \frac{U}{R+r} \qquad (2\text{-}2)$$

The efficiency of the battery can be described as followed:

$$\eta = \frac{I^2 R}{I^2 R + I^2 r} = \frac{R}{R+r} \qquad (2\text{-}3)$$

By (3-3), we know that the battery itself needs to reduce the internal resistance r to improve the efficiency.

From the perspective of battery heat balance, there is the following relationship:

978-1-5090-0064-7/15 $31.00 © 2015 IEEE

$$I^2 r - h(T - T_{amb})S_{sur} = 0 \qquad (2\text{-}4)$$

where h is the integrated heat conduction coefficient from the core of the battery to the outer surface environment, T is the temperature of the core of the battery, T_{amb} is the ambient temperature of the battery surface, and S_{sur} is the

heat dissipation area of the battery surface and the environment. When the type (3-2) substitutes type (3-4), and puts T on the left of the equation, we can get the following equation,

$$T = T_{sur} + \frac{U^2 r}{hS_{sur}(R+r)^2} \qquad (2\text{-}5)$$

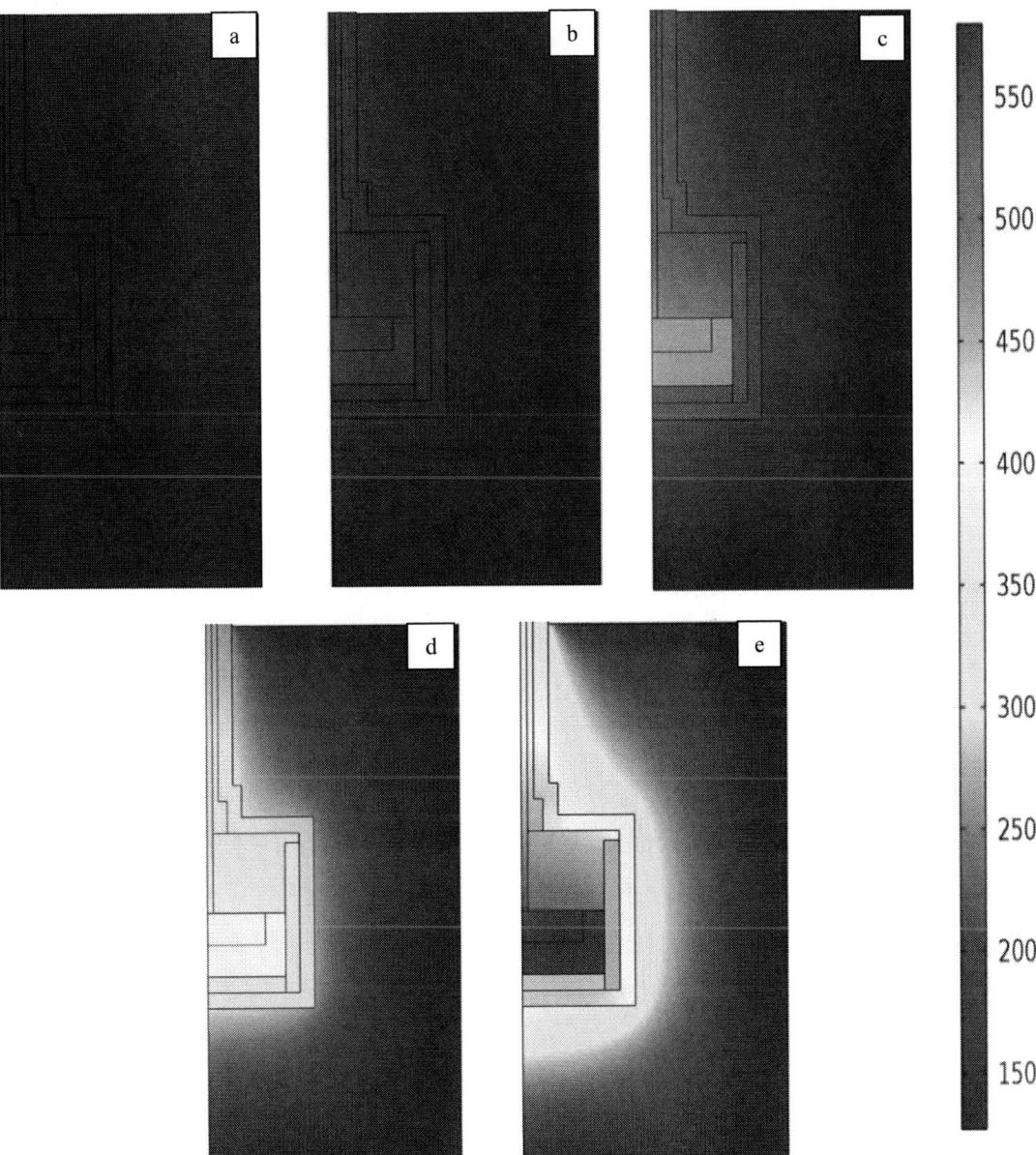

Fig. 5: The temperature distribution field of different current density in the electrolyte thickness of 10 mm

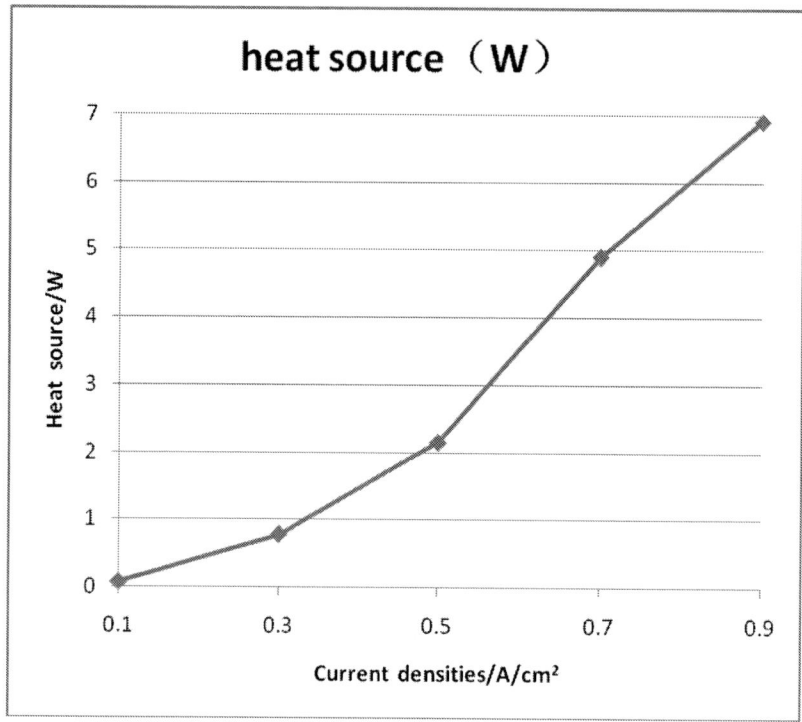

Fig. 6: Relationship between the heat generation and the current density in 10mm electrolyte thickness

where T is core temperature of the battery. As the core temperature of the liquid metal battery needs to reach a certain degree (above 400 degree) to work properly, assume that the critical working temperature of the battery is Tc, and then there is the following formula:

$$T_{sur} + \frac{U^2 r}{h S_{sur}(R+r)^2} > T_c \qquad (2\text{-}6)$$

We put r of the type (3-6) on the left, there is the following formula:

$$r > \frac{h S_{sur}(T_c - T_{sur})}{I^2} \qquad (2\text{-}7)$$

According to the type (3-3) and type (3-7), in the premise of the normal working of the liquid metal battery, we try to improve the energy efficiency of the battery as much as possible, the battery internal resistance r must satisfy type (3-7) and the r value is as small as possible. In other words, when the r value equals the right side of type (3-7), .the battery energy efficiency reaches the highest.

To be further analysis, if the internal resistance of the liquid metal battery r is represented by the internal parameters of the battery, namely r can be expressed as followed:

$$r = f(\rho, l, S),$$
$$r \propto \rho, r \propto l, r \propto \frac{1}{S} \qquad (2\text{-}8)$$

where ρ is the equivalent resistivity, l is the height of the battery, S is cross section area of the battery.

The internal resistance of the battery can also be expressed as followed:

$$r = \frac{f_1(\rho) f_2(l)}{f_3(S)} \qquad (2\text{-}9)$$

where $f_1(\rho)$ is the function of ρ, $f_2(l)$ is the function of l, $f_3(S)$ is the function of S.

When the type (3-9) substitutes type (3-7), there is the following formula:

$$\frac{f_1(\rho) f_2(l)}{f_3(S)} > \frac{h S_{sur}(T_c - T_{sur})}{I^2} \qquad (2\text{-}10)$$

By type (3-10), we know that it satisfies a certain constraint relationship between the battery temperature and the battery electrolyte thickness in the premise of satisfying the normal work of the battery. , that is, formula (3-10) should be satisfied.

When the current density is a certain value (namely, S is a certain value), the battery size should satisfy the following formula:

$$f_2(l) > \frac{h f_3(S) S_{sur}(T_c - T_{sur})}{f_1(\rho) I^2} \qquad (2\text{-}11)$$

When the battery size is a certain value (namely, l is a certain alue), the battery current density j (j=I/f3(S)) should satisfy the following formula:

$$j = \frac{I}{f_3(S)} > \frac{hS_{sur}(T_c - T_{sur})}{f_1(\rho)f_2(l)I} \quad \text{(2-12)}$$

Combined the type (3-11) and type (3-12), we know that the battery size and the current density should be met the type (3-11) and the type (3-12) respectively, which confirm the above simulation results from the perspective of mathematical derivation.

When we need to improve the battery thermal efficiency as much as possible in the premise of guaranteeing the normal work of the battery, the battery internal resistance should satisfy the type (3-7), and the battery internal resistance r is smaller, the thermal efficiency of the battery is high, namely when the value of the internal resistance r equals to the type on the right side of type (3-7), the thermal efficiency of the battery reaches the highest.

Combined the type (3-2) and type (3-3),we know that when the value of the load resistance R is larger, it may cause the battery unable to work as the current density may be lower to its lower limit value; and when the value of the load resistance R is smaller, it may lead to a lower energy efficiency of the LMBs, therefore there is a maximum load (at this time the current density reaches the lower limit value)for a battery , which makes the battery work in highest efficiency.

5. Experiment verification

The simulation results and the mathematical deviation can be verified by the experiment. Fig.7 shows the relationship between the energy efficiency and the current density by experiment. From fig.7, we know that the higher the current density, the lower the energy efficiency of the liquid metal battery. And when the current density changes from 0.025 A/cm2 to 0.05 A/cm2, the energy efficiency is almost the same. And the current density cannot be so small that it cannot reach the working temperature, therefore the current density has a lower limit value. This also verifies the above simulation results and mathematical deviation.

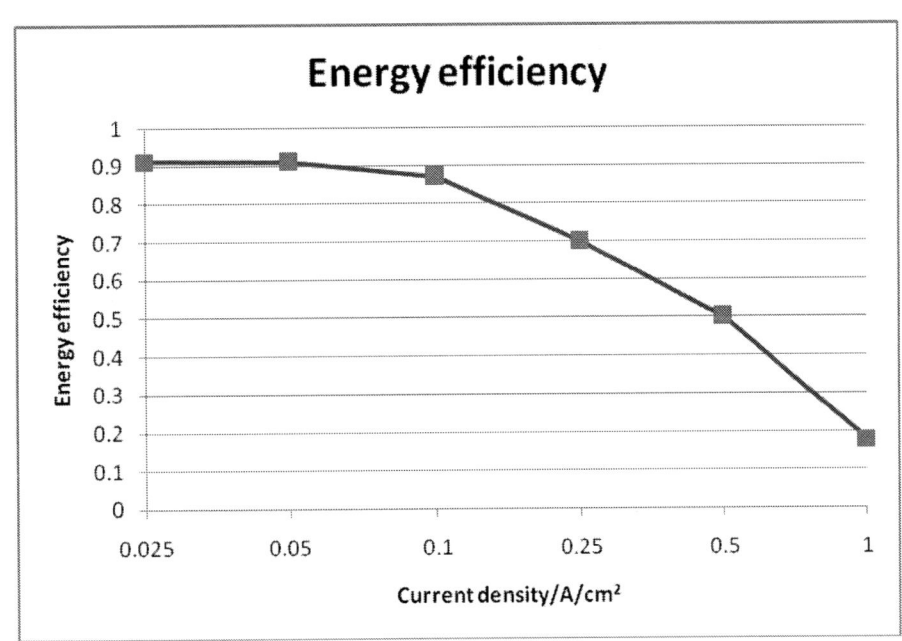

Fig. 7: Experimental relationship between the current density and the energy efficiency

IV. CONCLUSIONS

1) The high temperature of the liquid metal battery distributes in the center area of the battery, ant the temperature decreases gradually when it closes to the edge of the battery, due to the axial thermal conductivity of cell is less than the radial thermal conductivity, which makes heat in the diameter direction conduct less than the axial direction.

2) There is a similar temperature distribution in the center of the battery under different electrolyte thickness. From the mathematical derivation, we know that the

electrolyte thickness has a critical value under a certain current density in the premise of satisfying the normal working temperature of the LMBs. And when it reaches the critical value, the battery reaches the highest thermal efficiency in a certain current density. The simulation results also confirm the conclusion.

3) There is also a roughly the same temperature distribution in the center of the battery under different current densities. From the mathematical derivation, we know that the current density has a critical value under a certain electrolyte thickness in the premise of satisfying the normal working temperature of the LMBs. And when

it reaches the critical value, the battery reaches the highest thermal efficiency in a certain electrolyte thickness. The simulation results also confirm the conclusion.

4) There is a minimum value of the internal resistance to achieve the LMB's normal working in a certain temperature. And the smaller the internal resistance, the higher the thermal efficiency will be under this condition. When the internal resistance reaches the lower limit value, the thermal efficiency reaches the highest in this condition.

5) To a certain liquid metal battery (the internal resistance of the battery is a certain value), there is a maximum load resistance (at this time the current density reaches the lower limit value), which makes the battery work in highest efficiency. And the lower the resistance of the load, the lower the battery energy efficiency.

REFERENCES

[1] Weber N, Galindo V, Stefani F, et al. Current-driven flow instabilities in large-scale liquid metal batteries, and how to tame them[J]. Journal of Power Sources, 2014, 265: 166-173.

[2] Wang K, Jiang K, Chung B, et al. Lithium-antimony-lead liquid metal battery for grid-level energy storage[J]. Nature, 2014, 514(7522):348-350.

[3] Cairns E J, Shimotake H. Recent advances in fuel cells and their application to new hybrid systems [M]. 1969.

[4] Agruss B. The Thermally Regenerative Liquid-Metal Cell [J]. Journal of The Electrochemical Society, 1963, 110(11): 1097-1103.

[5] Cairns E J, Shimotake H. High-temperature batteries [J]. Science, 1969, 164(3886): 1347-1355.

[6] William H. Electrolytically-refined aluminum and articles made therefrom: U.S. Patent 1,534,315[P]. 1925-4-21.

[7] Bradwell D J, Kim H, Sirk A H C, et al. Magnesium–Antimony Liquid Metal Battery for Stationary Energy Storage[J]. Journal of the American Chemical Society, 2012, 134(4): 1895-1897.

[8] Kim H, Boysen D A, Newhouse J M, et al. Liquid metal batteries: Past, present, and future[J]. Chemical reviews, 2012, 113(3): 2075-2099.

[9] Díaz-González F, Sumper A, Gomis-Bellmunt O, et al. A review of energy storage technologies for wind power applications [J]. Renewable and Sustainable Energy Reviews, 2012, 16(4): 2154-2171.

[10] Sudworth J, Tiley A R. Sodium Sulphur Battery [M]. Springer Science & Business Media, 1985.

[11] Hamlen R P, Atwater T B, Linden D, et al. Handbook of Batteries [J]. McGraw-Hill, New York, USA, 2002: 38.1.

[12] Okuyama, R.; Nomura, E. J. Power Sources 1999, 77, 164

[13] Sato N. Thermal behavior analysis of lithium-ion batteries for electric and hybrid vehicles [J]. Journal of Power Sources, 2001, 99(1): 70-77.

[14] Botte G G, Johnson B A, White R E. Influence of Some Design Variables on the Thermal Behavior of a Lithium-Ion Cell [J]. Journal of the Electrochemical Society, 1999, 146(3): 914-923.

[15] Doyle M, Newman J, Gozdz A S, et al. Comparison of modeling predictions with experimental data from plastic lithium ion cells[J]. Journal of the Electrochemical Society, 1996, 143(6): 1890-1903.

[16] Kim G H, Smith K, Lee K J, et al. Multi-domain modeling of lithium-ion batteries encompassing multi-physics in varied length scales [J]. Journal of The Electrochemical Society, 2011, 158(8): A955-A969.

[17] Masset P, Guidotti R A. Thermal activated (thermal) battery technology: Part II. Molten salt electrolytes [J]. Journal of power sources, 2007, 164(1): 397-414.

[18] CHEN Ze-shao, GE Xin-shi, GU Yu-qubg. Measure technology of heat & thermophysics [M]. Hefei: University of Science and Technology of China Press, 1990: 18−62.

Comparison of Detection Methods for Power Quality in Micro-grid

Wang Huihui[1,2]   Wang Ping[1]

[1] School of Electrical Engineering and Automation, Tianjin University, Tianjin 300072, China;
E-mail: pingw@tju.edu.cn
[2] School of Control and Mechanical Engineering, Tianjin Chengjian University, Tianjin 300384, China;
E-mail: tjwanghuihui2015@126.com

Abstract–The micro-grid brings benefits for the economic development, but the transient power quality disturbances and harmonics generated by distributed power and grid-connected are inhibited to the development of micro-grid system. Detection and identification of these transient power quality disturbances are important. To circumvent this problem, the paper discusses the methods of power quality detection in frequency domain, mainly the Fourier transform, wavelet transform, HHT, S-transform and atomic decomposition as the main methods. It compares the advantages and disadvantages of these methods and analysis the power quality disturbance problems in micro-grid. By theoretical analysis and MATLAB simulation, it can be concluded that S-transform has high accuracy and well noise immunity in power quality detection, which provides an important reference to solve the problem of the micro-grid power quality disturbances.

Keywords– Disturbance detection, Micro-grid, Power quality, S-transform, Wavelet transform

I. INTRODUCTION

In recent years, long distance transmission by large power system is expensive and not reliable. On the other hand new energy including wind energy or solar energy has been developed rapidly. Consequently, the micro-grid system consisting of distributed generation has been widely applied. Micro-grid is a nearest small power system form the user. The wide application of micro-grid improves the reliability of power supply and the utilization rate of distributed generation system, which benefits the environment and economy. At the same time, it causes power quality disturber problems. The typical micro-grid system contains distributed power, energy storage devices, control devices and load. The instability of distributed power may cause voltage amplitude and frequency deviation of micro-grid. Energy storage devices and power control devices are mostly consist of power electronic devices which bring prolifically harmonic and inter-harmonic to the micro-grid. The on-off of high-power load may cause impact on micro-grid, while causing voltage sags at the same time. The islanded state and grid-connected state operation of micro-grid have different power quality disturbance. In addition, popularization of computer, precision instruments and other digital electronic devices needs high power quality [1], thus the problem of power quality disturbances in micro-grid is becoming more and more serious [2].

In order to solve these problems, firstly, the optimization of detection methods, rapid capture of the instantaneous change of the waveform and sampling on waveform characteristics with high sampling rate are needed. Secondly, accurate detection and identification on the sampled disturbance type are important. Finally, the establishment of analysis system which can reflect the characteristics and variation of power quality index is also important [3]. In this paper, disturbance detection method on micro-grid power quality is the research object, The Fourier transform, wavelet transform, HHT transform, S-transform and other major power quality disturbance detection methods are discussed. The good noise immunity and accurate detection and analysis method of power quality on micro-grid is obtained through simulation.

II. POWER QUALITY DISTURBANCES IN MICRO-GRID

Micro-grid has the advantages including safety, reliability and self-healing. It mainly consists by the control center, distributed power supply, power users, energy storage devices and power grid with the function of fault reconstruction. Micro-grid is designed to be a seamless access for distributed power supply and to provide high quality electric power to loads. But because of the harmonic source like micro sources in micro-grid, small capacity of micro grid, weak self-adjustment ability, difference on control mode of grid-connected and islanded operation, output power of micro sources easily affected by environment, lead to the power quality disturbance problems as the following aspects.

In micro-grid, voltage fluctuation and flicker are caused by active power output and the load reactive power. Active power is mainly generated by micro sources, such as photo-voltaic power and wind power. It's output power is influenced by weather and wind. Micro-grid's change between different operation modes also causes voltage fluctuations in the grid. In grid-connected state, the voltage is relatively stable, while it is easily affected in islanded. Because capacity of micro-grid is small and self-adjustment ability is weak, high power load and capacitors' on-off generates more serious voltage fluctuation. There is over 5% voltage fluctuation when micro-grid converts between grid-connected and islanded operation mode [4], which has a certain effect on the sensitive loads.

Harmonics and inter-harmonics mainly caused by the nonlinear loads and the power electronic interface in energy storage devices. As shown in Fig.1, harmonics and inter-harmonics generated by multi grid-connected inverters may extend to PCC, which after superposition and amplification may affect power quality in large power system or may damage the load devices [5-8]. Harmonic level in the distribution girds ascent gradually with the increase of permeability of micro-grid.

Zero sequence or negative sequence current is formed because of the unbalance of the three-phase voltage or the unbalanced three-phase load in three-phase four wire micro-grid system, resulting in unbalanced three-phase voltage in PCC. Single phase micro-source leads directly to three-phase unbalance in micro-grid. As shown in Fig.1, circulation is formed within the multiple micro-sources and within the parallel inverters in micro-grid. Besides voltage tracking precision of each micro-source is not identical. For the above reasons, power quality disturbances in micro-grid are unique.

There is always frequency fluctuation in micro-grid whether in grid-connected operation or islanded operation. In grid-connected mode, micro-grid's stable frequency is achieved by the frequency modulation of main power grid. If it operates in islanded mode, micro-grid's frequency can be guaranteed stable by primary and secondary sides adjust frequency. When micro-grid operates in island mode, the AC power generated by distributed generator could not be transmitted to the public power grid. The amplitude and frequency in islanding are unable to control, so the frequency fluctuation problem should be considered carefully if micro-grid is in island operation [9].

To sum up, the distributed power which in micro-grid has the characteristics including the intermittence, complexity, diversity and instability, distributed loads in micro-grid and influenced by switching between grid-connected and island. The disturbance signal of power quality is rapid, unstable and complex. Consequently the difficulty of accurate detection increases and the detection algorithm needs to be more sophisticated.

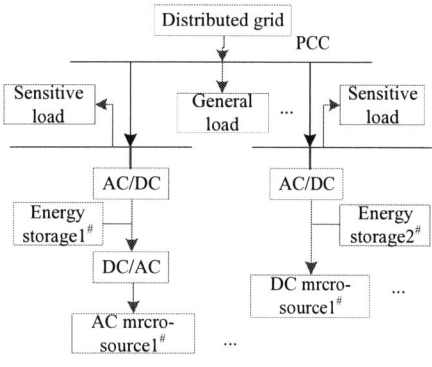

Fig. 1: Micro-grid system

III. THE METHOD OF POWER QUALITY DISTURBANCES DETECTION

1. Time domain analysis methods

Time domain analysis methods mainly refers to various vector transformation and the theory of instantaneous reactive power detection methods for power quality, mainly including $\alpha\beta$ transform based on the stator coordinate, dq transform based on rotor and symmetrical component method. Time domain analysis method will simplify the problems to be solved, mainly because that the measurements are dealt with mathematical transformation. The instantaneous power theory is mainly used for APF harmonic and reactive power detection, dq

transform is mainly used for voltage sag detection. Reference [10] proposed a voltage sag detection algorithm by using the combination of *dq* transform and morphological and forming a low pass filter, which can provide real-time detection for the start-end time of voltage sag, amplitude and phase shift. The experiments prove that the algorithm has good dynamic characteristics. Voltage sag detection principle block diagram is shown in Fig.2. Literature [11] studies the *dq* theory in calculating required compensation voltage of dynamic voltage compensator. This method shows accuracy in simulation experiments.

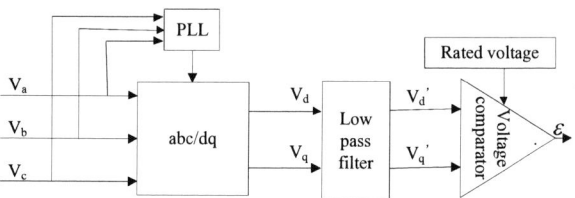

Fig. 2: The schematic of detection for sag

2. Analysis method in frequency domain

2.1. Fourier Transform

Fourier transform as a classical signal analysis method, has been widely used in signal analysis field. The discrete Fourier transform and the fast Fourier transform have been applied to detect stable signal of power quality, which have the advantages of orthogonally and completeness etc. FFT algorithm is used for accurate harmonic detection, but it has limitations in the detection of the non-stationary signal, such as the detection of inter-harmonics. Because it is difficult to achieve synchronous sampling, the application of FFT algorithm will produce the picket fence effect and spectrum leakage, use of windowed, interpolation and spectrum correction improved algorithm [12] can eliminate the problems mentioned above. But the choice of window function is more complex, the time window function selected too wide will result in poor real-time performance. Reference [13] researched the selection of window function of short time Fourier transforms in power quality harmonic analysis.

2.2. The modern spectrum analysis method

The modern spectrum analysis method using signal knowledge to forecast the finite signal intercepted by window function, it can improve the spectral resolution. This method is mainly used in the detection of harmonic and inter-harmonic. Typical modern spectrum analysis method for the detection of harmonics and inter-harmonics are Prony and MUSIC algorithm. Literature [14] distinguished harmonics and inter-harmonics which have two or a plurality of spectral lines very close to each other based on high resolution and accurate Prony algorithm. Prony algorithm is used to decompose fundamental and inter-harmonics, and then Hilbert or other algorithm is used to detect power quality disturbance caused by inter-harmonics. This method required less sampling data and is more practicable, but the anti-interference and anti-noise is poor. Literature [15] puts forward multiple signal classification (MUSIC) algorithm combined with genetic algorithm to estimate harmonics and inter-harmonics spectrum. The simulation results show that the algorithm is good at de-noising.

3. Time frequency analysis method

The transient power quality disturbances have serious problems. The time domain or frequency domain analysis method alone cannot achieve the expected effect, resulting in the time-frequency analysis. The time-frequency analysis can reflect signal characteristics in both time domain and frequency domain, and can also reflect localization information of time-frequency signal. Typical time-frequency analysis methods used in power quality disturbance analysis are short time Fourier transforms wavelet analysis, S-transform and Hilbert Huang transform etc.

3.1. Short time Fourier transform

Short time Fourier transform is designed to solve non-stationary power signal which the Fourier transform cannot solved. Short time is achieved by adding window function. The non-stationary signal to be sampled is made stationary in the window function period. That is to make the non-stationary sampling signal into a series of smooth windowed signal. Then Fourier transform is used to analyze these stationary signals. Given the signal $f(t)$ by using window function $g(t)$ to intercept signal $f(t)$ in the time domain. Move the window function to get the continuous short-time Fourier transform in different time, so that the signal in different limited time of width is stationary.

$$F(t,f) = \int_{-\infty}^{\infty} f(t)g(t-\tau)e^{-j2\pi f\tau} dt \tag{1}$$

Short time Fourier transform has a single resolution. If the time width of window function is reduced, the time resolution is improved, but the frequency resolution will be reduced. And if we want to improve the frequency resolution to make the window function become wider, the similarity of short-time stationary assumption is reduced, which is not suitable for the analysis of multi-scale or rapid change signals. The literature [16] proposed a multi-resolution with short-time Fourier transform method. Firstly, wavelet determines the main frequency range of the signal and then determines the window function. After that the corresponding short time Fourier transform is used to estimate the frequency and amplitude of the signal. Literature [17] researched short-time Fourier transform method for the detection of harmonic and frequency. For signals with amplitude containing noise and time-varying frequency, this method could give a real-time estimation of fundamental frequency/amplitude and harmonic frequency/phase of the signal. Literature [18] studies short time Fourier transform in voltage sag detection, such as detecting amplitude changes, start-end time and the disturbance source location under the disturbance voltage. This method has better detection accuracy and anti-noise ability.

3.2. Wavelet transform

Wavelet transform is a new tool of time-frequency analysis. It adds signal with a time-frequency variable window, which can automatically adjust the window size according to the frequency. Wavelet transform has high frequency resolution and low time resolution in the low frequency part, low frequency resolution and high time resolution in the high frequency part. It is widely used in the field of transient power quality disturbance analysis. Flow char of wavelet transform in transient power quality disturbance detection are shown in Fig.3.

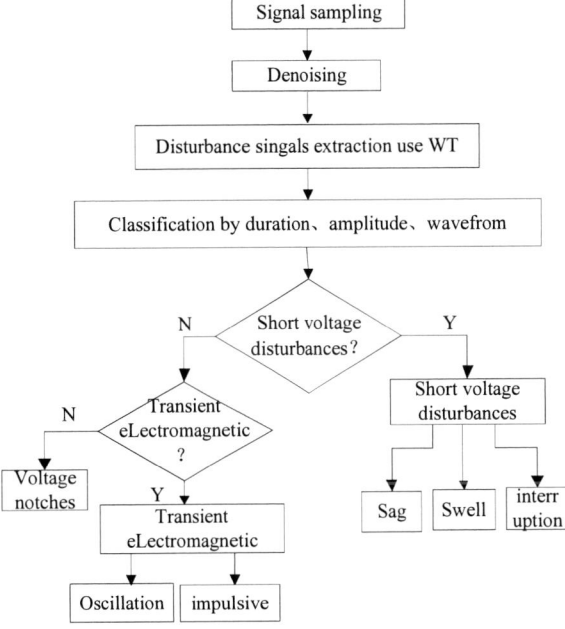

Fig. 3: The flowchart of detection for transient PQ

Noise pollution is serious in the actual signal sampling. Noise not only weaken the detection ability of power quality monitoring system based on wavelet transform, but also hinder the recovery of important information such as disturbance location and classify of the interception signal [19]. In the case of the low SNR, it's hard to extract accurate disturbance signal due to the interference of the noise [20]. The signal is usually singular in the mutation point, and the singularity of signal is often described by local maxima modulus of the wavelet transform. Local maxima modulus of wavelet transform attenuation rapidly along with the increasing of scale, according to which we can distinguish signal mutation and noise in order to eliminate the influence of noise [11]. The literature [19,20] respectively proposed different de-noising methods. According to the noise signal, a reasonable choice of noise threshold is important, which can make the wavelet transform method more accurate in detecting and locating the disturbance. But the de-noising method based on threshold have the problem of waveform distortion in the actual use, so some scholars put forward the de-noising method of choosing different wavelet coefficients [10]. Direct wavelet coefficient method solves the problem of de-noise limitation compared with threshold method, but it is time-consuming. The literature [20] proposed de-noising and extraction characteristic values in the use of discrete wavelet transform method, in the case of various SNR still can quickly detect power quality disturbances, has good real-time performance and robustness.

978-1-5090-0064-7/15 $31.00 © 2015 IEEE

Detection of power quality disturbances is affected by the signal-to-noise ratio. Due to the presence of the disturbance is not single, the detection algorithm is more complicated. The literature [21-24] presents improved wavelet transform algorithm, which can improve detection accuracy of power quality disturbance and reduce the detection time in multi-disturbances circumstance, enhancing robustness of detection algorithm in different SNR conditions. S-transform or wavelet transforms combined recursive algorithm, support vector machine and the neural network together achieving good effect. The literature [25] get the wavelet transform and Hilbert together for power quality disturbance detection, analysis, and extraction of the signal by Hilbert, eliminating the spectral leakage and detect the flick accurately. Some scholars combines the wavelet transform with neural network, where wavelet transform is used to detect and extract disturbance feature value and the neural network method is used for identification of power quality disturbance type. Literature [26] used the new classifier improves the ability of classification. Wavelet transform with neural network are combined to detect power quality disturbances. The literature [21, 27] proposed the detection of harmonic and inter-harmonic based on wavelet packet transform. Use the wavelet packet transform defined power quality indexes in frequency domain of three-phase system, the steady and unsteady system, balance and imbalance, in sinusoidal and non-sinusoidal signal under the condition of based on Wavelet packet power quality index quantification values are close to the real value.

In short, wavelet transform has many advantages. Combined with the intelligent algorithm, it can be flexibly applied to analyze disturbance signals of power quality in time or frequency domain to detect power quality disturbance with accuracy. But there are also disadvantages of wavelet transform. Wavelet transform is poor in noise immunity especially in the detection of power quality disturbance signal.

3.3.S-transform

S-transform based on the Gaussian window function, is a method of time-frequency analysis developed from the continuous wavelet transform and short-time Fourier transform. It is a continuous wavelet with particular wavelet basis multiple by a phase shift factor, the shift factor is equivalent to a kind of phase revision on continues wavelet transform. The continuous wavelet transform of signal $f(x)$ is

$$S_x(t,f) = \int_{-\infty}^{\infty} x(\tau) |f| e^{-\pi f^2 (t-f)^2} e^{-j2\pi f\tau} d\tau \qquad (2)$$

The window function is the Gaussian window function

$$w(t,f) = |f| e^{-\pi t^2 f^2} \qquad (3)$$

When $f \to \dfrac{n}{NT}, \tau \to kT$, where as T is the sampling interval, N is the total number of sampling points, the discrete S-transform can be expressed as where.

$$S[\frac{n}{NT}, kT] = \begin{cases} \sum_{m=0}^{N-1} X[\frac{m+n}{NT}] e^{\frac{-2\pi^2 m^2}{n^2}} e^{\frac{j2\pi mk}{N}} & n \neq 0 \\ \frac{1}{N} \sum_{m=0}^{N-1} X[\frac{m}{NT}] & n = 0 \end{cases}$$

$$k, m, n = 0, 1, ..., N-1 \qquad (4)$$

S-transform has good time-frequency resolution characteristics which makes it widely used in disturbance signal analysis of power quality. Compared with the wavelet transform, S-transform has obvious advantages in detection speed, accuracy and locating [28]. For example the voltage sag can be detected accurately [29]. When signal to be detected is Non-stationary and contains noise, S-transform can detect power quality disturbance and the islanding in systems containing distributed power more accurately which compared with wavelet transform [30]. In view of the fact that S-transform does well in denoising and time-frequency resolution, S-transform has been widely used in the detection of power quality disturbances in recent years.

3.4. Hilbert-Huang transform（HHT）

The signal is first processed by the empirical mode decomposition, then after decomposition the intrinsic mode functions is transformed by the Hilbert transformation to obtain the instantaneous frequency etc. Empirical mode decomposition is used to decompose the signal into intrinsic mode functions according to combination conditions. Intrinsic mode functions have two conditions; firstly the number of local maxima together with local minima is equal to or one less than the number of zero crossing points. That is to say an extreme value must be immediately followed by a zero crossing point. Secondly, for the signal at any one point, the average value of upper envelope defined by a local maxima and lower envelope defined by the local minima should be close to zero. Hilbert Huang transform compared with Fourier transform, short-time Fourier transform and wavelet transform has excellent adaptability. It is not restricted by the Heisenberg uncertainty principle, suitable for processing mutations signal. It does not need to pre-select basis function, but with the help of Hilbert transform to get the phase function. And then the local instantaneous frequency is generated by derivation of the phase function. Fourier transform has global frequency, while wavelet transform has regional frequency.

Reference [31] proposed variable frequency wavelet decomposition algorithm based on Hilbert-Huang transform which is used for the detection of power quality. This method compensates the frequency spectrum leakage compared with discrete wavelet packet and can accurately measure parameters like RMS, total harmonic distortion and flicker. The literature [32, 33] proposed an application of Hilbert-Huang transform combined with mathematical morphology which is used for power quality disturbance detection and location. For Hilbert-Huang transform is recently brought forward, its application still needs further research.

3.5. Atomic Decomposition algorithm

978-1-5090-0064-7/15 $31.00 © 2015 IEEE

Atomic decomposition method is widely used in signal processing applications in recent years, such as radar signals, voice signals, et al, it has a good effect in modeling, feature extraction. The idea of atomic decomposition is on over-complete decomposition, selecting the best matching atom linear representation of the signal, it has high freedom of choice atoms, and can adaptively select the expanded function according to the signal characteristics, thus atomic decomposition greatly increasing flexibility and simplicity. Selected atomic library is the key in the atomic decomposition algorithm, choice of the atom library depending on the disturbance feature [34], but the critical question is the most difficulty problem in atom decomposition algorithm, since characteristic disturbance is more complex, and therefore Select the library and atomic parameters is a difficult problem. Figure 4 below is a flowchart of atomic decomposition algorithm to detect and classify PQ disturbances.

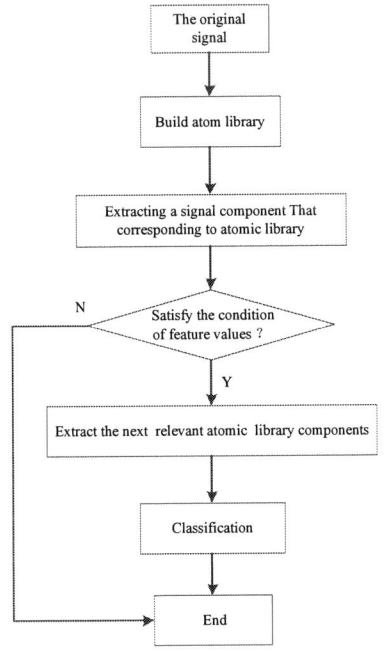

Fig.4: The flowchart of atomic decomposition algorithm

IV. SIMULATIONS

According to the analysis of transient power quality disturbance in second section, transient power quality disturbance signal in micro-grid has the characteristics of rapidness, non-stationary and high complexity. According to the analysis and research on the algorithm in third section, S-transform is accurate and real-time in non-stability, mixed and transient disturbance signal detection. According to transform information of S-transform modular matrix, transient power quality disturbance signal can be detected from the amplitude, frequency and phase. Wavelet transform also played a crucial role in power quality detection in time-frequency domain analysis, but it has certain limitation compared with S-transform. Parts of micro-grid transient disturbance signal are simulated using both the wavelet transform and S-transform detection algorithm. The conclusion is drowning with the simulation data.

The actual disturbance signal contains noise signal, thus the simulation signal model are composed of a standard model and an additive random noise. The simulation software is MATLAB. The fundamental frequency of signal is 50Hz. In addition to transient vibration and high frequency harmonic the frequency of signal is not higher than 3kHz. According to Shannon sampling theorem, the sampling rate is taken to be equal to 3.2kHz. Thus there are 64 sampling points in each cycle of the waveform lasting for 8 cycles. 6 kinds of single disturbance and 2 kinds of composite disturbance are given as an example. The 6 kinds of single disturbance include sag, swell, momentary interruption, flicks, harmonics, transient oscillations, and 2 kinds of composite disturbance include swell with harmonic and sag with harmonic. The start-end time and peak value of the disturbance signal are detected. Wavelet transform and S-transform are adopted as the detection algorithm to detect the three kinds of single disturbances signal and one composite disturbance signal. Thus verify the accuracy of the two algorithms for disturbance detection.

If S-transform is used for detection of power quality disturbance, the disturbance signal occurred time can be detected by the time amplitude envelope, the mean of amplitude square curve and frequency amplitude envelope in S-transform modulus Matrix(STM).

When wavelet transform is used for the detection of power quality disturbance, disturbance signal singularity and detected noise have different wavelet transform modulus maxima in different wavelet scales. First singularity of signal is analyzed at different scales using these characteristics. Then the start-end time of disturbance and mutation is detected using db3 wavelet decomposition. Comparison of the simulation results and theoretical values are shown in Table I. The simulation waveforms containing sags, swells and harmonics are given as an example, shown in Fig.5-Fig.8.

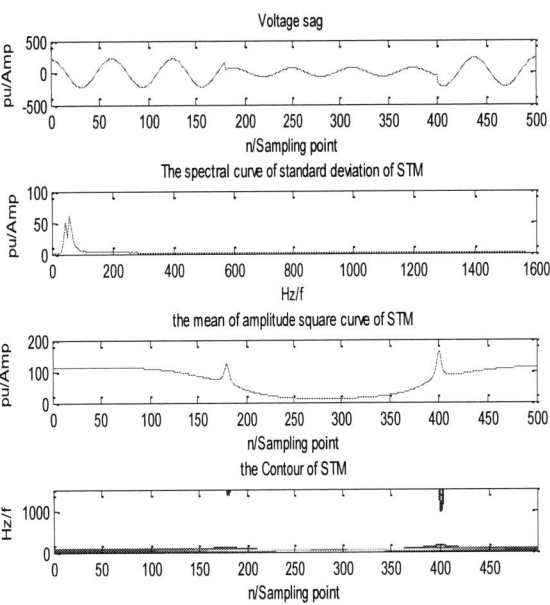

Fig. 5: Voltage Sag detection by ST

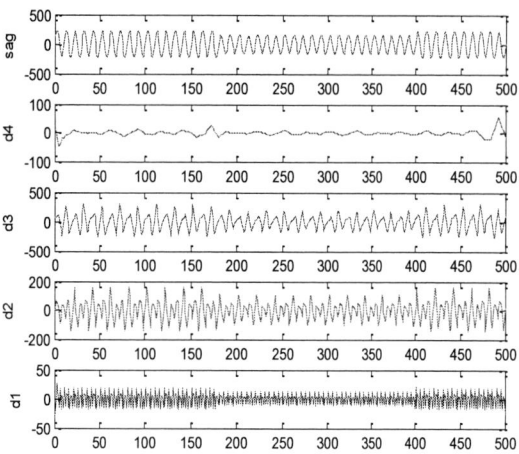

Fig. 6: Voltage sag detection by WT

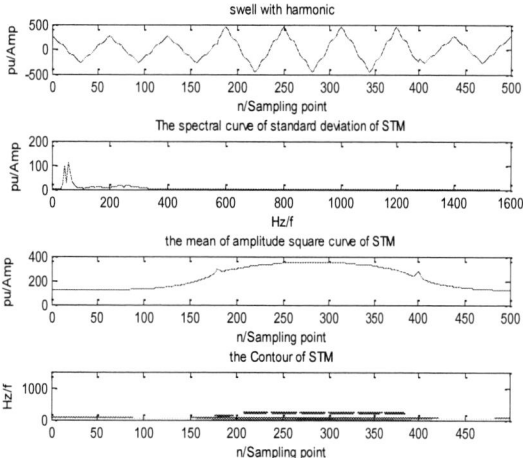

Fig.7: Swell with harmonic detection by ST

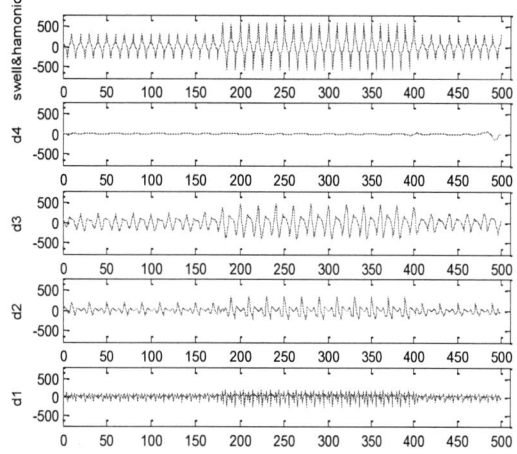

Fig.8: Swell with harmonic detection by WT

Table 1: The comparison of WT and ST detection accuracy for PQ disturbance

	Simulation result/ms(WT)	Simulation result/ms(ST)	Theoretical value(ms)
sag	(56.3,	(58.6,	(57.6,
	129.6)	128.0)	128.0)
Swell	(49.7,	(49.9,	(50.0,
	9.8)	10.2)	10.0)
Transient Oscillation	35.8	35.8	35.2
Swell with harmonics	(58.5,	(64.0,	(57.6,
	127.2)	131.2)	128.0)

Both S-transform and wavelet transform extract different characteristic values of disturbance signals, according to the characteristic value to detect the disturbance time. The results according to Table I and figure 5-figure 8 show that the wavelet transform to detect start-end time of sags and swells has an error interval of 0.2%~4.8%. The error interval of S-transform is 0.2%~2%. It is the same for transient oscillation. For the compound disturbance like swell with harmonic, wavelet transform's detection error interval is 2.5%~11.1% on start-end time, while S-transform's error interval is 0.6%~9.7% on start-end time. Analysis shows that S-transform is better than the wavelet transform in detection of disturbance signal and mutation point location.

V. CONCLUSION

This paper discusses several kinds of detection and identification methods for the transient power quality disturbance. Analysis the similarities and differences of transient power quality disturbances in micro-grid and power system, the result provide an important reference to transient power quality disturbance detection and identification in micro-grid.

Using S-transform and wavelet transform to detect the standard disturbance signals (sag, swell, transient oscillation, swell with harmonics) are simulated on the MATLAB platform. As can be seen from Fig.5-Fig.8 and Table1, S-transform performs well in detection of disturbance signal, anti-noise and accuracy, compared with wavelet transform.

Summarizing the previous research work, after analysis, comparison and simulation, conclusions are drawn in this paper. S-transform has its unique advantages in detecting the disturbance signal, which provides reference for transient power quality disturbance detection and identification problem in micro-grid. But the application of detection using phase element in S-transform modulus matrix (STM) has to be developed.

REFERENCES

[1] Yuan Siyuan, AI Qian, Huang Qidong.the power quality problems in smart grid[J]. Modern Architecture Electric. 2012, 23（11）:30-35.

[2] Yang Xinfa,Su Jian,Lv Zhipeng,et al. Overview on Micro-grid Technology [J].Proceedings of the CSEE.2014,35（1）：57-70.

[3] Lin Xuehai. Basic problems of modern power quality[J]. Power System Technology, 2001, 25(10) :5-13.

[4] Liu Jinxin,Li Peng,Cui Hongfen,et al. Power quality problems and control measures in micro-grid) [J].China Power.2012,45(3):38-43.

[5] Lei Zhang, Linchuan Li,Wei Cui, etal. Transient Disturbance Detection for Micro Grid Power Quality based on Neural Network Adaptive Control [J]. ICNC, 2013,297-301.

[6] M Amin Zamani, Amirnaser Yazdani, Tarlochan S Sidhu. A Control Strategy for Enhanced Operation of Inverter-Based Micro-grids Under Transient Disturbances and Network Faults[J]. IEEE TRANSACTIONS ON POWER DELIVERY, 2012, 27(4):1737-1748.

[7] Peng Li, Liu Jinxin, Li Xiaochun,etal. Detection of Power Quality Disturbances in Micro-Grid Based on Generalized Morphological Filter and Backward Difference[J]. IEEE,2011,1663-1667.

[8] Huang Dongdong,Wu Zaijun,Dou Xiaobo,et al. A power quality composite control strategy based on large-scale grid-connected photovoltaic power generation [J]. Power System Protection and Control,2015,43（3）:107-112.

[9] Li Mohsen Hamzeh, Amin Ghazanfari, Hossein Mokhtari, etal.Integrating Hybrid Power Source Into an Islanded MV Microgrid Using CHB Multilevel Inverter Under Unbalanced and Nonlinear Load Conditions[J].IEEE TRANSACTIONS ON ENERGY CONVERSION,2013,28(3):643-652.

[10] U D Dwivedi, S N Singh.Enhanced Detection of Power-Quality Events Using Intra and Interscale Dependencies of Wavelet Coefficients[J]. IEEE TRANSACTIONS ON POWER DELIVERY, 2010, 25(1):358-367.

[11] Xu Yonghai，Xiao Xiangning, Yang Yihan, Chen Xueyong. Wavelet transform in power quality analysis of electric power systems[J]. Electric Power Systems, 1999.23（23）：55-59.

[12] Quanming Zhang, Huijun Liu, Hongkun Chen,etal. A Precise and Adaptive Algorithm for Inter-harmonics Measurement Based on Iterative DFT[J]. IEEE TRANSACTIONS ON POWER DELIVERY,2008.

[13] Puneet Singla,Tarunraj Singh. Desired Order Continuous Polynomial Time Window Functions for Harmonic Analysis[J]. IEEE TRANSACTIONS ON INSTRUMENTATION AND MEASUREMENT, 2010.

[14] Gary W Chang, Cheng-I Chen. An Accurate Time-Domain Procedure for Harmonics and Inter-harmonics Detection[J]. IEEE TRANSACTIONS ON POWER DELIVERY,2010.

[15] Gao Peisheng, Gu Xiangwen, Wu Weilin. Inter-harmonic/harmonic spectrum estimation based Roots of multiple signal classification and genetic algorithms[J]. TRANSACTIONS OF CHINA ELECTROTENICAL SOCIETY, 2008,23(6):109-114.

[16] Xu Yonghai, Zhao Yan. Identification of Power Quality Disturbance Based on Short-Term Fourier Transform and Disturbance Time Orientation by Singular Value Decomposition[J].PowerSystemTechnology,2011,35(8):174-180.

[17] Elisabetta Lavopa, Pericle Zanchetta, Francesco Cupertino. Real-Time Estimation of Fundamental Frequency and Harmonics for Active Shunt Power Filters in Aircraft Electrical Systems[J]. IEEE TRANSACTIONS ON INDUSTRIAL ELECTRONICS,2009.

[18] Zhao Fengzhan, Yang Rengang. Voltage sag disturbance detection based on short time Fourier transform[J]. Proceedings of the CSEE,2007, 27(10):28-36.

[19] U D D wivedi, S N Singh. Denoising Techniques With Change-Point Approach for Wavelet-Based Power Quality Monitoring[J]. IEEE TRANSACTIONS ON POWER DELIVERY, 2009,24(3):1719-1728.

[20] M.A.S.Masoum, S.Jamali N. Ghaffarzadeh. Detection and classification of power quality disturbances using discrete wavelet transform and wavelet networks[J].IET.Sci.Meas.Technol,2010(4):193–205.

[21] Flavio B. Costa. Boundary Wavelet Coefficients for Real-Time Detection of Transients Induced by Faults and Power-Quality Disturbances [J]. IEEE TRANSACTIONS ON POWER DELIVERY,2014, 29(6):2674-2687.

[22] H Eristi, Y Demir. Automatic classification of power quality events and disturbances using wavelet transform and support vector machines[J]. IET Gener. Transm. Distrib, 2012,10（6）：968-976.

[23] M A S Masoum, S Jamali, N Ghaffarzadeh. Detection and classification of power quality disturbances using discrete wavelet transform and wavelet networks[J]. Published in IET Science，Measurement and Technology, 2010, 4(4):193-205.

[24] S Mishra, C N Bhende, B K Panigrahi. Detection and Classification of Power Quality Disturbances Using S-Transform and Probabilistic Neural Network[J]. IEEE TRANSACTIONS ON POWER DELIVERY,2008, 23(1):280-288.

[25] Norman C. F. Tse , John Y. C. Chan , Wing-Hong Lau,et al. Hybrid Wavelet and Hilbert Transform With Frequency-Shifting Decomposition for Power Quality Analysis[J].IEEE TRANSACTIONS ON INSTRUMENTATION AND MEASUREMENT,2012,61(12): 3225-3233.

[26] Walid G Morsi, Mohamed E EI-Hawary. Time-frequency single-phase power components measurements for harmonics and inter-harmonics distortion based on Wavelet Packet transform；Part I: Mathematical formulation[J]. CAN. J. ELECT. COMPUT. ENG.2010, 35(1):1-7.

[27] P K Dash, B K Panigrahi,and G Panda. Power Quality Analysis Using S–Transform[J]. IEEE TRANSACTIONS ON POWER DELIVERY,2003.

[28] Milan Biswal, P.K.Dash. Detection and characterization of multiple power quality disturbances with a fast S-transform and decision tree based classifier [J]. Digital Signal Processing,2013,23:1071–1083.

[29] Prakash K Ray,Nand Kishor, Soumya R Mohanty. Islanding and Power Quality Disturbance Detection in Grid-Connected Hybrid Power System Using Wavelet and –Transform[J]. IEEE TRANSACTIONS ON SMART GRID,2012,3（3）:1082-1095.

[30] Norman C F,John Y C Chan,Wing-Hong Lau etal. Hybrid Wavelet and Hilbert Transform With Frequency-Shifting Decomposition for Power Quality Analysis[J]. IEEE TRANSACTIONS ON INSTRUMENTATION AND MEASUREMENT ,2012,61 (12):3225-3234.

[31] Huang Yong, Liu Yongqiang, Qi Zhiping. Detection and Location of Power Quality Disturbances Based on Mathematical Morphology and Hilbert-Huang Transform[J]. The Ninth International Conference on Electronic Measurement & Instruments, 2009,2：319-325.

[32] Chen Huali, Sun Yunlian, Cheng Yan. Harmonic and Inter-harmonic Detection of Grid-connected Distributed Generation Based on Modified Mathematical Morphology Filter and Hilbert-Huang Transformation[J]. IPEMC, 2009,1155-1161.

[33] Song Haijun, Huang Chuanjin, Liu Hongchao,et al. A New Power Quality Disturbance Detection Method Based on the Improved LMD [J].Proceedings of the CSEE,2014,34(10):1700-1708.

[34] Wang Ning,Li Linchuan,Jia Qingquan,Dong Haiyan. Classification of Power Quality Disturbance Signals Using Atomic Decomposition Method [J]. Proceedings of the CSEE,2011,31(4):51-57.

978-1-5090-0064-7/15 $31.00 © 2015 IEEE

BIOGRAPHIES

 **Wang Huihui**, She was born in Gansu, China in 1986, obtained her Bachelor's and master's degree both from Tianjin Chengjian University in 2008 and 2010, she is now the lecturer in Tianjin Chengjian University and currently completing a Ph.D in electrical engineering in Tianjin university, research direction is Power electronic conversion technology and power quality analysis.

 **Wang Ping.** She was born in Tianjin, China in 1959. Now she is a professor and doctorial tutor in Tianjin University whose research interests include power electronic transformation technology and its control, intelligent detection and control as well as electronic circuits and systems, etc. Some items of science and technology have been worked on and many papers have been published in some important core journals.

Design and Development of Closed Loop Model of an Adjustable Speed Permanent Magnet Synchronous Motor Drive Using PI Controller

Pabitra Kumar Biswas[1]   Atanu Banerjee[2]   Chiranjit Sain[3]

[1]Electrical & Electronics Engineering Department, NIT Mizoram,796012, India
E-mail: pabitra.biswas2009@gmail.com
[2]Electrical Engineering Department, NIT Meghalaya, 793003, India
E-mail: atanu_banerjee@nitm.ac.in
[3]Electrical Engineering Department, Siliguri Institute of Technology, Siliguri, 734009, India,
E-mail:sain.aec@gmail.com

Abstract-This paper deals with the development of closed loop model of an adjustable speed Permanent Magnet Synchronous Motor (PMSM) drive fed from a voltage source inverter operating under 180° conduction mode, self synchronized with the rotor position information is represented. An automated closed loop adjustable speed PMSM drive can be achieved by employing a speed controller, to which the information about the reference speed and actual speed should be available. The incorporation of Sinusoidal Pulse Width Modulation (SPWM) strategy establishes near sinusoidal armature phase currents and comparatively less torque ripples without sacrificing torque/weight ratio. In this closed loop model of PMSM drive, the information about reference speed is provided to a speed controller, to ensure that actual drive speed tracks the reference speed with ideally zero steady state speed error. Design of the speed controller and current controller will be incorporated in order to achieve the desired performance of the closed loop model of PMSM drive accordingly. Therefore classical PI controller is chosen for such a closed loop adjustable speed PMSM drive for the performance optimization.

Keywords- Closed Loop Model, PMSM drive, PI controller, Voltage Source Inverter (VSI),

I. INTRODUCTION

The Permanent Magnet Synchronous Motor (PMSM) is a rotating electrical machine where the stator is a classic three phase stator like that of an induction motor and the rotor has surface mounted permanent magnets. In this regard, The Permanent Magnet Synchronous Motor is equivalent to an induction motor where the air gap magnetic field is produced by a permanent magnet [1]. Thus with the development of permanent materials and control technology the PMSM is mostly used due to high torque/inertia ratio, high power density, high efficiency, reliability and easy for maintenance in different industrial applications. The schematic block diagram of closed loop model of adjustable speed PMSM drive has been represented in Fig.1. This model basically involves development of model of PMSM i.e. for machines having sinusoidal air gap flux distribution. The PMSM, therefore, has a sinusoidal induced emf and requires sinusoidal currents to produce constant torque. The rotor position information is very crucial for field oriented control. For widespread industrial applications, such as high performance motor drives, accurate motor speed control is required in which regardless of sudden load changes and parameter variations [2-3]. Hence, the control system must

be designed very carefully as it required to ensure the optimum speed operation under the environmental variations, load variations and structural perturbations. In this paper, the model of a complete closed loop adjustable speed Permanent Magnet Synchronous Motor drive is developed using conventional Proportional Integral (PI) controller, where a three phase two-level voltage source inverter (VSI) feeds the PMSM armature and the VSI is switched according to a sinusoidal pulse width modulation (SPWM) strategy.

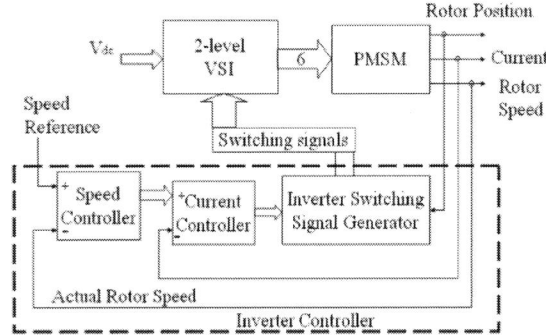

Fig. 1: Schematic block diagram of closed loop model of adjustable speed PMSM drive

II. SYSTEM DESCRIPTION

The complete set-up of the developed model has been represented in Fig 2, The system comprising of four (05) necessary components like SPWM based Voltage Source Inverter (VSI), PMSM motor, position sensor, controller (Speed and current) and the mechanical block. Here an absolute position encoder is mounted on the rotor, which is assumed capable of providing the rotor position information at each instant of time. The inverter is assumed powered from the DC side by a constant DC voltage source, V_{dc} which is not varied. The controller realization starts with the adjustable speed reference, at which the drive is intended to run, irrespective of the load torque variation within a feasible range [4-5]. The information of this reference speed is provided to a speed controller, which is a PI controller in this model in order to track the reference speed [6]. The speed controller output forms the torque or current reference, which is fed to next controller which is the current controller. The current controller output is fed to the block responsible for the generation of the switching signals of the six power

978-1-5090-0064-7/15 $31.00 © 2015 IEEE

electronic devices of the VSI. The speed control arrangement consists of two loop control system with an outer speed loop and inner current loop. The current controller in this case is also a PI controller. It starts with an adjustable speed reference, with which the actual machine speed, obtained by integrating the rotor position information provided by the absolute position encoder is compared [2]. The Permanent Magnet Synchronous Machine is analyzed on the basis of "D-Q axes rotor reference frame theory" and the control on the machine is

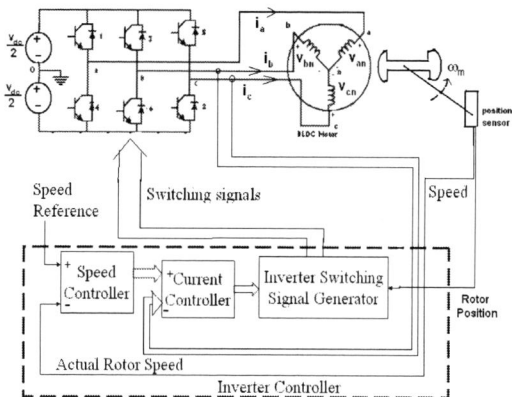

Fig 2: The scheme of the closed loop adjustable-speed PMSM drive with the three phase voltage source inverter.

exercised by controlling the q-axis component of the armature current.

II. DESIGN AND MATHEMATICAL ANALYSIS OF CLOSED LOOP ADJUSTABLE SPEED PMSM DRIVE

The design of the speed controller is important from the point of view of imparting desired transient and steady state characteristics to the speed-controlled PMSM drive system. Selection of gain and time constants of such a controller by using the symmetric-optimum principle is straightforward if the d axis stator current is assumed to be zero [3]. As the closed loop system yields a two loop control structure i.e. outer loop is a speed controller loop and inner loop is a current controller loop. The reset time and gain of the outer speed PI controller should be critically found out to ensure that the dynamic response of the drive is satisfactory, zero steady state speed error is ensured and the system remains stable. For the closed loop model, block diagram of complete drive system is represented in Fig. 3. The transfer functions of all the blocks are determined; afterwards the gain and time constants are calculated in order to design the proposed closed loop model of an adjustable speed PMSM drive [4]. The speed PI controller output is treated as the q axis current reference i_{qs}, which is responsible for generation of electromagnetic torque, for $v_{ds} = 0$, as the approximate analysis of the PMSM suggests. This current reference is compared with actual 'q'-component of armature current (i_{qs}) and the current error is fed to the proportional integral controller with unity gain. The inner current loop is provided for controlled yet fast dynamic of current with least sudden overshoot [6].

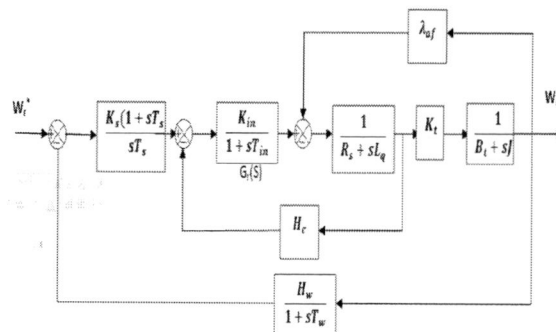

Fig. 3: Block diagram of the speed-controlled PMSM drive

The motor q axis voltage equation with d-axis current being zero becomes [1]

$$v_{qs} = (R_s + L_q p) i_{qs} + \omega_r \lambda_{af} \quad (1)$$

And the electromechanical equation is

$$\frac{P}{2}(T_e - T_l) = Jp\,\omega_r + B_1\omega_r \quad (2)$$

Where the electromagnetic torque is given by

$$T_e = \frac{3}{2} \cdot \frac{P}{2} \lambda_{af} i_{qs} \quad (3)$$

If the load is assumed to be frictional, then

$$T_l = B_l \omega_m \quad (4)$$

This upon substitution gives the electromechanical equation as[1]

$$(J_p + B_t)\omega_r = \left\{ \frac{3}{2}\left(\frac{P}{2}\right)^2 \cdot \lambda_{af} \right\} i_{qs} = K_t \cdot i_{qs} \quad (5)$$

where,

$$B_t = \frac{P}{2} B_l + B_1 \quad (6)$$

$$K_t = \frac{3}{2}\left(\frac{P}{2}\right)^2 \cdot \lambda_{af} \quad (7)$$

The inverter is modelled as gain with a time lag by

$$G_r(s) = \frac{K_{in}}{1 + s\,T_{in}} \quad (8)$$

where,

$$K_{in} = 0.65\,\frac{V_{dc}}{V_{cm}} \qquad T_{in} = \frac{1}{2f_c}$$

Where V_{dc} is the dc-link voltage to the inverter, V_{cm} is the maximum control voltage and f_c is the switching (carrier) frequency of the inverter.

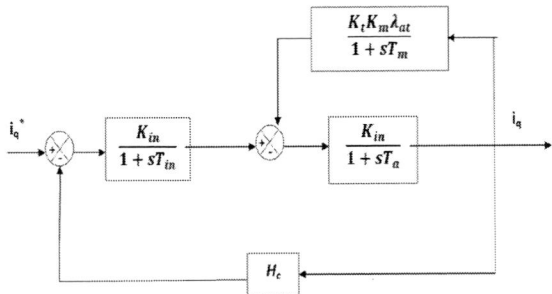

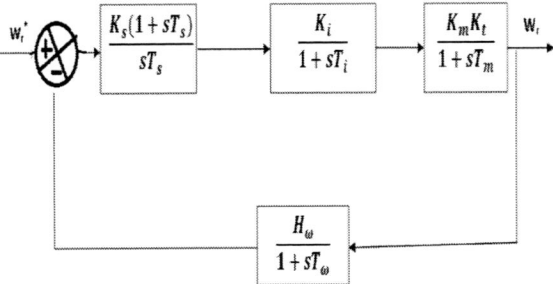

Fig.4: Current control loop

The induced emf loop crosses the q-axis current loop and it could be simplified by moving the pick –off point for the induced emf loop from speed to current output point [1].This gives the current-loop transfer function from fig.4

$$G_i(s) = \frac{K_{in}K_a(1+sT_m)}{H_cK_{in}K_a(1+sT_m)+(1+sT_{in})\{K_bK_a+(1+sT_a)(1+sT_m)\}} \quad (9)$$

where,

$$K_a = \frac{1}{R_s}, T_a = \frac{L_q}{R_s}, K_m = \frac{1}{B_t}, T_m = \frac{J}{B_t}, K_b = K_t K_m \lambda_{af}$$

The following approximations are valid near the vicinity of crossover frequency

$$1 + sT_r \cong 1$$
$$1 + sT_m \cong sT_m$$
$$(1+sT_a)(1+sT_{in}) \cong 1 + s(T_a + T_{in}) \cong 1 + sT_{ar}$$

where $T_{ar} = T_a + T_{in}$

With this the current loop transfer function is approximated as[1]

$$G_i(s) \cong \frac{(K_{in}K_aT_m)s}{K_bK_a+(T_m+H_cK_{in}K_aT_m)s+(T_{ar}T_m)s^2} \quad (10)$$

$$G_i(s) \cong (\frac{T_mK_{in}}{K_b})\frac{s}{(1+sT_1)(1+sT_2)} \quad (11)$$

It is found that $T_1 < T_2 < T_m$ hence on further approximation $(1+sT_2) \cong sT_2$ .the approximate current loop transfer function is then given by[1]

$$G_i(s) \cong \frac{K_i}{(1+sT_i)} \quad (12)$$

where $K_i = (\frac{T_mK_{in}}{T_2K_b})$,

This simplified current loop transfer function is substituted in the design of the speed controller as follows.

Fig.5: Simplified speed control loop

The speed loop with the simplified current loop is shown in fig.5. Near the vicinity of cross over frequency, the following approximations are valid [1]

$$(1+sT_m) \cong sT_m$$
$$(1+sT_i)(1+sT_\omega) \cong 1+sT_{\omega i}$$

$$(1+sT_\omega) \cong 1$$

The speed loop transfer function with the approximate is given by[1]

$$G_s(s) \cong \frac{K_iK_mK_tH_wK_s(1+sT_s)}{T_mT_ss^2(1+sT_{wi})} \quad (13)$$

IV. DISCUSSION AND PERFORMANCE ANALYSIS OF CLOSED LOOP PMSM DRIVE USING PI CONTROLLER

The entire model of PMSM closed loop drive is divided into two loops, inner loop current and outer loop speed. In the first case, current control loop is determined using mathematical block diagram representation, using PI controller. By getting the simplified and exact current loop transfer functions of current controller, it is added to the speed control loop i.e. outer loop using PI controller. Simplified current loop transfer function is coupled with simplified speed loop and exact current control loop is coupled with exact speed control loop. Hence a typical performance study is introduced to familiar with the different performance indices of the closed loop system corresponding to time domain and frequency domain specifications. The time domain specifications of overall closed loop system like peak time, rise time, settling time, steady state value etc are determined analytically which in turn helps to familiar with the performance of PI controlled PMSM drive accordingly. The frequency domain specifications of the system like gain margin, phase margin, gain and phase cross over frequencies are also executed by bode- diagram representation respectively. The stability analysis of overall speed and current loop representation of closed loop adjustable speed PMSM drive can be determined by nyquist diagram representation. Therefore different performance indices can be taken for the determination of dynamic as well as steady state response of closed loop PMSM drive [7-9].
PI control is a name commonly given to two-term control; P stands for Proportional term and I for Integral term. P-I controller is mainly used to eliminate the steady state error resulting from P controller. The characteristics of PI control actions are (i) steady state accuracy improves (ii)

978-1-5090-0064-7/15 $31.00 © 2015 IEEE

rise time increases (iii) bandwidth decreases (iv) overshoot reduces [10-12].

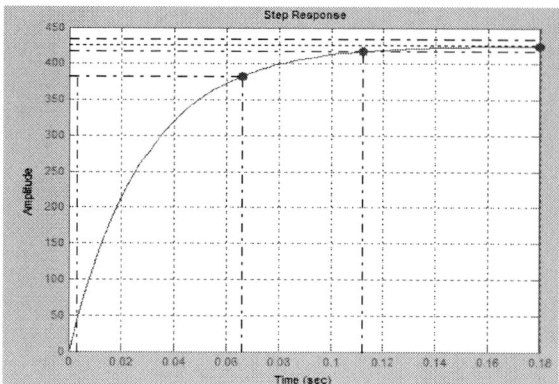

Fig. 6: Step response representation of simplified current loop

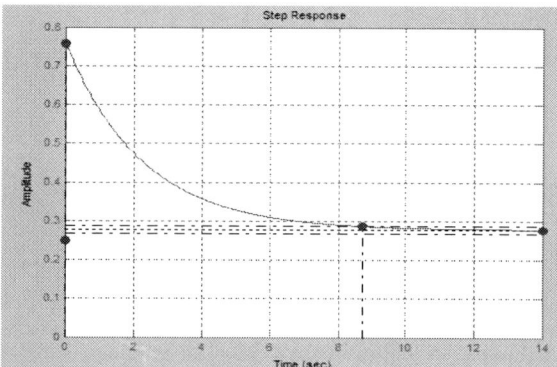

Fig. 7: Step response representation of exact current loop

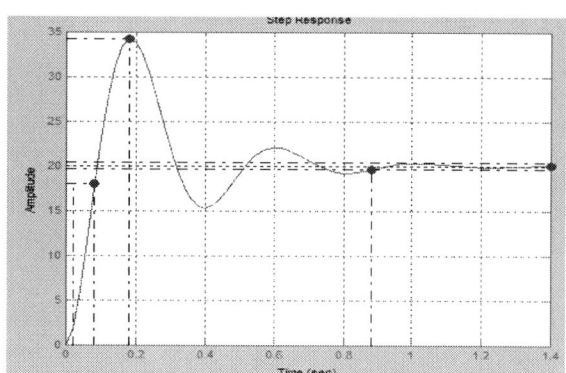

Fig. 8: Step response representation of simplified speed control loop with PI controller

Table 1: Time Domain Specifications

Type of Controller	Peak Amplitude	Rise Time	Settling Time	Peak overshoot (%)	Steady state value (%)
PI	22	4.54 sec	14.5 sec	10.2	20

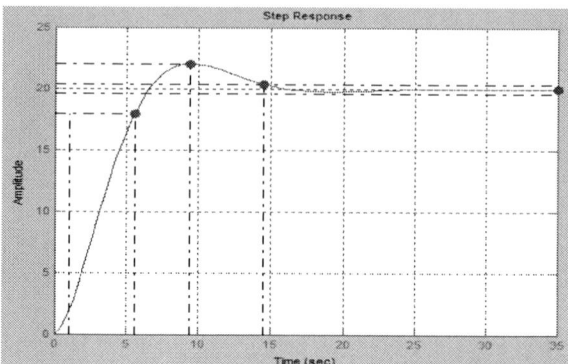

Fig. 9: Step response characteristics of overall speed and current loop using PI controller

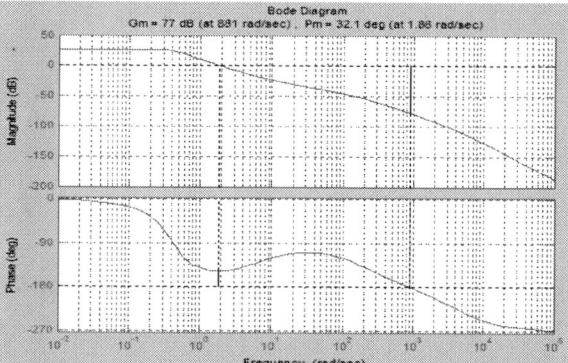

Fig.10: Bode diagram representation of overall speed and current loop using PI controller

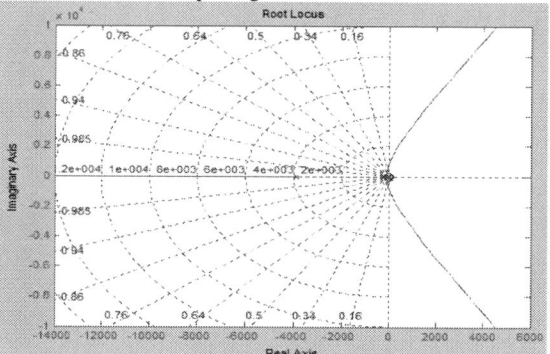

Fig. 11: Root locus representation of overall speed and current loop using PI controller

V. CONCLUSION

This paper significantly describes the performance of overall closed loop model of an adjustable speed PMSM drive using PI controller. The different time domain specifications of the system like settling time, rise time, peak overshoot, steady state value and frequency domain specifications like gain margin, phase margin etc. are determined analytically. Therefore this model can predict dynamic as well as steady state behaviour of the system respectively. Hence a stable performance with improved overshoot can be gained by designing a PI controlled PMSM drive. By taking the two loops control structure consisting of outer loop speed and inner loop current, the overall structure can be implemented using different classical controllers and soft computing controllers in future. Thus a comparative study will be carried out by taking different estimations of the controller gains for performance optimization.

APPENDIX

The parameters of the Permanent Magnet Synchronous Machine (PMSM), on which the studies are made in this paper, are: Number of pole P=4, armature resistance R_s= 3.2 ohm, combined moment of inertia J= 0.061 kg-m^2, damping co-efficient F_n=0.45 Nm-s/radian, d-axis inductance L_d=0.053H, q-axis inductance Lq= 0.041H, proportional gain (K_p) =0.048, integral gain (K_i) = 0.28

REFERENCES

[1] R. Krishnan," Electric Motor Drives- Modelling, Analysis and Control", PHI, New Jersey, 2001

[2] Bimal K. Bose, "Modern Power Electronics and AC Drives", Prentice Hall, 2001

[3] Katsuhiko Ogata,"Modern Control Engineering", PHI, Third Edition, 1998

[4] M. Gopal," Control Systems, Principles and Design", Tata McGraw Hill, Fourth Edition, 2012

[5] A.V.Santand K.R.Rajagopal,"PM synchronous motor speed control using hybrid fuzzy PI with novel switching functions," IEEE Trans. Mag., vol. 45, no. 10, pp. 4672–4675, Oct. 2009

[6] A. V. Sant and K. R. Rajagopal, "Novel hybrid fuzzy-PI controllers for the speed control of permanent magnet synchronous motors," presented at the IEEE Int. Magn.Conf., Sacramento ,CA,May4–8,2009

[7] H. M. Kamel, H. M. Hasanien , and H. E. Ibrahim, " Speed control of permanent magnet synchronous motor using fuzzy logic controller", Proc. IEEE Int. Conf.: Electric Machines and Drives, Miami, Florida, U.S.A., pp.1587-1591, 3-6 May 2009.

[8] P. Pillay, and R. Krishnan, "Modelling of permanent magnet motor drives", IEEE Trans. Industrial Electronics, vol. 35, no. 4, pp. 537 - 541, Nov. 1988

[9] Faa-Jeng Lin, Chih-Hong Lin, " A Permanent-Magnet Synchronous Motor Servo Drive Using Self-Constructing Fuzzy Neural Network Controller", IEEE Transactions on Energy Conversion, 2004, pp.66-72

[10] Kyeong-Hwa Kim, Myung-Joong Youn. A simple and robust digital current control technique of a PM synchronous motor using time delay control approach. IEEE Transactions on power electronics, vol. 16, pp.72-82, 2001

[11] T. M. Jahns and V. Blasko, "Recent advances in power electronics technology for industrial and traction machine drives," Proc. IEEE, vol. 89, pp. 963–975, June 2001

[12] Y. C. Hsu, et al., "A fuzzy adaptive variable structure controller with applications to robot manipulators," Systems, Man, and Cybernetics, Part B: Cybernetics, IEEE Transactions on, vol. 31, pp. 331-340, 2002

[13] Gupta .N.P, Gupta .P, " Performance Analysis of Direct Torque control of PMSM Drive Using SVPWM -Inverter" Power Electronics(IICPE), 2012 IEEE India International Conference on, pp. 1-6, 6-8 Dec.2012

[14] K.Ang, G.Chong, and Y.Li, "PID control system analysis, design and Technology," IEEE Trans. control system Technology, Vol.13, PP. 559- 576, July 2005

[15] Al-Nabi E, Wu B, Zargari NR, Sood V. "Input power factor compensation for high-power CSC fed PMSM drive using d-axis stator current control." IEEE Transactions on Industrial Electronics, pp. 59(2):752–761, 2012

[16] Marufuzzaman M, Reaz MBI, Ali MAM, Rahman LF. "Hardware approach of two way conversion of floating point to fixed point for current dq PI controller of FOC PMSM drive" Electronics and Electrical Engineering.;7(123): pp.79–82, 2012

BIOGRAPHIES

Pabitra Kumar Biswas was born in West Bengal, India in 1980. He completed his B.Tech from Asansol Engg. College, WBUT, India. He received his ME. Degree from Bengal Engineering and Science University, West Bengal, India and PhD. Degree in Electrical Engineering from National Institute of Technology, Durgapur, India. He is presently working as an Assistant Professor in Electrical Engineering in National Institute of Technology, Mizoram, India. He has published a numbers of research papers in National/International Conference Records/Journals. From 16.07.2007 to 05.02.2015, he served as an Assistant Professor in Electrical Engineering in Asansol Engineeng College, Asansol, India. His research interests include Electromagnetic Levitation System, Active Magnetic Bearing and Power electronics.

Atanu Banerjee was born in Asansol, West Bengal, India. He received his B.E. degree in Power Electronics Engg. from the Nagpur University in the year of 2001 and M.E in Electrical Engg. Department with specialization in Power Electronics & Drives in 2008 from Bengal Engineering & Science University, Shibpur (Now IIEST, Shibpur). He has completed his Ph.D in Electrical Engg from the Indian School of Mines, Dhanbad, India in 2013. He worked in industries for almost three years & has academic experience of more than 12 years. Presently he is in National Institute of Technology, Meghalaya as an assistant professor in the Electrical Engg. Department. His research interests include induction heating and high frequency switching in power electronic converters, adjustable speed drives. He has published few books & several journal/conference research papers. Also Dr. Banerjee has filed two patents to the Govt. of India. Currently he is guiding few research scholars for M-Tech & Ph.D

Chiranjit Sain received B.Tech in Electrical Engineering from Asansol Engineering College, Asansol,India and also received M.Tech from National Institute of Technical Teachers Training and Research,(NITTTR) Kolkata, India in the specialization of Mechatronics Engineering under Electrical Engineering department. Presently he is working as an Assistant Professor in the Electrical Engineering Department at Siliguri Institute of Technology, Siliguri India since 2010 and also pursuing Ph.D in the Electrical Engineering Department from National Institute of Technology, Meghalaya, India. His present research of interests includes Electrical Machine-Drives Systems, Power Electronics Converters, Soft-computing analysis etc.

Estimation of Vehicle States and Road Friction Based on DEKF

Ying Xu[1,2]　　Banglin Deng[1]*　　Gang Xu[1]

[1] Department of Mechatronics and Control Engineering, Shenzhen University, PRChina
[2] Department of Mechanical and Automotive Engineering, South China University of Technology, PRChina
*Email: dengbl@szu.edu.cn
Email: yxu@szu.edu.cn

Abstract–On vehicle active safety control systems, various types of state parameters of vehicle and road are needed to be estimated. This paper adopts information fusion technology, using of dual extended Kalman filter (DEKF) theory for rapid simulation and estimation of these parameters. Using different vehicles model and tire model, DEKF recursive estimation models are established and verified. Experimental results show that DEKF is designed by vehicle dynamic model based on three degrees of freedom(3-DOF), using the Highway Safety Research Institute tire model, not only accurately estimates the vehicle state parameters, but also estimates the road tire friction coefficient in real-time. In the DEKF, two recursive state and parameter estimation model exist in parallel, while they are dependent on each other, and have real-time interaction correction to forecast information, which quickly converges towards estimated true value for simulation. The accurate estimation of DEKF theory in the vehicle state and road information, make it possible that some of the parameters can be estimated, which is proved to be difficult to obtain, and it also provides the necessary conditions for the vehicle active control system. In the meantime, the validity and feasibility of this algorithm have been verified by CarSim and HIL driving simulator, and offers the possibility of application in real car in future.

Keywords–dual extended Kalman filter (DEKF), Tire Model, Vehicle Dynamic Models, Vehicle state.

I. INTRODUCTION

Nowadays, driving safety is particularly important. Therefore, vehicle active safety control system has been developed rapidly. The current active control systems, such as traction control system (TCS), Anti-lock braking system (ABS), and Electronic Stability Program (ESP), have been used in vehicle driving, which greatly improved the safety, performance and efficiency of road vehicles. Vehicle status (such as velocity and yaw rate) and the road surface friction coefficient which obtained difficultly and costly, it has great significance in the design of vehicle chassis electronic control systems [1]. In practice, vehicle longitudinal velocity, side slip angle and other statue signals should be collected in real time, but need to install expensive equipment. Meanwhile at the braking force control process, under the different road condition, the simulation estimation methods usually can't get the real-time value of road adhesion coefficient. So the entire control process will be unstable.

In order to reduce the costs of measurement devices in vehicle active safety control, various estimation methods are applied in estimating vehicle states and the road surface friction coefficient. Romberg is the basis of classical control theory to estimate, by appropriately adjusting parameters to match the goal of estimating the parameters of the vehicle [2]; Robust estimation used fault-tolerant first to deal with the data acquired, and then to the minimize interference of control system to realize the estimation [3]; Sliding mode estimation based on variable structure control to estimate, by observations and residuals of measurement estimates to calculate (easy to get side slip angle) [4]; Liapunov stimation, its constructed function built by Liapunov stability, using known vehicle state parameters, can deduced feedback gain of algorithm fastly to achieve the value of vehicle status [5]; Kalman filter is to build the model by noise and signal in status spatial, by observation and the latest state variables by previous estimating, then the current estimation value can determined. In the above estimation methods, Kalman filtering technique is most widely used in the estimation of the vehicle states and road friction coefficient [6-8], but only to estimate for certain types of vehicle parameters. Now, there is no model that can estimate both the vehicle states and road friction coefficient simultaneously. Classical Kalman filtering is only applicable to linear system, but vehicle process is non-linear system, so most estimations use Extended Kalman Filter (EKF). EKF estimation method applies only to the acquisition of the tire force, but lack of complex model on tire characteristics [9]. Therefore, in the vehicle active safety control process, real-time, effectiveness and accuracy ought to be emphasized within the estimation of vehicle states and road friction coefficient.

This paper uses Dual Extended Kalman Filter (DEKF) which is a crossing multi-model algorithm to estimate vehicle states and road friction coefficient accurately. DEKF estimator is designed based on vehicle dynamic model and HSRI (Highway Safety Research Institute) tire model; both the two models are built with EKF model, and are described in sections 2 and 3. The detailed design DEFK in section 4. In sections 5, joint simulation is designed which is composed of Carsim (vehicle engineering software) and matlab/simulink. Finally, in sections 7, conclusions are addressed.

II. TIRE MODEL

As complex tire model is for nonlinear system, this paper used HSRI tire model which presented by Michigan Highway Safety Research Institute, can adapt unsteady vehicle traveling simulation. The model is able to reflect the characteristics of the tire in the nonlinear region, and its structure is simple [10]. It is meaningful to the implementation of simulation estimates. Force coordinates of tire shown in Fig.1.

$$F_x = -c_s \frac{s}{1-s} f(L) \qquad (1)$$

$$F_y = -c_\alpha \frac{s}{1-s} f(L) \qquad (2)$$

where, F_x is the longitudinal tire force; F_y is the lateral tire force; μ is the Friction coefficient of tire; c_s is the longitudinal stiffness coefficient of tire; c_α is the cornering stiffness coefficient of tire; s is the slip rate of tire.

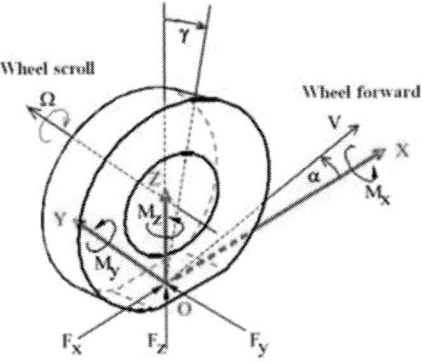

Fig.1: HSRI tire model coordinates

$$f(L) = \begin{cases} 1 & L \geq 1 \\ L(2-L) & L \prec 1 \end{cases} \qquad (3)$$

$$L = \frac{1}{2} \mu F_z (1-s)[(c_s s)^2 + (c_\alpha \tan^2 \alpha)^2]^{-\frac{1}{2}} \qquad (4)$$

$$\mu = \mu_0 (1 - AV) \qquad (5)$$

where, μ_0 is the Limiting friction coefficient of tire; A is the factor of vehicle velocity; F_z is the vertical tire force; V is the wheel velocity contacting on ground; α is the side slip angle.

Variables involved in the HSRI tire model are expressed as follows:

$$F_{zfl,fr} = (\frac{1}{2}mg \pm ma_y \frac{h}{t_f})\frac{b}{l} - \frac{1}{2}ma_x\frac{h}{l} \qquad (6)$$

$$F_{zrl,rr} = (\frac{1}{2}mg \pm ma_y \frac{h}{t_r})\frac{a}{l} + \frac{1}{2}ma_x\frac{h}{l} \qquad (7)$$

$$\alpha_{fl} = \delta_{fl} - \arctan\left(\frac{v_y + ar}{v_x + \frac{t_f}{2}r}\right) \qquad (8)$$

$$\alpha_{fr} = \delta_{fr} - \arctan\left(\frac{v_y + ar}{v_x - \frac{t_f}{2}r}\right) \qquad (9)$$

$$\alpha_{rl} = \delta_{rl} - \arctan\left(\frac{-v_y + br}{v_x + \frac{t_r}{2}r}\right) \qquad (10)$$

$$\alpha_{rr} = \delta_{rr} - \arctan\left(\frac{-v_y + br}{v_x - \frac{t_r}{2}r}\right) \qquad (11)$$

$$s_{ij} = 1 - \frac{w_{ij}r}{v_{ij}} \qquad (12)$$

$$v_{fl} = \sqrt{(v_x - \frac{t_f}{2}r)^2 + (v_y + ar)^2} \qquad (13)$$

$$v_{fr} = \sqrt{(v_x + \frac{t_f}{2}r)^2 + (v_y + ar)^2} \qquad (14)$$

$$v_{rl} = \sqrt{(v_x - \frac{t_r}{2}r)^2 + (v_y - br)^2} \qquad (15)$$

$$v_{rr} = \sqrt{(v_x + \frac{t_r}{2}r)^2 + (v_y - br)^2} \qquad (16)$$

$$V = \sqrt{V_x^2 + V_y^2} \qquad (17)$$

where, t_f is the front wheel track; t_r is the rare wheel track; δ_{ij} is the angle of wheel; w_{ij} is the rolling rate of wheel; s_{ij} is the slip ratio of wheel; α_{ij} is the side slip angle of wheel; v_{ij} is the velocity of wheel; F_{zij} is the vertical tire force of wheel. The subscript i denotes the front tire (f) or the rear one (r), and j denotes the left tire (l) or the right one (r).

III. VEHICLE MODEL

In order to study, estimator is built by three degrees of freedom-wheel vehicle dynamics model，which with the movements of longitudinal, lateral and yaw. The vehicle dynamics model is shown in Fig.2.

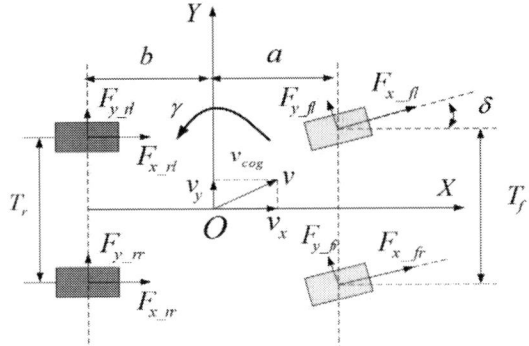

Fig.2: Three degrees of freedom vehicle model

The equations with respect to the longitudinal, lateral, and yaw dynamics are expressed as Eqs. (18) − (20), respectively:

$$v_x = a_x + v_y r \qquad (18)$$

$$v_y = a_y - v_x r \qquad (19)$$

$$r = \frac{1}{I_z}\Gamma \qquad (20)$$

$$\Gamma = a(F_{x\_fl}\sin\delta_{fl} + F_{y\_fl}\cos\delta_{fl}) - \frac{t_f}{2}(F_{x\_fl}\cos\delta_{fl} - F_{y\_fl}\sin\delta_{fl}) +$$

$$a(F_{x\_fr}\sin\delta_{fr} + F_{y\_fr}\cos\delta_{fr}) + \frac{t_f}{2}(F_{x\_fr}\cos\delta_{fr} - F_{y\_fr}\sin\delta_{fr}) \quad (21)$$

$$-b(F_{x\_rl}\cos\delta_{rl} + F_{y\_rl}\sin\delta_{rl}) - \frac{t_r}{2}(F_{x\_rl}\cos\delta_{rl} - F_{y\_rl}\sin\delta_{rl})$$

$$-b(F_{x\_rr}\cos\delta_{rr} + F_{y\_rr}\sin\delta_{rr}) + \frac{t_r}{2}(F_{x\_rr}\cos\delta_{rr} - F_{y\_rr}\sin\delta_{rr})$$

The expressions expressed the functional relationship among the longitudinal velocity, lateral velocity and yaw rate. Through these three variables, you can also get the expressions of a_x (longitudinal acceleration), a_y (lateral acceleration) and β (side slip angle of the vehicle). The expressions are shown as follows.

$$a_x = \frac{1}{m}(F_{x\_fl}\cos\delta_{fl} - F_{y\_fl}\sin\delta_{fl} + F_{x\_fr}\cos\delta_{fr}$$
$$- F_{y\_fr}\sin\delta_{fr} + F_{x\_rl}\cos\delta_{rl} - F_{y\_rl}\sin\delta_{rl} \quad (22)$$
$$+ F_{x\_rr}\cos\delta_{rr} + F_{y\_rr}\sin\delta_{rr})$$

$$a_y = \frac{1}{m}(F_{x\_fl}\sin\delta_{fl} + F_{y\_fl}\cos\delta_{fl} + F_{x\_fr}\sin\delta_{fr}$$
$$+ F_{y\_fr}\cos\delta_{fr} + F_{x\_rl}\cos\delta_{rl} + F_{y\_rl}\sin\delta_{rl} \quad (23)$$
$$+ F_{x\_rr}\cos\delta_{rr} + F_{y\_rr}\sin\delta_{rr})$$

$$\beta = \arctan\frac{v_y}{v_x} \qquad (24)$$

In section 2, it has been described in detail of HSRI nonlinear tire model. The 3-DOF vehicle model is designed by HSRI tire model. By using expressions 1 and 2, F_x and F_y can be expressed.

IV. ESTIMATION DESIGN OF DEKF

In the DEKF Algorithm, two recursive state and parameter estimator (vehicle states estimator and road friction coefficient estimator) exists in parallel [11], while they are dependent on each other, and have real-time interaction correction to forecast information, which quickly converges towards estimated true value for simulation, as shown in Fig. 3.

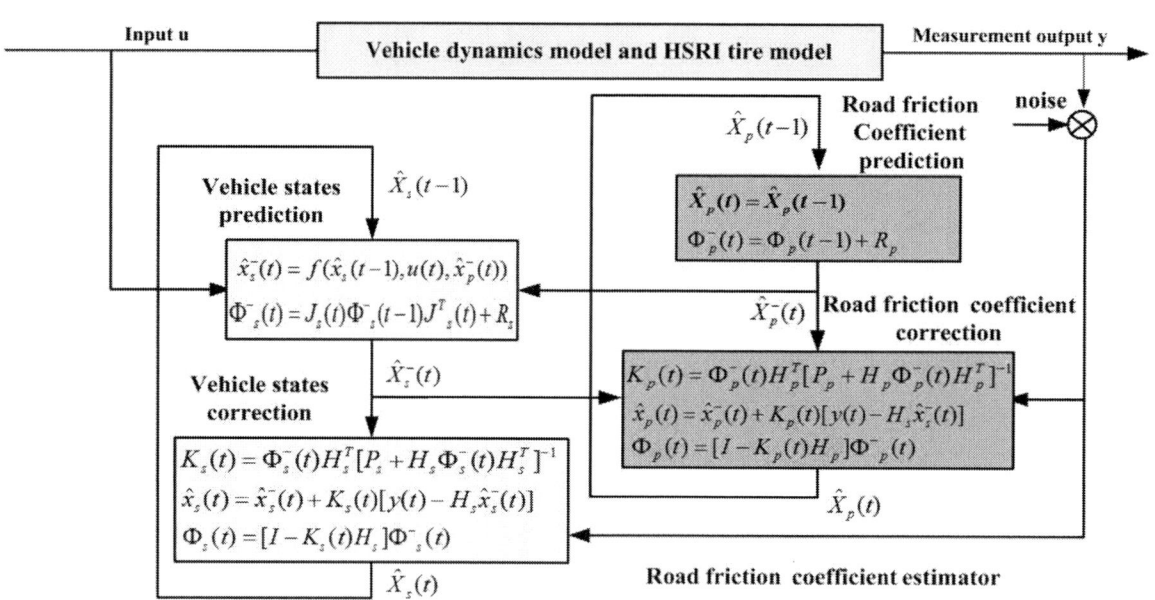

Fig.3: Dual Extended Kalman Filter Algorithm

Eqs. (1) – (24) are used to define the system equations of the filters $f(\cdot)$ and $h(\cdot)$ as follows:

$$\dot{x}_s(t) = f(x_s(t), x_p(t), u(t), w(t)) \qquad (25)$$

$$y(t) = h(x_s(t), x_p(t), v(t)) \qquad (26)$$

where, the road friction coefficient parameters vector $x_p(t) = \begin{bmatrix} \mu_{fl}, & \mu_{fr}, & \mu_{rl}, & \mu_{rr} \end{bmatrix}^T$; the vehicle states vector $\dot{x}_s(t) = [v_x, v_y, r, \Gamma, a_x, a_y, \alpha_{ij}, s_{ij}, F_{z\_ij}]^T$;

the input vector $u(t) = [\delta_{ij}, w_{fl}, w_{fr}, w_{rl}, w_{rr}]$, $w(t)$ is the progress noise matrix; the output of vehicle states estimator $y_1(t) = [a_y, r]$, the output of road friction coefficient estimator $y_2(t) = [a_x, a_y, \dot{r}]$.

Since the whole simulation process is nonlinear, the equations of vehicle states and road friction coefficient estimator are expanded on the preferred point by Taylor.

978-1-5090-0064-7/15 $31.00 © 2015 IEEE

The equations leave the first-order function only, so as to achieve the linearity of the model. Therefore, $F_s(t)$, $H_s(t)$, $H_p(t)$ are Jacobian matrices of $f(\cdot)$, $h(\cdot)$.

$$F_s(t) = \begin{bmatrix} \dfrac{\partial f_1}{\partial x_{s1}} & \cdots\cdots & \dfrac{\partial f_1}{\partial x_{sm}} \\ \cdots\cdots\cdots\cdots \\ \dfrac{\partial f_m}{\partial x_{s1}} & \cdots\cdots & \dfrac{\partial f_m}{\partial x_{sm}} \end{bmatrix} \quad (27)$$

$$= \begin{bmatrix} 0 & r & v_y & 0 & 1 & 0 & 0 & \cdots & 0 \\ -r & 0 & -v_x & 0 & 0 & 1 & 0 & \cdots & 0 \\ 0 & 0 & 0 & 1/I_z & 0 & 0 & 0 & \cdots & 0 \\ 0 & & & \cdots & & & & & 0 \\ \cdots & & & & & & & & \cdots \\ 0 & & & \cdots & & & & & 0 \end{bmatrix}$$

$$H_s(t) = \begin{bmatrix} \dfrac{\partial h_1}{\partial x_{s1}} & \cdots\cdots & \dfrac{\partial h_1}{\partial x_{sm}} \\ \cdots\cdots\cdots\cdots \\ \dfrac{\partial h_n}{\partial x_{s1}} & \cdots\cdots & \dfrac{\partial h_n}{\partial x_{sm}} \end{bmatrix} \quad (28)$$

$$= \begin{bmatrix} 0 & 0 & 0 & 0 & 0 & 1 & 0 & 0 & 0 & \cdots & 0 \\ 0 & 0 & 1 & 0 & 0 & 0 & 0 & 0 & 0 & \cdots & 0 \end{bmatrix}$$

$$H_p(t) = \begin{bmatrix} \dfrac{\partial h_1}{\partial x_{p1}} & \cdots\cdots & \dfrac{\partial h_1}{\partial x_{pn}} \\ \cdots\cdots\cdots\cdots \\ \dfrac{\partial h_m}{\partial x_{p1}} & \cdots\cdots & \dfrac{\partial h_m}{\partial x_{pn}} \end{bmatrix} = \begin{bmatrix} \dfrac{\partial h_1}{\partial \mu_{fl}} & \cdots\cdots & \dfrac{\partial h_1}{\partial \mu_{rr}} \\ \cdots\cdots\cdots\cdots \\ \dfrac{\partial h_m}{\partial \mu_{fl}} & \cdots\cdots & \dfrac{\partial h_m}{\partial \mu_{rr}} \end{bmatrix} \quad (29)$$

$$J_s(t) = e^{F_s(t)*\Delta t} \approx I + F_s(t)*\Delta t \quad (30)$$

where, J_s is the state transition matrix, and Δt is the sampling time, $\Delta t = 0.001s$.

As it shown in Fig.3, P and R are the covariance matrices of $v(t)$ and $w(t)$. $v(t)$ is gauss white Noise Measurement; and $w(t)$ is gauss white noise excitation; P_s and P_p are measurement noise covariance matrices of vehicle states and road friction coefficient estimators; R_s is progress noise covariance matrix of vehicle states estimator; R_p is progress noise covariance matrix of road friction coefficient estimator; Φ_s and Φ_p are estimation error covariance matrices of vehicle states and road friction coefficient estimators. In the real system, the matrices of measurement noise covariance and progress noise covariance will be changed by each iteration calculation. As we known, the parameters in those covariance matrices will influence the performance and convergence of estimator [12]. Generally, these parameters determination needs a lot of offline experimental data. Through previous experiments and analysis, the values of

these matrices are shown as follow: $R_s = 10000 \times I_{18\times18}$, $R_p = 0.0001 \times I_{4\times4}$, $P_s = I_{2\times2}$, $P_p = I_{3\times3}$.

V. APPLICATION AND VERIFICATION

The whole estimation algorithm is designed to achieve by matlab/simulink, Joint Simulation is built by Carsim vehicle model and matlab estimation model. Within Carsim brake test conditions, we set initial speed of 100km/h with the wheel cylinder pressure of 3Mpa. Meanwhile, the feature of road is designed as follow: the total length of road is 100m, the first 20m with the road adhesion coefficient of 0.8, and the rest with the road adhesion coefficient of 0.2. At 100km/h braking test, the initial value of the vehicle state variables and estimation error covariance matrix were set as following:

$\hat{x}_s^-(t_0) = [\ 80/3.6, 0, 0, 0, 0, 0, 0, 0, 0, 0, 0, 0, 0, 0, 1000, 1000, 1000, 1000\ ]^T$, $\hat{P}_s^-(t_0) = I_{18\times18}$, $\hat{x}_p^-(t_0) = [1,1,1,1]^T$, $\hat{P}_p^-(t_0) = I_{4\times4}$. The whole estimation process is divided into three sections, namely the input of control, the output of measurement and states of estimation.

Wherein the input of control comprises δ_{ij} (the angle of wheels) and w_{ij} (the angular velocity of wheels), the curves are shown in Fig.4.

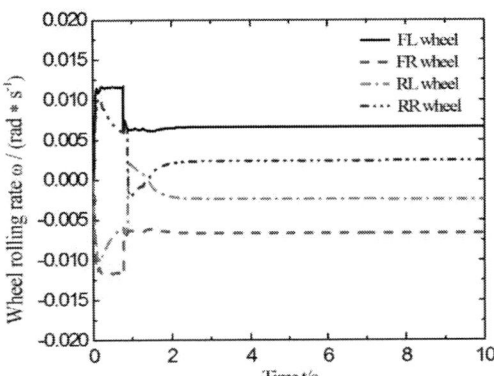

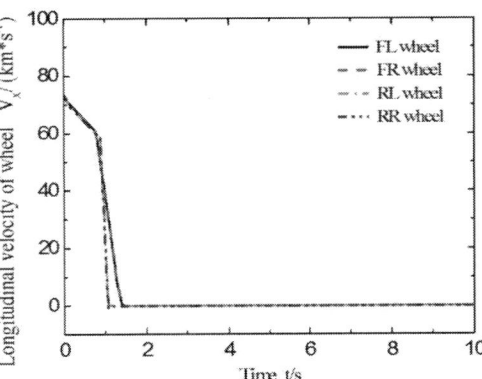

Fig.4: The input of DEKF control

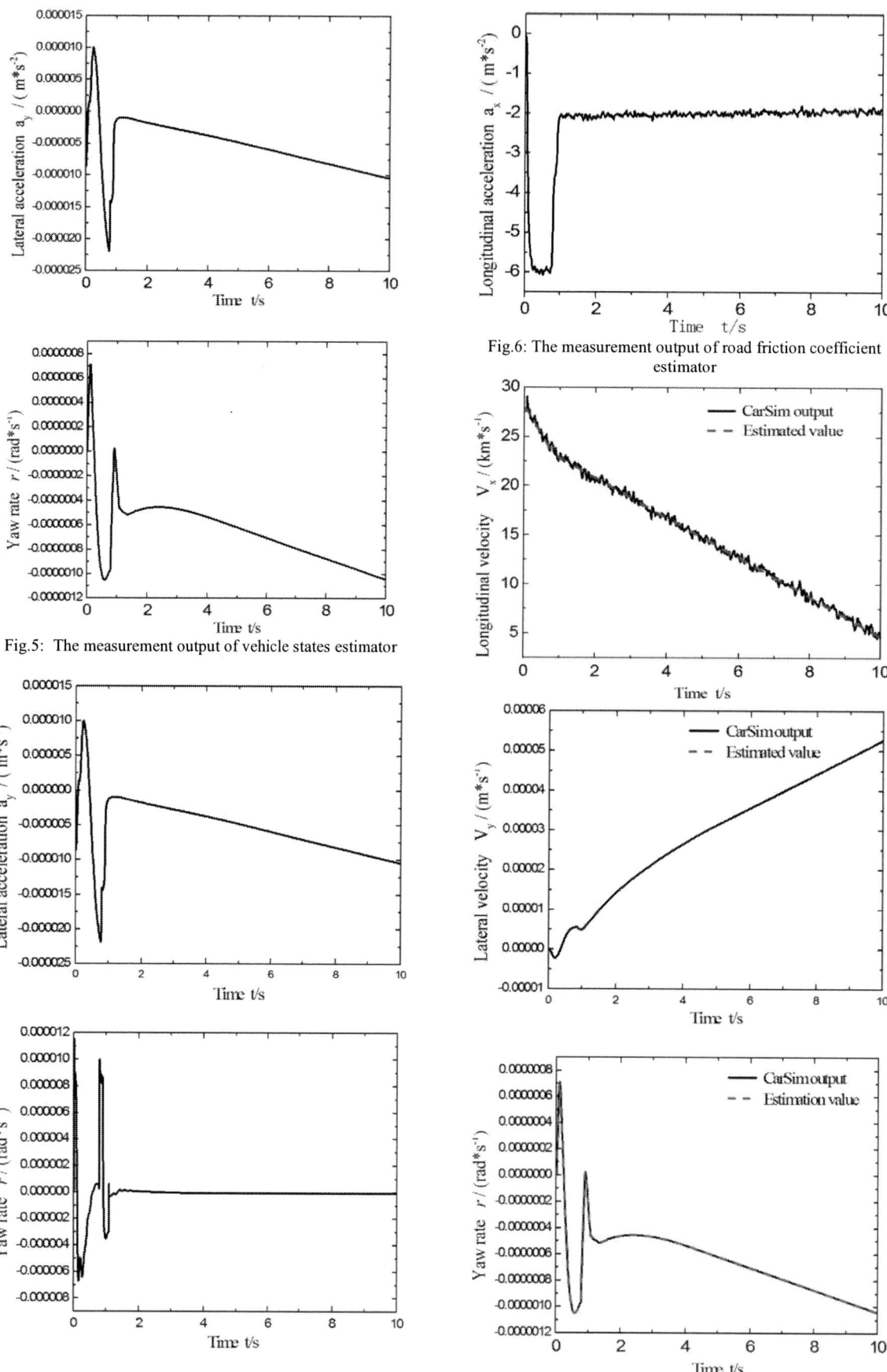

Fig.6: The measurement output of road friction coefficient estimator

Fig.5: The measurement output of vehicle states estimator

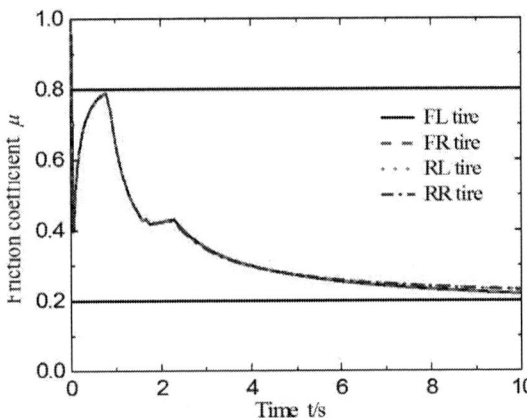

Fig.8: Estimated values of the road friction coefficient

VI. CONCLUSIONS

In this paper, an estimator of DEKF is designed, and the vehicle state parameters and road friction coefficient value show real-timely and accurately. By using simple HSRI tire model and three degrees of freedom vehicle dynamic model, the calculate time is reduced, and the simulation is more effective. The whole estimation is achieved by Matlab/ Simulink, which receives the passing parameters of Carsim. Compared with the output of Carsim, DEKF estimation parameters show favorable consistency, which indicates the DEKF estimation are accurate. When the road adhesion coefficient is changed, the DEKF estimation algorithm can accurately estimate the vehicle state. Meanwhile, the DEKF estimation method can estimate the value of road friction coefficient in real time. The application of the DEKF estimation can reduce the cost of the vehicle sensor, and achieve real-time vehicle parameters which are difficult to measure. So, it is possible to improve the active safety of the vehicle chassis.

ACKNOWLEDGMENT

This paper is financially supported by National Natural Science Foundation of China (No. 61403259, 51577120) and Science and Technology Research and Development Foundation of Shenzhen (No. JCYJ20140418182819128)

REFERENCES

[1] P. Lugner, H. Pacejka, M. Plochl, Recent advances in tyre models and testing procedures, Vehicle System Dynamics, vol. 43, no. 6- 7, 2005, P413- 436,.

[2] Sasaki H，Nishimaki T．A side-slip angle estimation using neural network for a wheeled vehicle[J]．SAE Paper 2000-01-0695

[3] Hai Wang；Zhihong Man; et al.Robust Sliding Mode Control for Steer-by-wire Systems with AC Motors in Road Vehicles［C］．2013 IEEE 8th Conference on Industrial Electronics and Applications (ICIEA 2013)，2013:P674-683

[4] Hiemer M，Vietinghoff A，Kiencke U．Determination of the vehicle body side slip angle with non-linear observer strategies[C]．SAE Paper 2005-01-0400

[5] Cherouat H，Braci M，Diop S．Vehicle velocity，side slip angles and yaw rate estimation[C]．IEEE International Symposium on Industrial Electronics，2005:P 349-355.

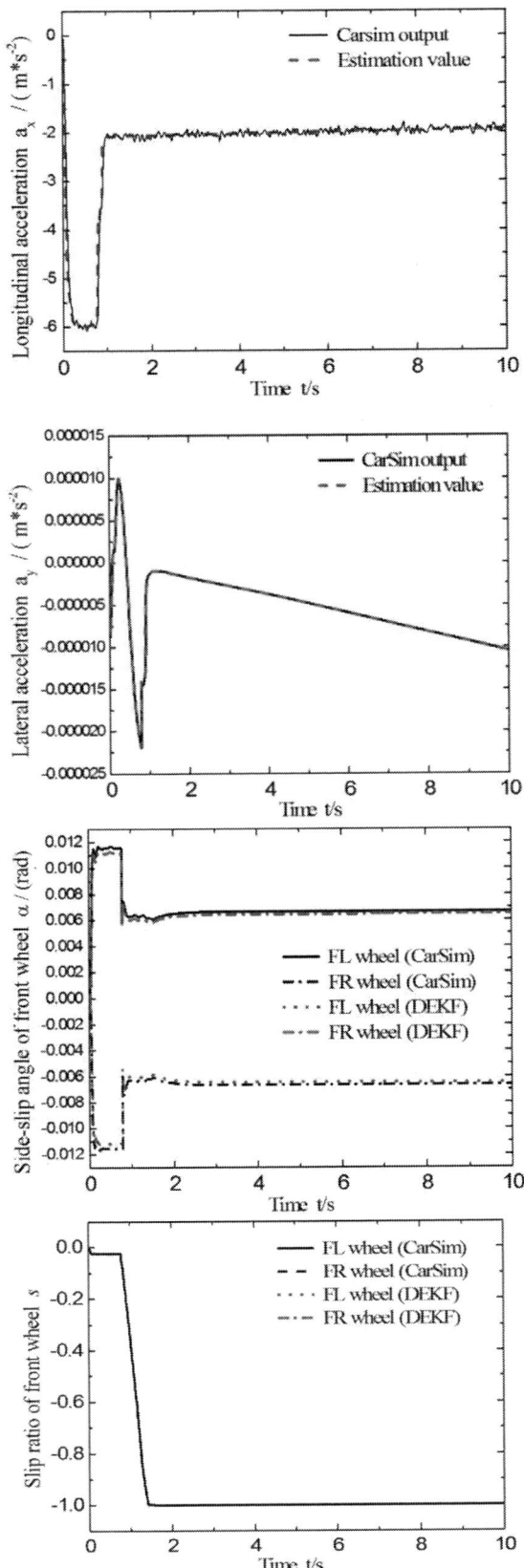

Fig.7: Compare the output of vehicle state parameters between DEKF and Carsim estimation

[6] Sanghyun Hong ; Smith, T.; Borrelli, F.; Hedrick, J.K. Vehicle inertial parameter identification using Extended and unscented Kalman Filters [C]. 16th International IEEE Conference on Intelligent Transportation Systems (ITSC 2013), 2013:P1436-1441

[7] Zong Chang-fu ; Hu Dan; Yang Xiao; et al.Vehicle driving state estimation based on extended Kalman filter [J]，Journal of Jilin University（Engineering and Technology Edition），Vol. 39, Iss.1, JAN 2009:P7-11.

[8] Rezaeian, Ayyoub ; Zarringhalam, Reza; et al.Cascaded Dual Extended Kalman Filter for combined vehicle state estimation and parameter identification [C] . SAE Technical Papers, SAE 2013 World Congress and ExhibitionVol. Vol. 2, 2013.

[9] M. C. Best, T. J. Gordon, and P. J. Dixon, "An Extended Kalman Filter for Real-time State Estimation of Vehicle Handling Dynamics" Vehicle System Dynamics, vol. 34, no. I, p. 57 75 July 2000.

[10] TIELKING J T, MITAL N K. A comparative evaluation of five traction tire models[R]. Highway Safety Research Institute Report, Interim Document 6, University of Michigan, January, 1974: P1–27.

[11] WAN E A, NELSON A T. Neural dual extended Kalman filtering: applications in speech enhancement and monaural blind signal separation[C]//Proceeding of Neural Networks for Signal Processing Workshop, IEEE, 1997: P466–475.

[12] WELCH G, BISHOP G. Introduction to Kalman filter[R]. Department of Computer Science, University of North Carolina at Chapel Hill, NC 27599-3175, 2001.

BIOGRAPHIES

Ying Xu received the B.S. degree in control engineering from the Jilin University of Technology in 1997, and M.E. and Ph.D. degrees in electrical and computer engineering from the Jilin University in 2005 and 2009, respectively. From 1997 to 2013, she was an Assistant Professor with the State Key Laboratory of Automotive Simulation and Control, Jilin University, Changchun, China. Since 2013, she has been teaching with the College of Mechatronics and Control Engineering, Shenzhen University, Shenzhen, China. Her current research interests include information fusion, computer simulation, vehicle dynamic modeling, and automotive electrical and electronic control technology.

B.L. Deng obtained his BE from Jilin University in 2004, and ME and PhD degrees both from the Hunan University in 2007 and 2014 respectively . During his pursuing PHD, he was invited to the California State University, Fresno for a one-year research program supported by China Scholarship Council. He is now the lecturer of College of Mechatronics and Control Engineering of Shenzhen University. His research interests include the related scientific issues during application of bio-fuels (bio-alcohols) in conventional internal combustion engines, such as combustion phenomenon and mechanism, the air pollutants' forming mechanism and the heat transfer process during engine working. He has published several papers in these fields in famous international journals (Applied Energy, Energy, Energy Conversion and Management, etc.).

Operation Strategy of Stand-alone Microgrid

Wenxiao WEI[1], Jinyong LEI[1], Xiaobin GUO[1], Ping YANG[2,3,4], Zhirong XU[2],
Yujia ZHANG[2], Shaoxiong ZHOU[5]

[1] Electric Power Research Institute, CSG, China
[2] School of Electric Power, South China University of Technology, China
[3] Guangdong Key Laboratory of Clean Energy Technology, South China University of Technology, China
[4] Guangdong Engineering Laboratory for wind Power Control and Integration Technology, China
[5] Guangdong Interwork Energy Technology Co., Ltd, Guangzhou, China

Abstract–In this paper, based on the situation of micropower sources and local loads, microgrid mathematical models for photovoltaic generation system, lithium battery energy storage system (ESS), vanadium redox flow battery energy storage system, wind power generation system and loads, are established. According to the microgrid electrical topology, energy management strategy of stand-alone microgrid are designed and used as theoretical guidance for microgrid stable operation. Finally, by using PSCAD simulation software, simulation verification is performed for the whole microgrid system models and microgrid system control strategies. The simulation results show that the designed microgrid control strategies can achieve the goal of system stability control, and verify the validity of system control strategies.

Keywords– Energy management, Hybrid energy storage, Microgrid, Operation in stand-alone mode, System control

I. INTRODUCTION

As the fossil fuels are increasingly depleted, wind power system and photovoltaic generation system making full use of renewable energy, are developing rapidly for their high efficient and clean features. Massive blackout spawned the birth of the distributed power which takes wind and photovoltaic power generation as the main energy. However, when the external environment changes, the output power of distributed generations would have a huge fluctuation. In order to make full use of the environmentally friendly advantage of distributed generations, maintain the stability of power grid, improve the quality of electric energy, microgrid which is a single control unit that can achieve self-control, protection and management, is combined with wind power , photovoltaic distributed generations, energy storage, local loads , protection devices and other monitoring [1,2]. In addition, it can operate in either grid-connected mode to exchange power with the grid, or in islanded mode [3].

In recent years, many technologies related to the system control and optimizations of microgrid with hybrid energy storage have been developed. On the one side, many scholars are doing the research of optimal design of microgrid system architecture, control of underlying devices, and computer simulation. On the other side, many other scholars are interested in improving the efficiency and stability of microgrid electricity supply in grid-connected mode [4].

From the perspective of the parts of microgrid study, the energy management study of stand-alone microgrid focuses

on MPPT control of photovoltaic array, wind turbine control and charge-discharge management of energy storage [5-8]. Besides the above mentioned study, the energy management study of grid-connected microgrid focuses on the management of distributed generations [9, 10]. From the perspective of the total system of microgrid study, it has two directions on energy management strategy study. On the one side, many researches are based on the features of wind power and photovoltaic generation microgrid, study the system design and control strategy of microgrid [11-14]. On the other side, some other studies are based on the multi-agent technology, consider the wind power and photovoltaic generations as distributed micropower sources, and use the agent of each micropower sources to study energy management [15-18].

According to the microgrid electrical topology, this paper establishes mathematical models for photovoltaic generation system, lithium battery energy storage system (ESS), vanadium redox flow battery energy storage system, wind power generation system and loads, designs energy management strategy of stand-alone microgrid which are used as theoretical guidance for microgrid stable operation. Finally, by using PSCAD simulation software, simulation verification is performed for the whole microgrid system models and microgrid system control strategies. The simulation results show that the designed microgrid control strategies can achieve the goal of system stable control, and verify the validity of system control strategies.

II. MICROGRID TOPOLOGY ANALYSIS

Microgrid which is one of the important forms of distributed generation, can parallel operating with power network through the distribution network, so to form a combined operating system of large network and small network, simultaneously, it can supply electricity for local loads independently. This mode improves the flexibility and reliability of power supply. What is more, the effects of connecting a large number of low power distributed generation to traditional power grid will be minimized by connecting the microgrid to the power grid through single point. Also, microgrid combines different types of disperse distributed generation together for power supply, which improves the efficiency of small distributed generation. Therefore, distributed generation will become a necessary part of the city smart distribution network and the combination of distributed generation and distributed network will be the future direction of smart grid.

Independent new energy generation system is an open system which can develop on the spot. System can make full use of local renewable resource(wind and solar energy, besides some fuel systems will aid when necessary). We could add fans and PV modules to expand in spite of the increasing demand for electricity. Simultaneously, the battery will balance loads, the stable voltage and the store energy. Under the background of current environment friendly era, independent new energy generation system is a kind of effective and economic power distribution scheme for remote area.

Microgrid combined with wind power, solar energy and energy storage technology, which only need one-time investment is especially suitable for stand-alone power system. Because of the features of convenient maintenance, low cost and long service life, there is a long-term benefit for users. This achievement can be widely applied to situations which have strict requirements for stand-alone power supply and stability of power system. It can be also applied to undersea installations, bases, mountains, islands, grasslands, railways or border posts for communication and lighting.

After preliminary investigation and communication, a trinity microgrid has been set initially on the base of a 50kW photovoltaic generation already built. Besides the photovoltaic generation, this stand-alone microgrid also has wind power generation, energy storage, monitoring system, and loads as well. The main system includes:

(1) A 50kW photovoltaic generation system (PV) (already contains photovoltaic array, needs photovoltaic inverters).

(2) Two 1kW wind power generations (WP)

(3) A 30kWh lithium iron phosphate battery energy storage system.

(4) A 20kWh vanadium redox flow battery energy storage system (VRB).

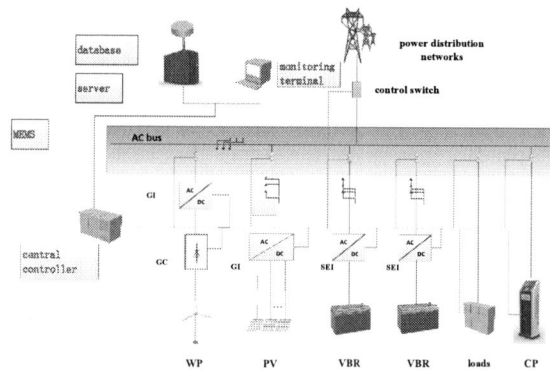

Fig.1: Microgrid Topological Graph

(5) A microgrid monitoring system.

(6) A microgrid control system.
(7) A microgrid energy management system (MEMS).

(8) Two charging piles (CP).

(9) A weather station.

(10) Several related loads (already set).

(11) Several grid-connected inverters (GI), stored energy inverters (SEI) and a grid-connected controller (GC).

1. The Scale of Construction
Considered the commonly used power supply mode and users' load characteristic, this paper proposes the construction scheme as follows.

Table 1: The Scale of Construction of Micro Power Sources

Project System	Specific Type	Phase	Power
Distributed Generation System	Photovoltaic Generation	Three	50kW
	Wind Power Generation	Single	2kW
Energy Storage	Lithium Battery	Three	30kW
	Vanadium redox flow battery System	Three	5kW
Load	AC Charging Pile	Single	14kW
	Air-conditioning	Single	<10kW

III. MICROGRID MODEL BUILDING

photovoltaic generation system, lithium battery energy storage system (ESS), vanadium redox flow battery energy storage system, wind power generation system and loads, and provides model foundation for algorithm design verification.

1. Microgrid Topological Structure
In the microgrid, micropower sources include photovoltaic generation, wind power generation, lithium battery system, and vanadium redox flow battery system, while loads include AC charging piles, air-conditioning, and computers, etc. The following sections specify different sub system models.

2. Photovoltaic Generation Model
Photovoltaic grid-connected inverter used for connecting photovoltaic generation with power grid, means that the inverter changes direct current (DC) to alternating current (AC) so that the power of photovoltaic generation obtain the same frequency and phase with power grid in order to connect the microgrid with the power grid. In this article, the photovoltaic grid-connected inverter use voltage-oriented control. The inner loop use current control under dq axis to realize the decoupled control of active current and inactive current. The outer loop uses dc voltage control to realize the fast tracking of dc voltage. The MPPT controller is used for PV array maximum power point tracking, in order to find out the voltage of the maximum power point as reference of dc voltage controller. The photovoltaic grid-connected system usually operates near unity power factor, so the reactive power is set to 0. The control structure of photovoltaic grid-connected inverter is shown as Fig.2.

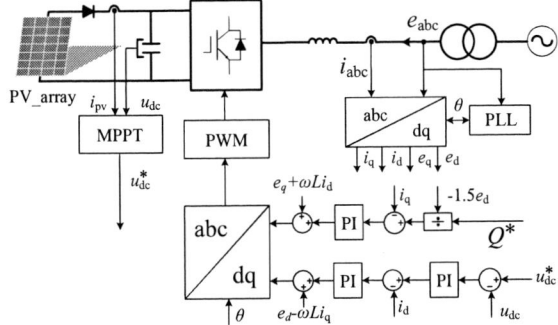

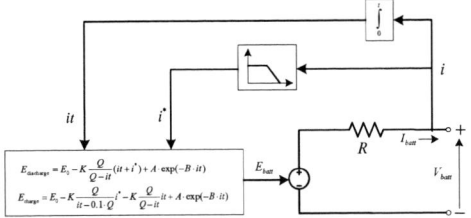

Fig.2: Control Structure of Photovoltaic Grid-connected Inverter

2.1. Inner loop current control

The inner loop current control system uses the PI controller to control the current of DQ Axis, to decoupling control active current and reactive current, the model of controller is shown as

$$\begin{cases} u_{cd} = K_{p\_id}\left(i_d^* - i_d\right) + K_{I\_id}\int\left(i_d^* - i_d\right)dt - \omega L i_q + e_d \\ u_{cq} = K_{p\_iq}\left(i_q^* - i_q\right) + K_{I\_iq}\int\left(i_q^* - i_q\right)dt + \omega L i_d + e_q \end{cases} \quad (1)$$

K_{p_id} d axis PI controller proportional coefficient

K_{I_id} d axis PI controller integral coefficient

e_d d axis system voltage

2.2. Direct voltage control

To realize the fast track control of direct voltage, we take PI controller as outer controller. The export is taken as D Axis setpoint. The control is shown as

$$i_d^* = K_{p3}(u_{dc}^* - u_{dc}) + K_{I3}\int (u_{dc}^* - u_{dc})dt \quad (2)$$

K_p proportional coefficient

2.2.1. Multiple Energy storage system

In this chapter the microgrid with multiple energy storage system is established, consisted of lithium battery and vanadium redox flow battery. Two models of storage battery describe and the storage control system are described as follows:

2.2.1.1. Lithium battery model

This paper uses the generalized equivalent circuit model which has a simple structure and considers battery internal nonlinear characteristics. The calculation methods of circuit parameters are simple. The generalized equivalent circuit model covers lead-acid battery, nickel-cadmium battery, nickel-metal hydride battery, and lithium battery. It has a high fitting degree in short-term dynamic simulation.

The generalized battery model is assembled in series connection by voltage source and constant resistance. The only state variable is the SOC of battery in charging or discharging status. The model is shown as Fig.3.

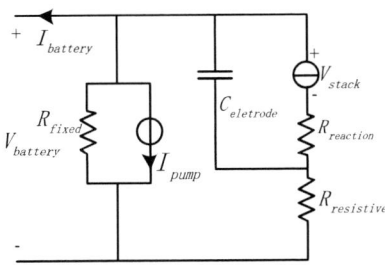

Fig.3: The lithium battery generalized model

According to the diagrammatic sketch above, the mathematical model of lithium battery can be expressed as follows:
The mathematical model of discharging:

$$V_{batt} = E_0 - R \cdot i - K\frac{Q}{Q - it}(it + i^*) + A \cdot \exp(-B \cdot it) \quad (3)$$

The mathematical model of charging:

$$V_{batt} = E_0 - K\frac{Q}{it - 0.1 \cdot Q}i^* - K\frac{Q}{Q - it} + A \cdot \exp(-B \cdot it) \quad (4)$$

V_{batt} is the voltage of battery (V);
E_0 is the constant voltage of battery (V);
K is the Polarization constants (V/(Ah));
Q is the battery capacity (Ah);
it is The actual battery charge(Ah);
A is the amplitude of area index;
B is the inverse of time constant in area index (Ah)-1;
R is the battery internal resistance (Ω);
i is the battery current (A);
i^* is the battery current after filtering (A);

2.2.1.2. Vanadium redox flow battery model

The vanadium redox flow battery model is shown as Fig.4. The stack voltage relevant to SOC uses the ideal voltage source V_{stack} to simulate. The pump loss can be simulated by I_{pump} and this controlled current source is relevant to SOC and I_{stack}. The dynamic response ability also is an important parameter of VRB, using capacitance to simulate. $R_{reavtion}$ and $R_{resistive}$ denote internal resistance of VRB. R_{fixed} and I_{pump} represents the external parasitic loss.

Fig.4: The vanadium redox flow battery model

2.2.1.3. Storage control system model

As shown in Fig.5 is circuit topology of battery energy system. As the main micropower source of the microgrid, lithium battery system work in PQ control mode when the microgrid is in grid connected mode. When the microgrid is in the islanded mode, the lithium work in VF control mode will provide voltage and frequency support. The Liquid flow battery whose control strategy uses the current loop control simply, only has an auxiliary microsource, In order to achieve the goal of active and reactive power control.

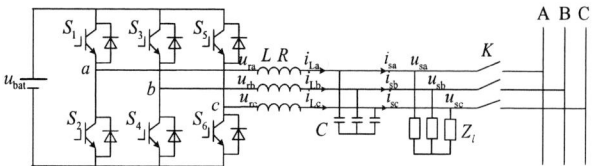

Fig.5: The circuit topology of lithium battery energy system

$$\begin{cases} U_{rd}^P = U_{sd}^P + L\dfrac{dI_{Ld}^P}{dt} - \omega L I_q^P \\[2mm] U_{rq}^P = U_{sq}^P + L\dfrac{dI_{Lq}^P}{dt} + \omega L I_d^P \\[2mm] U_{rd}^N = U_{sd}^N + L\dfrac{dI_{Ld}^N}{dt} + \omega L I_q^N \\[2mm] U_{rq}^N = U_{sq}^N + L\dfrac{dI_{Lq}^N}{dt} - \omega L I_d^N \end{cases} \quad (5)$$

According to the circuit topology, dq0 transformation and symmetrical component method, the dynamic equation of the inverter in the dq coordinate system can be expressed as equation (5).

The current loop control equation can be expressed as follows:

$$\begin{cases} U_{rd}^P = K_{pid}^P (I_{Ld}^{P*} - I_{Ld}^P) + K_{iid}^P \int (I_{Ld}^{P*} - I_{Ld}^P)dt + U_{sd}^P - \omega L I_q^P \\[2mm] U_{rq}^P = K_{piq}^P (I_{Lq}^{P*} - I_{Lq}^P) + K_{iiq}^P \int (I_{Lq}^{P*} - I_{Lq}^P)dt + U_{sq}^P + \omega L I_d^P \\[2mm] U_{rd}^N = K_{pid}^N (I_{Ld}^{N*} - I_{Ld}^N) + K_{iid}^N \int (I_{Ld}^{N*} - I_{Ld}^N)dt + U_{sd}^N + \omega L I_q^N \\[2mm] U_{rq}^N = K_{piq}^N (I_{Lq}^{N*} - I_{Lq}^N) + K_{iiq}^N \int (I_{Lq}^{N*} - I_{Lq}^N)dt + U_{sq}^N - \omega L I_d^N \end{cases} \quad (6)$$

U_{rd}^P d axis positive sequence voltage

U_{rd}^N d axis negative sequence voltage

K_{pid}^P positive sequence current controller proportional coefficient

K_{iid}^N negative-sequence current controller integral coefficient

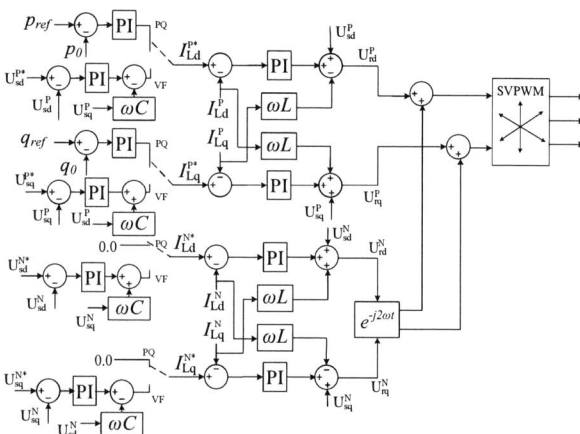

Fig.6: The whole energy storage system control structure diagram

The whole energy storage system control structure diagram is shown as Fig.6.

2.2.2. Wind Power Generation System Model

In this project, the wind turbine which is a single-phase generation with a small capacity is simulated with controlled current source in simulation.

Wind generator control in grid-connected is as follows:

We set the active power p_w and reactive power q_w, then

$$\begin{cases} S_w = \sqrt{p_w^2 + q_w^2} = e_{rms} i_{rms} \\ \text{phix} = \arctan(q_w / p_w) \end{cases} \quad (7)$$

set the grid voltage phase to phi1, so the instantaneous current in grid-connected is

$$i_{grid} = \sqrt{2} i_{rms} \cos(\text{phi1} + \text{phix}) = \frac{\sqrt{2} S_w}{e_{rms}} \cos(\text{phi1} + \text{phix}) \quad (8)$$

IV. MICROGRID OPERATION STRATEGY

This chapter explains the algorithm design of microgrid control system of stand-alone mode.

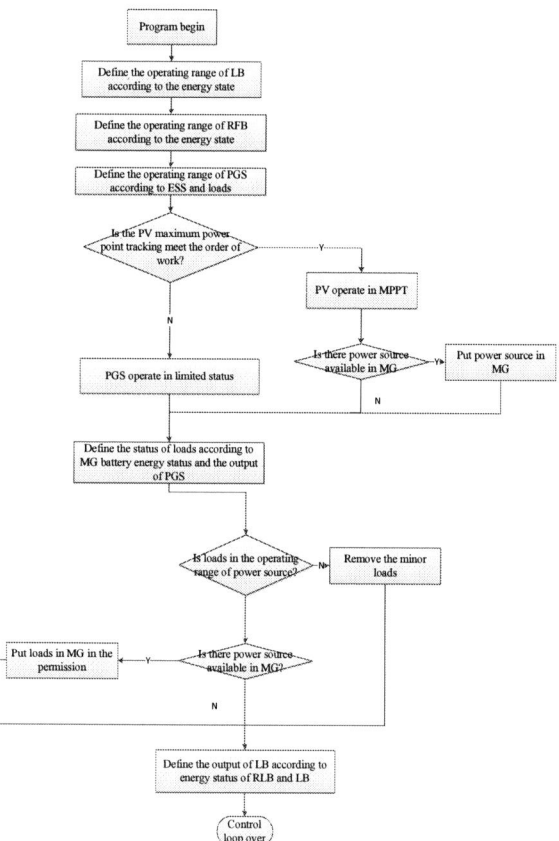

Fig.7: microgrid energy management system control flow in the stand-alone mode

1. Microgrid Energy Management System in the stand-alone Mode

The microgrid energy management system control flow in the stand-alone mode is shown as Fig.7.

1.1. Step 1: Confirm the main power source operating range

978-1-5090-0064-7/15 $31.00 © 2015 IEEE

Confirm the maximum charge-discharge power of main power source according to the energy status (SOC) of main power source, which means the operating range is set $[P_{ch\max,Li}, P_{dh\max,Li}]$. Next, according to the confirmed microgrid upper and lower reserve demand, the maximum and minimum output of voltage and frequency control unit (main power source) are limited to:

$$P_{Li,\min} = P_{ch\max,Li} + P_{down}$$
$$P_{Li,\max} = P_{dh\max,Li} - P_{up} \tag{9}$$

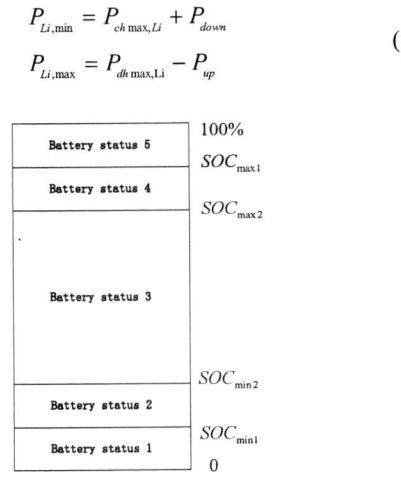

Fig.8: Energy Storage System Region Zoning Diagram

P_{up} and P_{down} are the microgrid upper and lower reserve demands, which can be confirmed by the maximum load or power source of microgrid, and can be set 5kW or 7kW. In the same way, the upper and lower limit of vanadium redox flow battery operation power can be confirmed according to the energy status of vanadium redox flow battery energy storage system, which is $[P_{\min,VR}, P_{\max,VR}]$.

The restriction of charge and discharge power of energy storage system is shown as Fig.8.

According to the battery characteristics of vanadium redox flow battery and lithium battery, several energy region and power restriction are shown as below.

The energy region zoning and power restriction of lithium battery is shown as below.

Table.2: Energy region zoning

Storage Battery Status	Minimum SOC (%)	Maximum SOC (%)	Discharge Power Restriction	Charge Power Restriction
Status 1	0	10	0	30kW
Status 2	10	20	$\min\{30\,\mathrm{kW}, \frac{Soc-10}{100\Delta t}C\}$	30kW
Status 3	20	80	30kW	30kW
Status 4	80	90	30kW	$\min\{\frac{90-Soc}{100\Delta t}C, 30\,\mathrm{kW}\}$
Status 5	90	100	30kW	0

The standard charge and discharge flow of storage battery is: use steady current of 0.3C to charge the storage battery until the rated power of the battery equals to rated voltage, then steady voltage charge by controlling the battery.

The energy storage system flow of lithium battery: discharge the storage battery at a steady current of 0.3C until the battery voltage equals to the cut-off voltage.

The energy region zoning and power restriction of vanadium redox flow battery is shown as below.

Table.3: energy region zoning

Storage Battery Status	Minimum SOC (%)	Maximum SOC (%)	Discharge Power Restriction	Charge Power Restriction
Status 1	0	0	0	5kW
Status 2	0	10	$\min\{5\,\mathrm{kW}, \frac{Soc}{100\Delta t}C\}$	5kW
Status 3	10	90	5kW	5kW
Status 4	90	100	5kW	$\min\{5\,\mathrm{kW}, \frac{100-Soc}{100\Delta t}C\}$
Status 5	100	100	5kW	0

1.2. Step2: Confirm the operation region of photovoltaic gene-ration

According to the charging region of vanadium redox flow battery and lithium battery, we confirm the charging region of photovoltaic generation:

$$P_{pv} = \min(P_{pv,mppt}, P_{net} - P_{Li,\min} - P_{VR,\min}) \tag{10}$$

Ensure the power of microgrid photovoltaic generation under the allowable value.

1.3. Step3: Confirm the switching instruction of microgrid loads

If cutting off loads is considered, cutting load quantity P_{cut} is set as follows:

$$P_{cut} = P_{net} - P_{pv,mppt} - P_{Li,\max} - P_{VR,\max} \tag{11}$$

If secondary loads are only two charging piles, then cut off these two charging piles one by one.

Step 4: Distribute net load power between lithium battery and vanadium redox flow battery

If the energy status of vanadium redox flow battery energy storage system is in the normal range, then use the vanadium redox flow battery energy storage system to carry the maximum net load of microgrid:

$$P_{VR} = \lim(P_{\min,VR}, P_{net} - P_{pv}, P_{\max,VR}) \tag{12}$$

If the energy status of lithium battery is too low, the energy status of lithium battery needs to be maintained by reducing the charge and discharging the power of vanadium redox flow battery:

$$P_{VR} = P_{\max,VR} \tag{13}$$

For long-term stand-alone microgrid, if the next period is predicted to be illumination insufficient, it will keep the power supply of important microgrid loads running normally by cutting off microgrid secondary loads.

The microgrid control flows in the stand-alone warning operation mode are shown as Fig.9 and Fig.10.

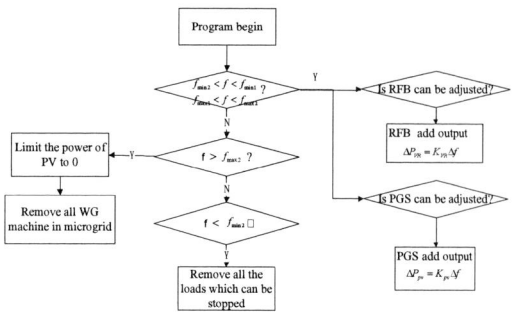

Fig.9: Microgrid control flow in the unusual frequency mode

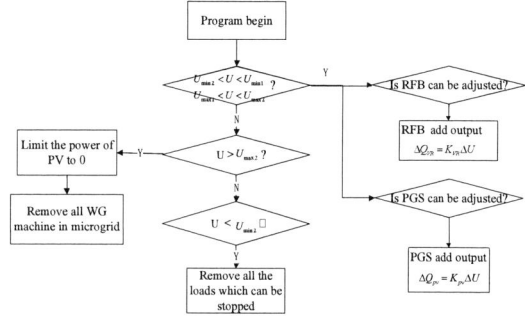

Fig.10: Microgrid control flow in the unusual voltage mode

V. PSCAD BASED MICROGRID CONTROL SYSTEM ALGORITHM PROOF

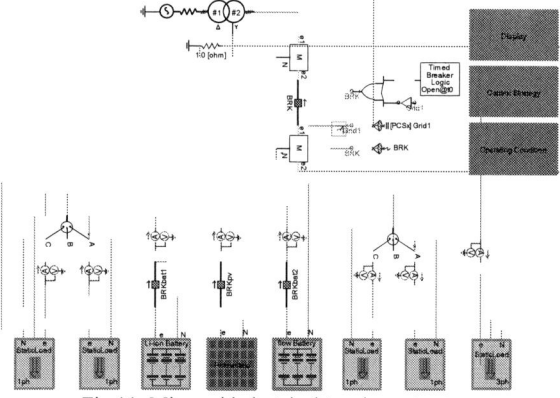

Fig.11: Microgrid electrical topology structure

According to chapter 2 microgrid system math models, this chapter establishes microgrid system model in PSCAD simulation software, simulating proof microgrid control algorithms in chapter 3 based on PSCAD simulation software, ensuring that the microgrid control algorithms are correct.

1. PSCAD Based Microgrid Simulation Model

According to microgrid electrical topology structure, we have established microgrid system model in PSCAD, which is shown as Fig.11.

The brief instruction of microgrid simulation system is shown as below:

(1)Microgrid simulation system has four components: microgrid main circuit, different operating simulation mode, microgrid control strategy, simulation result display.

(2)Microgrid main circuit includes a 30kWh lithium battery energy storage system, a 5kWh vanadium redox flow battery energy storage system, a 50kWp three-phase photovoltaic generation system, two 1kW single-phase wind power generation, two 7kW AC charging piles, and several loads as air-conditionings and computers.

(3)Operating simulation mode is used to simulate different boundary conditions and trigger control algorithm action.

(4)Microgrid control strategy mode is used to program microgrid control strategies in different operation mode, such as grid-connected operation mode control program, stand-alone operation mode control program, etc.

(5)Simulation result of display mode is used to display system response waveform under different control strategies, such as system voltage waveform, frequency changing waveform, and power changing waveform.

2. Microgrid System Simulation Operation Mode

Microgrid system simulation includes energy management system simulation and dynamic simulation. Dynamic simulation includes transient-status part and steady-status part. The former is used to prove stand-alone microgrid switching control strategy and steady control strategy while the latter is used to prove stand-alone microgrid energy management used to prove stand-alone microgrid energy management strategies.

2.1. Microgrid Energy Management System Simulation in stand-alone Mode

Needed simulation includes:

(1) Power mode of lithium battery energy storage system is normal, while the discharge power of lithium Battery is too large.

The initial status is as follow:
a. SOC of lithium battery energy storage system is 0.5, power generation is 0kW.

b. Vanadium redox flow battery energy storage system has no output, SOC=0.5, maximum output power is 5kW.

c. Photovoltaic generation system output power is 10kW, maximum output power is 10kW.

d. Wind power generation output power 2kW.
e. Loads are 12kW.

Simulate the microgrid by increasing the power of loads from 12kW to 17kW and 22kW.

(2) Power mode of lithium battery energy storage system is normal, but the charge power of lithium Battery too large.

The initial status is as follow:

a. SOC of lithium battery energy storage system is 0.5, power generation is 0kW (the maximum power limit is 20kW).

b. Vanadium redox flow battery energy storage system has no output, SOC=0.5.

c. Photovoltaic generation system output power is 20kW which is in maximum mode.

d. Wind power generation output power 2kW.

e. Loads are 22kW.

Simulate the microgrid by increasing the photovoltaic generation system output power to 40kW.

2.2. Simulation results in stand-alone Mode

(1) Mode 1

This result is system response of "Power mode of lithium battery energy storage system is normal, while the discharge power of lithium Battery is too large", in this simulation, 10s represent 1h.

Fig.12 is the result of mode 1, at the beginning, loads are powered by photovoltaic generation system and wind generation system, the power of lithium battery and vanadium redox flow battery is 0. When 5s, loads very from 12kW to 17kW. To keep the stable of system, vanadium redox flow battery output very from 5kW to loads, and the energy status decreases during this mode. When 10s, loads very from 17kW to 22kW, because the maximum power of vanadium redox flow battery is 5kW, to keep the stable of system, lithium battery output very from 5kW to loads. When 12.8s, the status of vanadium redox flow battery is very low, the loads are totally powered by lithium battery. When 17s, the status of the lithium battery reaches the limit, to keep the stable of system, we remove the minor load, system is powered by vanadium redox flow battery again.

(2) Mode 2

This result is system response of "Power mode of lithium battery energy storage system is normal, but the charge power of lithium Battery is too large", in this simulation, 10s represents 1h.

Fig.12 is the result of mode 2. At the beginning, loads are powered by photovoltaic generation system and wind generation system. At the time of 5s , photovoltaic generation system output very from 20kW to 30kW, now the vanadium redox flow battery is charging in maximum status, other energy is absorbed by lithium battery. At the time of 10s, the output of photovoltaic generation system turns from 30kW to 40kW,

Lithium battery is increasing immediately to keep the balance of system energy. When vanadium redox flow battery SOC=70%, charging power turns to lithium battery until lithium battery SOC=70%, then vanadium redox flow battery charges in maximum status. When the status of lithium battery reach 85%, we will limit charging power, when SOC=90%, we set charging power 0. If the microgrid power is too lager, we must remove the wind generation

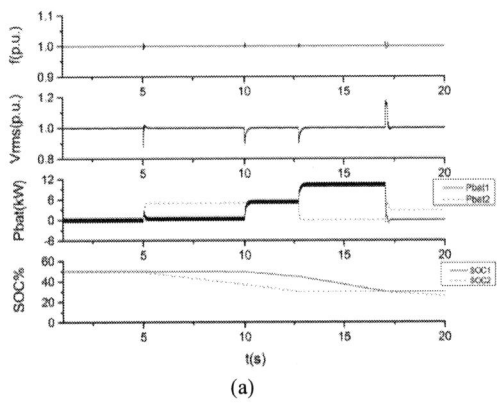

(a)

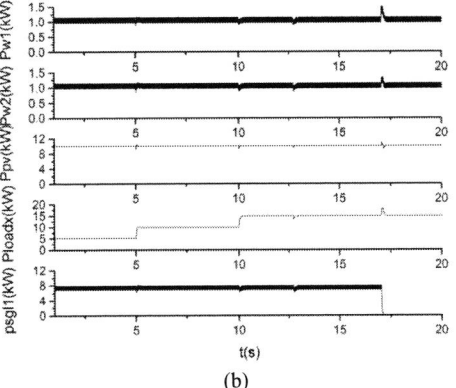

(b)

Fig.12 The result of mode 1

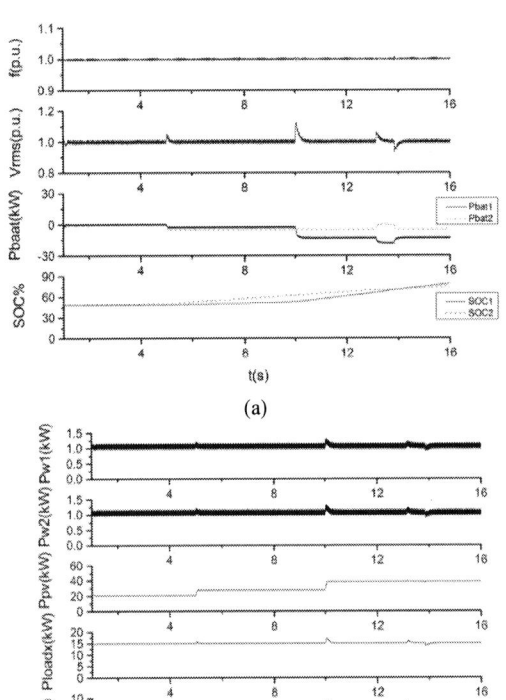

(a)

(b)

Fig.13 The result of mode 2

system. Then the charging power of vanadium redox flow battery and photovoltaic system will decrease. Until the energy status reach maximum, the simulation is done.

VI. CONCLUSION

This article based on the microgrid system electrical topology structure, finishes the model design and control strategy designed for microgrid, the main works are follow:

(1) Microgrid mathematical models for photovoltaic generation system, lithium battery energy storage system (ESS), vanadium redox flow battery energy storage system, wind power generation system and loads are established.

(2) According to the microgrid electrical topology, energy management strategies of stand-alone microgrid are presented and used as theoretical guidance for microgrid stable operation.

(3) Finally, by using PSCAD simulation software, simulation verification is done for the whole microgrid system models and microgrid system control strategies. The simulation results show that the designed microgrid control strategies can achieve the goal of system stable control, and verify the validity of system control strategies.

ACKNOWLEDGMENT

The authors would like to thank the support from the National High-tech R&D Program (863 Program) of China (2014AA052001), and Guangdong Strategic Emerging Industry Core Technology Research Projects (2012A032300001).

REFERENCES

[1] R. H. Lasseter, "MicroGrids," in Power Engineering Society Winter Meeting, 2002. IEEE: IEEE, 2002, pp. 305-308 vol.1.

[2] N. Hatziargyriou, H. Asano, R. Iravani, and C. Marnay, "Microgrids," IEEE POWER & ENERGY MAGAZINE, vol. 5, pp. 78-94, 2007.

[3] M. Ross, R. Hidalgo, C. Abbey, and G. Joos, "Energy storage system scheduling for an isolated microgrid," IET RENEWABLE POWER GENERATION, vol. 5, pp. 117-123, 2011.

[4] Z. Bo-Quan and Y. Yi-Min, "Wind and solar photovoltaic power generation situation and development trend," electric power, 2006.

[5] F. Valenciaga and P. F. Puleston, "Doubly fed induction generator model for transient stability analysis supervisor control for a stand-alone hybrid generation system using wind and photovoltaic energy," IEEE TRANSACTIONS ON ENERGY CONVERSION, vol. 20, pp. 398-405, 2005.

[6] Y. Ming, L. Geng-Yin, Z. Jian-Cheng, Z. Wei-Ran, X. Yi-Feng, "directly driven wind turbine with permanent magnet synchronous generators Modeling and Control Strategy," Power System Technology, 2007.

[7] M. Mei-Qin, Y. Shi-Jie and S. Jian-Hui, "Photovoltaic Array with MPPT functionality of Matlab general simulation model," Journal of System Simulation, 2005.

[8] C. Miao-Miao, "Capacity storage unit optimization scenery combined cycle power generation system design," Electrotechnical Application, 2006.

[9] C. Lin, "Some Problems of access to distributed generation power system," vol, 2007, p. 111.

[10] P. Wei, S. Kun, K. Li, Q. Zhi-Ping, " Influence and improve the distribution network for distributed power supply voltage quality," Proceedings of the CSEE, pp. 152-157, 2008-05-05 2008.

[11] W. Bo, W. Jie, Y. Jing-Ming, "A New Fuzzy Controller and Its Application," Microcomputer Information, 2004.

[12] Z. Miao, W. Jie, "Based on hierarchical hybrid power control system of fuzzy control solar wind," Acta Energiae Solaris Sinica, pp. 1208-1213, 2006-12-28 2006.

[13] Y. Huan-Ying, "Intelligent AC load management of independent wind and solar systems," vol, 2005, p. 65.

[14] T. Xi-Sheng, "stability of energy management in distributed generation super capacitor energy storage system," vol, 2006, p. 126.

[15] L. Dan, W. Jie, Z. Jun, G. Hong-Xia, Y. Xiao-Ming, " Decentralized generation dispatch planning Multi-Agent System," Control Theory & Applications, 2008.

[16] Z. Jun, W. Jie, L. Jun-Feng, G. La-Mei, L. Min, and IEEE, "An Agent-based Approach to Renewable Energy Management in Eco-building," in IEEE International Conference on Sustainable Energy Technologies Singapore, SINGAPORE, 2008, pp. 46-50.

[17] J. Chang, S. Y. Jia and IEEE, "Windy-solar Power Generation System based on Multi-agent System," in 7th International Conference on Machine Learning and Cybernetics Kunming, PEOPLES R CHINA, 2008, pp. 2446-2449.

[18] G. Lu, "Management controls and systems platform in wind / photovoltaic hybrid power systems research and development," vol, 2008, p. 70.

Design of MGCC for User-side Microgrid

Bing TIAN[1], Jinyong LEI[1], Xiaobin GUO[1], Ping YANG[2,3,4], Haozhe YUAN[2],
Zhirong XU[2], Shaoxiong ZHOU[5], Ting He[2]

[1] Electric Power Research Institute, CSG, China
[2] School of Electric Power, South China University of Technology, China
[3] Guangdong Key Laboratory of Clean Energy Technology, South China University of Technology, China
[4] Guangdong Engineering Laboratory for wind Power Control and Integration Technology, China
[5] Guangdong Intework Energy Technology Co., Ltd, China
E-mail:249064141@qq.com

Abstract–As the microgrid control centers, microgrid central controller can achieve coordinated control of various equipment of microgrid and maintain safe, reliable and economic operation. So, it receives wide attention. A microgrid central controller is proposed in this paper for high reliability, low cost, generic, compact design. Microgrid central controller uses modular software and hardware design based on embedded system. The design reduces the costs of controller, and improves the portability of the controller. The paper also presents a general method based on profiles which enhanced controller for microgrid with different types and adaptability of equipment from different manufacturers. After preliminary testing, microgrid central controller has fully functional, stable and reliable operation.

Keywords–Central Controller, Embedded Operating System, Generalized Design, Modular Design,

I. INTRODUCTION

Microgrid is made up of distributed generation, energy storage unit, load and the monitoring and protection device [1, 2]. Microgrid have a flexible mode of operation and schedulability. It can reduce adverse effects of intermittent distributed generation to distribution networks and maximize the use of output power of distributed generation. It realizes reliability of power supply through coordination between the relevant control device [3-5]. From the microcosmic perspective, microgrid can be seen as small power system, and from the macroscopic perspective, microgrid can be considered a "virtual" power supply or load of distribution network [6-9]. The important difference between microgrid and separate power supply of traditional distributed generation is that microgrid required coordination control for distributed generation, storage unit, load, to maintain the balance of microgrid active power and reactive power and realize the stable operation of microgrid [10, 11].

Microgrid central controller (MGCC) is core device of microgrid control system that it should has island detection, fault protection control, stable operation control, operation mode transformation, energy management and energy optimization. Based on operation of microgrid, central controller has flexible function configuration, milliseconds level stable operation control, minutes level operation optimization, and hours level energy-saving scheduling, ensuring the security and stable operation of microgrid [12]. Some papers have presented the designs of microgrid controller, but these designs are lack of controller details designs and has large device size that these design are lack of practical value [13-15].

Due to distributed photovoltaics (PV) and energy storage as the main power source in microgrid, the user-side microgrid has the advantages of closer to the load, low cost, easy to be promoted, and is concerned by national and social. Therefore, low cost, high reliability, universalization and miniaturization central controller is urgently needed for user-side microgrid.

The paper presents a design of microgrid central controller, which has hardware module and software task framework design, and uses the hardware platform based on stm32F407 chip and real-time operating system software platform based on µC/OS-II. According to the different requirements of software function module on real-time, the paper designs algorithm task switch control logic, and proposes generalized achieved method based on configuration file to adapt the different type and different manufacturers of bottom equipment in microgrid. It also proposes a user interface generalized design method to achieve configuration function of user interface. Preliminary tests show that the the microgrid central controller of design in this paper has stable and reliable operation.

II. MICROGRID TOPOLOGY ANALYSIS

Microgrid topology based on distributed PV is shown as figure 1. Microgrid 10kV bus is divided into I, II two segments. 0.4kV bus is connected to the II segment of 10kV bus through the public transformer #2 and grid-connected switch that the microgrid with 0.4kV bus has grid-connected and stand-alone operation mode. The microgrid with 0.4kV bus includes 0.5115MW PV system, 450kWh battery energy storage, reactive power compensation equipment and load. I segments of 10kV bus connects to three-way loads through three transformers and 2.3925MW PV system through two transformers. Microgrid with I segments of 10kV bus only has grid-connected operation mode.

In grid-connected operation mode, microgrid connects to the distribution network through a point of common coupling (PCC), and exchanges power with the distribution network under the coordinated scheduling. According to exchange power of PCC, microgrid has "source loading" double feature, which can be delivered power to the distribution network as a power point or absorb power from the distribution network as a load. Microgrid also can achieve 0 exchange power with

978-1-5090-0064-7/15 $31.00 © 2015 IEEE

distribution network. In grid-connected operation mode, microgrid as the controlled "source load" should be shown in 3 features in distribution network: 1, economic power distribution of load in normal operation; 2, output peaking power to peak load; 3, power support in emergency cases.

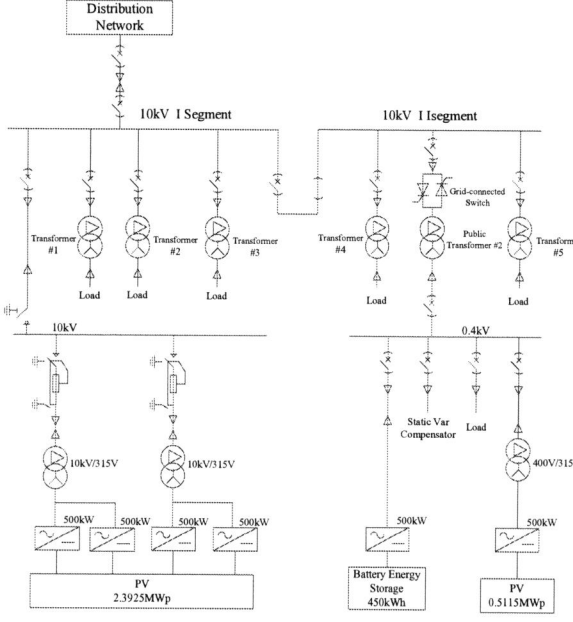

Fig. 1: Microgrid topology

When the energy storage system exceeds its range, or PV output fluctuations affects voltage fluctuation of PCC which is greater than 10%, or distribution network system occurs short-circuit fault, microgrid central controller disconnects grid-connected switch that microgrid run in stand-alone mode. Due to the limited accommodation capacity of distributed generation, variety of loads, limited storage capacity of energy storage, high power quality requirement of the components, and so on, microgrid accommodation ability does not necessarily meet the stability operation requirement in stand-alone mode.

Microgrid central controller need to comprehensively consider the state of charge (SOC) of energy storage system, load fluctuation, PV system output, and so on, and chooses microgrid power balance control measures to maintain stability operation of microgrid.

When distribution network is failure or power quality requirements are not met, microgrid central controller changes grid-connected mode to stand-alone mode. At this point, distribution network can not exchange power with microgrid through PCC. Microgrid occurs power imbalance. Due to the energy storage devices of microgrid is small capacity, and devices of microgrid contain electronic interface, microgrid is lack of capacity to regulate active and reactive power which results in voltage and frequency instability in the process of switching.

Before microgrid from the stand-alone mode switches to grid-connected mode, the microgrid central controller

sends grid-connected signals to energy storage device which adjust the voltage and frequency of its output based on the microgrid and distribution network voltage to achieve the synchronism period. When the microgrid power is in balance and microgrid voltage and frequency is synchronization with distribution network, microgrid central controller switches on the grid-connected switch.

In the process of microgrid mode switching, active and reactive power are provided by energy storage devices. Advantages of energy storage devices are stability output. In conditions and cost permit, super capacitors can be attached to improve the stability of microgrid when the mode switching.

III. MICROGRID CONTROL SYSTEM

In order to improve the reliability communication between the central controller and devices to achieve real-time and stability control, the paper designs communication architecture of microgrid control system based on data servers, Ethernet and RS485 serial communication.

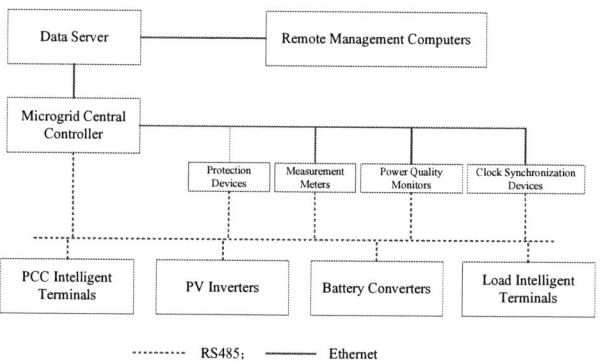

Fig. 2: Structure of microgrid control system

Central controller of microgrid connects PV inverters, battery converters and load intelligent terminals with RS485 serial communication, and it uses the corresponding operation control for primary side devices. Secondary side devices such as central controller, power quality monitors, measurement meters and protection devices is connected by Ethernet to achieve real-time data exchange and monitor microgrid operation. Controller will collect data into the data server. Server sends data to the remote management computers through Ethernet, so that operators can monitor microgrid in real-time. Operators can change the server's data through computers and control the microgrid central controller to change the operation mode of microgrid equipment.

IV. MICROGRID CENTRAL CONTROLLER

Microgrid central controller is a core device of microgrid control systems, microgrid control strategies are implemented through this device. Microgrid central controller analyzes real-time data of loads, distributed generations, grid-connected switch information and historical data such as loads and generating capacity.

According to the renewable energy most absorption, constant power operation and peak load shifting, microgrid central controller selects the control strategies and coordinating controls distributed generations, energy storages and loads to maintain microgrid stability.

1. Hardware Design of the Controller

The hardware platform of controller is used modular structure design. Controller includes the main control chip, human-machine interface module, terminal communication module, data storage module, device communication module. The controller has variety of communication interfaces and communication modes to achieve reliable communication with multiple devices.

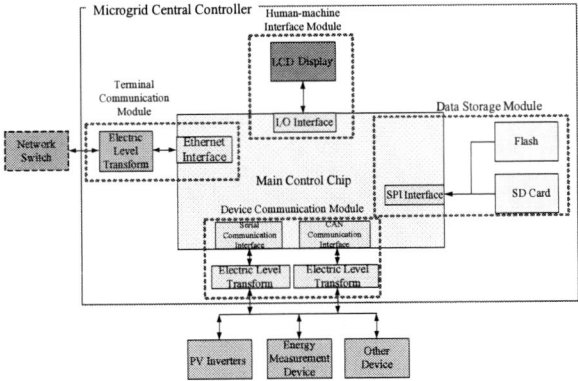

Fig. 3: Hardware design diagram of microgrid central controller

The description of the function of each modules are as follows:

The controller uses stm32F407 as the main control chip which has high-speed input/output (I/O) interface, Ethernet interface, serial peripheral interface (SPI), serial communication interface and controller area network (CAN) communication interface.

Human-machine interface module has liquid crystal display (LCD) to display equipment operation status and custom operation parameters and logs.

Terminal communication module is responsible for mutual communication between the central controller and terminals.

Device communication module provides interfaces for devices, and interconnect with devices which is not Ethernet-enabled such as PV inverters, energy measurement devices.

Data storage module stores the operation logs and key data in order to diagnosis and analysis system.

2. Software Design of Controller

2.1. Design of Software Framework

In order to make software more secure and simple, microgrid central controller uses µC/OS-II embedded system. µC/OS-II is preemptive multitasking real-time embedded operating system that it can be ported, cured and cropped, and applicable to a variety of microcontrollers and digital signal processing chips. Under µC/OS-II embedded operating systems, all operations are the handling of task. Device tasks can be roughly summarized as follows:

Algorithm tasks: according to models, the grid-connected photovoltaic inverter and user-side load of system are scheduling and optimization controlled.

Ethernet communication tasks: It is responsible for the communication between the central controller and the terminals, accessing to data and issued instructions, providing data for scheduling algorithm and the means of implementation.

RS485 communication tasks: It is responsible for RS485 communication between the central controller and the underlying devices, accessing to data and issuing instructions.

CAN communication tasks: It is responsible for CAN-bus communication between the central controller and the underlying devices, accessing to data and issuing instructions.

Data storage tasks: It is responsible for systems operation mission-critical data and reading or writing logs.

LCD display tasks: It is responsible for the LCD data display and user interface custom operations.

2.2. Algorithm Design

System software is used modular design. According to the "different time scales", software modules is divided. The priority as Figure 4 (a) shows, and software modules divided as follows:

a) Minute and hour level control module: 1, power recently forecast module; 2, power of real-time forecast module; 3, load power forecast module; 4, energy optimization module; 5, microgrid basic information underlying device settings module; 6, user-related settings module.

b) Second level control module: 1, microgrid system black-start module; 2, normal logic operation modules.

c) Millisecond level control module: 1, islanding detection control module; 2, protection control module; 3, grid-connected switch module; 4, frequency stability control module; 5, voltage stability control module.

Software switching logic as Figure 4(b) shown. In order to microgrid safe, stable and reliable operation, microgrid central controller detects microgrid is islanded or not. Then controller starts protection control module and microgrid system black-start module and judges whether the grid-connected switch is on.

After microgrid start-up, controller controls the microgrid voltage and frequency until stability, and then operates normal logic. In microgrid normal operation, according to basic settings of the user and devices, central controller adjusts the microgrid operation, implementing energy optimization, forecasting distributed generation power and load power, achieving microgrid optimization operation.

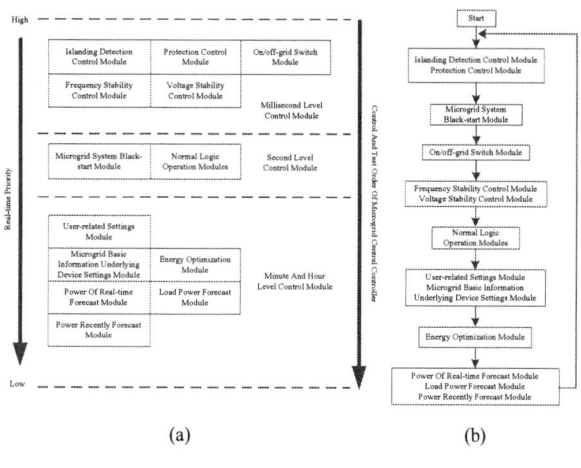

(a) Real-time Priority of Software Logic Module   (b) Software Switching Diagram

Fig. 4: Software module structure diagram

V. MONITORING INTERFACE OF MICROGRID CENTRAL CONTROLLER

Considering the microgrid central controller in different work environments, the number and type of equipments such as photovoltaic inverters, energy storage devices and loads can be configured by microgrid central controller. The human-machine interface is used generalized design that number and type of each devices which can be configured on operation and control interface of human-machine interface. Operation and control interface is used to control that the photovoltaic inverters, energy storage devices and loads is start or stop and displays the active and reactive power of photovoltaic inverters, energy storage devices and load, as well as voltage and current of photovoltaic inverters output, and SOC of energy storage devices. This interface allows operators to monitor PV inverters, energy storage devices and loads.

Real-time information of all interfaces is refreshed every 1 second to monitor devices in real-time. All interfaces display color can change. The number in interface displayed is green in normal operation. The number in interface will change color if the number is over ultra-limited and background highlights. If the number is higher than ultra-limited, number's color is red or orange. If the number is lower than ultra-limited, number's color is blue.

PV monitoring interface is mainly used for monitoring photovoltaic real-time information, including characterization of PV performance parameters such as temperature, efficiency, active power P and reactive power

Q, voltage V, current I, peak power, the daily generating capacity, and so on, drawing every day real-time curves and prediction curve of PV generating capacity.

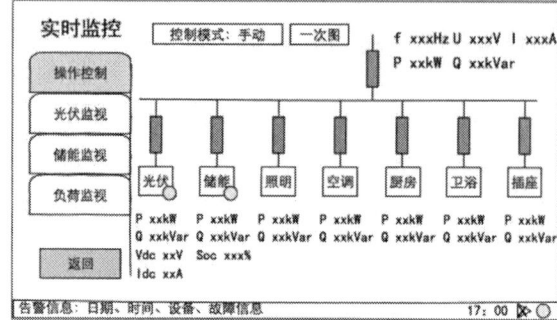

Fig. 5: Interface diagram of operation and control

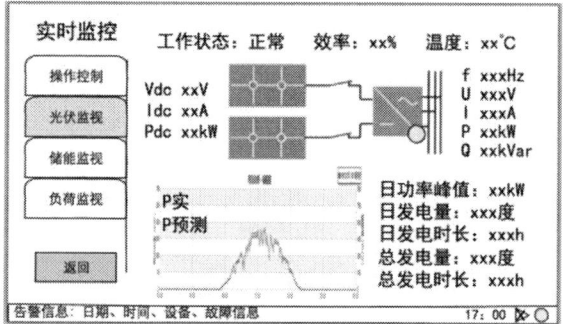

Fig. 6: Monitoring interface diagram of photovoltaic

Energy storage system monitoring interface is mainly used to monitor real-time energy information, including temperature, active power P and reactive power Q, voltage V and current I, discharge capacity, discharge duration and other performance parameters which can be used to characterize storage generation, drawing curves of every day energy storage output power and storage capacity.

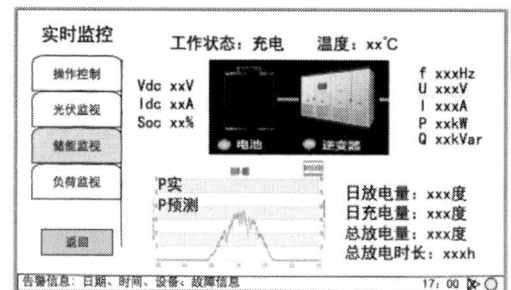

Fig. 7: Monitoring interface diagram of energy storage systems

Load monitoring interface is primarily used to monitor microgrid load information in real time, including various types of loads on electricity consumption proportion, electric quantity of distributed generation output which can be used to characterize the electrical parameters of microgrid users, drawing curves of real-time electric power of user in every day.

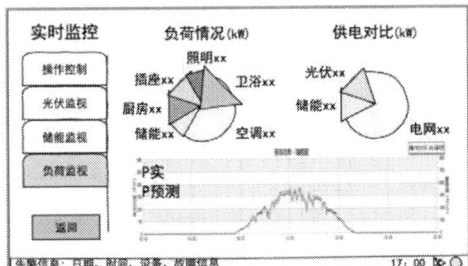

Fig. 8: Monitoring interface diagram of load

VI. MICROGRID CENTRAL CONTROLLER PROTOTYPE

The product structure design is considered the compact, beautiful, safe, durable, easy production, low cost and other factors. The development of hardware circuit is designed by professional structural engineers. External structure design of controller is used clean design style, and its middle part is the product logo and LCD, as shown in Figure 9. Microgrid central controller contains two power input lines and single-panel LCD as a human-machine interface which monitor equipment such as PV inverters, energy storage devices and loads and real-time display the microgrid and operation status. Structure of microgrid central controller is used simple design that its two power input lines have separate detection circuit, as shown in Figure 10. Controller has I/O interface, Ethernet interface, serial peripheral interface, serial communication interface and controller area network communication interface. These interfaces improve controller communication reliability and anti-interference ability. Module design of controller makes system easy for installation, commissioning, maintenance.

Fig. 10: Internal structure design of microgrid central controller

Fig. 9: External structure design of microgrid central controller

VII. CONCLUSION

This paper presents a modular, compact, universal microgrid central controller design, and has produced a prototype, achieving miniaturization and low cost. After initial test, prototype is more stable and reliable in operation. But it needs to carry out the following work before the large-scale application.

(1) Microgrid central controller needs dynamic simulation test based on low-cost embedded platform to test microgrid central controller whether can meet the microgrid control indicators or not.

(2) The demonstration application test function and reliability of microgrid central controller in actual projects.

ACKNOWLEDGMENT

The authors would like to thank the support from the National High-tech R&D Program (863 Program) of China (2014AA052001), Guangdong Technology Research Projects (2012B040303005), Science and technology project of Electric Power Research Institute, CSG (SEPRI-K143003), and Science and technology project of China Southern Power Grid (K-KY2014-009).

REFERENCES

[1] Waleed K A Najy, H H Zeineldin, Wei Lee Woon, "Optimal protection coordination for microgrids with grid-connected and islanded capability," IEEE TRANSACTIONS ON INDUSTRIAL ELECTRONICS, vol. 60, pp. 1668-1677, 2013.

[2] J. M. Guerrero, J. C. Vasquez, J. Matas, M. Castilla, and L. G. de Vicuna, "Control Strategy for Flexible Microgrid Based on Parallel Line-Interactive UPS Systems," in IEEE International Symposium on Industrial Electronics Vigo, SPAIN, 2009, pp. 726-736.

[3] E. Barklund, N. Pogaku, M. Prodanovic, C. Hernandez-Aramburo, and T. C. Green, "Energy Management in Autonomous Microgrid Using Stability-Constrained Droop Control of Inverters," IEEE TRANSACTIONS ON POWER ELECTRONICS, vol. 23, pp. 2346-2352, 2008.

[4] Wang S, Li Z, Wu L, et al, "New metrics for assessing the reliability and economics of microgrids in distribution system". IEEE Transactions on Power Systems, vol. 28, pp 2852-2861, 2013.

[5] A. L. Dimeas and N. D. Hatziargyriou, "Operation of a multiagent system for Microgrid control," IEEE TRANSACTIONS ON POWER SYSTEMS, vol. 20, pp. 1447-1455, 2005.

[6] Liu Zhenya, Electric "Power and Energy in China". *Beijing: China Electric Power Press*, pp. 174-174, 2012.

[7] Q. Jiang, M. Xue and G. Geng, "Energy Management of Microgrid in Grid-Connected and Stand-Alone Modes," *IEEE Transactions on Power Systems,* vol. 28, pp. 3380-3389, 2013.

[8] Wang Chengshan, Wu Zhen, Li Peng, "Research on Key Technologies of Microgrid," *Transactions of China Electrotechnical Society,* vol 29, pp. 1-12, 2014.

[9] SU Ling, ZHANG Jianhua, WANG Li, et al, "Study on some key problems and technique related to microgrid," *Power System Protection and Control*, vol 38, pp. 235-239, 2010.

[10] M. Marzband, A. Sumper, A. Ruiz-Álvarez, J. L. Domínguez-García, and B. Tomoiagă, "Experimental evaluation of a real time energy management system for stand-alone microgrids in day-ahead markets," *Applied Energy,* vol. 106, pp. 365-376, 2013.

[11] Wang Chengshan, Li Peng. "Development and challenges of distributed generation, the micro-grid and smart distribution system," *Automation of Electric Power Systems*, vol 34 pp. 10-14, 2010.

[12] R. H. Lasseter, "MicroGrids," in *Power Engineering Society Winter Meeting, 2002. IEEE*: IEEE, 2002, pp. 305-308 vol.1.

[13] LIU Xueping, LIU Tianqi, LI Xingyuan. "Power coordination strategies for hybrid isolated microgrid and its simulation," *Power System Technology*, vol 34, pp. 202-205, 2010.

[14] Y. W. Li, D. M. Vilathgamuwa and P. C. Loh, "Design, analysis, and real-time testing of a controller for multibus microgrid system," *IEEE TRANSACTIONS ON POWER ELECTRONICS,* vol. 19, pp. 1195-1204, 2004.

[15] IEEE Std "IEEE standard for interconnecting distributed resources with electric". *New York, USA: The Institute of Electrical and Electronics Engineers Inc*, pp. 1547-2003, 2003.

Study on Black Start Strategy of Multi-microgrids with PV and Energy Storage Systems Considering General Situations

Lei YU[1], Jinyong LEI[2], Xiaobin GUO[1], Ping YANG[2,3,4], Zhiji ZENG[2],
Zhirong XU[2,5], Qunru ZHENG[2]

[1] Electric Power Research Institute, CSG, China
[2] School of Electric Power, South China University of Technology, China
[3] Guangdong Key Laboratory of Clean Energy Technology, South China University of Technology, China
[4] Guangdong Engineering Laboratory for wind Power Control and Integration Technology, China
[5] National-Local Joint Engineering Laboratory for Wind Power Control and Integration Technology, South China University of Technology, China
E-mail:249064141@qq.com

Abstract–In recent years, with the rapid development of the microgrid, the multi-microgrids (MMG) has become a new type of power grids, which is comprised of multiple microgrids (MG). It's necessary to study a safe and effective black-start strategy for the MMG, because MMG is more complicated than MG not only in the architecture but also in the control mode. This paper puts forward a control strategy with 3-level structure, based on the hierarchical control theory. In addition, the paper designs a black start strategy based on the serial restoration strategy, for the MMG with the PV and energy storage systems. The results of the simulation on PSCAD/ EMTDC, verify the effectiveness and feasibility of the MMG black start strategy.

Keywords—black start strategy, energy storage, hierarchical control, microgrid, multi-microgrids, serial restoration.

I. INTRODUCTION

In recent years, with rapid economic development, the demand for energy is increasing sharply. The distributed generation technology receives extensive attention for its low pollution, high energy efficiency and high reliability. Nonetheless, there are some problems with distributed generation (DG), which are the high access cost of a single supply, the difficulty to control and the significant impact on the safe and stable operation of power system [1-2].

The concept of microgrid (MG) is put forward to harmonize contradiction between the grid and DGs. The microgrid system combines the distributed generations, loads, energy storage systems and control systems into a small power system [3]. The control strategy reduces the untoward impact when the distributed generations are connected to the system. The microgrids and the larger grids can support each other, and the flexibility of microgrids ensures the reliability of power supply, for the microgrid can be connected to the larger grid or work on the isolated situation [4].

The multi-microgrid (MMG) concept is developed within the framework of the More-Microgrids EU project. The MMG concept was created as a higher level structure similar to the MG but defined at the Medium Voltage (MV). The MMG consists of several MGs and DG units connected on several adjacent MV feeders, together with controllable MV loads. The coordination of several MG and other DG units requires the adoption of a hierarchical control scheme that enables the MMG to provide the flexibility and controllability necessary to support secure system operation [5].

The blackout will bring so huge impact to the economic development and people's daily life. The capability of black start is vital, in order to reduce the interruption time and the economic loss. The black start means a process of expanding the recovery scope of the power system, through the power supplies with black start capability driving the other power supplies without the capability. The black start is the last line of defense for safe and stable operation of the power system [3].

The conventional power system black-start process can be divided into three phases: the preparation phase and recovery phase, the load restoration stage. In the preparation phase, you determine the system partition, select grid recovery policy, select a recovery path, determined black start power supply; recovery phase complete transmission line charging and starts the appropriate black start unit, synchronization subsystem, the maintenance of active and reactive power balance, and load the recovery phase, need to load as soon as possible [6-10].

The black-start process of conventional power systems can be divided into three stages: the preparation stage, the recovery stage and the load restoration stage. In the preparation stage, the Power Dispatching Center should determine the system partition, select the restoration strategy, select a recovery path, determine black start power supplies, etc. In the recovery phase, the Power Dispatching Center should complete the charge of transmission line, start the appropriate black start unit, synchronize the subsystems, maintain the active and reactive power balance, etc. In the load restoration stage, the loads should be recovered as soon as possible [6-10].

As the characteristics of microsources is different from the characteristics of rotary electric machines, the selection of the main power supply of MG black start and the black start strategy is different [11]. The restoration strategies of the MG can be divided into parallel (or called Bottom-up) restoration strategy and serial (or called Top-down) restoration strategy. In a parallel restoration strategy, as the microsources with black start capability are started, the subsystems are built up, and they are connected to the MG through the synchronization devices. The parallel restoration strategy has the characteristics of short recovery time, complexity of the control system, necessity of pre-synchronization and the existence of several subsystems at the initial stage of black start [11-13]. In a serial restoration strategy, the main power supply establishes the stable voltage and frequency, and the other microsources start under the reference voltage and frequency. The serial restoration strategy has the characteristics of longer recovery time than the parallel strategy, and simple control structure [14].

The restoration strategies of the MMG can be also divided into parallel restoration strategy and serial restoration strategy. The [15-16] put forward a control strategy with 3-level structure, based on the hierarchical control theory. The first level control is controlled by the Central Autonomous Management Controller (CAMC). CAMC sends the control signals to the next level controllers according to the electrical information collected from DGs and loads. The second level control is controlled by the Microgrid Central Controller (MGCC). MGCC is responsible for the operation control of a single microgrid, the control of the MG is realized by the third level controller which receive the control signals from MGCC. The third level control is controlled by the bottom controllers, which include the Load Controller (LC) and the Microsource Controller (MC). In the [15-16], the serial restoration strategy is adopted.

This paper studies the structure of the MMG and its control system, and puts forward a black start strategy of the MMG based on a hierarchical control scheme. Compared with the other researchers, this paper shows a black start process with specific action criteria. In this strategy, the best energy storage system is selected as the main power supply of black start, and the black start strategy is based on a serial restoration strategy. The results of the simulation verify the effectiveness and feasibility of the MMG black start strategy.

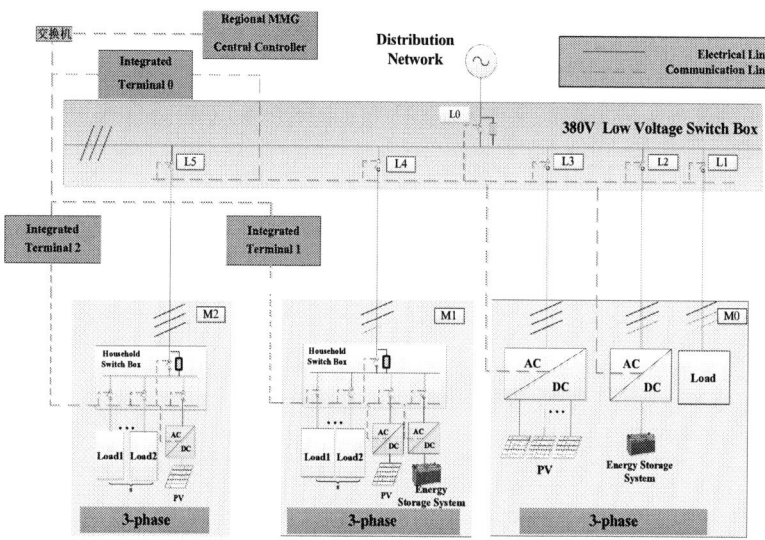

Fig.1: Multi-microgrid topology

II. MULTI-MICROGRID ARCHITECTURE

The architecture of a MMG is shown in Fig.1 .This multi-microgrid includes two layers. In Fig.1, M0 is the top-layer microgrid, M1 is the bottom-layer microgrid, and M2 is a kind of structure for DG interconnected.

The components of M0, M1 and M2 are as shown in Table 1.

Table 1 : Composition of the MMG

	Number	Composition
Top-layer microgrid	M0	PV energy storage system loads
Bottom-layer microgrid	M1	PV energy storage system loads
Structure for DG interconnected	M2	PV loads

The topology of the communication system is also shown in Fig.1. The MMG contains a regional MMG central controller and 3 integrated terminals.

The integrated terminal is a kind of equipment connecting the regional MMG central controller to LC and others controller in the MG, which can realize the islanding detection of the MG, switching between grid-connected situation and isolated situation, optimization and scheduling of the MMG, and failure protection, etc. It has the functions of bi-directional measurement, monitoring and energy management. It collects the electrical information from the PV, energy storage systems and loads, and is controlled by the regional MMG central controller.

The regional MMG central controller is responsible for the control and protection of the whole MMG, has the functions of the islanding detection, switching between grid-connected situation and isolated situation, optimization and scheduling of the MMG, and failure protection, etc.

III. THE HIERARCHICAL CONTROL THEORY OF THE MMG

The output power of the PV is influenced by the light, temperature and other environmental conditions, and has the obvious fluctuation and intermittence. The energy storage system can achieve the complement of the output power of the PV system. But due to the high cost of the storage device, the capability of the battery is limited, so it can only be the short-term energy supplement.

Considering the requirement of large-scale application and the characteristics of the MMG with PV and energy storage systems, a hierarchical control system is adopted in this paper. According to the response rate, time scales, and communication needs, the hierarchical control system is divided into three levels, as shown in Fig.2. Each layer is connected to the nearby layer by the communication cable.

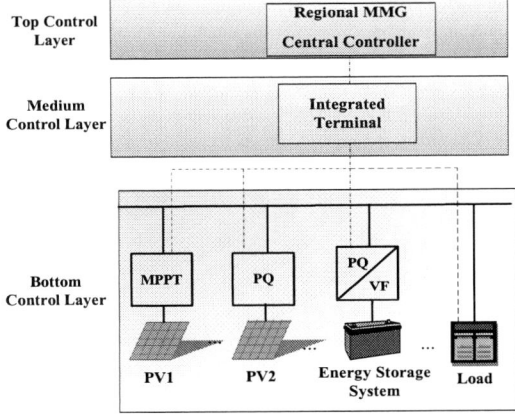

Fig.2: The control system of the MMG based on the hierarchical control

The first layer is the bottom control layer, which includes the load controllers, PV controllers and storage controllers.

The main functions are: (1) The LCs and MCs receive the control signals from the medium control layer, and realize the control and protection of the specific switch, the converters of the PV and energy storage devices and controllable loads, based on the local information without other communication with outside. (2) The LCs and MCs send the electrical information, such as the voltage, current , power and the SOC of the energy storage devices to the medium control layer, providing real-time data for the operation control of MG.

The second layer is the medium control layer, including the integrated terminals. The main functions are: (1) The medium control layer realizes the short-term and super short-term forecast based on light, and temperature and history data, and control the PV and loads. (2) The medium control layer receive the electrical information from the bottom control layer, and send the control signals to the bottom control layer, to maintain the single microgrid or the region (such as the M2 in Fig.1) stable. And the integrated terminal can realize optimization and scheduling, and power balance of the MG. (3) The

medium control layer receive the control signals from the regional MMG central controller, is responsible for the synchronization between a single microgrid and the distributed network. The integrated terminals adjust the power exchange between MG and MMG to reduce the impact of impulse current when MG connected.

The third layer is the top control layer, which is composed of the regional MMG central controller. The main functions are: (1) The regional MMG central controller receives and processes the data from the integrated terminals, and sends the control signals to the integrated terminals to maintain the MMG stable and realize optimization and scheduling, and power balance of the MMG. (2) The regional MMG central controller realizes the adjustment of the output power and voltage by setting a series of parameters of the integrated terminals, to make the MMG play a role of PV node in the bulk power grid.

IV. BLACK START STRATEGY OF MMG

Compared with the other researchers, this paper shows a black start process with specific action criteria.

1. Selection of the main power supply

The choice of main power supply is the key step of MMG black start. In view of the actual situation, the main power supply must meet the following conditions:

a) With the functions of charging and discharging.
b) With the capability of adjusting the voltage and frequency.
c) With sufficient reserve capacity.

For the MMG with PV and energy storage systems, the storage device with high SOC (Stage of Charge) and high capacity should be selected as the main power supply.

2. Procedure of the MMG black start

This paper presents a black start strategy for MMG with PV and energy storage systems, based on a serial restoration strategy. One of the microsources is selected as the main power supply to establish the stable voltage and frequency. After the battery systems (BS) are all connect, the loads and PV are connected in specific sequence with the BS adjusting the output power correspondingly.

The restoration sequences are based on the importance of the loads, BS, and PV. Based on the parameters of the microsources and load, and the importance of the MGs or regions, the restoration sequence can be obtained:

a) BS in M0 > BS in M1 > BS in M2
b) Load in M0 > Load in M1 > Load in M2
c) PV in M0 > PV in M1 > PV in M2

">" indicates that the former should be connected before the latter.

The black start program of the MMG with the PV and battery systems is as shown below:

1) Check whether the black start conditions are met:
   a) Removal of all loads;

b) The voltage and frequency of the MMG system are both 0;

c) Microsources with the capability of black start work properly;

d) Energy storage units with sufficient capacity;

2) Choose a BS as the main power supply and start it with V/f control mode.

3) Check whether the voltage and frequency are stable:

$$\begin{cases} f_{n\min} < f < f_{n\max} \\ U_{n\min} < U < U_{n\max} \end{cases} \quad (1)$$

Here f is the frequency in the MMG, and $f_{n\min}$ is the lower limits of the frequency, and $f_{n\max}$ is the upper limits of the frequency, and U is the voltage, and $U_{n\min}$ is the lower limits of the voltage, and $U_{n\max}$ is the upper limits of the voltage. If the condition is met, the program enters the next step.

4) Connect the next BS according to the restoration sequence mentioned above, until all of the BS is connected. The BS works with PQ control mode, except the main power supply.

5) Check the condition to connect the loads:

$$K_{L_j} \cdot P_{Lj} + P_{net} \leq \sum P_{BS,\max} \quad (2)$$

Here P_{Lj} is the active power of the j th load, and P_{net} is the net power of the MMG, and K_{L_j} is the impact coefficient of the load j when it's connected, and $P_{BS,\max}$ is the maximum power of the BS. If the condition is met, adjust the active power of the PQ adjustable microsources. Then, the load is connected to the microgrid.

6) Check the condition to connect the PV system:

$$P_{pvh} + \sum_{x=1}^{h-1} P_{pvx} + \sum P_{BS,\min} \leq \sum P_L \quad (3)$$

Here P_{pvh} is the output power of the h th PV to connect, and $\sum_{x=1}^{h-1} P_{pvx}$ is the total power of the PV connected. If the condition is met, adjust the active power of the PQ adjustable microsources. Then the PV system is connected.

7) Check whether there are loads and microsources unconnected. If so, return to step 5.

8) End.

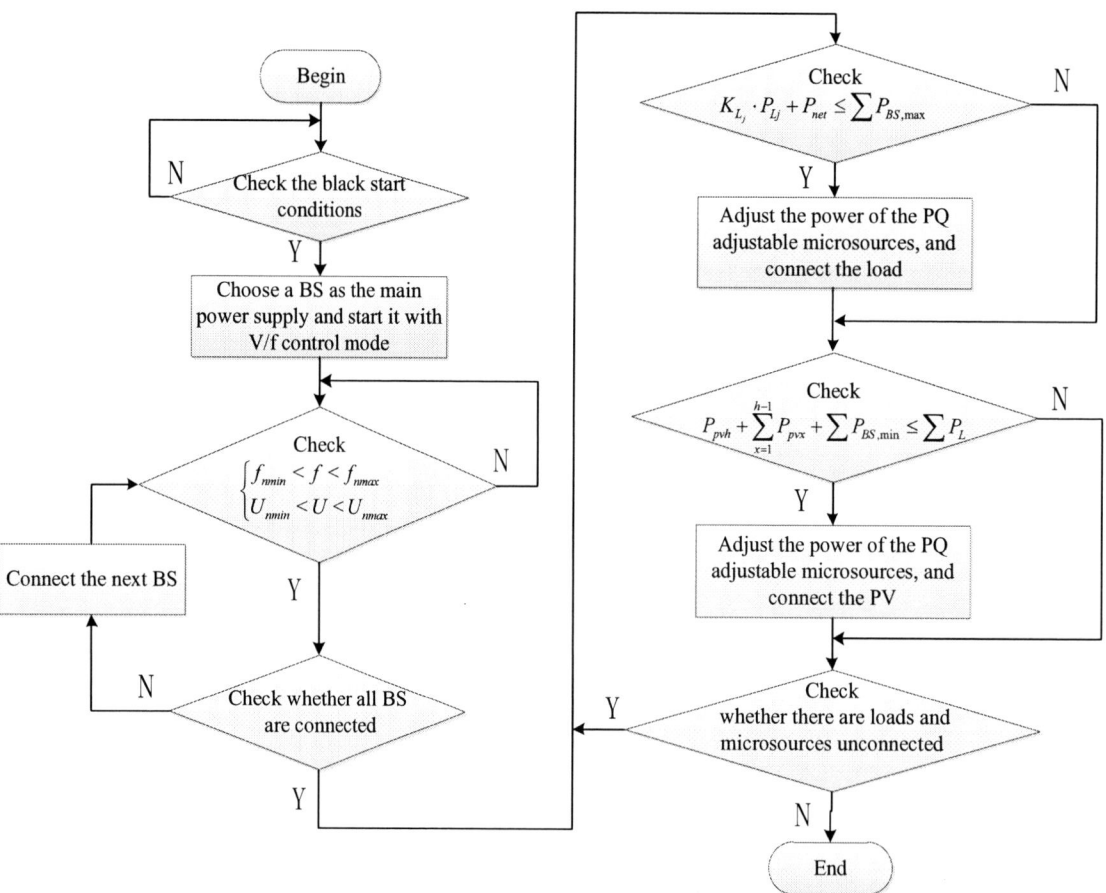

Fig. 3: Control process of the MMG black start

I. SIMULATION PLATFORM AND RESULTS

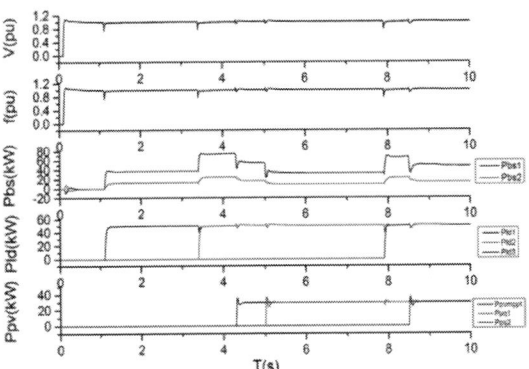

Fig.5: The simulation waveforms of the MMG black start

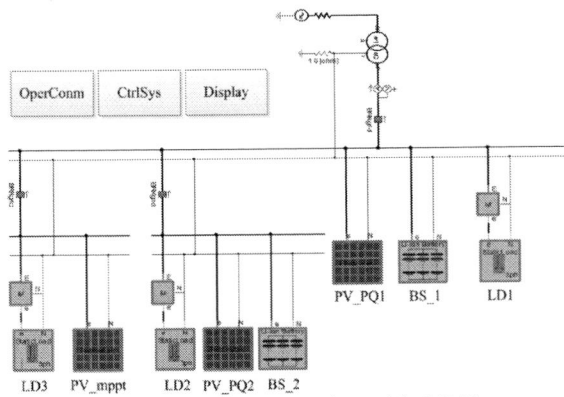

Fig.4: The simulation topology of the MMG

In order to verify the feasibility of the MMG black start strategy proposed in this paper, the simulation model is built on PSCAD/EMTDC simulation platform, whose structure is shown in Fig.4.

Table 2: Composition of the simulation system

Distributed generation systems	PV_mppt（30kW）
	PV_PQ1（30kW）
	PV_PQ2（30kW）
Energy storage systems	BS_1（90kWh）
	BS_2（30kWh）
Loads	LD1（50kW）
	LD2（50kW）
	LD3（50kW）

As shown in Fig.4, the MMG is composed of the PV, BS and loads. The distributed generation systems contain a set of 30kW PV system controlled by MPPT (Maximum Power Point Tracking) mode, and two sets of 30kW PV systems controlled by PQ control mode. The energy storage systems are comprised of a set of 90kWh main power supply and a set of 30kWh auxiliary power supply. The loads are comprised of three 50kW static loads.

This paper sets a typical condition to verify the feasibility of the MMG black start strategy proposed in this paper. The concrete details of the typical condition are as shown below:

Table 3: The concrete details of the typical condition

Weather information	Light intensity	1000W/m²	
	Temperature	25℃	
Battery information	Battery systems	BS_1	BS_2
	SOC	65%	60%
	Capacity	90kWh	30kWh

The simulation waveforms of the MMG black start is shown in Fig. 5.

When the black start conditions are met, the regional MMG central controller selects BS_1 to serve as the main power supply, which has higher SOC and capacity.

The regional MMG central controller sends the control signals to the specific integrated terminal, to start BS_1 with V/f mode. After the voltage and frequency are stablized, BS_2 is connected, with the PQ mode. At t=1.2s, the condition to connect LD1 is met, the regional MMG central controller sends the control signals to the specific integrated terminals, to adjust the output power of BS_2 and connect LD1. When the adjustment of the power of BS_2 is complete, the integrated terminal sends the connection signal to the LC controlling LD1.

At t=3.4s, the condition to connect LD2 is met, and then LD2 is connected. At t=4.3s, PV_mppt is connected. At t=5.0s, PV_PQ1 is connected. At t=7.9s, LD3 is connected. At t=8.5s, PV_PQ2 is connected.

At last, all of the loads and the microsources are connected, and the voltage and frequency of the MMG meet the stability conditions. The MMG black start is complete. The results of the simulation indicate that the MMG system recovers the stable operation, which verify the feasibility of the black start strategy.

II. CONCLUSION

This paper studies on the black start strategy of the MMG with the PV and energy storage systems. This paper puts forward a control strategy with 3-level structure, based on the hierarchical control theory. And based on the serial restoration strategy, this paper studies the control strategies of the microsources, and designs a black start strategy for the MMG with the PV and multiple energy storage systems. Compared with the other researchers, this paper shows a black start process with specific action criteria.

In the process of the MMG black start, the primary reference source with black start capability starts with V/f control mode to establish stable voltage and frequency. After that, loads and other microsources are connected in sequence. After all the loads and microsources are connected and the voltage and frequency are stablized, the MMG finally works in the stable operation.

At last, the simulation platform is built to verify the effectiveness of the MMG black start strategy proposed in this paper. The results of the simulation indicate that the MMG system recovers the stable operation under the control of the black start strategy, which verify the feasibility of the black start strategy.

ACKNOWLEDGMENT

This work was supported by the National High-tech R&D Program (863 Program) of China (2014AA052001), Science and Technology Planning Project of Guangdong Province, China (2012B040303005), Science and Technology Project of Science Academy of China Southern Power Grid (SEPRI-K143003), and Science and Technology Project of China Southern Power Grid Company (K-KY2014-009).

REFERENCES

[1] H. Wang, "Control strategy of microgrid with different DG types", Electric Power Automation Equipment, vol. 32, no. 5, pp. 19-23, 2012.
[2] C. S. Wang, "Development and Challenges of Distributed Generation, the Micro-grid and Smart Distribution System", Automation of Electric Power Systems, vol. 34, no. 2, pp10-14, 2012.
[3] Z. X. Lu, "Overview on Microgrid Research", Automation of Electric Power Systems, vol. 31, no.19, pp100-107, 2007.
[4] X.S. Tang, et al. "Control technologies of micro-grid operation based on energy storage", Electric Power Automation Equipment, vol. 32, no. 3, pp. 99-103, 2012.
[5] Rua. D, et al, "Impact of multi-Microgrid Communication systems in islanded operation", Innovative Smart Grid Technologies (ISGT Europe), 2011 2nd IEEE PES International Conference and Exhibition on IEEE, pp1-6, 2011.
[6] S. M. Xiong, "The Summary of Black-Start Restoration of Electrical Power System after Blackout", Proceedings of the EPSA, vol. 11, no. 3, pp. 12-17, 1999.
[7] D. M. Xia, "Analysis on Constraints during the Process of Power System Black Start-up", Northeast Electric Power Technology, vol. 32, pp40-44, 2009.
[8] X. Y. Fang, Z. Q. Zeng, "Study on Power System Black Start", Electric Power, vol. 33, no. 1, pp. 40-43, 2000.
[9] Y. Liu, "Research on Power System Black-Start Restoration and Its Decision Support Technique". Diss. North China Electric Power University, 2007.
[10] H. R. Zhong, "Research on Optimization Strategy of Black-start and Network Reconfiguration of Power Systems", Diss. North China Electric Power University, 2012.
[11] L.H. Mu, M.D. Xia, Z. Liu, "Research on black-start for microgrid", Power System Protection and Control, vol. 42, no. 22 pp. 32-37, 2014.
[12] Moreira C L , Resende F O , Lopes J, "Using low voltage microgrids for service restoration", Power Systems , IEEE Transactions on , vol. 22, no. 1 pp. 395-403, 2007.
[13] Q. Meng, L.H. Mu, X.F. Xu , "Black-start strategy of isolated microgrid", Electric Power Automation Equipment, vol. 34, no. 3, pp. 59-64, 2014.
[14] X. Huang, X.M. Jin, and L. Ma. "An Optimized Island Micro-Grid Black-Start Control Method." , Transactions of China Electrotechnical Society, vol. 28, no. 4, pp. 182-190, 2013.
[15] Resende F O , Gil N J , Lopes J A, "Service restoration on distribution systems using Multi-MicroGrids" . European Transactions on Electrical Power , vol. 21, no. 2, pp. 1327-1342, 2011.
[16] Hatziargyriou N, "Operation of Multi-Microgrids", Wiley-IEEE Press , 2013

Design Optimization of Linear Induction Motor

Abhay Kumar[1]   M.A.Hasan[2]   Md. Junaid Akhtar[2]   S.K.Parida[2]   R.K.Behera[2]

[1] Department of Electrical Engineering, Indian Institute of Technology, IIT Guwahati, Guwahati, India
[2] Department of Electrical Engineering, Indian Institute of Technology, IIT Patna, Patna, India
E-mail: skparida@iitp.ac.in

Abstract–This paper presents the design optimization of linear induction motor. Application of linear induction motor includes various industrial processes like conveyor belt, vertical movement, steel power plant, induction heating and automated material handling. All these applications require an efficient and high torque providing linear induction motor. Objective function of the optimization problem discussed in this paper includes efficiency, output thrust and machine weight. Various machine design parameters have been used as constraint variables. Optimization done with Quasi-Newton process shows significant improvement in machine efficiency and output torque compared to the results reported in literature. Machine design and optimization is carried out using RMxprt software.

Keywords–Linear Induction Motor, Design Optimization, Quassi-Newton Technique

I. INTRODUCTION

Certain merits of linear induction motor like high starting torque, simple structure, reduction in mechanical losses and smaller size make it an attractive choice as motor in motor driven systems [1]. There are various applications in industries where vertical motion of the machine is required. For example, robotic machines made for building constructions, rope-less lifts in super-skyscrapers, material transport for underground constructions etc. Linear induction motor can be used for all these purposes being able to facilitate linear upward and downward motion. One of the advantages of linear induction motor is its non-contact motion and quick response [2]. In steel power plants, for high quality and high productivity, linear induction motor has been researched and practically used for transporting thin steel plates during reheating and galvanization process [3]. Basic principle of operation of linear induction motor also facilitates eddy current generation by travelling magnetic field. This can help in induction heating [4].

In coal plants, conveyor belt is used to carry the coal blocks from one place to other for further processing. Conventional induction motors used for rotating conveyor belts provide forces at relatively smaller area. Due to stretching or mismatch of forces at two ends, belt faces danger of getting slipped. Linear induction motor can provide such a provision where a belt of conducting material passes between a pair of linear stator blocks carrying poly-phase system of coils [5].

Design of linear induction motor demands preparation of data sheet for stator and rotor dimensions. Number of stator and rotor slots, inner and outer dimensions, tooth and slot dimensions and conductor size requirement are some of the design parameters require to manufacture a linear induction motor [6]. RMxpert software provide platform to design various parameters of the electrical machine by providing performance characteristics of motor for a set of stator and rotor dimensions. By varying these dimensions, a suitable design can be obtained.

Optimization provides an optimum solution for design problem. An optimized design of the motor gives stator and rotor dimension values for which machine gives optimum performance. Various optimization techniques are available in the literature [7]. This paper uses Quasi-Newton optimization process to perform the design optimization.

Paper is organized as follows. Section 2 presents linear induction motor modeling. Section 3 discusses Quasi Newton optimization technique used in this paper. Results obtained for optimized machine is presented in section 4. Finally conclusion discusses the significance of presented work and results obtained.

II. LINEAR INDUCTION MOTOR MODELING

Fig. 1 presents the cross-sectional view of the rotor and stator section of the LIM. Non-uniform airgap causes an effective airgap g_e different from physical airgap g_m. Relation between physical and effective airgap is given in (1-2).

$$g_e = k_c g_o \qquad (1)$$

$$g_o = g_m + d \qquad (2)$$

where d is the thickness of the conducting layer and k_c is Carter's coefficient given by (3),

$$k_c = \frac{\lambda}{\lambda - \gamma g_o} \qquad (3)$$

Parameter λ is the slot pitch which is the distance between the centers of the two consecutive teeth. Stator slot depth h_s can be calculated from (4).

$$h_s = \frac{A_s}{W_s} \qquad (4)$$

where A_s is the cross sectional area of the slot and W_s is the slot width. A relation for the slot cross sectional area with number of turns per slot and conductor cross section area is given in (5).

$$A_s = \frac{10.N_c A_w}{7} \qquad (5)$$

978-1-5090-0064-7/15 $31.00 © 2015 IEEE

Here N_c is the number of turns per slot. Yoke height of the stator core h_y is the portion of the core below teeth. If it assumed that the flux in the yoke is one half of the flux in air gap, it can be expressed as,

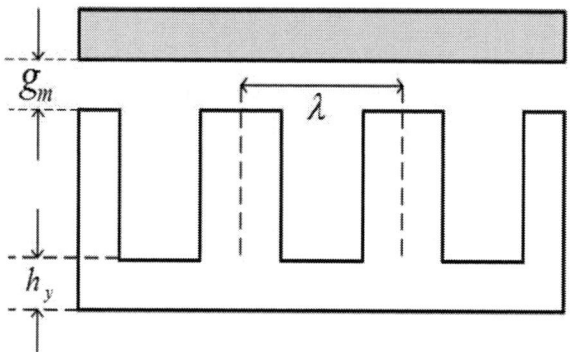

Fig. 1: Cross section of linear induction motor

$$h_y = \frac{\Phi_p}{2 B_{y\max} W_s} \quad (6)$$

A linear induction motor of the specification given in table I is designed using RMxpert software. Complete datasheet is presented in table II. Mathematical model is extended to obtain the electrical equivalent circuit model of LIM. Electrical equivalent circuit of LIM consists of series elements R_1 and X_1, core reactance component X_m in parallel with a variable resistance R_s/s representing mechanical load. Equations representing equivalent circuit components are given in (7-11).

$$R_1 = \frac{\rho_w l_w}{A_{wt}} \quad (7)$$

$$X_1 = \frac{2\mu_O \pi f \left[\left(\lambda_s \left(1 + \frac{3}{p}\right) + \lambda_d \right) \frac{w_s}{q_1} + \lambda_e I_{ce} \right] N_i^2}{p} \quad (8)$$

$$X_m = \frac{24\mu_o \pi f W_{se} k_w N_i^2 \tau}{\pi^2 p g_e} \quad (9)$$

$$R_2 = \frac{X_m}{g_e} \quad (10)$$

$$G = \frac{2\mu_o f \tau^2}{\pi g_e} \quad (11)$$

where ρ_w, l_w and A_{wt} are volume resistivity, length and cross sectional area of stator winding copper wire. K_p is pitch factor, K_w is winding factor and g_e is equivalent air gap. Based on the electrical equivalent circuit, performance parameters of LIM like electromagnetic torque, power output and efficiency can be obtained as given in (12-14).

$$T = \frac{mI_2^2 R_2}{V_s s} \quad (12)$$

$$P_o = mI_2^2 R_2 \frac{1-s}{s} \quad (13)$$

$$\eta = \frac{mI_2^2 R_2 (1-s)}{mI_2^2 R_2 + mI_1^2 R_1 s} \quad (14)$$

III. DESIGN OPTIMIZATION

Solution of an optimization problem follows an iterative process which starts with an initial point x_o, producing a sequence of points x_k that converges at an optimum point x^*. If x is a vector of design variables with constraints $x_{min} < x < x_{max}$ and f is a function to be optimized, then second order Taylor expansion around x_k is given by (15)

$$m_k(p) = f_k + p^T \nabla f_k + \frac{1}{2} p^T B_K P . \quad (15)$$

Where $p = x - x_k$, and B_k denotes the Hessian matrix (second order partial derivatives of x). In this paper following design variable constraints have been used.

Slot width: 10 mm < W_s < 25 mm
Slot height: 20 mm < h_s < 35 mm
Primary height: 45 mm < h_y < 60 mm
Secondary sheet thickness: 2.5 mm < d < 7.5 mm
Air gap: 0 mm < g_m < 7 mm
Current density: 3 A/mm^2 < J_1 < 6 A/mm^2

Objective function for the optimization used in this paper includes efficiency, electromagnetic torque and weight of the machine. Objective function is built so as to maximize efficiency and torque and minimize weight of the machine. Objective function is given by (16).

$$f(x) = (T(x) * \eta(x)) / W(x) \quad (16)$$

Where $T(x)$, $\acute{\eta}(x)$ and $W(x)$ represent torque, efficiency and weight function respectively.

Table 1: Linear induction motor specification

Specification	Value
Rated Power	3.73 kW
Rated Voltage	440 V
Number of poles	4
Rated frequency	50 Hz
Rated speed	1428

Table 2: Complete Datasheet

Design Parameter	Design value
Number of stator slots	36
Stator outer diameter	213 mm
Stator inner diameter	119 mm
Number of rotor slots	30
Inner diameter of rotor	61.8 mm

IV. SIMULATION AND RESULTS

Electrical machine design is an iterative process. Search of a machine design with desired output require repetitive calculations to satisfy all the constraints. Help of software based platform in this work is highly appreciated. ANSYS RMxprt is a design tool which calculates the performance of a hypothetical machine based on electrical equivalent circuit. It helps in making decision regarding machine dimensions and material selection. User provides a set of initial machine dimensions. Software does rigorous electromagnetic transient analysis to produce possible performance of the machine.

Table 3: Design of LIM

Design parameter	Design value (in mm)
Stator top tooth width	5.98
Stator bottom tooth width	5.96
Length of the stator core	140
Diameter of the conductor	1.151
Stator resistance	4.12 ohms
Air gap	0.3
Length of the rotor	140
Rotor stacking factor	0.96

Table 4: Optimum Design of LIM

Design parameter	Design value (in mm)
Primary current density	5.6
Primary width	140
Primary height	49.5
Secondary sheet thickness	5
Air gap	5
Motor length	1090
Tooth width	12
Slot width	18
Efficiency	71.47
Weight	50 kg
Output torque	43.74

Based on iterative process, a machine design for the specification given in table 1 and 2 is prepared. Table 3 presents complete stator and rotor dimensions for the LIM design. Performance of this machine is given in Table 4. ANSYS RMxprt also generates the magnetic field and flux density distribution in the air gap. Magnitude of air gap flux and magnetic field is shown through different color lines. Fig. 2-3 gives magnetic field and flux representation.

As given in table 3 and 4, a theoretical design of Linear Induction motor of efficiency 67 percent, power factor 0.87 and electromagnetic torque of 24 N-m is produced. Flux density and magnetic field distribution shows the end effect in LIM. Red color closed loop lines at extreme left end shows that due to end effect, flux density is more than elsewhere.

As we increase the load, the load current changes. Corresponding change in speed and input power is given in fig. 4-5. For full load condition, full load current of approx. 10 A is drawn at rated speed of 1428 rpm. As load is increased, output power requirement increases and hence secondary current increases. Since magnetizing reactance remains constant, primary side current increases. This characteristic is shown in fig. 4-5. As we increase speed of the motor, input power has to be increased if constant torque operation is desired. This is presented by fig. 6. In low speed range, torque is directly proportional to the speed. As speed increases, slip decreases. When speed is increased beyond full load speed, voltage drop across magnetizing reactance becomes significant. If the load is increased beyond breakdown point, decrease in rotor power factor becomes significant and torque decreases. This is presented by fig. 7-8.

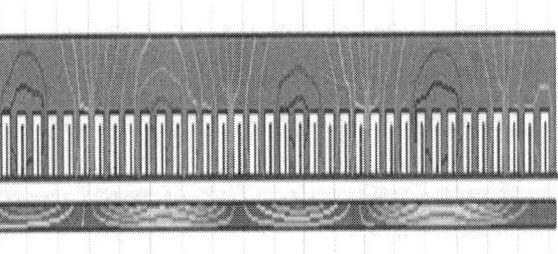

Fig. 2: Flux density distribution

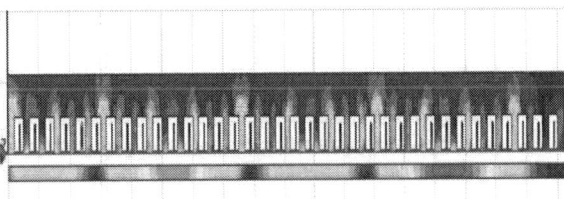

Fig. 3: Magnetic field distribution

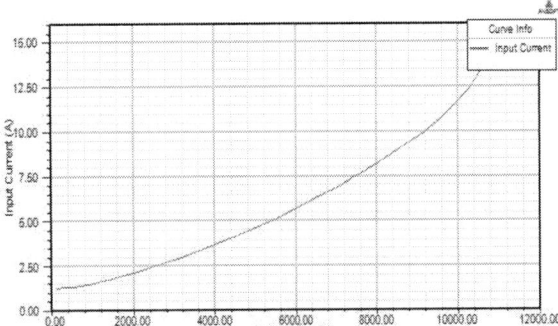

Fig. 4: Current Vs output power

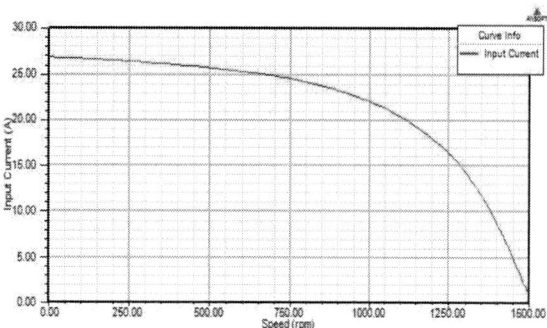

Fig. 5: Current Vs speed

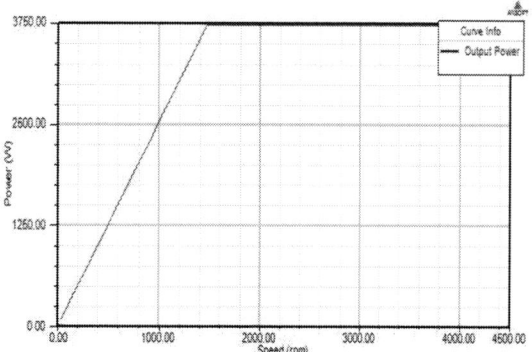

Fig. 6: Power Vs speed

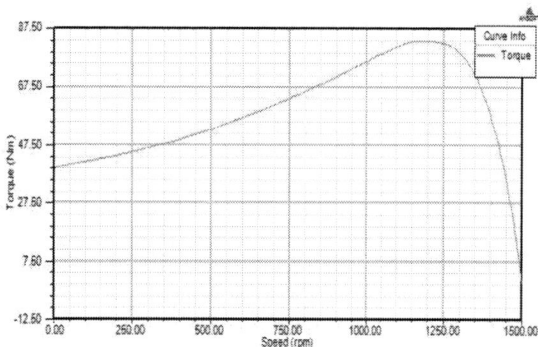

Fig. 7: Torque Vs speed

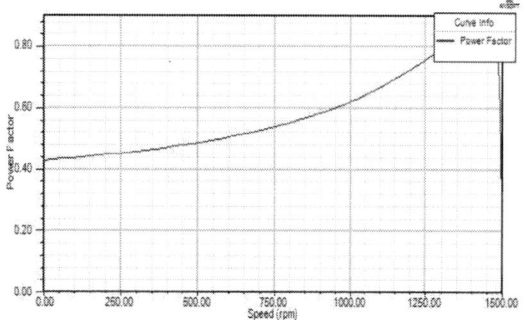

Fig. 8: Power factor Vs speed

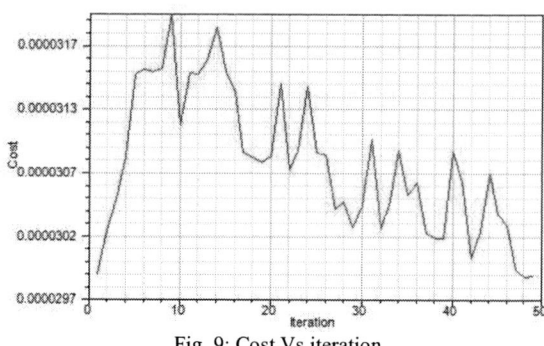

Fig. 9: Cost Vs iteration

Above discussed theoretical machine has a very high efficiency but lower electromagnetic torque. For applications discussed in this paper where LIM's are employed, there is need of high starting and full load torque. Also a low weight machine is preferred for dynamic operation. In order to achieve these requirements, optimization is done. An initial setting of efficiency, weight and torque is fed to Quasi-Newton algorithm. Based on results obtained and cost Vs iteration relation, initial settings may be required to be changed. Finally an optimized design is obtained.

Accuracy and closeness of desired output in optimization varies with number of iterations. Fig. 9 presents the cost Vs iteration curve. Cost signifies the closeness of result to the desired value. For higher accuracy, more number of iteration is required. Optimization presented in this paper use 50 iterations. Table 4 presents the optimized machine design. It was found that different design constraints affect motor performance in different ways. Study of independent effect of each constraint is analyzed by varying a particular constraint keeping others constant.

An increase in secondary sheet thickness increases the primary weight and decreases the torque. Power factor and efficiency increases first then starts decreasing. Thus design optimization requires a compromise. In this work power factor is compromised which allows a higher value of air gap.

An increase in current density keeping output torque constant decreases conductor wire diameter and increases the winding resistance. This results in reduction in weight and efficiency. Since both weight and efficiency optimization is desired, so the results may vary with different number of iterations. This work has presented the result obtained for 50 iterations. Objective function is optimized for a value of $3.1*10^{-5}$.

V. CONCLUSION

This paper has presented an optimized design of linear induction motor for better efficiency, weight and torque. A conventional design based on electrical equivalent circuit is prepared in RMxprt software. Optimization of the machine design has been carried out using Quasi-Newton optimization technique. All design parameters affect performance of the machine in a different way. Their independent effects have been discussed in this paper.

Optimized machine achieves better desired output as compared to the conventional design

REFERENCES

[1] Sadler, G.V.; Davey, A.W., "Applications of linear induction motors in industry," in *Electrical Engineers, Proceedings of the Institution of*, vol.118, no.6, pp.765-776, June 1971

[2] Morizane, T.; Masada, E., "Study on the feasibility of application of linear induction motor for vertical movement," in *Magnetics, IEEE Transactions on*, vol.29, no.6, pp.2938-2940, Nov 1993

[3] Fujisaki, K., "Application of electromagnetic force to thin steel plate," in *Industry Applications Conference, 2002. 37th IAS Annual Meeting. Conference Record of the*, vol.2, no., pp.864-870 vol.2, 13-18 Oct. 2002

[4] Yamada, Takahiro; Fujisaki, K., "Basic Characteristic of Electromagnetic Force in Induction Heating Application of Linear Induction Motor," in *Magnetics, IEEE Transactions on*, vol.44, no.11, pp.4070-4073, Nov. 2008

[5] Laithwaite, E.R.; Tipping, D.; Hesmondhalgh, D.E., "The application of linear induction motors to conveyors," in *Proceedings of the IEE - Part A: Power Engineering*, vol.107, no.33, pp.284-294, June 1960

[6] Shiri, A.; Shoulaie, A., "Design Optimization and Analysis of Single-Sided Linear Induction Motor, Considering All Phenomena," in *Energy Conversion, IEEE Transactions on*, vol.27, no.2, pp.516-525, June 2012

[7] Xian Liu; Wilsun Xu, "A Global Optimization Approach for Electrical Machine Designs," in *Power Engineering Society General Meeting, 2007. IEEE*, vol., no., pp.1-8, 24-28 June 2007

Design Technique of Coupled Inductor Filter for Suppressing Switching Ripples in PWM Converters

Hyeon-gyu Choi   Jung-Ik Ha

Department of Electrical Engineering, Seoul National University, Seoul, Korea
E-mail: nadcha@snu.ac.kr, Jungikha@snu.ac.kr

Abstract–This paper proposes a design method of a coupled inductor filter for suppressing switching ripples in PWM converters. High attenuation performance in the ripples is achieved by adjusting positions of poles and zeros. The previous researches only focused on a double zero and a double pole, which determine notching and bandwidth frequencies of the filter. However, the proposed filter gives the increased attenuation in high frequency region by placing another double pole between bandwidth and notching frequencies. For this property, this paper proposes the design procedure of a coupled inductor filter. In addition, control strategies to find the resonance frequency and minimize the ripples are presented in the full paper. Furthermore, an uncoupled general form circuit, named LLC-LC filter is proposed as a filter topology.

Keywords–Coupled inductor filter, Design optimization, Power filters

I. INTRODUCTION

With the development of pulse width modulation (PWM) converters, demand for high quality energy conversions are increasing. Especially in grid-connected system or high precision servo motor drive system, high attenuation property in the ripples generated by PWM is essential. For these reasons, many filter topologies have been studied for a long time such as L, LC, and LCL filters, etc. A coupled inductor filter is one of the powerful conventional concepts [1]. It is composed of one core, two coils, and two capacitors. Compared with 4th order cascaded LC filter (two cores, two coils, and two capacitors), it has smaller number of magnetic components and shows better harmonic attenuation characteristics. As a result, total size of the circuits can be reduced.

For these reasons, many papers dealt with the coupled inductor filters. However, most of them focused only on its application not on a coupled inductor filter itself. Balog and Krein summarized the characteristics of the coupled inductor filter and treated it as a filter block itself [2]. However, the design method was not considered.

This paper gives more precise approaches to the coupled inductor filter to improve the attenuation. First, the basic principles are introduced. Second, with bode plot construction technique, positions of the poles and zeros are considered. From this result, a design method is proposed. Finally, simulation results are presented to verify the proposed design.

II. PRINCIPLES OF COUPLED INDUCTOR FILTER

1. Ideal model

A basic operation principle comes from an ideal model with perfect coupled inductor and infinite capacitor in Fig.

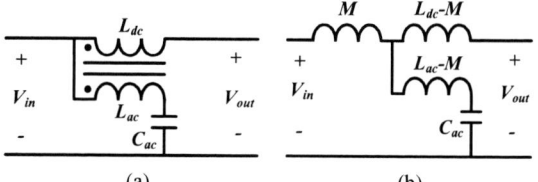

(a)   (b)

Fig. 1: Coupled inductor filter. (a) circuit diagram. (b) T-equivalent model.

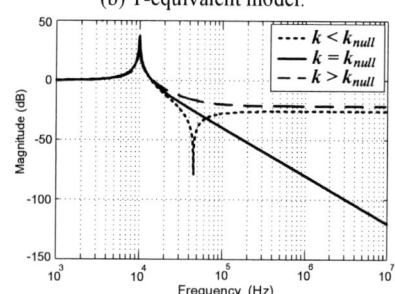

Fig. 2: Bode plot of the realistic model.

1(a). A branch which is composed of L_{ac} and C_{ac} is shown as a short circuit for ac components of the current and an open circuit for dc components. Thus, total ac components of the current flow through this branch. As a result, only dc components of the voltage are shown at the output port. This ripple steering property is the basic concept of the coupled inductor filter [3].

2. Realistic model

The imperfect coupling of the transformer and the finite value of the capacitor are considered in realistic model. In this sense, a T-equivalent model can be drawn as Fig. 1(b). The transfer function is as follows:

$$H(s) = \frac{V_{out}(s)}{V_{in}(s)} = \frac{1 + s^2 C_{ac}(L_{ac} - M)}{1 + s^2 C_{ac} L_{ac}}, \quad (1)$$

where $M = k\sqrt{L_{ac}L_{dc}}$ and $k_{null} = \sqrt{L_{ac}/L_{dc}}$. The bode plots are depicted in Fig. 2. If coupling coefficient k equals k_{null}, the transfer function has two imaginary poles. It is typical 2nd order low pass filter. If k is larger than k_{null}, the transfer function has two imaginary poles and two real zeros. It shows flat property at high frequency. Thus, it is meaningless for filtering performance. And if k is smaller than k_{null}, it has not only two imaginary poles but also two imaginary zeros. It appears notch characteristic which is useful nature for PWM converter. Generally, major portion of the ripple component occurs at multiples of the switching frequency. By placing zeros at main switching ripple frequency, high attenuation is achieved.

Low pass and notch filtering properties cannot be obtained

978-1-5090-0064-7/15 $31.00 © 2015 IEEE

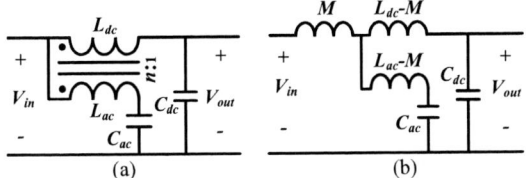

Fig. 3: Coupled inductor filter (a) with output capacitor (b) equivalent circuit.

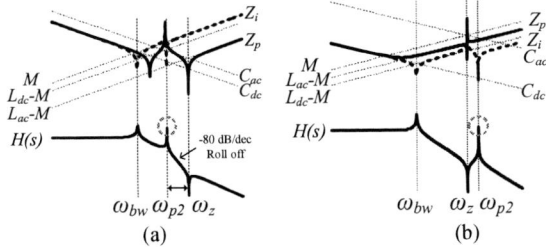

Fig. 4: Bode plot construction of coupled inductor with output capacitor (a) with $L_{dc} - M > L_{ac} - M$ (b) with $L_{dc} - M < L_{ac} - M$.

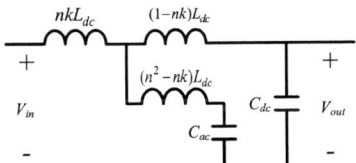

Fig. 5: T-equivalent model represented by alternative parameters; n, k, and L_{dc}

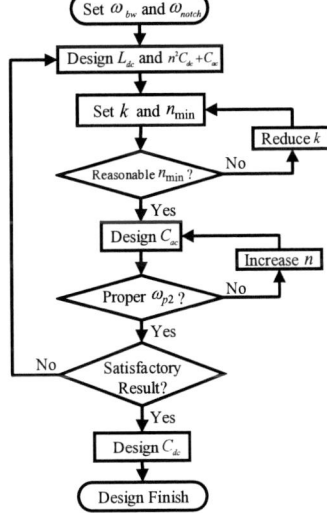

Fig. 6: Design flow chart

simultaneously using the circuit shown in Fig. 1. And the leakage inductance of dc side inductor (L_{dc}-M) does not participate in poles and zeros location. In other words, there is a redundant design factor.

3. Coupled inductor filter with output capacitor

To present both low pass and notch filter characteristics, another capacitor (C_{dc}) is added [2]. Fig. 3(a) and Fig. 3(b) shows the coupled inductor with output capacitor and its equivalent circuit. Transfer function is as follows:

$$H(s) = \frac{1 + s^2 C_{ac}(L_{ac} - M)}{1 + s^2(C_{ac}L_{ac} + C_{dc}L_{dc}) + s^4 C_{ac}C_{dc}(L_{ac}L_{dc} - M^2)}. \quad (2)$$

There are two double poles and one double zero as shown in Fig. 4. Most previous researches focused on a double zero (ω_{notch}) which determines notching frequency and one double pole (ω_{bw}) which determines the bandwidth of the filter. They did not consider second double pole (ω_{p2}) because of the complexity of (2). Fei, et al obtained the attenuation with a LCL-LC filter [4]. In the filter design proposed in this paper, ω_{p2} is located between ω_{bw} and ω_z to obtain the higher attenuation with the filer as shown in Fig. 4(a).

III. DESIGN PROCEDURE

To analyze the complicated circuits, bode plot construction method as in [5] is used. The position of the second double pole which is highlighted by dotted circle in Fig. 4 can be changed by the ratio of L_{dc}-M and L_{ac}-M. In case of Fig. 4(b) where L_{dc}-M is smaller than L_{ac}-M, ω_{p2} is located at the right side of ω_z. It definitely makes the filter performance worse because the attenuation rate at high frequency decreases. On the other hand, ω_{p2} is located between ω_{bw} and ω_z in Fig. 4(a), where L_{dc}-M is larger than L_{dc}-M. Higher attenuation is obtained because -80 dB/dec roll off region is enlarged as shown in Fig. 4(a). As a result, the attenuation rate increases for the frequencies higher than ω_{p2}. Thus, the second double pole position should be considered in design procedure.

To derive the practical design method for a coupled inductor filter, it is better to change circuit parameters M, L_{ac}, and L_{dc} to n, k, and L_{dc}. With a little calculation, parameters are obtained as shown in Fig. 5 and its transfer function is

$$H(s) = \frac{1 + s^2(n^2 - nk)L_{dc}C_{dc}}{1 + s^2 L_{dc}(n^2 C_{ac} + C_{dc}) + s^4 n^2(1 - k^2)L_{dc}^2 C_{dc}C_{ac}}. \quad (3)$$

There are two design criteria. First, n should be smaller than 1 because the second double pole is located between ω_{bw} and ω_z (L_{dc}-M>L_{ac}-M). Second, n should be larger than k to get a notch characteristic (L_{ac}-M>0). To sum up, design boundary condition is

$$k < n < 1. \quad (4)$$

1. Design of L_{dc} and $n^2 C_{ac} + C_{dc}$

From (3), poles are calculated as

$$\omega_{bw} = \pm \sqrt{\frac{2}{L_{dc}\left[\sqrt{(n^2 C_{ac} + C_{dc})^2 + 4n^2(k^2-1)C_{dc}C_{ac}} + (n^2 C_{ac} + C_{dc})\right]}},$$
$$\omega_{p2} = \pm \sqrt{\frac{2}{L_{dc}\left[-\sqrt{(n^2 C_{ac} + C_{dc})^2 + 4n^2(k^2-1)C_{dc}C_{ac}} + (n^2 C_{ac} + C_{dc})\right]}} \quad (5)$$

If k is close to 1, ω_{bw} is approximated as

$$\omega_{bw} \approx \sqrt{\frac{1}{L_{dc}(n^2 C_{ac} + C_{dc})}}. \quad (6)$$

From (6), the bandwidth of the filter is determined by self-inductance of dc side transformer L_{dc} and equivalent

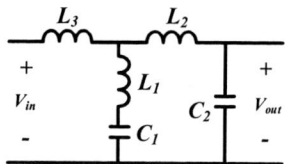

Fig. 7. LLC-LC filter.

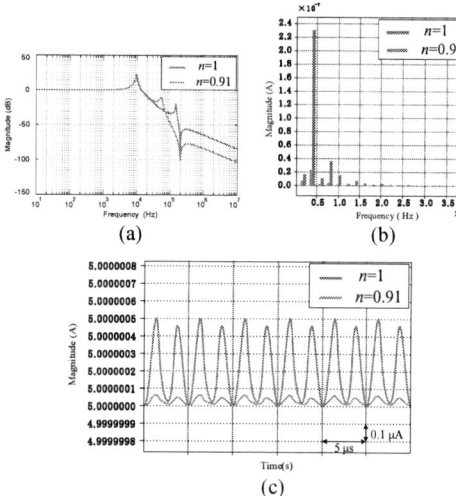

Fig. 8: Simulation result of the simulation model.
(a) Bode plot (b) frequency spectrum (c) output current.

capacitor C_{eq} ($=n^2C_{ac}+C_{dc}$). These parameters can be designed in a similar way of LC filter [6].

2. Design of n and k

The position of second double pole is important in performance of the filter as aforementioned. However, unwanted peak point is accompanied. In a usual PWM modulation, ripple currents are concentrated on multiples of switching frequency. Therefore, it is better to place the second pole between bandwidth pole and notching zero.

From (5), ω_{p2} decreases by reducing n and k. At first, k is selected to feasible maximum value temporarily, e.g. 0.95. From (4), minimum n is selected, e.g. 0.96. If it is hard to implement such turn ratio because of small L_{dc}, which means number of turns is relative small, reduce k and select minimum n again until reasonable values are obtained.

3. Design of capacitors, C_{ac} and C_{dc}

Once n and k are determined, C_{ac} can be designed as

$$C_{ac} = \frac{1}{\omega_{notch}^2(n^2 - nk)L_{dc}}. \tag{7}$$

From the parameters obtained with above procedures, the position of the second double pole can be calculated by (5). If it is smaller than expected value, increase n and calculate C_{ac} again. If it is larger than expected one even with minimum n, reduce k and repeat above steps again. If proper k and n are selected, C_{dc} is obtained as

$$C_{dc} = C_{eq} - n^2C_{ac}. \tag{8}$$

Flowchart diagram in Fig. 6 shows overall steps of designing a coupled inductor filter.

The bode plot can be drawn using parameters obtained above. And then, the input ripple current should be checked. Because current and voltage stress on the switching devices increases if the input ripple current is severe.

If the obtained parameters are too subtle or not implementable, an uncoupled type filter LLC-LC, shown in Fig. 6, can be one solution. The size of L_1, C_1, and C_2 are not bulky because only small ripple current flows through them. Also L_2 is not bulky because it has small inductance. By adding these small components, dramatic filtering effect can be achieved. Compared with LCL-LC filter [4], better PWM harmonic attenuation can be achieved. A design method and control techniques for LLC-LC filter will be discussed in a full paper.

V. SIMULATION RESULTS

Simulation is accomplished on 3 phase voltage source inverter system. Rating power is 600 W and switching frequency is 100 kHz. Bandwidth of the filter is set to 10 kHz. The modulation method is space vector PWM. The notching frequency is 200 kHz because twice of switching frequency is produced by space vector PWM.

Table 1: Designed parameters

Parameters	Values
L_{dc}	112 μH
k	0.9
n	0.91
C_{dc}	1.6 μF
C_{ac}	0.57 μF

The simulation results are compared with the coupled inductor filter which do not consider the position of the second pole, $n = 1$. Designed transfer functions are shown in Fig. 8(a). Because of the second double pole position, dramatic attenuation difference is obtained. Fig 8(c) shows wave form of the simulation result when current reference is 5 A. The magnitude of the current ripple is only 8.3×10^{-6} % of the rated current with the proposed filter. High order harmonics are reduced significantly as shown Fourier analysis in Fig. 8(b).

VI. CONCLUSION

In this paper, a novel design method of a coupled inductor filter is presented. The position of the second double pole is adjusted by two design factor; turn ratio (n) and coupling coefficient (k). In this way, the ripple steering ability is dramatically enhanced. The proposed designed filter is verified by simulation. Experimental results and more analysis will be added in a full paper.

REFERENCES

[1] G. B. Crouse, "Filtering electric currents," US Patent 1804859, May 12, 1931.

978-1-5090-0064-7/15 $31.00 © 2015 IEEE

[2] Balog, Robert S., and Philip T. Krein, "Coupled-inductor filter: A basic filter building block." IEEE Trans. Power Electron, vol. 28, no. 1, pp. 537-546, Jan. 2013.

[3] R. P. Severns and G. Bloom, Modern DC-to-DC Switchmode Power Converter Circuits, New York: Van Nostrand Reinhold, 1985, pp. 266–324.

[4] Li, Fei, et al. "An LCL-LC Filter for Grid-Connected Converter: Topology, Parameter, and Analysis." IEEE Trans Power Electron, vol. 30, no. 9, pp. 5067-5077, Sep. 2015.

[5] B. Choi, Pulsewidth Modulated DC-to-DC Power Conversion: Circuits, Dynamics, and Control Designs, New Jersey: John Wiley & Sons, 2013, pp. 257-260.

[6] Ahmad, A. Ale, et al. "A new design procedure for output LC filter of single phase inverters," the 3rd Int. Conf. Power Electron. Intell. Transp. Syst., Shenzhen, China, Nov. 2010.

Analysis of the Influence of Photovoltaic Sources Integrated into Distribution Network in Different Ways

Xiyuan MA[1], Jinyong LEI[1], Xiaobin GUO[1], Ping YANG[2,3,4], Jiajun Peng[2],
Zhirong XU[2], Shaoxiong ZHOU[5]

[1] Electric Power Research Institute, CSG, Guangzhou, China
[2] School of Electric Power, South China University of Technology, China
[3] Guangdong Key Laboratory of Clean Energy Technology, South China University of Technology, China
[4] Guangdong Engineering Laboratory for wind Power Control and Integration Technology, China
[5] Guangdong Intework Energy Technology Co., Ltd, Guangzhou, China
E-mail: 243109541@qq.com

Abstract–Analysis of the impact of photovoltaic sources massively connect to distribution network will be helpful to take measures in advance to improve the stability of the power grid. According to the voltage changes of distribution network under both ways of the photovoltaic sources single-point and multi-points access to distribution network, this paper puts forward three quantitative indicators, which include of voltage difference change rate, static voltage indicator and net loss variation. In view of practical engineering network topology, the quantitative indicators are calculated to analyze the influence of photovoltaic sources integrated into the distribution network, and the rationality of photovoltaic sources access scheme of the demonstration project is verified. It proves that these quantitative indicators have practical engineering applications value.

Keywords–Distribution network, High permeability, Photovoltaic, Quantitative indicators.

I. INTRODUCTION

As distributed energy massively integrated into distribution network, its clean, efficient and environmentally friendly features are shown adequately[1-4]. Meanwhile, the trend of the distribution characteristics and each node voltage will have corresponding change, resulting in voltage fluctuation and flicker, which will has a variety effects on the stability of the distribution network[5-8].

Aiming at the effects on each point voltage distribution feeders when photovoltaic sources integrated into distribution network, the paper[9] establishes the deviation model of medium voltage distribution network feeder voltage at each point, to study the effect of photovoltaic sources integrated into distribution network and the photovoltaic allowed capacity that does not occur overvoltage. But this paper does not analyze the network loss caused by photovoltaic sources integrating into the distribution network. The paper[10] proposes the voltage-reactive power sensitivity algorithm in order to analyze the influence of photovoltaic sources accessing. But the algorithm only considers the distribution network voltage sensitivity to the change of reactive power, and actually active power of distributed power supply to the distribution network can't be ignored. According to the specific distribution network topology, the paper[11] calculates the allowed photovoltaic sources access capacity of each user by the method of simulation. But it is difficult to draw general conclusions in typical topologies because this method needs to build models according to the specific application objects, and it's difficult to build

models when the topology of the distribution network is complex.

The voltage deviation and fluctuation of the distribution network is the key factor limiting the acceptable capacity of photovoltaic sources integrating into distribution network. So aiming at the problem of voltage rise, this paper puts forward three quantitative indicators to assess the influence of high permeability photovoltaic sources access to distribution network in different ways, and the rationality of photovoltaic sources access scheme of demonstration projects is verified. It proves that these quantitative indicators have practical engineering applications value.

II. TYPICAL TOPOLOGIES ANALYSIS IN DIFFERENT WAYS

1. Photovoltaic sources single-point access to distribution network

As shown in fig.1 is the typical low voltage distribution network line load distribution when the photovoltaic sources single-point access. There are N users on the line, the apparent power of the user N is $P_n + Q_n (n = 1, 2, \cdots N)$. The voltage of the head of line is U_0 and set the value remains the same. The voltage of the user N is $U_n (n = 1, 2, \cdots N)$. The line impedance between the user $N-1$ and user N is $R_n + jX_n = l_n (r + jx)$, where l_n is the line length between the user $N-1$ and user N, r and x are resistance and reactance of the line unit length respectively. The photovoltaic sources capacity of the user p is P_V[9].

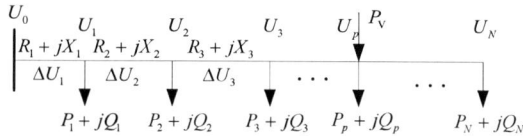

Fig. 1: Load distribution of low voltage line in distribution network for single photovoltaic sources access

Due to the large customer side power factor and low voltage line reactance is small, so this paper ignores the effect of reactive power.

(1) If users in front of the photovoltaic sources access point $0 < m < p$, the voltage of the m user can be shown as follows:

$$U_m = U_0 - \sum_{k=1}^{m} \Delta U_k = U_0 - \sum_{k=1}^{m} \frac{(\sum_{n=k}^{N} P_n - P_V) r l_k}{U_{k-1}} \quad (1)$$

After the photovoltaic sources access to distribution network, the user side voltage rise in front of the photovoltaic sources access point, and it related to line parameter, load size, photovoltaic output power and the location of photovoltaic sources access.

The voltage difference between m and $m-1$ is shown as follows:

$$U_m - U_{m-1} = -\frac{(\sum_{n=m}^{N} P_n - P_V) r l_m}{U_{m-1}} \quad (2)$$

$\sum_{n=m}^{N} P_n > P_V$ means the sum of the active power of the user load behind the m user(includes the m user) is greater than the photovoltaic active power. In this situation, the line voltage would reduce, $U_m - U_{m-1} < 0$.

$\sum_{n=m}^{N} P_n < P_V$ means the sum of the active power of the user load behind the m user(includes the m user) is less than the photovoltaic active power. In this situation, the line voltage would rise, $U_m - U_{m-1} > 0$.

(2) If users behind the photovoltaic sources access point $p < m < N$, the voltage of the m user can be shown as follows:

$$U_m = U_0 - \sum_{k=1}^{p} \frac{(\sum_{n=k}^{N} P_n - P_V) r l_k}{U_{k-1}} - \sum_{k=p+1}^{m} \frac{\sum_{n=k}^{N} P_n r l_k}{U_{k-1}} \quad (3)$$

The voltage difference between m and $m-1$ is shown as follows:

$$U_m - U_{m-1} = -\frac{\sum_{n=m}^{N} P_n r l_m}{U_{m-1}} \quad (4)$$

The voltage of point m is always less than the voltage of point $m-1$.

In conclusion, if the head of line voltage remains the same, when photovoltaic sources single-point access to distribution network, as photovoltaic power gradually increase, the line voltage change trend has the following three conditions:

(1) decreasing;
(2) first decreasing and then increasing again decreasing;
(3) first increasing and then decreasing.

2. Photovoltaic sources multi-point access to distribution network

As shown in fig.2 is the typical low voltage distribution network line load distribution when the photovoltaic sources multi-points access.

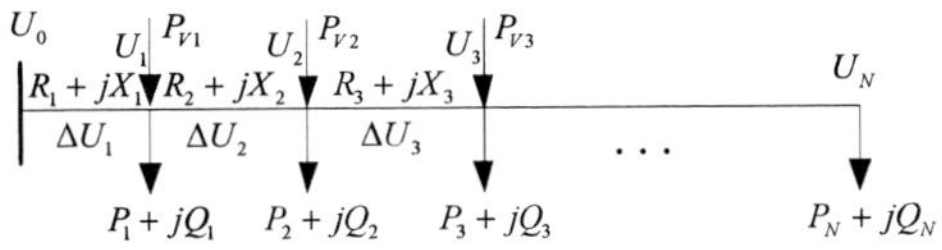

Fig. 2: Load distribution of low voltage line in distribution network for multiple photovoltaic sources access

The voltage of point m is shown as follows:

$$U_m = U_0 - \sum_{k=1}^{m} \frac{\sum_{n=k}^{N} (P_n - P_{Vn}) r l_k}{U_{k-1}} \quad (5)$$

The voltage difference between m and $m-1$ is shown as follows:

$$U_m - U_{m-1} = -\frac{\sum_{n=m}^{N} (P_n - P_{Vn}) r l_m}{U_{m-1}} \quad (6)$$

where P_{Vn} is the photovoltaic sources access capacity of the user n.

In conclusion, when photovoltaic sources multi-points access to distribution network, the difference between m and $m-1$ may reduce. If the photovoltaic sources access capacity greater enough, the voltage of point $m-1$ would even greater than the voltage of point m.

III. DEFINITION OF THREE QUANTITATIVE INDICATORS

1. Voltage difference change rate indicator

In order to determine the voltage difference changes between m and $m-1$ when the photovoltaic sources access capacity changes. This paper defines the voltage difference between m and $m-1$ derivative of the photovoltaic sources access capacity as the voltage difference change rate indicator. Consider the most complex condition of the photovoltaic sources single-point access to the distribution network, the expression of voltage difference change rate is shown below:

$$a_m = \frac{\partial(\Delta U_m)}{\partial(P_V)} = \frac{R_m}{U_{m-1}} + \frac{(\sum_{n=m}^{N} P_n - P_V) R_m}{U_{m-1}^2} (\sum_{k=1}^{m-1} \frac{\partial \Delta U_k}{\partial P_V}) \quad (7)$$

If users behind the photovoltaic sources access point $(p < m < N)$, $P_V = 0$.

After simplified (7), voltage difference change rate is a geometric progression, and the common ratio is $(\sum_{n=m}^{N} P_n - P_V) R_m / U_{m-1}^2 + 1$. Set the line voltage turning point is M, the photovoltaic sources access point is p. The conclusions are listed as follows:

(1) $1 < m < M$: The common ratio larger than 1, so the geometric progression is an increasing progression. With the increase of P_V, the voltage difference ΔU_m that closer to the turning point M would be larger.

(2) $M < m < p$: The common ratio smaller than 1, so the geometric progression is a decreasing progression. With the increase of P_V, the voltage difference ΔU_m that closer to the point p would be smaller.

(3) $\forall m > p$: The common ratio larger than 1, so the geometric progression is an increasing progression. With the increase of P_V, the voltage difference ΔU_m that closer to the end of the line would be larger.

2. Static voltage indicator
According to the national standard *GB12325-90 Power quality-Deviation of supply voltage*, the admissible deviation of supply voltage is $|\Delta U| \leq 7\%$.

$$|\Delta U| = \left| \frac{U_m^* - U_M}{U_M} \right| \leq 7\% \qquad (8)$$

Where the U_M is the rated voltage of distribution network. In this paper, $U_0 = U_M$; U_m^* is the actual voltage of point m.

The static voltage indicator is defined as follows:

$$S_m = \frac{U_m^* - U_M}{U_M} \qquad (9)$$

For different situations, the static voltage indicator has different expression:

(1) photovoltaic sources single-point access & in front of the user:

$$S_m = \frac{1}{U_0} \sum_{k=1}^{m} \frac{(\sum_{n=k}^{N} P_n - P_V) r l_k}{U_{k-1}} \qquad (10)$$

(2) photovoltaic sources single-point access & behind the users:

$$S_m = \frac{1}{U_0} \left[\sum_{k=1}^{P} \frac{(\sum_{n=k}^{N} P_n - P_V) r l_k}{U_{k-1}} + \sum_{k=p+1}^{m} \frac{r l_k \sum_{n=k}^{N} P_n}{U_{k-1}} \right] \qquad (11)$$

(3) photovoltaic sources multi -point access:

$$S_m = \frac{1}{U_0} \sum_{k=1}^{m} \frac{r l_k (\sum_{n=k}^{N} P_n - P_{Vn})}{U_{k-1}} \qquad (12)$$

In order to stabilize the distribution network voltage of point m, the static voltage indicator of point m should satisfy this condition: $-7\% \leq S_m \leq 7\%$.

3. Net loss variation indicator
Through the reasonable design access scheme, distributed energy access to distribution network is helpful to reducing line loss, improving the transmission efficiency of the distribution network.

Before the photovoltaic sources access, the line loss of the distribution network S_B is shown as follows:

$$S_B = \sum_{k=1}^{N} \frac{\Delta U_{kb}^2}{R_k} \qquad (13)$$

Where ΔU_{kb} is the voltage difference between point k and $k-1$ before the photovoltaic sources access; R_k is the line resistance between point k and $k-1$. $k = 1, 2, 3, \cdots, N$.
After the photovoltaic sources access, the line loss of the distribution network S_A is shown as follows:

$$S_A = \sum_{k=1}^{N} \frac{\Delta U_{ka}^2}{R_k} \qquad (14)$$

Where ΔU_{ka} is the voltage difference between point k and $k-1$ after the photovoltaic sources access.
The net loss variation indicator is defined as follows:

$$\Delta S = S_A - S_B \qquad (15)$$

If $\Delta S < 0$, the photovoltaic sources access is helpful to reducing line loss; else if $\Delta S > 0$, the photovoltaic sources access would increase line loss.

4. The comprehensive indicator
For the sake of comprehensively reflecting the influence of photovoltaic sources integrated into distribution network under three proposed indicators, there is a way to combined them in a weighted manner. The first problem should be solved is that the different dimension among the three indicators. Set the maximum allowed value of voltage difference change rate indicator is a_{max}, the maximum allowed value static voltage indicator of is S_{max}, the minimum value of net loss variation indicator is ΔS_{min}. The definition of the comprehensive indicator is showed as followed:

$$Y = \alpha \frac{a_{max} - a_m}{a_{max}} + \beta \frac{S_{max} - S_m}{S_{max}} + \gamma \frac{\Delta S - \Delta S_{min}}{\Delta S} \qquad (16)$$

When the α, β and γ is the weight coefficients for the three proposed indicators, and $\alpha + \beta + \gamma = 1$. The comprehensive indicator reflects influence of photovoltaic sources integrated into distribution network.

IV. EXAMPLE

Taking a glass factory demonstration project as example, this paper calculates three quantitative indicators in order to analyze the influence of photovoltaic sources integrated into distribution network in different ways.

1. Photovoltaic sources single-point access to distribution network

Select a workshop distribution station of the glass factory as the calculating object. The photovoltaic sources access capacity is 2.3925MWp. The topology of the distribution station is shown as follows:

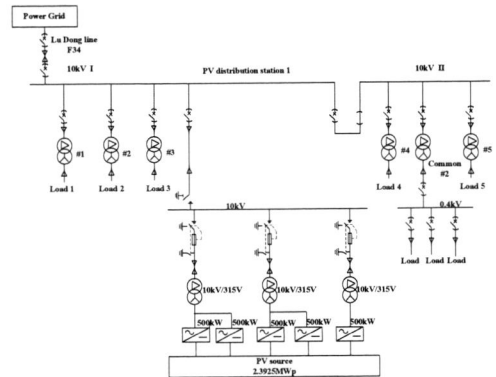

Fig. 3: The topology of workshop distribution station

The load and line parameters are shown in the following table:

Table 1: Load and line parameters of distribution station in a workshop   Statistic units: $kW\ km$

Loads	Power	Lines	Length
Load 1	P1=350	F34~#1	L01=0.2
Load 2	P2=350	#1~#2	L12=0.2
Load 3	P3=450	#2~#3	L23=0.2
Load 4	P4=500	#3~#4	L34=0.6
Load 5	P5=400	#4~Common #2	L4g2=0.3
Common Load	Pg2=500	Common #2~#5	Lg25=0.5

10kV bus line resister rate is $\rho = 0.25\Omega/km$, $U_0 = 10kV$.

(1) According to the topology of the distribution station, load and line parameters, calculating the line voltage of each point. The voltage distribution diagram is shown as follows:

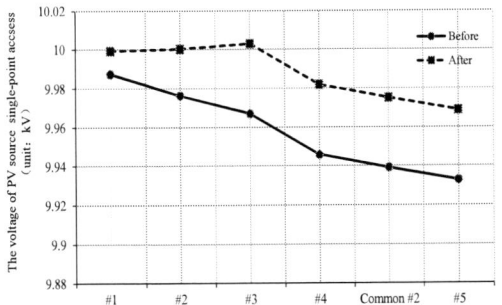

Fig. 4: The line voltage distribution diagram of PV single-point access in the workshop

(2) According to the formulas above, calculating the voltage difference change rate indicator. The results is shown as follows:

Table 2: The voltage difference change rate of PV single-point access in the workshop   Statistic units: A^{-1}

Nodes	Voltage difference change rate	Nodes	Voltage difference change rate
#1	5.00×10-6	#2	5.00×10-6
#3	5.00×10-6	#4	4.15×10-8
#5	1.02×10-8	Common #2	9.45×10-9

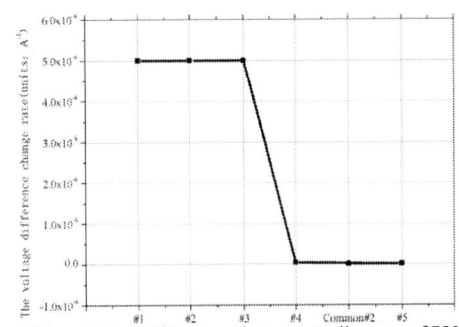

Fig. 5: The voltage difference change rate diagram of PV single-point access in the workshop

(3) According to the formulas above, calculating the static voltage indicator. The results is shown as follows:

Table 3: The static voltage indicator of PV single-access in the workshop

Nodes	Static voltage indicator	Nodes	Static voltage indicator
#1	0.00788%	#2	-0.00175%
#3	-0.0289%	#4	0.181%
#5	0.311%	Common #2	0.249%

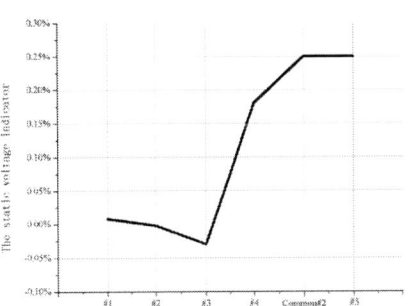

Fig. 6: The static voltage indicator of PV single-access in a workshop

(4) According to the formulas above, calculating the net loss variation indicator. The results is shown as follows:

Table 4: The net loss variation indicator of PV single-access in the workshop   statistic units: kW

Before PV access	After PV access	Net loss variation indicator	Net loss of the distribution network
11.287	4.04	-7.247	Reduction

978-1-5090-0064-7/15 $31.00 © 2015 IEEE

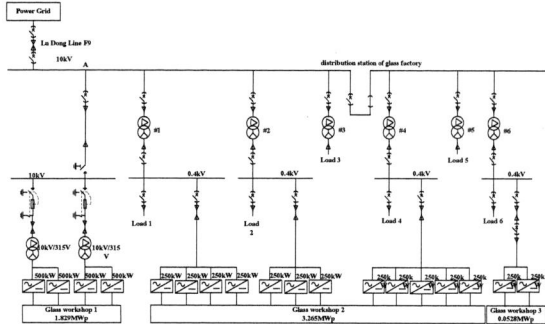

Fig. 7: The topology of glass factory PV distribution station

2. *Photovoltaic sources multi-point access to distribution network*

Select distribution station of the glass one factory as the calculating object. The photovoltaic sources access capacity of glass workshop 1 is 1.829MWp, capacity of glass workshop 2 is 3.265MWp and the capacity of glass workshop 3 is 0.528MWp. The topology of the distribution station is shown as follows:

The load and line parameters are shown in the following table:

Table 5: Load and line parameters of glass factory PV distribution station   Statistic units: *kW km*

Loads	Power	Lines	Length
Load 1	P1=900	F9~Point A	L0A=0.1
Load 2	P2=500	Point A~#1	LA1=0.2
Load 3	P3=600	#1~#2	L12=0.2
Load 4	P4=1000	#2~#3	L23=0.3
Load 5	P5=300	#3~#4	L34=0.5
Load 6	P6=600	#4~#5	L45=0.3
		#5~#6	L56=0.3

10kV bus line resister rate is $\rho = 0.25\Omega/km$, $U_0 = 10kV$. The photovoltaic sources access capacity of each point are listed as follows: $P_{VA} = 1.829MW$, $P_{V1} = 1.000MW$, $P_{V2} = 0.750MW$, $P_{V4} = 1.250MW$, $P_{V6} = 0.5280MW$.

(1) According to the topology of the distribution station, load and line parameters, calculating the line voltage of each point. The voltage distribution diagram is shown as follows:

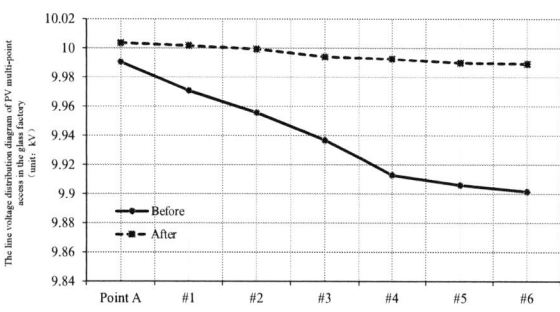

Fig. 8: The line voltage distribution diagram of PV multi-points access in the glass factory

(2) According to the formula above, calculating the voltage difference change rate indicator. The results is shown as follows:

Table 6: The voltage difference change rate of PV multi-points access in the glass factory   Statistic units: A^{-1}

Nodes	Difference change rate	Nodes	Difference change rate
#1	5.00×10-6	#2	5.00×10-6
#3	7.51×10-6	#4	1.25×10-5
#5	7.52×10-6	#6	7.51×10-6
Point A	2.50×10-6		

According to the table above, the variation of the photovoltaic sources access capacity have a relatively remarkable influence on # 4 and the derivative of point #4 is the biggest.

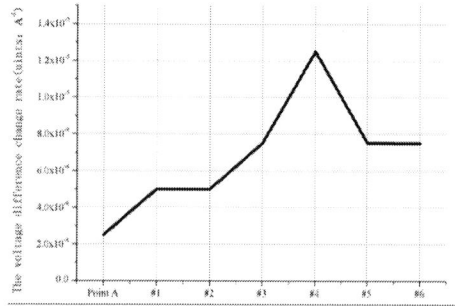

Fig. 9: The voltage difference change rate of PV multiple point access in the glass factory

(3) According to the formula above, calculating the static voltage indicator. The results is shown as follows:

Table 7: The static voltage indicator of PV multi-points access in the glass factory

Nodes	Static voltage indicator	Nodes	Static voltage indicator
#1	-0.0178%	#2	-0.134%
#3	-0.254%	#4	-0.46%
#5	-0.495%	#6	-0.531%
Point A	-0.0364%		

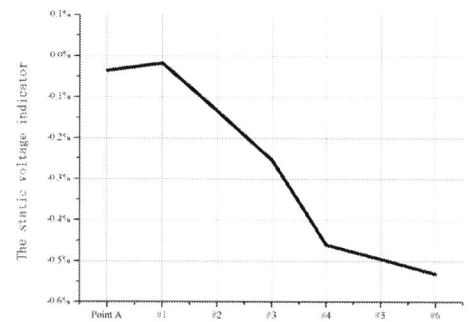

Fig. 10: The static voltage indicator of PV multi-access in the glass factory

According to the chart above, the voltage of each point within the limit of the admissible deviation of supply voltage.

(4) According to the formula above, calculating the net loss variation indicator. The results is shown as follows:

Table 8: The net loss variation indicator of PV multi-points access in the glass factory   Statistic units: kW

Before PV access	After PV access	Net loss variation indicator	Net loss of the distribution network
26.142	1.229	-24.913	Reduction

V. CONCLUSION

According to the two Ways of photovoltaic sources single-point and multi-point access to distribution network, this paper analyzes the topology of photovoltaic sources integrated into distribution network. Combined with the actual project, this paper puts forward three quantitative indicators with actual engineering application value to analyze the influence of photovoltaic sources integrated into distribution network, which include of voltage difference change rate, static voltage indicator and net loss variation, and the three quantitative indicators are provided in details in the thesis. Finally, aiming at the actual project, the quantitative indicators are calculated to analyze the influence of photovoltaic sources integrated into the distribution network, and the rationality of photovoltaic sources access scheme of the demonstration projects is verified.

ACKNOWLEDGMENT

The authors would like to thank the support from the National High-tech R&D Program (863 Program) of China (2014AA052001), Science and Technology Planning Project of Guangdong Province, China (2012B040303005), Science and Technology Project of Science Academy of China Southern Power Grid (SEPRI-K143003), and Science and Technology Project of China Southern Power Grid Company (K-KY2014-009).

REFERENCES

[1] DJAPIC P, RAMSAY C, RUDJIANTO D, et al, "Taking an active approach," IEEE Power and Energy Magazine,2007.

[2] W.LIU, D.PENG, G.Q.BU et al," A survey on system problems in smart distribution network with grid-connected photovoltaic generation," Power System Technology,2009.

[3] WALLING R A, SAINT R, DUGAN R C, et al, "Summary of distributed resources impact on power delivery systems", IEEE Trans on Power Delivery,2008.

[4] Y.YOU, D.LIU, W.P.YU, et al, "Technology and its trends of active distribution network," Automation of Electric Power Systems ,2012.

[5] NUROGLU F M, ARSOY A B, "Voltage profile and short circuit analysis in distribution systems with DG," Proceedings of the IEEE Canada Electric Power Conference(EPEC'08) ,2008.

[6] K ASHEM M A, LEDWICH G, "Distributed generation as voltage support for singlewire earth return systems," IEEE Trans on Power Delivery,2004.

[7] A.Y.HAN, X.DENG, M.H.WEN, et al, "Strategy of Large Power System Coping with Accession of Microgrid with High Penetration," Automation of Electric Power Systems , 2010.

[8] B.ZHAO, X.S.ZHANG, B.W.HONG, "Energy penetration of large-scale distributed photovoltaic sources integrated into smart distribution network," Electric power automation equipment, 2012.

[9] J.LIU, X.Q.TONG, Z.M.PAN, et al, "The maximum power of distributed PV generation according to over-voltage in distribution network," Power System Protection and Control, 2014.

[10] B.LI, T.Q.LIU, X.Y.LI, "Impact of distributed generation on power system voltage stability," Power System Technology,2009.

[11] TONKOSKI R, TURCOTTE D, EL-FOULY T H M, "Impact of high PV penetration on voltage profiles in residential neighborhoods," IEEE Trans on Sustainable Energy, 2012.

Design Considerations for High Frequency DCM Flyback Converter

Jun Lee   Kyung-Hwan Lee   Jung-Ik Ha

Department of Electrical and Computer Engineering, Seoul National University, Seoul, Korea
E-mail: leejun1672@snu.ac.kr, kyugahsal@snu.ac.kr, jungikha@snu.ac.kr

Abstract– This paper presents design considerations for a high frequency (≥1 MHz) Discontinuous Conduction Mode (DCM) flyback converter. As switching frequency increases, the problems such as switching loss and undesirable effects due to the parasitic capacitance of switching device are exacerbated. Considering these issues, design instructions are proposed about DC link capacitance, PCB transformer, snubber, magnetizing inductance, switch, output capacitance and turn ratio. Following the instructions, 1-MHz 65-W flyback converter was implemented. Its output voltage was regulated to 19 V in DC 100-373 V input voltage. The experimental results and loss breakdown analysis are given. To observe the effect of variation in switching device capacitance, Si and SiC MOSFET switches are tested. The switching waveforms and the efficiency of the two cases are compared.

Keywords–DCM, Flyback converter, high frequency switching, SiC MOSFET

I. INTRODUCTION

Flyback converters are used widely as cost-effective isolated power converters [1], [2]. Switching frequencies of power converters can be higher as switching devices and transformers are developed [3], [4]. Higher switching frequency reduces the maximum flux linkage and allows a converter to be smaller. Also, ripple on the output voltage or output capacitance can be reduced because energy used per period is decreased.

Appearance and enhancement of GaN transistors offer chances to choose much higher switching frequency, and a 5-MHz 30-W converter was designed with 87% efficiency [5]. However, the application had narrow input voltage range, and there are demands of higher output power. In this paper, design considerations for a high frequency flyback converter and comparisons between design options are covered. For verification, a DCM flyback converter (whose circuit is drawn in Fig. 1) is designed for DC 100V~373 V (AC 100V~264V) input voltage, 19 V output voltage and 65 W output power system, for 1MHz switching frequency.

II. DESIGN CONSIDERATIONS FOR USING HIGH SWITCHING FREQUENCY

Design considerations for high switching frequency flyback converter is actually not quite different from those with lower switching frequencies. By increasing the switching frequency, the importance or priority of considerations changes. In this section, variations in using high switching frequency are analyzed, and corresponding design ways are suggested.

1. Minimum input voltage and DC link capacitance C_{dc}

There are two duties for DCM operation of a flyback converter; switch turn-on duty, D_1, and flux-removing duty, D_2. A flyback converter should be designed to keep D_1 larger than proper time to use the switch normally while satisfying $D_1 + D_2 < 1$.

D_1 becomes $D_{1,max}$ when the input voltage is its minimum value. For AC input voltage source, $V_{dc,min}$ can be obtained with a numerical iteration with an equation,

$$\frac{1}{2}C_{dc}V_i^2 - \frac{1}{2}C_{dc}V_f^2 = P_{in}\Delta t , \qquad (1)$$

assuming that $P_{in} = P_o/\eta$ is constant with constant efficiency η, 0.8. Fig. 2 shows the iteration result for the test board. Using 100uF input capacitor, $V_{dc,min}$ is designed to be DC 100V for AC 100V input voltage.

Smaller C_{dc} makes the ratio between $V_{dc,max}$ and $V_{dc,min}$ larger. If V_{dc} varies widely, so does D_1. As a converter is designed with high frequency, the switch may not be driven normally with small D_1 (<0.05) or short turn-on time(<50 ns). Since $V_{dc,max}$ is specified by the grid, ratio between $D_{1,max}$ and $D_{1,min}$ only can be regulated with $V_{dc,min}$. After $V_{dc,min}$ is set, following equation can be used to know C_{dc}, approximately.

$$C_{dc} = 2P_{in}(\sin^{-1}(\frac{V_{dc,min}}{V_{dc,max}}) + \frac{1}{4f_{grid}})/(V_{dc,max}^2 - V_{dc,min}^2) . \quad (2)$$

What needs extra attention is that the turn-on and turn-off delays depend on input and output capacitances of the switch, C_{iss} and C_{oss}. C_{oss} may vary more than 100 times as voltage across the switch changes. If turn-off delay exists, input voltage is applied on the inductor longer. So,

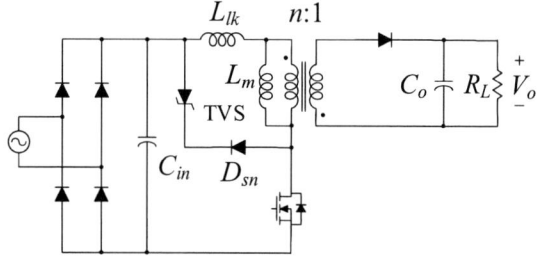

Fig. 1: Flyback converter with TVS clamping circuit.

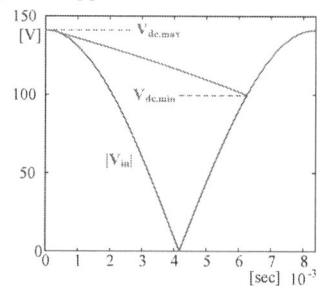

Fig. 2: Rectified AC input voltage and V_{dc} calculated with iteration.

978-1-5090-0064-7/15 $31.00 © 2015 IEEE

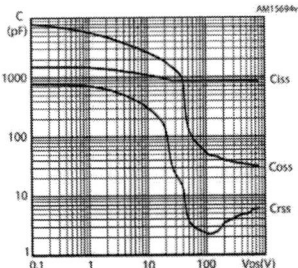

Fig. 3: Capacitance variations of Si MOSFET, STB13N80K5 [6].

'effective' duty becomes higher than the reference. Hence the operation range of D_1 must be designed with enough margin. The variation of capacitances used for the Si MOSFET test board is shown in Fig. 3.

One may wish to make the size of a DC link capacitor smaller keeping the operation range properly, but resonance between the grid inductance comes to be another issue. Voltage ripple on C_{dc} with higher frequency may degrade the performance of output voltage regulation.

2. Transformer

Peak voltage stress on the switch is determined by the leakage energy of the transformer. The leakage energy is related with the leakage inductance, L_{lk}. To reduce its value to about 1% of the magnetizing inductance, L_m, a PCB

Fig. 4: Transformer made for the 1 MHz test board (26mm * 20mm * 12mm).

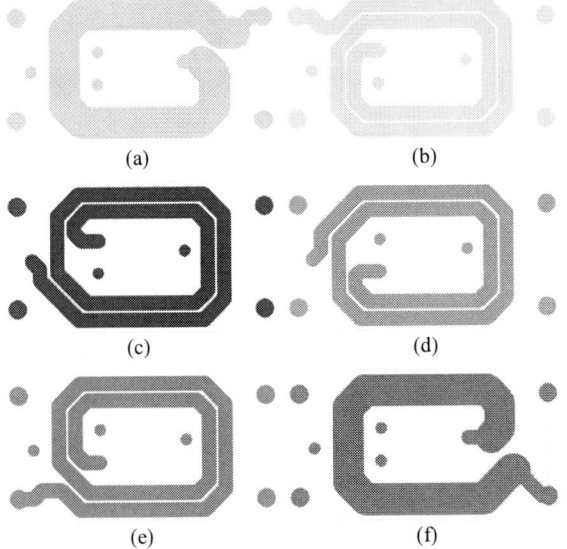

(a)   (b)

(c)   (d)

(e)   (f)

Fig. 5: (a)-(f) Layers of the PCB transformer in order; (b)-(e) are the primary winding, and (a), (f) are the secondary winding.

transformer is used for the test board. For the test board, 6-layer PCB was made with ML95S core of Hitachi Metals. Fig. 4 is the transformer made for the test board. Its volume is about 30% of transformers for under-100-kHz switching frequency.

Fig. 5 shows the PCB windings each layer. The transformer has 4 layers of 2-turn primary wire and 2 layers of 1-turn secondary wire. There are many combinations for order of the layers, and it is better to choice one with smaller leakage inductance by running a magnetic field simulation. The secondary wire should have much thicker width than the primary wire due to higher current.

If the core loss is more dominant than the conduction loss, increasing the number of turn can decrease the overall loss. Flux variation on the transformer is reversely proportional to the number of turn, and the conduction loss is proportional to it. A designer may choose the optimal turn number with following equation,

$$-\frac{dP_{core}}{dN_p} = \frac{dP_{cond}}{dN_p} . \qquad (3)$$

However, the leakage power remains large due to the high switching frequency; because the leakage energy is generated every period. So a snubber circuit becomes necessary.

3. Primary snubber

A diode for the snubber must be chosen carefully. To effectively clamp the rising voltage of the resonance of L_{lk} and C_{oss}, the snubber diode, D_{sn}, must turn on sufficiently fast. For faster snubber operation, a TVS (transient voltage suppressor) can be considered to replace an RCD snubber. When input voltage is low, a TVS clamping circuit would not be activated and would not make any loss. The designer should be aware of this because rest of the leakage energy might be consumed in the other parts of the circuit; circuit resistance or devices. Fig.1 is a flyback circuit which used a TVS in its snubber.

The Zener voltage of TVS, V_{TVS}, should have some margin, satisfying

$$V_{TVS} < V_{sw,rate} - V_{dc,max}, \qquad (4)$$

while $V_{sw,rate}$ is the rated voltage of the switch.

4. Peak and rms currents

Primary and secondary peak and rms currents are important for converter design. Many losses are related to these, and there are some devices have current limits.

Maximum peak currents and rms currents are calculated as follow.

$$I_{p,pk} = \frac{V_{dc,min}D_{max}}{L_m f_{sw}}, \qquad (5)$$

$$I_{p,rms} = I_{p,pk}\sqrt{\frac{D_{max}}{3}}, \qquad (6)$$

$$I_{s,pk} = nI_{p,pk}, \tag{7}$$

$$I_{s,rms} = \sqrt{\frac{2nI_oI_{p,pk}}{3}}. \tag{8}$$

Above equations show that peak and rms currents will not be affected by f_{sw} if L_m is selected reversely proportional to f_{sw}. To transfer the rated power at the worst case, L_m is determined.

$$L_m = \frac{(V_{dc,min}D_{max})^2}{2P_{in}f_{sw}}. \tag{9}$$

By using this magnetizing inductance, considerations related to currents become not far from those at lower switching frequency.

5. Switch

Two main loss generated by the switch is turn-off loss and conduction loss. As the switching frequency getting higher, turn-off loss of the switch becomes more dominant than the conduction loss; it is calculated and shown in Fig. 9. It would be better to focus on the input and output capacitance of the switch rather than the on-resistance.

6. Output capacitance C_o

Magnitude of the output voltage ripple can be written as

$$\Delta V_o = \frac{I_o(I_{p,pk} - I_o/n)^2}{I_{p,pk}f_{sw}C_o}, \tag{10}$$

where $I_o = P_o/V_o$ is assumed to be constant. Keeping the ripple in the same magnitude, C_o can be reduced reversely proportionally to f_{sw} due to fixed $I_{p,pk}$ from (5) and (9).

7. Turn ratio n

The turn ratio of a flyback converter is selected to get proper duty ratio range and voltage stress on the switch, hence it depends only on $V_{dc,max}$ and $V_{dc,min}$. While transformer and other parts would be changed in accordance with the switching frequency, the turn ratio has no relation with it.

III. EXAMPLE EXPERIMENTS AND RESULTS

1. Experimental set up

Following above steps, 1-MHz, 65-W flyback converter prototype of Fig.1 was implemented. The parameters are summarized in Table 1. The experiment was done with DC input voltage source. Si MOSFET STB13N80K5 and SiC MOSFET C2M0160120D are used for the switch S_1. The output voltage was controlled to be 19V by PI controller with a DSP, TMS320C28346, and the duty is calculated every switching period.

Table 1: Parameters for experiment

Parameters	Value
f_{sw}	1 [MHz]
V_{dc}	100~373 [V]
C_{dc}	100 [uF]
V_o	19 [V]
P_o	65 [W]
N_1:N_2	8:2
L_m	11.4 [uH]
L_{lk}	76 [nH]
C_o	32 [µF]

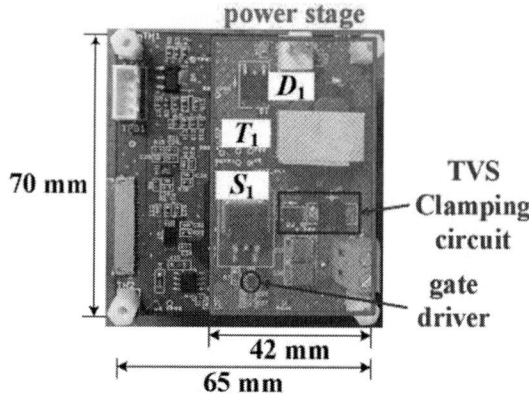

Fig. 6: Test board. Power stage = 70mm x 42mm x 15mm.

2. Experimental results

The switching waveforms of two switches for 373 V input voltage and 100% load cases are shown in Fig. 7. Results of 100 V input voltage and 50% load cases are shown in Fig. 8.

Two waveforms had very similar shape although the duties were different. This is due to the difference of turn-off delay which is related with C_{iss} and C_{oss}. Shorter turn-off delay enables a designer to use a smaller input capacitor or wider input voltage range. The C_{iss} of the Si and SiC MOSFET are 870 pF and 527 pF according to [6] and [7].

Efficiencies of the SiC flyback converter of various input voltages were measured almost always higher than that of the Si flyback converter as shown in Fig. 9.

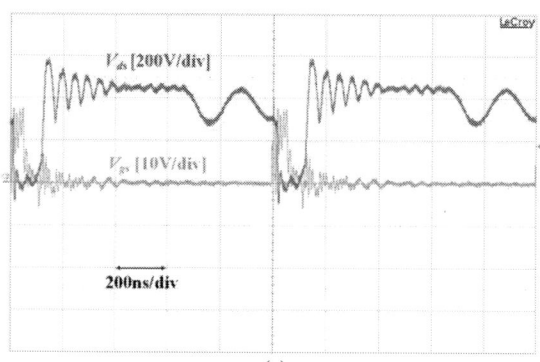

(a)

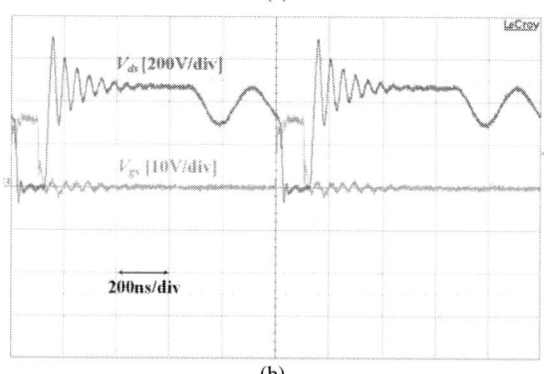

(b)

Fig. 7: V_{ds} and V_{gs} @ V_{dc}=373 V, 100% load with (a) Si MOSFET, (b) SiC MOSFET.

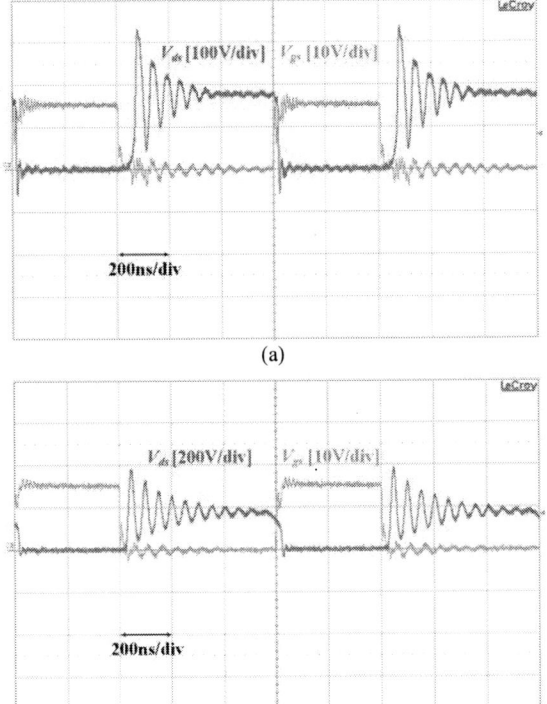

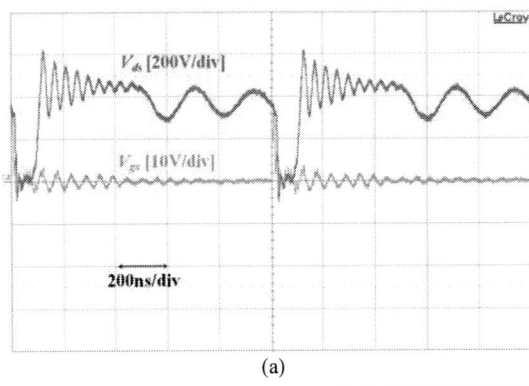

(a)

(a)

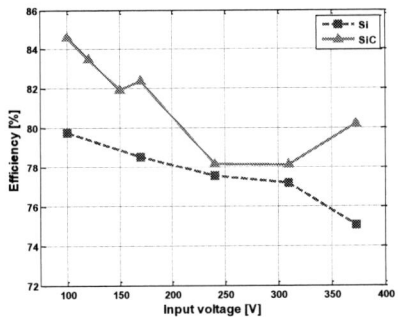

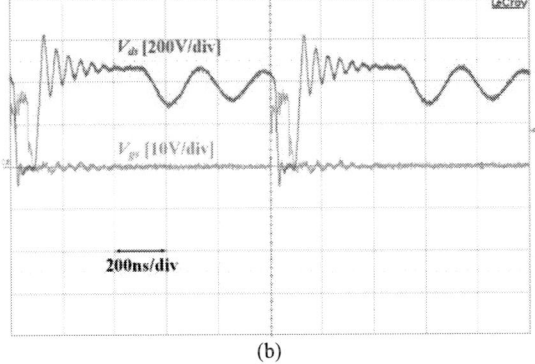

(b)

(b)

Fig. 8: V_{ds} and V_{gs} @ V_{dc}=100 V, 100% load with (a) Si MOSFET, (b) SiC MOSFET.

Fig. 10: V_{ds} and V_{gs} @ V_{dc}=373 V, 50% load with (a) Si MOSFET, (b) SiC MOSFET.

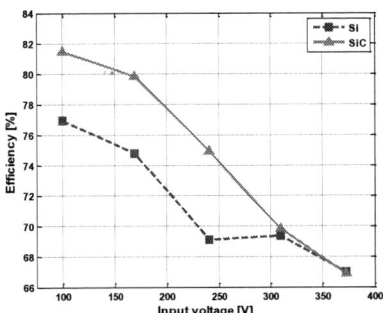

Fig. 9: Efficiency comparison between Si and SiC MOSFTET test boards at 100% load.

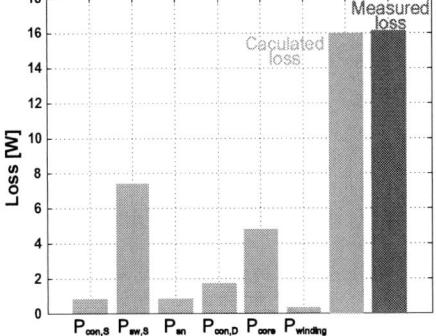

Fig. 11: Efficiency comparison between Si and SiC MOSFET test boards at 50 % load.

Fig. 10 shows waveforms of 50% load at 373V input voltage. The result shows that the turn-on duty of the Si flyback converter has big error from the effective turn-on duty. Fig. 11 compares the efficiency between Si and SiC MOSFET test boards at 50% load. SiC MOSFET test board had higher efficiency also for all 50% load cases. This can be explained with following three reasons. C_{oss} of the SiC MOSFET is about 1/5 of that of Si MOSFET when V_{ds} is low, according to the datasheets. This means that the SiC MOSFET stores less energy than the Si MOSFET. The power loss of the stored energy is

$$P_{store,S} = E_{store,S}f_{sw}, \tag{11}$$

hence SiC MOSFET has smaller capacitive loss. Cross section between V_{ds} and I_p is actually the dominant part in turn-on and turn-off loss. Smaller C_{oss} makes the cross section smaller. On-resistance of the SiC MOSFET is also smaller; so is conduction loss.

Fig. 12: Loss breakdown analysis for Si flyback converter @ V_{dc}=100 V, 100% load.

Loss breakdown graph for Si MOSFET flyback with experimental result values at 100 V input voltage, 100% load condition is shown in Fig. 12. The figure shows 45% of the total loss is the switching loss.

978-1-5090-0064-7/15 $31.00 © 2015 IEEE

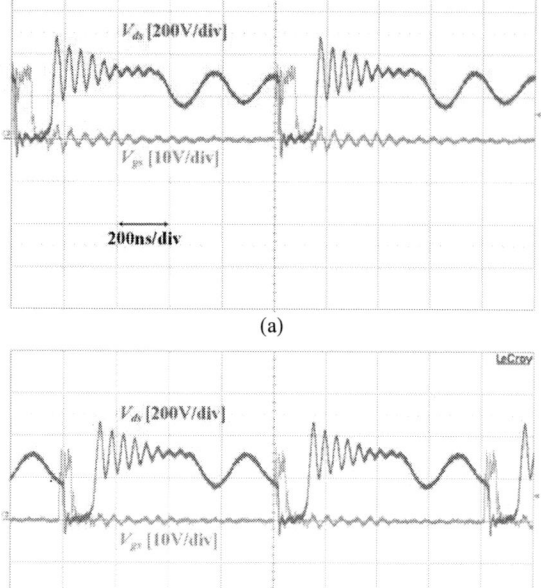

(a)

(b)

Fig. 13: V_{ds} and V_{gs} @ V_{dc}=240 V, 100% load
with Si MOSFET at
(a) 1 MHz, (b) 1.23 MHz switching frequency.

At 240 V input voltage case in Fig. 11, Si MOSFET test board had lower efficiency than the expected one from other cases. This is because V_{ds} is at the peak point of the resonance, which is shown in Fig. 13(a). From Fig. 12, it is expected that reducing the switching loss may increase the efficiency significantly. The final consideration for high frequency converter is QR(quasi-resonant) operation. A designer may pick a switching frequency arbitrarily. However, if the input voltage and load condition are fixed, there is relevant benefit to pick a frequency by solving

$$T_{sw} = \left(1 + \frac{V_{dc}}{nV_o}\right)\sqrt{\frac{2P_oL_mT_{sw}}{\eta V_{dc}^2}} + (2m-1)\pi\sqrt{(L_m + L_{lk})C_{oss}} \ . \ (12)$$

The first term of the right side is time that the magnetizing flux takes to be zero. The second term is time to get m-th valley in the latter resonance. This approach has two difficulties to use in the first design; a designer should expect the efficiency of the converter quite accurately, and C_{oss} is assumed to be constant. Depending on the experimental result, f_{sw}=1.23 MHz is picked for a controlled experiment, and its result is shown in Fig. 13(b).

By using the higher switching frequency, 1.23 MHz, the switch was turned off at the valley of the voltage resonance. The efficiency is increased to 73.7% from 69% although the switching loss might be increased because it is proportional to the switching frequency. The increased efficiency also can be checked with D_l, which is decreased as shown in waveforms.

IV. CONCLUSION

In this paper, main design considerations in a high frequency flyback converter are covered. The problems due to high frequency are pointed and several solutions are presented. Following the suggested instructions, an 1-MHz, 65-W, 100~373-V-to-19-V flyback converter prototype was implemented. Experimental results of various loads and input voltages are shown for Si and SiC MOSFETs. The efficiency of Si MOSFET converter was about 75%, and the switching loss was calculated as 45% of the total losses. The SiC MOSFET converter has the decreased switching loss, and its efficiency was about 80%, which is higher than that of Si MOSFET converter. From parameters estimated from the result waveforms, new switching frequency for QR operation was calculated, and the efficiency was increased by 4.7%.

ACKNOWLEDGEMENT

This work was supported by Research Resettlement Fund for the new faculty of SNU, the Brain Korea 21 Plus Project in 2015.

REFERENCES

[1] Hsing-Fu Liu and Lon-Kou Chang, "Flexible and low cost design for a flyback AC/DC converter with harmonic current correction," *IEEE Trans. Power Electron.*, vol. 20, no. 1, pp.17-24, Jan. 2005.
[2] Y. Li and R. Oruganti, "A low cost flyback CCM inverter for AC module application," *IEEE Trans. Power Electron.*, vol. 27, no. 3, pp.1295-1303, March 2012.
[3] I. Barbi, J. C. Fagundes, and E. V. Kassick, "A compact AC/AC voltage regulator based on an AC/AC high frequency flyback converter," *IEEE-PESC '91 Record*, pp. 846-852, 1991.
[4] A. D. Sagneri, D. I. Anderson, and D. J. Perreault, "Transformer Synthesis for VHF Converter," in *Proc. Int. Power Electron. Conf.*, Jun. 2010, pp. 2345-2353.
[5] Z. Zhang, K. D. T. Ngo, and J. L. Nilles, "A 30-W flyback converter operating at 5 MHz," in *Proc. Applied Power Electronics Conference and Exposition (APEC)*, Fort Worth, TX, USA, Mar. 2014, pp. 1415–1421.
[6] STMicroelectronics, "N-channel 800 V, 0.37 Ohm typ., 12 A MDmesh K5 Power MOSFET in D2PAK package," STB13N80K5 datasheet, April, 2013.
[7] CREE, "Silicon Carbide Power MOSFET Z-FET™ MOSFET," C2M0160120D datasheet, 2013.

BIOGRAPHIES

Jun Lee was born in Seoul, Korea, in 1993. He received the B.S. degree in electrical engineering from Seoul National University, Seoul, Korea, in 2015, where he is currently pursuing the M.S. degree in electrical engineering. His current research interests are electric energy conversion, power electronics and magnetic manipulation system.

Kyung-Hwan Lee (S'14) was born in Seoul, Korea, in 1991. He received the B.S. degree in electrical engineering from Seoul National University, Seoul, Korea, in 2013 where he is currently pursuing the Ph.D. degree in electrical engineering. His current research interests include AC-DC and DC-DC power conversion and high-frequency power converters.

Jung-Ik Ha (S'97–M'01–SM'12) was born in Korea in 1971. He received the B.S., M.S., and Ph.D. degrees from Seoul National University, Seoul, Korea, in 1995, 1997, and 2001, respectively, all in electrical engineering.

From 2001 to 2002, he was a Researcher with Yaskawa Electric Company, Kitakyushu, Japan. From 2003 to 2008, he was a Senior and Principal Engineer with Samsung Electronics Company, Suwon, Korea. From 2009 to 2010, he was a Chief Technology Officer with LS Mecapion Company, Seoul. Since 2010, he has been with the Department of Electrical and Computer Engineering, Seoul National University, Seoul, where he is currently an Associate Professor. His research interests are on circuits and control in high-efficiency and integrated electric energy conversions for various industrial fields.

Control Method Considering Current and Voltage Limits in Magnetic Manipulation Systems

Jun Lee and Jung-Ik Ha

Department of Electrical and Computer Engineering, Seoul National University, Seoul, Korea
E-mail: leejun1672@snu.ac.kr, jungikha@snu.ac.kr

Abstract–This paper proposes a control method for magnetic manipulation system which has voltage and current limits on its actuator circuits. The proposed method modifies current and voltage references making the largest ones bounded in the rated values for delicate attitude control. When current or voltage larger than the rated value is required for a force-torque command, the method generates the modified currents and voltages. Modified references make force and torque with the same direction and decreased magnitude with the original command. An actuator system with the proposed method may have slower control performance but achieves delicate attitude control. Simulative comparison between the proposed method and a simple limiting method, which simply cutoffs references, is done.

Keywords– Control method, Current limit, Magnetic manipulation, Micro-robot, Voltage limit,

I. INTRODUCTION

There have been researches about control of small magnetic-material-body using magnetic manipulation system [1], [2]. From shape of the robot to multi-body control, various topics have been covered [3], [4]. Most of them tried to maximize magnetic field and its gradient by using proper coil configuration. These approaches are useful and meaningful when commands (position, attitude, force, etc.) are fixed. A system which includes an inverter circuit has rated current and voltage conditions with it, hence commands requiring high current or voltage cannot be accomplished. Considering that many medical applications are suggested with magnetic manipulation systems, current and voltage limits come to be severe problems.

In this paper, a control method, proportional limiting method which can deal with over-limit commands is suggested. The method generates proportionally limited current and voltage references when magnetic field and force commands are given. Firstly, control faults due to currents and voltages are shown. By applying the proposed method, the faults are modified or removed. The effectiveness of the suggested method is verified with physical and electrical states in simulations. An example of magnetic manipulating system is shown in Fig. 1.

II. FUNDAMENTALS OF MAGNETIC MANIPULATION SYSTEM

Magnetic manipulation is about controlling position and attitude of a magnetic material body, a robot. For better control performance, the robot is usually used after magnetization, which induces bigger force and torque.

Magnetic moment of a robot is written as

$$\mathbf{M} = \mathbf{M_v} V \qquad (1)$$

Force and torque applied to the robot are

$$F = (M \cdot \nabla) B, \qquad (2)$$

$$T = M \times B, \qquad (3)$$

$$= \mathrm{Sk}(M) B, \qquad (4)$$

where B is the applied magnetics flux density at the location of the robot, P. B is the sum of individual magnetic fields made by currents flowing on the coils surrounding the control region and is expressed as

$$B(P) = \begin{bmatrix} \tilde{B}_1(P) & \cdots & \tilde{B}_1(P) \end{bmatrix} \begin{bmatrix} i_1 \\ \vdots \\ i_n \end{bmatrix} = \beta(P) I. \qquad (5)$$

A command of torque and force can be written as

$$\beta_e(P) = \frac{\partial \beta(P)}{\partial e}, \qquad (6)$$

$$\begin{bmatrix} T \\ F \end{bmatrix} = \begin{bmatrix} \mathrm{Sk}(M) \\ M^T \beta_x(P) \\ M^T \beta_y(P) \\ M^T \beta_z(P) \end{bmatrix} \begin{bmatrix} i_1 \\ \vdots \\ i_n \end{bmatrix} = A_{T,F}(M, P) I. \qquad (7)$$

Using these equations, the required currents are calculated with pseudo inverse of the actuation matrix,

$$I = A_{T,F}(M, P)^+ \begin{bmatrix} T_{ref} \\ F_{ref} \end{bmatrix}. \qquad (8)$$

Since the torque applied on the robot tends to make the robot be aligned with the magnetic flux applied on it. So a more simple command, (9), is used often.

$$I = A_{B,F}(M, P)^+ \begin{bmatrix} B_{ref} \\ F_{ref} \end{bmatrix}. \qquad (9)$$

To get a solution for arbitrary reference, the actuation matrix should have 6 or more columns. The fat actuation matrix has its pseudo inverse matrix if it has full row rank. By using 8 coils, a magnetic manipulation system comes to have more chance to have full row rank.

III. CONTROL METHOD UNDER CURRENT AND VOLTAGE LIMITS

If magnetic manipulating system is used in its safe region, there is no need for proper limiting method. But if a command is given without any consideration, the robot may not work following the command due to current and voltage limits. Moreover, (7)-(9) indicates that the actuation matrix changes according to the position and attitude of the robot, so it may be hard, even to the designer, to determine safe command region for a system. Therefore, control method that ensures safety is necessary.

1. Safe alternative reference
Before further argument, safe operation under current or voltage limits are defined. If any current solution obtained from (9) is larger than the current limit, it means that given magnetic field or force command is achievable in no way. Proper countermeasures may differ for various systems according to main goals. In this paper, keeping the same directions of magnetic field and its gradient (force) is set as the main goal. This condition makes the robot not move or aligns along wrong direction.

2. Proportional limiting method
Since magnetic fields depend on currents flowing on coils, steady state and transient errors are all caused by current errors. But relationship between currents and voltages,

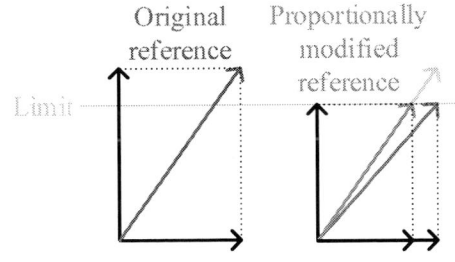

Fig. 2: Proportional limiting method

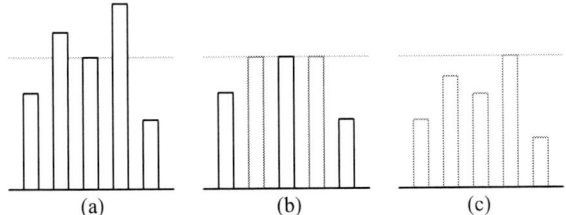

Fig. 3: Difference between limiting methods;
(a) original current or voltage references and limit value (yellow line),
(b) simple limiting method,
(c) proportional limiting method

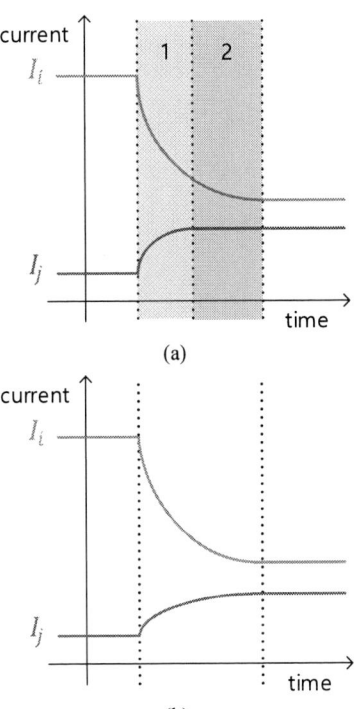

(a)

(b)

Fig. 4: Two currents in transient state using (a) simple, (b) proportional voltage limiting method

$$\frac{di_n}{dt} = \frac{V_n}{L_n} \qquad (10)$$

shows that transient errors are actually caused by voltage limit.

Once a new command is given to the system, the goal currents and voltages are calculated. References should be dealt with limits appropriately before they are inputted to the controller. Just cutting-off the references larger than the rated value makes the direction of the net reference different from the original reference.

To achieve the main goal in the prior section, the proportional limiting method is proposed. Fig. 2 depicts the main concept of the method. By multiplying the same ratio to all references, the direction of new net reference is aligned with the original reference. Difference between the proportional limiting method and the simple limiting method is shown in Fig. 3, applying to five-reference case.

3. Current limit case – steady state error
Because each magnetic field is proportional to corresponding current, if all currents are scaled with the same ratio, the direction of the sum of magnetic fields is not changed. Since there might be multiple current references bigger than the rated value, currents should be scaled depending on the biggest current. By multiplying (*rated current*)/(*the largest current reference*) to all reference currents, the condition can be satisfied.

4. Voltage limit case - transient state error
Using P controllers with the same gain, each coil takes the

978-1-5090-0064-7/15 $31.00 © 2015 IEEE

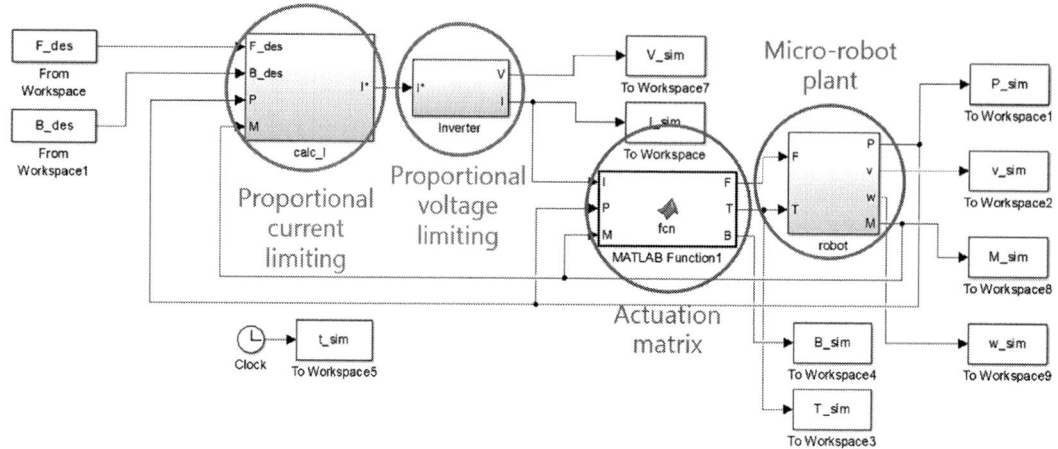

Fig.5: Simulink® simulation block diagram

Table 1: Parameters for simulation

Parameters	Value
d (distance between coil and center)	5 [cm]
L_{coil}	50 [mH]
R_{coil}	0.1 [Ω]
M_{robot}	10^5[A/m]
x_{robot}	2 [mm]
y_{robot}, z_{robot}	1 [mm]
m_{robot}	8.9 x 10^{-6}[kg]
J_{robot}	8.3 x 10^{-10}[kg m²]
$|B_{ref}|$	5 [mT]
a_{max}	19.6 [m/s²]
ω_c (controller cutoff frequency)	1000 [rad/s]
b_{drag}	0.02 [N/(m/s)]
b_{rot_drag}	5 x 10^{-8}[N m/(rad/s)]
I_{lim}	30 [A]
V_{lim}	100 [V]

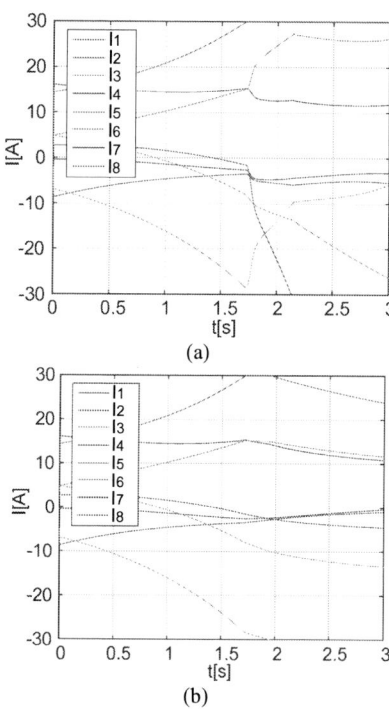

(a)

(b)

Fig.7. Currents on coils with
(a) simple, (b) proportional limiting method

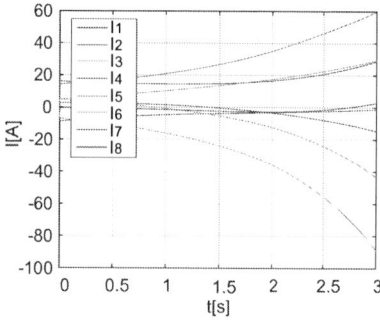

Fig. 6: Currents with no limit

equal duration to reach to a new reference. If voltages larger than DC link voltage is simply cut-offed, currents stay in the transient states for different durations and unexpected net magnetic field would be obtained. Its graphical explanation is shown in Fig. 4(a). Note that the

robot would operate unexpectedly not only in duration 2 but also in duration 1.

A new voltage reference is generated in a similar way to the current reference. By multiplying (*rated voltage*)/(*the largest voltage reference*) to all reference voltages, transient times of coils become same, like Fig. 4(b). This figure also explains that all currents would be controlled with the same speed of the worst one.

IV. SIMULATION RESULTS

A simulation was set with parameters in Table 1 using the coil arrangement scaled from Fig. 1. The inductance of the coil was also calculated from simulation. A coil with the inductance makes 50mT at the center of control region if 30A current flows on it. Magnetic field made by coils in the control region was calculated by regarding each coil as a point dipole. The Simulink® simulation block diagram is shown in Fig. 5.

Since current reference depends on position, magnetic moment and aligned direction of the robot, *P* and *M* were predicted during the simulation by implementing a plant block.

1. Current limit case

978-1-5090-0064-7/15 $31.00 © 2015 IEEE   74

Because magnetic field in the control region generated by a coil differs at every point, current references changes according to the position of the robot even if the command is kept constantly. Fig. 6 shows currents flowed on coils for command

$$\begin{bmatrix} B_{ref} & 0 & 0 & \dfrac{F_{max}}{\sqrt{5}} & 0 & \dfrac{2F_{max}}{\sqrt{5}} \end{bmatrix}^{\mathrm{T}}$$

with no current limit for 3 seconds. Fig. 7(a) shows the currents with simple limiting method which cutoffs currents over 30A. Currents after about 1.7s have awkward shapes because current references were generated for wrong attitude (and position) of the robot. Fig. 8(a) and Fig. 9(a) shows that the robot moved and headed wrongly. Aligned direction of the robot is (M_x, M_y, M_z).

The results showed that the proposed method kept the direction of the magnetic field. The robot received no force along y-axis and the ratio between F_x and F_z was also maintained to be 1:2. It means that the only difference from no-limit case is the scale of the magnetic field magnitude and the gradient.

2. Voltage limit case

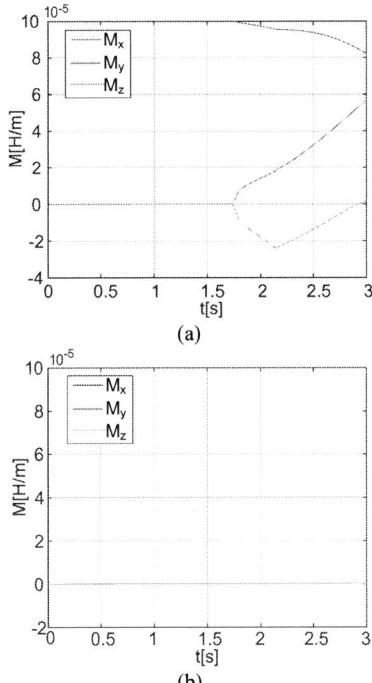

Fig. 9: Attitude of the robot with
(a) simple, (b) proportional limiting method

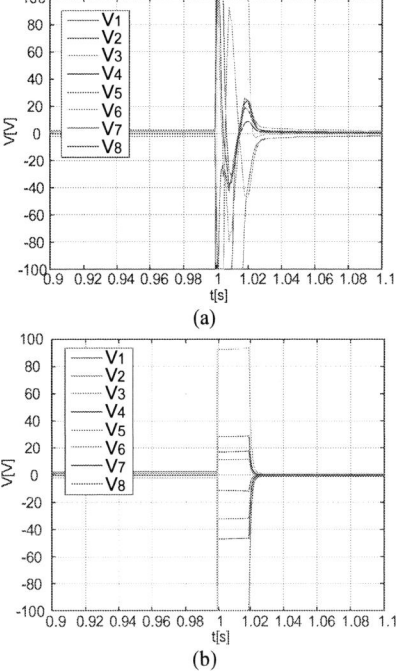

Fig. 10: Voltages with
(a) simple, (b) proportional limiting method

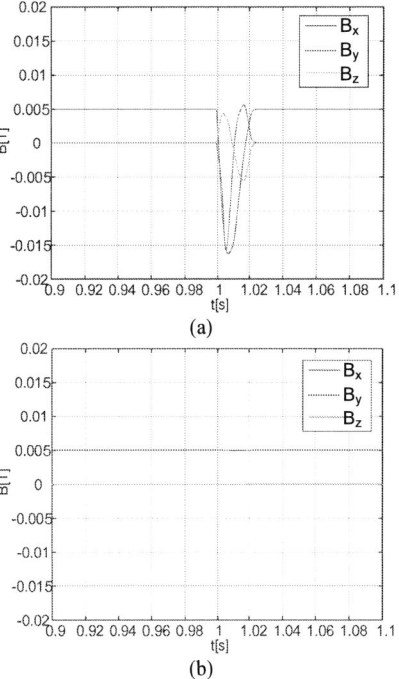

Fig. 11: Magnetic flux on the robot with
(a) simple, (b) proportional limiting method

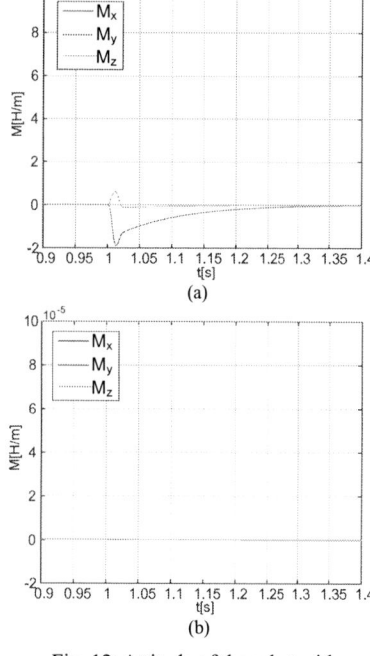

Fig. 12: Attitude of the robot with
(a) simple, (b) proportional limiting method

For a command which changes from
$$\begin{bmatrix} B_{ref} & 0 & 0 & \dfrac{F_{\max}}{\sqrt{5}} & 0 & \dfrac{2F_{\max}}{\sqrt{5}} \end{bmatrix}^T \text{ to } \begin{bmatrix} B_{ref} & 0 & 0 & 0 & \dfrac{F_{\max}}{\sqrt{5}} & -\dfrac{2F_{\max}}{\sqrt{5}} \end{bmatrix}^T$$
after 1 second, two limiting methods were applied and the results are shown in Fig. 10 to Fig. 12. In Fig. 12(a), the robot is misaligned, hence the voltage references to achieve the command are fluctuating. Similar to the current limit case, voltage references for coils fluctuate a lot due to the misalignment.

Applying the proportional limiting method, misalignment and fluctuation of the voltage references are disappeared.

V. Conclusion

The proposed proportional limiting method enhanced both steady and transient state operations, under current and voltage limits. Magnetic field and its gradient were scaled keeping the directions for delicate control. Force and torque applied on the magnetized robot were modified to satisfy the given limit, and unexpected force or torque was not occurred. The method also makes the actuator circuit more reliable since it reduces other currents or voltages under the limit values when one reference is limited, while the conventional simple limiting method could have multiple currents or voltages with the limited value.

Acknowledgement

This work was supported by the Robot industry fusion core technology development project of the Ministry of Trade, Industry & Energy of KOREA(10052980).

References

[1] C. Pawashe, S. Floyd and M. Sitti, "Modeling and experimental characterization of an untethered magnetic micro-robot," *The Int. Journal of Robotics Research*, pp.1077-1094, Aug. 2009.

[2] J. J. Abbott, O. Ergeneman, M. P. Kummer, A. M. Hirt and B. J. Nelson, "Modeling magnetic torque and force for controlled manipulation of soft-magnetic bodies," *IEEE Trans. Robotics*, vol. 23, no. 6, pp.1247-1252, Dec. 2007.

[3] E. Diller, J. Giltinan, G. Z. Lum, Z. Ye, and, M. Sitti, "Six-degrees-of-freedom remote actuation of magnetic microrobots," in Robotics: *Science and Systems Conf.*, 2014.

[4] E. Diller, J. Giltinan and M. Sitti, "Independent control of multiple magnetic microrobots," *The Int. Journal of Robotics Research* 32(5), pp. 614–631.

[5] S. Schuerle, S. Erni, M. Flink, B. E. Kratochvil, and, B. J. Nelson, "Three-dimensional magnetic manipulation of micro- and nanostructures for applications in life sciences, " *IEEE Trans. Magnetics*, vol. 49, no. 3, pp. 321-330, Jan. 2013.

Biographies

Jun Lee was born in Seoul, Korea, in 1993. He received the B.S. degree in electrical engineering from Seoul National University, Seoul, Korea, in 2015, where he is currently pursuing the M.S. degree in electrical engineering. His current research interests are electric energy conversion, power electronics and magnetic manipulation system.

Jung-Ik Ha (S'97–M'01–SM'12) was born in Korea in 1971. He received the B.S., M.S., and Ph.D. degrees from Seoul National University, Seoul, Korea, in 1995, 1997, and 2001, respectively, all in electrical engineering.
From 2001 to 2002, he was a Researcher with Yaskawa Electric Company, Kitakyushu, Japan. From 2003 to 2008, he was a Senior and Principal Engineer with Samsung Electronics Company, Suwon, Korea. From 2009 to 2010, he was a Chief Technology Officer with LS Mecapion Company, Seoul. Since 2010, he has been with the Department of Electrical and Computer Engineering, Seoul National University, Seoul, where he is currently an Associate Professor. His research interests are on circuits and control in high-efficiency and integrated electric energy conversions for various industrial fields.

The Research of Battery Model and Intelligent Charging in the Rail Transport System

Qin Zhuqian   Huang Shenghua   Zou Xunhao   Kan Guangqiang

State Key Laboratory Ratory of Advanced Electromagnetic Engineering and Technology
Huazhong University of Science and Technology, Wuhan, 430074, R.P. China
E-mail:qinzhuqian@126.com

Abstract—This paper introduces a third-order dynamic model for lead acid battery whose parameters vary with the state of charge (Soc), charge current and the electrolyte temperature(θ). According to the practical dynamic model, this paper also proposes two control structure for charging battery applying to constant current and constant voltage mode: i) a innovative third closed loops control strategy composed of innermost inductance current loop, inner battery current loop and outer battery voltage loop based on the DC/DC converter, ii) a novel double closed loops control strategy composed of inner current loop and outer either battery current loop under constant current mode or battery voltage loop under constant voltage mode based on the three-phase voltage-sourced PWM rectifier. Moreover, the variable proportional-integral (PI) controller and the power feed-forward control strategy are used on PWM rectifier. The simulation and experiments are carried out to prove the performance of the battery model that shows a good compromise between complexity and precision and the feasibility of the system.

Keyword—Battery Model, Charging Control Strategy

I. INTRODUCTION

The lead acid battery is studied because of the reliable technology and the low cost. So the battery model is very pivotal section of an electrical system in the simulation. In addition, the model is better to be simple and accurate to achieve reasonable simulation results. There exist many battery models, such as electrochemical models, equivalent circuit models, black box model, neural network model and adaptive structure model. Among them, equivalent circuit models are composed of electromotive forces, resistors, capacitors and other elements. Considering the connection ways of the resistors and capacitors in the circuit, the equivalent circuit model can be further grouped into four categories: Rint, radio control (RC), Thevenin, and Partnership for a New Generation of Vehicles models. The Thevenin model was widely used and studied. An equivalent circuit model which takes into account the non-linear characteristics due to the polarization effect within the battery is proposed in [1]. But the model is so complicated and there is not a complete structure to charge battery. The mathematical model proposed in this paper is a modified Thevenin model. The models commendably represented battery behavior and easily used in simulation software. Meanwhile, Efficient charging of the battery is critical to maintain good battery life, safety and reliability. In [2] the power converter for battery is carried out. But the structure is cumbersome. The strategy of the PWM rectifier and a Buck-Boost circuit respectively linking to the battery for bidirectional power converter is proposed in this paper.

II. BATTERY MODEL AND PROPOSED SYSTEM CONTROL STRUCTURE

The modeling and model parameter determination is complicated by the battery's dynamic behavior. The difference between the electromotive force (EMF) and the terminal voltage of the battery is caused by the polarization effect, which consists of ohmic voltage drop and overvoltage effects. The ohmic voltage drop is cause by the resistances of the active materials, the supportive grids within the electrodes and the porous separators. On the other hand, the overvoltage effects represent the extra energy needed to force the electrochemical reaction to proceed. Consequently, the terminal voltage of the battery is always less than the EMF during discharge, which higher during charging.

1. Simple Battery Model

The simplest battery model uses a series connection of an ideal voltage source and a simple resistor, as shown in Fig.1.

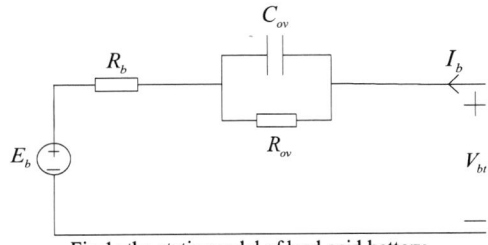

Fig.1: the static model of lead acid battery

In addition to the series connection of the voltage source E_b and the ohmic resistance R_b, the Thevenin battery model includes a parallel resistor-capacitor network in order to describe the overvoltage effects. However the components are constant, whereas in reality all these parameters are the function of stage-of-charge, current, temperature

2. Third-order Dynamic Battery Model

The model is shown in fig.2.

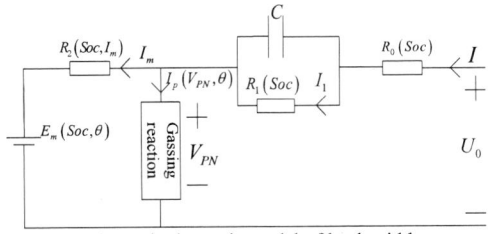

Fig.2: Thevenin dynamic model of lead acid battery

The model composed of the main reaction branch which approximated the battery dynamics under most conditions and the parasitic reaction branch which accounted for the gas evolution reaction of battery at the end of a charge are based on variable electrical parameters.

In fig.2, E_m is the electromotive force of battery, R_0 is the terminal resistance, I_p is the parasitic branch current which takes into account gas evolution reaction, R_1, R_2 are the main branch resistances which think about battery polarization phenomenon .

The third-order dynamic model parameters are the non-linear function of the state of charge, charge current and the electrolyte temperature. Some of the parameters are defined as follows.

$$C(I,T) = \frac{k_c C_0 \left(1 - \frac{T}{T_f}\right)^{\varepsilon}}{1 + (k_c - 1)\left(\frac{I}{I^*}\right)^{\delta}} \qquad (1)$$

where C is the battery total capacity, k_c, δ and ε are constant for a given battery, I^* is the nominal battery current, I is the charge current, T is the electrolyte temperature, T_f is the electrolyte freezing temperature, C_0 is the battery capacity at 0 °C.

$$Soc = \frac{Q_e}{C(0,T)} \qquad (2)$$

where $Q_e = \int_0^t I_m(\tau)d\tau$ Q_e is the battery remaining capacity.

$$\frac{dT}{dt} = \frac{1}{C_\theta}\left[P_s - \frac{T - T_a}{R_\theta}\right] \qquad (3)$$

where C_θ is the battery thermal capacitance, R_θ is the battery thermal resistance, T_a is the temperature of the environment, P_s is the heat that is generated internally in the battery.

$$E_m = E_{m0} - K_E(273 + \theta)(1 - Soc) \qquad (4)$$

$$R_1 = -R_{10}\ln(Soc) \qquad (5)$$

$$R_0 = R_{00}\left[1 + A_0(1 - Soc)\right] \qquad (6)$$

$$R_2 = R_{20}\frac{\exp\left[A_{21}(1 - Soc)\right]}{1 + \exp\left(A_{22}\frac{I_m}{I^*}\right)} \qquad (7)$$

where E_{m0}, K_E, R_{10}, R_{00}, A_0, R_{20}, A_{21}, A_{22} are constant for a particular battery.

$$I_p = V_{PN}G_{P0}\exp\left[\frac{V_{PN}}{V_{P0}} + A_p\left(1 - \frac{T}{T_f}\right)\right] \qquad (8)$$

where I_p is the current loss in the parasitic branch, V_{PN}

is the voltage at the parasitic branch, G_{P0}, V_{P0}, A_p are the constant.

3. Control Structure

The whole scheme including every part of the control system is shown in the Fig.3. The system configuration is composed of a PWM controlled three phase full bridge converter, a battery pack and a DC/DC converter.

The black part is a PWM rectifier control strategy, using variable PI regulator to control the battery voltage and current fluctuation when switches among constant charging mode and constant voltage charging mode.

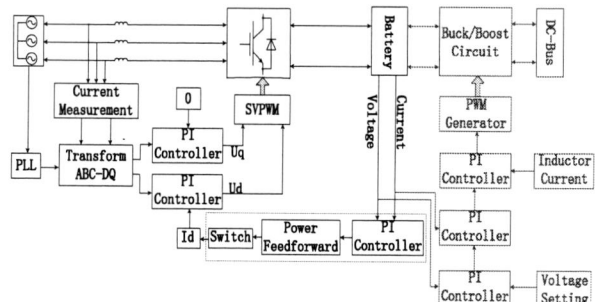

Fig.3: The proposed system control structure

To overcome the power network volatility and the effect of load disturbance, the power feed-forward control strategy is also applying. Neglect the loss of three-phase bridge circuit and switching devices, there is as follows:

$$\frac{3}{2}\left(e_d i_d + e_q i_q\right) = U_{dc}\left[C\frac{dU_{dc}}{dt} + i_{bat}\right] \qquad (9)$$

Use the voltage oriented vector control, there is $e_q = 0$. At steady state, and we have the equation $\frac{dU_{dc}}{dt} = 0$. The equation (9) can been expressed as follows:

$$\frac{3}{2}e_d i_d = U_{dc}i_{bat} \qquad (10)$$

If reference current is followed as $i_d^* = K_f\frac{i_{bat}}{e_d}$, take it into (10).

$$\frac{3}{2}K_f = U_{dc} \qquad (11)$$

It has been shown that when battery voltage is steady, it is independent of power voltage and load. Feed forward signal tracks with incoming current rapidly, which can keep a power balance between input and output. So battery voltage can keep constant.

If reference current is followed as $i_d^* = K_f\frac{U_{dc}}{e_d}$, take it into (10).

$$\frac{3}{2}K_f = i_{bat} \qquad (12)$$

It has been shown that when battery current is steady, it is independent of power current and load.

The blue part is a DC/DC converter control strategy in which voltage PI regulator output saturation value as constant current setting equaled to double closed loop control under constant current mode and quit saturation realizing third closed loop to control the PWM duty ratio directly based on voltage feedback under constant voltage mode.

In addition, the PI regulator will reduce the damping and deteriorate the stability. A method to find the most appropriate parameters of PI controller which is critical to system will be carried on in this paper.

III. SIMULATION RESULTS

Simulations of the third-order dynamic battery model are performed using Matalb/Simulink. The model functioning in the charging state is illustrated in Fig.4.The constant current setting is 100A. The test of charge and discharge processes is carried out and the waveform is shown in Fig.5.

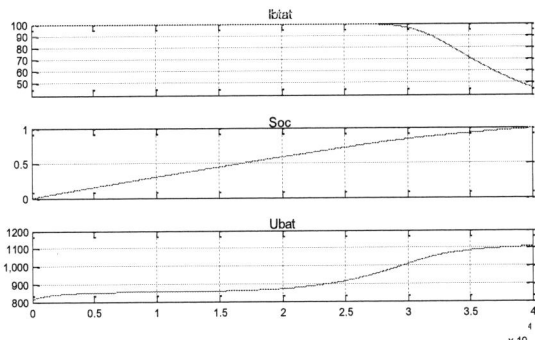

Fig.4: Battery state in constant current mode

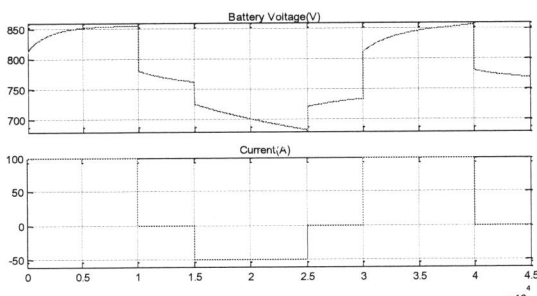

Fig.5: Battery state in constant current charge or discharge mode

From in Fig.4, The simulation of charge process of the third-order dynamic model shows that the voltage, current and Soc change are reasonable.

From in Fig.5, Constant current discharge begins on 15000 second, the drop is caused by the internal resistance of the battery. The smooth portion of the drop is determined by the capacitance and internal resistance functions. On 30000 second, constant current charge begins with jump due to internal resistance before it starts the smooth portion of the climb which results from the

capacitance.

Simulation of constant current constant voltage charger based on the battery model is also essential to validate the function of supposed structure, define the parameters of controller and observe the performance of system. All the element's parameters are strictly consistent with the hardware configuration. The charging current is set to 100A and the charging voltage is set to 850V, the battery waveforms are shown in Fig.6 to Fig.7.

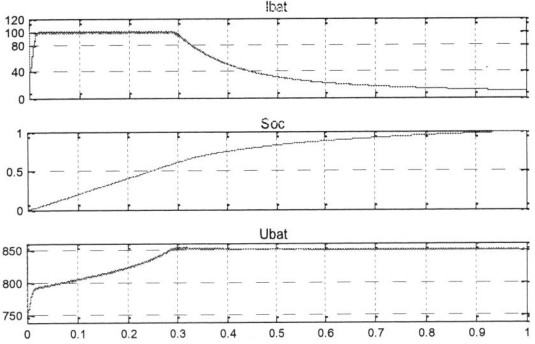

Fig.6: Battery state in constant current and constant voltage mode

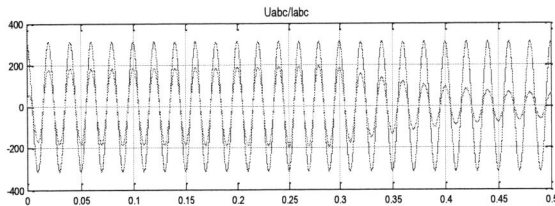

Fig.7: Grid state in constant current and constant voltage mode

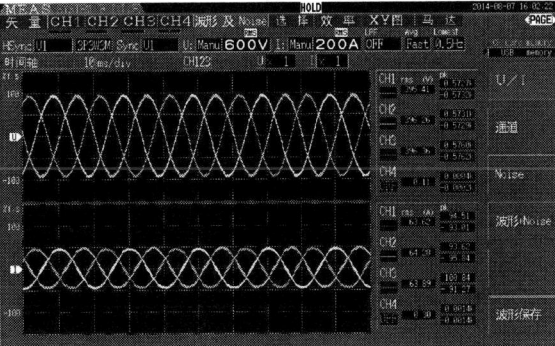

Fig.8: Grid voltage and current in constant current

From in Fig.6, the charging current could reach the setting value without overshoot, and the ripple of current after stabilizing is limited within 1A. the charging voltage could reach the setting value with overshoot and ripple.

From in Fig.7, The phase of grid voltage is keeping with the phase of grid current. The experiment waveforms are shown in Fig.8. Because of the grid transformer, The phase of grid voltage isn't in accordance with the phase of grid current. It demonstrates the precision of the battery model and the effectiveness of the proposed control strategy.

IV. CONCLUSION

In this paper, The third-order dynamic model of lead acid battery could show the battery behavior well and truly. Moreover, there is a thermal model in the battery model, which has important significance for refining the battery management system. In addition ,a bidirectional power converter for lead acid battery is proposed. A three phase rectifier and a bidirectional DC-DC converter are designed which is aimed at controlling the current and voltage of battery. The variable PI controller could control the charging current and charging voltage precisely. The simulation and experiment verify the effectiveness of system. One drawback of the control strategy is the stability of system when switches among the two charging modes. Future work will focus on fuzzy control to improve the system robustness.

ACKNOWLEDGMENT

The author would like to thank Lin Zhenjun for his guidance during the time on the field of power electronics, This work is supported by the National Natural Science Foundation of China (Grant No. 51277053).

REFERENCES

[1] Massimo Ceraolo. New Dynamical Models of Lead-Acid Batteries.IEEE[J].Transaction on power Systems,2000,15(4):1184-1190

[2] Chen Zheng, Nie Ziling, Fu Yuhong, and C. C. Mi, "A bidirectional power converter for battery of plug-in hybrid electric vehicles," the 36th Annual Conference on IEEE Industrial Electronics Society (IECON 2010), pp. 3049-3053, 7-10 Nov. 2010

[3] Stefano Barsal, Massimo Ceraolo. "Dynamical Models of Lead-Acid Batteries: Implementation Issues," IEEE Transactions on Energy Conversion. Vol.17, No.1, IEEE, March 2002.

[4] Shuo Tian, Munan Hong, Minggao Ouyang. "An Experimental Study and Nonlinear Modeling of Discharge I-V Behavior of Valve-Regulated Lead-Acid Batteries," Energy Conversion, IEEE Transactions on, Vol.24, May. 2009, pp:452-458.

[5] Olivier Tremblay, Louis-A. Dessaint, Abdel-Illah Dekkiche. "A Generic Battery Model for the Dynamic Simulation of Hybrid Electric Vehicle," Vehicle Power and Propulsion Conference, 2007.IEEE, Sep. 2007.

[6] S. Mischie and D. Stoiciu. A new and improved model of a lead acid battery. FACTA UNIVERSITATIS (NIS), 20(2):187–202, Aug. 2007.

[7] B. Singh, B. N. Singh, A. Chandra, et al. "A Review of Three-Phase Improved Power Quality AC-DC Converters,"IEEE Transaction on Industrial Electronics, Vol. 51, No. 3,pp.641-660, 2004.

[8] V. Blasko and V. Kaura, "A New Mathematical Model and Control of a Three-phase Voltage Source Converter," IEEE Trans. Power Electronics, Vol.22, No.1, pp.116-123, Jan. 1997.

[9] Cichowlas M, Malinowski M, Kazmierkowsk M P, et al., "Direct power control for three phase PWM rectifier with active filtering function,"IEEE Conference on International Symposium on Industrial Electronics,Rio de Janeiro: IEEE, pp. 913-918, 2003.

[10] B. Schweighofer, K. M. Raab, and G. Brasseur, "Modeling of high power automotive batteries by the use of an automated test system," IEEE Trans. on Instrument Measurement, vol. 52, no. 4, pp. 1087–1091,Aug. 2003.

[11] C. Zhan, X.C. Wu, S. Kromlidis, V. Ramachandaramurthy, M. Barnes,N. Jenkins, and A.J. Ruddell. Two electrical models of the lead-acid battery used in a dynamic voltage restorer. In Generation, Transmissionand Distribution, IEE Proceedings-, volume 150, pages 175–182, March.2003.

[12] J. Zhang, S. Ci, H. Sharif, and M. Alahmad, "An enhanced circuit-based model for single-cell battery," in Applied Power Electronics Conference and Exposition (APEC), 2010, pp. 672-675.

[13] Aden Seaman,Thanh-Son Dao and John McPhee , A survey of mathmematics-based equivalent-circuit and electrochemical battery models for hybrid and electric vehicle simulation,Journal of Power Sources,vol.256,no.0,pp.410-423,2014.

[14] Baogong Ji, Benfeng Gao, Kuangcheng Li. "The Modeling and Simulation of Lead-acid Batteries Based on the Third-order Model," Journal of Aemored Force Eengineering Institute, Vol. 17, Sep. 2003, pp:75-78.

[15] R. Giglioli, A. Buonarota, P. Menga, and M. Ceraolo, "Charge and discharge fourth order dynamic model of the lead-acid battery," in Proc. 10th Int. Elect. Vehicle Symp., Hong Kong, China, Dec. 1990.

[16] D.C.Lee, G.M.Lee,and K.D.Lee,"DC-bus voltage control of three-phase AC DC PWM converters using feedback linearization,"IEEETrans.onIndustryApplications,Vol.36,No .3,May/June 200.

BIOGRAPHIES

Qin Zhuqian was born in China in 1989.He obtained his B.E. degree of electrical engineering from China University of Mining and Technology.

He is now studying the master degree in HuaZhong University of Science and Technology, and his research direction is power electronics and power drives.

Huang Shenghua is now the professor of Huazhong University of Science and Technology. His main research interests include new special motor and its control system, power electronic devices and systems and the applications of power electronics in Power system.

Zou Xunhao obtained his BSc degrees in the college of electrical engineering from Zhejiang University in 2013. He is now working towards his master degree in Huazhong University of Science and Technology.His mainresearch interests are control of PM synchrounous motors and generators.

Kan Guangqiang obtained his BSc degrees in the college of electrical engineering from southwest jiaotong University in 2014. He is now working towards his master degree in Huazhong University of Science and Technology. His main research interests are control of PM synchrounous motors and power electronics.

978-1-5090-0064-7/15 $31.00 © 2015 IEEE

A Hierarchical Group Control Method of Electrical Loads in Smart Home

Zhaojing Yin[1] Yanbo Che[1] Wei He[2]

[1]Key Laboratory of Smart Grid of Ministry of Education, Tianjin University, China
E-mail: lab538@163.com
[2] State Grid Jiangxi Electric Power Research Institute, Nanchang, China
E-mail: lanlyhw@163.com

Abstract–With the development of smart home, the amount of household electricity loads is greatly increasing. How to effectively control these loads has been one of the most important issues in the research of smart home system. Keeping the total electricity loads under limitation would ensure the safety of users' electricity consumption. However, limiting the amount of electricity load or simply cutting the electricity off when it exceeds the limit will bring inconvenience to users. This paper proposes a hierarchical group control method of household facility, which classifies household electricity loads according to their importance and formulates a rule of cutting off when the total load exceeds the limitation. Based on the MATLAB software, calculating and simulating the appliances' working state by original Particle Swarm Optimization (PSO) has verified the correctness and effectiveness of the proposed group method. This method can be applied to current smart home control unit to improve the security and convenience of the system.

Keywords–Electrical safety, group control, load hierarchy, power system, smart home

I. INTRODUCTION

Smart home is one of the emerging application domains of the Internet of Things which is the third wave of the global information industry. The smart home is a concept of the pervasive computing, and it gradually becomes essential for the people living in information age [1-2]. In literature [3-6], several kinds of smart home system were introduced. Smart home can provide services to make household life more comfortable, safer, and more energy-saving.

Nowadays, rapid development of smart home creates the complexity of the household facility in citizens' home. Home Energy Management System (HEMS) was recently researched and developed to solve this question. In [7-11], HEMS and related knowledge was introduced. In traditional home, home capacity is generally greater than the total load. But all-electric smart home may cause the home capacity sufficient for all the loads. If the home overloaded frequently, it will affect the stability of power system and lead to a serious of security threats, it will also affect the users' safety. Overloading can harm the circuit, even catch a fire. Limiting the amount of electricity load or simply cutting the electricity off when it exceeds the limitation will be inconvenient for users. So how to control the load orderly and effectively has been one of the most important issue in the research of the smart home system.

To solve this problem, a household electricity load control method was proposed, which formulates a rule for cutting the load off when the total load exceeds the limitation. The household electrical load was rated into hierarchy so that the HEMS can group control the household electricity load through this rule. At last, the Particle Swarm Optimization (PSO) was used to verify the correctness of the hierarchical group control method. The literature [12] introduced the PSO, the literature [13-16] introduced the PSO and related improved algorithm's application in smart home.

II. CONTROL METHOD DESIGN

1. Theoretical background

Maslow's hierarchy of needs is a theory on psychology proposed by Abraham Maslow in 1943 in the paper "A Theory of Human Motivation" in Psychological Review. This theory was introduced in literature [17-19]. This theory accurately describes many realities of personal experiences. Maslow is a humanistic psychologist. Humanists do not believe that human beings are pushed and pulled by mechanical forces and they focus upon potentials. They believe that humans strive for an upper level of capabilities. Humans seek the frontiers of creativity, the highest reaches of consciousness and wisdom. This has been labeled "fully functioning person", "healthy personality", or as Maslow calls this level, "self-actualizing person." Maslow has set up a hierarchic theory of needs. All of his basic needs are instinctual, equivalent of instincts in animals. According to Maslow, there are five levels of basic needs. Beyond these needs, higher levels of needs exist. These include needs for understanding, esthetic appreciation and purely spiritual needs. In the levels of the five basic needs, the person does not feel the second need until the demands of the first have been satisfied or the third until the second has been satisfied, and so on.

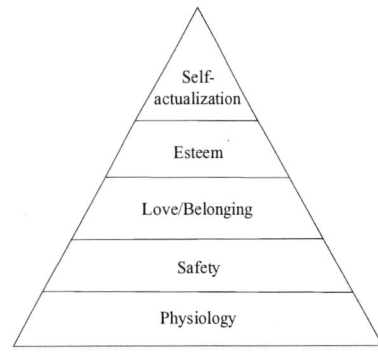

Fig. 1: Maslow's hierarchy of needs

Maslow's basic needs are as follows: Physiological Needs, Safety Needs, Needs of Love, Affection and Belongingness, Needs for Esteem and Needs for Self-Actualization. The first level is biological needs, such as needs for oxygen, food, water, and a relatively constant body temperature. When all physiological needs controlling thoughts and behaviors are satisfied, the needs for security then become active. When the needs for safety and for physiological well-being are satisfied, the next class of needs for love, affection and belongingness can emerge. When the first three classes of needs are satisfied, the needs for esteem will become dominant. When all of the foregoing needs are satisfied, then and only then are the needs for self-actualization activated.

The smart home system is just like a person. Of course, this system contains the users. The household appliances were regarded as the person's needs, so they can be rated into hierarchy to control based on the Maslow's hierarchy of needs theory.

2. Implementation

As mentioned above, this paper proposed a method for the HEMS to control the household electricity load. Our theoretical background is the Maslow's hierarchy of needs theory. Now, in this section, how to rate the electricity load based on the Maslow's hierarchy of needs will be introduced.

The smart home is just like a person whose needs should be considered. As the theory indicates that the basic need for a person is the physiological needs, so only when a person makes himself alive, he can pursue the next level. For a smart home, there must be a method which satisfies residents' needs for being alive to control the appliances. . So the first step is to get the system alive, the basic electric is the most important for the system which can ensure the normal operation of the system. So the first group of the electricity load can be determined as: the load which can ensure the smart home system work normally, which is essential to the residents or which requires continuous supply, these load's power outage or an extended power outage will cause serious impact on residents, even endanger property and lives. The appliances in this group include washing machines, microwave ovens, water heaters, clothes dryer, etc.

The second group doesn't need to consider the person's basic living (the person is a part of the smart home).When the needs for living are satisfied, the next need group should be considered, which just like the need for safety in the Maslow's hierarchy of needs theory. With the first group, our system can work normally, human in the smart home and the smart home itself can be alive. But the person may still be bothered by the trivia related to food and clothing which belong to the first group. Because sometimes, a person may finish something difficultly by the appliances only belong to the first group. They were limited by the trivia. So the second group can be determined as: the load which can provide convenience for the basic need. The appliances in this group include washing machines, microwave ovens, water heaters, clothes dryer, etc.

The first and the second group have already solved the basic needs and made life convenient. So now the next level of need should be considered. In the Maslow's hierarchy of needs theory, these needs are love/

belongingness, affection and self-actualization. Based on the basic need, human can enjoy themselves. So the third group can be determined as: the load which beyond basic needs and help to improve the quality of life. The appliances in this group include TV, computer, sound, air conditioning, etc.

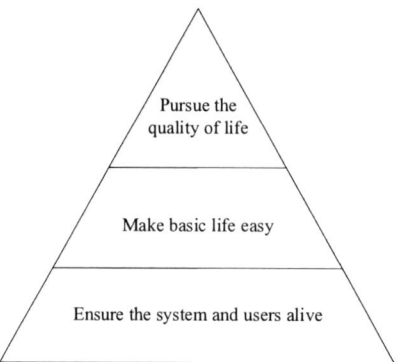

Fig. 2: The hierarchy of household

Now, the three hierarchies have been rated, named group I, group II, and group III respectively. So their total power respectively are P_I, P_{II}, P_{III}, P_Σ, the home capacity is P_N. During the control process:

$$P_\Sigma = P_I + P_{II} + P_{III} \qquad (1)$$
$$P_\Sigma < P_N \qquad (2)$$

In the home energy management system, the Central Control System control the electricity load, all the load state can be changed. Keeping the $P_\Sigma < P_N$ is the control goal. Once the users' consumption behavior makes

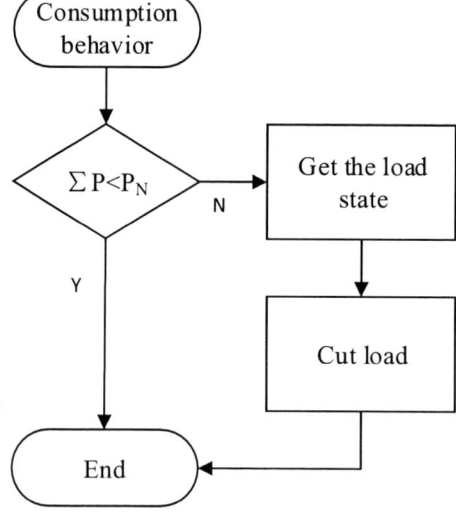

Fig. 3: HEMS Control System

$P_\Sigma \geq P_N$, the Central Control System should response relevantly in a short time limit by changing the load state. As the rule formulated, the system cuts the group III first, then cuts the II and keeps the group I working all the time. Rating the household appliances into hierarchy was just for formulating this rule. The Fig. 3 represents the control process.

III. ANALYSIS OF EXAMPLE

1. Example Introduction

For a smart home with capacity of 7kW, some appliances' hierarchy and power are listed in Table 1. The total power of appliances is 11.8kw, greater than the home capacity.

Table 1: Appliances' Hierarchy and Power

Appliances	Hierarchy	P/kW
General Lighting	I	0.3
Control System	I	0.1
Refrigerator	I	0.3
Rice Cookers	I	1.5
Electric Pan	I	1.5
Washing Machine	II	0.5
Clothes Dryer	II	0.5
Water Heater	II	2.0
Microwave	II	1.0
Hoods	II	0.8
Air Conditioner	III	2.0
Decorative Lighting	III	0.3
TV	III	0.4
Computer	III	0.3
Loudspeaker	III	0.3

In the MATLAB environment, a data file was created to save the appliances' state, power and their hierarchies.

$$P = \begin{bmatrix} 0.3 & 0.1 & 0.3 & 1.5 & 1.5 \\ 0.5 & 0.5 & 2.0 & 1.0 & 0.8 \\ 1.8 & 0.3 & 0.4 & 0.3 & 0.3 \end{bmatrix}$$

$$S = \begin{bmatrix} 1 & 1 & 1 \\ 1 & 1 & 1 \\ 1 & 1 & 1 \\ 1 & 1 & 1 \\ 1 & 1 & 1 \end{bmatrix}$$

The P is the power of the appliances. The S is the state of the appliances. The different row in P and different column in S represents the different group, from I to III. The $P \cdot S = P_\Sigma$, so the state should fulfill the $P \cdot S < P_N$, under this condition, the best case is the maximum number of 1 in the S. The original PSO was used to solve this optimization problem. In the original PSO, the particle change the velocity and position of according to the following equation:

$$V^{k+1} = \omega V^k + c_1 r_1 \left(P_{id}^k - X^k \right) + c_2 r_2 \left(P_{gd}^k - X^k \right) \quad (3)$$

$$X^{k+1} = V^{k+1} + X^k \quad (4)$$

X^k : denotes the current position of the k-th particle;

V^k : denotes the velocity of the k-th particle;

P_{id}^k : the best position found by the k-th particle so far;

P_{gd}^k : the best position found from the particle's neighborhood;

r_1 and r_2 are two vectors of random numbers uniformly chosen from [0, 1];

c_1 and c_2 are acceleration coefficients.

ω : inertia weight

The figure 4 represents the process of the PSO.

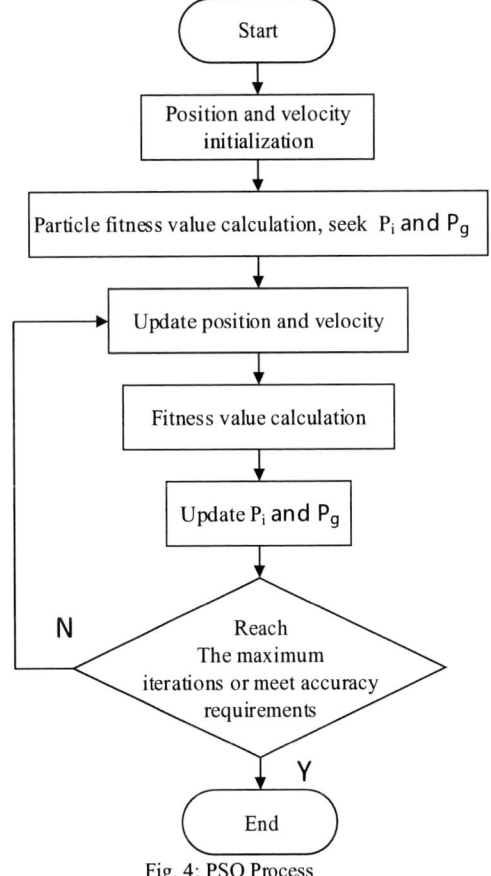

Fig. 4: PSO Process

In this case,

$$c_1 = c_1 = 0.8, \; \omega_{min} = 0.1, \; \omega_{max} = 1.2$$

Set the S as the X in the equations so that the PSO can output the state of every appliance. In order to reflect the hierarchical group control, S was as a 5x2 matrix (S_1) during the optimization algorithm so that the first column of the S state can't be changed by the optimization algorithm. The second and third group take part in the load control by the optimization algorithm. In the three situations simulated in this section, the system worked entirely or partly respectively setting priority or not for group III and group.

2. Situation A

Turn on all the appliances and observe the optimization of the system.
Now:

978-1-5090-0064-7/15 $31.00 © 2015 IEEE

$$S_1 = \begin{bmatrix} 1 & 1 & 1 \\ 1 & 1 & 1 \\ 1 & 1 & 1 \\ 1 & 1 & 1 \\ 1 & 1 & 1 \end{bmatrix}$$

This mean that all the appliances are working. The system worked not entirely: the Water Heater, Microwave and air conditioning should be cut so that the P_Σ will less than P_N and the number of 1 in the S can get the maximum. Then the system worked partly and the solution was got:

$$S_1' = \begin{bmatrix} 1 & 1 & 0 \\ 1 & 1 & 1 \\ 1 & 0 & 1 \\ 1 & 0 & 1 \\ 1 & 1 & 1 \end{bmatrix}$$

$$\mathbf{P} \cdot S_1' = 6.8\text{kW} < P_N$$

The PSO successfully cut the correct load. Let the hierarchical control system worked entirely, after the PSO, the S should be:

$$S_1'' = \begin{bmatrix} 1 & 1 & 0 \\ 1 & 1 & 0 \\ 1 & 0 & 0 \\ 1 & 1 & 0 \\ 1 & 1 & 0 \end{bmatrix}$$

$$\mathbf{P} \cdot S_1'' = 6.5\text{kW} < P_N$$

This is the correct state which cutting the group III first, and then cutting group II.

3. Situation B
Now set the state to S_2:

$$S_2 = \begin{bmatrix} 1 & 1 & 1 \\ 1 & 1 & 0 \\ 1 & 0 & 0 \\ 1 & 1 & 1 \\ 0 & 1 & 0 \end{bmatrix}$$

The system worked partly and the six solutions were got:

$$S_2' = \begin{bmatrix} 1 & 0 & 1 \\ 1 & 1 & 0 \\ 1 & 0 & 0 \\ 1 & 1 & 1 \\ 0 & 1 & 0 \end{bmatrix} \quad S_2' = \begin{bmatrix} 1 & 1 & 1 \\ 1 & 0 & 0 \\ 1 & 0 & 0 \\ 1 & 1 & 1 \\ 0 & 1 & 0 \end{bmatrix}$$

$$S_2' = \begin{bmatrix} 1 & 1 & 1 \\ 1 & 1 & 0 \\ 1 & 0 & 0 \\ 1 & 0 & 1 \\ 0 & 1 & 0 \end{bmatrix} \quad S_2' = \begin{bmatrix} 1 & 1 & 1 \\ 1 & 1 & 0 \\ 1 & 0 & 0 \\ 1 & 1 & 1 \\ 0 & 0 & 0 \end{bmatrix}$$

$$S_2' = \begin{bmatrix} 1 & 1 & 0 \\ 1 & 1 & 0 \\ 1 & 0 & 0 \\ 1 & 0 & 1 \\ 0 & 1 & 0 \end{bmatrix} \quad S_2' = \begin{bmatrix} 1 & 1 & 1 \\ 1 & 1 & 0 \\ 1 & 0 & 0 \\ 1 & 0 & 0 \\ 0 & 1 & 0 \end{bmatrix}$$

Either solution of S_2' fits:

$$\mathbf{P} \cdot S_2' < P_N$$

Let the hierarchical control system worked entirely, there will only be the last two solutions of S_2'.

4. Situation C
Now set the state to S_3:

$$S_3 = \begin{bmatrix} 1 & 1 & 0 \\ 1 & 1 & 0 \\ 1 & 0 & 1 \\ 1 & 1 & 1 \\ 1 & 1 & 1 \end{bmatrix}$$

The system worked partly and then the two solutions were got:

$$S_3' = \begin{bmatrix} 1 & 1 & 0 \\ 1 & 1 & 0 \\ 1 & 0 & 1 \\ 1 & 0 & 1 \\ 1 & 1 & 1 \end{bmatrix} \quad S_3' = \begin{bmatrix} 1 & 1 & 0 \\ 1 & 1 & 0 \\ 1 & 0 & 1 \\ 1 & 1 & 1 \\ 1 & 0 & 1 \end{bmatrix}$$

Either solution of S_3' fits:

$$\mathbf{P} \cdot S_3' < P_N$$

Let the hierarchical control system worked entirely, and the solutions are another two solutions:

$$S_3'' = \begin{bmatrix} 1 & 1 & 0 \\ 1 & 1 & 0 \\ 1 & 0 & 0 \\ 1 & 1 & 0 \\ 1 & 1 & 1 \end{bmatrix} \quad S_3'' = \begin{bmatrix} 1 & 1 & 0 \\ 1 & 1 & 0 \\ 1 & 0 & 0 \\ 1 & 1 & 1 \\ 1 & 1 & 0 \end{bmatrix}$$

$$S_3^* = \begin{bmatrix} 1 & 1 & 0 \\ 1 & 1 & 0 \\ 1 & 0 & 1 \\ 1 & 1 & 0 \\ 1 & 1 & 0 \end{bmatrix}$$

IV. DISCUSSION

In this section, whether the system working entirely or partly, the group I can keep the original state and enable to be changed. Comparing the entirely working system and partly working system, the partly working system can cut the less appliances than the entirely working system, it's almost like cutting load random until the total power less than the home capacity, but it didn't consider the person's needs.

In A and C, the solutions of entirely working system are different from the solutions of partly working system. In B, the solutions of partly working system contain the solutions of entirely working system. But most solutions of the partly working system changed the state of appliances in group II before group III, and the entirely working system doesn't do this. So the entirely working system can make the users' life easier than the partly working system. In a word, the control process can ensure the sum of the working load less than the home capacity. And the system cut the load in the rule which was created above: the system cut the group III first, then cut the group II and keep the group I working all the time, under this premise, the system cut the minimal number of load.

V. CONCLUSION

Several situations of the appliances' state were simulated and the correctness of the hierarchical group control method was verified in the simulation. .And the method can be used into the HEMS or only the smart home system. It can effectively keep the load state optimal from overloading and benefit the power system.

ACKNOWLEDGMENT

This work is supported by State Grid Corporation of China, science and technology projects on Study on the key technologies of all-electric smart home for electric power alteration and applications. (NO.521820150007)

REFERENCES

[1] Boon L T, Husin M H, Zaaba Z F, et al , "A proposed automated smart home control system for detecting human emotions through facial detection", Information and Communication Technology for The Muslim World (ICT4M), 2014 The 5th International Conference on. IEEE, 2014: 1-4.

[2] Ye X, Huang J, "A framework for cloud-based smart home", Computer Science and Network Technology (ICCSNT), 2011 International Conference on. IEEE, 2011, 2: 894-897.

[3] Li M, Lin H J, "Design and implementation of smart home control system based on wireless sensor networks", IEEE Transaction on Industrial Electronics, vol. pp, 2014 (99): 1-12.

[4] Akinyemi L A, Shoewu O O, Makanjuola N T, et al, "Design and Development of an Automated Home Control System Using Mobile Phone", World Journal Control Science and Engineering, 2014, 2(1): 6-11.

[5] Brito J, Gomes T, Miranda J, et al, "An intelligent home automation control system based on a novel heat pump and Wireless Sensor Networks", Industrial Electronics (ISIE), 2014 IEEE 23rd International Symposium on. IEEE, 2014: 1448-1453.

[6] Pang D F, Lu S L, Zhu Q Y, "Design of Intelligent Home Control System Based on KNX/EIB Bus Network", Wireless Communication and Sensor Network (WCSN), 2014 International Conference on. IEEE, 2014: 330-333.

[7] Chaudhari R B, Dhande D P, Chaudhari A P, "Home Energy Management System", International Journal of Advanced Electronics and Communication Systems, 2014, 3(3).

[8] Han D M, Lim J H, "Smart home energy management system using IEEE 802.15. 4 and ZigBee", Consumer Electronics, IEEE Transactions on, 2010, 56(3): 1403-1410.

[9] Han D M, Lim J H, "Design and implementation of smart home energy management systems based on ZigBee", Consumer Electronics, IEEE Transactions on, 2010, 56(3): 1417-1425.

[10] Han J, Choi C S, Lee I, "More efficient home energy management system based on ZigBee communication and infrared remote controls", Consumer Electronics, IEEE Transactions on, 2011, 57(1): 85-89.

[11] Bouhafs F, Mackay M, Merabti M, "Home Energy Management Systems", Communication Challenges and Solutions in the Smart Grid. Springer New York, 2014: 53-67.

[12] Kennedy J, "Particle swarm optimization", Encyclopedia of Machine Learning. Springer US, 2010: 760-766.

[13] Lugo-Cordero H M, Fuentes-Rivera A, Guha R K, et al, "Particle Swarm Optimization for load balancing in green smart homes", Evolutionary Computation (CEC), 2011 IEEE Congress on IEEE, 2011:715-720.

[14] Huang C M, Yang S P, Yang H T, et al, "Combined particle swarm optimization and heuristic fuzzy inference systems for a smart home one-step-ahead load forecasting", Journal of the Chinese Institute of Engineers, 2014, 37(1): 44-53.

[15] Liu T, Xiao X, Ying D, "Kiln Landscape Evolution Simulation Based on Particle Swarm Optimization and Cellular Automata Model", International Journal of Smart Home, 2015, 9(1): 141-150.

[16] Jiang B, Fei Y, "Smart Home in Smart Micro-grid: A Cost-Effective Energy Ecosystem with Intelligent Hierarchical Agents", Smart Grid, IEEE Transactions on, 2015, 6(1): 3-13.

[17] McLeod S, "Maslow's hierarchy of needs", Simply Psychology, 2007, 1.

[18] Simons J A, Irwin D B, Drinnien B A, "Maslow's hierarchy of needs", Retrieved October, 1987, 9: 2009.

[19] Gawel J E, "Herzberg's theory of motivation and Maslow's hierarchy of needs", ERIC Clearinghouse on Assessment and Evaluation, 1997.

BIOGRAPHIES

Zhaojing Yin was born in Liaoning, China . He received his B.S. egree from Nanjing University of Science and Technology. He is presently a student and pursuing for master degree in the School of Electrical Engineering and Automation at TianjinUniversity.

Yanbo Che was born in Shandong, Chian. He received his B.S. degree from Zhejiang University, Hangzhou, China, in 1993. He received his M.S. and Ph.D. degrees fromTianjin University, Tianjin, China, in 1996 and 2002, respectively. Since 1996, he has been engaged in teaching and scientific research of power electronic technology and power systems. He is presently an Associate Professor in the School of Electrical Engineering and Automation at TianjinUniversity. His current research interests include power electronics, new energy and micro-grids.

Wei He was born in Hubei, China. He received his B.S. degree from Wuhan University, Wuhan, China, in 2007. He received his M.S. degree from Wuhan University, Wuhan, China, in 2009. He received his Ph.D. degree from North China Electric Power University, Baoding, China, in 2013. He is presently work st State Grid Jiangxi Electric Power Research Institute. He has been engaged in new energy generation and power system since 2007.

A Research of the Strategy of Electric Vehicle Ordered Charging based on the Demand side Response

Jidong Wang[1], Yuhao Yang[1] and Wei He[2]

[1]Key Laboratory of Smart Grid of Ministry of Education, Tianjin University, China, Email: jidongwang@tju.edu.cn
[2]State Grid Jiangxi Electric Power Research Institute, Nanchang, China, Email: lanlyhw@163.com

Abstract-The rapid development of electric vehicle as well as photovoltaic generation leads measures to combine their advantages in urgent need. The appliance of the TOU price also brings about new challenge to the schedule of electric vehicle. This paper studies the method to minimize the electric cost of the users by the response to the photovoltaic generation and TOU price. Firstly, the model of the photovoltaic generation is proposed. Secondly, the model of optimization of electric vehicle is suggested. Lastly, optimization algorithm is utilized in this model. Simulation verifies the valid of the model.

Keywords-electric vehicle; demand side response; photovoltaic generation; optimization

I. INTRODUCTION

The technology of electric vehicle [1-2] as well as photovoltaic generation [3-4] is becoming more and more popular. However, the unordered charging of electric vehicle leads to the uncertainty of its charging time. As a result, the electric vehicle is unable to respond to the photovoltaic generation system sufficiently. That is to say, the advantages of them cannot combine. To correct the electric vehicle charging disorder, it is a valid approach to change the electric vehicle charging time by participating in demand side response[5]. By this method, the electric vehicle will be able to respond to the curve of PV power systems and the TOU price. Furthermore, the users will cost less and the solar Resource will be made full use of.

This paper studies the strategies of the electric vehicle participating in demand side response in which the PV system is included. The charging process of electric vehicles will be in an orderly state that the user total load, PV system power curve and the price curve allows the user to achieve minimal costs. The strategy makes full use of the distributed photovoltaic energy and the price ladder so the advantages of electric vehicle and PV system combine sufficiently.

The specific tasks include: (1) Study the characteristics of the photovoltaic power generation system and electric vehicle charging characteristics to establish photovoltaic system model and the electric vehicle charging model. (2) Analyze the behavior characteristics of residential users and the other loads of the users to give the objective function and constraints of demand response in which the PV system is included, and further develop the optimization model of such demand response. (3) Study the correlation optimization algorithm and choose the binary tree pruning method as the algorithm to obtain electric vehicle charging time of the optimal solution and develop strategies of the electric vehicle to participate in demand response containing photovoltaic system based on the optimal solution.

II. MODEL OF DEVICES

The optimization involves photovoltaic generation, electric vehicle and other loads. This section mainly discusses the model of such devices.

1. Model of photovoltaic generation
The model of photovoltaic generation can be shown as Fig. 1:

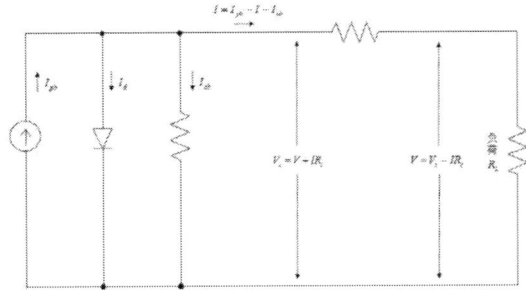

Fig. 1: the Model of photovoltaic generation

In the model of photovoltaic generation，I and V represents the current and voltage of the generation, I'_{sc} and V'_{oc} represents the short circuit current and open circuit voltage. I'_m and V'_m represents the current and voltage when the power is at its maximum value，T , T_{air} and T_{ref} represent the temperature of photovoltaic generation, air and reference temperature. S and S_{ref} represent the real and reference intensity of solar radiation. K represent the temperature factor of the photovoltaic. Then,

$$I'_{sc} = I_{sc}\left(\frac{S}{S_{ref}}\right)(1+aT) \tag{1}$$

$$V'_{oc} = V_{oc}(1-c\Delta T)(1+b\Delta S) \tag{2}$$

In equation (1) and (2), I_{sc} and V_{oc} represent the short circuit current and open circuit voltage under rated condition. a , b , c are equal to 0.0025,-0.1949 and 0.0029, ΔT and ΔS are the variable quantity of the temperature and intensity of solar radiation：

$$\Delta T = T - T_{ref} \tag{3}$$

$$\Delta S = \frac{S}{S_{ref}} \text{-}1 \tag{4}$$

978-1-5090-0064-7/15 $31.00 © 2015 IEEE

I'_m and V'_m could be calculated by equation (5) and (6):

$$I'_m = I_m \left(\frac{S}{S_{ref}} \right)(1 + a\Delta T) \qquad (5)$$

$$V'_m = V_m (1 - c\Delta T)(1 + b\Delta S) \qquad (6)$$

In equation (5) and (6), I_m and V_m represent the current and voltage when the power is equal to its maximum value under rated condition.

I could be calculated by equation (7):

$$I = I'_{sc} \left(1 - C_1 \left\{ \exp\left[\frac{V}{C_2 V'_{oc}} \right] - 1 \right\} \right) \qquad (7)$$

In equation (7), coefficients C_1 and C_2 could be calculated by equation (8) and equation (9):

$$C_1 = \left(1 - I'_m / I'_{sc} \right) e^{-V'_m/(C_2 V'_{oc})} \qquad (8)$$

$$C_2 = \left(\frac{V'_m}{V'_{oc}} - 1 \right) \left[\ln\left(1 - I'_m / I'_{sc} \right) \right]^{-1} \qquad (9)$$

In conclusion, the relationship between the the current and voltage of the photovoltaic generation is shown as Fig. 2:

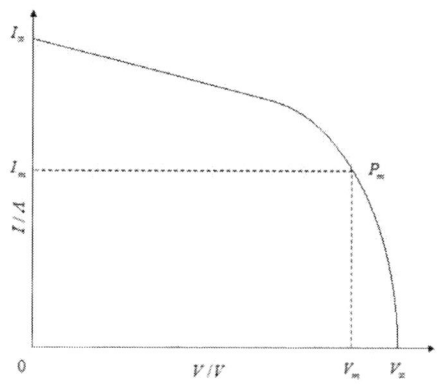

Fig. 2: relationship between the the current and voltage of the photovoltaic generation

2. Model of loads and electric vehicle

In this model, loads except electric vehicle are regarded as a whole. It is assumed that a day is divided into N time slots and at the kth time slots the power of loads is $x(k)$, then:

$$\overline{x_{min}} \leq \frac{\sum_{k=1}^{N} x(k)}{N} \leq \overline{x_{max}} \qquad (10)$$

$$x_{min}(k) \leq x(k) \leq x_{max}(k) \qquad (11)$$

In equation (10) and equation (11), $\overline{x_{min}}$ and $\overline{x_{max}}$ represent the minimum and maximum value of the average power. $x_{min}(k)$ and $x_{max}(k)$ represent the minimum and the maximum of the power at the kth time slot.

In order to the electric vehicle, it is assumed that P_0 represents the power of the electric vehicle while charging. $E(k)$ represents the energy stored by the electric vehicle at the kth time slot. $y(k)$ represents the state of the electric vehicle. $y(k) = 1$ represent at the kth time slot the electric vehicle is in charge state while $y(k) = 0$ represent at the kth time slot the electric vehicle is not in charge state. Therefore, when the electric vehicle is charging,

$$E(k+1) = E(k) + P_0 y(k)\Delta t \qquad (12)$$

In equation (12), Δt represents the time interval of each time slot. If the electric vehicle is in use, the state of the vehicle will follow equation (13)

$$\begin{cases} E(k+1) = E(k) - \Delta E_s(k) \\ y(j) = 0, \quad j = k, k+1, \ldots k_N \end{cases} \qquad (13)$$

In equation (12), ΔE_s represents the consumption of energy and k_N represents the last time slot that the electric vehicle in use.

III. MODEL OF OPTIMIZATION

In order to the residents, the objective function ought to be the electricity costs. Therefore, the objective function is presented by equation (14):

$$z = \sum_{k=1}^{N} p(k)u(x_\Sigma(k))x_\Sigma(k)\Delta t \qquad (14)$$

In equation (14), $x_\Sigma(k)$ represents the power sent by the grids, function $u(t)$ is unit step function. $p(k)$ represents the electrical price at the kth time slot.

$$x_\Sigma(k) = x(k) + y(k)P_0 - x_{DG}(k) \qquad (15)$$

$x_{DG}(k)$ represents the power of photovoltaic generation at the kth time slot.

In terms of the restrictions, the power of the loads ought to obey equation (10) and equation (11). In addition, the energy is supposed to satisfy the need of the utilization of the vehicle.

$$E(k) \geq \Delta E_s(k) \qquad (16)$$

IV. MODEL SOLUTION

In the model, the decision variables are 0-1 variables. Therefore, the method in paper [6] is introduced in the solution of this model [6]. All the feasible solutions of the electric vehicle commitment problem can be described by a binary tree shown in Fig. 3. Because of this, we might as well call this binary tree "solution tree". In the solution tree, the node of the tree represents the state of the electric vehicle The edge between two nodes of the tree represents the status of the electric vehicle and its weight represents the on/off of the charging system in the electric vehicle (0 —off, 1—on).

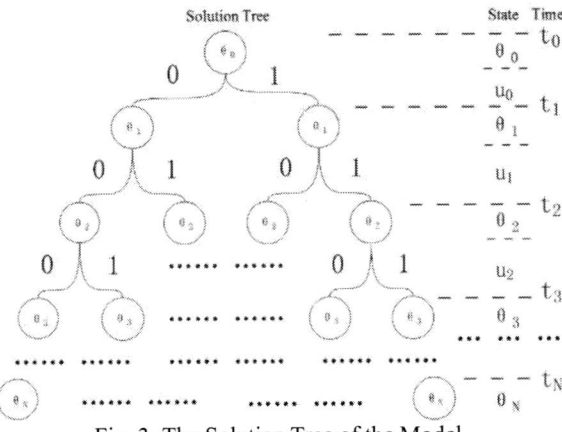

Fig. 3: The Solution Tree of the Model

With the definition of the solution tree, all the feasible solutions of the problem can be obtained after a breadth-first and preorder traversal of the solution tree.

Lots of the nodes violate the constraints. Because of this, it is necessary for us to check whether the value of a node satisfies the comfort constraints in order to obtain the feasible solutions of the problem subjecting to the constraints. We should compare the value with the limits set by the following steps:

(a)if the value of the node satisfies the comfort constraints, go on traversing to the next node;

(b)if the value of the node violates the constraints, remove the whole subtree whose root node is this node from the solution tree such that no further traversal will be performed to this node's child nodes.

By this method, the solutions that violate the comfort constraints are removed from the feasible solution set during BPTST. We can call this method violation pruning.

Through the steps above, we are already able to obtain the exact optimal solution of the optimization problem. However, this kind of traversal costs so much time that it is not practical to apply this method to real life. Because of this, we come up with an "Inferior Pruning" operation for the traversal process to greatly shorten the traversal time and keep the solution as optimal as possible.

It can be found that in many cases the value of a node in the solution tree is very close to its cousin nodes' values but at the same time the cost of this node is significantly lower than that of its cousin nodes', that is to say, this node is significantly superior to its cousin nodes because

the schedule that this node represents offers the similar energy service with the least electricity cost. In these cases, if we remove all the inferior cousin nodes and their child nodes from the solution tree and just go on traversing to the superior node's child nodes, the computation cost will be greatly reduced and the time needed will be significantly shortened. This is the basic idea of inferior pruning.

The steps of "Inferior Pruning" can be summarized as following: while travelling to the nth layer, all nodes will be categorized in accordance with the weights from root to the node and the node with equal objective function will be introduced into the same class. In each class, the node with the minimum objective function will be selected and the other nodes will be pruned.

In conclusion, the steps of the optimization can be described as following:

(a)Traversal the original binary of the optimization problem solution layer by layer and obtained all possible electricity plan with violation pruning and inferior pruning.

(b)After traversing, calculate the path with the minimum objective function from the root of the Nth layer to last layer and select this path as the optimal solution.

V. EXAMPLE ANALYSE

1. Parameter Setting of the Model
An example of a resident house with an electric vehicle, a photovoltaic generation and other loads is selected. In the example, it is assumed that the electricity price is 0.31 Yuan/kW*h between 00:00 to 6:00 and 23:00 to 24:00, and the price is 0.61 Yuan/kW*h on the other time.

In terms of the photovoltaic generation, it is assumed that V_m, I_m, V_{oc} and I_{sc} are equal to 34.4V, 4.51A, 4.9V and 43.2A. In addition, the illumination intensity in one day is shown in Fig. 1. The curve of the loads except the electric vehicle is shown in Fig. 2

2. Simulation Results
The state of the electric vehicle after optimization is shown in Fig. 6.

Before optimization, the user's cost of electricity is 11.54 yuan. However, after optimization, the cost turns into 9.79 yuan. Therefore, the amount of cost decrease is 1.75 yuan.

VI. CONCLUSION

Based on the photovoltaic generation characteristics and combined with the increasing integration of distributed generations, this paper presents a comprehensive consideration of distributed generation, and other loads to optimize electric vehicle operation. Compared to non-optimized energy management, the proposed strategy has taken advantage of local distributed energy, improved local consumption proportion of renewable energy sources, and improved the economic efficiency of the users.

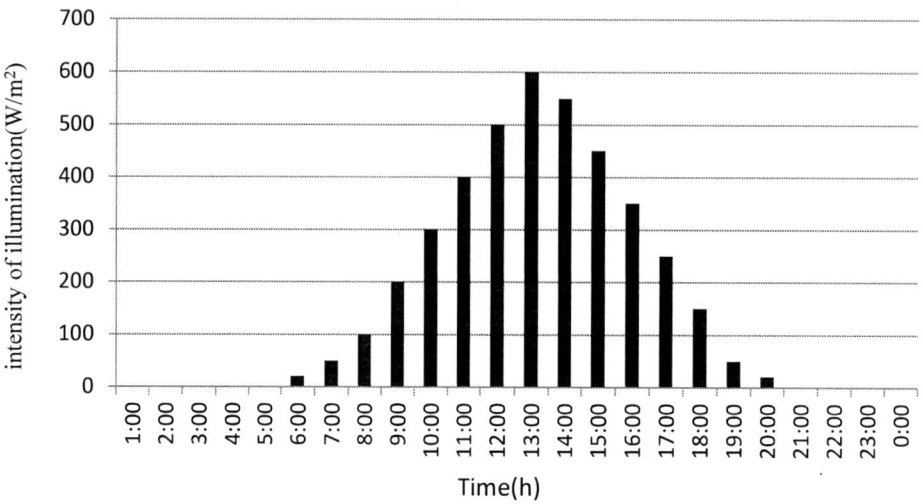

Fig. 4: The illumination intensity in one day

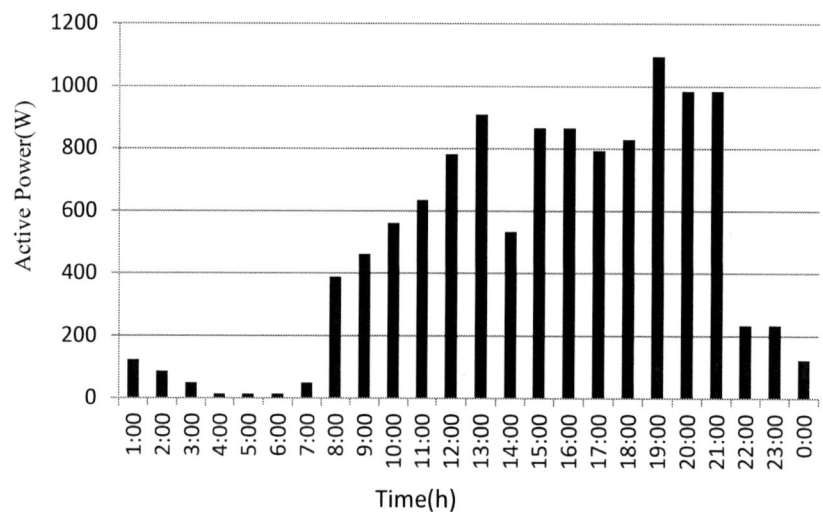

Fig. 5: The curve of the loads except the electric vehicle

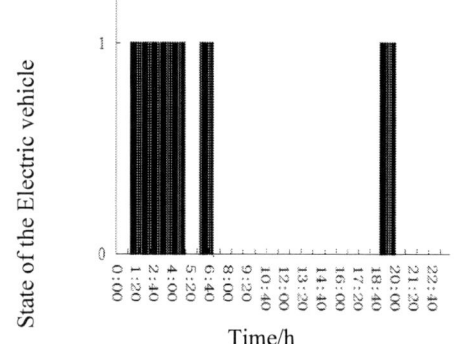

Fig. 6: The state of electric vehicle before optimization

For user-side energy management issues, the toughest problem is not to obtain optimal load schedule, but to accurately perceive and access the users' needs. Therefore, future work stresses on how to analyze the influence of large amounts of load on the energy management system,

how to improve the optimization results and how to realize a flexible and interactive smart power utilization in the environment of the smart grid.

ACKNOWLEDGEMENT

This work is supported by National Natural Science Foundation of China (NSFC) (51477111) and Smart Grid Corporation of China science and technology projects on Study on the key technologies of all-electric smart home for electric power alteration and applications. (NO.521820150007).

REFERENCES

[1] L. Gan, X-N. Zhang, "Modeling of plug-in hvbrid electric vehicle charging demand in probabilistic power flow caculations", IEEE Transactions on Smart Grid, Vol. 3, No.1, 2012, pp: 492-499.

[2] Ashrari, E. Bibeau, S. Shahidinejad, et al, "EV charging profile prediction and analysis based on vehicle usage data", IEEE Transactions on Smart Grid, Vol. 3, No.1, 2012, pp: 341-350.

978-1-5090-0064-7/15 $31.00 © 2015 IEEE

[3] M-D Tabone, D-S. Callaway, "Modeling Variability and Uncertainty of Photovoltaic Generation: A Hidden State Spatial Statistical Approach", IEEE Transaction on Power Systems, Vol. 30, 2015, pp:2965-2973.

[4] M. Zhang, J. Chen, "Islanding and Scheduling of Power Distribution Systems With Distributed Generation", IEEE Transaction on Power Systems, Vol. 30, 2015, pp:3120-3129.

[5] B. Chakrabarti, D. Bullen, C. Edwards, et al, "Demand Response in the New Zealand Electricity Market," IEEE PES Transmission and Distribution Conference and Exposition, 2012.

[6] C-S. Wang, Y. Zhou, J-D Wang, et al, "A novel Traversal-and-Pruning algorithm for household load scheduling", Applied Energy, Vol. 102, 2013, pp:1430-1438.

A research for the interactive behavior between Electric Vehicle and Residential Energy Management System based on Probability Distribution

Jidong Wang[1]   Kaijie Fang[1]   Wei He[2]

[1]Key Laboratory of Smart Grid of Ministry of Education, Tianjin University, China
[2]State Grid Jiangxi Electric Power Research Institute

Abstract-With the human living environment getting rapid deterioration, Electric Vehicle(EV) is gaining more and more attention as a kind of potential new energy vehicle. Besides, Electric Vehicle plays a significant role in the smart home of the future, the strategies of its changing and discharging and intelligent access to the Residential Energy Management System are under pretty hot study. This paper, based on peak-valley electricity price, builds a residential energy management system which includes the renewable energy like solar energy and electric vehicle, makes the most optimal schedule for users through optimal algorithm. Specially, this paper studies the probability distribution of independent variables like daily trip miles, return time and the spending time in the charging behavior. Then, this paper adapts this result of probability distribution to whole energy management system. In addition, the influence brought by the charging behavior has been considered into the energy system, the system can reschedule for the appliances and Photovoltaic system output.

Keyword: Electric Vehicle, Genetic Algorithm, Home Energy Management System, Probability Distribution

I. INTRODUCTION

Compared with the traditional petroleum vehicles, electric vehicles (EV) indeed have some comprehensive advantages in the areas of pollution emission reduction, energy saving and nation's security on imported petroleum. As a result, electric vehicles have being included in the crucial development of automotive industries. Research [1], has shown that under the medium growing speed, the rate of electric vehicles in amount of total cars will rise up to 35%, 51% and 62%,till 2020,2030 and 2050 in U.S.A. And electric vehicles are divided into three kinds based on different ways of using the source energy, pure electric vehicle (PEV), plug-in hybrid electric vehicle (PHEV) and fuel cell electric vehicle(FCEV)[2]. Many automotive manufactures, like Toyota and BMW, have pushed some which suit for household into the market. One of these EV products called "TESLA Model S" gain lots of fame and reputation. Limited by the lack of charging-piles and the user experience, charging using distribution grid at home plays an important role in charging methods for household electric vehicle. The charging load brought by charging behaviors changes the initial home daily load curve. And due to the uncertainty of return time and daily traveling miles of household vehicles, the charging load is highly uncertain. So does the influence caused by electric vehicle charging load.

The largest proportion in the user side of smart grid is common residential users. And with smart appliances being popular, the Home Energy Management System (HEMS) becomes possible. Work[3] states that HEMS can improve the energy efficiency, users'comfort and save electricity bills through controlling all appliances and managing all energy.

This paper brings the pure electric vehicle into HEMS and studies the uncertain charging load of PEV based its probability distribution. In addition, this paper gain daily charging load expectation in different times and the daily charging load curve of PEV by using the Monte Carlo stochastic simulation method. The charging load expectation and curve is applied to the HEMS. In the part of photovoltaic system in HEMS, this paper considers uncertainty of the output of photovoltaic array and connects the array to battery device. In the part of the objective function of HEMS, this paper comes up with an objective function of comprehensive satisfaction, considering user tariffs and comfort. At the same time, this paper constructs a nonlinear program problem with constraints which are based on the balance of power and the constraints of different loads and its character of probability distribution. At the last section, this paper adopt Genetic Algorithm to solve the nonlinear program problem and gain the schedule for different appliance.

II. PROBABILITY MODEL FOR PEV

Considering various types of electric vehicle, this paper adopt the pure electric vehicle as household daily transportation. The return time and the daily traveling miles are directly relevant to the users 'driving habits, driving characters and the charging mode for one pure electric car. And the statistical data shown in Fig.1 and Fig.2 comes from national household travel survey(NHTS) which is published by Department of Transportation U.S in 2001[4]. The two bar charts obviously indicate that the return time probabilities and the daily traveling miles probabilities suit for some special probability density function.

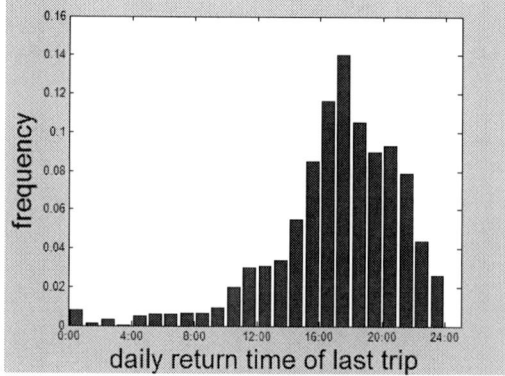

Fig.1: The different frequencies of daily return time per residential vehicle

978-1-5090-0064-7/15 $31.00 © 2015 IEEE

Obviously, the return time mainly converges on the period between 15:00 and 22:00 where it reaches the peak at about 18:00. Supposing the pure electric vehicle start charging immediately after returning home, referring to the peak-valley price of grid in china shown in Fig.3, this electric vehicle charging load has pretty high probability to meet the high price of grid. And according to Fig.1 and Fig.2, the return time and daily traveling miles approximately meet the normal distribution and normal logarithmic distribution[5].

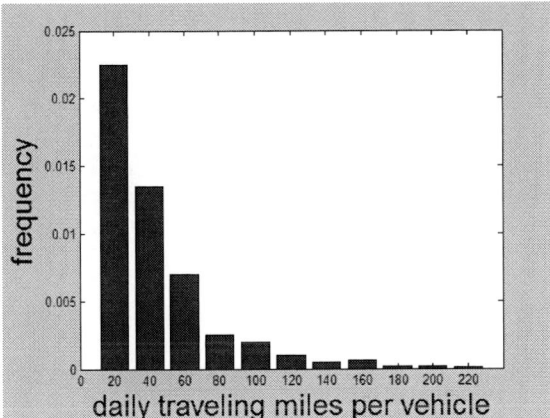

Fig.2: The different frequencies of daily traveling miles per residential vehicle

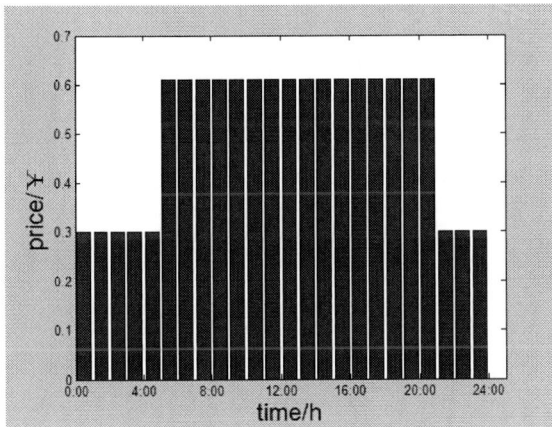

Fig.3: The peak-valley price of grid

This paper adopt maximum likelihood estimation to gain the parameters of normal distribution and normal logarithmic distribution based on the data from NHTS. And the start time of charging is equal to the return time of last trip based on above assumptions.

The probability density function for start charging time is as follow:

$$f_{es}(t) = \begin{cases} \dfrac{1}{\sigma\sqrt{2\pi}}\exp\left[-\dfrac{(t-\mu)^2}{2\sigma^2}\right] & (\mu-12) < t \le 24 \\ \dfrac{1}{\sigma\sqrt{2\pi}}\exp\left[-\dfrac{(t-(\mu-24))^2}{2\sigma^2}\right] & 0 < t \le (\mu-12) \end{cases} \quad (1)$$

The likelihood function can be figured out:

$$L(\theta) = \prod_{i=1}^{n} f(t_i; \mu, \sigma^2)$$
$$= (\frac{1}{\sigma\sqrt{2\pi}})^n \cdot \exp[-\frac{1}{2\sigma^2}\sum_{(\mu-12)<t_i\le 24}(t_i-(\mu-24))^2]\cdot\exp[-\frac{1}{2\sigma^2}\sum_{0<t_i\le(\mu-12)}(t_i-\mu)^2] \quad (2)$$

We can get the parameters through solving the function in which the partial derivative is zero.

$$\begin{cases} \dfrac{\partial\ln(L)}{\partial\mu} = 0 \Rightarrow \mu = \dfrac{\displaystyle\sum_{i=1}^{n}t_i + \sum_{(\mu-12)<t_i\le 24}(24)}{n} \\ \dfrac{\partial\ln(L)}{\partial\sigma^2} = 0 \Rightarrow \sigma^2 = \dfrac{\displaystyle\sum_{(\mu-12)<t_i\le 24}(t_i-(\mu-24))^2 + \sum_{0<t_i\le(\mu-12)}(t_i-\mu)^2}{n} \end{cases} \quad (3)$$

In formulas(1)~(3), t and ti respectively represent for the start charging time and No.i sampled data, while mu and sigma are the parameters we need. Take the data from NHTS into these functions: $\mu = 17.6; \sigma = 3.4$

This paper deal with the daily traveling miles with the same method and gain the probability density function as follow:

$$f_{em}(x) = \frac{1}{\sigma_1\sqrt{2\pi}}\exp\left[-\frac{(\ln x - \mu_1)^2}{2\sigma_1^2}\right] \quad (4)$$

Take the data from NHTS into same process: $\mu_1 = 3.2; \sigma_1 = 0.88$

The probability density function and cumulative density function are shown in Fig.4 and Fig.5.

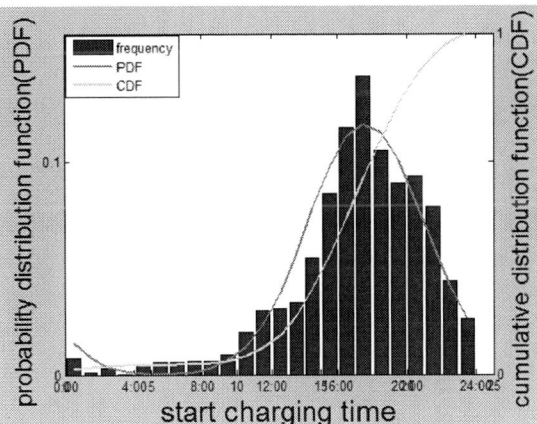

Fig.4: The PDF and CDF curve for start charging time

However, the charging behavior is influenced by not only the start charging time, but also the spending time for charging. Furthermore, the spending time for charging one PEV is relevant to remaining battery capacity, charging power and charging mode. Research[6] instruct that for the charging mode various batteries almost adopt the constant voltage-constant current method. And study[7] shows that the charging process can be simplified into a constant power charging process. The charging power used in this paper is just the simplified constant power curve which is shown in Fig.6.

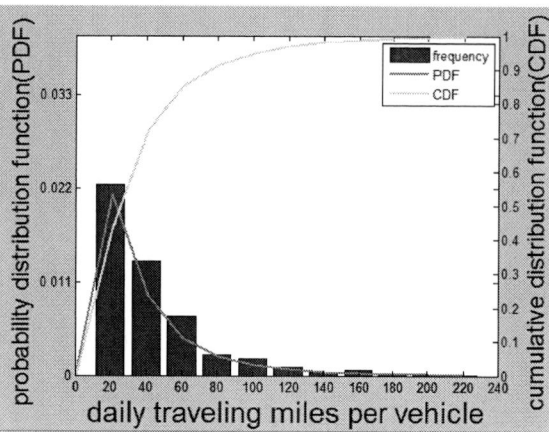

Fig.5: The PDF and CDF curve for daily traveling miles

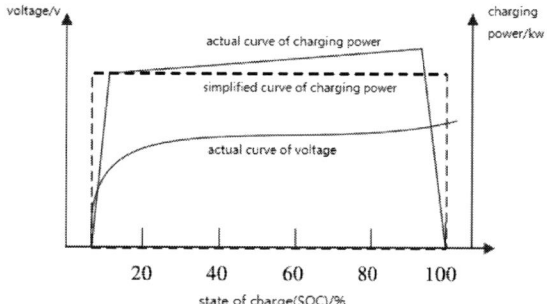

Fig.6: Charging power curve

Remaining battery capacity can be regarded as a linear function which is related to the daily traveling miles. Assuming that the household PEV start its everyday travel after fully charging the battery, then the remaining state of capacity(SOC) can be calculated as follow:

$$soc = (1 - \frac{E_c}{E_n}) \times 100\% \qquad (5)$$

$$E_c = \frac{hx}{\eta} \qquad (6)$$

In formulas(5)~(6), E_c represents the energy consumption caused by daily travel, E_n represents rated energy capacity, h and η represent the energy consumption per mile and the constant about vehicle efficiency, respectively.

Then we can get the spending time for charging:

$$\int_0^T p(t)dt = E_c \Rightarrow T = \frac{E_c}{p} = kx, \quad k = \frac{h}{p\eta} \qquad (7)$$

In formulas (5)~(7), these parameters, like k and η, refer to the studies[6,8]. Actually, the spending time for charging is just the value of function of random variable where the PDF has been figured out in formula (4). We can easily gain the PDF of the spending time for charging.

$$f_{e_c}(T) = \frac{1}{k} f_{em}(\frac{T}{k}) = \frac{1}{k\sigma_1 \sqrt{2\pi}} \exp\left[-\frac{(\ln(T/k) - \mu_1)^2}{2\sigma_1^2}\right] \qquad (8)$$

In order to get the charging load, the expectation of

charging power for each time in one day should be obtained which is based on the PDF (1) and (8). This paper use the Monte Carlo stochastic simulation method in the following five steps:

Step1.create two arrays named STT and SPT sized N in which elements should satisfy the PDF of start time and spending time for charging and should be arranged randomly.

Step2.create one array named JUT sized 24*N which is used to judge whether the PEV is under charging or not in each hour. And the elements are 0 or 1 where the rules are defined as follow:

$$\begin{cases} JUT_k(i) = 1, STT + SPT > JUT_k(i) > STT \\ \quad or \quad STT + SPT - 24 > JUT_k(i) > STT - 24 \\ JUT_k(i) = 0, else \\ k = 1 \cdots 24 \end{cases} \qquad (9)$$

Step3. Calculate the number (the sum is named n) of elements that their value must be one in array JUT. And the rate of n/N is just probability of charging for time k in one day.

Step4. Calculate the charging load expectation for period k:

$$E(JUT_k) = \frac{n}{N} * P \qquad (10)$$

Step5. Due to pretty uncertainty of Monte Carlo method, the expectation should be averaged after several above experiments. And the charging load curve is shown in Fig.7.

The expectation of the charging load for single PEV do not have specific physical meaning. And the load curve cannot represent the real charging load which is obviously sectionalized for explicit one day instead of being continuous. However, through the Monte Carlo method, the continuous load curve implies the probability information in the long time term. And we can regard this simulated load as a virtual load. Applying this virtual load to the HEMS as one regular load, like light load, the system can schedule other appliances and photovoltaic system to reduce the grid fee and improve users' comfort.

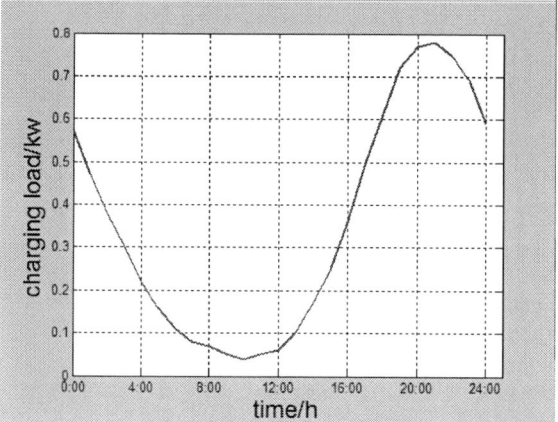

Fig.7: The charging load curve

Although the virtual load cannot reflect real one day PEV load, but the virtual load indeed can affect the schedule and plan for appliances and PV system. Because, high load means pretty high probability charging in one specific day. That is to say, the schedule should avoid this period. So introducing the virtual load of PEV is meaningful for HEMS.

III. PHOTOVOLTAIC SYSTEM

Renewable energy like solar energy and wind energy is extremely significant part of smart home. And many smart houses have been equipped with PV system and wind generators. The research [9] raises an engineering mathematics model for PV array:

$$I = I'_{SC}(1 - C_1\{\exp[\frac{v}{c_2 v_{oc}}] - 1\}) \quad (11)$$

This I-V curve can be determined while four photovoltaic battery parameters are given by manufacturers and the temperature and light intensity are obtained by sensors. After gaining This I-V curve, we can draw the power output curve of the PV array afterwards. Fig.8 is one power output curve of the PV array someday.

Though Fig.8, we can find that the output of PV array is pretty undulant instead of being smooth at some value. Our residential system cannot accept this low-quality power output. In order to stabilize the output, the PV array is directly connected to the battery device where this paper takes the lead-acid battery as the energy-storage device. The solution is that the PV array charges the battery device and the battery offer the stable power to the home. The rules of charging and discharging for the lead-acid battery are as follow:

While the array charging the battery in period [k, k+1]:

$$Q_B(k+1) = Q_B(k) + \gamma_c \cdot P_c(k) \cdot \delta; \quad (12)$$

While the battery offering home power in period [k, k+1]:

$$Q_B(k+1) = Q_B(k) - P_d(k) \cdot \delta / \gamma_d; \quad (13)$$

While the battery self-damping in the capacity in period [k, k+1]:

$$Q_B(k+1) = Q_B(k) - Q_B(\delta); \quad (14)$$

In formula (12)-(14), $Q_B(k)$ means the capacity of the lead-acid battery which value unit is KWH. $P_c(k)$ And $P_d(k)$ represents the charging power of the array and the constant output of the battery. γ_c and γ_d is the efficiency rate for the process of charging battery and offering home power. δ Is the session time of period [k, k+1]. And in this paper, the period time is set as one hour.

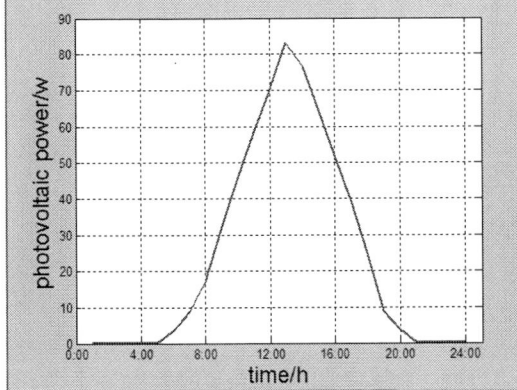

Fig.8: The output of PV array

The schedules for PV system are just the choices of charging the battery or offering power or self-damping for the lead-acid battery in different period. And HEMS should intelligently select suit period to charging battery. So do the behavior of offering home power and self-damping.

The PV array's parameters are given:
Vm=34.4V,Im=4.51A,Isc=4.9A,Voc=43.2V

The battery of PV system is the lead-acid battery. And the upper and lower limit of the capacity is 12KWH and 2KWH. γ_c and γ_d are all 90%. The initial capacity of battery is 7KWH. The charging power comes from the Real-time array output and discharging power is constant set as 0.8kw.

IV. HOME ENERGY MANAGMENT SYSTEM

The HEMS in this paper consist of photovoltaic system with its battery, one PEV and three smart appliances which can be planed for its start time by system in order to reduce the grid fee.

The objective function of HEMS considers not only the economics but also the user's comfort where it comes up with two parted function separately named as economic function and comfort function.

The economic function:

$$\min C_1 = C_{grid} * P_{grid} + C_{DG}$$
$$C_{DG} = k * sum_{DG} \quad (15)$$

In formula(15), C_1 represents the whole cost of the grid while C_{grid} is the peak-valley price and P_{grid} is the power offered by the grid. At the same time, k is a constant which is relevant to the wastage rate of PV system and its battery, sum_{DG} means the whole number of switching the device.

The comfort function:

$$\min C_2 = \sum C(i) = \sum |ST_{real}(i) - ST_{set}(i)| * f(i) \quad (16)$$

In formula(16), $ST_{real}(i)$ and $ST_{set}(i)$ represents the real start time and set start time for No.i appliance. And f(i) is defined the penalty cost for No.i appliance. The penalty cost is a kind of punishment for changing the user's habit in using household appliance while considering user's comfort.

The general function:

$$\min C = C_1 + C_2 \quad (17)$$

The constraints of HEMS mainly consist of the balance of the power and the limit of PV system.

The balance of the power:

$$P_{grid}(k) + P_{DG}(k) = P_{load}(k) + P_{ev}(k) \qquad (18)$$

The limits of PV system:

$$Q_B^L \leq Q_B \leq Q_B^U \qquad (19)$$

In formula(20),the $P_{ev}(k)$ is the virtual load of PEV which is mentioned while calculating the PEV's charging load curve.

Besides, this paper chooses three different appliance washing machine, electromagnetic oven and computer. Three appliances' situations are given:

Washing machine:7:00 is the set time and work for 2 hours. The power is 1.6kw and penalty value is 0.
Electromagnetic oven:18:00 is the set time and work for one hour. The power is 2kw and penalty value is 3.
Computer:8:00 is the set time and work for 4 hours. The power is 0.2kw and penalty value is 2.

So this paper has constructed a nonlinear program problem with constraints which are based on the balance of power and the constraints of PV devices. .The solution is the schedule or plan for the three appliances and the PV system. Obviously, we should use special intelligent algorithms.

V. GENETIC ALGORITHM FOR SOLUTION

The genetic algorithm simulates the phenomenons of crossover and mutation in the process of the natural selection. And the schedules and plans are the compositions of the gene of one individual. The algorithm creates next population which fits the fitness function more and consists of different kinds of individual starting from initial population after the option of crossover and mutation. With the evolving one generation by one generation, the average fitness value of the population become better and better and the value will get the convergence at last.

This paper mixes the binary and floating-point coding method while coding the gene. The start time for the appliances uses the binary where the time is converted to five bits binary like ST(i)=[x,x,x,x,x]; And the schedules for PV system adopt the floating-point coding method where the schedules for each hour's choices are set as [y1,......y24] and each element can be zero, one and two which stand for self-damping, charging battery and the power output of the battery. So the gene's length will be 39 bits. And this paper's fitness function adopts the max function. So we use big constant like 100 to subtract the general function:

$$\max C' = C_{con} - C \qquad (20)$$

In addition, each generation should meet the constraints of the HEMS. Constraints should be added into the algorithm and exclude the nonconforming individuals. Notably, the roulette method has some probability to miss the nice individual and this problem becomes serious while getting closed to the convergence. Because the individuals become similar and have more and more little differences. So this paper selects the individual which has highest

fitness value in one population and puts the individual directly into next generation passing the options of crossover and mutation. Fig.9 shows the whole process of the algorithm.

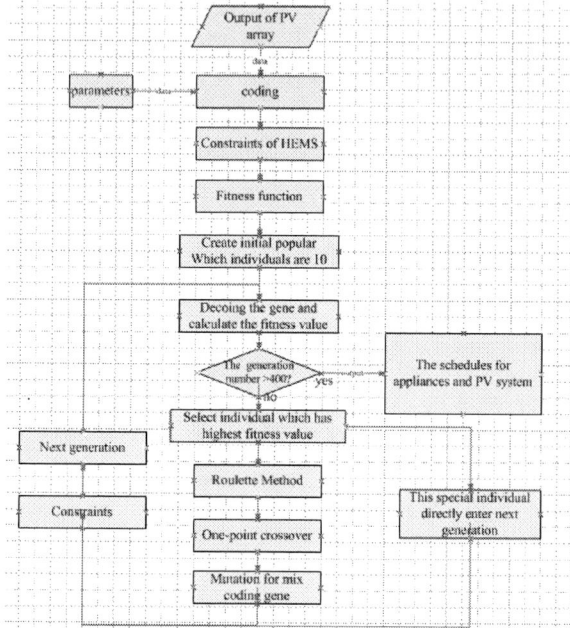

Fig.9: The whole process of GA

After the genetic algorithm solving and optimizing, the schedules for three appliance and PV system are given as follow:

Scheduled time for washing machine was 00:00 which is pushed advanced to meet the low price of grid.

Scheduled time for electromagnetic oven and computer is still set time for each device. The result indicates that the algorithm indeed takes the user's comfort into consider.

The schedules for PV system are shown in Tab.1Period k means time of o'clock k-1 to o'clock k. And state of 0,1,2 is explained in float-point coding for PV system.
And the state of capacity(SOC) is shown in Fig.10

Table 1: The schedules for PV system

period	1	2	3	4	5	6	7	8
state	2	2	1	1	1	0	0	2
period	9	10	11	12	13	14	15	16
state	2	2	1	1	1	1	2	2
period	17	18	19	20	21	22	23	24
state	0	2	0	2	0	0	1	1

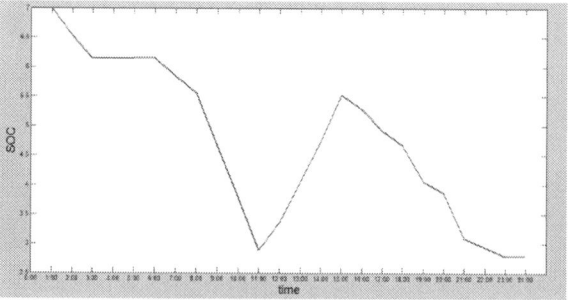

Fig.10: The SOC daily curve

From the schedules, we can find that the algorithm arrange the PV array to charge the battery while the output of PV array is quite high and arrange the battery to offer power while the virtual load of PEV is quite high. This arrangement is just like carrying the high output of PV array to offer the virtual load of PEV in a different period. So this arrangement can save the grid fee.

Finally, the average and max fitness value per generation are shown in Fig.11. With the evolving one generation by one generation, the average and max fitness value go up to one stable value. Obviously, the algorithm has pretty good convergence after about four hundred generation. The average value increase to 92.4 from about 91.7 and the max value increase to 92.4 from about 92.2. The average value and max value are identical at last. But the average will still fluctuate due to the option of crossover and mutation.

REFERENCES

[1] Duvall M, Knipping E, Alexander M, et al. Environmental assessment of plug-in hybrid electric vehicles. Volume 1: Nationwide greenhouse gas emissions[R],Palo Alto, CA：Electric Power Research   Institute,:2007,1015325.

[2] Taylor M J, Alexander A. Evaluation of the impact of plug-in electric vehicle loading on distribution system operations[C]//IEEE Power & Energy Society General Meeting, Calgary, Canada, 2009: 1-6.

[3] WANG Jidong, SUN Zhiqing, ZHOU Yue and DAI Jiaqiang, "Optimal dispatching model of smart home energy management system," in Proc. IEEE Innovative Smart Grid Technologies-Asia, Tianjin, China, 2012,pp. 1-5.

[4] A. Vyas, D. Santini, "Use of National surveys for estimating "Full" PHEV potential for Oil Use Reduction," presented at the Plug-In Conference, San Jose CA, July 2008

[5] TIAN Liting,SHI Shuanglong,JIA Zhuo.A Statistical Model for Charging Power Demand of Electric Vehicles[J].Power System Technology.2010,34(11):

[6] XIONG Xicong.Study of the Probability Simulation of Electric Vehicle Charging Load and its Impact to the Loss-of-life of Distribution Transformer[D] .Chon-qing:School of Electrical Engineering of Chongqing University, Chongqing, China,2014: 10-15.

[7] SONG Yonghua,YANG Yuexi,HU Zechun.Present Status and Development Trend of Batteries for Electric Vehicles[J].Power System Technology,2011, 35(4): 1~7.

[8] Shidore N, Bohn T, Duoba M, et al. PHEV "All electric range" and fuel economy in charge sustaining   mode   for low   SOC   operation   of   the   JCS   VL41M   Li-ion battery   using   battery HIL[C] . Proceedings of 23rd International Electric Vehicle Symposium, Anaheim, USA, 2007: 1419-1431.

[9] WANG   Lianfen,XU   Shubai.Analytic   Hierarchy Process[M]. The China Renmin University Press:1990.

A New Method to Suppress the Commutation Torque Ripple for BLDC Motor Based on ZETA Converter

Zhen Chen[1], Hancheng Zhang[2], Xiangdong Liu, Hengzai Hu, Jing Zhao, Congzhe Gao

School of Automation, Beijing Institute of Technology, Beijing, 100081, Beijing, China
[1]E-mail: chenzhen76@bit.edu.cn
[2]E-mail: 20091631@bit.edu.cn

Abstract-The commutation torque ripple generated in the commutation interval is one of the main drawbacks of Brushless DC (BLDC) motor and restricts its applications. In this paper, in order to reduce the commutation torque ripple, a novel circuit topology including a zero energy thermonuclear assembly (ZETA) converter and a dc-link voltage selection circuit is proposed and investigated. Firstly, the cause of commutation is analyzed. Then the desired DC link voltage is calculated for the novel circuit topology to make the incoming and outgoing phase currents change at the same rate during commutation, and to effectively reduce the commutation torque ripple. Finally, simulation results show that the proposed topology can effectively suppress the commutation torque ripple under both low-speed and high-speed operation.

Keywords- BLDC motor, commutation torque ripple, ZETA converter.

I. INTRODUCTION

Brushless DC (BLDC) motor is widely used in industrial applications due to its high efficiency, sample structure and easy-to-control features [1]-[2]. However, torque ripple commonly existing in the BLDC motor restricts its application in high precision drive system. The torque ripple in BLDC motor is caused by many reasons, especially the commutation torque ripple. Hence, suppression of commutation torque ripple becomes the key to reduce the BLDCM torque ripple.

So far, many methods have been proposed to suppress the torque ripple. such as the overlap of the phase change method [3], the hysteresis current method[4], the pulse width modulation (PWM) method[5], the current predictive control method[6], the direct torque control [7] and so on, but those methods inevitably exist the problems of over or under compensation, which may affect the effect of the torque ripple suppression. A dual switching logic is presented in [8] to reduce the commutation torque ripple of the BLDC motor. In this method, the inverter is operated at a 120° electrical conduction mode in the low-speed range, and a 180° electrical conduction mode in the high-speed range. The method presented in [9] uses a single dc-link current sensor to regulate the incoming current and outgoing current to minimize the toque ripple. In [10], a buck converter is used in front of the three-phase bridge to generate the suitable dc-link voltage which is proportional to the motor speed and the torque ripple is greatly reduced at lower speed. In [11], A super-lift Luo-converter instead of a buck converter is employed, and the torque ripple is effectively suppressed at high speed. A new topology of drive circuit for brushless DC motors control based on a quasi-Z-source net is proposed in [12]. In [13], a new circuit topology

using a single-ended primary-inductor converter (SEPIC) converter is presented. In this paper, a novel circuit topology including a zero energy thermonuclear assembly (ZETA) converter and a dc-link voltage selection circuit is proposed and employed in a BLDC motor drive system to reduce the commutation torque ripple. In a BLDC motor, the commutation torque ripple arises due to the incoming and outgoing phase currents changing at different rate during commutation. The control strategy is to generate the desired commutation voltage which is proportional to the motor speed to let the incoming and outgoing phase currents change at the same rate during commutation, then the commutation torque ripple will be decreased. To reduce the commutation torque ripple, the commutation torque ripple and its influence on the steady characteristic of the BLDC motor are firstly analyzed. Secondly, the operation principle of the novel circuit is analyzed. Thirdly, the optimum duty ratio of the ZETA converter is obtained under different speed and load conditions due to the ZETA converter can generate higher or lower voltage than bus voltage. Finally, simulation results show that the proposed topology can suppress the commutation torque ripple effectively under both low-speed and high-speed operation.

II. MATHEMATICAL ANALYSIS

A typical circuit diagram for BLDC motor drive system is shown in Fig. 1.

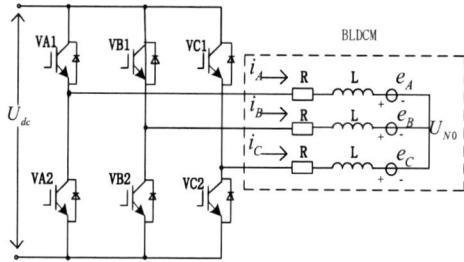

Fig. 1:
Circuit diagram for BLDC Motor drive system

The BLDC Motor voltage equation of three windings with phase variables is

$$
\begin{bmatrix} u_A \\ u_B \\ u_C \end{bmatrix} = \begin{bmatrix} R & 0 & 0 \\ 0 & R & 0 \\ 0 & 0 & R \end{bmatrix} \begin{bmatrix} i_A \\ i_B \\ i_C \end{bmatrix}
$$
$$
+ \begin{bmatrix} L & 0 & 0 \\ 0 & L & 0 \\ 0 & 0 & L \end{bmatrix} \frac{d}{dt} \begin{bmatrix} i_A \\ i_B \\ i_C \end{bmatrix} + \begin{bmatrix} e_A \\ e_B \\ e_C \end{bmatrix} + \begin{bmatrix} U_{N0} \\ U_{N0} \\ U_{N0} \end{bmatrix} \quad (1)
$$

and the electromagnetic torque is given by

$$T_e = \frac{e_A i_A + e_B i_B + e_C i_C}{\omega} \qquad (2)$$

where U_{N0} is the neutral point-to-ground voltage, u_A, u_B, and u_C are the terminal phase voltages with respect to the power ground, i_A, i_B, and i_C are phase currents, e_A, e_B and e_C are trapezoidal back EMFs, $L=L_s-M$ is the equivalent inductance of phase windings, L_s and M are self-inductance and mutual inductance, respectively, R is the resistance of the phase windings, and ω is the speed of the rotor.

In this paper, the commutation of the current from phase B to phase C is considered, and is assumed to be constant and equal to I_m. The initial values $i_b = -i_a = I_m$, $i_c = 0$ are known at the beginning of commutation. The voltage initial values at the beginning of commutation can be drawn as follows: $e_A = -E_m$, $e_B = e_C = E_m$, $u_A = u_B = 0$ and $u_C = U_{dc}$ .where E_m is the back-EMF that is constant during commutation. Then the electromagnetic torque at the beginning of commutation is

$$T_e = \frac{2E_m I_m}{\omega} \qquad (3)$$

And the voltage equation (1) can be rewritten as

$$u_A = 0 = Ri_A + L\frac{di_A}{dt} + e_A + U_{N0}$$

$$u_B = 0 = Ri_B + L\frac{di_B}{dt} + e_B + U_{N0} \qquad (4)$$

$$u_C = U_{dc} = Ri_C + L\frac{di_C}{dt} + e_C + U_{N0}$$

Then, we can get

$$u_{AB} = u_A - u_B = 0$$
$$= R(i_A - i_B) + L(\frac{di_A}{dt} - \frac{di_B}{dt}) + e_A - e_B$$
$$u_{AC} = u_A - u_C = -U_{dc} \qquad (5)$$
$$= R(i_A - i_C) + L(\frac{di_A}{dt} - \frac{di_C}{dt}) + e_A - e_C$$

On the other hand, we have

$$i_A + i_B + i_C = 0 \qquad (6)$$

Then, from equations (5) and (6), we can obtain that

$$3Ri_A + 3L\frac{di_A}{dt} - 4E_m = -U_{dc} \qquad (7)$$

As the frequency of PWM is very high and the PWM period is much shorter than the electrical time constant L/R, the effect of R can be neglected, the phase currents can be obtained as

$$i_A = -I_m + \frac{4E_m - U_{dc}}{3L}t$$

$$i_B = I_m + \frac{-U_{dc} - 2E_m}{3L}t \qquad (8)$$

$$i_C = \frac{2U_{dc} - 2E_m}{3L}t$$

According to equations (2) and (8), the electromagnetic torque during commutation can be calculated as

$$T_e = \frac{2E_m I_m}{\omega} + \frac{2E_m}{3L\omega}(U_{dc} - 4E_m)t \qquad (9)$$

Then the commutation torque ripple can be expressed as

$$\Delta T_e = 2\frac{U_{dc} - 4E_m}{3L\omega}E_m t \qquad (10)$$

According to equation (8), the time that i_B decreasing from I_m to 0 is

$$t_1 = \frac{3LI_m}{U_{dc} + 2E_m} \qquad (11)$$

And the time that i_c increasing from 0 to I_m is

$$t_2 = \frac{3LI_m}{2U_{dc} - 2E_m} \qquad (12)$$

According to equations (8), (11) and (12), different current behaviors under different speeds are obtained, as shown in Fig.2, and the following conclusions can be also drawn:

1) If $U_{dc} < 4E_m$, then $t_1 < t_2$, and the torque keeps decreasing during commutation.

2) If $U_{dc} > 4E_m$, then $t_1 > t_2$, and the torque keeps increasing during commutation.

3) If $U_{dc} = 4E_m$, then $t_1 = t_2$, and the torque is constant during commutation.

From equation (10), it can obtained that if $U_{dc} = 4E_m$ during the commutation, the commutation torque ripple will be reduced. From equations (11) and (12), the commutation interval during the commutation can be expressed as

$$t = \frac{L I_m}{2 E_m} \quad (13)$$

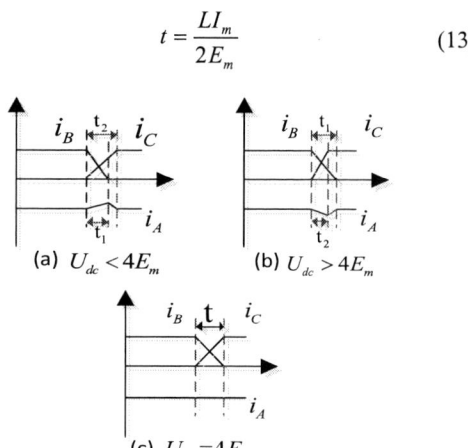

(a) $U_{dc} < 4E_m$ (b) $U_{dc} > 4E_m$

(c) $U_{dc} = 4E_m$

Fig. 2: Current behaviors during commutation.

III. PROPOSED CONTROL STRATEGY

From equation (10), it can be seen that when the speed of the motor is relatively low, the bus voltage U_{dc} should be reduced to satisfy $U_{dc} = 4E_m$. And when the speed of the motor is relatively high, the DC link voltage U_{dc} should be boosted to satisfy $U_{dc} = 4E_m$. Hence, a ZETA converter which can buck/boost the bus voltage is employed to supply the proper voltage during commutation, as shown in Fig. 3.

The basic working principle of a ZETA converter is as follows[14]: when V1 is switched on, electric source charges to the inductor L1, at the same time, the load's power supply is meted by the electric source and the capacitor C1, as shown in Fig.4 (a). When V1 is switched off, the inductor L1 charges to the capacitor C1 through the diode D1, and also the current flowing through inductor L2 will flow through the diode D1. as shown in Fig.4 (b).

The output voltage of the ZETA converter can be expressed as

$$V_O = \frac{D}{1-D} V_{CC} \quad (14)$$

where D is the duty ratio of V1. As E_m is proportional to speed, so the duty ratio of V1 can be calculated by

$$D = \frac{4 k_e \omega}{V_{CC} + 4 k_e \omega} \quad (15)$$

where k_e is the back EMF coefficient. To achieve an immediate change of the input voltage of inverter, S1 and S2 are required to be complementary to each other. By switching off S1 and switching on S2 during commutation, the commutation torque ripple can be suppressed effectively. After the commutation process, the S1 is turned on and the S2 is turned off.

Some parameters of the simulation system are listed in Table.1.

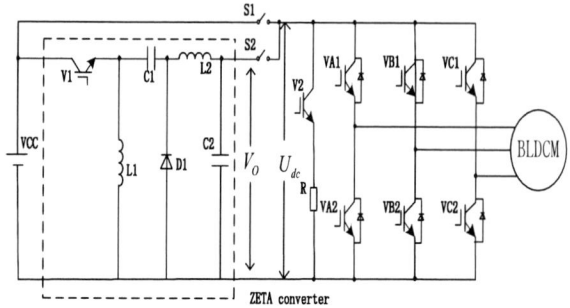

Fig. 3: Configuration of BLDC Motor driving system with a ZETA converter.

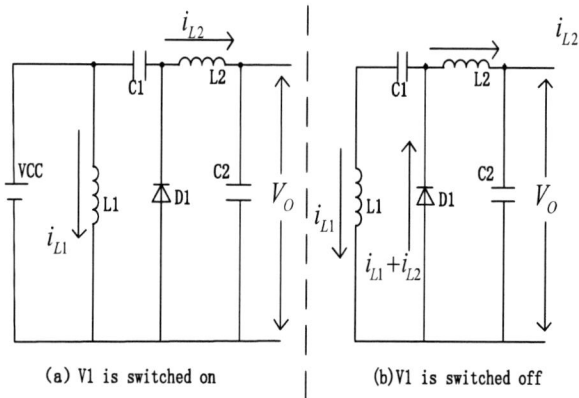

(a) V1 is switched on   (b)V1 is switched off

Fig. 4: The two working states of ZETA converter.(a) V1 is switched on;(b)V1 is switched off;

Table 1: Parameters of the simulation system.

Rated voltage(V)	10
Rated speed(rpm)	1500
Pole Pairs	8
Phase resistance(Ω)	0.434
Phase inductance(mH)	0.82
Back-EMF coefficient(V/rpm)	0.0059
PWM switch frequency(KHz)	20

IV. EXPERIMENT RESULTS

To verify the effectiveness of the proposed strategy, simulations are carried out. The simulation model of the system is shown in Fig.5.

Under the condition that the reference speed is 500r/min and the load torque is 0.02 N.m, some simulated results are shown from Fig.6 to Fig.11. The speed response is shown in Fig.6. The simulated three-phase currents with and without a ZETA converter are shown in Fig.7 and Fig.8, respectively. Comparing Fig.7 and Fig.8, we can

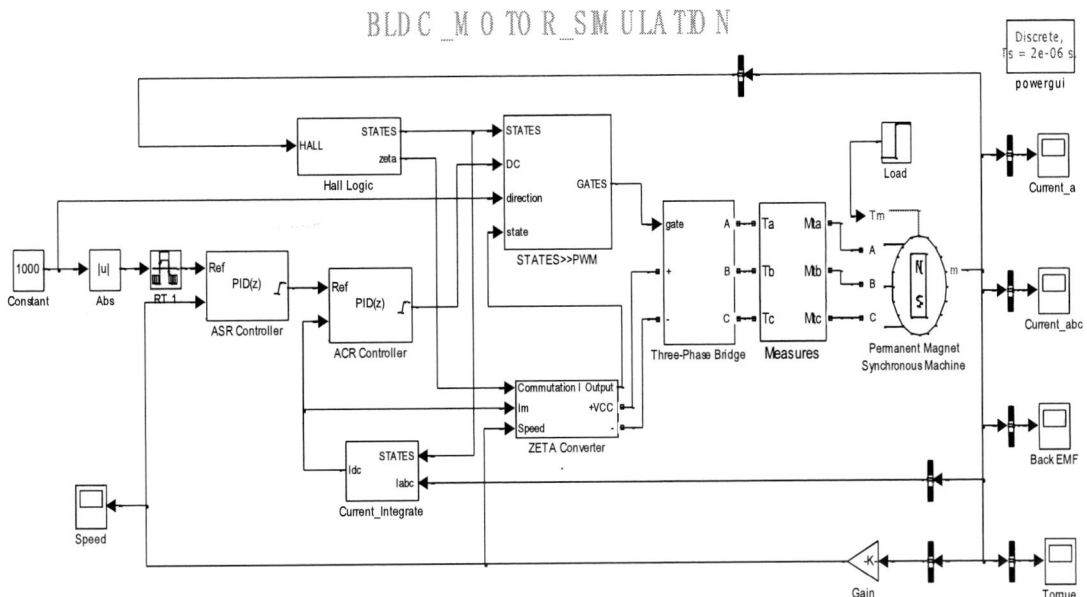

Fig. 5: Simulation model of the system.

get that the incoming and outgoing phase currents change at the nearly same rate during commutation. The simulated phase current with and without a ZETA converter is shown in Fig.9 (a) and Fig.9 (b), respectively. From Fig.7 and Fig.9 (a), it can be seen that the non-commutation phase current becomes smooth during commutation when a ZETA converter is used. On the contra, without the ZETA converter, the non-commutation phase current will increase or decrease during the commutation process. Fig.10 shows the simulated electromagnetic torque. As the non-commutation phase current's change will result in electromagnetic torque's change, so, from the comparison of Fig.10 (a) and Fig.10 (b), it can be seen that the commutation torque ripple is effectively suppressed when the ZETA converter has been employed. The switch signals and the DC link voltage are shown in Fig.11. From Fig.11, it can be seen that the switch signals for S1 and S2 are complementary to each other, and the DC link voltage at the beginning of commutation

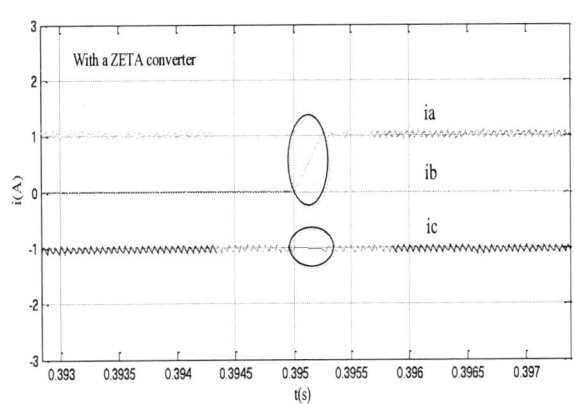

Fig. 7: Simulated three phases currents with a ZETA converter at n = 500 r/min.

is switched to be the ZETA converter's output voltage to meet the equation $U_{dc} = 4E_m$.

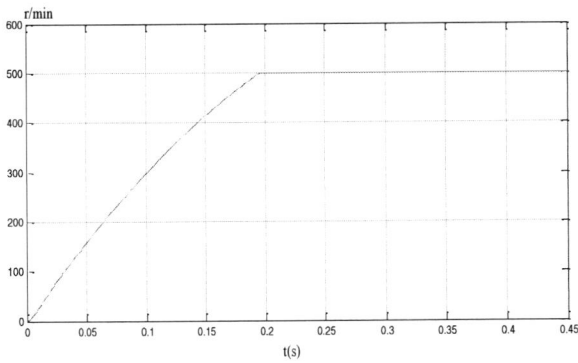

Fig. 6: Speed response when the reference speed n= 500rpm.

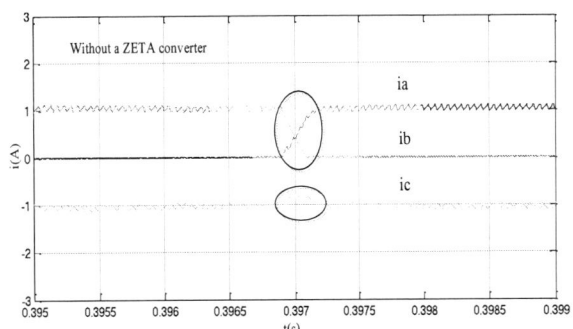

Fig. 8: Simulated three phases currents without a ZETA converter at n = 500 r/min.

978-1-5090-0064-7/15 $31.00 © 2015 IEEE

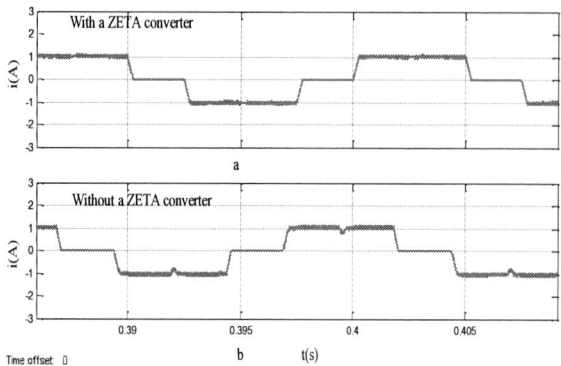

Fig. 9: Simulated phase currents at n = 500 r/min. (a) with ZETA converter; (b) without a ZETA converter.

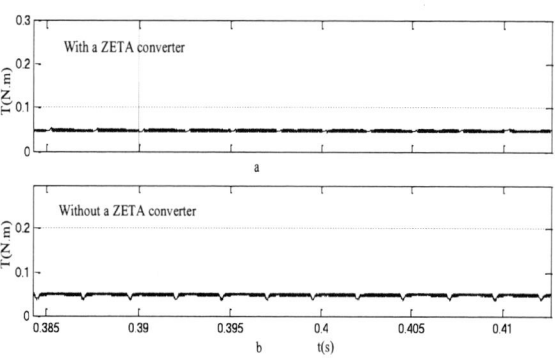

Fig. 10: Simulated electromagnetic torque at n = 500 r/min. (a) with ZETA converter; (b) without a ZETA converter.

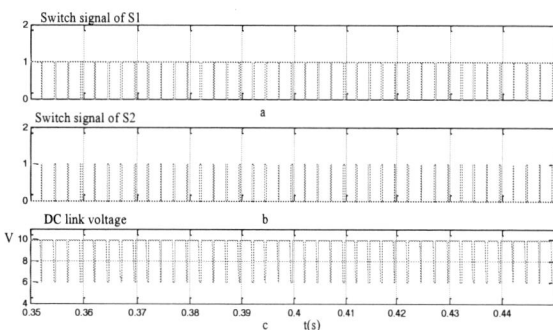

Fig. 11: The switch signal and the DC link voltage U_{dc} when n=500rpm.(a) Switch signal of S1. (b)Switch signal of S2. (c) The DC link voltage U_{dc}.

Under the condition that the reference speed is 1000 r/min and the load torque is 0.02 N.m, some simulated results are shown from Fig.12 to Fig.17. Fig.12 shows the speed response. The simulated three-phase currents are shown in Fig.13 and Fig.14. The simulated phase current is shown in Fig.15. Fig.16 shows the simulated electromagnetic torque, and Fig.17 shows the switch signals and the DC link voltage.

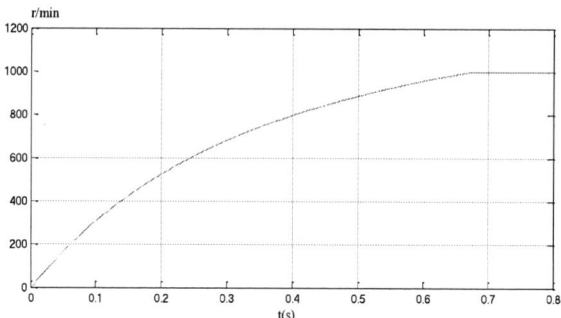

Fig. 12: Speed response when the reference speed n= 1000rpm.

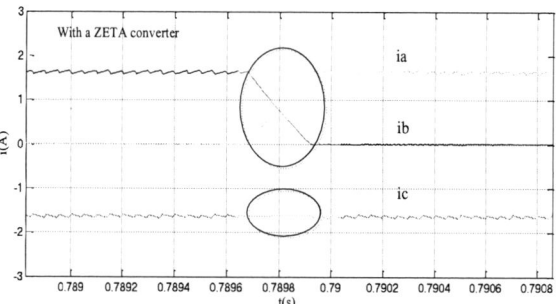

Fig. 13: Simulated three phases currents with a ZETA converter at n = 1000 r/min.

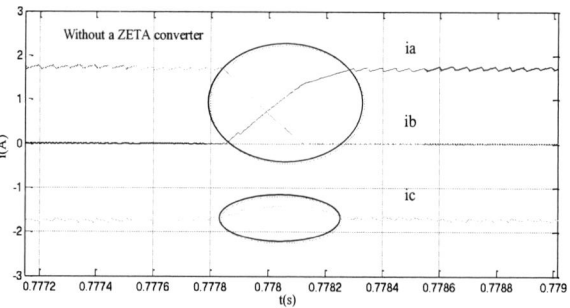

Fig. 14: Simulated three phases currents without a ZETA converter at n = 1000 r/min.

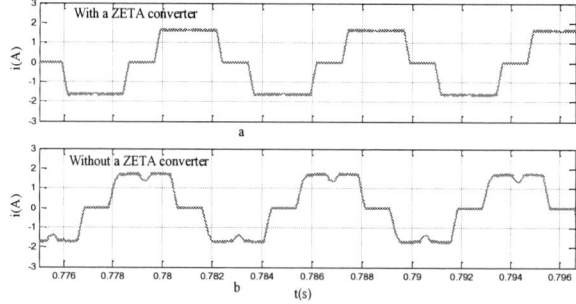

Fig. 15: Simulated phase currents at n = 1000 r/min. (a) with ZETA converter; (b) without a ZETA converter.

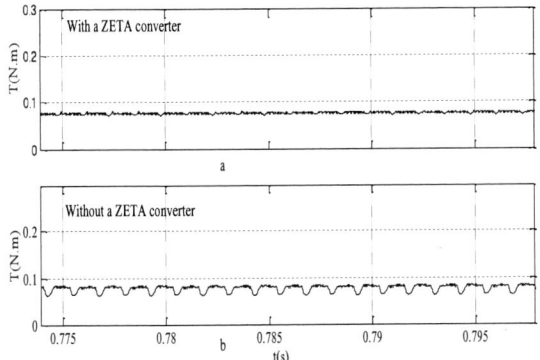

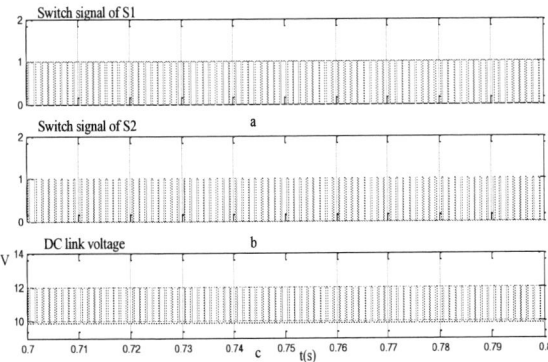

Fig. 16: Simulated electromagnetic torque at n = 1000 r/min. (a) with ZETA converter; (b) without a ZETA converter.

Fig. 17: The switch signal and the DC link voltage U_{dc} when n=1000rpm.(a) Switch signal of S1. (b)Switch signal of S2. (c) The DC link voltage U_{dc} .

From Fig.13, Fig.15 (a) and Fig.16, the similar simulated conclusion that the commutation torque ripple can be greatly reduced under the effect of the ZETA converter. And from the simulated results, it can be obtained that the proposed topology can suppress the commutation torque ripple effectively under both low-speed and high-speed operation.

V. CONCLUSION

In this paper, a novel circuit topology including a zero energy thermonuclear assembly (ZETA) converter and a dc-link voltage selection circuit is proposed and employed in BLDC motor. The zeta converter is employed at the front of the three-phase bridge and the optimum duty ratio of the ZETA converter is obtained under different speed and load conditions. Simulation results show that this method can effectively reduce commutation torque ripple.

REFERENCES

[1] A. Sathyan, N. Milivojevic, Y.-J. Lee, M. Krishnamurthy, and A. Emadi. An FPGA-based novel digital PWM control scheme for BLDC motor drives, IEEE Trans. Ind. Electron, vol. 56, no. 8, pp. 3040–3049, Aug. 2009.

[2] G. J. Su and J. W. Mckeever. Low-cost sensorless control of brushless DC motors with improved speed range, IEEE Trans. Power Electron, vol. 19, no. 2, pp. 296–302, Mar. 2004.

[3] Murai Y, Kawase Y, Ohashi K, et al. Torque Ripple Improvement for Brushless DC miniature motors [J]. IEEE Transactions on Industry Applications, 1989,25(3):441-450.

[4] Kim Gwang Heon, Kang Seog Joo, Won JongSoo. Analysis of the Commutation Torque Ripple Effect for BLDC-Motor fed by HCRPWM VSI. Applied Power Electronics Conference and Exposition. Boston, America,1992:277-284.

[5] Zhang Xiangjun, Chen Boshi. The different influences of four PWM modes on commutation torque ripples in brushless DC motor control system[J]．Electric Machines and Control，2003，7(2)：87-91(in Chinese).

[6] Holtz J, Springob L. Identification and Compensation of Torque Ripple in High-Precision Permanent Magnet Motor Drives[j]．IEEE Transactions on Industry Electronics, 1996, 4:309-320.

[7] Y. Liu, Z.Q. Zhu, D.Howe. Direct Torque Control of Brushless DC Drives with reduced Torque RIPPLE[J]. IEEE Transactions on Industry. Application,2005,41(2):599-608.

[8] Bharatkar, S.S, Yanamshetti, R, Chatterjee, D, Ganguli, A.K. Dual-mode switching technique for reduction of commutation torque ripple of brushless dc motor. IET Electr. Power Appl, 2011, 5, (1), pp. 193–202.

[9] Joong-Ho, S, Choy, I. Commutation torque ripple reduction in brushless DC motor drives using a single DC current sensor. IEEE Trans. Power Electron., 2004, 19, (2), pp. 312–319.

[10] Zhang, X.F., Lu, Z.Y. A new BLDC motor drives method based on BUCK converter for torque ripple reduction. Proc. IEEE Power Electron. Motion Control Conf, 2006, pp. 1–4.

[11] Chen, W, Xia, C.L. A torque ripple suppression circuit for brushless DC motors based on power dc/dc converters. Proc. IEEE Industrial Electronics Applications Conf., 2008, pp. 1453–1457.

[12] Xia Kun, Zhu Linling, Zeng Yanneng, Xu Xinyue, Yang Yihua. Researches on the Method of Suppressing Commutation Torque Ripple for Brushless DC Motors Based on a Quasi-Z-Source Net. Proceedings of the CSEE,35(4).

[13] Shi Tingna, Guo Yuntao, Song Peng, Changliang Xia. A new approach of minimizing commutation torque ripple for brushless DC motor based on DC–DC Converter. IEEE Trans. Ind. Electron, 2010, 57, (10), pp. 3483–3490.

[14] Zhaoan Wang, Jinjun Liu. Power electronic technology[M]. Beijing：Mechanical Industry Press，2009(in Chinese).

Modeling and Control of DAB Converter for Solar Micro-grid Application

Ranjan K. Behera[1]   Olorunfemi Ojo[2]

[1] Department of Electrical Engineering, Indian Institute of Technology Patna, Patna-800013, INDIA
Email: rkb@iitp.ac.in
[2] Center for Energy Systems Research, Department of Electrical and Computer Engineering, Tennessee Tech University, TN, USA.
And Eskom Centre of Excellence in HVDC Engineering, School of Engineering, University of KwaZulu-Natal, Durban 4041, Republic of South Africa
Email: JOjo@tntech.edu

Abstract– In this paper, the modeling and control of the DC sub-grid, fed from Solar panels and integrated into the grid using dual active bridge (DAB) converters, is studied. The dynamic performance of the DAB converter depends on accurate modeling of the system. A small signal (SS) modeling of DAB converter is developed for the closed loop control and its dynamics are studied for different operating conditions. The simulation results of the proposed system are compared and it is found that the proposed system gives better dynamical response as compared to classical control system. An AC-DC hybrid micro-grid structure is modeled and its performance is studied. The performance of the dynamic system is simulated and the complete control algorithm is implemented in the laboratory.

Keywords– Dual Active Bridge converter, micro grid, solar energy.

I. INTRODUCTION

The micro-grid traditionally generates alternating current (AC) power serving AC loads; recently there is an increasing interest in exploring direct current (DC) microgrid structures due to many benefits, which includes higher efficiency, more robust system operation, no impedance matching issue, no synchronization, and simple waveforms. So, DC micro-grid, DC sources power DC loads and this is becoming dominant in houses, on-board DC grids in ships, aero-planes, some cities and datacenters where the loads are intrinsically DC. A more recent innovation is the introduction of hybrid AC-DC micro-grid power systems in which the traditional AC and DC micro-grids are interfaced using bi-directional converters, enabling the transfer of power from the AC to the DC subsystems and vice versa. This configuration should result in improved system stability and optimal utilization of energy resources. There is an emerging research on hybrid AC-DC micro-grids and the potential for providing premium power for the various types of micro-grids is very high. The DC sub-grid is fed from Solar panels and integrated into the grid using dual active bridge (DAB) converters [1]-[7].

In order to realize power distribution between energy generation systems and storage systems in micro grids, various bidirectional DC–DC converters (BDCs) have been proposed as an everlasting key component to interface between a high-voltage bus, where an energy generation system such as a fuel cell stack or a photovoltaic array is installed, and a low-voltage bus, where usually an energy storage system such as a battery or a super capacitor is implemented, as shown in Fig. 1 [8].

Generally, BDC is divided into non-isolated type [9], [10] and isolated type [8], [11], and galvanic isolation for BDC is required for flexibility of system reconfiguration and meeting safety standards [11]. State-of-the-art isolated bidirectional DC–DC converter (IBDC) is based on the single-phase and H-bridge topology with a high-frequency isolation transformer. Fig. 2 depicts a typical configuration of IBDC. Compared to traditional DC–DC converter circuits, this converter has many advantages, such as electrical isolation, high reliability, easy to realize soft switching control, and bidirectional energy flow [11]–[13]. In this paper, an all solar DC-AC micro-grid as shown schematically in Fig. 1 is proposed and a closed loop control system is designed for constant load voltage control. Initially the modeling of DAB is carried out and a small signal analysis is developed. A closed loop controller is designed for improving the dynamic performance of DAB converter system. The simulated and experimental results are presented for verifying the dynamic performance of the proposed system

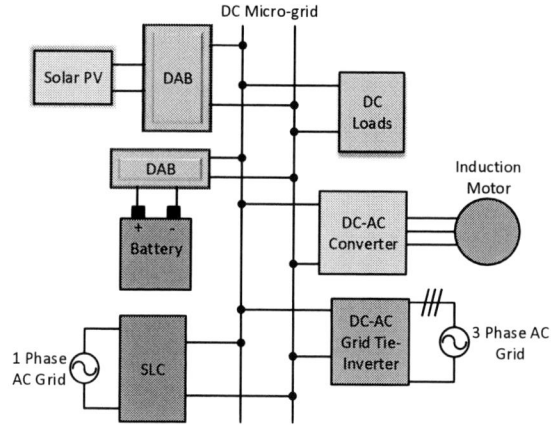

Fig. 1: Proposed AC-DC Micro-grid

II. MODELING AND CONTROL OF DAB

A single phase Dual Active Bridge (DAB) converter topology as shown in Fig. 2 is considered most promising alternative because of high converter efficiency and the high power density. It is due to the low number of inductors and due to the employed capacitive filters on the HV side and on the LV side. Fig. 2 shows the configuration of a full-bridge isolated bidirectional dc–dc

978-1-5090-0064-7/15 $31.00 © 2015 IEEE

converter, where V_1 and V_2 are input and output voltage, respectively, n is the turns ratio of the transformer. There are two H-bridge converter linked with a HF transformer. The first H-bridge provides square wave AC voltage with duty ratio of 50% to the primary winding of the high-frequency transformer. This bridge consists of four power semiconductor switches S_1, S_2, S_3, and S_4 commonly insulated gate bipolar transistors (IGBTs) for high power applications and metal semiconductor field effect transistor (MOSFET) for low power application.

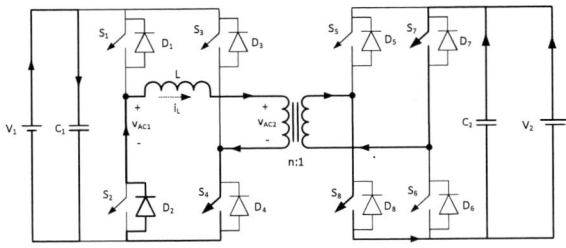

Fig. 2: Single phase Dual Active Bridge (DAB) converter

In the port2, the second H-bridge consists of four power semiconductor switches, S_5, S_6, S_7 and S_8, are connected to the secondary winding of the HF transformer, and operates in the boost mode by means of 'phase-shift' control. This DC-DC converter can operate in bidirectional mode. Hence, the H-bridge in port1 and port2 can be considered as primary or secondary depending on the direction of power flow. The analysis of the DAB converter is simplified by reflecting the model to one side of the transformer and considering the transformer magnetizing inductance much larger than the leakage inductance. The DAB converter is, therefore, represented by two active H-bridges linked by the transformer leakage inductance L. Consider the switching function (1) and (2) of port1 converter is u_1 and of port2 is u_2. The switching function u can be defined as

$$u_1(\tau) = \begin{cases} 1 & 0 \leq \tau \leq T_s/2 \\ -1 & T_s/2 \leq \tau \leq T_s \end{cases} \quad (1)$$

where $T_S = 1/f_s$ and f_s is the switching frequency.

$$v_{AC1}(\tau) = u_1(\tau) \bullet v_1(\tau)$$

$$u_2(\tau) = \begin{cases} 1 & 0 \leq \tau \leq T_s/2 \\ -1 & T_s/2 \leq \tau \leq T_s \end{cases} \quad (2)$$

$$v_{AC2}(\tau) = u_2(\tau) \bullet v_2(\tau)$$

Considering the inductor current (i_L) and output voltage (v_0) as the state variable. The dynamical equations of DAB can be written as

$$\frac{di_L(\tau)}{dt} = -\frac{R}{L}i_L(\tau) + \frac{u_1(\tau)}{L}v_i(\tau) - \frac{u_2(\tau)}{L}v_o(\tau) \quad (3)$$

$$\frac{dv_o(\tau)}{dt} = -\frac{1}{R_L C_2}v_o(\tau) + \frac{u_2(\tau)}{C_2}i_L(\tau) \quad (4)$$

The state space modeling of the DAB converter is simplified by reflecting the model to one side of the transformer and considering the transformer magnetizing inductance much larger than the leakage inductance. The DAB converter is, therefore, represented by two active H-bridges linked by the transformer leakage inductance L with internal resistance of R. With the HV side full bridge, three different voltage levels are possible for $v_{AC1}(t)$

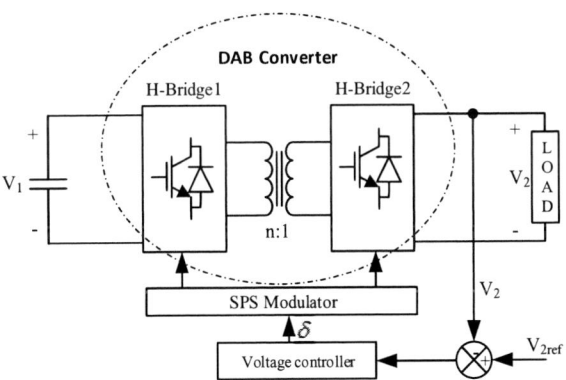

Fig. 3: Closed-loop control of DAB converter

The state space modeling can be written as [16],

$$\frac{dX}{dt} = AX + Bu \quad (5)$$

Controller design and stability analysis for a power converter require the linearized model of power converters. The derivation of small signal control-to-output transfer function, which is the dynamic response of a converter from a small perturbation in the control signal. The output voltage needed to be controlled, hence the state space small signal model of the DAB can be written as [16], [17].

$$\frac{d}{dt}\begin{bmatrix} \Delta v_{20} \\ \Delta i_{L1R} \\ \Delta i_{L1I} \end{bmatrix} = \begin{bmatrix} -\dfrac{1}{R_L C_2} & -\dfrac{4\sin(D\pi)}{\pi C_2} & -\dfrac{4\cos(D\pi)}{\pi C_2} \\ \dfrac{2\sin(D\pi)}{\pi L} & -\dfrac{R}{L} & \omega_s \\ \dfrac{2\cos(D\pi)}{\pi L} & -\omega_s & -\dfrac{R}{L} \end{bmatrix}\begin{bmatrix} \Delta v_{20} \\ \Delta i_{L1R} \\ \Delta i_{L1I} \end{bmatrix} + \quad (6)$$

$$\begin{bmatrix} \dfrac{4}{C_2}(I_{0I}\sin(\pi D) - I_{0R}\cos(\pi D)) \\ \dfrac{2V_{20}}{L}\cos(\pi D) \\ -\dfrac{2V_{20}}{L}\sin(\pi D) \end{bmatrix}[\Delta d]$$

Now the transfer function of the DAB model can be written as

$$G_{v2d} = C(sI - A)^{-1}B \quad (7)$$

978-1-5090-0064-7/15 $31.00 © 2015 IEEE

Putting the circuit parameters in (7) from Table I,

$$G_{v2d} = \frac{1.25 X 10^5 s^2 + 2.5 X 10^8 s - 4.37 X 10^{16}}{s^3 + 4.33 X 10^3 s^2 + 1.67 X 10^{10} s + 3.63 X 10^3} \quad (8)$$

Using the transfer function (8), the overall block diagram of the converter is shown in Fig. 4. The Bode plot of the controller is shown in Fig. 5. It is shown that the closed loop system is stable and bounded.

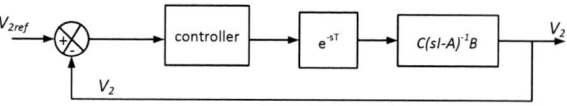

Fig. 4: Closed loop control block diagram.

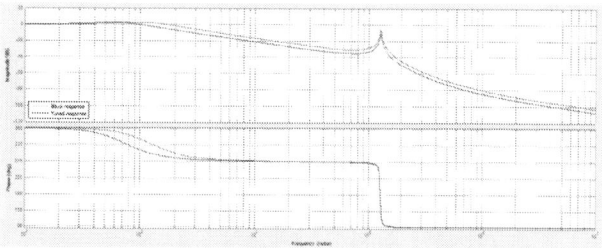

Fig. 5: The Bode Plot of closed loop control of DAB system

III. DISCUSSION ON SIMULATION AND EXPERIMENTAL RESULTS

Initially the complete closed loop control of DAB converter has been carried out using MATLAB/SIMULINK environment. A phase shift (PS) switching scheme is used for the controlling the power flow from port1 to port2 or vice versa.

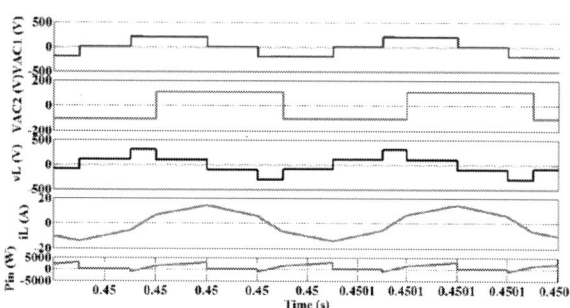

Fig. 6: Simulation results of V_{AC1}, V_{AC2}, v_L, i_L, and p_{in}.

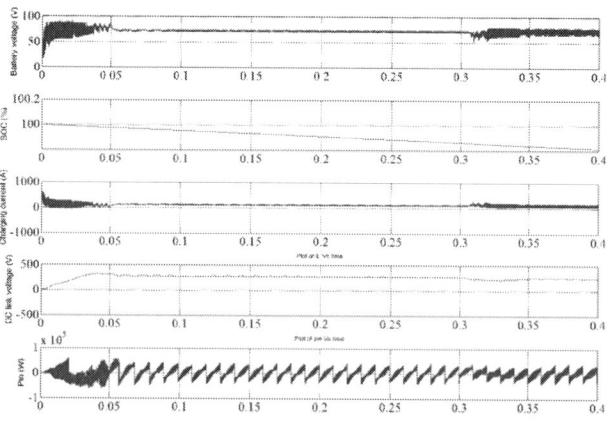

Fig. 7: Simulation results of step change of voltage (V_2)

Fig. 6 shows the simulation results for of DAB converter with PS method. Fig. 6 shows the simulation results of V_{AC1}, V_{AC2}, v_L, i_L, and p_{in} of the closed loop system. Fig. 7 shows the dynamics of voltage (V_2) tracking of the port2 side. A step change in voltage (V_2) 200 V is applied at 1 sec, it is observed that the voltge is tracking perfectly according to reference value. Small over shoot in the voltage may be rectified. Fig. 8 shows the waveforms of voltage reacking error. Fig. 9 shows the dynamics of voltage V_1. A step change of port1 voltage (V_1) of 80 V is applied at 0.3 s and the voltage tracking to reference is perfect. There is a small dip in the port2 voltage (V_2). However, the closed loop system stabilizes the voltage. This voltage change can be observed in the port2 current (I_2). Fig. 10 shows simulation results of the battery charging system in microgrid. It is shown that the microgrid control structure is operating very efficiently under all adverse conditions.

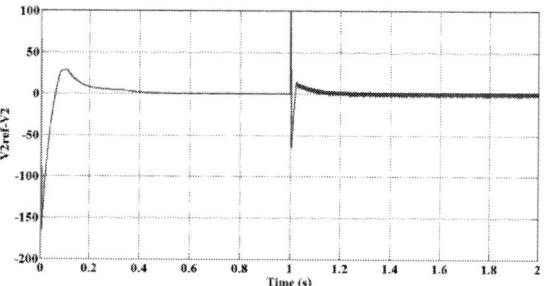

Fig. 8: Voltage tracking error

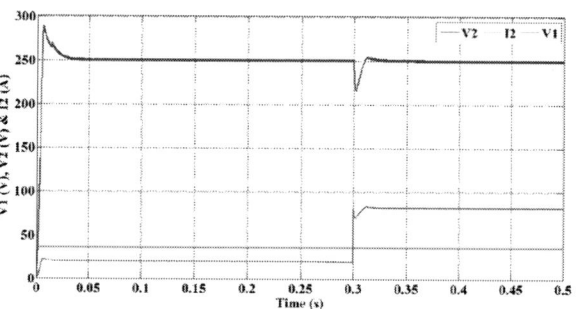

Fig. 9: Simulation results of step change of voltage (V_1), I_2 and V_2.

Fig. 10: Simulation results of the battery charging system in microgrid

The experimental verification of the proposed switching schemes is obtained through a prototype experimental setup as shown in Fig. 11. A digital signal processor that runs on dSPACE control board (DS1105) is used for controlling of DAB converter. The experimental setup is developed. Fig. 12 shows control signal of port1 converter.

Fig. 11: Experimental Setup

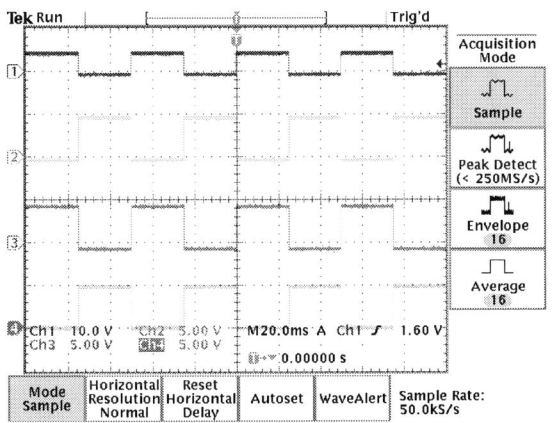

Fig. 12: Control signal of port1 converter

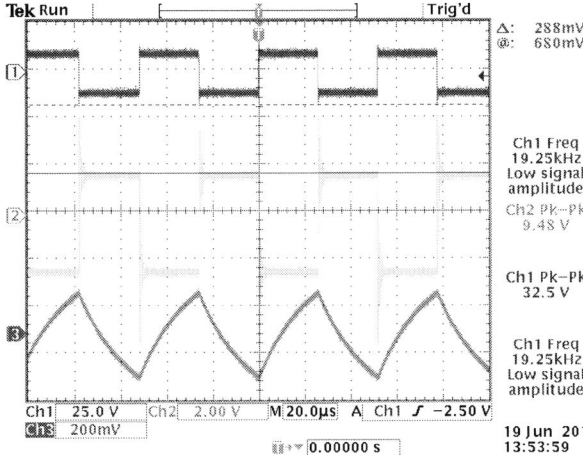

Fig. 13: Experimental results of port1, port2 voltage and inductor currents

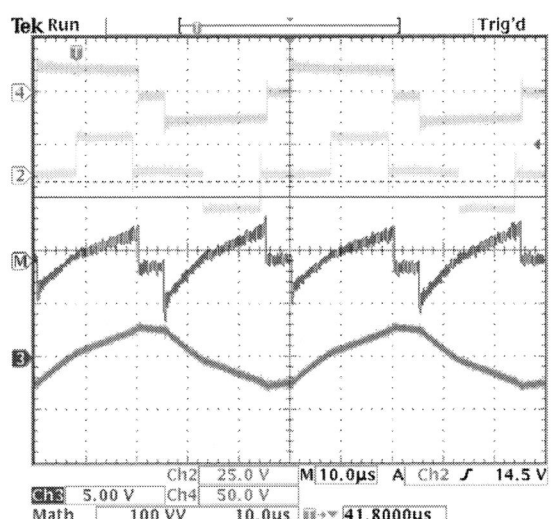

Fig. 14: Experimental results of the steady state operation of the DAB converter; showing from top (a) Input ac voltage v_{AC1} (50V/div); (b) Output voltage v_{AC2} (50V/div); (c) Input instantaneous power $P1$ (625W/div); (d) Input ac current i_L (10V/div)

Table 1: Circuit parameters of the dab converter

Rated Power	3 kW
Rated DC voltage	360 V
Transformer turns ratio	1:1
Transformer leakage inductance	1.6 μH
Transformer winding resistance	17 mΩ
Auxiliary inductor	20 μH
Inductor winding resistance	20 mΩ
Switching frequency	20 kHz
Load resistance	12.96 Ω
Filter capacitance	33 μF

Experimental results of port1, a port2 voltage and inductor current of DAB is shown in Fig. 13. With both duty cycle variations and phase shift change, the magnitude of the voltages and reactive power can be regulated. Fig. 14 shows the primary ac voltage v_{AC1}, secondary ac voltage v_{AC2}, input power and inductor current. It is regulated based on the desired specifications.

IV. CONCLUSION

This paper presented a modeling, control and experimental realization of a single DAB converter used in all solar micro-grid system. The small signal modeling of DAB converter system is carried out and closed loop controller is designed for tracking the output voltage for a battery charging applications. To verify the proposed system few experimental results are presented

ACKNOWLEDGMENT

The author thankful to Indo-U.S. Science and Technology Forum (IUSSTF) and Department of Science and Technology, Govt. of India, for providing financial support under Bhaskara Advanced Solar Energy (BASE) Fellowship Program. Under this fellowship, it was possible to do research at Tennessee Technological University, Tennessee, USA for a period of 3 months on advanced research on Grid Interaction including Smart Grids and System Development and Integration.

Authors are thankful to Dr. Jianfu Fu for using his experimental setup for verification of proposed control strategy.

REFERENCES

[1] S. Bifaretti, P. Zanchetta, A. Watson, L. Tarisciotti, and J. C. Clare, "Advanced power electronic conversion and control system for universal and flexible power management," *IEEE Trans. Smart Grid*, vol. 2, no. 2, pp. 231–243, June 2011.

[2] J. Rocabert, A. Luna, F. Blaabjerg, and P. Rodriguez, "Control of power converters in AC microgrids," *IEEE Trans. Power Electronics*, vol. 27, no. 11, pp. 4734–4749, November 2012.

[3] G. Pepermans, J. Driesen, D. Haeseldonckx, R. Belmans, and W. D'haeseleer, "Distributed generation: Definition, benefits and issues," *Energy Policy*, vol. 33, no. 6, pp. 787–798, 2005.

[4] T. C. Green and M. Prodanovic, "Control of inverter-based micro-grids," *Electr. Power Syst. Res. Distrib. Generation*, vol. 77, no. 9, pp. 1204–1213, 2007.

[5] F. Katiraei, R. Iravani, N. Hatziargyriou, and A. Dimeas, "Microgrids management," *IEEE Power Energy Mag.*, vol. 6, no. 3, pp. 54–65, May/Jun. 2008.

[6] A. Ipakchi and F. Albuyeh, "Grid of the future," IEEE Power Energy Mag., vol. 7, no. 2, pp. 52--62, 2009.

[7] K. Moslehi and R. Kumar, "A reliability perspective of the smart grid," IEEE Trans. Smart Grid, vol. 1, no. 1, pp. 57--64, 2010.

[8] W. Chen, P. Rong, and Z. Y. Lu, "Snubberless bidirectional DC–DC converter with new CLLC resonant tank featuring minimized switching loss," IEEE Trans. Ind. Electron., vol. 57, no. 9, pp. 3075–3086, Sep. 2010.

[9] F. H. Khan and L. M. Tolbert, "Bi-directional power management and fault tolerant feature in a 5-kWmultilevel DC–DC converter with modular architecture," IET Power Electron., vol. 2, no. 5, pp. 595–604, Jul. 2009.

[10] C. M.Wang, C. H. Lin, and T. C. Yang, "High-power-factor soft-switched DC power supply system," IEEE Trans. Power Electron., vol. 26, no. 2, pp. 647–654, Feb. 2011.

[11] P. Das, S. A. Mousavi, and G. Moschopoulos, "Analysis and design of a nonisolated bidirectional ZVS-PWM DC–DC converter with coupled inductors," IEEE Trans. Power Electron., vol. 25, no. 10, pp. 2630–2641, Oct. 2010.

[12] S. Inoue and H. Akagi, "A bidirectional DC–DC converter for an energy storage system with galvanic isolation," *IEEE Trans. Power Electron., v*ol. 22, no. 6, pp. 2299–2306, Nov. 2007.

[13] W. H. Li, W. C. Li, Y. Deng, and X. N. He, "Single-stage single-phase high-step-up ZVT boost converter for fuel-cell microgrid system," IEEE Trans. Power Electron., vol. 25, no. 12, pp. 3057–3065, Dec. 2010.

[14] B. Zhao, Q. Song, W. Liu, and Y. Sun, "Overview of dual-active-bridge isolated bidirectional DC–DC converter for high-frequency-link power-conversion system," IEEE Trans. Power Electron., vol. 29, no. 8, pp. 4091– 4106, August 2014.

[15] B. Zhao, Q. Yu, and W. Sun, "Extended-phase-shift control of isolated bidirectional DC–DC converter for power distribution in microgrid," IEEE Trans. Power Electron., vol. 27, no. 11, pp. 4667-4680, November 2012.

[16] H. Qin, J. W. Kimball, "Generalized Average Modeling of Dual Active Bridge DC–DC Converter," in *Power Electronics, IEEE Transactions on* , vol.27, no.4, pp.2078-2084, April 2012.

[17] Jianfu Fu, "Multiport High Frequency Transformer Coupled Bidirectional DC-DC Converters for Hybrid Renewable Energy System," *PhD Thesis*, the Faculty of the Graduate School, Tennessee Technological University, USA.

Propulsion System Design of Electric Vehicle

Md. Junaid Akhtar[1] R. K. Behera[2] S. K. Parida[3]

[1, 2, 3] Department of Electrical Engineering, Indian Institute of Technology, Patna, India
[1]E-mail: junaid@iitp.ac.in
[2]E-mail: rkb@iitp.ac.in
[3]E-mail: skparida@iitp.ac.in

Abstract—This paper presents the systematic design methodology for the design of electric vehicle propulsion system. Power rating of three phase induction motor is calculated to achieve the pre-specified vehicle dynamic characteristics. The effect on electric motor power rating with variation of vehicle maximum speed is analyzed. Simulation results for vehicle dynamics are shown and validated. Furthermore, the obtained result is utilized to design a three wheeler electric vehicle for maximum speed of 50 km/hr.

Keywords–electric vehicle, induction motor, tractive power, road load.

I. INTRODUCTION

The hazardous gasses emitted by internal combustion engines (ICE) are the major cause of global warming and air-pollution. Moreover, the petroleum resources will also be facing the risk of getting depleted in the near future. Therefore, replacing ICE with electric motors can be a feasible solution for environmental pollution and petroleum fuel dependency. The idea of the hybrid electric vehicle (HEV) and electric vehicle (EV) was perceived in the middle of the 20th century [1]. After the introduction of the ICE, HEV's and EV's continued to dominate the automobile market side by side with the ICE for many years. The energy density of gasoline is far more than what the electrochemical battery could offer. In the near future, it is predicted by many that fuel cells and batteries will replace ICE completely.

EV's are the only automobiles group that has zero emission. It has the ability to provide emission free transportation [2]. These vehicles use an electric motor for propulsion of the vehicle and batteries as electrical-energy storage devices to power the electric motor. EV's also reduce the noise pollution level. Electric motor in EV should have high efficiency, higher power density, small in size and low weight [3]. It should also have wide operating range and wide speed constant power range [4]. EVs are mainly comprised of three different blocks: (i) an energy source, (ii) an electric drive and (iii) an electric motor. The electric motor is connected to the drive wheels of the vehicle through the mechanical transmission [5]. The key components of EV are shown in Fig. 1. The control inputs are given to the electronic controller through the brake or accelerator. Then the electronic controller sends the control signals to the three phase inverter and hence the three phase induction motor (IM) is driven. The IM is coupled to the driving wheels through the gearbox and differential.

In this work, a design procedure for the EV propulsion system is presented based on the vehicle dynamics and power rating of the three phase IM is calculated. This paper is organized as follows. Section II discusses the different EV configuration. Design steps to calculate the electric motor power rating are discussed in Section III whereas simulations and experimental results are shown in Section IV. Conclusions are discussed in Section V.

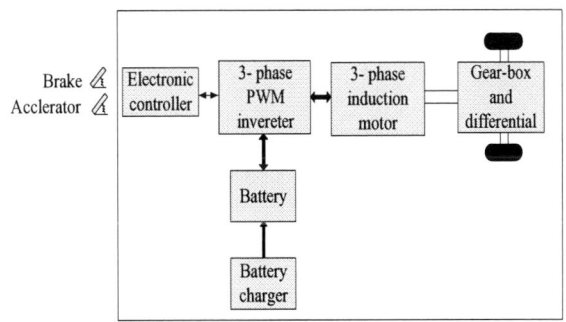

Fig. 1: General electric vehicle configuration

II. CONFIGURATION OF ELECTRIC VEHICLE

There are various configuration of EV due to arrangement of electric propulsion and energy sources [6]. These configurations are shown in Fig. 2(a)-(f). Fig. 2(a) shows the EV configuration. In this type of configuration the motor is connected to driving wheels through the clutch, gear box and differential. The clutch and gearbox helps the driver to shift the gear ratios and the torque is transmitted to the wheels. High torque is obtained in the lower gears at low speed and low torque is obtained at higher gears at high speed. The differential is used to drive the wheel at different speed when the vehicle is turning. Different EV configuration is obtained by replacing the gearbox with fixed gearing and hence removing the clutch as shown in Fig. 2(b). In this configuration, the weight and size of the mechanical transmission is reduced. Fig. 2(c) shows the third type of EV configuration. In this configuration, the electric motor, fixed gearing and differential are assembled together. Fig. 2(d) shows the fourth type of EV configuration. In this, two electric motors are used. These motor separately drive the driving wheels with fixed gear. In Fig. 2(e), the electric motor is placed inside the wheel to shorten the mechanical transmission path from the electric motor to the driven wheels. This is called in-wheel drive system. In this type of configuration, fixed planetary gear is used. In the last type of EV configuration, the mechanical   gear   is   removed

978-1-5090-0064-7/15 $31.00 © 2015 IEEE

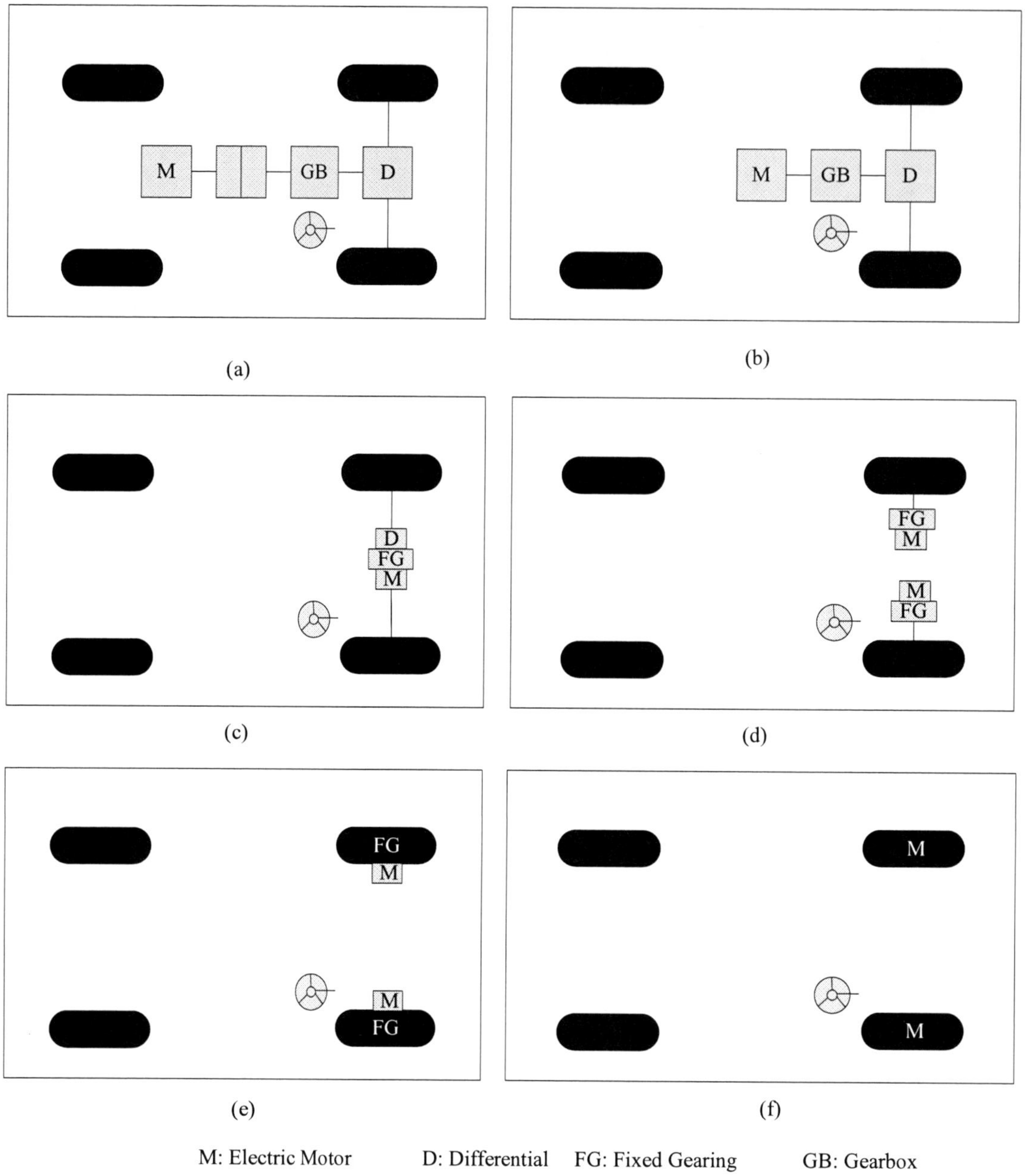

(a)

(b)

(c)

(d)

(e)

(f)

M: Electric Motor   D: Differential   FG: Fixed Gearing   GB: Gearbox

Fig. 2: Electric vehicle configuration due to variations in electric propulsions

completely. A low speed outer rotor electric motor is installed inside the wheel like an in-wheel drive system as show in Fig. 2(f). In this type of configuration gearbox is not used. The speed control of the electric motor is same as the wheel speed and hence the vehicle speed.

III. ELECTRIC VEHICLE DESIGN

In designing the EV various variables are to be considered. These design variables include:

    a) electric motor rated speed.
    b) electric motor power rating.
    c) maximum speed of the electric motor.
    d) constant power region beyond the rated speed.
    e) gear ratio between the wheel shaft and motor

shaft.

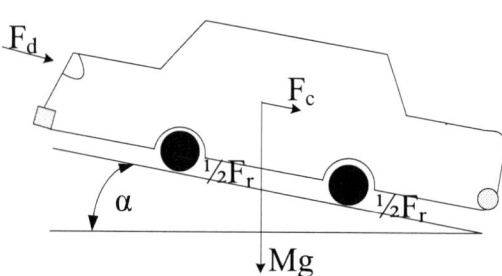

Fig. 3: Forces acting on electric vehicle

The vehicle has to overcome several forces acting on it while travelling is known as road load [7] - [9] as shown in Fig. 3. This road load F_l consist of three components: (a)

978-1-5090-0064-7/15 $31.00 © 2015 IEEE

aerodynamic drag force F_d, (b) rolling resistance force F_r and (c) climbing force F_c. The road load is given by the following equation.

$$F_l = F_d + F_r + F_c \tag{1}$$

The aerodynamic drag force is the force acting on the vehicle body when it is travelling through air. It consist of three components: the skin friction drag (due to air flow in the boundary layer), normal pressure drag (depends on the vehicle frontal area and speed), induced drag (due to trailing vortices). The expression for aerodynamic drag is given by

$$F_d = 0.5 \rho C_d A (v + v_o)^2 \tag{2}$$

where; C_d is the aerodynamic drag coefficient, ρ is the air density in kg/m^3, A is the frontal area in m^2, v is the head wind velocity in m/s. In general, ρ is taken as 1.23 kg/m^3, C_d lies between 0.2 and 1.5.

The rolling resistance force is due to the deformation on the wheel and road surface. Some of the factors affecting the rolling resistance are the pressure and temperature of the tire, speed of the vehicle and material of the tire. The rolling resistance is given by (3).

$$F_r = MgC_r \tag{3}$$

where, M is the vehicle mass in Kg, g is acceleration due to gravity and C_r is the coefficient of the rolling resistance. The value of C_r lies around 0.013.

The climbing force is the force required by the vehicle to climb up the inclined surface. The climbing force is given by the (4).

$$F_c = Mg \sin \alpha \tag{4}$$

where, α is the angle of incline surface.

The road load F_l is overcome by the motive force F available at the wheels so the vehicle is driven with the acceleration a which is given by (5).

$$a = \frac{(F - F_l)}{k_r M} \tag{5}$$

where, k_r is correction factor because there is an increase in vehicle mass due to inertia of rotational masses.

The tractive power [10] is expressed as given below.

$$P_t = \frac{k_r M}{2 t_a} \left(V_f^2 + V_b^2 \right) \tag{6}$$

where, P_t is the tractive power, t_a is the acceleration time, V_f is the vehicle final speed and V_b is the vehicle base speed.

The power rating obtained by (6) is only used for accelerating the vehicle. The power consumed to overcome rolling resistance and dynamic drag should also

be considered. During acceleration, the average drag power is given by (7).

$$\bar{P}_{drag} = \frac{1}{t_a} \int_0^{t_a} \left(Mgf_r V + \frac{1}{2} \rho C_d A V^3 \right) dt \tag{7}$$

where, V is the vehicle speed.

The acceleration time for an EV is given by (8).

$$t_a = \int_0^{V_b} \frac{k_r M}{P_t / V_b - Mgf_r - 1/2 \rho C_d A V^2} dV$$
$$+ \int_{V_b}^{V_f} \frac{k_r M}{P_t / V - Mgf_r - 1/2 \rho C_d A V^2} dV \tag{8}$$

The vehicle speed is expressed in terms of time t, as

$$V = V_f \sqrt{\frac{t}{t_a}} \tag{9}$$

Substituting (9) in (7) and integrating we get

$$\bar{P}_{drag} = \frac{2}{3} MgC_r V_f + \frac{1}{5} \rho C_d A V_f^3 \tag{10}$$

The total tractive power for accelerating the vehicle from zero to speed V_f in t_a seconds is given as

$$P_t = \frac{k_r M}{2 t_a} \left(V_f^2 + V_b^2 \right) + \frac{2}{3} MgC_r V_f + \frac{1}{5} \rho C_d A V_f^3 \tag{11}$$

From (11) it is observed that for a given acceleration performance, small power rating of the motor can be obtained if vehicle base speed is kept low. However, the rate of change of power rating to the vehicle base speed reduction is not identical. Differentiating (11) with respect to vehicle base speed V_b, we get

$$\frac{dP_t}{dV_b} = \frac{k_r M}{t_a} V_b \tag{12}$$

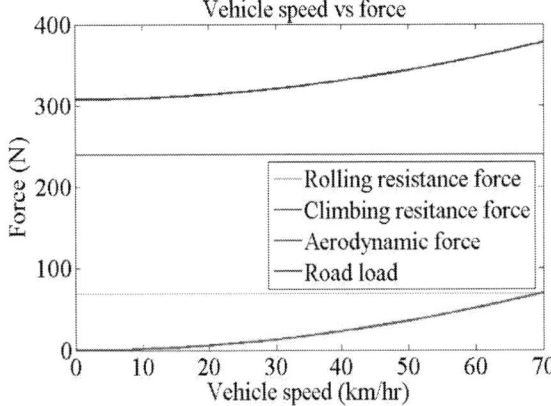

Fig. 4: Road load characteristics as a function of speed.

978-1-5090-0064-7/15 $31.00 © 2015 IEEE

IV. SIMULATIONS AND RESULTS

To obtain the road load characteristics, the following assumptions are made:
  1) velocity is independent of rolling resistance.
  2) head wind velocity is zero.
The following vehicle parameters are chosen:
- Mass of the vehicle 700 kg.
- Acceleration time of 30 sec.
- Vehicle base speed 25 km/hr.
- Vehicle final speed 50 km/hr.
- Acceleration due to gravity 9.8 m/s^2.
- Coefficient of rolling resistance 0.01.
- Air density 1.23 kg/m^3.

The resistive forces given in (1)-(4) are plotted versus vehicle speed as shown in Fig. 4. It is observed from Fig. 4 that the rolling resistance force and climbing force are constant whereas the aerodynamic force increases as the vehicle speed increases. The total force i.e. road load is the sum of all the three forces and it also increases as the vehicle speed increases.

The tractive power rating of an electric motor is calculated using (11) and shown in Fig. 5. It is observed that a low x (high vehicle base speed) the power requirement is more whereas if the vehicle base speed is decreased then the power requirement also decreases gradually. But with high x (low value of vehicle base speed), if the vehicle speed is decreased, the power requirement does not change significantly. The power rating decline rate to the vehicle speed reduction (dP_t / dV_b) versus the speed factor is calculated using (12) and is shown in Fig. 6. It is observed that for high vehicle base speed, the decline rate of change of power rating w.r.t. vehicle base speed is large. However, for low vehicle base speed, the decline rate of change of power rating w.r.t. vehicle base speed is small.

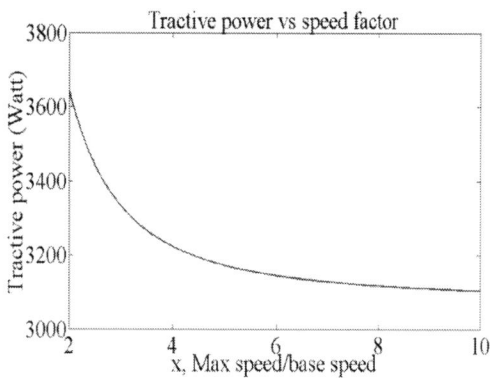

Fig. 5: Power rating vs speed factor

A three phase IM is designed to fulfill the vehicle parameters given above using (11). The IM power rating is calculated as 3.64 kW. So according to this, power rating of 5 hp three phase squirrel cage IM is designed for EV. The base speed of the IM is 2400 rpm so it has wide speed operating range. Fig. 7 shows the fabricated IM. The three wheeler electric auto rickshaw is shown in Fig. 8. The fabricated IM is then fitted to the three wheeler auto-rickshaw as shown in Fig. 9.

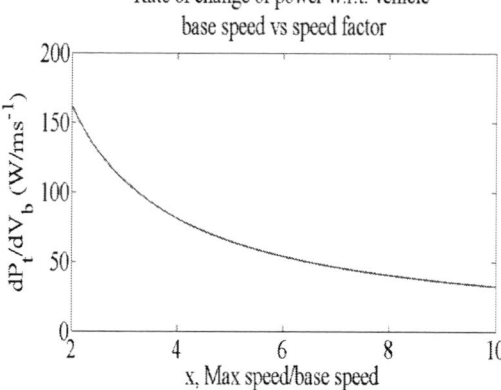

Fig. 6: Rate of change of power w.r.t. vehicle base speed vs speed factor.

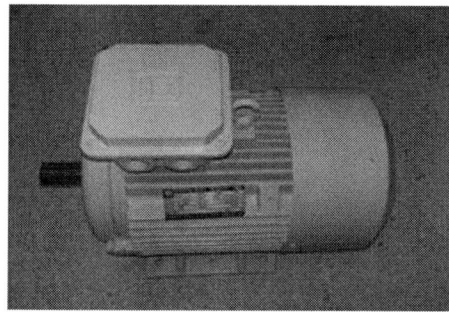

Fig. 7: Fabricated 5 hp induction motor

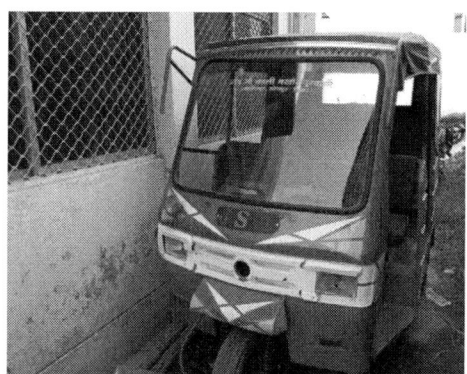

Fig. 8: Three wheeler auto-rickshaw

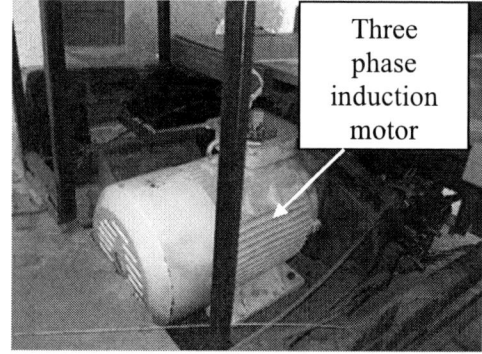

Fig. 9: Three phase induction motor fitted in electric auto rickshaw.

V. CONCLUSION

A design procedure for the EV propulsion system is presented based on the vehicle dynamics. This methodology helps to calculate the motor power rating according to the vehicle dynamics. It is observed that as the speed factor (i.e. ratio of vehicle maximum speed by vehicle base speed) is small, the power rating requirement is high and as the vehicle base speed is decreased for the same vehicle maximum speed the power rating decreases significantly. But for higher value of the speed factor, when the vehicle base speed is increased for the same vehicle maximum speed, the power rating does not change significantly. A three wheeler EV is developed for maximum speed of 50 km/hr. A three phase 5 hp, IM is designed and fabricated. This fabricated IM is then installed in the electric auto rickshaw.

ACKNOWLEDGMENT

Authors acknowledge the support provided by the National Mission on Power Electronics Technology (NAMPET) project by Ministry of Communications and Information Technology, Govt. of India and Centre for Development of Advanced Computing (CDAC), Thiruvananthapuram. No-R&D/SP/EE/CDAC/MDF/2012-13/45.

REFERENCES

[1] C. C. Chan and K. T. Chau, 'Modern electric vehicle technology'(Oxford University Press, 2001).

[2] C. C. Chan, "The State of the Art of Electric, Hybrid, and Fuel Cell Vehicles," Proceedings of the IEEE , vol.95, no.4, pp.704,718, April 2007.

[3] Emadi, "Transportation 2.0," Power and Energy Magazine, IEEE , vol.9, no.4, pp.18,29, July-Aug. 2011.

[4] S. S. Williamson, S. M. Lukic, A. Emadi, "Comprehensive drive train efficiency analysis of hybrid electric and fuel cell vehicles based on motor-controller efficiency modeling," Power Electronics, IEEE Transactions on , vol.21, no.3, pp.730,740, May 2006

[5] J. Larminie and J. Lowry, 'Electric vehicle technology explained'(John Wiley & Sons, 2004).

[6] M. Ehsani, Y. Gao and A. Emadi, 'Modern electric, hybrid electric, and fuel cell vehicles' (CRC Press, 2010).

[7] M. Ehsani, K. M. Rahman, H. A. Toliyat, "Propulsion system design of electric and hybrid vehicles," Industrial Electronics, IEEE Transactions on , vol.44, no.1, pp.19,27, Feb 1997.

[8] M. Ehsani, K. M. Rahman, H. A. Toliyat, "Propulsion system design of electric vehicles, "Industrial Electronics, Control, and Instrumentation, 1996., Proceedings of the 1996 IEEE IECON, 22nd International Conference on , vol.1, no., pp.7,13 vol.1, 5-10 Aug 1996

[9] B. Tabbache, A. Kheloui, M. E. H. Benbouzid, "Design and control of the induction motor propulsion of an Electric Vehicle," Vehicle Power and Propulsion Conference (VPPC), 2010 IEEE , pp.1,6, 1-3 Sept. 2010.

[10] M. Ehsani, Y Gao, S. Gay, "Characterization of electric motor drives for traction applications," Industrial Electronics Society, 2003. IECON '03. The 29th Annual Conference of the IEEE , vol.1, no., pp.891,896 vol.1, 2-6 Nov. 2003.

Comparative Analysis of 2-Level and Multi-Level Inverter Fed Coupled IM Drives Based on V/f and DTC Techniques

Shrikant Misal[1]*, Swetha T[1], Mamta Wandalkar[2], Chang Yan Tai[3]

[1]Department of Electrical Engineering, M.S. Ramaiah University of Applied Sciences, Bangalore, India
[2]Department of Electrical Engineering, CMR Institute of Technology, Bangalore, India
[3]Department of Electrical Engineering, The Hong Kong Polytechnic University, Hong Kong
*Corresponding Author: shrik88@gmail.com

Abstract—For industrial sectors involving coupled induction motor drives, control of speed/torque is an utmost critical issue especially in dynamic conditions. Considering this aspect, there is a dire need of a flexible and efficient control which could best fit in the scheme of coupled drive operation. In this paper an effort has been made to explore different facets of speed/toque control for coupled induction motors based on traditional scalar (Volt/Hz) and advanced vector (Direct Torque) techniques. Also in order to facilitate improvements in terms of harmonic contents and power losses, it has been proposed to replace traditional 2-Level Inverter in Induction Motor drive with an advanced Reversing Voltage Multi Level Inverter (MLI) topology providing 5-Levels of output phase voltage. Based on the simulation study for both the control strategies in MATLAB/Simulink, it has been shown that harmonic distortions present in the fundamental output voltage of a coupled drive fed by Multilevel Inverters are comparatively less when compared to its 2-Level counterpart. These reduced distortions are in-turn responsible for toning down filtering requirements at the input of motor, thereby enhancing the efficiency of drive system.

Keywords-Direct Torque Control, Multi-Level Inverter, Induction Motor, V/f control

I. INTRODUCTION

In recent years, the power demand from industrial sector has shot up drastically and it is estimated to double up in near future considering the pace at which it is progressing. At present almost 60% of the load demands are in process industries where heavy works are carried out using Induction Motor (IM) drives. To put a tab on this high power consumption, control of motors using an efficient and flexible technique with varying load pattern plays a critical role. Although it may be convenient to say that the speed or torque control can be achieved by mere variation of stator voltage and frequency of supply but in actual sense it is highly impossible to do so without sacrificing the drive efficiency. In simple terms an increase in loading proportionally decreases the efficiency of operation. Hence to account for this aspect, advanced scalar and vector control techniques [1-4][8] are coming up of late to make the process of individual or simultaneous speed/torque control quite amicable.

For load sharing applications like cranes, hoists etc. where there is a presence of coupled motor drives, the only ambiguity of control is with synchronization of different part movements. Thus in present industrial scenarios, to facilitate a precise variation in speed and torque individually, the control techniques are preferred to be independent. This type of control scheme is coherently termed as Scalar or V/f control which basically deals with steady state condition wherein only magnitude and frequency of voltage or current are controlled neglecting coupling effect in the machine. Some typical advantages of this control scheme are simple, conventional and cost effective speed control for load sharing applications. The accuracy that can be achieved using V/f Control [1] for both speed and torque control is around ±3%. But owing to constrain of individual speed/toque control circuits for coupled motors, there are healthy chances of the drive getting slowed down. This is highly undesirable from the dynamic operation point of view. On similar lines another control scheme which has garnered considerable attention in recent times is the advanced vector control technique also called as Direct Torque Control (DTC). This scheme [2] mainly deals with dynamic state of IM motor drive wherein instantaneous positions of voltage, current and flux linkage space vector can be controlled amicably by hexagonal (normally) unit space displacement. The potential advantages which work in favor of DTC [4,5] are its higher accuracy, precision, response time of around 5ms which is 10 times faster than the scalar control. These factors are certainly responsible for blessing this scheme with staggering reliability and flexibility of speed/torque control which is the foremost need of a coupled drive. The only compromise which is required here is with the complexity of control.

An extensive research has already been made in [2,3] regarding the scalar and vector control strategies with individual drive operations. Several critical issues involving stability, efficiency and reliability pertaining to open loop and closed loop operation of IM drive have been effective discussed with their probable solution. Also the Voltage Source Inverter (VSI) topology implemented for the coupled drive operation is basically a 2-Level Inverter having voltage levels $V_{dc}/3$ and $2V_{dc}/3$. Even though 2-Level VSI seems to be satisfactory for coupled drives, there will obviously a considerable presence of Total Harmonic Distortions in the phase voltage waveform responsible for dampening the system stability. Thus several MLI configurations like diode clamped, capacitor clamped, cascaded bridge, reversing voltage, hybrid etc. discussed in [6,8] have been successfully tested for single motor drive operation with Field Oriented Control (FOC). The above discussion clearly hints that there is a potential room for research in this field to explore the application of scalar and vector controls in coupled drive system which is hardly discussed in any of the referred literatures.

1. Objectives of the paper
The main scope of this paper is to propose the application of V/f and DTC techniques in coupled motor drives to

achieve simultaneous torque control with a single controller and carry out their comparative analysis based on their respective performance with 2-Level and Multi-Level VSI drives.

2. Organization of the paper

V/f and DTC configurations for coupled motors are introduced in II, selection and design of induction motor parameters in III, simulation circuits and control strategies in IV, simulation results and analysis in V followed by concluding remarks in VI.

II. CONFIGURATIONS OF VOLT/HZ AND DIRECT TORQUE CONTROLLERS FOR COUPLED IM DRIVES

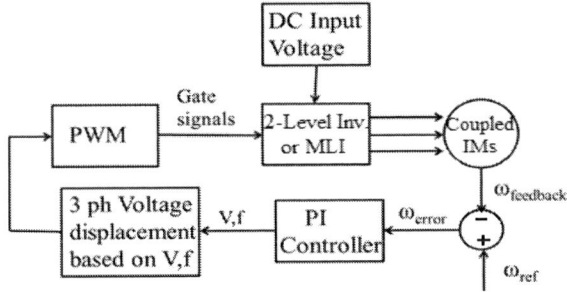

Fig. 1: Block Diagram of V/f Control

Fig.1 shows the block diagram of a coupled induction motor drive with V/f control. Here the VSI can be a 2-Level or Multi-Level depending upon the output requirements. The speed from IM is sensed and fed back to the comparator circuit wherein the reference and feedback speeds are compared and an error signal is generated. The error signal is passed through a PI controller which is responsible for fine tuning it by mitigation of overshoots. This speed voltage error is converted into 3 phase voltage signals which are 120° phase shifted with respect to each other. Based on these 3 phase voltage signals, the PWM generator gives out requisite gate pulses in order to initiate the triggering action for inverter switches. For load sharing in multiple motors, speed reference of two motors is desired to be kept same since both are driven by a common drive technology and the torque needs to be equally divided among both motors based on their ratings. This is the reason for requirement of both speed and torque control in coupled drives.

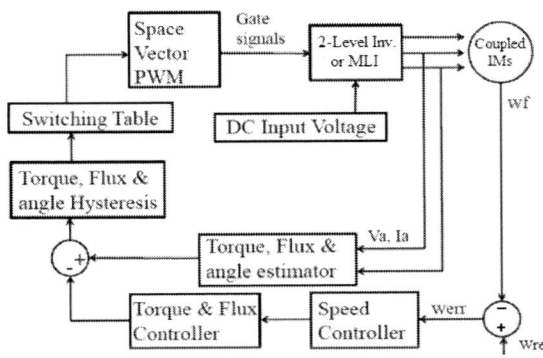

Fig. 2: Block Diagram of DTC Model

Block diagram in Fig.2 highlights the basic schematic representation of DTC. Similar to Fig.(1), VSI drive can have 2-Level or Multi-Level output. The electromagnetic torque and stator flux are calculated from primary motor inputs like stator voltage and current. The control initiates with direct selection of stator voltage vectors according to torque and flux errors. These errors are nothing but the difference between references of torque and stator flux linkage with their actual values. The optimum vector selection for inverter switching is made with an intention to restrict the torque and flux errors within pre-specified hysteresis bands. The speed controller block is nothing but a comparator circuit where the reference speed and sensed drive speed are compared to give an error signal. The error signal is processed with the help of a PI controller. From the speed output the torque and flux is estimated using the torque and flux controller.

III. SELECTION AND DESIGN OF MOTOR PARAMETERS

1. IM Drive Selection

Consider that for the movement of heavy goods or equipments, a coupled drive has been implemented in a crane application [9]. Here the selection of coupled drive is governed by its load sharing capability to perform simultaneous different operations at a constant torque. Thus motor parameters can have a significant impact on the design of variable frequency and torque controller. In context of the given situation it is desired to meet front end high power requirements with simultaneous control of speed and torque. Taking this aspect into mind, IMs rated for 20 HP and 50HP have been selected for coupling. Detailed technical specifications of the chosen motors are as shown in Table I which will be further used in calculation of performance parameters.

2. Design of Performance Parameters for IM Drive

From specifications provided in Table I, design steps have been carried out in this section for obtaining performance parameters to be fed in simulation models. The design process in [7,10] involves calculation of losses (stator, rotor, auxiliary), calculation of motor resistances and reactances (stator, rotor, magnetizing). Table II highlights the designed parameters for 20HP and 50HP IMs respectively based on the design equations.

Table 1: Specifications of Induction Motors Used for Coupling

Power Rating	20 HP (15kW)	50HP (37kW)
Voltage (V)	415	415
Frequency (Hz)	50	50
Current (A)	27.5	63
Speed (rpm)	1470	1480
Power Factor	0.84	0.87
Efficiency @ F.L (%)	91%	94%
No. of Poles	4	4

Table 2: Designed Parameters for Induction Machine Used in Coupled Drive

IM Power Rating	20 HP = 15 kW	50 HP = 37 kW
Stator Input	16.603 kW	39.396 kW
Total Power Loss	1.494 kW	2.36376 kW
Slip	0.02	0.013
Rotor Copper Loss	306.12 W	487.3 W
Stray Loss	332.06 W	787.92 W
Friction and Windage Loss	74.7 W	118.15W
Maximum Torque	97.491 Nm	238.6 Nm
Rotor Input	15306 W	37484.615 W
Rotor Output	14925.18 W	36878.55 W
Rotor Current	24.715 A	60.36 A
Rotor Resistance	0.16725 Ω	0.0445 Ω
Stator Copper Loss	781.12 W	970.33 W
Stator Resistance	0.34429 Ω	0.08149 Ω
Stator & Rotor Reactance	0.4569 Ω	0.2383 Ω
Mutual Inductance	0.03088 H	0.0126 H

IV. SIMULATION CIRCUITS AND CONTROL STRATEGIES

1. Simulation circuit for proposed V/f control with 2-Level and Multi-Level VSI fed coupled drive system

A scheme of coupled motors implementing 3-Phase, 2-Level (basic) and 5-Level reversing voltage inverter topologies for V/f control is as shown in Fig.3.

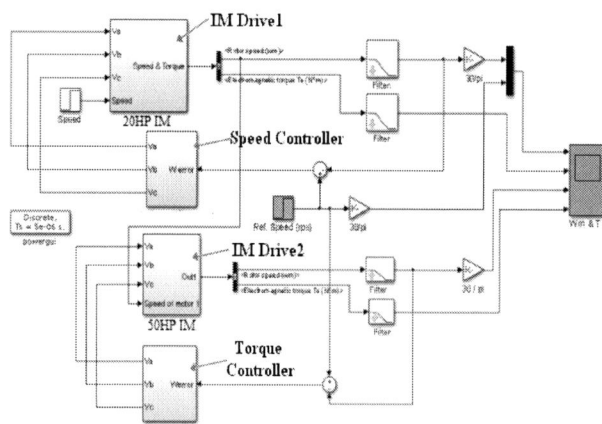

Fig. 3: Simulink Model for V/f Control of Coupled Motors

The chosen IMs are of 20HP and 50HP with a rated speed of 1470rpm. Control of speed and torque has been achieved with an individual speed and torque control scheme. Motors are coupled in such a fashion that output speed/torque of one form the driving source of other. Here switching action for VSI is initiated using a reliable Sinusoidal PWM technique.

2. Controller and drive circuits for proposed V/f coupled drive system

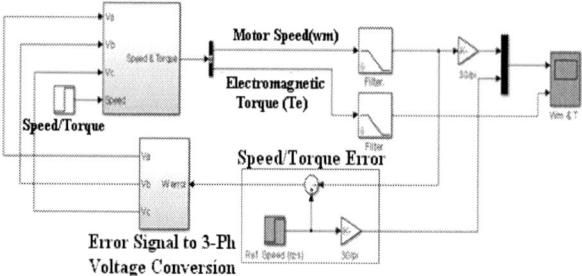

Fig. 4: Simulink Model for V/f Control of Coupled Motors

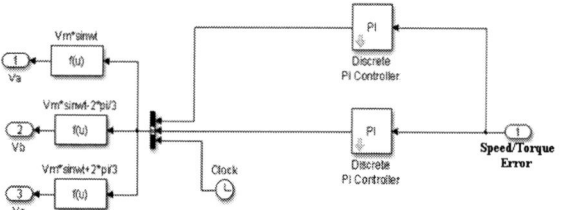

Fig. 5: Error Signal to Voltage Conversion Block

The basic schematic of control circuit typically used for speed torque variations in coupled drives is as shown in Fig.4.

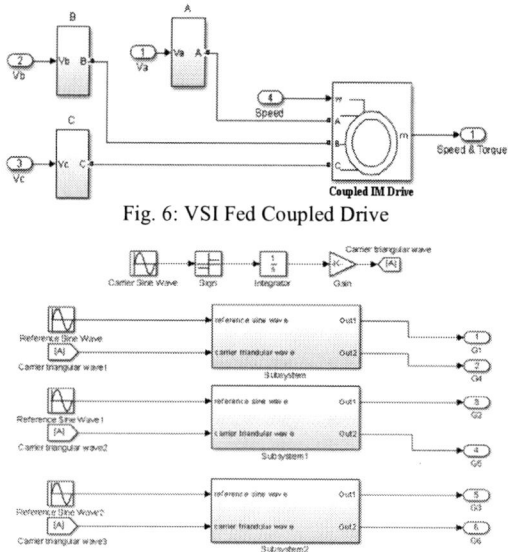

Fig. 6: VSI Fed Coupled Drive

Fig. 7: Switching Scheme for 3-Ph. Reversing Voltage MLI

Based on the output speed/torque requirements, the reference signal is made to vary in steps for generating a desired error signal. As seen from Fig.5, for a chosen time instant this signal is processed through different PI controllers to dampen the overshoot that percolates into the system because of instantaneous voltage and frequency variations. Lastly the output from PI controller is passed through a voltage conversion block wherein based on the specified time instance and processed speed error, a near to perfect sinusoidal magnitude will be generated. This sine wave as evident from Fig.7 is fed as a reference signal to the control circuit of VSI for comparison with a repeating ramp signal to give out a train of PWM pulses.

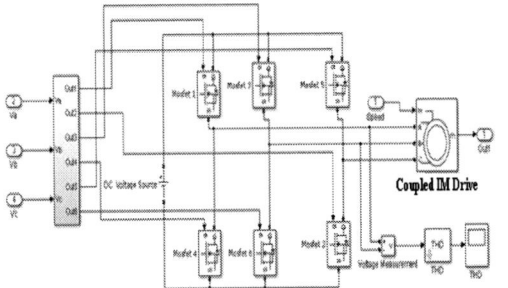

Fig. 8: Two-Level VSI with PWM Scheme for Coupled Drives

These pulses depending upon the topology of VSI have certain variations in switching patterns as seen from Figures 8 and 9. Here the choice of a particular reversing voltage topology for MLI is governed by the fact that it creates a multilevel stepped voltage half-wave with only positive values by using a simple inverter configuration having less number of DC sources. This factor particularly contributes towards improvement of system response.

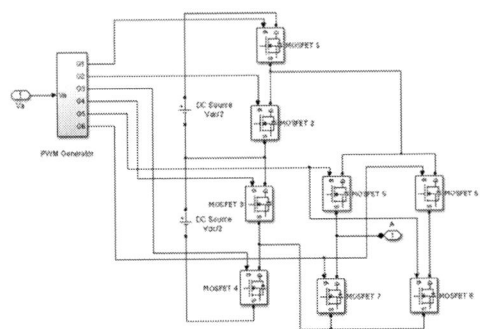

Fig. 9: Five-Level Reversing Voltage Inverter with PWM Scheme for Coupled Drives

3. Simulation circuit for proposed DTC control with 2-Level and Multi-Level VSI fed coupled drive system

A DTC basically involves connection of different subsystems working in tandem to make it capable of providing a faster and precise torque control. The subsystems involved are speed/torque controller, torque and flux estimator, torque and flux hysteresis controller, vector switching table, switching control and sector selection.

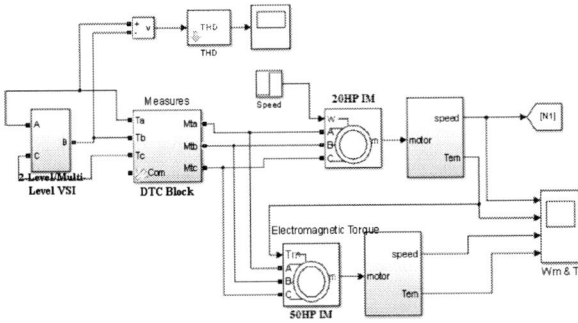

Fig. 10: Simulink Model for DTC of Coupled Motors

These components have been extensively discussed in this section for analyzing the DTC control. Fig.10 represents a generalized scheme of DTC for implementation in coupled motor drives.

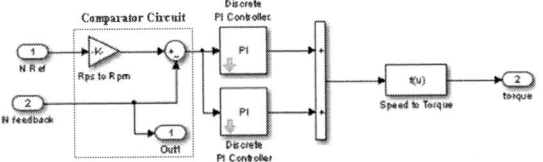

Fig. 11: Speed and Torque Controller Block

As seen from Fig.11 the functioning of speed/torque controller block resembles pretty close to that for V/f technique. Only difference observed will be the speed error signal is converted to equivalent torque value using the formula $T_{max} = \dfrac{974 \times kW}{N(rpm)}$. Thus the maximum torque at which the motor must be operating should be computed from speed feedback signal.

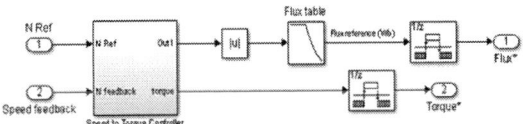

Fig. 12: Torque and Flux Controller Block

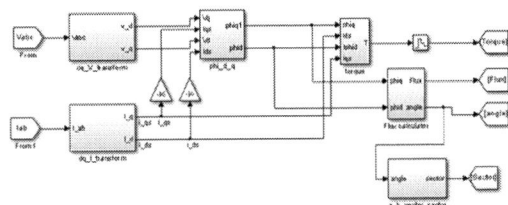

Fig. 13: Torque and Flux Estimator Block

By computation of speed feedback signal obtained from flux table, the requisite flux at which coupled motors will work can be efficiently determined with the help of scheme highlighted in Fig.12. The control of flux and torque require a certain level of feasibility and that is only possible with their pre-estimation to accomplish successive 3-phase to 2-phase transformation as shown in Fig.13.

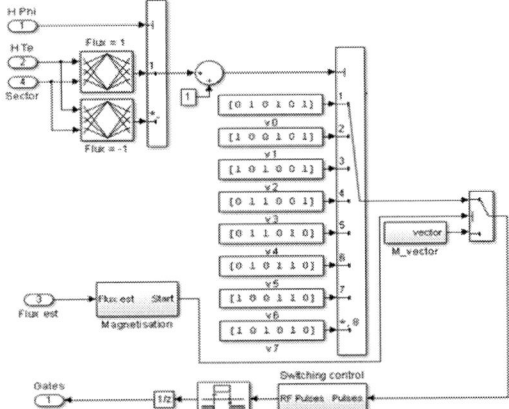

Fig. 14: Switching Control Block

The voltage and current values measured from inverter output serve as an input to Torque and Flux estimator block. These values which are sensed in 'abc' reference frame require transformation to 'dq' frame using abc to dq transformation. This step is initiated in order to express the obtained torque and flux with a two dimensional orientation as unit vectors which can possibly be controlled by initializing active switching states for VSI shown in Fig.14. The output sector is selected based on the flux angle as it forms an input to switching control block responsible for generating gate pulses.

The output sector is selected based on the flux angle as it forms an input to switching control block responsible for generating gate pulses. This block compares flux and torque, based on which switching action for VSI to compensate the torque value is initiated. Also here the precision of the torque and flux waveforms is maintained by using a flexible hysteresis controller which can vary the operating window as per requirements. The choice and switching scheme of 2-Level and Multi-Level inverters will remain the same as in V/f technique of Fig.6.

V. SIMULATION RESULTS AND ANALYSIS

4. For proposed V/f control with 2-Level and Multi-Level VSI fed coupled drive system

From Fig.15 implementing a 5-Level reversing voltage inverter drive it is observed that both motors gradually start their run at an initial speed of 1380rpm until the build-up of required air-gap flux upto 1.5sec. Once the desired flux builds up of flux, there comes the need to maintain the frequency in order to avoid saturation effect. Hence by variation of supply voltage the desired stability is pitched in to attain a constant speed of 1450rpm from 1.5sec onwards. The torque at which first motor is operating is 88N-m which is slightly at a lower scale compared to the rated torque. On similar lines with speed variation, induction of V/f control enhances the torque of 20HP motor to about 92N-m which is nothing but the rated value.

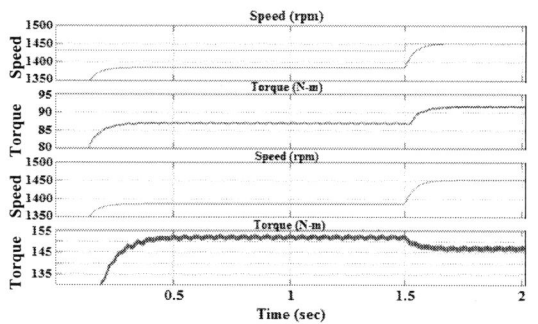

Fig. 15: Output Speed and Torque Waveforms from Two Coupled Motors using Reversing Voltage MLI

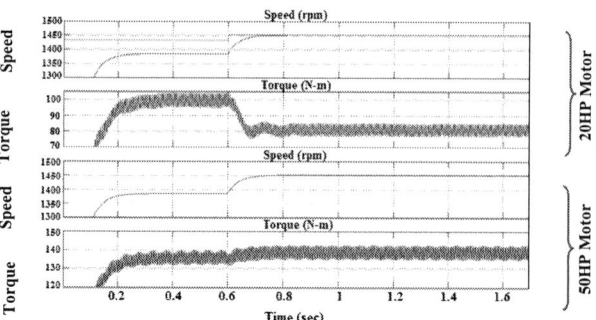

Fig. 16: Output Speed and Torque Waveforms from Two Coupled Motors using 2-Level Inverter

In this process the torque associated with coupled 50HP motor shoots up to almost 1.5 times the one obtained from first motor i.e. approximately 145N-m considering the mechanical coupling which gives the effect of a single shaft. Here the discussed operation for V/f control with MLI holds good even for a 2-Level VSI except for the few degradation in performance parameters. As seen from Fig.16 similar to previous operation, both motors are running at a speed of 1370rpm initially upto 0.6sec for initial flux build-up and thereafter from 0.6sec onwards V/f control takes over the mantle to deliver a constant speed of 1450 rpm. Similarly torque achieved for first motor is 98N-m at critical speed of 1370rpm which is slightly on a higher note than the rated one. After 0.6sec the torque of motor decreases and attains a rated value of about 83N-m. Because of single shaft sharing the net torque at which second motor should operate is around 140N-m.

Since the motor is been fed with VSI drive having either 2-Level or Multi-Level (5-Level) operation, analyzing its performance based on the magnitude of fundamental and harmonic distortions is utmost critical aspect. These parameters in-turn are responsible for determining the drive losses and hence the system efficiency. Figures 17 and 18 show the output phase voltage waveform for 3-Ph, 2-Level and 5-Level VSIs. For both the cases magnitude of voltage is maintained at 375V except for the number of steps involved. With more discretization of voltage the THD content can be potentially toned down.

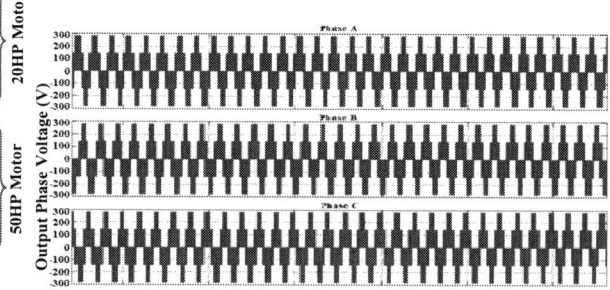

Fig. 17: 3-Ph. Output Voltage Waveforms for 2-Level Inverter

978-1-5090-0064-7/15 $31.00 © 2015 IEEE   118

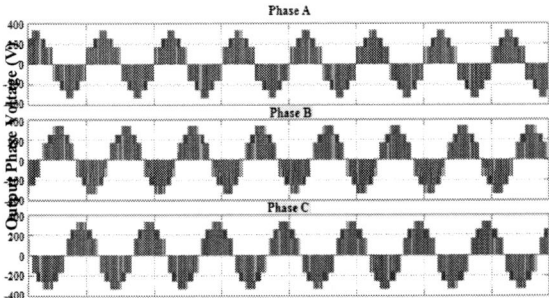

Fig. 18: 3-Ph. Output Voltage Waveforms for 5-Level Inverter

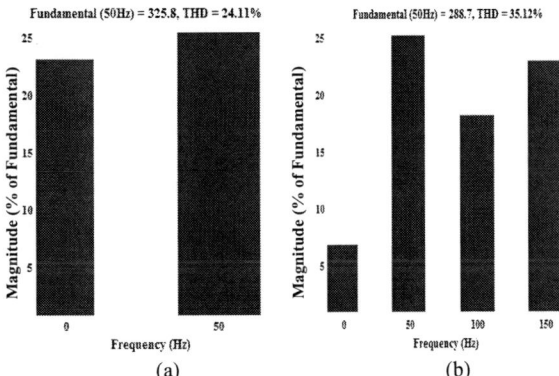

Fig. 19: THD in Output Voltage for (a) Multi Level Inverter and (b) 2-Level Inverter Topologies

Fig.19 shows the percentage of THD magnitude present in the phase voltage in context of the fundamental frequency of operation. For MLI, the THD magnitude is 24.11% of the fundamental voltage as against 35.12% for 2-Level inverter which is pretty high. Another important parameter which plays a significant role in determining the overall system efficiency is determination of losses associated with switching components of VSI. Considering that the switching element used here is MOSFET, the power losses can be categorized into switching and conduction losses. Since the switching frequency used here is nothing but the supply frequency component of 50Hz, switching losses can be neglected to directly obtain conduction losses. The conduction losses for the MOSFET is calculated using the formula $P_{CM} = R_{DSON}*(I_{rms})^2$, where R_{DSON} is drain to source on state resistance and I_{rms} is RMS value of MOSFET on state current. Since ON state resistance for MOSFET is readily available in the component datasheet, it becomes convenient to obtain the conduction losses from measured RMS value of drain current. So the magnitude of conduction losses considering 6 switches (for 2-Level) and 24 switches (for 5-Level) comes out to be 312W and 1248W respectively during active state of operation. Thus knowing the drive conduction losses and motor operational losses, the overall power output of complete drive system has been amicably computed. The nature of this output power as seen from Figures 20 and 21 is a constant value of 13280W and 13090W for respective 2-Level and Multi-Level inverter fed drives only because motor torque is maintained constant.

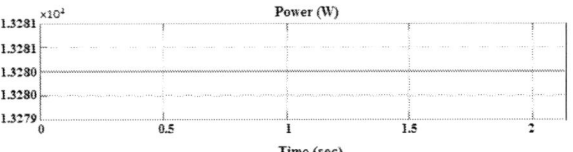

Fig. 20: Output Power of Induction Motor using V/f Control for 5-Level Reversing Voltage Inverter Topology

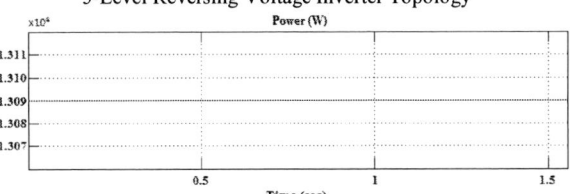

Fig. 21: Output Power of Induction Motor using V/f Control for 2-Level Inverter Topology

The data obtained from above analysis has been effectively summarized in Table III such that a comparison has been made for V/f operation based on 2-Level and 5-Level Inverter drive.

Table 3: Obtained Output Parameters for Different Inverter Topologies of V/f Control

Parameters	MLI	2 Level Inverter
THD	24.11%	35.12%
Fundamental Voltage	345 V	288 V
Conduction Losses	1248 W	312 W
Power Output	13280 W	13090 W
Efficiency	89%	87%

5. For proposed DTC control with 2-Level and Multi-Level VSI fed coupled drive system

As discussed in V/f control, DTC also deals with the control of IM torque by varying frequency of operation, ultimately to have a tab on the speed of motor. This is achieved by requisite estimation of motor's magnetic flux and torque in terms of measured voltage and current quantities as discussed in Section IV.

As seen from Figures 22 and 23, the behavior of torque speed curve for DTC remains the same as with V/f control technique but a prominent difference which separates them is the speed of operation. The transient period is comparatively reduced in this control scheme from 1.5sec to 0.5sec for MLI operation and from 0.6sec to 0.5sec for 2-Level Inverter operation.

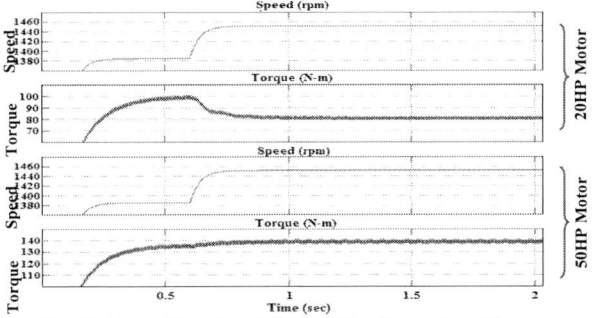

Fig. 22: Output Speed and Torque Waveforms from Two Coupled Motors using Reversing Voltage MLI

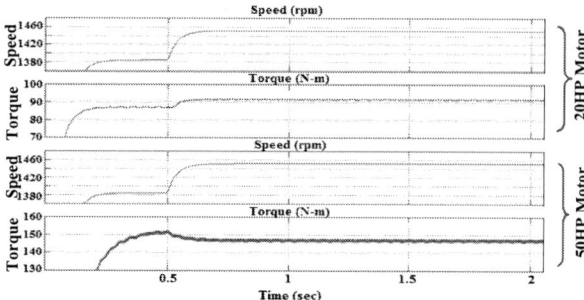

Fig. 23: Output Speed and Torque Waveforms from Two Coupled Motors using 2-Level Inverter

The nature of phase voltage will be unchanged for the chosen inverter topologies except for the magnitude 305V and 293V respectively. For the change in fundamental output voltage, the THD content is bound to change with 29.27% and 37.88% magnitudes as shown in Fig.24. It must also be noted that for the increased feasibility of output power capability upto 13860W, the drive has to compensate with increased filtering requirements since hysteretic current controller may induce some flux errors. Even then the superiority of this control scheme is justified by the fact that overall system efficiency gets a considerable boost with same losses as with V/f control. Table IV summarizes the obtained parameters.

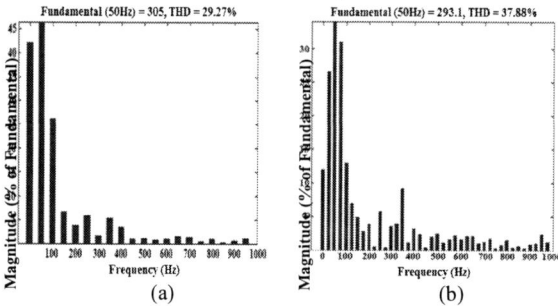

Fig. 24: THD in Output Voltage for (a) Multi Level Inverter and (b) 2-Level Inverter Topologies

Table 4: Obtained Output Parameters for Different Inverter Topologies of DTC Control

Parameters	MLI	2 Level Inverter
THD	29.27%	37.88%
Fundamental Voltage	305V	293V
Conduction Losses	1248 W	312 W
Power Output	13860 W	13000 W
Efficiency	92.89%	87.13%

VI. CONCLUSION

The suggested V/f and Direct Torque control techniques for coupled IM drives with 2-Level/Multi-Level Inverter have been implemented and analyzed in MATLAB Simulink. From the obtained simulation results an extensive comparative study has been carried out for the two schemes based on performance parameters. The comparative analysis clearly depicts that choice of VSI for coupled motor drive has a tremendous influence on its harmonic distortions and output efficiency. Even though a 2-Level Inverter performs decently for coupled drive operation but it is over-shadowed by the superior

performance of 5-Level Inverter where improvements in THD by 11%, fundamental magnitude by 57V, output efficiency by 2% in case of V/f control and THD by 9%, fundamental magnitude by 12V, output efficiency by 6% for Direct Torque Control have been achieved. From the compared magnitudes of performance parameters it is pretty much evident that MLI forms the best choice for implementation in coupled drives with both V/f and Direct Torque controls. Also another aspect which has been brought to fore is regarding the choice of control. It is clear from the simulation study that even though DTC offers 2% (with 2-Level Inverter) and 3% (with 5-Level Inverter) slightly more THD as compared to V/f control but it potentially lifts up the overall system efficiency by almost 3% with corresponding improvement of 60% in transient response time. Thus it is very critical for a designer to strike an efficient balance between the choice of VSI and control strategy keeping in mind the system abnormalities. This simulation study intends to aid control designer in selection of appropriate inverter drive for coupled induction motors to maximize reliability and flexibility of the chosen crane application.

ACKNOWLEDGMENT

Authors would like to sincerely thank respective Managements of MSRUAS, CMRIT and HK PolyU for providing all facilities required to carry out this research work.

REFERENCES

[1] M. Amiri, M. Feyzi and H. Saberi, "A Modified Torque Control Approach for Load Sharing Application using V/F Induction Motor Drives", Proc. IEEE Power Electronics, Drive Systems and Technologies Conference (PEDSTC) Conference, pp.1-6, Feb.2013.

[2] D. Casadei, F. Profumo, G. Serra, A. Tani, "FOC and DTC: Two Viable Schemes for Induction Motors Torque Control", IEEE Transaction on Power Electronics, vol.17, no.5, pp.779-787, Nov. 2002.

[3] Iyer J., Tabarraee K., Chiniforoosh S. and Jatskevich J., "An Improved V/f Control Scheme for Symmetric Load Sharing of Multi-Machine Conference on Electrical and Computer Engineering (CCECE), pp.1487-1490, May.2011.

[4] Aarniovuori L., Laurila L., Niemela, M., Pyrhonen, J., "Measurements and Simulation of DTC Voltage Source Converter and Induction Motor Losses", IEEE Transaction on Industrial Electronics, vol.59, no.5, pp.2277-2287, May 2012.

[5] Mishra R., Singh S., Singh B., Kumar D., "Investigation of Transient Performance of VSI fed Induction motor Drives Using Volt/Hz and Vector Control Techniques", Proc. IEEE Power, Control and Embedded Systems Conference (ICPCES), pp.1-6, Dec.2012.

[6] Baby T., Sabah V., Lall K., Chitra A., "Multi-level Inverter Fed Induction Motor Drive for Pumping Application", Proc. IEEE Int. (TAP Energy), pp.85-92, June 2015.

[7] Curiac R., Li H., "Specific Design Considerations for AC Induction Motors Connected to Adjustable Frequency Drives", Proc. IEEE Pulp6, June 2010.

[8] H. Joshi, P. N. Tekwani, A. Hinduja, "Multilevel Inverter for Induction Motor Drives: Implementation using Reversing Voltage Topology", Proc. IEEE Industrial Power Electronics Conference (IPEC), pp.181-186, Oct 2010.

[9] D. Hartanto, G.H. Vaupel, W. Gansel, G. L. Fischer, "Comparison of Conventional and Frequency Converter Driven Overhead Bridge Crane Control Systems", Proc.

IEEE Power Electronics and Motion Control Conference (PEMCC), pp.940-945, Aug.2000.

[10] I. Daut, K. Anayet, M. Irwanto, N. Gomesh, M. Muzhar and M. Asri, Syatirah, "Parameters Calculation of 5 HP AC Induction Motor", Proc. International Conference on Applications and Design in Mechanical Engineering (ICADME), pp.12B1-12B5, Oct. 2009.

The Comparative Study between Different Performance Indices of a Permanent Magnet Synchronous Motor Drive on Variable Sensor Angle

Atanu Banerjee[1]   Pabitra Kumar Biswas[2]   Chiranjit Sain[3]

1. Electrical Engineering Department, NIT Meghalaya,793003, India
E-mail: atanu_banerjee@nitm.ac.in
2. Electrical & Electronics Engineering Department, NIT Mizoram,796012, India
E-mail: pabitra.biswas2009@gmail.com
3. Electrical Engineering Department, Siliguri Institute of Technology, Siliguri, 734009, India
E-mail: sain.aec@gmail.com

Abstract- **In this paper the development of a model for self-controlled Permanent Magnet Synchronous Motor drive (PMSM) fed from a three phase voltage source bridge inverter which operates under self-control mode with the rotor position information is presented. The rotor position is monitored online via an absolute position encoder and this information is passed on to the inverter controller block for generating switching signals. It is found that with some settings of the position sensor, near-sinusoidal armature phase currents result, which therefore give rise to less torque ripple. The different performance indices of the system like electrical speed, electromagnetic torque, phase voltage, phase current and rotor position are modelled and subsequently simulated for different sensor angle with varying load. Results obtained through such process are compared and analyzed for optimization of sensor angle.**

Keywords- Permanent Magnet Synchronous Motor (PMSM) Drive, Rotor Position, Sensor angle, Voltage Source Inverter (VSI), 180⁰ conduction mode.

I. INTRODUCTION

Permanent Magnet Synchronous Machine (PMSM) is a rotating electrical machine with a balanced three phase armature in its stator and a permanent magnet material in its rotor. Due to the absence of any slip ring-brush arrangement, the maintenance problem does not exist and the ruggedness also increases [1]. Generally, in a PMSM, the nature of air gap flux density distribution and the induced excitation voltages in the stator phase windings produced by the permanent magnet material employed, exhibit sinusoidal waveforms. In this paper, a model is developed for PMSM, when its armature is fed by a three phase two-level transistorized voltage source inverter (VSI), self-synchronized with the rotor position information and operating under 180° conduction of the inverter switches. Permanent Magnet Synchronous motors are increasingly used in variable speed industrial drives [2, 3]. New developments and applications have been greatly accelerated by improvements in permanent magnet materials, especially rare earth magnets. In open loop manner the speed control of such a machine can be done in a way similar to that of a conventional dc

machine-by changing the equivalent conceptual" brush" position by varying the sensor position with respect to the rotor frame, characterized by the variable 'sang' in the model [4].

II. THE SYSTEM DESCRIPTION

Fig. 1 shows the drive system for which the proposed model is developed.

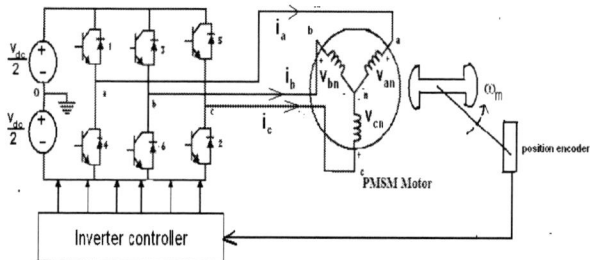

Fig.1: The PMSM drive operating with the inverter operating under 180° conduction

Here, the PMSM three phase armature is fed from a three phase voltage source bridge inverter (operating under 180° conduction) consisting of six self-controlled switches viz. IGBT's with anti parallel diodes. The inverter devices are switched in synchronism with the rotor position information. The rotor position information is assumed available to the inverter by placing, on the rotor, a gray-coded disc. Additionally a set of three infrared emitting diodes (IRED) and a corresponding set of three receptor photodiodes are placed on a stationary frame. The position sensor system, thus formed, is assumed to finally yield three digital electrical pulses, which can be considered as a three bit binary word which goes on changing after each 60°, as the rotor rotates. Thus, the rotor position information after every 60° can be identified [5, 6]. Simple decoder logic uses this information to yield the switching signals of the six self-controlled inverter devices as per the 180° conduction logic. The inverter is conceived to be powered from the DC side from a controllable DC voltage source, V_{dc}, which may be varied to achieve speed control in an open-loop manner, in the same way as armature voltage control, done in a conventional DC machine with mechanical commutator.

III. THE SENSOR POSITION

The concept of the sensor position is all about positioning the gray-coded disc on the rotor and positioning the IREDs with the receptor photodiodes on a stationary frame. Once set, the instant of switching on of a particular armature phase through an inverter device can be synchronized with

978-1-5090-0064-7/15 $31.00 © 2015 IEEE

a particular position of the rotor permanent magnet in space. The sensor position is defined through an angle variable 'sang' which is defined to be zero as shown in Fig. 2 when the averaged permanent magnet field MMF space vector (averaging done over each 60° interval for which a particular set of three inverter devices are on as per 180° conduction logic) is in space quadrature with respect to the armature voltage space vector [7, 8].

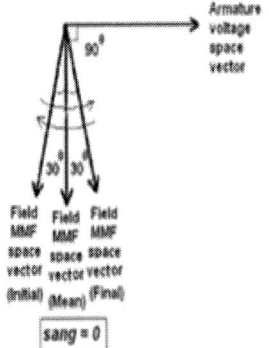

Fig. 2: Illustration of the concept of the sensor position for the sensor angle (sang) = 0°.

The corresponding phasor diagram for that instant of inverter switching is shown in Fig. 3.

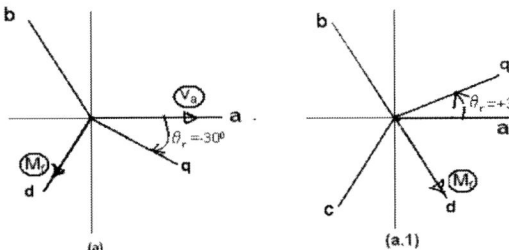

Fig 3 : Illustration of the Field MMF (Mf) and the armature voltage (Va) space vectors with different switching combinations (a) when 6, 1,2 are ON (5 has just become off and 6 has just become on) and rotor position θr=-30° (a.1) When 6,1,2 are still ON (5 is off and 6 is on) and rotor position θr=+30°

The switching of the inverter would thereafter continue in a similar way after each 60° interval from the last rotor position with phase displacement of 30° electrical. This would have been possible by adjusting the rotor position encoder system. The value for sang, the sensor lead angle, for this case would therefore be defined as 30° lead or −30° as depicted in the Fig. 4 and the corresponding phasor diagram is also shown in Fig.5.

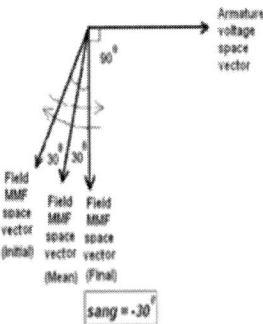

Fig. 4: Illustration of the concept of the sensor position for the sensor angle (sang) = -30°.

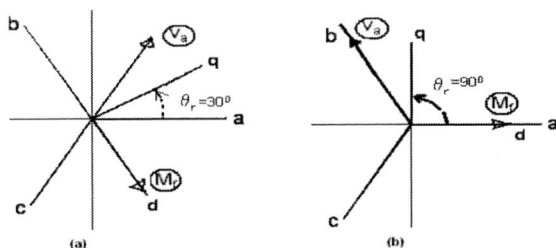

Fig 5: Illustration of the Field MMF (Mf) and the armature voltage (Va) space vectors with different switching combinations (a) when 1,2,3 are ON (6 just off and 3 just on) and rotor position θr = 30° (b) When 2, 3, 4are ON (1 just off and 4 just on) and rotor position θr =90°

In a similar way, any value of sang may be realized (negative means 'leading values' and positive means 'lagging values'). It can be concluded that, as the average space angle for different sensor lead angles are going to be different, the torque profiles would be different for each sensor lead angle. Because the inverter operates under 180° conduction mode, the actual instantaneous field MMF with respect to the armature voltage space vector, however, oscillates within ±30° about its own average space position. In the developed model here, the performance of the PMSM drive can be tested for different values of 'sang'.

IV. THE DEVELOPED MODEL OF PMSM DRIVE

The overall structure of PMSM drive can comprise of three (03) main components i.e. The Inverter block, abc-dq transformation block and the machine block. This model of PMSM can be derived by assuming the inverter to be operated 180° conduction or under sinusoidal pulse width modulation (SPWM).

1. 3-Phase Voltage Source Inverter

In voltage source inverter, a dc voltage source with very small internal impedance is used as input of the inverter. The dc side terminal voltage is constant, but the ac side output voltage may be constant or variable irrespective of load current. For providing adjustable frequency power to industrial applications, three phase inverter are more used than single phase inverter [9]. It uses a minimum of six thyristors. For one cycle of 360°, each step should be of 60° interval for six step inverter. In inverter terminology, a step is defined as a change in the firing from one thyristor to the next thyristor in proper sequence. There are two patterns of gating the thyristor. In one pattern, each thyristor conducts for 180° and in the other, each thyristor conducts for 120°. But in both these patterns, gating signals are applied and removed at 60° intervals of the output voltage waveform [3]. The rotor position is monitored online via an absolute position encoder and this information is passed on to the inverter controller block. This inverter controller block process this information and accordingly generates the switching signals for the inverter [10, 11].

978-1-5090-0064-7/15 $31.00 © 2015 IEEE

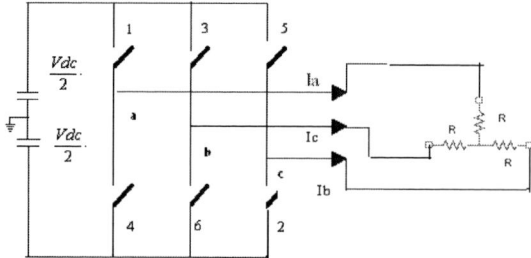

Fig. 6: Schematic Diagram of Three Phase VSI

It is to be noted that whenever the inverter operates under 180° conduction or under SPWM at each time instant the following are true:

i)   Either one switch of the upper three inverter devices and two switches out of the lower three inverter devices are ON simultaneously.

ii)   OR two switches of the upper three inverter devices and one switch of the lower three inverter devices are ON simultaneously.

2. The abc-dq Transformation Block
The Permanent Magnet Synchronous Machine with sinusoidal back EMF is analysed and modelled using d-q reference frame theory [12-15]. The d-q reference frame theory is well-known, where a three phase system (called a-b-c frame of reference) of current or voltage or flux linkage is transformed into an equivalent two phase system (called d-q reference frame, actually d-q-0 frame but zero sequence neglected finally, assuming balanced operation) of current or voltage or flux linkage, by means of a transformation relationship, known as Park's transformation.

$$i_a = \left(i_{qs} \cos \theta_r + i_{ds} \sin \theta_r \right) \tag{1}$$

$$i_b = \left[i_{qs} \cos(\theta_r - 2\pi/3) + i_{ds} \sin(\theta_r - 2\pi/3) \right] \tag{2}$$

$$i_c = \left[i_{qs} \cos(\theta_r + 2\pi/3) + i_{ds} \sin(\theta_r + 2\pi/3) \right] \tag{3}$$

$$\begin{bmatrix} V_d \\ V_q \\ V_0 \end{bmatrix} =$$

$$\sqrt{\frac{2}{3}} \begin{bmatrix} \cos \theta_r & \cos\left(\theta_r - \frac{2\pi}{3}\right) & \cos\left(\theta_r + \frac{2\pi}{3}\right) \\ -\sin \theta_r & -\sin\left(\theta_r - \frac{2\pi}{3}\right) & -\sin\left(\theta_r + \frac{2\pi}{3}\right) \\ \frac{1}{\sqrt{2}} & \frac{1}{\sqrt{2}} & \frac{1}{\sqrt{2}} \end{bmatrix} \begin{bmatrix} V_{an} \\ V_{bn} \\ V_{cn} \end{bmatrix} \tag{4}$$

3. The Machine
The Permanent Magnet Synchronous Machine is analyzed on the basis of "D-Q axes rotor reference frame theory". The 'Machine' block of the developed model accepts the D-axis (V_{ds}) and Q-axis (V_{qs}) components of the PMSM's stator (armature) voltage, the instantaneous rotor position and the load torque details (T_l, f_n) as inputs and computes the stator currents and the stator flux linkages in d-q frame the electromagnetic torque (T_e), the rotor speed (θ_r). Fig 4 represents the geometrical orientation of a PMSM

motor having sinusoidal air gap flux distribution along with direction of flux and torque producing component [16,17].

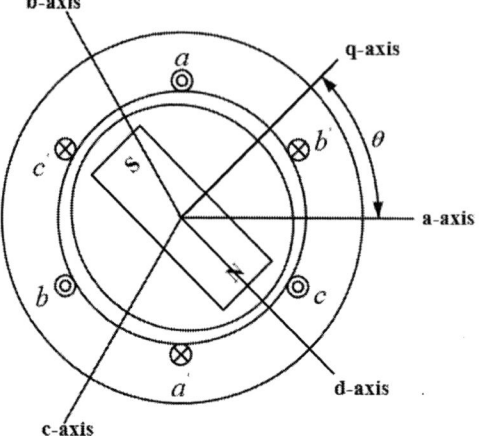

Fig. 7: Permanent magnet synchronous motor

The necessary machine equations are summarized as follows, where symbols have their usual meanings:

$$\omega_r = \frac{P}{2} \omega_m \tag{5}$$

$$\theta_r = \frac{P}{2} \theta_m \tag{6}$$

$$T_{em} = \frac{d\omega_m}{dt} + f_n \omega_m + T_l \tag{7}$$

$$\theta_r = \int_0^t \omega_r (\xi) d\xi + \theta_r(0) \tag{8}$$

V. THE PERFORMANCE EVALUATION

Performance of the PMSM can be checked for deferent values of sensor angles by changing 'sang'. The sensor angle is used as a node of PMSM template which has been taken as a main variable. The different performance indices of the PMSM drive i.e. electrical speed, electromagnetic torque, phase voltage, phase current, q-axis current and the rotor position information have been represented for different values of sensor angles by changing 'sang'. The results shown here for sang=0° and sang= -30°.

1. Comparative Performance Analysis (No-Load Operation)
The following responses are obtained for the analysis of PMSM drive under different operating conditions. A step voltage of 144V is applied at the dc link of the VSI of the PMSM drive for a sensor lead angle is 0° and -30° keeping all other conditions unaltered. In Fig. 3 with sensor angle "sang"=0°, the electrical speed is found to settle finally to a steady state value after starting from zero and no tendency of instability is observed. In the second case i.e. when 'sang'= -30°, the magnitude of electrical speed increases with respect to time and desired steady state is reached faster than previous case. This variation of sensor

978-1-5090-0064-7/15 $31.00 © 2015 IEEE   124

angle is an important parameter to control the speed of the machine in open loop manner.

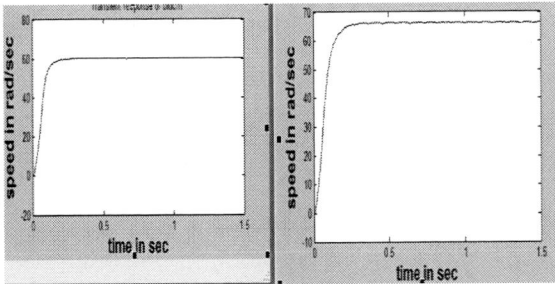

Fig. 8: Speed Vs time response when 144V is applied with "sang"=0° and -30°, T_l =0, Fn =0.45 Nm-sec/rad

The electromagnetic torque in Fig. 4 with sensor angle 'sang'=0°, is initially found higher for high currents but finally settles to a steady pattern having ripples with a frequency corresponding to the six times of the rotor speed. It is found that when sensor lead angle "sang" is changed from 0° to -30°, the average electromagnetic torque increases manifold, which causes the machine to run at increased steady state speed. This type of response decreases the peak to peak torque ripples as a percentage of the average electromagnetic torque at steady state also getting reduced, which is advantageous.

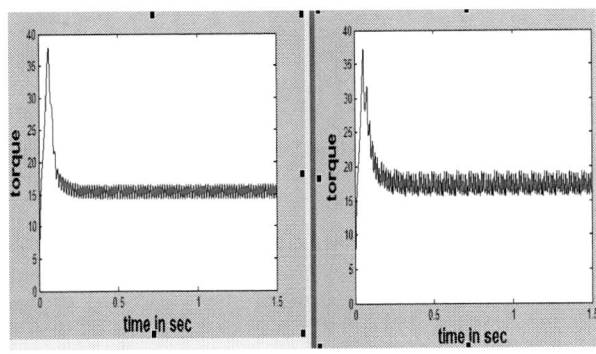

Fig. 9: Electromagnetic torque Vs time response when 144 V is applied with "sang"=0° and -30°, T_l =0. F_n =0.45 Nm-sec/rad

The q-axis current response in Fig. 5 resembles almost similar characteristics like electromagnetic torque or interaction torque developed by the machine. It is observed that with the change in sensor angle, the quadrature axis current tries to communicate with the new value of sensor angle. In this manner the ripple component is also getting reduced as compared to the previous one, which is also a sign of improved operation.

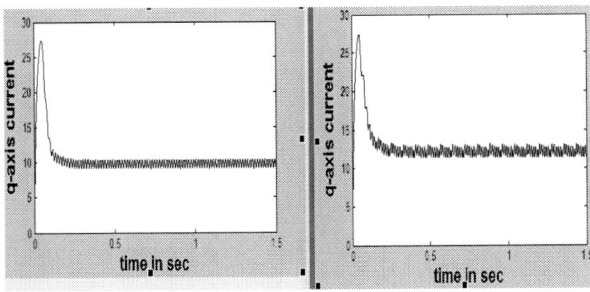

Fig. 10: q-axis Current Vs time response when 144 V is applied with "sang"=0° and -30°, T_l =0, F_n =0.45 Nm-sec/rad

It is also observed in Fig. 6 that with the different switching combination of the inverter, space angle between the filed mmf and the armature mmf is getting changed so that the rotor is very much sensitive with the position information. Hence it can be concluded that with the change in sensor lead angle from 0° to -30°, due to the different torque profiles, the rotor positions can be controlled which is advantageous for adjustable speed drives applications.

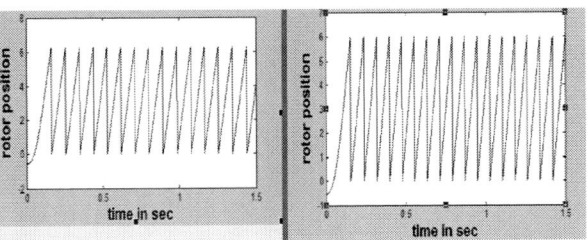

Fig. 11: Rotor position Vs time response when 144 V is applied with "sang"=0° and -30°, T_l =0, F_n =0.45 Nm-sec/rad

2. Comparative Performance Analysis (On-Load Operation)

The following responses are obtained for the analysis of PMSM drive under different operating conditions i.e. when a step voltage of 144V is applied at the dc link of the VSI of the PMSM drive for a sensor lead angle is 0° and -30° and a load torque of 5 N-m is applied to the machine keeping all other conditions unaltered. In this study different performance characteristics of the machine like speed, torque, phase voltage, phase current, q-axis current, rotor position are analyzed for different values of sensor lead angle with sudden load variation. Thus a comparative study may be carried out on PMSM drive with the load variation with the corresponding changes in the sensor lead angle. It is found that when the sensor lead angle is changed from 0° to −30°, with the load applied the armature phase current wave shape is drastically improved, resembling a pure sinusoid much more. The average electromagnetic torque therefore increases manifold, which causes the machine to run at an increased steady state speed compared to the previous case. For the same reason the peak to peak torque ripple as a percentage of the average electromagnetic torque at steady state also decreases, which is an advantageous feature. The rotor position information with corresponding changes in the sensor lead angle, on-load operation is also sensitive in this case. Hence a typical control algorithm may be implemented on PMSM drive to monitor its performance characteristics under different operating conditions.

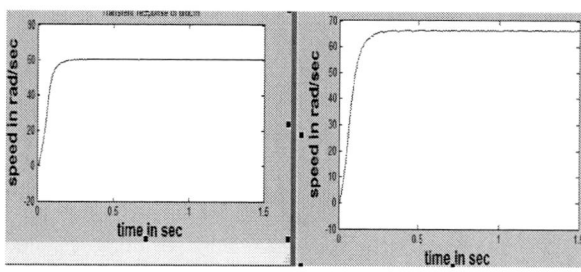

Fig. 12: Speed Vs time response when 144V is applied with 'sang'=0° and -30°, T$_l$ =5N-m, F$_n$ =0.45 Nm-sec/rad

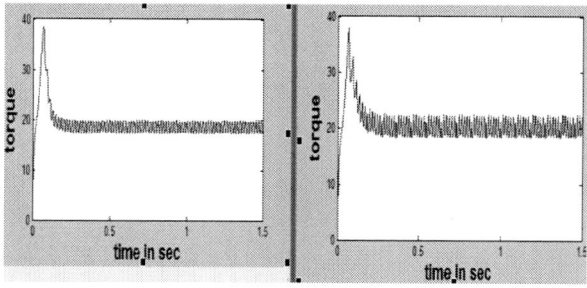

Fig. 13: Electromagnetic torque Vs time response when 144 V is applied with 'sang'=0° and -30°, T$_l$ =5N-m. F$_n$ =0.45 Nm-sec/rad

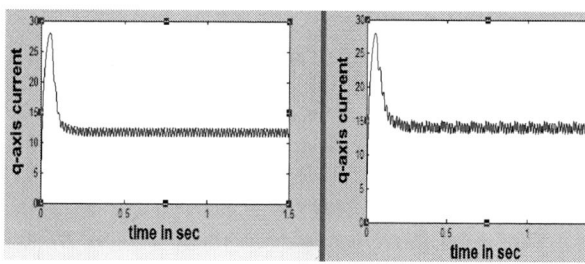

Fig. 14: q-axis Current Vs time response when 144 V is applied with "sang"=0° and -30°, T$_l$ =5N-m, F$_n$ =0.45 Nm-sec/rad

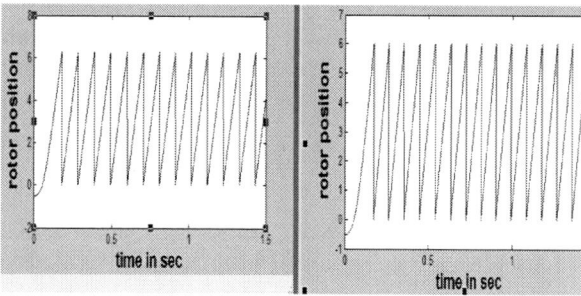

Fig. 15: Rotor position Vs time response when 144 V is applied with 'sang'=0° and -30°, T$_l$ =5N-m, F$_n$ =0.45 Nm-sec/rad

VI. CONCLUSION

In this manuscript an extensive comparative study between different parameters of a permanent magnet synchronous motor drive operated through self synchronous 180° conduction mode bridge inverter has been significantly described. Effect of change in sensor angle "sang" an important factor on which performance of such drive will be responsible. Setting of position sensor to achieve this particular "sang", for which the armature currents are near

sinusoidal reduces the torque to weight ratio a bit, which in turn helps to develop the simple control circuit of the inverter. Changing of sensor angle on PMSM drive is tested and monitored on no load and on load operation. Hence the PMSM drive is very much sensitive with respect to the change in sensor angle with the corresponding load variation. With the help of this concept closed loop model of the adjustable speed PMSM drive has been tested and verified for performance optimization.

APPENDIX

The PMSM for which the model is developed in this paper has the following parameters: p=4, Rs = 3.2 ohm, J= 0.061 kg-m^2, damping co-efficient F$_n$= 0.45 Nm-s/radian, L$_d$=0.053H, Lq= 0.041H, Load torque T$_l$=0 and 5 N-m

REFERENCES

[1] R. Krishnan," Electric Motor Drives- Modeling, Analysis and Control" , PHI, New Jersey, 2001
[2] Bimal K. Bose, " Modern Power Electronics and AC Drives", Prentice Hall, 2001
[3] M.H Rashid, Power Electronics circuits, Device and Applications, prentice hall, 1993.
[4] P. C. Krause, "Analysis of Electric Machinery", McGraw Hill, New York, 1987
[5] F. De Belie, J. Melkebeek" Seamless Integration of a Low-Speed Position Estimator for IPMSM in a Current-Controlled Voltage-Source Inverter" IEEE Int. Conf., Sensorless Control for Electrical Drives (SLED), pp.50-55, 9-10 July 2010
[6] Gunawan Dewantoro and Yong-lin Kuo," Robust speed control of Permanent Magnet Synchronous Motor drive using fuzzy logic controller" Proc. IEEE, pp. 27-30, June 2011
[7] T. J. Vyncke, S. Thielemans, M. Jacxsens, and J. A. Melkebeek," Analysis of some design choices in model based predictive control of ying-capacitor inverters, COMPEL": The International Journal for Computation and Mathematics in Electrical and Electronic Engineering, vol. 31, no. 2, p. 619-635, Feb. 2012
[8] Merzoug M.S. and Naceri F., "Comparison of Field Control and Direct Torque Control for PMSM, Proceedings of World Academy of Science, Engineering and Technology, Vol. 35, Nov 2008
[9] Rahideh A., Rahideh A., Karimi M., Shakeri A., Azadi M., "High Performance Direct Torque Control of a PMSM using Fuzzy Logic and Genetic Algorithm", IEEE International Conference on Electric Machines and Drives System, pp 932 – 937, May, 2007
[10] Jong-Sun Ko, Hyunsik Kim, and Seong-Hyun Yang, "Precision Speed Control of PMSM Using Neural Network Disturbance Observer on Forced Nominal Plant", IEEE Transactions of Electrical Engineering, Vol. 3, pp-1746-1752, June 2004
[11] X. Lin-Shi, F. Morel, A.M. Llor, B. Allard, and J.M. Rtif, "Implementation of Hybrid Control for Motor Drives," IEEE Trans. on Ind. Electron., Vol. 54, No. 4, Aug. 2007
[12] M. Kojima, K. Hirabayashi, Y. Kawabata, E.C. Ejiogu, T. Kawabata, "Novel vector control system using Deadbeat controlled PWM inverter with output LC filter," IEEE Transactions on Industry Applications, vol. 40, no. 1, pp. 162-169, 2004
[13] Al-Nabi E, Wu B, Zargari NR, Sood V. "Input power factor compensation for high-power CSC fed PMSM drive using

d-axis stator current control." IEEE Transactions on Industrial Electronics, pp. 59(2):752–761, 2012.

[14] Bose Bimal K., "Neural Network Applications in Power Electronics and Motor Drives-An Introduction and Perspective", IEEE Transactions on Power Electronics, Vol. 54, No. 1, Feb 2007

[15] Bhim Singh Fiete, B.P. Singh and Sanjeet Dwivedi, "DSP Based Implementation of Vector Control Scheme For Permanent Magnet Synchronous Motor Drive" IETE Journal of Research, Volume 53, Issue 2, pages 153-164,March 2007.

[16] J. J. H. Paulides, G. W. Jewell, and D. Howe, "An Evaluation of Alternative Stator Lamination Materials for a High – Speed, 1.5 MW, Permanent Magnet Generator", IEEE Trans. On Magnetics, Vol. 40, No. 4, pp. 2041 – 2043, July 2004

[17] Marufuzzaman M, Reaz MBI, Ali MAM, Rahman LF. "Hardware approach of two way conversion of floating point to fixed point for current dq PI controller of FOC PMSM drive" Electronics and Electrical Engineering. 7(123): pp.79–82, 2012.

Machine-Drives Systems, Power Electronics Converters, Soft-computing analysis etc.

BIOGRAPHIES

Atanu Banerjee was born in Asansol, West Bengal, India. He received his B.E. degree in Power Electronics Engg. from the Nagpur University in the year of 2001 and M.E in Electrical Engg. Department with specialization in Power Electronics & Drives in 2008 from Bengal Engineering & Science University, Shibpur (Now IIEST, Shibpur). He has completed his Ph.D in Electrical Engg from the Indian School of Mines, Dhanbad, India in 2013. He worked in industries for almost three years & has academic experience of more than 12 years. Presently he is in National Institute of Technology, Meghalaya as an assistant professor in the Electrical Engg. Department. His research interests include induction heating and high frequency switching in power electronic converters, adjustable speed drives. He has published few books & several journal/conference research papers. Also Dr. Banerjee has filed two patents to the Govt. of India. Currently he is guiding few research scholars for M-Tech & Ph.D

Pabitra Kumar Biswas was born in West Bengal, India in 1980. He completed his B.Tech from Asansol Engg. College, WBUT, India. He received his ME. Degree from Bengal Engineering and Science University, West Bengal, India and PhD. Degree in Electrical Engineering from National Institute of Technology, Durgapur, India. He is presently working as an Assistant Professor in Electrical Engineering in National Institute of Technology, Mizoram, India. He has published a numbers of research papers in National/International Conference Records/Journals. From 16.07.2007 to 05.02.2015, he served as an Assistant Professor in Electrical Engineering in Asansol Engineeng College, Asansol, India. His research interests include Electromagnetic Levitation System, Active Magnetic Bearing and Power electronics.

Chiranjit Sain received B.Tech in Electrical Engineering from Asansol Engineering College, Asansol, India and also received M.Tech from National Institute of Technical Teachers Training and Research, (NITTTR) Kolkata, India in the specialization of Mechatronics Engineering under Electrical Engineering department. Presently he is working as an Assistant Professor in the Electrical Engineering Department at Siliguri Institute of Technology, Siliguri India since 2010 and also pursuing Ph.D in the Electrical Engineering Department from National Institute of Technology, Meghalaya, India. His present research of interests includes Electrical

High Power High Voltage Gain Interleaved DC-DC Boost Converter Application

Jiexun Liu   Dawei Gao   Yue Wang

State Key Laboratory of Automotive Safety and Energy
Tsinghua University, Beijing, China
Email: liu-jx12@mails.tsinghua.edu.cn

Abstract—DC-DC converter is usually an important part in many industrial applications. Especially, high power and high voltage gain DC-DC converter is now required in Fuel Cell Vehicle applications. This paper presents a high power high voltage gain interleaved DC-DC converter. It is a kind of non-isolated boost converter, which will raise a DC voltage of 120 V to a voltage of 700 V. The rating power of the converter may be at 60 kW. There are four modules of DC boost converter in this converter topology. All four modules are connected in parallel and controlled by 90 phase shifting degree to each other. The converter topology will be simulated by MATLAB/SIMULINK. The classical PI controller is selected for each current loop. Experiments on this converter are conducted based on a novel test bench, which uses a boost converter and a buck converter to form an energy cycle. Then experimental results will be studied in the lab for this converter.

Keywords—DC-DC Converter, Fuel Cell Vehicle, High Power, High Voltage Gain

I. INTRODUCTION

The polymer electrolyte membrane fuel cell (PEMFC) is now a hot research point since it is regarded as one of the most important future energy sources. DC-DC converter is a key part of Fuel Cell Vehicles [1], because it controls the output power of the Fuel Cell fast and smoothly. Besides the basic functions, high-reliability, high-efficiency, high-power density, and high-stability of the converter are also common requirements for all applications. Therefore, design of DC-DC converters that have step-up or step-down characteristics and meet the necessary requirements has important practical significance.

The basic topologies of DC-DC converters are classified as isolated and non-isolated [2]. The isolated DC-DC converter cannot achieve as high efficiency as non-isolated DC-DC converter because of the transformer in the converter. Therefore, non-isolated DC-DC converter is more often used in the industrial applications. The two basic topologies for non-isolated DC-DC converter are Boost and Buck converter. Others are mostly based on these two topologies. For high power high voltage gain applications, since the classical boost converter cannot reach so high voltage gain, dual boost DC-DC converter [3-6] is considered in this paper. To reduce the current ripple at the input side and the voltage ripple at the output side, interleaving technique is used in this dual boost

DC-DC converter. The phase shifting control should also be considered according to the interleaving technique. Here, a 4-phase interleaved step-up converter is studied. Moreover, classical PI controller is selected in the current loop. The control is implemented by Digital Signal Processing (DSP, here is TMS320F28335) Board to

control the duty cycle of the Pulse Wide Modulation (PWM) signal. Finally, conclusions on the high power high voltage gain DC-DC converter are made based on simulation by MATLAB/SIMULINK and experimental results in the lab.

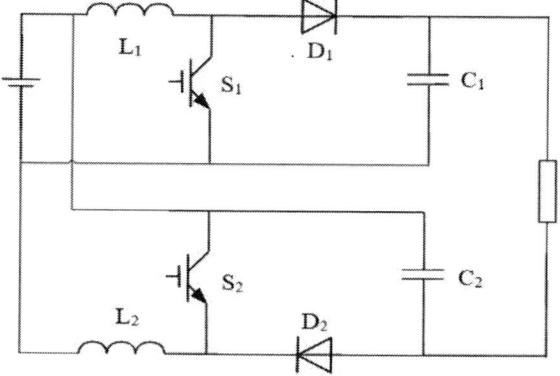

Fig. 1: Dual boost converter

II. INTERLEAVED DUAL BOOST CONVERTER

Classical boost converter can step up input voltage, but it cannot meet the demand of this experiment, which is to step up the input voltage from 120 V to the output voltage of 700 V. Because the IGBTs cannot work in totally open state in boost converter, that is to say, there is a limit for the duty cycle of PWM signal. Therefore, the dual boost converter is considered to achieve the high voltage gain.

1. Circuit Structure

The dual boost converter topology is shown in Fig. 1. The module above is composed of inductor L_1, switching device S_1 (here is IGBT FF450R12KT4), diode D_1 and capacitor C_1. And the module below is composed of L_2, S_2, D_2 and C_2. The input side of the converter connect in parallel sharing the same power supply, which is a DC source with the input voltage of 120 V. The output side of the converter connect in series, so the overall output voltage is equal to the summation of the output voltages of the two boost converters minus the input voltage.

To reduce the component size, reduce the current and voltage ripple and increase the converter efficiency, interleaving technique is usually used in DC-DC converter [7-9]. The high power high voltage gain DC-DC boost converter proposed in this paper is a combination of two 2 phase interleaved boost converter together as shown in Fig. 2. Such circuit is called as 4 phase interleaved DC-DC boost converter [1]. The four switching devices are controlled by 90 phase shifting degree to each other

978-1-5090-0064-7/15 $31.00 © 2015 IEEE   128

simultaneously, which is called the interleaving technique method, in order to smooth output current ripple, raise rating power and efficiency as described above [10-11].

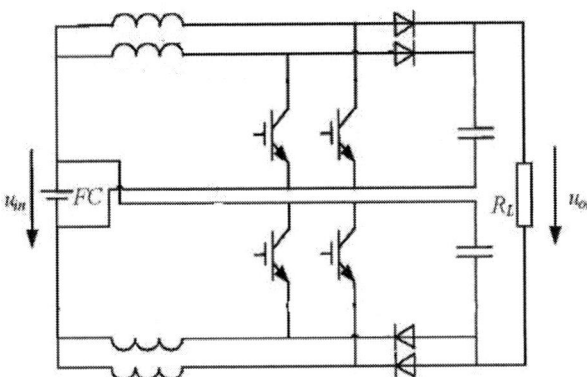

Fig. 2: Interleaved dual boost converter

2. Basic relations

Based on Fig. 1, the voltage of C_1 can be calculated as:

$$U_{C1} = U_{in} \cdot \frac{1}{1-D} \qquad (1)$$

In the same way, the voltage of C_2 can be calculated as:

$$U_{C2} = U_{in} \cdot \frac{1}{1-D} \qquad (2)$$

The relation between the input voltage, the output voltage and the voltage of C_1 and C_2 is:

$$U_{C1} + U_{C2} = U_{in} + U_{out} \qquad (3)$$

Take (1) and (2) into (3), then the relation between the input voltage and the output voltage can be achieved:

$$U_{out} = U_{in} \cdot \frac{1+D}{1-D} \qquad (4)$$

where U_{in} is the input voltage, U_{out} is the output voltage, U_{C1} is the voltage of capacitor C_1 and U_{C2} is the voltage of capacitor C_2. D is the duty cycle of PWM signal.

The inductor size can influence the input current ripple, but it is determined by many factors of the converter. With a maximum current ripple, the inductance can be determined as follows:

$$L = \frac{U_{in} \cdot D}{4 \cdot \Delta I \cdot f} \qquad (5)$$

where L is the inductance, ΔI is the maximum input current ripple and f is the frequency of the switching devices, here is the IGBT.

The output voltage ripple depends on the size of capacitor. However, there are two capacitors that connect in series, and both of them have an effect on the output voltage ripple. The capacitance depends on output current, duty cycle and also the output voltage, switching frequency. The equation of capacitance is as follows:

$$C = \frac{I_{out} \cdot D}{2 \cdot \Delta U_{out} \cdot f} \qquad (6)$$

where I_{out} is the output current of the converter, and ΔU_{out} is ripple of the output voltage.

The inductor current is the feedback signal of the control system, and the input current of the converter is the target signal of the control system. Relationship between these two currents can be obtained as follows:

$$I_L = \frac{I_{out} + I_{in}}{N} \qquad (7)$$

where I_L is the inductor current, and N is the number of the single cell converter in the proposed converter topology.

III. CONTROL SYSTEM DESIGN

1. Control Scheme

The overall control scheme of interleaved dual boost converter is shown in Fig. 3. The inductor current of the converter measured by Hall sensor is chosen as feedback signal, and it is sent to DSP to be converted to digital signal. The input current of the converter is considered as the target signal in this paper. Inductor target current is calculated by (7). Error signal e which is used as the input of the PI controller can be achieved through comparing the feedback signal with the inductor target signal. Then the duty cycle of PWM signal is obtained from the result calculated by PI controller. So the drive signals of the IGBTs are finally obtained. Since the interleaving technique is used in the converter, the drive signals of the IGBTs should have 90 phase shifting degree to each other, but the duty cycle of PWM signal for the four IGBTs should be the same.

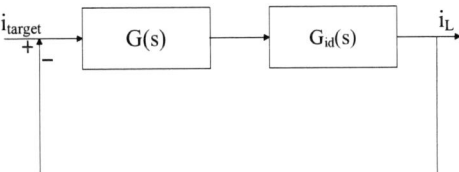

Fig. 3: Control scheme of the converter

2. Control System Design

In order to simplify the analysis of the control system of the dual boost converter, several necessary assumptions are needed. There is no time delay for switch devices to turn on or off and no conductive losses. The inductor is an ideal element and work in unsaturated regions. Moreover, the parasitic resistance and leakage inductance are assumed to be zero. The capacitor equivalent series resistance is also assumed to be zero and the ratio of ripple to output voltage is extremely small so that the steady-state capacitor voltage can represent the dynamic voltage. And the path transmission loss is also ignored.

978-1-5090-0064-7/15 $31.00 © 2015 IEEE   129

Based on these assumptions, the state-space average model of the dual boost converter with time averaging method can be described by continuous functions, which is linearized to equations as follows:

$$L_1 \frac{d\hat{i}_{L1}(t)}{dt} = u_{in}(t) - (1-D)\hat{u}_{C1} + U_{C1}\hat{d}(t) \quad (8)$$

$$L_2 \frac{d\hat{i}_{L2}(t)}{dt} = u_{in}(t) - (1-D)\hat{u}_{C1} + U_{C1}\hat{d}(t) \quad (9)$$

$$L_3 \frac{d\hat{i}_{L3}(t)}{dt} = u_{in}(t) - (1-D)\hat{u}_{C2} + U_{C2}\hat{d}(t) \quad (10)$$

$$L_4 \frac{d\hat{i}_{L4}(t)}{dt} = u_{in}(t) - (1-D)\hat{u}_{C2} + U_{C2}\hat{d}(t) \quad (11)$$

$$C_1 \frac{d\hat{u}_{C1}(t)}{dt} = (1-D)\hat{i}_{L1} - I_{L1}\hat{d}(t) + (1-D)\hat{i}_{L2} - I_{L2}\hat{d}(t) - \hat{i}_o(t) \quad (12)$$

$$C_2 \frac{d\hat{u}_{C2}(t)}{dt} = (1-D)\hat{i}_{L3} - I_{L3}\hat{d}(t) + (1-D)\hat{i}_{L4} - I_{L4}\hat{d}(t) - \hat{i}_o(t) \quad (13)$$

where X represents nominal parameters, and $\hat{X}$ represents parameter variations.

Based on the equations above, control system of the converter in this paper can be designed in more details and finally transfer function of the control system can be obtained which is useful for design of PI controller.

In this control system, parameters in every single cell of the dual boost converter are assumed to be the same value, that is as follows:

$$L_1 = L_2 = L_3 = L_4 = L \quad (14)$$

$$C_1 = C_2 = C \quad (15)$$

Based on these assumptions, the voltage on capacitor C_1 and C_2 can be considered the same, the symmetry of the converter ensures that:

$$U_{C1} = U_{C2} = U_C \quad (16)$$

Using the same current reference for the current of each single cell, i.e.

$$i_{L1} = i_{L2} = i_{L3} = i_{L4} = i_L \quad (17)$$

Also, duty cycle in each single cell is the same, i.e.

$$d_1 = d_2 = d_3 = d_4 = d \quad (18)$$

From (8) – (18), the relation between duty cycle of PWM signal and inductor current can be obtained. So the transfer function of the control system is as follows [1]:

$$\frac{\hat{i}_L(s)}{\hat{i}_{LREF}(s)} = \frac{G(s)G_{id}(s)}{1 + G(s)G_{id}(s)} \quad (19)$$

where

$$G(s) = K_p + \frac{K_i}{s} \quad (20)$$

$$G_{id}(s) = \frac{\hat{i}_L(s)}{\hat{d}} = \frac{K_{id}(T_Z s + 1)}{\left(\dfrac{s}{\omega_n}\right)^2 + \dfrac{2\zeta}{\omega_n}s + 1} \quad (21)$$

And:

$$\begin{cases} K_{id} = \dfrac{I_L}{(1-D)} \\ T_Z = \dfrac{U_C \cdot C}{(1-D)I_L} \end{cases} \quad and \quad \omega_n = \sqrt{\dfrac{(1-D)^2}{L \cdot C}} \quad (22)$$

K_i and K_p are the parameters in PI controller, which are set in the experiment to obtain a steady output of the Fuel Cell.

IV. EXPERIMENTS

1. Prototype Design

A 60 kW prototype of interleaved dual boost converter is developed. It is used for the power system of Fuel Cell Vehicle. The input of the converter is Fuel Cells. The rated power of Fuel Cells is about 60 kW, and its output voltage is

Fig. 4: 3D Prototype of the converter

120 V. The output of the converter connects with batteries and motors. The batteries are used as auxiliary power source. When the output voltage is higher than batteries, batteries are charged. When the output voltage is lower

than batteries, batteries are discharged and both the fuel cells and batteries work as power sources. The motor is used to drive the vehicle. The details of the specifications are presented in Table I. Fig. 4 shows a photograph of 3D prototype of the converter and Fig. 5 shows the prototype of the converter including the DSP control board. TMS320F28335 is chosen as the control chip and the control system can provide four drive outputs for IGBTs.

Table 1: Prototype specifications

Input Voltage	120 V
Output Voltage	700 V
Power	60 kW
Switches and Diodes	FF450R12KT4
Inductors	L=350 µH
Capacitors	C=140 µF/1000 V

2.Test Bench

The large current in this experiment exceed the range of the equipment in our lab. Also to save the energy used in this experiment, a novel test bench is designed in this experiment. There are two converters in this experiment. One is the interleaved dual boost converter, the other is a buck converter. The input DC voltage is stepped up to an output voltage of 700 V through the interleaved dual boost converter, and then the output side of boost converter is connected to the input

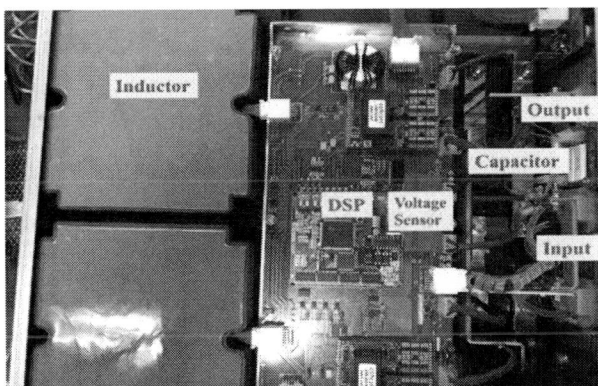

Fig. 5: Prototype of the interleaved dual boost converter

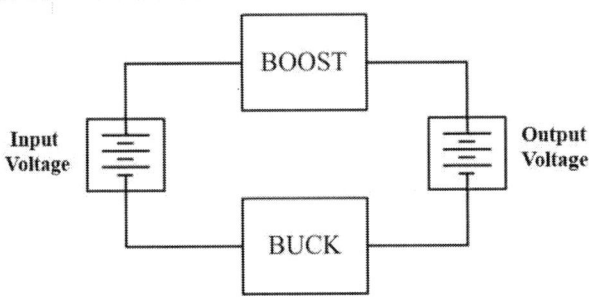

Fig. 6: Scheme of test bench

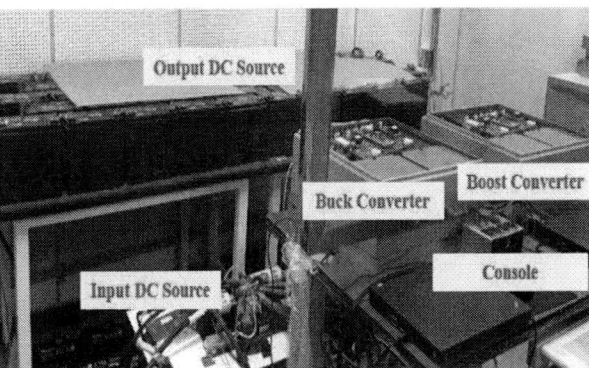

Fig. 7: Prototype of the test bench

side of the buck converter, and the input side of boost converter is connected to the output side of buck converter. So an energy cycle is formed and large current can be obtained in this experiment. The scheme of the test bench is shown in Fig. 6 and the prototype of the test bench is shown in Fig. 7.

The boost converter is controlled based on the input current, and the buck converter is controlled based on the output voltage. This is because the input voltage of the boost converter, should be steady. Both of the two converters are using PI controllers, but the values of PI controller parameters are set respectively. All the values of parameters and state of the converter in the experiment can be obtained through the

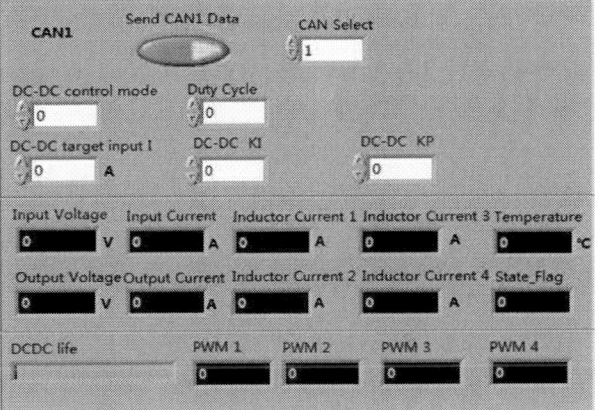

Fig. 8: Control window of the experiment

control window which is shown in Fig. 8. The purpose for PI controller is to reduce the ripple of the current and voltage to obtain a steady output power of the converter.

According to the dual boost converter topology, parameters of the components in the converter can be calculated based on the circuit principle. And to improve the efficiency of the converter and reduce the current and voltage ripple, the parameters of the components can be determined eventually.

V. PERFORMANCE EVALUATION

Firstly, the converter prototype is implemented by MATLAB/SIMULINK. Fig. 9 shows output voltage and capacitor voltage of the converter. It indicates that voltages of C_1 and C_2 are the same. The summation of the two capacitor voltages equals to the summation of input and output voltage. The ripple of output voltage is smaller than the capacitor voltage. This is because of the interleaving technique.

To compare the inductor current with the input and output current of the converter, different current curves are plotted in Fig. 10, in which the relation between input current, output current and inductor current is shown. It indicates that the inductor current in four cells has 90 phase shifting degree to each other. The frequency of switching device is 20 kHz. The input current ripple shown in Fig. 10 is smaller than the ripple of inductor current. This is also because of the interleaving technique. Fig. 11 shows the inductor current in the experiment. It demonstrates that the converter works in Continuous Conduction Mode (CCM) and the inductor current ripple is in a normal range. Experimental result matches the implemented results.

Fig. 12 shows change of the inductor current when the target input current of the converter step up from 200A to 300A. As can be seen, the inductor current does not have an obvious overshoot, and it has a relatively fast response in a short time. This is because the property of the Fuel Cell needs for this application. The Fuel Cell demands that an overshoot should absolutely be avoided and a slow response can be accepted. That is to say, values of PI controller parameters is suitable for this control system.

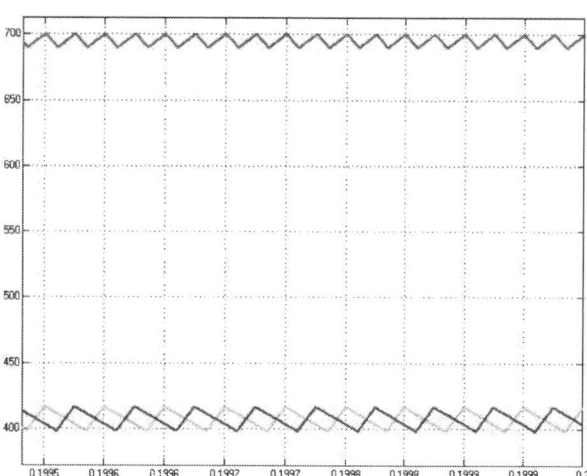

Fig. 9: Ouput voltage and voltage of the capacitor

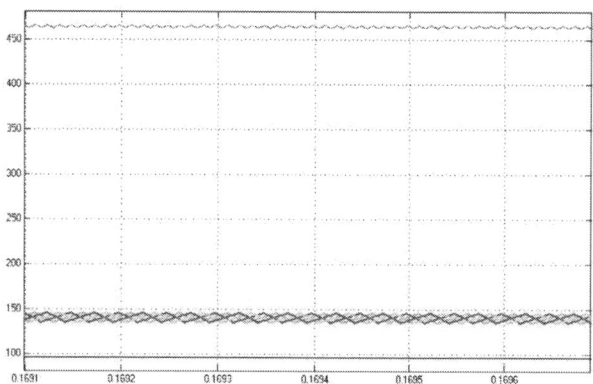

Fig. 10: Input current, output current and inductor current

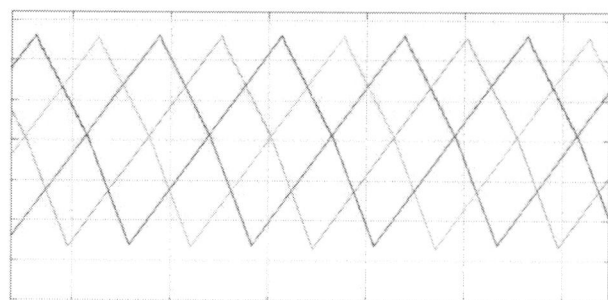

Fig. 11: Experimental waveforms of inductor current

Efficiency of the converter versus target input current is shown in Fig. 13. It indicates that the converter has a higher efficiency at low target input current condition. Efficiency decreases as the target input current increases. However, the efficiency is higher than 91% at all working conditions, which is a quite a high value for a DC-DC converter.

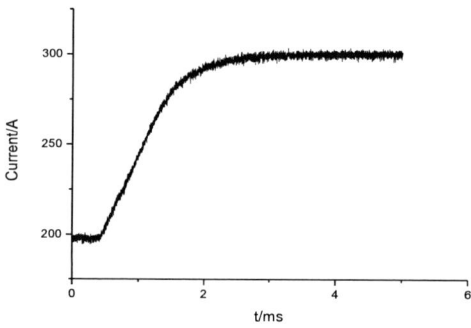

Fig. 12: Change of input current with a target input current step

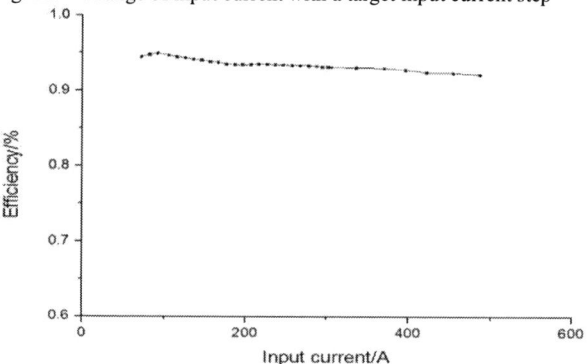

Fig. 13: Efficiency of the converter

VI. CONCLUSION

This paper demonstrates the interleaved dual boost converter, which can not only achieve high voltage gain but also have a high output power with interleaving technique. Parallel cells in the converter share the large current when it works in high power condition, so that the inductor current can be kept in a steady state.

A 60 kW prototype of the converter as well as a novel test bench is designed and developed to satisfy a high power condition. With a test cycle in the experiment, large current flows between boost and buck converter, which has less effect on the DC source. Ripple of the input current and the inductor current is reduced due to the interleaving technique. The converter has a relatively fast response to the change of load and target input current without an obvious overshoot. This totally meets the demand of application in Fuel Cell Vehicles.

REFERENCES

[1] Xuefeng Hu; Chunying Gong, "A High Gain Input-Parallel Output-Series DC/DC Converter With Dual Coupled Inductors," Power Electronics, IEEE Transactions on , vol.30, no.3, pp.1306,1317, March 2015

[2] Qing Du; Bojin Qi; Tao Wang; Tao Zhang; Xiao Li, "A High-Power Input-Parallel Output-Series Buck and Half-Bridge Converter and Control Methods," Power Electronics, IEEE Transactions on , vol.27, no.6, pp.2703,2715, June 2012

[3] Garcia, F.S.; Pomilio, J.A.; Spiazzi, G., "Modeling and Control Design of the Interleaved Double Dual Boost Converter," Industrial Electronics, IEEE Transactions on , vol.60, no.8, pp.3283,3290, Aug. 2013

[4] Lembeye, Y.; Ferrieux, J.P.; Barbaroux, J.; Avenas, Y., "New high power - high ratio non isolated DC-DC boost converter for Fuel cell applications," Power Electronics Specialists Conference, 2006. PESC '06. 37th IEEE , vol., no., pp.1,7, 18-22 June 2006

[5] Weerachat Khadmun, Wanchai Subsingha, "High Voltage Gain Interleaved DC Boost Converter Application for Photovoltaic Generation System," Energy Procedia, Volume 34, 2013, Pages 390-398

[6] Wang, J.; Reinhard, M.; Peng, F.Z.; Zhaoming Qian, "Design guideline of the isolated DC-DC converter in green power applications," Power Electronics and Motion Control Conference, 2004. IPEMC 2004. The 4th International , vol.3, no., pp.1756,1761 Vol.3, 14-16 Aug. 2004.

[7] P. Thounthong, P. Sethakul, S. Raël, and B. Davat, "Design and Implementation of 2-Phase Interleaved Boost Converter for Fuel Cell Power Source," in Proc. IET-PEMD 2008, 2 - 4 April 2008, York-UK, pp. 91–95

[8] P. Thounthong, P. Sethakul, S. Raël, and B. Davat, "Modeling and control of a fuel cell current control loop of a 4-phase interleaved stepup converter for dc distributed system," in Proc. 39th IEEE-PESC, 15-19 June 2008, Island of Rhodes-Greece, pp. 230–236.

[9] H. Xu, E. Qiao, X. Guo, X. Wen, and L. Kong, "Analysis and design of high power interleaved boost converters for fuel cell distributed generation system," in Proc. 36th IEEE-PESC, 11-14 Sept. 2005, pp.140-145.

[10] Coutellier, D.; Agelidis, V.G.; Sewan Choi, "Experimental verification of floating-output interleaved-input DC-DC high-gain transformer-less converter topologies," Power Electronics Specialists Conference, 2008. PESC 2008. IEEE , vol., no., pp.562,568, 15-19 June 2008

[11] Kajangpan, K.; Neammanee, B., "High gain double interleave technique with maximum peak power tracking for wind turbine converter," Electrical Engineering/Electronics, Computer, Telecommunications and Information Technology, 2009. ECTI-CON 2009. 6th International Conference on , vol.01, no., pp.292,295, 6-9 May 2009.

BIOGRAPHY

Jiexun Liu obtained his BSc degree in Automotive Engineering from Tsinghua University, Beijing, China, in 2012. And now he is working towards hie MSc degree in Tsinghua University. Since June 2012, he is involved in automotive power electronics, especially in DC-DC converter. His current research is focused on topologies and control methods of DC-DC converter.

A Bidirectional Flyback Cell Equalizer for Series-connected Lithium Iron Phosphate Batteries

Daiming Yang[1]   Shengyong Li[1]   Guoguang Qi[2]

[1] Group of Energy Storage Technology, Zhangjiagang Smart Grid Research Institute, Jiangsu, China
[2] Department of Computer Science, Tsinghua University, Beijing, China
E-mail: yangdaiming@aliyun.com

Abstract–Cell equalization is a critical technology to enlarge battery pack capacity and prolong battery cycle life. A cell equalizer based on an isolated bidirectional flyback converter is introduced in this paper. Energy can be transferred from one cell to the pack or from the pack to some cell through a double winding transformer. An n-MOSFET and a milliohm resistor for current sampling are connected to each winding in series. According to the sampling current, the n-MOSFETs in both sides are switched on alternately. One is for inverting, and the other is for rectifying. Therefore, the converter achieves synchronous rectification, which contributes to higher energy efficiency. Meanwhile, the average balancing current can be set based on the sampling resistors. By the application of a controller IC for multi-cell battery stacks balancing, the implementation of the cell equalizer is simplified, and the reliability of the circuit is improved. A prototype is designed for a twelve-cell series connected lithium iron phosphate battery pack. Every cell can be charged or discharged individually at the same time, which accelerates the balancing process. Experiments have been conducted to validate the balancing circuit.

Keywords–Bidirectional flyback converter, Cell equalizer, lithium iron phosphate battery, synchronous rectification;

I. INTRODUCTION

A lithium battery energy storage system (BESS) is formed by hundreds or thousands of cells. To enlarge the system capacity, cells with low capacities are often connected in parallel. Meanwhile, cells in series are to achieve a higher voltage. As is known to us, the performance of a serial system is determined by the weakest part. The capacity of a battery pack is limited to the minimum capacity of cells. Compared with lead acid batteries, lithium batteries cannot automatically balance by controlled overcharging [1]. A lithium-ion cell may be damaged irreversibly, sometimes even on fire or explosion, when over charge or over discharge. Thus, not only a real-time voltage monitor is needed to ensure cell safety, but also an auxiliary circuit for cell capacity or voltage equalization is necessary.

The simple and common implementation of battery equalization is dissipating cell energy by a power resistor. The voltages of all cells are kept to the lowest level [1], [2]. The balance circuit is simple but not advocated recently.

Nowadays, numbers of cell equalizers composed of power electronic devices are proposed [3]-[8]. The equalizers transfer energy through cells with little dissipation. Based on the energy storage components in equalizers, there are three major classifications. They are capacitor method, inductor method and transformer method. The capacitor method transfers the charge from one cell to another by a

capacitor connected with two switches or more [3]. The balancing current is related to the voltage difference of the two balancing cells. As the balancing process going on, the smaller and smaller cell voltage difference results in a declining balancing current. The equalizer should be turned off when the balancing current is so small that the circuit power loss is larger than the transferred power. However, a small voltage difference may cause a large state-of-charge (SOC) difference for lithium iron phosphate batteries [9]. Moreover, the uniformity of cell voltages is complicated considering the cell internal resistance and capacitance during charging or discharging [10]. It is not an optimal criterion of cell voltage for choosing the cells to equalize [11]. A more complex capacitor method, such as a resonant switched capacitor converter, can be used to increase the balancing current and the efficiency [12]. But it is difficult to integrate and control such an equalizer for a dozen cells. The inductor method overcomes the shortage that the balancing current is correlated positively to the voltage difference [6], [7]. The charge can be transferred bi-directionally between two adjacent cells by an inductor. The direction and the value of the balancing current are easily controlled. However, if the two balancing cells are not adjacent, the cells between them also need to get into balancing, which leads to a low efficiency. The transformer method is based on the topology of isolated DC/DC or DC/AC converters [4], [5], [8]. Multi-winding or double winding transformers are applied. In some applications, the current flowing through the windings of the transformer is uncontrollable [4], [8]. The balancing current is determined by the voltage difference, which is similar to the capacitor method. Some proposed circuits deliver the charge from the pack to cells in a unidirectional way [5]. It may take a long time to reach an equilibrium state if only one or two cells hold a fairly high voltage. Since a battery pack consists of multi cells, a multi-winding transformer can be used to connect the pack and all cells [4]. While only one cell is allowed to balance with the pack at one moment to ensure the balancing current controllable.

This paper introduces a cell equalizer based on isolated bidirectional flyback converter topology. Every cell connects to the battery pack by an independent double winding transformer, which makes it possible to balance each cell individually at the same time. The equalizer is controlled by a multi-cell balancing controller IC, which simplifies the implementation of the prototype. The balancing current can be adjusted by changing some components in the circuit. A lithium iron phosphate battery pack containing 12 cells in series is applied in the experiments for verification of the usefulness of the equalizer.

II. CELL EQUALIZER ANALYSIS

1. Circuit Topology

The main components of the introduced balancing circuit are a double winding transformer and two switches. As shown in Fig.1, the topology of an equalizer for n cells contains n balancing units, namely bidirectional flyback converters. All of the units are the same, and operate independently. The converter operates in the forward and reverse modes [13]. Energy is transferred from cells to the pack in the forward mode, displayed as the solid arrow lines in Fig.1. While in the reverse mode, energy is transferred from the pack to cells following the path of the dotted arrow lines. The principles of the two operation modes are the same due to the symmetrical circuits at the primary and secondary sides.

A simplified equivalent circuit is derived from the bidirectional flyback converter for analysis, as shown in Fig.2. The transformer in the flyback converter works as an inductor. L_e is the equivalent inductance of the transformer. When $Switch1$ is on and $Switch2$ is off, $Battery1$ charges the inductor, and the current through the inductor is increasing. Then $Switch1$ is turned off and $Switch2$ is turned on, the current through the inductor cannot turn inversed immediately so that it charges $Battery2$. $Switch2$ is turned off and $Switch1$ is turn on just the moment that the current decreases to zero, that means a new balance cycle starts. The charge transfers from $Battery1$ to $Battery2$. This process is the forward operation mode. If the current flowing through the inductor turns inversed, the charge transfers from $Battery2$ to $Battery1$, which is named as the reverse mode.

On one hand, the two switches cannot be switched on simultaneously for avoiding short circuit faults. On the other hand, either of them must be keeping on if the inductor current is not zero. Practically, the switches are power electronics devices like MOSFETs, and a body diode of the MOSFET offers a path for the current. Therefore, only the switch in the primary side is needed to control in the discharge balancing process, so is the switch in the secondary side in the charge balancing process.

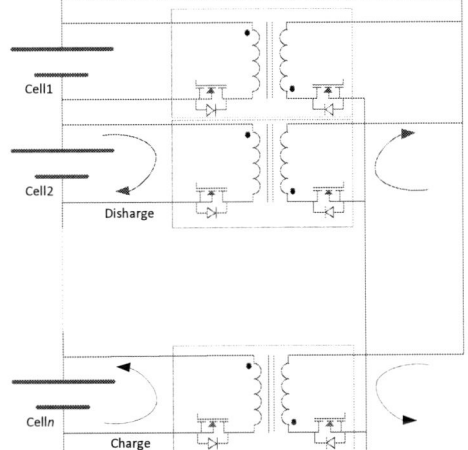

Fig. 1: Schematic of the equalizer circuit among n cells

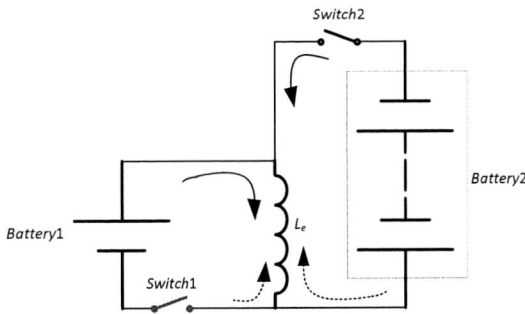

Fig. 2: Equivalent circuit of a bidirectional flyback converter

2. Operating Principle

The discharge balancing process of one balancing unit is shown in Fig.3 (a) and (b).

Firstly, the current of the primary side increases when the primary n-MOSFET is turned on. The energy stored in the primary inductance of the transformer is increasing, so is the magnetic flux density of the transformer magnetic core. Because of the reversed polarity on the secondary side, the current cannot flow through the secondary diode.

Secondly, at the moment turning the primary n-MOSFET off, the magnetic flux density decreases so that the induced electromotive force direction inverses. The current through the secondary diode is ascending. The energy that stored in the primary inductance releases through the secondary electric circuit, not the primary one. The transformer is for both isolation and energy storage. That is how one unit operates in a discharge balancing. The analogous process of cell charge balancing is to control the secondary n-MOSFET.

Since the cell voltage is low, for lithium iron phosphate battery, about 2.5V to 3.6V, and the forward voltage drop of a diode is 0.4V to 0.7V, the rectification by a diode is quite inefficiency. The energy dissipation of the diode is about 10% to 20% of the whole that transferred to the secondary side. The synchronous rectification technology is usually taken to improve efficiency by switching on the n-MOSFET in the rectifier side when current flows through the diode. The drain-source on-state resistance of an n-MOSFET can be low to several milliohms. The voltage on the switch is lower than 0.1V.

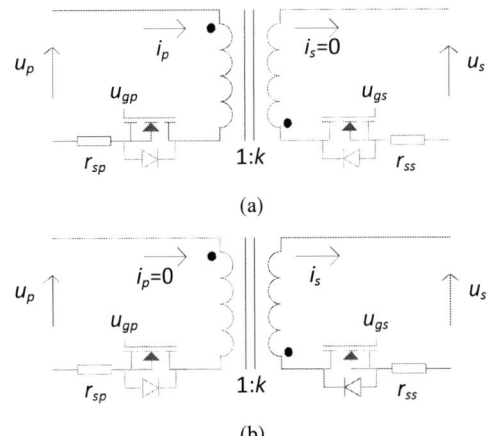

Fig. 3: Operating principle of a balancing unit

978-1-5090-0064-7/15 $31.00 © 2015 IEEE   135

Assume that the transformer with the ratio of 1:k is ideal and the resistances of all components are negligible. The primary current in Fig.3 (a) $i_p(t)$ is

$$i_p(t) = \begin{cases} \dfrac{u_p}{L_p} \cdot (t-t_0), t_0 \leq t < t_1 \\ 0, t_1 \leq t < t_2 \end{cases}. \qquad (1)$$

And the conduction duty ratio D is

$$D = \frac{t_1 - t_0}{T_s} = \frac{L_p \cdot i_{p(max)}}{u_p \cdot T_s}, \qquad (2)$$

where during interval $t_0 \sim t_1$ the primary switch is on, and during interval $t_1 \sim t_2$ the primary switch is off. T_s is switching period and equal to (t_2-t_0). L_p is the inductance of the primary side. $i_{p(max)}$ is the maximum current flowing through the primary side.

The average primary current I_p is

$$I_p = \frac{i_{p(max)}}{2} \cdot \frac{t_1 - t_0}{t_2 - t_0} = \frac{i_{p(max)} \cdot D}{2T_s}, \qquad (3)$$

The secondary current in Fig.3 (b) $i_s(t)$ is

$$i_s(t) = \begin{cases} 0, t_0 \leq t < t_1 \\ i_{s(max)} - \dfrac{u_s}{L_s} \cdot (t-t_1), t_1 \leq t < t_2 \end{cases}, \qquad (4)$$

where $i_{s(max)}$ is the maximum current flowing through the secondary side.

When t=t_2, $i_s(t_2)$=0, then

$$t_2 - t_1 = \frac{i_{s(max)} \cdot L_s}{u_s}. \qquad (5)$$

The average secondary current I_s is

$$I_s = \frac{i_{s(max)}}{2} \cdot \frac{t_2 - t_1}{t_2 - t_0} = \frac{i_{s(max)} \cdot (1-D)}{2T_s}, \qquad (6)$$

and the switching period T_s is

$$T_s = (t_2 - t_1) + (t_1 - t_0) = \frac{i_{s(max)} \cdot L_s}{u_s} + \frac{i_{p(max)} \cdot L_p}{u_p}. \qquad (7)$$

Assume that the transfer efficiency of the transformer from the primary side to the secondary is η_d. According to the relationship of the transfer power, the maximum secondary current can be determined using the following equations:

$$P_{out} = \eta_d \cdot P_{in}, \qquad (8)$$

$$\frac{1}{2} L_s \cdot i_{s(max)}^2 = \eta_d \cdot \frac{1}{2} L_p \cdot i_{p(max)}^2, \qquad (9)$$

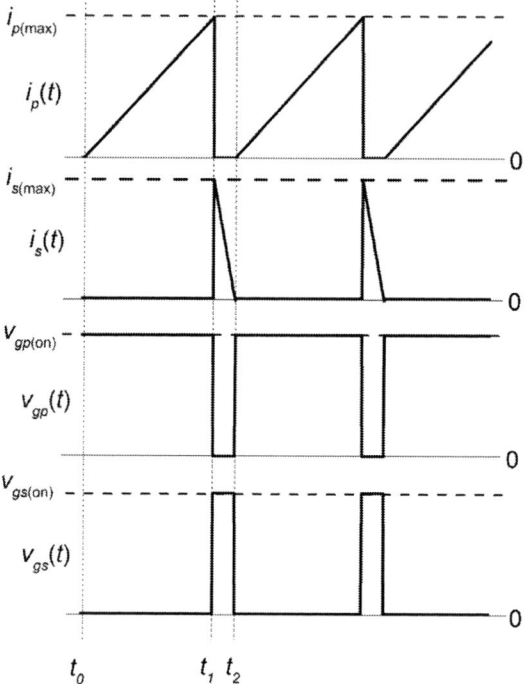

Fig. 4: Typical waveforms during discharge balancing

$$i_{s(max)} = i_{p(max)} \cdot \sqrt{\frac{\eta_d \cdot L_p}{L_s}}, \qquad (10)$$

as for an ideal transformer, the leakage inductance is zero, and the coupling coefficient is 1. The following equation is generated:

$$\frac{L_p}{L_s} = \frac{1}{k^2}, \qquad (11)$$

so (10) can be written as

$$i_{s(max)} = \frac{i_{p(max)} \cdot \sqrt{\eta_d}}{k}, \qquad (12)$$

The waveforms of the primary current $i_p(t)$, the secondary current $i_s(t)$, the gate-source voltage of the primary n-MOSFET $v_{gp}(t)$, and the gate-source voltage of the secondary n-MOSFET $v_{gs}(t)$ are shown in Fig.4. When the primary n-MOSFET is switched on or off is determined by the primary current $i_p(t)$ or the secondary current $i_s(t)$. At the moment of t_0 and t_2, the secondary current $i_s(t)$ decreases to zero, the primary n-MOSFET is switched on. At the moment of t_1, the primary current $i_p(t)$ increases to $i_{p(max)}$, the primary n-MOSFET is switched off. The on or off state of the secondary n-MOSFET is opposite to that of the primary one. The two n-MOSFETs cannot be switched on at the same time.

To measure the current flowing through the transformer, two milliohm resistors serial-connected to the n-MOSFETs are added respectively. $i_p(t)$ and $i_s(t)$ can be calculated by the voltages of the sampling resistors.

The charge balancing process is similar to the discharge one. During charge balancing process, the energy is transferred from the pack to cells. Therefore, the equalizer is a bidirectional buck-boost converter.

III. IMPLEMENTATION AND EXPERIMENT

1. Controller and Driver IC

For a battery pack with n cells, there are $2n$ n-MOSFETs needing to control and drive. That means a circuit module with $2n$ pins, which can drive n-MOSFETs individually is necessary. The driving circuit must deal with the problem of voltage differences among different cells. Meanwhile, the module should be available to detect the voltage on the milliohm resistors for sampling. It is difficult to build such a circuit module by discrete components. Fortunately, Linear Technology Corporation launched an active balancing controller IC named LTC3300 in 2013. It is the most suitable IC to control the introduced cell equalizer. Each IC can control balance for six cells, and it is convenient for more cells by cascading ICs. Control algorithm can run in a micro controller unit (MCU), and the control instructions for which n-MOSFETs are switch on or off are transmitted to the IC by serial communication.

2. Prototype cell equalizer

A printed circuit board (PCB) has been designed to verify the prosed cell equalizer, as shown in Fig.5. The size of the PCB is about 260mm*130mm. The prototype is suitable for as many as twelve cells balancing simultaneously and independently. Some components should be carefully selected. The n-MOSFETs with small drain-source on-state resistance are preferable. The primary inductance of the transformer should be lower than 10uH so that the switching frequency can be up to about 100kHz even though the primary voltage is several volts.

3. Experiment results

Twelve lithium iron phosphate cells with the nominal capacity of 10Ah are chosen for experiments. The rated voltage is 3.2V, and the voltage ranges from 2.5V to 3.6V. The cells are connected in series. A cell voltage monitor based on LTC6804 is applied in the experiments.

Fig.6 shows the voltages of the chosen cells during a balancing process about 1hour. At the first balancing stage (0.11~0.42h), the 3rd cell gets charge from the whole pack. The voltage of the 3rd cell increases, and those of others decrease. At the second balancing stage (0.5~0.76h), all cells get into balancing process to test the usability of all units.

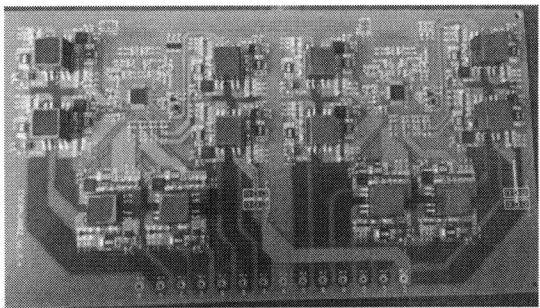

Fig. 5: Prototype cell equalizer

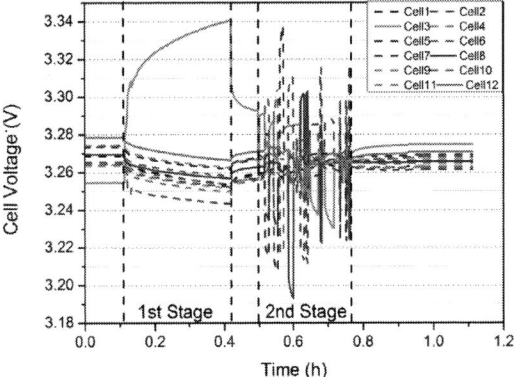

Fig. 6: Voltages of all cells during balancing process

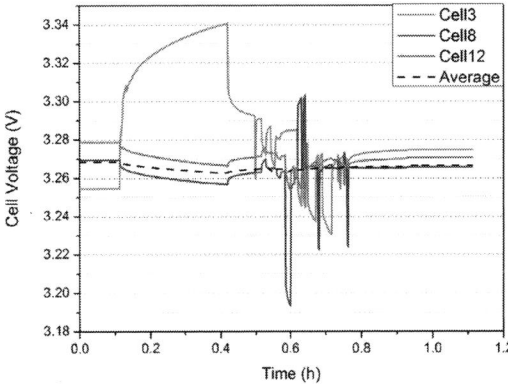

Fig. 7: Voltages of three selected cells and the average value

Three special cells among the twelve ones are selected for a further analysis. They are the 3rd cell, the one with the lowest voltage at the beginning, the 12th cell, the one with the highest voltage, and the 8th cell, the one with the voltage near the average value. The voltages of the three cells and the average value of all cells are shown in Fig.7.

At the end of the first balancing stage, the voltage of the 3rd cell increases to the highest one. However, the voltages of others decrease, so is the average value. During the second balancing stage, all cells are balances. For example, the 3rd cell discharges irregularly, and its voltage decreases at the end of the second balancing stage. The 8th cell discharges or charges randomly, and its final voltage is near to the average value again. The 12th cell doesn't get into the balancing as an individual cell. The change of its voltage is due to the voltage variation of the whole pack.

Besides the balancing results shown in the figure, the polarization effect of the cell voltage is obvious. It makes

978-1-5090-0064-7/15 $31.00 © 2015 IEEE   137

a difference in the detection of cells unbalance. It means that balancing cells just based on the voltages is difficult to achieve a good equalized state, or might lead to a more unbalanced result. Other factors, such as SOC, should be taken into consideration.

To display the effect of the balancing operation, some voltage and SOC characteristics of the battery pack are shown in Table I. The SOC value is calculated based on the OCV-SOC curve proposed in [14]. After the balancing, the cell average voltage decreases a little, which might be caused by the energy consumption of the cell equalizer. The maximum cell voltage difference decreases obviously, which leads to a smaller SOC difference among cells. The maximum cell SOC decreases and the minimum cell SOC increases so that the battery pack can get more charge or output more energy. It seems that the usable capacity of the pack is increased.

Table 1: Voltage ond SOC Characteristics of the Pack

	At the beginning	At the end
Cell average voltage (V)	3.268	3.266
Maximum cell voltage difference (mV)	24.1	13.3
Cell average SOC (%)	23.44	22.93
Maximum cell SOC (%)	26.88	25.54
Minimum cell SOC (%)	20.07	21.78

IV. CONCLUSION

In this paper, a cell equalizer based on a bidirectional flyback converter has been introduced. The operating principle is analyzed by the equivalent circuit. Energy can be transferred from one cell to the pack or in reverse direction through the same balancing path. A balancing controller IC is applied to control and drive the prototype for twelve cells balancing. Each cell can get into balancing for charge or discharge independently. Experiment results shows that the equalizer works well for a lithium iron phosphate battery pack.

ACKNOWLEDGMENT

This work was supported by the Jiangsu Province Science and Technology Plan (BE2014006-3), China.

REFERENCES

[1] S. W. Moore, and P. J. Schneider, "A review of cell equalization methods for lithium ion and lithium polymer battery systems," No. 2001-01-0959. *SAE Technical Paper*, 2001.

[2] M. Daowd, N. Omar, P. Van Den Bossche, and J. V. Mierlo, "Passive and active battery balancing comparison based on MATLAB simulation," in *Vehicle Power and Propulsion Conference (VPPC), 2011 IEEE* , vol., no., pp.1-7, 6-9 Sept. 2011.

[3] A. C. Baughman, and M. Ferdowsi, "Double-Tiered Switched-Capacitor Battery Charge Equalization Technique," in *Industrial Electronics, IEEE Transactions on* , vol.55, no.6, pp.2277-2285, June 2008.

[4] S. Li, C. C. Mi, M. Zhang, "A High-Efficiency Active Battery-Balancing Circuit Using Multiwinding Transformer," in *Industry Applications, IEEE Transactions on* , vol.49, no.1, pp.198-207, Jan.-Feb. 2013.

[5] C. H. Kim, M. Y. Kim, Y. D. Kim, and G. W. Moon, "A Modularized Charge Equalizer Using a Battery Monitoring IC for Series-Connected Li-Ion Battery Strings in Electric Vehicles," in *Power Electronics, IEEE Transactions on* , vol.28, no.8, pp.3779-3787, Aug. 2013.

[6] X. Lu, W. Qian, and F. Z. Peng, "Modularized buck-boost + Cuk converter for high voltage series connected battery cells," in *Applied Power Electronics Conference and Exposition (APEC), 2012 Twenty-Seventh Annual IEEE* , vol., no., pp.2272-2278, 5-9 Feb. 2012.

[7] Y. S. Lee, and G. T. Cheng, "Quasi-Resonant Zero-Current-Switching Bidirectional Converter for Battery Equalization Applications," in *Power Electronics, IEEE Transactions on* , vol.21, no.5, pp.1213-1224, Sept. 2006.

[8] B. Zhang, Z. Lu, M. Chen, D. Yang, et al, "A high efficiency DC/AC inverter for energy balancing of battery modules," in *Industrial Electronics and Applications (ICIEA), 2014 IEEE 9th Conference on* , vol., no., pp.1609-1614, 9-11 June 2014.

[9] D. Yang, J. Liu, Y. Wang, M. Chen, et al, "State-of-charge estimation using a self-adaptive noise extended Kalman filter for lithium batteries," in *Power and Energy Engineering Conference (APPEEC), 2014 IEEE PES Asia-Pacific* , vol., no., pp.1-5, 7-10 Dec. 2014.

[10] M. Chen, B. Zhang, Y. Li, G. Qi, et al, "Performance of inconsistency in lithium-ion battery packs for battery energy storage systems," in *Power and Energy Engineering Conference (APPEEC), 2014 IEEE PES Asia-Pacific* , vol., no., pp.1-5, 7-10 Dec. 2014.

[11] M. Einhorn, F. V. Conte, and J. Fleig, "Improving of active cell balancing by equalizing the cell energy instead of the cell voltage," in *World Electric Vehicle Journal*, vol. 4, no. 1, pp.400-404, 2011.

[12] M. Shoyama, T. Naka, and T. Ninomiya, "Resonant switched capacitor converter with high efficiency," in *Power Electronics Specialists Conference, 2004. PESC 04. 2004 IEEE 35th Annual* , vol.5, no., pp.3780-3786 Vol.5, 20-25 June 2004.

[13] H. S. H. Chung, W. L. Cheung, and K. S. Tang, "A ZCS bidirectional flyback DC/DC converter," in *Power Electronics, IEEE Transactions on* , vol.19, no.6, pp.1426-1434, Nov. 2004

[14] D. Yang, G. Qi, X. Li, "State-of-charge estimation of LiFePO4/C battery based on extended Kalman filter," in *Power and Energy Engineering Conference (APPEEC), 2013 IEEE PES Asia-Pacific* , vol., no., pp.1-5, 8-11 Dec. 2013

BIOGRAPHY

Daiming Yang obtained his B.S. and M.S. degrees both from the Tsinghua University, Beijing, China, in 2011 and 2013 respectively. He is currently a reseacher in Group of Energy Storage Technology, Zhangjiagang Smart Grid Research Institute, Jiangsu Province, China.

Since November 2010, he has been studied on battery management system (BMS) for large lithium-ion battery packs in battery energy storage system (BESS). His research interests include battery modeling, SOC estimation, BMS design and battery equalization technologies.

A Control Scheme for a High Speed Railway Traction System Based on High Power PMSM

Zhao S. F.[1] Huang X. Y.[1] Fang Y. T.[1] Li J.[2]

[1]Research Center for High Speed Railway, College of Electrical Engineering, Zhejiang University, Hangzhou, China
E-mail: 3090101746@zju.edu.cn
[2]University of Nottingham, Nottingham, UK
E-mail: jing.li@nott.ac.uk

Abstract–The high power permanent magnet (PM) traction drive has recently been highlighted due to its promising qualities such as high efficiency and power density. For better performance, a robust, efficient and reliable control and driving scheme needs to be designed. In this Paper a control scheme is designed which based on Most Torque Per Ampere (MTPA) strategy to obtain high power density and is specially designed to achieve smooth transition between constant torque mode and constant power (field weakening) mode. In addition, a novel low-frequency pulse width modulation (PWM) scheme is designed to achieve smooth transitions between various fundamental-carrier wave frequency ratios while maintaining the synchronization between stator and rotor flux throughout the entire operation range. The validity of the scheme has been proved by simulation results and experiment is underway.

Keywords–High Speed Railway, PMSM control, low frequency PWM.

I. INTRODUCTION

High Speed Railway (HSR) systems are developing fast globally. So far most systems employ induction motors as drive machines. In order to achieve higher efficiency and power density, a new PM traction motor has been designed and built in China. This new model exhibits high performance as shown in Table 1. However, compared to induction traction motors, the PM traction motor has several issues to be dealt with, especially the following:

- The rotor field flux is fixed, as a result, in high speed range field weakening need to be implemented and, specifically for the PM, the large current induced could cause the risk of demagnetization.

- During operation, the stator and rotor flux should be in constant synchronization. In the concerned case this should be especially guaranteed considering the high flux density and power.

Table 1: Specifications of the high power PM traction motor

Items (Units)	nomenclature	value
Rated Power (KW)	P_n	635
Rated Speed (RPM)	N_n	4200
Rated Torque (N·m)	T_n	1360
Maximum Torque ((N·m)	T_m	3500
Efficiency (%)	η	98
Rated line voltage (V)	V_{n-l}	2730
Rated phase current (A)	I_{n-p}	141
Maximum phase current (A)	I_{m-p}	350
Number of phases	M	3
Pole pares	P	3

In addition, due to the high voltage and current, the switching frequency of the IGBTs in the inverter should be kept low (in this case at 500 Hz or below) to reduce the temperature rise due to the switching loss. Meanwhile, the inverter adopts a simple three-phase-two-level topology for reliability reason. Considering these facts, and without the intrinsic low pass filter feature of the large rotor time constant in an induction motor, the PWM scheme should be specifically designed to mitigate torque ripple and current harmonics.

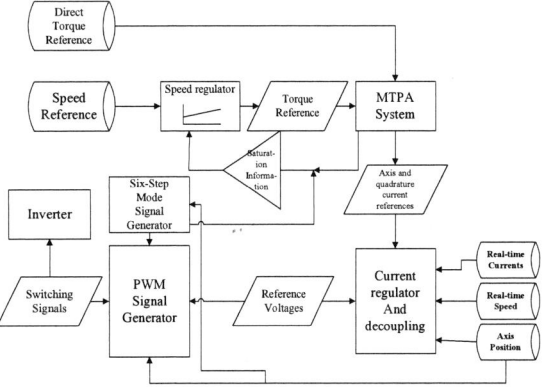

Fig. 1: Control scheme block diagram

In this paper, the MTPA, field weakening and the PWM designs are mainly addressed corresponding respectively to the issues mention above. A brief block diagram of the scheme is shown in Fig. 1. In Section 2 and 4 the existing MTPA/field weakening and PWM techniques are introduced and compared respectively, after which most feasible strategies are designed. In Section 3 a decoupling feedback current control mechanism is introduced which achieves fast and accurate decoupled current components control. In Section 5 the whole design is summarized.

II. MTPA/FIELD WEAKENING SCHEME

1. MTPA principle

According to the MTPA principle, for a reference torque command, the axis and quadrature current are decided by solving the following equations:

$$T_e = \frac{3}{2} \cdot P \cdot (\Psi_f \cdot i_q - (L_d - L_q) \cdot i_d) \quad (1)$$

$$\frac{d\,T_e}{d\sqrt{i_d{}^2 + i_q{}^2}} = 0 \quad (2)$$

where

T_e: Reference Torque;
i_d : axis current;
i_q : quadrature current.

Solving these equations is overwhelming for any embedded processor, so conventionally it is done off-line. For various reference torques, an i_d-T_e and an i_q-T_e curve can be drawn to form a pair of 1-dimention look-up table. Current references are generated according to the table until the voltage limit or the current limit for the inverter is reached. Another approach is to find a group of fitting curves and do calculation on-line, which may lower data storage cost but increase calculation work.

2. Field weakening scheme

2.1. Field weakening scheme in PWM mode
The current limits mentioned above are variable depending on the motor speed. Generally all field weakening control approaches can be categorized into two groups. The first group engages a feedback structure, through which the magnitude of the reference voltage vector and a set magnitude limit is compared and the difference is processed by a field weakening regulator (in most circumstances PI regulator) whose output is a adjustment quantity which is added to the axis current reference and also adjust the quadrature current reference through another regulator or simple calculation. The other group utilizes a group of target current curves which are usually formed in advance, and feed the current references directly forward in operation corresponding to the curve, instantaneous motor electrical angular speed and torque reference. This group of methods need a data look-up process also as the typical MTPA approach introduced above, hence they can be naturally done simultaneously. There are a few proposed approaches which are based on both with some adaptations. For example, in [5] the proposed strategy refers to the target current curve but without directly using a look-up table, instead it decides the status of operation by calculating the relative position and direction of the instantaneous operation point with regard to the current curve, and subsequently gives a current reference adjustment value.

As mentioned in Section 1, in our case the field weakening strategy requires to be especially fast, accurate and reliable to prevent demagnetization and minimize current transients. Methods featuring feedback loops are easy to apply and without the expense of data storage, yet with a feedback structure and in-loop regulators, its accuracy and fast response through the whole speed range is hard to guarantee. The strategy proposed in [5] demonstrated high accuracy and responding speed, but it involves large amount of extensive calculation which may bring great challenge to the processor. To sum up, a curve look-up based strategy is most desirable here.

In the proposed scheme the entire speed range is considered to form a 2-dimension look-up table (see Fig. 2), and the various voltage and current limits are taken into account. In general, the curves are bound to follow a MTPA—voltage limit ellipse arc—current arc trajectory. It should be noted that when the operation point moves along the voltage limit arc (which means the MTPA is not achieved but maximum torque output is maintained), the current vector magnitude is monotonically increasing, suggesting the current reference is still a smallest possible value, all the way until the current limit is reached (torque output hits a limit).

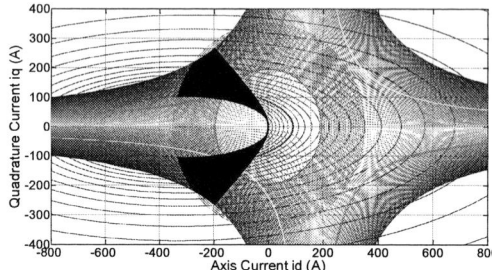

Fig. 2: 2-dimension look-up table generation plot
(Red: current limit circle; blue: voltage limit circle; green: positive equal-torque lines; orange: negative equal-torque lines; black: MTPA curves)

In many curve look-up strategies, as the motor electrical angular speed increases, the operation point moves along the trajectory which indicates the opposite gradient of the limit curve group. Here the torque limit of each curve can be easily obtained by introducing a set of limit indices (See Fig. 3), which are generated to mark the torque value

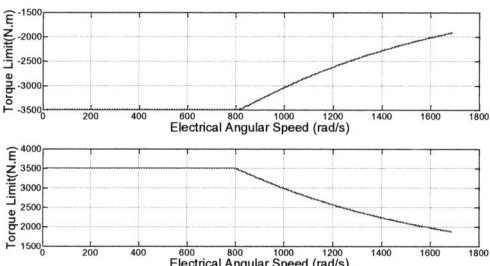

Fig. 3: Maximum torque limit indices (upper: negative; lower: positive)

beyond which no viable roots exist when solving the curve equations. These indices are fed back to where the torque reference is firstly generated and adjust the reference value correspondingly. Proper margins are always kept to maintain the capability of the controller. In such way, the MTPA and field-weakening process can always be executed simultaneously and instantly, and safety and fast responding speed can thus be achieved.

2.2. Current control in six-step mode
The Current control in PWM mode is finally implemented by inverter through a linear modulated voltage feed-forward mechanism, which generate reference voltage vectors (explained in Section 3). Yet in six-step mode, no vectors can be formed as in PWM, thus instead a power angle control is carried out. In this mode specific power angle of the fundamental component of reference voltage is calculated, and the phase of the six-step voltage wave is locked to that of the rotor according this angle. The process of angle calculation is virtually the same as the reference current curves forming process, except for a following additional procedure in which the power angle corresponding to each reference current value is calculated according to motor phasor relationships. It should be noted

978-1-5090-0064-7/15 $31.00 © 2015 IEEE

that in this mode MTPA can no longer be achieve due to the fact that the fundamental voltage magnitude is fixed along with the DC bus voltage. The current reference and then the power angle are functions of this fixed voltage value and torque command. Similar as in the field weakening scheme, a set of indices are introduced and in operation fed back to the origin of torque command to make adjustments. It is shown in Fig. 4.

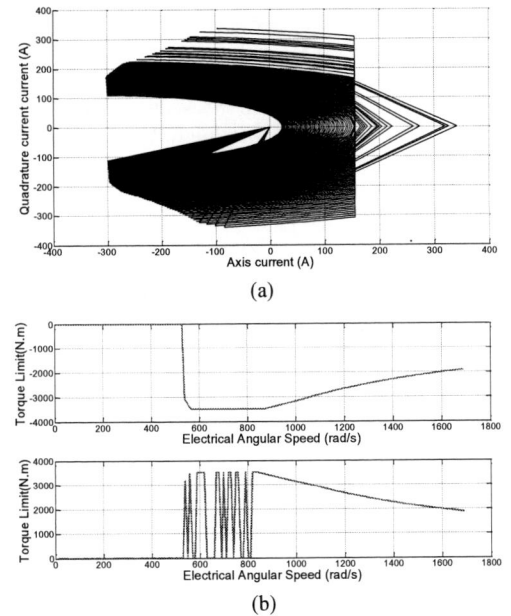

(a)

(b)

Fig. 4: Six-step mode current control: (a) current look-up table; (b) maximum torque limit indices (upper: negative; lower: positive)

III. DECOUPLING CURRENT CONTROL

The concerned PM traction motor employs an interior magnets structure in order to improve rotor strength and introduce high reluctance torque for better low speed acceleration. Due to this structure the quadrature inductance of the motor is significantly higher than the axis inductance. Thus, an effective decoupling current control scheme is required to achieve accurate and low transient current control.

There exist several kinds of decoupling methods, including feed-forward decoupling, feedback decoupling and PI regulator decoupling. In [6] the close-loop transfer functions for different methods are derived, and their pole-zero distribution at different speed analyzed. It is demonstrated that a PI decoupling structure (see Fig. 5)

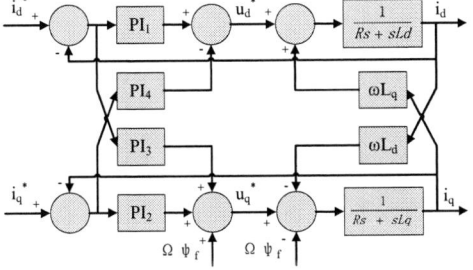

Fig. 5: PI decoupling structure

With their parameter setting satisfies the following relations:

$$PI_3 = \frac{\omega_e L_d}{R_s + s L_d} PI_1 \, , \quad PI_4 = \frac{\omega_e L_q}{R_s + s L_q} PI_2 \qquad (3)$$

where
ω_e : fundamental electrical angular speed;
R_s : stator phase resistance;
L_d : axis inductance;
L_q : quadrature inductance

can achieve effective decoupling for all speed and is insensitive to motor parameter change. In practice this structure can be realized by establishing difference models and is thus adopted in our case.

It should be noted that the stator resistance is function of temperature, and inductance is function of frequency. In actual application a lookup table of these functions is employed to make adjustment to main regulator parameter settings.

IV. PWM SCHEME

1. General PWM scheme design

In our case, a two-level voltage source inverter (VSI) is employed. The maximum switching frequency of high voltage level and high power capacity IGBTs is usually from a few hundred Hertz to over 1k Hertz. It is shown in PWM theory that a higher switching frequency brings lower harmonic loss in the motor windings. Yet, it has been proved that the switching loss significantly increases with the increase of current [8-10]. In a railway traction system the inverters are often very compactly built due to limited room, which increases the difficulty of heat dissipation, consequently the switching frequency has to be limited to prevent overheating caused by extensive switching loss. In a weighted conclusion of our case, the switching frequency should be kept no higher than 500 Hz. Thus, through most of the speed range, the fundamental-carrier wave frequency ratio is rather low and finally six-step mode is required in high speed range. As a result, synchronous modulation should be adopted in most of the speed range to eliminate low-frequency harmonics.

PWM strategy has been the subject of extensive research. The principle and harmonic analysis of various PWM schemes are well developed in many literatures. There are mainly two types of PWM schemes, namely carrier-based PWM and carrier-less PWM [1]. Among the former type space vector modulation (SVM) is widely applied due to its high DC bus voltage utilization, good harmonic performance and relative simplicity to apply. Yet theoretically the latter type can achieve the least harmonics by solving off-line various transcendental equations.

First, the traction system specifications should be analyzed in order to select a most desirable PWM scheme.

The rated electrical frequency of the traction PMSM is 210 Hz. Before going into six-step mode, the inverter should be working in synchronous PWM mode. Conventionally, fundamental-carrier wave frequency ratio in a carrier-based synchronous PWM scheme (or the pulse number per fundamental wave period in a carrier-less PWM scheme) is set to be an odd integer to achieve a central-symmetrical waveform within a fundamental period (in order to eliminate undesirable even-order harmonics). Thus, the lowest ratio besides six-step is 3. Considering the maximum switching frequency of 500 Hz, the highest fundamental frequency in 3-pulse mode is 500/3 = 167 Hz, which means the inverters go into the six-step mode rather early. Hence most of the time the inverter works in six-step mode, and the major objective of previous modulation process is to achieve a fast and smooth transition.

The actual DC bus voltage in practice is above 3600 V. It can be derived that the inverter maintains linear capacity even under maximum current and torque output condition before six-step mode. Consequently, over-modulation scheme, which is usually of great concern in general PWM strategy design, is not necessary in this case.

Then the characteristics of SVM of carrier-based PWM and off-line optimization sequence of carrier-less PWM should be summarized.

SVM: (1) easy to digitally apply; (2) has superior performance through a wide modulation-index range (but when very high) [2]; (3) relatively easy for different carrier ratio transition; (4) relatively complicate in over-modulation range.

Carrier-less modulation (typically SHEPWM (selected harmonics elimination PWM) and CHMPWM (current harmonics minimum PWM)): (1) superior Weighted THD performance with <u>high pulse number;</u> (2) intrinsically easy over-modulation; (3) relatively difficult for different pulse number transition. (There are researchers who proposed some methods which to a great extent overcome this difficulty, yet these methods requires extensive computation and rely particularly heavily on the accuracy of current sampling, which may bring difficulty under sophisticated traction environment, and so far no experiments have been done on a large PMSM.)

It now can be concluded that the need for fast and smooth transitions, simplicity to apply and no requirement for over-modulation scheme in the traction system favors SVM as the modulation strategy.

In order to better utilize the switching frequency of the inverter while implementing SVM scheme, a combined modulation strategy is designed which integrates asynchronous SVM scheme, synchronous SVM schemes of different fundamental-carrier wave frequency ratio and finally six-step mode. Fig. 6 shows the way in which different schemes are integrated.

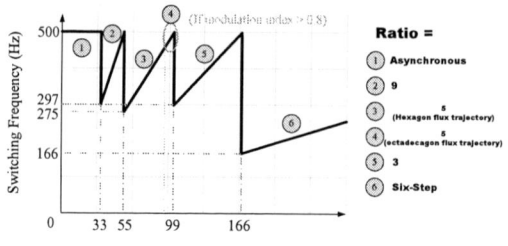

Fig. 6: SVM mode combination

As for the issue of transition between different ratios, theoretically the harmonic distortion can be minimized as possible if the ratio utilizing the most DC bus voltage is always chosen, i.e., ratio = 3, 5, 7, 9, 11, …(if needed) in an opposite order. However, the current and torque transients during the transitions are also considerable, which cause extra loss and even system instability if transitions are too frequent. In a weighted consideration, the combination in Fig. 6 is adopted.

The choice of these synchronous ratios corresponding to their different fundamental frequency sections refers to the conclusions of similar researches [3-4],[13],[16]. The major objective is to always keep the current and electromagnetic torque ripple on a modest level and to bring low transients during transitions. It has been demonstrated that by choosing appropriate positions of the voltage vector on its fundamental period trajectory low transient magnitudes can be achieved [8],[11],[13]. The principle is to select the reference voltage vector position where the phases of fundamental components of both ratio modes are the same, and execute the transition when the actual voltage vector reaches the same position. If the adjacent patterns of transition is chosen so that the stator flux trajectories of the last fundamental period of the former pattern and the first fundamental period of the latter have most of their sides overlapping and the corresponding voltages of either one of the three phases reach their peaks in both patterns on the selected transition position, the current transitions of both fundamental and harmonic components can be eliminated. However, this situation cannot always be reached considering the aforementioned specifications in our case. So the prior task is to guarantee the phase continuity of the fundamental voltages vectors, and make minor position adjustment of the transition point if necessary to lower the harmonic transients, which will be specified later.

2. PWM scheme design specifications

2. 1. Smooth transition

The aforementioned analysis of transition position choosing is often based on post-processing, i.e. to use the frequency domain analysis of the voltage and current data to derive an optimum transition point, and to summarize a discipline for the point choosing. Yet, such predetermined discipline can be sensitive to unpredictable influence factors in the close current control, which undermine its effectiveness, unless a real-time current component analysis is carried out which is not very practical. Thus, a relatively simple and robust rule to make adjustment during transition is needed.

- Adjacent stator flux trajectory overlapping.

When involved SVM patterns are chosen such that the adjacent flux trajectories are naturally overlapping on most of their sides, the fundamental voltage phases and magnitudes are continuous. Yet when not, there may exist a magnitude step at the instant of transition. Under these circumstances an elaborate minor additional vector should be added such that the adjacent trajectories are concentric (see Fig. 12). Theoretically the additional vector changes the instantaneous flux magnitude and thus brings minor change to torque, but it helps the next several vectors given by close-loop current controller maintain even angular speed and magnitude, which is essential to keep the transients fast and modest

- Current sampling instants.

The zero vector distribution within a SVM period is decisive to the frequency spectral pattern of the output waveform. This is reflected in the specific periodic pattern of the axis and quadrature current in time domain. In steady state the average value of each period represents the fundamental magnitude. In this aspect it is convenient to conceive that by always putting the current sampling point around the average value, a steady output by the current controller can be obtained (see Fig. 7). (At the first

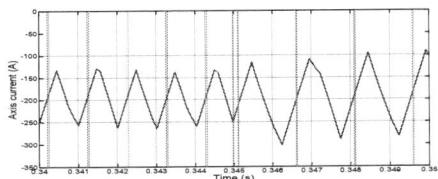

Fig. 7: Current sampling instants around transition

vector duration calculation and updating instant after the transition, the duration can be calculated using the vector magnitude of the previous instant to further prevent steer change.) This can be achieved by high frequency current sampling and following of the periodic current pattern. As the sampling point in SVM scheme is evenly distributed, the adjustment of zero vector duration during the transition can be utilized. The key is to keep the average angular velocity of the stator flux steady during transition, in which the zero vector duration is critical.

- Angle compensation

At low switching frequency, the reference voltage actually sent to SVM module must be compensated with an angle corresponding to half of the SVM period, so that the equivalent voltage vector is perpendicular to the average equivalent stator flux vector (which is a circle). After the transition, this compensation value experiences a step change dueto the change of the fundamental-carrier wave frequency ratio. This is problematic because due to the inertia of the current controller, the reference voltage

magnitude is virtually continuous, but with the step change of angle, the vector durations calculated with SVM principle also experience a step change (see Fig. 8), which may cause flux trajectory distortion and subsequently current and torque ripple. To handle this problem, the magnitude of the first few reference voltage vectors should be modified by a coefficient, which derived from:

$$k = \frac{|U_{ref}| \cdot \sin\alpha + |U_{ref}| \cdot \sin(\frac{\pi}{3} - \alpha)}{|U_{ref}| \cdot \sin(\alpha + \theta_{c1} - \theta_{c1}) + |U_{ref}| \cdot \sin[\frac{\pi}{3} - (\alpha + \theta_{c1} - \theta_{c1})]} \quad (4)$$

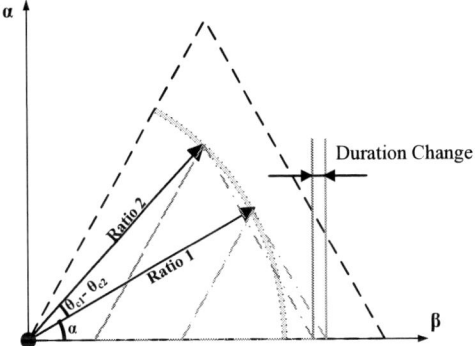

Fig. 8: Active vector duration step change

where

$|U_{ref}|$: reference voltage magnitude;

α: relative reference voltage angle in its sector;

θ_{c1}, θ_{c1}: compensation angles corresponding to the SVM pattern on left and right side of the transition respectively

The modifying coefficient is a temporary value and should not be kept on. It can be set to decay as the time passes on and finally set to unity so that after the transient the reference voltage vector magnitudes again comply with the current controller outputs.

2. 2. An integrated carrier synchronization strategy

So far, extensive analysis has been done on various synchronous SVM schemes. But most of the analysis focus on the principles of different pulse distribution strategies and their corresponding harmonic performances, meanwhile few has elaborated the specific way of carrier synchronization. According to the basic principle of SVM, it can be inferred that in most cases, the fundamental voltage frequency within each carrier cycle is constant, and the carrier durations are based on these piecewise constant values at each sampling point. This is practically valid considering the fact that the motor speed and the reference voltage angular velocity (which is given by current controller) do not change fast with respect to the carrier cycle. Yet, also out of practical consideration, there are still facts that will undermine the performance of this method. First, during very steep torque reference change, the angular velocity of the reference voltage vector can change very fast, under which circumstance the presumption of a constant fundamental voltage vector angular velocity within a carrier cycle may be acceptable but far from precise, which can bring considerable error. Second, in steady state the fundamental vector velocity should indeed be constant, but the reference voltage vector can hardly be truly smooth and rotate very evenly due to various influential factors inflicted on the current control loop. If the sampling for deciding a carrier cycle is executed when there happens to be a considerable disturbance on the current controller, the sampled fundamental vector velocity may significantly differ from the actual value and thus a distortion in flux occurs.

To handle this issue, the ordinary principle of the old analog SPWM can be referred to in which the switching signal is generated when the actual carrier wave and reference wave intersect (natural sampling). If the carrier wave can always be synchronized with the fundamental reference signal, the output voltage will also be well synchronous. The carrier waves are often triangle or saw-tooth, which can be conceived as the integration of a constant value, and the fundamental vector trajectory is also the integration of its instantaneous velocity. So if the two integrations are done always according to the same instantaneous velocity, the carrier and the fundamental wave will be synchronous. And the sampling instants, which are at fix position of the carrier wave, is also synchronized. This integration scheme can intrinsically eliminate the influences of most high frequency disturbances which can be problematic if otherwise, as explained above.

Now that the traction PMSM has a resolver installed and thus the precise position and rotational speed of the rotor are available all the time, it is convenient to use the speed as integrand for the carrier waves. The shapes of carrier wave corresponding to different SVM pattern can be easily designed by introducing specific integrand coefficients and reset/reversing thresholds as shown in Fig. 9.

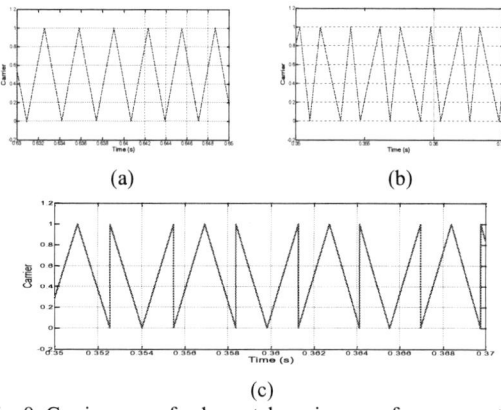

(c)

Fig. 9: Carrier wave: fundamental-carrier wave frequency ratio = (a) 9 or 3; (b) 5, hexagon flux trajectory; (c) 5, octadecagon flux trajectory.

In steady state, the fundamental voltage vector is strictly in synchronization with the rotor electrical angle, and this scheme works well. Still, with the transient effect, there may be a steady error. For instance, in the ratio = 3 unsaturated SVM mode, the sampling points should be at when the angle of reference voltage vector are 30°, 90°, 150°, 210°, 270°, 330° to generate a optimum waveform. If the sampled positions are 33°, 93°, 153°, 213°, 273°, 333°, it can be inferred that the carrier drags 3° in phase. In steady state this difference can be compensated by

adding a minor value to the carrier phase angle every cycle until it comes to zero.

Of course, the use of rotor electrical speed is invalid when the torque command and the reference voltage vector speed is changing since the rotor does not accelerate as the speed of voltage vector. Under this condition, the integrand for the carrier wave is set to be the instantaneous reference voltage vector angular velocity. Yet, in simulation the problem of the high frequency harmonics on the vector speed value cannot be ignored, in which even after integration the carrier wave is still not smooth enough, cause the sampling intervals uneven. To handle this issue, a special conditioning work needs to be done on the reference voltage vector angular velocity wave, which trims its significant harmonics and disturbances without introducing any time delay (see Fig. 10) (as of an ordinary LPF). This work involves some wave pattern recognition.

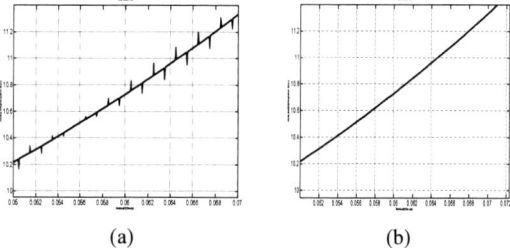

(a)   (b)

Fig. 10: Reference voltage vector angular velocity integration waveform: (a) before conditioning; (b) after conditioning.

When the steady state is recovered, the integrand can shift back to rotor electrical angular speed.

V. CONCLUSION

The specific design technique details are elaborated above. The whole system model is established on Matlab/Simulink platform. In Fig. 11 The phase current, stator flux trajectory, line voltage (U_{ab}) and electromagnetic torque behaviors around transitions are presented. A steady and smooth torque output is achieved.

In practical use, there still one difficulty that may be concerned with. A few detailed strategies introduced above require very high sample rate capability of the hardware system. For a basic DSP based system, the task can be challenging. Yet in our case a FPGA based sampling and computing module is considered. FPGAs can achieve very high sample rate and are very efficient in fixed-pattern computing such as SVM signal generation and conventional PI regulation. In fact the application of FPGA in motion control system is becoming a mainstream. With its help, the control specifications proposed in this paper can be realized without any major obstacles. The hardware control system based on DSP/FPGA is being constructed, and experiments will be carried out.

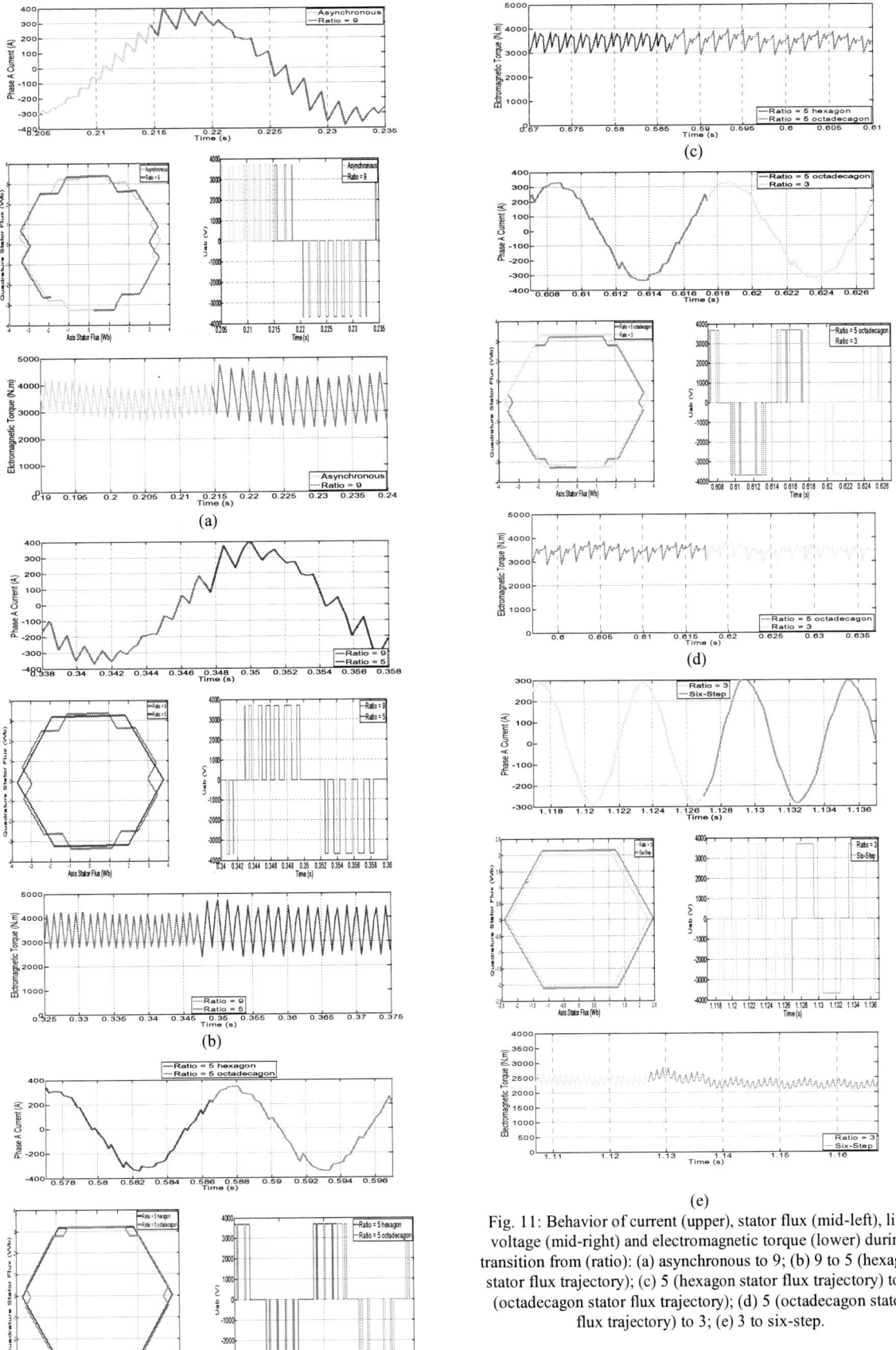

Fig. 11: Behavior of current (upper), stator flux (mid-left), line voltage (mid-right) and electromagnetic torque (lower) during transition from (ratio): (a) asynchronous to 9; (b) 9 to 5 (hexagon stator flux trajectory); (c) 5 (hexagon stator flux trajectory) to 5 (octadecagon stator flux trajectory); (d) 5 (octadecagon stator flux trajectory) to 3; (e) 3 to six-step.

REFERENCES

[1] J. Holtz, "Pulsewidth Modulation - A Survey", IEEE Transactions on Industrial Electronics, Vol. 39, No. 5, 1995, pp. 410-420.

[2] V. Blasko, "Analysis of a Hybrid PWM Based on Modified Space Vector and Triangle Comparison Methods", IEEE Transactions on Industrial Applications, Vol. 33, No. 3, 1997, pp. 756-764.

[3] Y. P. He, Y. L. Wen, J. F. Xu, J. H. Feng, "High-Power Permanent Magnet Flux-Weakening Strategy Based on Multi-Mode SVPWM", Transactions of China Electrotechnical Society, Vol. 27, No. 3, 2012, pp. 92-99.

[4] L. J. Liu, Y. Liu, W. Guo, W. T. Wang, Y. T. Wang, "An Algorithm of PWM Modulation under Low Switching Frequency", High Power Converter Technology, Vol. 1, 2013, pp. 7-11.

[5] Y. F. Sheng, S. Y. Yu, W. H. Gui, Z. N. Hong, "Field Weakening Operation Control Strategies of Permanent Magnet Synchronous Motor for Railway Vehicles", Proceedings of the CSEE, Vol. 30, No. 9, 2010, pp. 74-79.

[6] K. Z. Lin, "Research on Control Method for Permanent Magnet Motor Under Low Switching Frequency", M.E. thesis, Beijing Jiaotong University, 2014.

[7] D. G. Holmes, T. A. Lipo, 'Pulse Width Modulation for Power Converters: Principles and Practice' (WILEY, 2003).

[8] D. N. Sun, "Research on Key Control Technologies of Electric Traction Drive System for Metro Cars", Ph.D. thesis, Beijing Jiaotong University, 2012.

[9] K. Dong, "Research on Key Control Technologies and Performance Optimization of Traction Drive System for EMUs", Ph.D. thesis, Beijing Jiaotong University, 2015.

[10] K. K. Wei, "Study on Digital Control for Railway Traction Inverter", Ph.D. thesis, Beijing Jiaotong University, 2012.

[11] R. Sun, "The Study of Multi-mode Modulation Algorithm and Carrier Ratio Change Concussion Restrain", M. E. thesis, Beijing Jiaotong University, 2008.

[12] Y. Zhang, "Design and Field-Weakening Control of EV-Used PM Motor", M. E. thesis, Zhejiang University, 2014.

[13] K. Wang, X. J. You, C. C. Wang, M. L. Zhou, "Research on Synchronized SVPWM Strategies Under Low Switching Frequency", Proceedings of the CSEE, Vol. 35, No. 16, 2015, pp. 4175-4183.

[14] K. Wang, X. J. You, C. C. Wang, M. L. Zhou, "Research on the Comparison of Synchronized Modulation of SHEPWM and SVPWM Under Low Switching Frequency", Transactions of China Electrotechnical Society, Vol. 30, No. 14, 2015, pp. 333-341.

[15] D. H. Wu, X. H. Xia, Z. Y. Zhang, C. Li, "A SVPWM Overmodulation Method Based on Three-Phase Bridge Arm Coordinates", Transactions of China Electrotechnical Society, Vol. 30, No. 1, 2015, pp. 150-158.

[16] Z. W. Ma, Z. H. Yin, W. Jiang, W. Li, L. Zhang, "Research on Multi-mode Space Vector PWM Algorithm for High-power Two-level Inverters", Electric Drive for Locomotives, No.4, 2010, pp. 17-24.

[17] B. J. Chalmers, R. Akmese, L. Musaba, "Design and field-weakening performance of permanent-magnet/reluctance motor with two-part rotor", IEE Proceedings of Electric Power Applications, Vol. 145, No. 2, 1998, pp. 133-139.

[18] J. Holtz, B. Beyer, "The trajectory tracking approach-a new method for minimum distortion PWM in dynamic high-power drives", IEEE Transactions on Industry Applications, Vol. 30, No. 4, 1994, pp. 1048-1054

[19] J. Holtz, N. Oikonomos, "Synchronous Optimal Pulsewidth Modulation and Stator Flux Trajectory Control for Medium-Voltage Drives", Vol. 43, No. 2, 2007, pp. 600-608.

[20] A. M. Hava, R. J. Kerkman, T. A. Lipo, "Simple Analytical and Graphical Methods for Carrier-Based PWM-VSI Drives", IEEE Transactions on Power Electronics, Vol.14, No. 1, 1999, pp. 49-61.

[21] P. P. Robert, M. Gautier, C. Bergmann, "A Frequency Approach for Current Loop Modeling With a PWM Converter", IEEE Transactions on Industry Applications, Vol. 34, No. 5, 1998, pp. 1000-1014.

[22] T. A. Sakharuk, A. M. Stankovic, G. Tadmor, G. Eirea, "Modeling of PWM Inverter-Supplied AC Drives at Low Switching Frequencies", IEEE Transactions on Circuits and Systems – I: Fundamental Theory and Applications, Vol. 49, No. 5, 2002, pp. 621-631.

[23] B. J. Seibel, T. M. Rowan, R. J. Kerkman, "Field-Oriented Control of an Induction Machine in the Field-Weakening Region With DC-Link and Load Disturbance Rejection", IEEE Transactions on Industry Applications, Vol. 33, No. 6, 1997, pp. 1578-1584.

BIOGRAPHIES

Shuofeng Zhao received his B.E. degree in electrical engineering from Zhejiang University, Hangzhou, China in 2013. He is currently working towards Ph.D. degree in electrical machines and drives in Zhejiang University, Hangzhou, China.
His research interests include control and drive system design for PMSM in the applications of transportation and clean energy.

Xiaoyan Huang received the B.E. degree, from Zhejiang University, Hangzhou, China, in 2003, and received the Ph.D. degree in electrical machines and drives from the University of Nottingham, Nottingham, U.K., in 2008. From 2008 to 2009, she was a Research Fellow with the University of Nottingham. Currently, she is an associate professor with the College of Electrical Engineering, Zhejiang University, China, where she is working on electrical machines and drives. Her research interests are PM machines and drives for aerospace and traction applications, and generator system for urban networks.

Youtong Fang received the B.S. degree and Ph.D. degree in electrical engineering from Hebei University of Technology, Hebei, China, in 1984 and 2001 respectively. Currently, he is a professor with the College of Electrical Engineering, Zhejiang University, China. His research interests include the application, control, and design of electrical machines.

Jing Li received her MS.c (Distinction) and B.Eng. (Hons.) from Beijing Institute of Technology, obtained the Ph.D. degree from the University of Nottingham in 2010. She is currently a research fellow in the Power electronics, Machine and Control group, University of Nottingham. Her research interests are condition monitoring for motor drive systems and power distribution systems, advanced control and design of motor drive systems, and insulation study for electrical machines.

State-of-Charge Estimation of Lithium Iron Phosphate Battery Using Extreme Learning Machine

Zhihao Wang[1]   Daiming Yang[2]

[1] School of Electrical Engineering, Guangxi University, Guangxi, China
[2] Group of Energy Storage Technology, Zhangjiagang Smart Grid Research Institute, Jiangsu, China
E-mail: yangdaiming@aliyun.com

Abstract–Tracking the state-of-charge (SOC) is important for battery applications. A novel SOC estimation algorithm based on extreme learning machine (ELM) is proposed. Compared with the traditional neural network method, the ELM simplifies the computation procedure and shortens the learning time. A typical model of ELM is built for an 180Ah/3.2V · lithium iron phosphate battery, and data acquired from cell discharge experiments under a series of current are contributed to model training and predicting. To evaluate the effect of ELM on SOC estimation, the back propagation (BP) neural network and the support vector machine (SVM) are taken into comparison. The results show that the ELM method provides a more accurate SOC. What is more important is that the speed of network training is improved greatly.

Keywords–Extreme learning machine, lithium iron phosphate battery, neural network, State-of-charge (SOC) estimation.

I. INTRODUCTION

State-of-charge (SOC) estimation technology is important for the development of lithium-ion battery energy storage systems (BESSs) [1], [2]. An accurate SOC is the principal element for safety running of battery packs. It is also a reference basis for cell equalizing control. The function of a cell balancer is to keep all cells equalized during charging and discharging to maximize the pack capacity and protect cells from over charge or over discharge. It may lead to permanent damage when one cell is over-charged or over-discharged. Therefore, improving the precision of the estimated SOC makes difference in mitigating damage of cell, prolonging cycle life and decreasing maintenance cost.

There are some common methods of SOC estimation for lithium iron phosphate battery, including ampere-hour integration, open-circuit voltage mapping, Kalman filter, neural network, and so on [3]-[8]. Some parameters, such as cell voltage, current and/or temperature are inputs for these methods. According to different methods, functions with output of SOC are established. However, on one hand, because of the complicated structure of battery, it is impossible to track the real-time inner chemical reaction exactly. On the other hand, the relationship between SOC and parameters is non-linear, and the SOC is influenced by many other factors. As a result, there is no advanced SOC estimation algorithm which is suitable for the whole cycle lifetime and under different operation situations. Generally, in accordance with various requirements and situations, one or two preferable methods are applied.

Since the neural network hold features like multiple input/output support, highly nonlinear, robustness and fault tolerance, the prospect of this method for battery SOC estimation in BESSs is good [4], [8]. Error back propagation (BP) algorithm is one of the most popular feed forward neural network methods. The learning process consists of forward propagation of signal and back propagation of error. Any nonlinear function can be approximated in arbitrary precision by the algorithm. Practically, some obvious shortages of BP algorithm exist. It is complex in mechanism, requiring large computing resource and learning slowly. When dealing with small-amount samples, the generation is poor. While mass training data might lead to a local minimum easily [9]. Support vector machine (SVM) is a machine learning approach based on statistics [10]. It uses the minimization principle of the structural risk. When dealing with small samples, nonlinear and high pattern recognition issues, this method works well. However, it needs to set many parameters such as kernel function, error control parameters and penalty coefficient, which wastes a lot of time to adjust every parameter, making it difficult in practical applications.

Recently, based on the traditional neural networks, a simple and efficient single-hidden layer feed forward neural network (SLFN) named as extreme learning machine (ELM) is developed [11]. This method has some advantages like simple structure, learning fast, easy to adjust the parameters and not falling into local minimum. The ELM simply calculates the output weights, while the hidden layer parameters are obtained randomly [12]. By this way, the training efficiency has been significantly improved, and the performance of generalization is excellent.

In this paper, the ELM method is adopted for SOC estimation modeling of an 180Ah/3.2V lithium iron phosphate battery. Compared with models based on the BP network and the SVM method, the accuracy and learning time are analyzed. The results validate the superiority of ELM on SOC estimation.

II. EXTREME LEARNING MACHINE

1. Single-Hidden Layer Feed Forward Neural Network

Suppose N known samples (x_k, t_k), $k=1\sim N$, in which the input vector is $x_k = [x_{k1}, x_{k2}, ..., x_{kn}]^T \in R^n$, the output vector is $t_k = [t_{k1}, t_{k2}, ..., t_{km}]^T \in R^m$. The standard mathematical model of SLFNs with $\tilde{N}$ hidden layer nodes and excitation function $g(x)$ is

$$\sum_{i=1}^{\tilde{N}} \beta_i g_i(x_k) = \sum_{i=1}^{\tilde{N}} \beta_i g(\omega_i \cdot x_k + b_i) = o_k , \qquad (1)$$

where $k = 1,...,N$. $\omega_i = \left[\omega_{i1}, \omega_{i2},...,\omega_{i\tilde{N}}\right]^T$ is the weight vector connecting the ith hidden layer nodes and the input nodes. $\beta_i = \left[\beta_{i1}, \beta_{i2},...,\beta_{i\tilde{N}}\right]^T$ is the weight vector connecting the ith hidden layer nodes and the output nodes. b_i is the threshold for the ith hidden layer nodes.

To ensure the zero error output close to the expected value, that means

$$\sum_{k=1}^{N} \|o_k - t_k\| = 0 , \text{ so}$$

$$\sum_{i=1}^{\tilde{N}} \beta_i g(\omega_i \cdot x_k + b_i) = t_k , k = 1,...,N . \qquad (2)$$

The above equation can be simplified as

$$H\beta = T , \qquad (3)$$

where H is the output matrix of the hidden layer, $H(\omega_1,...,\omega_{\tilde{N}},b_1,...,b_{\tilde{N}},x_1,...,x_N)$

$$= \begin{bmatrix} g(\omega_1 \cdot x_1 + b_1) & \cdots & g(\omega_{\tilde{N}} \cdot x_1 + b_{\tilde{N}}) \\ \cdot & & \cdot \\ \cdot & \cdots & \cdot \\ \cdot & & \cdot \\ g(\omega_1 \cdot x_N + b_1) & \cdots & g(\omega_{\tilde{N}} \cdot x_N + b_{\tilde{N}}) \end{bmatrix}_{N \times \tilde{N}} ,$$

$$\beta = \begin{bmatrix} \beta_1^T \\ \cdot \\ \cdot \\ \cdot \\ \beta_{\tilde{N}}^T \end{bmatrix}_{\tilde{N} \times m} , T = \begin{bmatrix} t_1^T \\ \cdot \\ \cdot \\ \cdot \\ t_N^T \end{bmatrix}_{N \times m} .$$

If H is a invertible square matrix, which means $\tilde{N} = N$, SLFNs can approach to the training samples by zero error. Generally, the number of hidden layer nodes is much smaller than that of the training samples so that H is not a square matrix. Therefore, $\omega_i, b_i, \beta_i (i = 1,...,\tilde{N})$ that are suitable for $H\beta = T$ might not always exist. One solution is to adjust $\hat{\omega}_i, \hat{b}_i, \hat{\beta}_i (i = 1,...,\tilde{N})$ and reduce the error by error gradient reduction method. The parameter set of weight vectors ω_i, β_i and the threshold b_i, which named as vector W, can be adjusted by iteration as

$$\Delta W_k = -\eta \frac{\partial E}{\partial W} , \qquad (4)$$

where $\eta \in (0,1)$ is the proportional coefficient, that reflects the speed of learning in training. E is the output error,

$$E = \sum_{j=1}^{N} \left(\sum_{i=1}^{\tilde{N}} \beta_i g(\omega_i \cdot x_j + b_i) - t_j \right)^2 . \qquad (5)$$

2. Extreme Learning Machine Algorithm

Different with traditional function approximation theories, ELM algorithm produces the connected weights for input layer and hidden layer and the thresholds of the hidden layer nerve cells randomly. The weights and thresholds do not need to adjust. A good output is achieved only by setting the number of the hidden layer nerve cells.

As the input weights ω_i and hidden layer thresholds b_i are definite, training SLFN is equal to finding the least square solution $\hat{\beta}$.

$$\left\| H(\omega_1,...,\omega_{\tilde{N}},b_1,...,b_{\tilde{N}})\hat{\beta} - T \right\|$$

$$= \min_{\beta} \left\| H(\omega_1,...,\omega_{\tilde{N}},b_1,...,b_{\tilde{N}})\beta - T \right\| \qquad (6)$$

According to the Moore-Penrose Generalized Inverse definition, the minimum norm least square solution of the above linear system is

$$\hat{\beta} = H^+ T \qquad (7)$$

where H^+ is the Moore-Penrose Generalized Inverse of matrix H.

III. SOC ESTIMATION MODELING

The object of SOC estimation modeling is a lithium iron phosphate battery. The ELM neural network is used for SOC estimation.

The SOC is related to the voltage, current and temperature of the battery. During the charge and discharge process, the temperature changes little so that it can be neglected. The cell voltage and current are the model inputs, and the SOC is output. That is

$$\begin{cases} x_k = [u_k, i_k] \\ t_k = SOC_k \end{cases} , \qquad (8)$$

where k means the kth inputs or output.
The model structure is shown in Fig.1.

IV. SIMULATION RESULTS AND ANALYSIS

Experiments for an 180Ah/3.2V lithium iron phosphate battery have been conducted by the battery test system BTS-5V/200A, produced by Shenzhen Neware Tech. Limited Co. Ltd. Under the environment temperature of 25℃, the battery is discharged from the full-charge state to the voltage reaching 2.5V at several constant currents, ranging from 20A to 160A. The sample cycle for voltage and current is 1s, and the SOC is calculated by ampere-hour (Ah) integration. When the voltage decreases to 2.5V, it is named as the empty-discharge state, and the SOC is defined as 0. For example, test data under discharge currents of 60A and 90A is shown in Table I.

978-1-5090-0064-7/15 $31.00 © 2015 IEEE

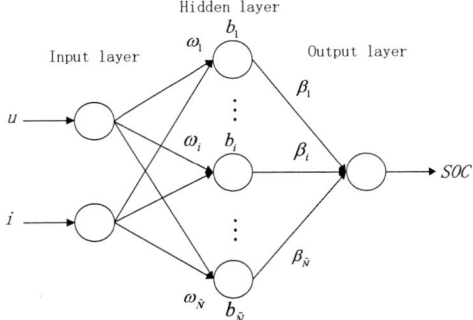

Fig. 1: ELM model structure for SOC estimation

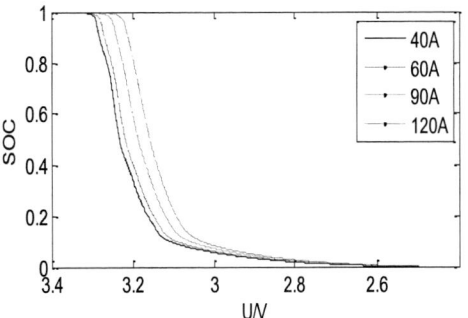

Fig. 2: SOC versus voltage at different discharge currents

Table 1: Voltage and SOC Characteristics of the Pack

Current (A)	Voltage (V)	SOC	Current (A)	Voltage (V)	SOC
59.93	3.30	0.993	89.93	3.30	0.998
59.93	3.26	0.849	89.94	3.26	0.986
⋮	⋮	⋮	⋮	⋮	⋮
59.94	3.20	0.401	89.94	3.20	0.597
59.94	3.10	0.107	89.94	3.10	0.151
59.94	3.00	0.061	89.94	3.00	0.072
⋮	⋮	⋮	⋮	⋮	⋮
59.94	2.90	0.036	89.94	2.90	0.042
59.93	2.70	0.010	89.94	2.70	0.012
⋮	⋮	⋮	⋮	⋮	⋮
59.93	2.50	0.000	89.94	2.50	0.000

The curves of SOC and voltage at the discharge currents of 40A/60A/90A/120A are displayed in Fig.2.

The terminal voltages and currents collected during the charge and discharge experiments vary greatly. If the data are taken as training samples directly, it might make a bad effect on the determination of the model structure and the output weights. Therefore, it is better to introduce data normalization process to eliminate this effect, and help to improve the convergence speed of the network training. The data normalization equation is

$$\bar{x}_i = \frac{x_i - x_{\min}}{x_{\max} - x_{\min}}, \tag{9}$$

where x_i is for input or output data. $x_{\min}$ means the minimum value of the data, and $x_{\max}$ is the maximum value. By the application of data normalization, all input and output data are limited in [0, 1].

Take the sampling voltages and currents under different discharge currents as the training data set. By the help of MATLAB, the Sigmoidal function is taken as the excitation function, and the number of the hidden layer nodes is set 40. Input the normalized data to the network model and then train the model.

The experiment data under discharge current of 60A is used for testing the generalization performance of the neural network. The SOC estimation results by Ah integration and the proposed ELM are compared in Fig.3. The errors between the two methods are shown in Fig.4.

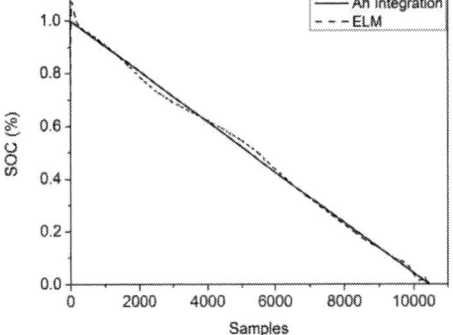

Fig. 3: SOC estimation comparison under current of 60A

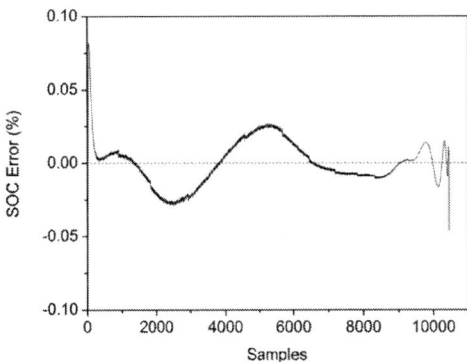

Fig. 4: SOC error under current of 60A

Then, the experiment data under discharge current of 90A is used for testing the generalization performance of the neural network. The SOC estimation results by Ah integration and the proposed ELM are compared in Fig.5. The errors between the two methods are shown in Fig.6.

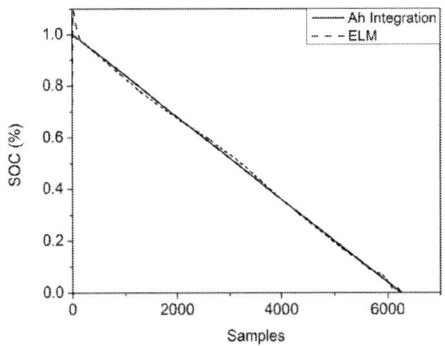

Fig. 5: SOC estimation comparison under current of 90A

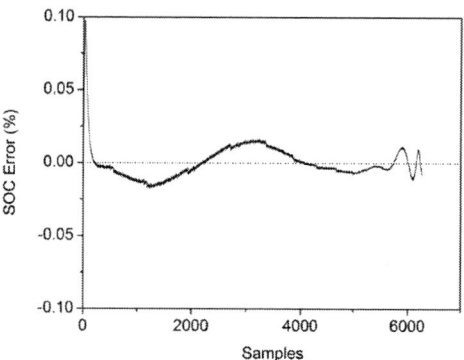

Fig. 6: SOC error under current of 90A

According to the experimental results, the ELM-based SOC estimation model proposed in this paper can accurately predict the SOC of lithium iron phosphate batteries. The maximum estimation error is less than 4%. In the SOC range of 0~0.95, the results by ELM almost coincide with that by Ah integration method. While in the range of 0.95~1, the errors between the two methods seem slightly large. That mainly due to the large change of voltage in a short time, which is caused by the polarization effect of the lithium iron phosphate battery at the beginning of the discharge. The voltage changes are related to the value of current and time. After the duration, the battery gets into a relatively stable state, and the SOC estimation errors reduce. For a more precise estimation by ELM at the beginning, a more sophisticated model matched the dynamic process is wanted.

BP neural network and BVM are taken into comparison with the proposed method to test the computing resource consumption. The result is shown in Table II. The RMSE is the root mean square error. The nodes of ELM are far fewer than that of SVM. The training speed of ELM is faster than both BP and SVM obviously.

Table 2: Comparison of computing resource consumption

Method	Training		Testing RMSE	Node Number
	Time/s	RMSE		
ELM	0.8916	0.0312	0.0344	40
BP	24.982	0.0298	0.0852	40
SVM	3284.6	0.0315	0.0873	12618

Table II shows that under the same nodes in the hidden layer and similar to the case of training accuracy, the learning time of ELM is less than 4% of that for BP neural network. The training speed is significantly accelerated. The estimation accuracy by ELM is better than that by BP method.

Compared with the SVM, the ELM needs much less nodes and achieves a higher accuracy. The node number of ELM is 40, about 0.3% of the support vectors for SVM. Meanwhile, to accomplish the calculation of the above SVM, a training time up to 3284.6s, about 55min, is needed. The results show that the ELM obtains an advantage on the training speed and generalization performance over the BP and SVM.

V. CONCLUSION

In this paper, a SOC estimation method based on ELM for lithium iron phosphate battery has been introduced. The performance of ELM in SOC prediction is evaluated, and the superiority is verified by compared with BP neural network and SVM. Traditional neural network learning algorithms are complicated in mechanism, need to set up a large number of training parameters, require intensive computing resources, and easily produce local optima. ELM algorithm does not need to adjust the connection weights between the input layer and the hidden layer, neither the threshold value of the hidden layer neurons. A unique optimal solution can be got by only setting the node number of hidden layer.

Comparative experiments among ELM, BP neural network and SVM show that parameters setting of ELM are easier, and the less calculation with a quicker learning speed and the better generalization performance may lead to better prospects in the SOC estimation.

ACKNOWLEDGMENT
This work was supported by the Jiangsu Province Science and Technology Plan (BE2014006-3), China.

REFERENCES

[1] H. Rahimi-Eichi, U. Ojha, F. Baronti, and M. Chow, "Battery Management System: An Overview of Its Application in the Smart Grid and Electric Vehicles," in *Industrial Electronics Magazine, IEEE* , vol.7, no.2, pp.4-16, June 2013.

[2] K. W. E. Cheng, B. P. Divakar, H. Wu, K. Ding, and H. F. Ho, "Battery-Management System (BMS) and SOC Development for Electrical Vehicles," in *Vehicular Technology, IEEE Transactions on* , vol.60, no.1, pp.76-88, Jan. 2011.

[3] H. He, R. Xiong, X. Zhang, F. Sun, and J. Fan, "State-of-Charge Estimation of the Lithium-Ion Battery Using an Adaptive Extended Kalman Filter Based on an Improved Thevenin Model," in *Vehicular Technology, IEEE Transactions on* , vol.60, no.4, pp.1461-1469, May 2011.

[4] M. Charkhgard, and M. Farrokhi, "State-of-Charge Estimation for Lithium-Ion Batteries Using Neural Networks and EKF," in *Industrial Electronics, IEEE Transactions on* , vol.57, no.12, pp.4178-4187, Dec. 2010.

[5] J. Kim, and B. H. Cho, "State-of-Charge Estimation and State-of-Health Prediction of a Li-Ion Degraded Battery Based on an EKF Combined With a Per-Unit System," in *Vehicular Technology, IEEE Transactions on* , vol.60, no.9, pp.4249-4260, Nov. 2011.

[6] S. Lee, J. Kim, J. Lee, and B. H. Cho, "State-of-charge and capacity estimation of lithium-ion battery using a new open-circuit voltage versus state-of-charge," in *Journal of Power Sources* , vol.185, no.2, pp.1367-1373, Dec. 2008.

[7] J. Xu, M. Gao, Z. He, Q. Han, and X. Wang, "State of Charge Estimation Online Based on EKF-Ah Method for Lithium-Ion Power Battery," in *Image and Signal Processing, 2009. CISP '09. 2nd International Congress on* , vol., no., pp.1-5, 17-19 Oct. 2009.

[8] I. H. Li, W. Y. Wang, S. F. Su, and L. Yuang-Shung, "A Merged Fuzzy Neural Network and Its Applications in Battery State-of-Charge Estimation," in *Energy Conversion, IEEE Transactions on* , vol.22, no.3, pp.697-708, Sept. 2007.

[9] G. B. Huang, H. Zhou, X. Ding, and R. Zhang, " Extreme Learning Machine: Theory and Applications," in *Neurocomputing*, vol.70, no.1-3, pp.489-501, Dec. 2006.

[10] V. N. Vapnik, "An overview of statistical learning theory," in *Neural Networks, IEEE Transactions on* , vol.10, no.5, pp.988-999, Sep 1999.

[11] G. B. Huang, Q. Y. Zhu, and C. K. Siew, "Extreme learning machine: a new learning scheme of feedforward neural networks," in *Neural Networks, 2004. Proceedings. 2004*

978-1-5090-0064-7/15 $31.00 © 2015 IEEE

IEEE International Joint Conference on , vol.2, no., pp.985-990 vol.2, 25-29 July 2004.

[12] G. B. Huang, H. Zhou, X. Ding, and R. Zhang, "Extreme Learning Machine for Regression and Multiclass Classification," in *Systems, Man, and Cybernetics, Part B: Cybernetics, IEEE Transactions on* , vol.42, no.2, pp.513-529, April 2012.

BIOGRAPHY

Zhihao Wang received the B.S. degree in electronic and information engineering from Zhejiang Sci-Tech University in 2012.He is currently working toward the M.S. degree in control engineering in GuangXi University.

His current research interests are in the areas of design and control of battery management system, charge equalization converter,and low-voltage high-current dc/dc converters.

A Sensorless Vector Strategy for the PMSM Using Improved Sliding Mode Observer and Fuzzy PI Speed Controller

Hui Deng[1] Guang-Zhong Cao[1] Su-Dan Huang[1,2] Lai-Juan Shi[3] Zhi-Ming He[4]

[1] Shenzhen Key Laboratory of Electromagnetic Control, Shenzhen University, Shenzhen, 518060, China
E-mail: denghui@email.szu.edu.cn; gzcao@szu.edu.cn
[2] College of Electrical Engineering, Southwest Jiaotong University, Chengdu, 610031, China
E-mail: hsdsudan@gmail.com
[3] China International Marine Containers Intelligent Technology CO., LTD., Shenzhen, 518067, China
E-mail: laijuan.shi@cimc.com
[4] Shenzhen Hpmont Technology CO., LTD., Shenzhen, 518055, China
E-mail: hezhiming@hpmont.com

Abstract–To reduce the cost and obtain accurate rotor position of the permanent magnet synchronous motor (PMSM), a sensorless vector strategy for the PMSM using an improved sliding mode observer (SMO) and a fuzzy PI speed controller is proposed. The mathematical model of the PMSM is firstly presented. Then, a sigmoid function is applied to the improved SMO for overcoming chattering, and the observer of back electromotive force instead of the low-pass filter is introduced to the SMO. The stability of the improved SMO is further proved with the Lyapunov function. Additionally, the fuzzy PI speed controller is designed. The simulation is performed based on MATLAB, and the experiment is carried out under the developed test rig based on DSP TMS320F2808. The effectiveness of the proposed strategy is finally verified.

Keywords–Fuzzy PI speed controller, permanent magnet synchronous motor (PMSM), sliding mode observer (SMO), sensorless vector control.

I. INTRODUCTION

Permanent magnet synchronous motor (PMSM) has been widely applied in industrial applications, owning to its inherent advantages of high ratio of torque, high power density, high efficiency, and easy control. The high performance speed control and position control of the PMSM require accurate rotor position and rotor velocity for synchronizing the excitation pulses of phases to the rotor position. However, the sensors are frequently used in the PMSM, and they result in reduced reliability, susceptibility to noise, additional cost, big bulk, and increased systematic complexity. The sensorless strategy of the PMSM can overcome this issue.

Several methods have been proposed for sensorless strategy of the PMSM [1]-[3], such as the flux linkage estimation method, the model reference adaptive method, the signal injection method, the Kalman filter method, and the sliding mode observer (SMO) method. To solve the aforementioned problem, an improved SMO is proposed to achieve the sensorless vector control of the PMSM. The conventional switch function is replaced by the sigmoid function to reduce chattering. To obtain the accurate rotor position and rotor velocity, an observer is created based on the back electromotive force (back-EMF).

For the vector control of the PMSM, the outer speed loop greatly affects the system performance. The proportional integral (PI) controller is usually applied in the outer speed loop. Due to the fixed proportional and integral gains of the PI controller, the system performance is easily affected

by the parameter variation, load disturbance, and speed variation [4]-[6]. Hence, an effective speed controller is of great importance for improving the system performance. Fuzzy control does not depend on the mathematic model of the system, and its algorithm is easy to design. The fuzzy PI regulator not only has the advantages of the PI controller, such as simple, stability, and reliability, but also has the merits of the fuzzy controller, such as quick response and small overshoot. Thus, a fuzzy PI speed controller is employed for the PMSM in this paper.

In this paper, a sensorless vector strategy for the PMSM using an improved SMO and a fuzzy PI speed controller is proposed, and the simulation and experiment are carried out to verify the proposed strategy.

II. MATHEMATICAL MODEL OF THE PMSM

The mathematical model of the PMSM in the frame (α, β) is given by [7], [8]

$$\frac{di_s}{dt} = -\frac{R}{L_s}i_s - \frac{1}{L_s}e_s + \frac{1}{L_s}u_s \tag{1}$$

$$i_S = \begin{bmatrix} i_\alpha & i_\beta \end{bmatrix}^T \tag{2}$$

$$u_S = \begin{bmatrix} u_\alpha & u_\beta \end{bmatrix}^T \tag{3}$$

$$e_S = \begin{bmatrix} e_\alpha \\ e_\beta \end{bmatrix} = \begin{bmatrix} -K_E\omega\sin\theta \\ K_E\omega\cos\theta \end{bmatrix} \tag{4}$$

where i_α and i_β, u_α and u_β, e_α and e_β are the phase currents, phase voltages, and back-EMF in the stationary reference frame (α, β), respectively. R is the stator phase resistance, L_s is the stator phase inductance, K_E is the flux linkage of the permanent magnet, and ω is the electrical angular velocity.

From (4), it is clear that the back-EMF contains the information of rotor position. Therefore, the information of the rotor position and rotor speed can be obtained through the estimated back-EMF with the observer.

III. THE IMPROVED SMO

The block diagram of the improved SMO is presented in Fig. 1. According to the theory of sliding mode variable structure [9], [10], a sliding mode observer of the PMSM can be established.

978-1-5090-0064-7/15 $31.00 © 2015 IEEE

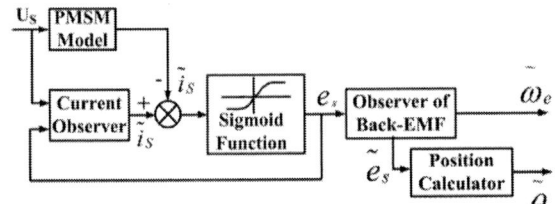

Fig. 1: Block diagram of the improved SMO

The sliding mode surface is given as

$$S = \overline{i}_\alpha = \tilde{i}_\alpha - i_\alpha = 0 \qquad (5)$$

where $\tilde{i}_\alpha$ is the observing current of the SMO, and i_α is the actual current. The sliding mode observer based on the mathematical model of the PMSM in the α-β coordinate frame is expressed as

$$\frac{d\tilde{i}_s}{dt} = -\frac{R}{L_S}\tilde{i}_s + \frac{1}{L_S}(u_s - Z_s) \qquad (6)$$

$$\tilde{i}_s = \begin{bmatrix} \tilde{i}_\alpha \\ \tilde{i}_\beta \end{bmatrix} \qquad (7)$$

$$Z_S = \begin{bmatrix} Z_\alpha \\ Z_\beta \end{bmatrix} = \begin{bmatrix} k*F(\tilde{i}_\alpha - i_\alpha) \\ k*F(\tilde{i}_\beta - i_\beta) \end{bmatrix} \qquad (8)$$

where k is a constant current observer gain. The sgn function of the conventional SMO is replaced by a sigmoid function defined as

$$F(x) = \frac{2}{1 + e^{-ax}} - 1 \qquad (9)$$

where a is the adjustable parameter. Subtracting (1) from (6), the error equation of the estimating current is given as

$$\begin{cases} \dfrac{d\hat{i}_\alpha}{dt} = -\dfrac{R}{L_S}\hat{i}_\alpha + \dfrac{e_\alpha}{L_s} - \dfrac{k}{L_S}F(\hat{i}_\alpha) \\ \dfrac{d\hat{i}_\beta}{dt} = -\dfrac{R}{L_S}\hat{i}_\beta + \dfrac{e_\beta}{L_s} - \dfrac{k}{L_S}F(\hat{i}_\beta) \end{cases} \qquad (10)$$

In order to determine the stability of the designated observer, the Lyapunov function is constructed as

$$V = \frac{1}{2}S(X)^T S(X) \qquad (11)$$

where

$$S(X) = \begin{bmatrix} S_\alpha(X) \\ S_\beta(X) \end{bmatrix} = \begin{bmatrix} \tilde{i}_\alpha - i_\alpha \\ \tilde{i}_\beta - i_\beta \end{bmatrix} . \qquad (12)$$

By calculation and analysis, the result is expressed as

$$\dot{V} \le 0 \quad . \qquad (13)$$

According to (13), the sliding motion exists and the state observer (6) is unanimous asymptotic stability in the global scope while k is large enough, that is

$$k > \max(|e_\alpha|, |e_\beta|) . \qquad (14)$$

If the system is in the state of sliding mode, $\hat{i}_\alpha = 0$, $\hat{i}_\beta = 0$, $\dfrac{d\hat{i}_\alpha}{dt} = 0$ and $\dfrac{d\hat{i}_\beta}{dt} = 0$, then

$$\begin{cases} u_{eq\alpha} = [kF(\hat{i}_\alpha)]_{eq} = \tilde{e}_\alpha \\ u_{eq\beta} = [kF(\hat{i}_\beta)]_{eq} = \tilde{e}_\beta \end{cases} . \qquad (15)$$

The rotor position and rotor speed of the PMSM can be obtained from (15) as follows

$$\tilde{\theta}_e = -\tan^{-1}\left(\frac{\tilde{e}_\alpha}{\tilde{e}_\beta}\right) \qquad (16)$$

$$\tilde{\omega}_e = \frac{d\tilde{\theta}_e}{dt} . \qquad (17)$$

The block diagram of the proposed sensorless control system of the PMSM is shown in Fig. 2. The rotor position and rotor speed is estimated through the improved SMO. The fuzzy PI controller is adopted in the speed loop instead of the conventional PI controller, and the PI controller is used in the current loop.

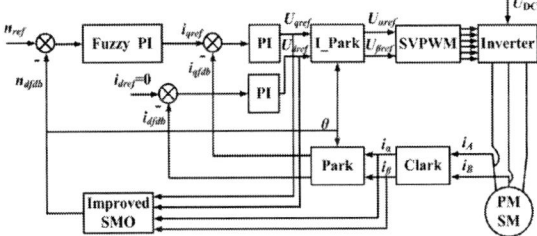

Fig.2: Block diagram of the proposed sensorless control system

IV. THE FUZZY PI SPEED CONTROLLER

A fuzzy logic controller (FLC) is free of accurate mathematical model and is based on the linguistic rules formed from the experience with the system. Compared to the PI controller, the FLC involves approximations, increased complexity, more computations, and higher memory requirements [11], [12]. The performance of the FLC is superior only under transient conditions, and the performance of the PI controller is superior under the steady state condition. A fuzzy PI controller is used, and it is based on the fuzzy rule table set, so the parameters of PI controller can be adjusted online, satisfied the self-adaption control.

The schematic diagram of the fuzzy PI controller is shown in Fig. 3. K_P and K_I are the parameters of PI controller, which can be adjusted through ΔK_P and ΔK_I according to external changes, as depicted in (18).

$$\begin{cases} K'_{\mathrm{p}} = K_{\mathrm{p}} + \Delta K_{\mathrm{p}} \\ K'_{\mathrm{I}} = K_{\mathrm{I}} + \Delta K_{\mathrm{I}} \end{cases} \qquad (18)$$

where K'_{P} and K'_{I} are the adjusted K_{P} and K_{I}, respectively.

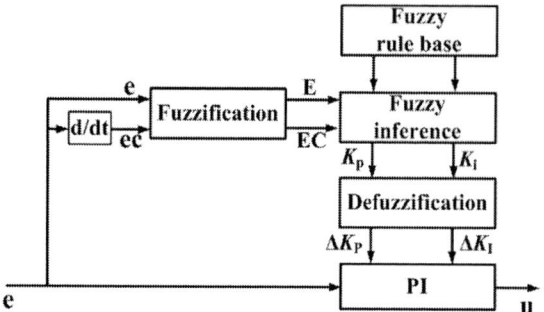

Fig. 3: Schematic diagram of the fuzzy PI controller

The FLC is a double-input and double-output system. Its inputs are error (e) and rate of error change (ec), and its outputs are ΔK_{P} and ΔK_{I}, which can be attained by the rules based on fuzzy mathematics. It is consist of four parts, i.e. fuzzification, fuzzy rule base, fuzzy inference, and defuzzification.

1. Fuzzification

In Fig. 3, E and EC are fuzzy variables of e and ec, K_{p} and K_{i} are fuzzy variables of ΔK_{P} and ΔK_{I}. The range of inputs and outputs are given. E and EC= {-3, -2.5, -2, -1.5,-1,-0.5, 0, 0.5, 1, 1.5, 2, 2.5, 3}, ΔK_{P} and ΔK_{I} = {-0.3, -0.25, -0.2,-0.15,-0.1,-0.05, 0, 0.05, 0.1, 0.15, 0.2, 0.25, 0.3}. The quantum factors K_{e} and K_{c} are used for transforming the inputs into the fuzzy variables E and EC, respectively. They can be described as

$$\begin{cases} E = e * K_e \\ EC = ec * K_c \end{cases} \cdot \qquad (19)$$

2. Fuzzy rule base

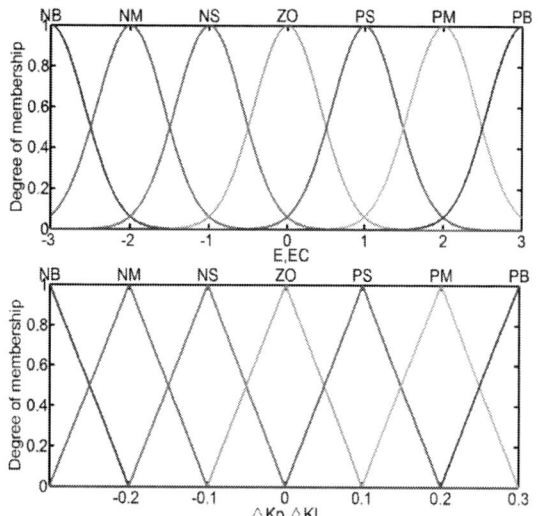

Fig.4: Membership functions of inputs and outputs

The fuzzy sets of the fuzzy variables are defined as {NB (negative big), NM (negative middle), NS (negative small), ZO (zero), PS (positive small), PM (positive middle), PB

(positive big)}. Their membership functions are illustrated in Fig. 4. Based on the common control engineering knowledge [13], the inputs of membership functions are chosen as gauss membership function, and the outputs of membership are selected as triangular membership functions, due to their high computational efficiency.

3. Fuzzy inference

The fuzzy logic rules of K_{p} and K_{i} are listed in Tables 1 and 2, and the surfaces of K_{p} and K_{i} are shown in Figs. 5 and 6, respectively. The Mamdani fuzzy inference is used. According to the expert experiment in the PMSM servo system and simulation analysis of the system, there are usually 49 rules in the system.

Table 1: Fuzzy logic rules of K_{p}

		EC						
		NB	NM	NS	ZO	PS	PM	PB
E	NB	PB	PB	PM	PM	PS	ZO	ZO
	NM	PB	PB	PM	PS	PS	ZO	NS
	NS	PM	PM	PM	PS	ZO	NS	NS
	ZO	PM	PM	PS	ZO	NS	NM	NM
	PS	PS	PS	ZO	NS	NS	NM	NM
	PM	PS	ZO	NS	NM	NM	NM	NB
	PB	ZO	ZO	NM	NM	NM	NB	NB

Table 2: Fuzzy logic rules of K_{i}

		EC						
		NB	NM	NS	ZO	PS	PM	PB
E	NB	NB	NB	NM	NM	NS	ZO	ZO
	NM	NB	NB	NM	NS	NS	ZO	ZO
	NS	NB	NM	NS	NS	ZO	PS	PS
	ZO	NM	NM	NS	ZO	PS	PM	PM
	PS	NM	NS	ZO	PS	PS	PM	PB
	PM	ZO	ZO	PS	PS	PM	PB	PB
	PB	ZO	ZO	PS	PM	PM	PB	PB

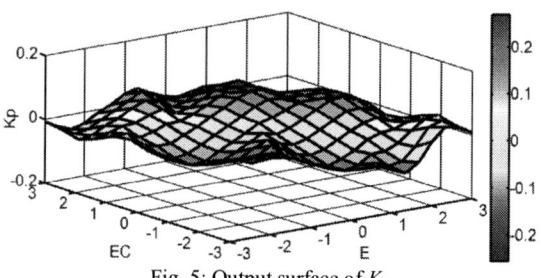

Fig. 5: Output surface of K_{p}

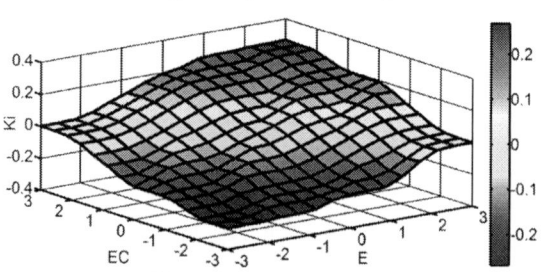

Fig. 6: Output surface of K_{i}

4. Defuzzification

The max-min composition & centroid method is employed in defuzzification. According to this rule, K_{p} and K_{i} are calculated, but they are fuzzy valves. The scaling factors y_1 and y_2 are used for transforming them into accurate valves (ΔK_P and ΔK_I), as given by

$$\begin{cases} \Delta K_p = K_p * y_1 \\ \Delta K_i = K_i * y_2 \end{cases}. \quad (20)$$

V. SIMULATION AND EXPERIMENTAL RESULTS

1. Simulation results

For a sensorless vector control system of the PMSM with the improved SMO and the fuzzy PI controller, simulation is performed based on MATLAB/Simulink. The parameters of the PMSM are listed in Table 3.

Table 3: Parameters of the PMSM

Parameters	Value	Unit
Rated power P_N	350	W
Rated voltage U_N	220	V
Rated speed n_N	3000	rpm
Rated torque T_N	1.1	N. m
Stator resistance R_s	3.1	Ω
Stator inductance $L_d=L_q$	8.5	mH
Rotor inertia J	0.9×10^{-4}	kg $\cdot$ m^2
Number of poles n_P	4	\
Flux ψ	0.25	Wb

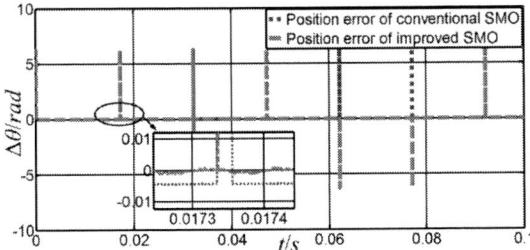

Fig.7: Position error of the conventional SMO and the improved SMO

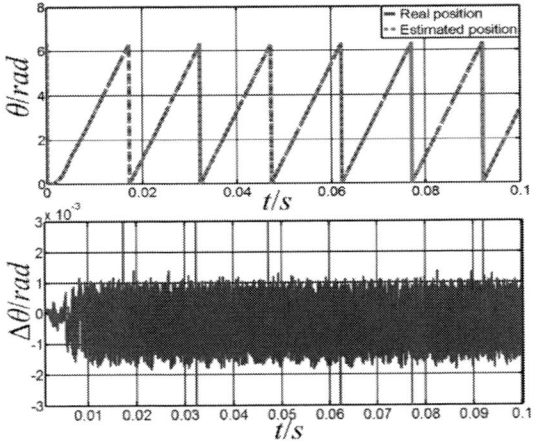

Fig. 8: Estimated position and position error

According to the sinusoidal current in the current loop, the parameters of PI controller are achieved by trial and error. The position error of the conventional SMO and improved SMO is shown in Fig. 7. From the partial enlarged detail, the position error of the improved SMO is smaller than the position error of the conventional SMO. The estimated

position and position error of the PMSM with the improved SMO are shown in Fig. 8, and the position error is $\pm1.5\times10^{-3}\ rad$ from (21).

$$\Delta\theta = \pm\frac{2\pi}{6.2}\times1.5\times10^{-3} = \pm1.5*10^{-3}(rad). \quad (21)$$

Fig. 9 shows the velocity response with load disturbance at 0.03s. Compared to the velocity response of PI controller, the overshoot is smaller and settling time is shorter with the fuzzy PI controller.

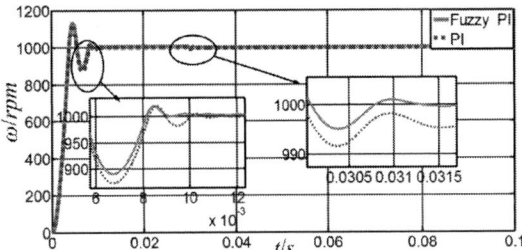

Fig. 9: Velocity response with load disturbance at 0.03s

The characteristic of velocity is shown in Fig. 10, the PMSM can be controlled sensorlessly from 300 to 1500 rpm with fixed load, and the maximum velocity control error is less than 7% of the reference velocity.

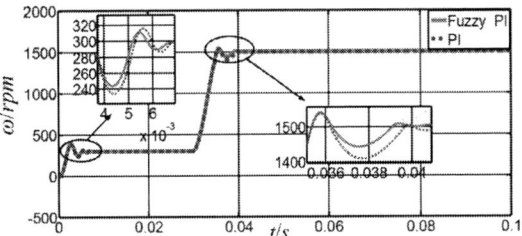

Fig. 10: Velocity response from 300 to 1500 rpm with fixed load

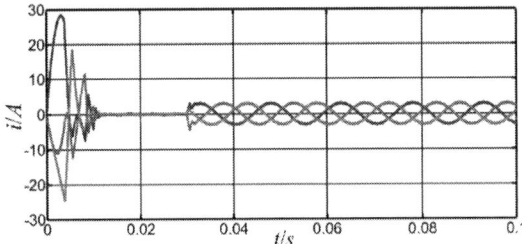

Fig. 11: Three-phase current response

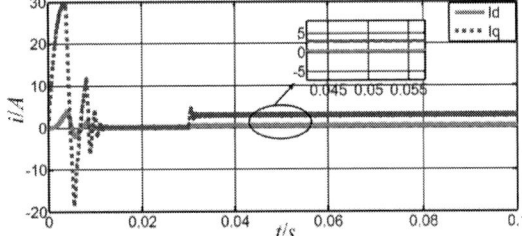

Fig. 12: Current response of I_d and I_q

The three-phase current response of the PMSM is shown in Fig. 11, which is fairly sinusoidal in the three cases. Fig. 12 shows the current response of I_d and I_q, I_d=0 and I_q=2A coincide with the principle of vector control.

978-1-5090-0064-7/15 $31.00 © 2015 IEEE

2. Experimental results

The experimental setup of the sensorless vector control system of the PMSM is shown in Fig.13, which is composed of the PMSM, a drive, a digital signal processor (DSP) controller, a DSP emulator, a current clamp, and an oscilloscope. The control algorithm has been implemented under the developed test rig based on DSP TMS320F2808. The reference velocity is set to 1000 rpm.

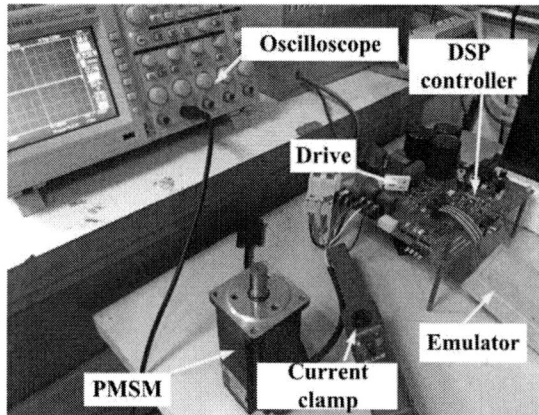

Fig. 13: Experimental setup for sensorless vector control system of the PMSM

Figs. 14 and 15 show that there are no large ripples at the estimated currents, the estimated back-EMF has a clear sinusoidal waveform, and the steady-state error is reduced. The error of the estimated position is close to zero, and thus the estimated position is approximately equal to the actual position. The current response of I_d and I_q show the vector control has a good performance.

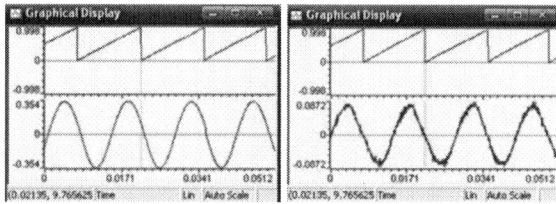

Fig. 14: Estimated position, back-EMF, and current

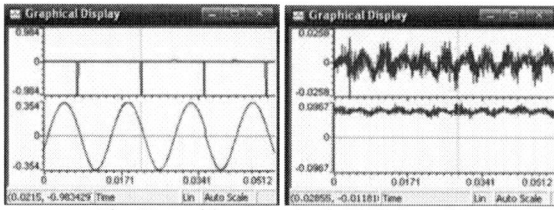

Fig. 15: Position error, back-EMF, I_d, and I_q

VI. CONCLUSION

The sensorless vector control strategy of the PMSM has been presented using an improved SMO and a fuzzy PI speed controller in this paper. The improved SMO is used to estimate the rotor position and rotor speed of the PMSM. Compared to the conventional SMO, the improved SMO has features of simpler structure and better estimated performance. The fuzzy PI speed controller is applied to adjust the rotor velocity, and the rotor velocity response is much quicker and smaller overshoot compared with the PI controller. The validity and superior performance of the proposed strategy are verified via the simulation and experiment.

ACKNOWLEDGMENT

The authors would like to thank the National Natural Science Foundation of China under Grant NSFC51275312, the National Key Technology R&D Program under Grant 2014BAH23F04, and Shenzhen Government Fund under Grant KC2014JSJS0001A.

REFERENCES

[1] G. L. Wang, R. F. Yang, and D. G. Xu, "DSP-based control of sensorless IPMSM drives for wide speed range operation", IEEE Trans. Ind. Electron., Vol. 60, No. 2, 2013, pp. 720-727.

[2] T. Bernardes, V. Foletto Montagner, H. A. Grundling, and H.Pinheiro "Discrete time sliding mode Observer for sensorless vector control of permanent magnet synchronous machine", IEEE Trans. Ind. Electron., Vol. 61, No. 4, 2014, pp. 1679-1691.

[3] J. Rivera Domínguez, A. Navarrete, M. A. Meza, A. G. Loukianov and J. Canedo, "Digital sliding mode sensorless control for surface mounted PMSM" , IEEE Trans. Ind. Informat., Vol. 10, No. 1, 2014, pp. 137-151.

[4] R. H. Du, Y. F. Wu, and W. Chen, "Adaptive fuzzy speed control for permanent magnet synchronous motor servo systems", IEEE Electr., Power Compon., Syst., Vol. 42, No. 8, 2014, pp. 798-807.

[5] Q. L. Viet, H. H. Choi, and J.W. Jung, "Fuzzy sliding mode speed controller for PMSM with a load torque observer", IEEE Trans. Power Electron., Vol. 27, No. 3, 2012, pp. 1530-1539.

[6] C. C. Lee, "Fuzzy logic in control system: fuzzy logic controller-part I/II", IEEE Trans. Syst., Cybern, Vol. 20, No. 2, 1990, pp. 404-418.

[7] R. A. Walambe, V. A. Joshi, and A. A. Apte, "Study of sensorless control algorithms for a permanent magnet synchronous motor vector control drive", International Conf. on Industrial Instrumentation and Control (ICIC), 28-30 May 2015, pp. 423-428.

[8] Y. Zhao, W. Qiao and L. Wu, "Dead-Time effect analysis and compensation for a sliding mode position observer based sensorless IPMSM control system", IEEE Trans. Ind. Appl., Vol. 51, No. 3, 2015, pp. 2528-2535.

[9] O. Barambones, and P. Alkorta, "Position control of the induction motor using an adaptive sliding mode controller and observers", IEEE Trans. Ind. Electron., Vol. 61, No. 12, 2014, pp. 6556-6565.

[10] V. I. Utkin, "Sliding mode control design principles and applications to electric drives", IEEE Trans. Ind. Electron., Vol. 40, No. 1, 1993, pp. 23-36.

[11] H. H. Choi, H. M. Yun, and Y. Kim, "Implementation of evolutionary fuzzy PID speed controller for PM synchronous motor", IEEE Trans. Ind. Inform., Vol. 11, No. 2, 2015, pp. 540-547.

[12] M. H. Asemani and V. J. Majd, "A robust H_∞ Non-PDC design scheme for singularly perturbed T–S fuzzy systems with immeasurable state variables", IEEE Trans. Fuzzy Syst., Vol. 23, No. 3, 2015, pp. 525-541.

[13] J. W. Jung, Y.S. Choi, V. Q. Leu, H. H. Choi, "Fuzzy PI-type current controllers for permanent magnet synchronous motors," IEEE Trans. Electr. Power Appl., Vol. 5, No. 1, 2011, pp.143-152.

BIOGRAPHIES

Hui Deng received his B.Sc. degree in Huazhong University of Science and Technology Wuchang Branch, Wuhan, Hubei, China, in 2013.

He is currently a postgraduate with the Shenzhen Key Laboratory of Electromagnetic Control, Shenzhen University, Shenzhen, Guangdong, China. His research interests include motor control and robotics.

Guang-Zhong Cao received the B.Sc., M.Sc., and Ph.D. degrees in electrical engineering and automation from Xi'an Jiaotong University, Xi'an, Shanxi, China, in 1989, 1992, and 1996, respectively.

He is currently a Professor and Director with the Shenzhen Key Laboratory of Electromagnetic Control, Shenzhen University, Shenzhen, Guangdong, China. He has published more than 80 articles in refereed journals and conferences. His research interests include motor control, and control theory and its application.

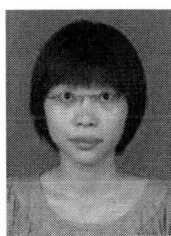

Su-Dan Huang received the B.Sc. and M.Sc. degrees in College of Mechatronics and Control Engineering from Shenzhen University, Shenzhen, Guangdong, China, in 2009 and 2012, respectively.

She is currently a joint PHD candidate with the Department of Electrical Engineering, Southwest Jiaotong University, Chengdu, Sichuan, China, and Shenzhen Key Laboratory of Electromagnetic Control, Shenzhen University, Shenzhen, Guangdong, China. Her research interests include design and control of planar switched reluctance motors, and control theory and its application.

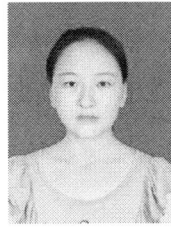

Lai-Juan Shi received the B.Sc. degree in electrical engineering from Zhengzhou Institute of Aeronautical Industry Management, Zhengzhou, China, in 2013.

She is currently an Engineer with the China International Marine Containers Intelligent Technology CO., LTD. and also a postgraduate with Shenzhen University, Shenzhen, Guangdong, China. Her research interests include intelligent control and automatic test.

Zhi-Ming He received the B.Sc. and M.Sc. degrees in electrical engineering and automation from Xi'an Jiaotong University, Xi'an, Shanxi, China, in 1997 and 2000, respectively.

He is currently an Engineer and the Director with the Shenzhen Hpmont Technology CO., LTD, Shenzhen, Guangdong, China. His research interests include motor control and industrial automation.

Development of the Three-Dimensional Scanning System Based on Monocular Vision

Yu-Xin Liang[1]   Guang-Zhong Cao[1]   Hong Qiu[1]   Su-Dan Huang[1, 2]   Shou-Qin Zhou[3]

[1] Shenzhen Key Laboratory of Electromagnetic Control, Shenzhen University, Shenzhen, 518060, China
E-mail: liangyuxin2@email.szu.edu.cn; gzcao@szu.edu.cn; sunday_qh@163.com
[2] College of Electrical Engineering, Southwest Jiaotong University, Chengdu, 610031, China
E-mail: hsdsudan@gmail.com
[3] China International Marine Containers Intelligent Technology CO., LTD., Shenzhen, 518067, China
E-mail: shouqin.zhou@cimc.com

Abstract–The emerging three-dimensional (3D) scanning technology based on computer vision is attracting more and more attention in industrial applications such as the 3D modeling of motors and automobiles. This paper develops a 3D scanning system using monocular vision to rapidly and automatically obtain the 3D dimension of the scanned object. The principle of the 3D scanning system is firstly introduced. The structure is then presented. The system consists of a camera, a line laser generator, and a rotation platform with a direct-current motor. The software of the system is developed based on C # and EmguCV library functions to realize image processing. Additionally, experiments are carried out through a fabricated prototype. Experimental results demonstrate that the developed system achieves the rapid acquisition for the 3D dimension of the scanned object and features simple structure, low cost, easy to extend, and rapid 3D reconstruction.

Keywords– Computer vision, linear laser, monocular vision, three-dimensional (3D) reconstruction, 3D scanner.

I. INTRODUCTION

The three-dimensional (3D) technology based on computer vision can obtain 3D dimension of the scanned object rapidly. It has brought great convenience in industrial applications, since the acquisition of the 3D stereo parameters and 3D model is a serious challenge in majority of the industrial applications, especially for the 3D modeling of motors, automobiles, and electron devices. The computer vision is replacing the traditional detection and measurement, which can improve the flexibility and automaticity of the production process [1], [2].

Currently, most of the existing 3D scanning devices have a complex structure, whereas the scanned results are only used to display the 3D vision in the plane. The actual size and the full range of the 3D dimension of the scanned object are not obtained via most of the existing 3D scanning devices, which are the necessary requirement of 3D technology. Thus, most of the existing 3D scanning devices cannot meet the need of 3D technology.

For the implementation of 3D scanning, there are various schemes, however, the contact 3D scanning and non-contact 3D scanning are the two main methods [3]. Contact 3D scanning requires physical contact to acquire the 3D information. The contact scanning exhibits high accuracy, nonetheless it usually time-consuming and may damage the scanned object. Non-contact 3D scanning calculates the 3D information by obtaining the reflected light of the scanned object in the image. The non-contact scan-

ing features faster scanning and simpler operation,–_compared with the contact 3D scanner.

To achieve the acquisition of the actual size and full range of the 3D dimension of the scanned object rapidly through a simple-structure 3D scanning system, a novel non-contact 3D scanning system based on monocular computer vision is developed in this paper. This system is different from the conventional 3D cameras. The mechanical platform of the system is mainly composed of a wide angle camera and an infrared ray line laser generator. The software of the system is developed based on C# and EmguCV library functions. The image processing is simplified, since an infrared ray is used as the line laser source. As the relative positions of the rotation platform, camera, and laser line generator are fixed, it is convenient to establish the mathematical model of the system.

II. THE PRINCIPLE OF THE MONOCULAR VISION

There are three coordinate systems in the monocular vision system, image coordinate system, world coordinate system and camera coordinate system. The diagram of the principle for the monocular vision system is shown in Fig. 1.

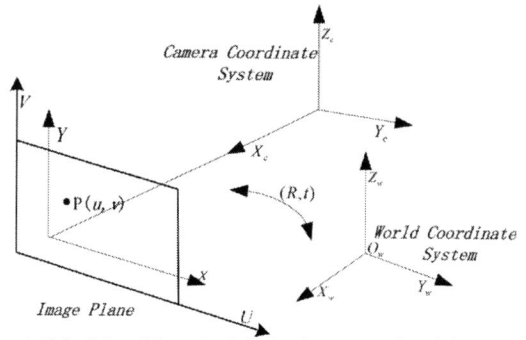

Fig. 1: Principle of the principle for the monocular vision system

The feature extraction point P is in the image coordinate system, and its coordinate is (u, v). The coordinates of P in the world coordinate system and camera coordinate system are (X_w, Y_w, Z_w) and (X_c, Y_c, Z_c). The relationship is represented as

$$\begin{bmatrix} X_c \\ Y_c \\ Z_c \\ 1 \end{bmatrix} = \begin{bmatrix} R & t \\ 0^T & 1 \end{bmatrix} \begin{bmatrix} X_w \\ Y_w \\ Z_w \\ 1 \end{bmatrix} \tag{1}$$

978-1-5090-0064-7/15 $31.00 © 2015 IEEE

where $[R\ t]$ is the rotating pan of the camera coordinate system and the target coordinate system transformation. The origin of the coordinate set of the image coordinate system is (u_0, v_0). The camera model is given as

$$Z_c = \begin{bmatrix} u \\ v \\ 1 \end{bmatrix} = \begin{bmatrix} f_x & 0 & u_0 & 0 \\ 0 & f_y & v_0 & 0 \\ 0 & 0 & 1 & 0 \end{bmatrix} \begin{bmatrix} R & t \\ 0^T & 1 \end{bmatrix} \begin{bmatrix} X_w \\ Y_w \\ Z_w \\ 1 \end{bmatrix} \quad (2)$$

where f_x is the amplification factor of the X-axis scale, and f_y is the amplification factor of Y-axis scale.

The targets are detected and identified firstly by the feature extraction and matching methods. Then, the internal parameters of the camera are obtained through the camera calibration. Finally, combined with internal parameters before and after obtaining, the dimension in the image can be calculated.

III. DESIGN OF THE 3D SCANNING SYSTEM

The 3D scanning system consists of the mechanical platform and software. For the mechanical platform, the camera and line laser generator are fixed on the mechanical platform. For the software for image processing, it is used to determine the software flow and algorithm of image processing.

1. Design of the mechanical platform

Because of the limit of the monocular vision, the acquisition of deep information from a single 2D image is limited. In order to obtain the deep information, the continuous target images acquired by a single camera in different time and space are needed. The target dimension and other parameters are calculated with the relationship of time and space among the continuous images.

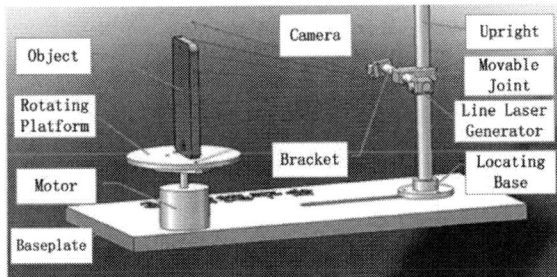

Fig.2: Design of the mechanical platform

The mechanical platform is presented in Fig. 2. The relative positions of the rotation platform, camera, and laser line generator are fixed. A rotating platform on the base is designed to place the scanned object. During the scanning process, the object rotates with the platform at a constant speed, and the camera records the images of different rotating degrees at the same time.

2. Design of the software

The workflow of the 3D scanning system is divided into four parts under the calibrated camera: image acquisition, image processing, feature extraction, and 3D reconstruction. According to the requirements of the 3D scanning system, the software flow is shown in Fig. 3.

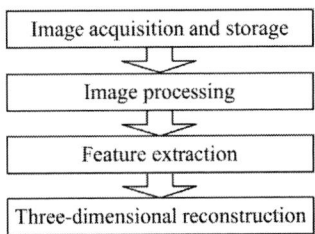

Fig. 3: Flow diagram of the software

3. Technical scheme of the software system

For the camera calibration method and steps, a lot of related researches have been performed. Quite a variety of effective methods are accessible. In this paper, the theory and the method of camera calibration are not introduced. However, the camera calibration is clarified. Zhang Zheng-You calibration method [4], [5] is adopted in this paper.

3.1. Image acquisition

Acquiring 2D image is the first step to obtain the 3D dimension of the scanned object. In order to obtain the actual size and the full range of information of the scanned object, a group of orderly 2D images is needed.

In this paper, the scanned object is put on the rotating platform. The camera images are transmitted to the computer through the USB port, and they are stored in the computer.

3.2. Pre-processing and feature extraction

The image acquired from the camera contains noise. To limit the usefulness of the information and improve the accuracy, the pre-process of the image is necessary, since the definition of images can be improved, the image processing can be easier with computer, and the feature analysis is more convenient. The laser provides a sharp contrast to the environment. Nevertheless, it is also difficult to find out which pixel is the laser and which pixel is the background in two images.

3.2.1. Factors affecting the laser line in the image

In order to separate the laser and ambient light, an infrared laser with a wavelength of 850 nm is selected. The main factors affecting the laser line in the image are described as follows.

(a) Width of laser source

For the ideal line laser, the line laser is approximately equal to the single pixel column in the image. There is a big gap between the ideal line laser generator and the realistic line laser generator. For the actual situation, the laser stripe has a certain width, especially when the angle of incidence is large.

(b) The surface of the scanned object

The surface material and properties are different among the different scanned objects. Different colors, materials, and surface smoothness may affect the reflection of laser light stripe on the surface of the scanned object. Especially in the case that the surface of the object has big change in curvature, and the images acquired by the camera may contain a lot of noise.

978-1-5090-0064-7/15 $31.00 © 2015 IEEE

(c)Influence of the environment

Even the infrared laser instead of visible laser is adopted. The image still contains a lot of environmental noise. These disturbances are random, complex, and non-estimated.

3.2.2. The extracting of the center of the line laser

There are several methods to extract the center of line laser, and two of them can be used compositionally. A gray weighted centroid algorithm is applied based on the extremum, which is mixed with two common algorithms to get better results [6]. Description of the algorithm is stated as follows.

- Step1: Scan the image by a pixel row, find the location of the pixel point having the largest gray value, and the found pixel point is recorded as X_{max};
- Step2: Select n pixel points range around the X_{max} on both left and right, and $2n+1$ point are selected;
- Step3: According to the gray weighted centroid algorithm, the calculation is shown as

$$\overline{x} = \sum_{i=X_{max}-n}^{X_{max}+n} i \cdot h_i \Big/ \sum_{i=X_{max}-n}^{X_{max}+n} h_i \qquad (3)$$

where h_i is the gray value, and $\overline{x}$ is the center position of the laser;

- Step4: Perform the same operation for the next line, and the remaining all lines are respectively performed the aforementioned operation in sequence.

4. Three-dimensional reconstruction

According to the structure of the hardware, point P on the scanned object is selected to be discussed. When the object performs a rotational motion, the intersection point of the laser and the scanned object is set as P. The horizontal plane of P is chosen to draw the top view, and the position model is built and depicted in Fig. 4.

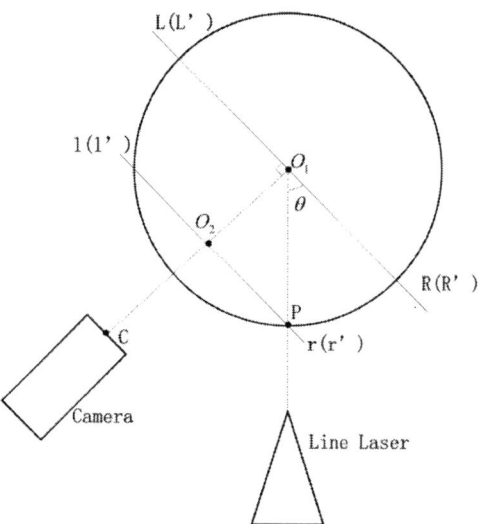

Fig. 4: Position model

Round O_1 is the moving track of the point. Set a polar coordinate system in this plane. Set the O_1R as the polar axis.

The lengths of segments are represented as l with different subscripts. The coordinates of P points on the plane can be expressed as $\left(l_{O_1P}, -\theta\right)$, According to the trigonometric function relationship in the position relation model, it can also be expressed as $\left(l_{O_2P}/\cos\theta, -\theta\right)$. The plane $ll'r'r$ is parallel to the image plane in the camera, l_{O_1P} is easy to obtain according to the principle of monocular vision. Extended to stereo space, the space can be seen as a superposition of lots of similar planes. By the same method, the coordinates of the same column with P can be obtained. The above work can be accomplished by data of one frame of the image.

The images are obtained clocked in the process of uniform rotation of the rotating platform. Assuming an initial angle of 0, n frames of image are obtained, the platform has rotated 360 degrees, and the degree of frame numbered m can be expressed as

$$rad_m = \frac{m \cdot 2\pi}{nt} \qquad . \qquad (4)$$

In order to make the preview of the 3D graphics facilitate, a new coordinate system (NCS) is built. The point P_m (the frame number is m, and the coordinates of P in the image is (u_p, v_p)). Simplified linear relationship coefficient as k, the coordinates (X_n, Y_n, Z_n) of P_m in the NCS are expressed as

$$\begin{cases} X_n = k l_{O_2P} \cdot \cos(rad_m) \\ Y_n = k l_{O_2P} \cdot \sin(rad_m) \\ Z_n = k v_p \end{cases} \qquad (5)$$

In (5), the l_{O_2P} can be calculated by the subtraction of the calibrated O_2 and P.

IV. IMPLEMENTATION OF THE PROTOTYPE AND EXPERIMENT

1. Implementation of the mechanical platform

The mechanical platform is composed of aluminum. The prototype of the platform is shown in Fig. 5, and the parameters of the platform are listed in Table 1.

Fig. 5: The mechanical platform

978-1-5090-0064-7/15 $31.00 © 2015 IEEE

Table 1: Parameters of the hardware

Parameters	Value	Unit
Maximum rotating speed	1	rads/s
Resolution of camera	640*480	Pixel
Maximum frame rate	30	fps
Wavelength of laser	850	nm

In order to filtrate the environmental light, a light filter is put before the lens of the camera.

2. Implementation of the software

The software of image processing is developed based on C # and EmguCV library functions [7], [8]. The software interface is manifested in Fig. 6.

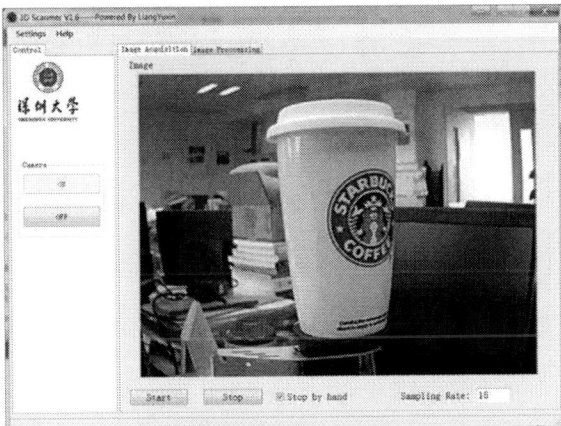

Fig. 6: The software interface

3. Experiment

Scanned results of a drink bottle and a yellow duck doll are presented in Figs. 7 and 8, respectively. The implementation of the scanning is less than 30s. The preliminary point clouds are drawn in MATLAB. The results demonstrate that the generated 3D model includes all 3D dimensions, and the developed system can rapidly achieve 3D reconstruction.

Fig. 7: Scan result of a drink bottle

Fig.8: Scan result of a yellow duck doll

V. CONCLUSION

In this paper, a novel non-contact 3D scanning system by using monocular computer vision has been presented to rapidly obtain the 3D dimension of the scanned object. The principle of the monocular vision, implementation scheme of the system, prototype, and experimental results have been given. The experimental results show that the generated 3D model includes all its 3D dimension; the developed system has advantages of low cost, simple structure, easy to extend, and rapidly 3D reconstruction; and the developed system is appropriate to the application of low cost, rapidly scanning, and insensitivity of high precision.

ACKNOWLEDGMENT

The authors would like to thank the National Natural Science Foundation of China under Grant NSFC51275312, the National Key Technology R&D Program under Grant 2014BAH23F04, and Shenzhen Government Fund under Grant JSGG20141015153303491 and KC2014JSJS025A.

REFERENCES

[1] Wikipedia. 2015. en.wikipedia.org/wiki/3D_scanner.

[2] W. H. Lei, "General Situation of Machine Vision", Journal of Applied Optics (In Chinese), No. 5, 2006, pp. 467-470.

[3] Z. Lv and Z. Zhang, "Build 3D Scanner System based on Binocular Stereo Vision", 2011 International Conf. on Intelligent Computation Technology and Automation (ICICTA), 28-29 Mar. 2011, pp.600-603.

[4] Z. Zhang, "A flexible new technique for camera calibration", IEEE Trans. Pattern Analysis and Machine Intelligence, Vol.22, No. 11, 2000, pp. 1330 - 1334.

[5] R. Y. Tsai, "A versatile camera calibration technique for high-accuracy 3D machine vision metrology using off-the-shelf TV cameras and lenses", IEEE Journal of Robotics and Automation, Vol.3, No.4, 1987, pp.323-344.

[6] Y. Chen , K. Wang, W. Zuo, and Q. Wu "Sub-Pixel Coded Structured Light Stripe Boundary Location Using Weighted Centroid Method", 2010 International Conf. Biomedical Engineering and Computer Science (ICBECS), 23-25 Apr. 2010, pp.1-5.

[7] A. Zelinsky, "Learning OpenCV - Computer Vision with the OpenCV Library", Robotics & Automation Magazine, Vol.16, No. 3, 2000, pp. 100 - 100.

[8] L. Zou and Y. Li, "A method of stereo vision matching based on OpenCV", 2010 International Conf. Audio Language and Image Processing (ICALIP), 23-25 Nov. 2010, pp.185-190.

BIOGRAPHIES

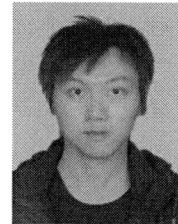

Yu-Xin Liang received his B.Sc. degree from Shenzhen University in 2015.

He is currently a postgraduate with College of Mechatronics and Control Engineering, Shenzhen University, Shenzhen. His research direction is computer vision.

Guang-Zhong Cao received the B.Sc., M.Sc., and Ph.D. degrees in electrical engineering and automation from Xi'an Jiaotong University, Xi'an, Shanxi, China, in 1989, 1992, and 1996, respectively.

He is currently a Professor and Director with the Shenzhen Key Laboratory of Electromagnetic Control, Shenzhen University, Shenzhen, Guangdong, China. He has published more than 80 articles in refereed journals and conferences. His research interests include motor control, and control theory and its application.

Hong Qiu received the B.Sc. and M.S. degrees from College of Mechatronics and Control Engineering, Shenzhen University, Guangdong Province, China, in 2007 and 2010, respectively.

He is currently a laboratory technician with the Shenzhen Key Laboratory of Electromagnetic Control, Shenzhen University, Shenzhen, Guangdong, China. His main research interest is embedded system development.

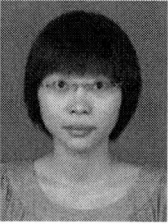

Su-Dan Huang received the B.Sc. and M.Sc. degrees in College of Mechatronics and Control Engineering from Shenzhen University, Shenzhen, Guangdong, China, in 2009 and 2012, respectively.

She is currently a joint PHD candidate with the Department of Electrical Engineering, Southwest Jiaotong University, Chengdu, Sichuan, China, and Shenzhen Key Laboratory of Electromagnetic Control, Shenzhen University, Shenzhen, Guangdong, China. Her research interests include design and control of planar switched reluctance motors, and control theory and its application.

Shou-Qin Zhou received the Ph.D. degree from Xi'an Jiaotong University, Xi'an, Shanxi, China, in 2001.

He is currently a senior engineer at professor grade and researcher with China International Marine Containers (CIMC) (Group) LTD., the director of CIMC Smart and Security Research Center (SSC), CIMC Intelligent Technology CO., LTD., General Manager, and research and development center of Shenzhen Intelligent container Engineering Technology, Shenzhen (national) leading professional and technical personnel, and professor gained Shenzhen government special allowance.

978-1-5090-0064-7/15 $31.00 © 2015 IEEE

Development of a 4-DOF SCARA Robot with 3R1P for Pick-and-Place Tasks

Wen-Bo Li[1]   Guang-Zhong Cao[1]   Xiao-Qin Guo[1]   Su-Dan Huang[1,2]

[1] Shenzhen Key Laboratory of Electromagnetic Control, Shenzhen University, Shenzhen, China
E-mail: liwenbochn@126.com; gzcao@szu.edu.cn; guoxq@szu.edu.cn;
2 College of Electrical Engineering, Southwest Jiaotong University, Chengdu, 610031, China
E-mail: hsdsudan@gmail.com

Abstract–The planar robot is very suitable for moving workpieces which are in high demand in industrial automation. This paper develops a 4 degree of freedom (4-DOF) selective compliance assembly robot arm (SCARA) robot with three rotary joints and one prismatic joint (3R1P) to realize pick-and-place tasks of the circular and rectangular workpieces. The structure of the robot is firstly presented. The kinematic model is then built, and the kinematic analysis is performed based on MATLAB. The trajectory planning is further implemented. A control interface is also designed via Visual C++ to control the robot for achieving pick-and-place tasks. The validity of the developed robot is finally verified through experimental results.

Keywords–Modeling, OpenCV, trajectory planning, SCARA, visual servoing.

I. INTRODUCTION

Planar robots play an increasingly important role in industrial automation, especially in the assembly industry. Industrial planar robots are developed for fast, accurate, and repetitive tasks [1]. As one kind of the planar robot, the selective compliance assembly robot arm (SCARA) robot is pliable in planar and rigid in Z axis. The SCARA robot features simpler structure, lighter mass, faster response, more precise positioning accuracy compared to most of robots. Thus, the SCARA robot is widely applied to assembly industry. Additionally, pick-and-place tasks are basic operation of assembly processes, such as inserting an edge connector socket [2] into a printed circuit board, and classifying workpieces of sorting systems [3]. Currently, one of the world's fastest SCARA robots is Adept 1. Adept 1 has several times of velocity of other joints robots, which can reach 10 m/s of the end effector and achieve less than ±0.02 mm repeated accuracy. The research of the SCARA robot focuses on accurate modeling [4], visual servo [5], human-robot collaboration [6], and trajectory planning [7], etc. In this paper, a 4 degree of freedom (4-DOF) SCARA robot with three rotary joints and one prismatic joint (3R1P) is developed for pick-and-place tasks by using visual servo control.

This paper is organized as follows. The hardware of the SCARA robot is presented in Section II. In Section III, the kinematic model of the SCARA robot is derived with the denavit-hartenberg (D-H) method, and the forward kinematic and inverse kinematic [8] are analyzed based on MATLAB. In Section IV, a straight line and an arc line for trajectory motion are planned in joint coordinates via Robotics Toolbox of MATLAB [9]. Section V discusses how to locate workpieces for pick up tasks by using position from visual servo. Conclusion remarks are given in Section VII.

II. SYSTEM STRUCTURE

The prototype of the 4-DOF SCARA robot with 3R1P is shown in Fig. 1. Axis 1, 2 and 4 are the rotational joints, and Axis 3 is the prismatic joint. An Electromagnet clamp is installed on Axis 4, which is applied as the end effector. And a camera is installed on the electromagnet clamp.

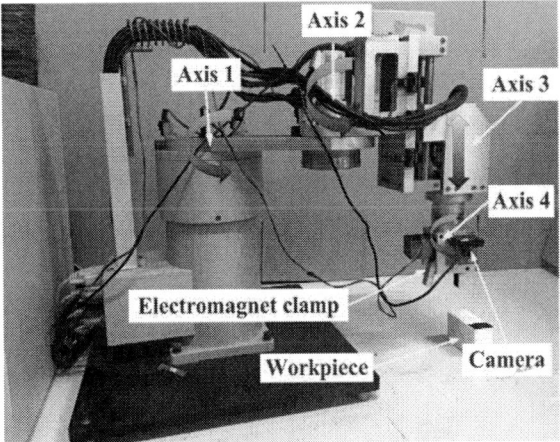

Fig. 1: Prototype of the 4-DOF SCARA robot with 3R1P

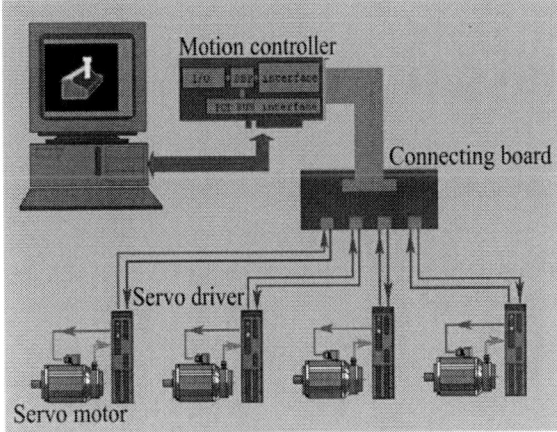

Fig. 2: Connection schematic diagram

Fig. 2 presents the connection of each part of the system. Each axis is driven by a Panasonic alternating current (AC) servo motor controlled by its Servo Driver. The parameters of the AC servo motor are given in Table 1. The PC+DSP control method is used in this system, and it is one of the most efficient methods to control robots in assembly industry. A Motion controller is installed on PC using PCI bus interface, its core is composed of an ADSP2181 DSP and a FPGA, and it can realize high-efficient calculation for control.

978-1-5090-0064-7/15 $31.00 © 2015 IEEE

Table 1: Parameters of the AC servo motor

Parameters	Value
Input	3Φ AC 42 V 1.0 A
Rated output	0.05 KW
Rated frequency	200 Hz
Rated REV	3000 r/min

III. KINEMATIC MODEL

Kinematic Modeling is analyzed in this section, and it consists of forward and inverse kinematic modeling. The D-H parameters of the robot are listed in the Table 2.

Table 2: D-H parameters of the SCARA robot

Axis	θ	d	a	α	Range
1	θ_1	0	l_1=200 mm	0	-10° ~109°
2	θ_2	0	l_2=200 mm	0	-8° ~102°
3	0	d_3	0	0	-48 mm~48 mm
4	θ_4	0	0	0	-180° ~180°

1. Forward kinematic

When the robot movement of each joint is known, the process to solve the pose of the end effector is denoted as the forward kinematic of the robot. The D-H transformation matrix between link *i-1* and *i* is represented as (1). For the SCARA robot, after confirm the link coordinate, the transformation matrix from base coordinate to Axis 4 can be obtained and given in (2).

$$T_i = \begin{bmatrix} \cos\theta_i & -\sin\theta_i\cos\alpha_i & \sin\theta_i\sin\alpha_i & a_i\cos\theta_i \\ \sin\theta_i & \cos\theta_i\cos\alpha_i & -\cos\theta_i\sin\alpha_i & a_i\sin\theta_i \\ 0 & \sin\alpha_i & \cos\alpha_i & d_i \\ 0 & 0 & 0 & 1 \end{bmatrix} \quad (1)$$

$$^0T_4 = {}^0T_1\,{}^1T_2\,{}^2T_3\,{}^3T_4 \quad (2)$$

where $^{i-1}T_i$ is the transformation matrix from coordinate n to m. Substituting D-H parameters into (1), the transformation matrix between links can be expressed as

$$^0T_1 = \begin{bmatrix} \cos\theta_1 & -\sin\theta_1 & 0 & l_1\cos\theta_1 \\ \sin\theta_1 & \cos\theta_1 & 0 & l_1\sin\theta_1 \\ 0 & 0 & 1 & 0 \\ 0 & 0 & 0 & 1 \end{bmatrix} \quad (3)$$

$$^1T_2 = \begin{bmatrix} \cos\theta_2 & \sin\theta_2 & 0 & l_2\cos\theta_2 \\ \sin\theta_2 & -\cos\theta_2 & 0 & l_2\sin\theta_2 \\ 0 & 0 & 1 & 0 \\ 0 & 0 & 0 & 1 \end{bmatrix} \quad (4)$$

$$^2T_3 = \begin{bmatrix} 1 & 0 & 0 & 0 \\ 0 & 1 & 0 & 0 \\ 0 & 0 & 1 & d_3 \\ 0 & 0 & 0 & 1 \end{bmatrix} \quad (5)$$

$$^3T_4 = \begin{bmatrix} \cos\theta_4 & -\sin\theta_4 & 0 & 0 \\ \sin\theta_4 & \cos\theta_4 & 0 & 0 \\ 0 & 0 & 1 & 0 \\ 0 & 0 & 0 & 1 \end{bmatrix} \quad (6)$$

Substituting (3), (4), (5), and (6) into (2), the transformation matrix of the end effector coordinate can be represented as

$$^0T_4 =$$
$$\begin{bmatrix} \cos(\theta_1+\theta_2-\theta_4) & \sin(\theta_1+\theta_2-\theta_4) & 0 & l_1\cos(\theta_1)+l_2\cos(\theta_1+\theta_2) \\ \sin(\theta_1+\theta_2-\theta_4) & -\cos(\theta_1+\theta_2-\theta_4) & 0 & l_1\sin(\theta_1)+l_2\sin(\theta_1+\theta_2) \\ 0 & 0 & 1 & d_3 \\ 0 & 0 & 0 & 1 \end{bmatrix} \quad (7)$$

According to the kinematic theory of robots, the solution of kinematic can be given by

$$\begin{cases} n_x = \cos(\theta_1+\theta_2-\theta_4) \\ n_y = \sin(\theta_1+\theta_2-\theta_4) \\ n_z = 0 \\ o_x = \sin(\theta_1+\theta_2-\theta_4)\cdot \\ o_y = \cos(\theta_1+\theta_2-\theta_4) \\ o_z = 0 \\ a_x = 0 \\ a_y = 0 \\ a_z = 1 \\ p_x = l_1\cos(\theta_1)+l_2\cos(\theta_1+\theta_2) \\ p_y = l_1\sin(\theta_1)+l_2\sin(\theta_1+\theta_2) \\ p_z = d_3 \end{cases} \quad (8)$$

2. Inverse kinematic

In most situations, the movement of each joint from point A to point B is required. Inverse kinematic analysis is employed to solve this problem. There are two methods to solve the inverse kinematic problem, which are the closed-form solution and numerical solution [10]. However, in practice, due to the iterative of the numerical solution, the efficient of the numerical solution is less than closed-form solution. Therefore, the closed-form solution is chosen for the inverse kinematic.

$$\left(^0T_1\right)^{-1}\,{}^0T_4 = {}^1T_2\,{}^2T_3\,{}^3T_4 \quad (9)$$

$$\begin{bmatrix} \cos\theta_1 & \sin\theta_1 & 0 & -l_1 \\ -\sin\theta_1 & \cos\theta_1 & 0 & 0 \\ 0 & 0 & 1 & 0 \\ 0 & 0 & 0 & 1 \end{bmatrix}\begin{bmatrix} n_x & o_x & a_x & p_x \\ n_y & o_y & a_y & p_y \\ n_z & o_z & a_z & p_z \\ 0 & 0 & 0 & 1 \end{bmatrix} =$$

$$\begin{bmatrix} \cos(\theta_2+\theta_4) & \sin(\theta_2+\theta_4) & 0 & l_2\cos\theta_2 \\ \sin(\theta_2+\theta_4) & -\cos(\theta_2+\theta_4) & 0 & l_2\sin\theta_2 \\ 0 & 0 & 1 & d_3 \\ 0 & 0 & 0 & 1 \end{bmatrix} \quad (10)$$

The closed-form solution can be divided into the algebraic method and geometric method in terms of the solve method. Because the SCARA robot has relatively simple configuration, so the algebraic method is used. Equation (2) can be deformed as (9), compared the elements on both sides of (10), the solution of inverse kinematic can be derived as

$$\begin{cases}
\theta_1 = \arctan\left(\dfrac{A}{\pm\sqrt{1-A^2}}\right) - \phi \\[2mm]
\theta_2 = \arccos\left(\dfrac{r\sin(\theta_1+\phi)-l_1}{l_2}\right) \\[2mm]
d_3 = p_z \\[2mm]
\theta_4 = \arcsin\left(n_y\cos(\theta_1)-n_x\sin(\theta_1)\right)-\theta_2 \\[2mm]
A = \dfrac{l_1^2+p_x^2+p_y^2-l_2^2}{2l_1\sqrt{p_x^2+p_y^2}} \\[2mm]
\phi = \arctan\left(\dfrac{p_x}{p_y}\right) \\[2mm]
r = \sqrt{p_x^2+p_y^2}
\end{cases} \qquad (11)$$

3. Modeling based on MATLAB

The Robotics Toolbox of MATLAB is developed by Professor Peter Corke. It provides a series function of kinematic and path planning to research of robots, and it is widely applied in the robot development. Meanwhile, the toolbox can also perform the image simulation based on robots. In Robotics Toolbox, it is important to construct the joints model. Function Link and SerialLink can be used in modeling. Fig. 3 illustrates the simulation model in 3-D space.

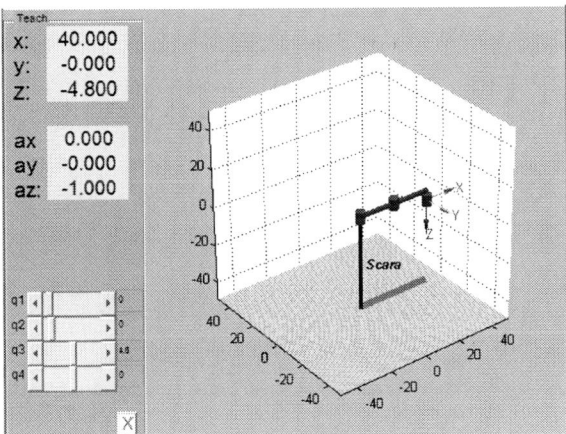

Fig. 3: Simulation model of the SCARA robot

IV. Path Planning

1. Path planning in joint space

The final pose can be solved by using the inverse kinematic while the initial pose is known, however, the path planning of each joint need calculate, respectively. Assumed a joint has an initial angle θ_i, when the motions begin and the final angle θ_f is in the motions end. If the joint motions during the rotation need to be smooth, at least 4 constraints of the joints trajectory functions are required. Meanwhile, assumed V_i and V_f are velocity of the joint at the begin and the end of the motion, then a unique cubic polynomial (12) can be solved by using the constraints mentioned above. Equation (13) is the solution of (12).

$$\theta(t) = c_0 + c_1 t + c_2 t^2 + c_3 t^3 \qquad (12)$$

$$\begin{cases}
c_0 = \theta_i \\[2mm]
c_1 = v_i \\[2mm]
c_2 = -\dfrac{3\theta_i - 3\theta_f + 2v_i t + v_f t}{t^2} \\[2mm]
c_3 = \dfrac{2\theta_i - 2\theta_f + v_i t + v_f t}{t^3}
\end{cases} \qquad (13)$$

In practice, the cubic polynomial will cause mutation of acceleration. To solve this problem, the acceleration of the motion should also be constrained. Assumed a_i and a_f are acceleration of the joint at the begin and the end of the motion, then a unique quintic polynomial equation can be obtained as

$$\theta(t) = c_0 + c_1 t + c_2 t^2 + c_3 t^3 + c_4 t^4 + c_5 t^5 \qquad (14)$$

$$\begin{cases}
c_0 = \theta_i \\[2mm]
c_1 = v_i \\[2mm]
c_2 = \dfrac{a_i}{2} \\[2mm]
c_3 = \dfrac{\left[20\theta_f - 20\theta_0 - (8v_f + 12v_i)t_f - (3a_i - a_f)t_f^2\right]}{2t_f^3} \\[2mm]
c_4 = \dfrac{\left[30\theta_0 - 30\theta_f + (14v_f + 16v_i)t_f - (3a_i - 2a_f)t_f^2\right]}{2t_f^4} \\[2mm]
c_5 = \dfrac{\left[12\theta_f - 12\theta_0 - (6v_f + 6v_i)t_f - (a_i - a_f)t_f^2\right]}{2t_f^5}
\end{cases} \qquad (15)$$

2. Path planning in Cartesian space

In most situations, the path of end effector is required to a specific curve. For example, the motion path of the end effector should be a straight line or arc line, when the initial point and end point is known. The interpolation method can achieve this purpose. Equation (16) can be used to transform Cartesian coordinate to joint coordinate, if all the interpolating points are in task space.

$$\begin{cases}
\theta_1 = \pi - \arccos\left(\dfrac{l_1^2 + l_{21}^2 - (x^2 + y^2)}{2l_1 l_2}\right) \\[2mm]
\theta_2 = \arctan\left(\dfrac{y}{x}\right) - \arctan\left(\dfrac{l_2\sin(\theta_2)}{l_1 + l_2\cos(\theta_2)}\right)
\end{cases} \qquad (16)$$

V. Position Control

This section clarifies how to calibrate workpiece by the camera. In Visual C++ Compiler Environment, OpenCV has been installed to perform the camera calibration and image process.

In computer vision, it is meaningful to confirm the relationship between real location and image of the workpiece. For this goal, the geometry model of the camera is needed. It contains internal camera parameters and extrinsic camera parameters. Zhang Zheng-you method is adopted to calibrate the camera. The calibration results are given as

978-1-5090-0064-7/15 $31.00 © 2015 IEEE

$$M_1 = \begin{bmatrix} 703.6503 & 0 & 320.9820 & 0 \\ 0 & 701.5517 & 287.6760 & 0 \\ 0 & 0 & 1 & 0 \end{bmatrix} \quad (17)$$

$$M_2 = \begin{bmatrix} 0.0026 & 0.9998 & 0.0195 & 97.9088 \\ 1 & -0.0025 & 0.0069 & -26.6459 \\ 0.0069 & 0.0195 & -0.9998 & -27.6573 \\ 0 & 0 & 0 & 1 \end{bmatrix} \quad (18)$$

Fig. 4 depicts the flowchart of the position control method. The flow chart is the workflow of pick up workpiece. All of the procedure is performed on control software designed by Visual C++ 6.0 environment.

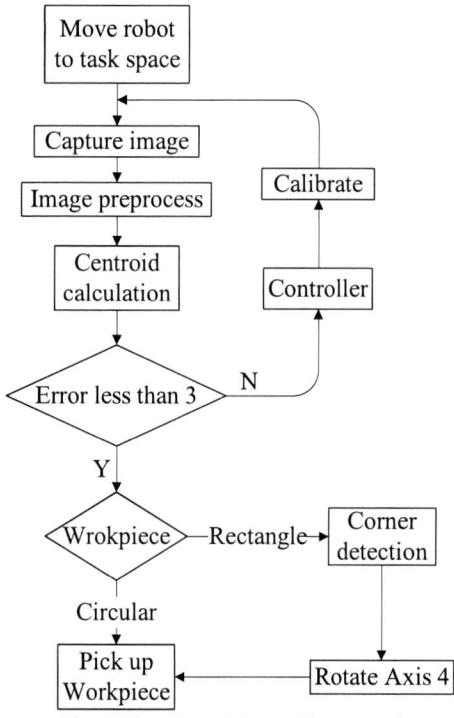

Fig. 4: Flowchart of the position control

VI. EXPERIMENTAL RESULTS

Tables 3 and 4 list the experimental results of the circular and rectangle workpieces, respectively. The workpiece is needed 3 times to calibration generally. The desired centroid is the centroid of the workpieces in the camera coordinate, when the workpiece is placed under the electromagnet clamp. As the number of the calibration is increased, the absolute errors of X and Y positioning are decreased. The absolute errors of positioning for the circular and rectangle workpieces are satisfying, which are less than 3 pixels and 1 pixel, respectively.

Table 3: Experimental result of the circular workpiece

Calibration times	Desired Centroid	Actual Centroid	X absolute error	Y absolute error
1	(261,80)	(182,264)	79	180
2	(261,80)	(269,85)	8	5
3	(261,80)	(264,81)	3	1

Table 4: Experimental result of the rectangle workpiece

Calibration times	Desired Centroid	Actual Centroid	X absolute error	Y absolute error
1	(246,204)	(181,380)	65	176
2	(246,204)	(241,203)	5	1
3	(246,204)	(245,203)	1	1

VII. CONCLUSION

A 4-DOF SCARA robot with 3R1P has been developed in this paper. The kinematic model, the straight line path and arc line path, the visual servo, and the experimental results of the robot have been presented. Experimental results demonstrate that the robot can calibrate the workpieces with less than 3 pixels error for the circular and rectangle workpieces, and the developed robot is feasible and valid.

ACKNOWLEDGMENT

The authors would like to thank the National Natural Science Foundation of China under Grant NSFC51275312, the National Key Technology R&D Program under Grant 2014BAH23F04, and Shenzhen Government Fund under Grant JSGG20141015153303491 and KC2014JSJS025A.

REFERENCES

[1] L. E. G. Moctezuma, A. Lobov and J. L. M. Lastra, "Free Shape Paths in Industrial Robots", IECON 2012 - 38th Annual Conference on IEEE Industrial Electronics Society, 25-28 Oct 2012, pp. 3739-3743

[2] H. Chen and Y. Liu, "Robotic assembly automation using robust compliant control", Robotics and Computer-Integrated Manufacturing, Vol. 29, No. 2, 2013, pp.293-300

[3] C. Pop, S. M. Grigorescu and A. Davidescu, "Colored object detection algorithm for visual-servoing application", Optimization of Electrical and Electronic Equipment (OPTIM), 24-26 May 2012, pp. 1539-1544

[4] L. U. Odhner and A. M. Dollar, "The Smooth Curvature Model: An Efficient Representation of Euler–Bernoulli Flexures as Robot Joints", IEEE Transactions on Robotics, Vol. 28, No.4, 2012, pp. 761-772

[5] M. A. Pérez and M. Bueno, "3D Visual Servoing Control for Robot Manipulators Without Parametric Identification", IEEE Latin America Transactions, Vol. 13, No. 3, 2015, pp. 569-577

[6] Y. Li and S. S. Ge, "Human–Robot Collaboration Based on Motion Intention Estimation", IEEE/ASME Transactions on Mechatronics, Vol. 19, No. 3, 2014, pp. 1007-1014

[7] L. M. Capisani and A.F. Ferrara, "Trajectory Planning and Second-Order Sliding Mode Motion/Interaction Control for Robot Manipulators in Unknown Environments", IEEE Transactions on Industrial Electronics, Vol. 59, No. 8, 2012, pp. 3189-3198

[8] D. Manocha and J. F. Canny. "Efficient inverse kinematics for general 6R manipulators", IEEE Transactions on Robotics and Automation, Vol. 5, No. 10, 1994, pp. 648-658

[9] P. I. Corke. "A Robotics Toolbox for Matlab", IEEE Robotics and Automation Magazine, Vol.3, No.1, 1966, pp. 24-32

[10] D. Martins and R. Guenther. "Hierarchical kinematic analysis of robots", Mechanism & Machine Theory, Vol. 38, No. 6, 2003, pp. 497-518

978-1-5090-0064-7/15 $31.00 © 2015 IEEE

BIOGRAPHIE

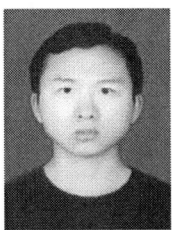

Wen-Bo Li received his B.Sc. degree from Shenzhen University in 2015.

He is currently a postgraduate with College of Mechatronics and Control Engineering, Shenzhen University, Shenzhen. His research direction is motor control.

Guang-Zhong Cao received the B.Sc., M.Sc., and Ph.D. degrees in electrical engineering and automation from Xi'an Jiaotong University, Xi'an, Shanxi, China, in 1989, 1992, and 1996, respectively.

He is currently a Professor and Director with the Shenzhen Key Laboratory of Electromagnetic Control, Shenzhen University, Shenzhen, Guangdong, China. He has published more than 80 articles in refereed journals and conferences. His research interests include motor control, and control theory and its application.

Xiao-Qin Guo received the B.Sc., M.Sc. degree in electrical engineering and automation from Northwestern polytechnical University Shanxi Province, China.

She is currently an associate professor with College of Mechatronics and Control Engineer, Shenzhen University. She has published over 30 papers in refereed Journals and conferences. Her research interests are control theory and its application, robot and computer vision.

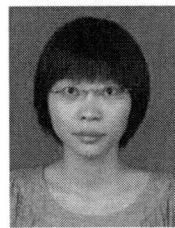

Su-Dan Huang received the B.Sc. and M.Sc. degrees in College of Mechatronics and Control Engineering from Shenzhen University, Shenzhen, Guangdong, China, in 2009 and 2012, respectively.

She is currently a joint PHD candidate with the Department of Electrical Engineering, Southwest Jiaotong University, Chengdu, Sichuan, China, and Shenzhen Key Laboratory of Electromagnetic Control, Shenzhen University, Shenzhen, Guangdong, China. Her research interests include design and control of planar switched reluctance motors, and control theory and its application.

Yaw Angle Control of a Boxfish-like Robot Based on Cascade PID Control Algorithm

Hanbo Deng[1]   Wei Wang[2]   Wenguang Luo[1]   Guangming Xie[3]

[1]The Guangxi University of Science and Technology，Liu Zhou, Guangxi, China
E-mail: hanbo_deng@foxmail.com
[2,3]The Peking University，Beijing, China
E-mail: wangweiw4y4@pku.edu.cn
[3]E-mail: xiegming@pku.edu.cn

Abstract–Because of the complexity of water environment, it is difficult to precisely model the fish robot system. Therefore, model-based attitude control algorithm is difficult to apply to the robotic fish. Cascade PID control uses the measurements of controlled variable and its derivative to constitute a double feedback loop. It can adjust the interference that influence the intermediate variable in advance and improve the dynamic quality and working frequency of the whole system. It is better than the traditional PID controller in anti-interference performance, rapidity, adaptation and stability. This article adopts the cascade PID algorithm to control the yaw angle of a robotic fish for the first time. This article innovatively uses a Kalman filter to eliminate the system angle error caused by periodic oscillation of robotic fish. A yaw angle control framework was proposed for biomimetic robotic fish. Based on the proposed framework, systematically yaw angle control experiments with the robotic fish were carried out in situations where the reference inputs were step signal, square wave and sine wave. Experiments showed that the robotic fish can accurately track the reference of the yaw angle in real time, verifying the effectiveness of the proposed algorithm.

Keywords–Attitude control, Cascade PID control, Kalman filter, robotic fish

I. INTRODUCTION

In recent years, in order to meet the shortage of land resources, marine resources are constantly to be exploited. Due to the special nature and complexity of the marine environment, autonomous underwater vehicle (AUV) became an indispensable tool for seabed exploration, marine environment monitoring, and even for maritime military strategy. As a branch of autonomous underwater vehicle, robotic fish has been received widespread concern for its efficiency, hidden and highly maneuverable underwater athletic ability when it swims and has made some achievements [1-8]. Especially in the special water environment, the robotic fish can swim through the pathway in a particular attitude, narrow gap, etc.

Fish use strong muscle to drive their own bodies to swim. Therefore robotic fish need to design reasonable driving system to get enough thrust. The motor drive is the most common way of driving at present. The tail of the robotic fish swings through the rotation of the motor drives the robotic fish joints that generates propulsion. We generally adopt PID controller to control the propulsion system of robotic fish in order to achieve better promoting effect.

PID control is by far the most common method of control by which the majority of the feedback loop or less distortion

used to control [9]. PID controller has a simple structure, it has the advantage of robustness against the model error and easy to operate. Robotic fish attitude control requires high precision, however the traditional PID controller cannot meet the performance requirements. Cascade PID control uses the measurements of controlled variable and its derivative to constitute a double feedback loop. It can adjust the interference that influence the intermediate variable in advance and improve the dynamic quality and working frequency of the whole system [10]. It is better than the traditional PID controller in anti-interference performance, rapidity, adaptation and stability.

In this paper, the author uses an autonomous robotic fish entity to conduct the experiments and adopts the cascade PID control algorithm to the control the yaw angle of the robotic fish for the first time. This article innovatively uses a Kalman filter to eliminate the system angle error caused by periodic oscillation of robotic fish. Finally, a yaw angle control framework was proposed for biomimetic robotic fish. Based on the proposed framework, systematically yaw angle control experiments with the robotic fish were carried out when the reference inputs were step signal, square wave and sine wave. Experiments showed that the robotic fish can accurately track the reference of the yaw angle in real time, verifying the effectiveness of the proposed algorithm.

II. BOXFISH-LIKE ROBOT

The boxfish-like robot has a similar appearance with the real boxfish. It has a pair of pectoral fins and a tail. The boxfish-like robot can flexibly implement various swimming modes such as forward swimming, turning, downward and upward swimming, and rolling [11]. As shown in Fig. 1, it is the configurations of the boxfish-like robot

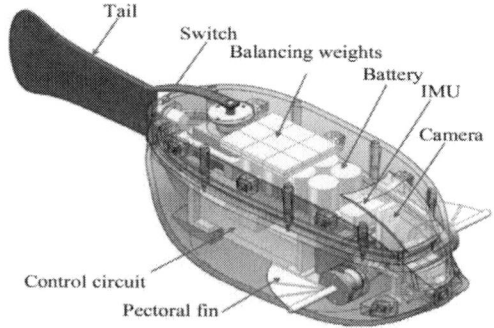

Fig. 1: Configurations of the boxfish-like robot

The robot uses Raspberry Pi as the main controller, equipped with the Linux operating system, and two STM32 microcontrollers as the secondary controller. It can achieve two control modes: manual mode and autonomous mode. In manual mode, the robotic fish that can be controlled by the host computer system though wireless communication network (WiFi), and monitor various states of boxfish-like robot (voltage and current, sensor information, etc.); In autonomous mode, the boxfish-like robot can automatically swim through the sensor, Inertial measurement unit (IMU), video cameras in order to sense of its surroundings and complete the set task and record the results.

Inertial measurement unit (IMU) as the important sensor of feedback attitude information. It is constituted by a triaxial accelerometer, a three-axis gyroscope and a three-axis electronic compass. It is arranged on the axis of the robot to monitor the yaw, roll and pitch angles of the robot. The sampling frequency is 50Hz. The precision of the yaw angle is $1°\sim 3°$.

We designed an artificial central pattern generator (CPG) to control the basic locomotion of the robot [12]. The control parameters of the CPG model are the desired frequency, amplitude, offset and phase difference of the oscillations. The CPG model shown in Fig. 2, it can be divided input saturation functions, the coupled neural oscillators and output transition function [13].

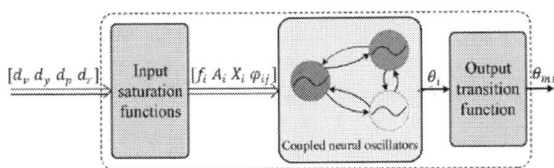

Fig.2: A schematic diagram of robotic fish model

Input saturation functions received from the upper controller can be divided into four command inputs, namely speed d_v, yaw angle d_y, pitch angle d_p and roll angle d_r, turn the upper control command to the CPG model control input parameters. CPG model input parameters include the desired frequency f_i, amplitude A_i, offset X_i and phase difference φ_{ij} of the oscillations. Based on the coupled neural oscillator , these control parameters can generate the stability of the CPG signals θ_i, but these signals can not be delivered directly to the steering engine, the output must be converted by output transition function, translate these CPG signal into the drive signal which can be identified to the steering engine. Among them, the coupled linear oscillators take the forms:

$$\dot{a}_i = \alpha_i(A_i - a_i) \qquad (1)$$
$$\dot{x}_i = \beta_i(X_i - x_i) \qquad (2)$$
$$\dot{\emptyset}_i = 2\pi f_i + \sum_{j \in T_i} u_{ij}(\emptyset_i - \emptyset_i - \varphi_{ij}) \qquad (3)$$
$$\theta_i = x_i + a_i \cos \emptyset_i \qquad (4)$$

Among them, where α_i , x_i , $\emptyset_i$ are the state variables representing the amplitude, offset and phase of the ith oscillator, and the variable θ_i is its output., which is the last swing angle of the steering engine. u_{ij} and φ_{ij} determine the coupling weight and phase bias of the jth

oscillator to the ith oscillator. Where f_i, A_i, X_i are the desired frequency, amplitude, offset of the oscillations. α and β is the constant proportion coefficient. Therefore, controlling the parameters of the tail fin and the pectoral fins α_i , x_i , f_i, the direction and speed of robotic fish can be controlled. φ_{ij} can be used to match each servo movement between steering engines, so that the boxfish-like robot produces a variety of swimming modes.

Yaw angle is mainly controlled by the tail fin of the robot with a servo motor. We adopt the servo motion control system to control the motion of the tail fin in the robotic fish propulsion system with a single joint of the mechanical mechanism. The controller (STM32) of the robotic fish can recognize each step instructions of the CPG to produce PWM wave with different duty ratio to control the motor position. It can realize the motion of the tail fin. Forces generated by the tail can drive the robot to move forwards. Moreover, the yaw angle of the robot can be controlled by changing the offset of the tail fin.

In this paper, we mainly controlled the yaw angle of a robotic fish using a Cascade PID control algorithm for the first time.

III. DESIGN OF CASCADE PID CONTROL SYSTEM

Because of the complexity of the water environment, it is difficult to precisely model the fish robot system. Therefore, model-based attitude control algorithm is difficult to apply to the robotic fish. Through state response experiments, we found that the robot propelling by its tail is a second-order system. Therefore, it can be controlled using a PID controller [14-15].

1. Cascade Control System
Single-loop PID control algorithm:

$$\mathbf{u}(t) = K_p[e(t) + \frac{1}{T_I}\int_0^t e(t)dt + T_D\frac{de(t)}{dt}] \qquad (5)$$

Where K_p is proportionality coefficient, T_I is integral time coefficient, T_D is Derivative time coefficient.In application, the PID controller needs to be discretized to be suitable to program in a microcontroller. The discrete form of the PID controller is:

$$\mathbf{u}(k) = K_p e(k) + K_I \sum_{i=0}^{k} e(i) + K_D[e(k) + e(k-1)] \qquad (6)$$

Where $K_I = K_p \frac{T}{T_I}$ is integral coefficient; $K_D = K_p \frac{T_D}{T}$ is differential coefficient. T is the sampling period, k is the sample number, u(k) is output value of the kth sampling time, e(k) is error value of the input for the kth sampling time. Here we normally use u(k) as an output value of the PID controller to directly control actuators. However the Single-loop PID controller can't meet our performance requirements. Therefore the cascade PID control system is introduced.

Cascade PID control system has two adjusters in series, one of the output of adjusters can work as another reference adjuster, thus adding a secondary loop control.

978-1-5090-0064-7/15 $31.00 © 2015 IEEE   169

Cascade PID control algorithm can effectively control the time-delay characteristics of the controlled object and thus improve the dynamic response of the system.

For our robot, the PID controller can directly control the yaw angle of the robot. However, the change of the yaw angle is caused by the angular velocity. Therefore, we design a cascade PID controller as shown in Fig. 3.

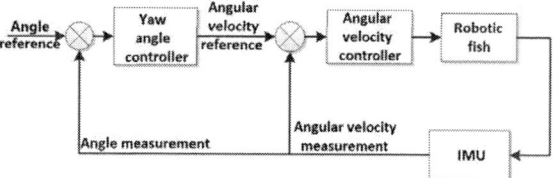

Fig.3: Cascade PID control plan

In this system, the main controller (yaw angle controller) control yaw angle by PD controller that is the outer loop. The secondary controller (angular velocity controller) control angular velocity by PID controller that is the inner loop. The error between the yaw angle measurement value and the yaw angle reference value is the input of the outer loop. The output of the outer loop is the angular velocity reference. The error between the angular velocity measurement value and the angular velocity reference value is the input of the inner loop. The output of the inner loop controls angular velocity of the robot, namely it controls the steering engine to change the angular velocity, thus controls the yaw angle to change direction. To meet the faster inner loop control, we adopt the same sampling period for the inner loop and the outer loop which are 50Hz.

2. Kalman Filtering
In experiments, we found that the output of cascade PID controller, namely the offset of the tail fin oscillated largely. This is caused by the oscillation of the robot while swimming. The oscillation resulted in unstable control of the yaw angle. Moreover, the oscillation of the PID output makes the robot consume large energy. Therefore, Kalman filter is introduced to solve the oscillation signals. Specifically, we filter the error of the outer loop and inner loop to reduce its oscillation amplitude. After filtering, the offset of tail fin decreased significantly.

IV. EXPERIMENTS AND RESULTS

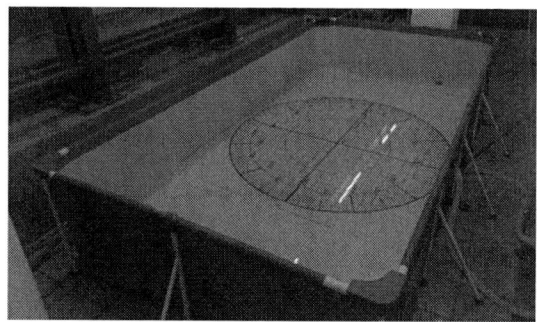

Fig. 4: The Experimental site

We use the boxlike-fish robot to conduct the experiments. Experimental site is a 4 meter-long, 2 meter-wide and 0.8

meter-deep pool, as shown in Fig. 4. In the experiments, the robot began to swim from a stationary attitude and recorded the yaw angle and tail fin offset in real time, as shown in Fig. 5. For the convenience of the experiment, we fixed tail fin amplitude and frequency, so that the robot will swim at the same linear speed.

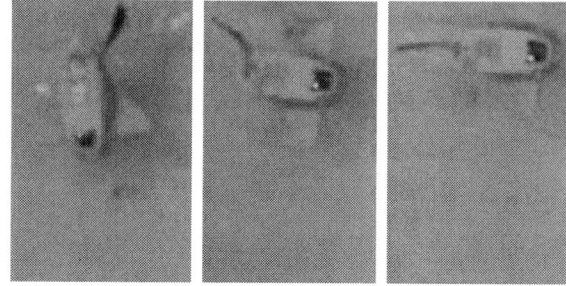

Fig. 5: The control process of the robot swimming

We first adopt the single-loop PID controller to control the yaw angle. But we found that it had a great overshoot and the dynamic response was too slow. As shown in Fig. 6. Therefore, the system cannot respond quickly to meet our control requirements.

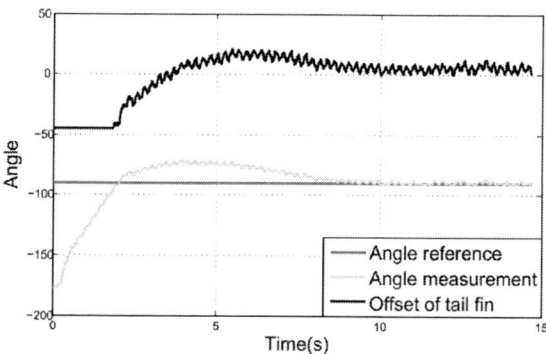

Fig. 6: The Single-loop PID control effect

When we adopt the cascade PID controller, the performance is significantly better than single-loop PID controller. Compared with the single-loop PID controller, the cascade PID controller is faster in dynamic response and shorter in adjustment time. Moreover, it only has a small overshoot. As shown in Fig. 7. But it can be found that the offset of the actuator (tail fin) is large.

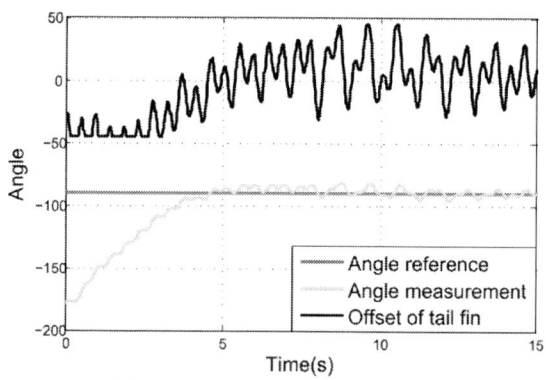

Fig. 7: The cascade PID control effect

The large offset is not expected. Kalman filter can obviously reduce the amplitude of the error of the inner loop. It leads to significantly reduce the offfset of the

978-1-5090-0064-7/15 $31.00 © 2015 IEEE   170

actuator (tail fin), smooth the oscillation of the signals. As shown in Fig. 8. After Kalman filter, the system controls better results, as shown in Fig. 9.

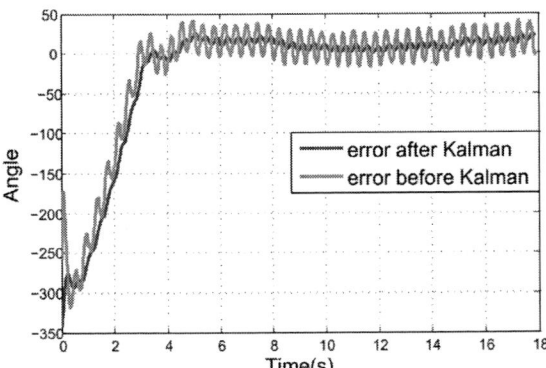

Fig. 8: Error curve before and after Kalman filter

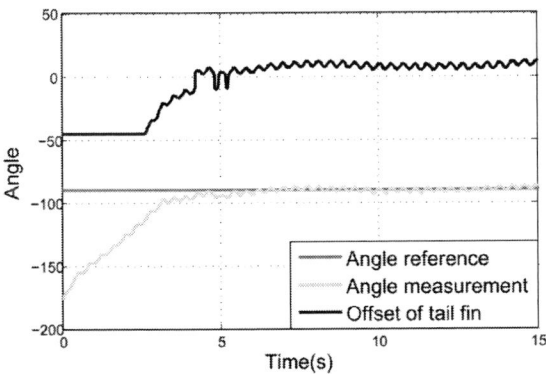

Fig. 9: The cascade PID control effect after Kalman filter

When the more error between angle reference and angle measurement, the greater the offset of tail. We limit a maximum for the offset of tail to protect physial mechanical properties.When we need to turn robot continuously by changing yaw angle, it can fleetly adjust to the specified yaw angle at 6 seconds. In addition to this, it doesn't produce overshoot and accurately keep expected yaw angle. As shown in Fig. 10.

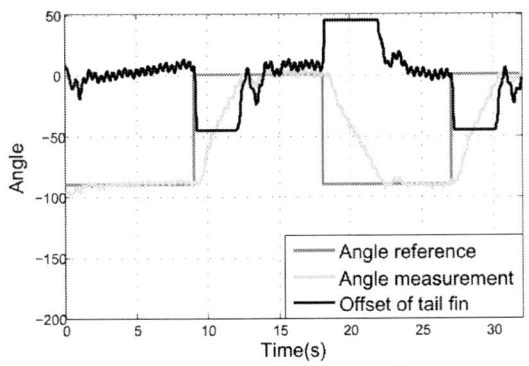

Fig. 10: Tracking square wave trajectory

When the robot tracks the sinusoidal trajectory, it is accurate, and have only a small time lag. It illustrates that the robot can not only maintain steady state, but also has the certain real-time performance. As shown in Fig. 11.

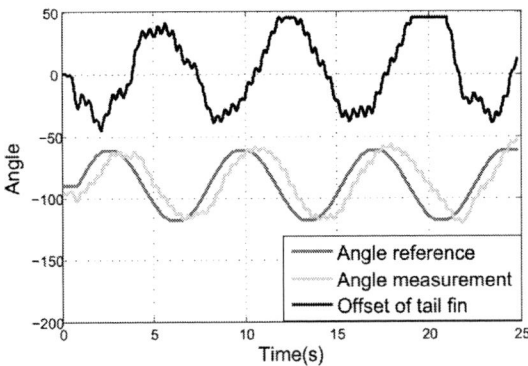

Fig. 11: Tracking sinusoidal trajectory

V. CONCLUSION AND FUTURE WORK

In this paper, we studied yaw angle control of a boxfish-like robot. According to the characteristics of fish swimming, we used a cascade PID controller to control the yaw angle of the robot. And introduced a Kalman filter to smooth the oscillation of the signals. The actual swimming performance indicated that this method has a good dynamic tracking ability of yaw angle of the robot. In the future, we will realize the control of roll angle and pitching angle.

REFERENCES

[1] Triantafyllou, Michael S., and George S. Triantafyllou. "An efficient swimming machine." Scientific american 272.3 (1995): 64-71.

[2] Wei Wang, Xingxing Zhang, Jianwei Zhao, Guangming Xie, "Sensing the Neighbouring Robot by the Artificial Lateral Line of a Bio-inspired Robotic Fish", The 2015 IEEE/RSJ International Conference on Intelligent Robots and Systems (IROS), accepted.

[3] Wei Wang, Hanbo Deng, Jianwei Zhao, Guangming Xie,"Electrode Size Affects Underwater Electric Current Communication Between Two Fish Models", The 34th Chinese Control Conference (CCC), accepted.

[4] Wei Wang, Guangming Xie, Hong Shi, "Dynamic Modeling of an Ostraciiform Robotic Fish Based on Angle of Attack Theory", The IEEE World Congress on Computational Intelligence (WCCI), pp. 3944 – 3949, 2014.

[5] Wei Wang, Jianwei Zhao, Wei Xiong, Fayang Cao, Guangming Xie, "Underwater Electric Current Communication of Robotic Fish: Design and Experimental Results", International Conference on Robotics and Automation (ICRA), pp. 1166 – 1171, 2015.

[6] Wei Wang, Guangming Xie, "An Adaptive and Online Underwater Image Processing Algorithm Implemented on Miniature Biomimetic Robotic Fish", The 19th World Congress of the International Federation of Automatic Control (IFAC), pp. 7598-7603, 2014.

[7] Yu, Junzhi, et al. "Amphibious Pattern Design of a Robotic Fish with Wheel-propeller-fin Mechanisms." Journal of Field Robotics30.5(2013):702-716.

[8] Crespi, Alessandro, et al. "Controlling swimming and crawling in a fish robot using a central pattern generator." Autonomous Robots25.1-2(2008):3-13.

[9] Wang, Weibing, et al. "Bio-inspired design and realization of a novel multimode amphibious robot." Automation and Logistics, 2009. ICAL'09. IEEE International Conference on. IEEE, 2009.

[10] Xian-lun, T. A. N. G., et al. "Implementation of PID algorithm based on MATLAB in cascade control system."

978-1-5090-0064-7/15 $31.00 © 2015 IEEE

Journal of Chongqiang University (Natural Science Edition) 28 (2005): 61-63.

[11] Wei Wang, Guangming Xie, "Online High-Precision Probabilistic Localization of Robotic Fish Using Visual and Inertial Cues", IEEE Transactions on Industrial Electronics. 62(2), pp. 1113-1124, 2015.

[12] Wei Wang, Guangming Xie, "CPG-based Locomotion Controller Design for a Boxfish-like Robot" International Journal of Advanced Robotic Systems. DIO: 10.5772/58564, 11(87), 2014.

[13] Wei Wang, Jiajie Guo, Zijian Wang, Guangming Xie, "Neural Controller for Swimming Modes and Gait Transition on an Ostraciiform Fish Robot", IEEE/ASME International Conference on Advanced Intelligent Mechatronics (AIM), pp. 1564-1569, 2013.

[14] Ijspeert, Auke Jan, et al. "From swimming to walking with a salamander robot driven by a spinal cord model." science 315.5817 (2007): 1416-1420.

[15] Kamimura, Akiya, et al. "Automatic locomotion design and experiments for a modular robotic system." Mechatronics, IEEE/ASME Transactions on 10.3 (2005): 314-325.

Genetic Algorithm Based Back-Propagation Neural Network Approach for Fault Diagnosis in Lithium-ion Battery System

Zuchang Gao[1], Cheng Siong Chin[1], Wai Lok Woo[1], Junbo Jia[2], Wei Da Toh[2]

[1] Faculty of Science Agriculture and Engineering, Newcastle University, Newcastle upon Tyne, NE1 7RU, United Kingdom
[2] Clean Energy Research Centre, School of Engineering, Temasek Polytechnic, 21 Tampines Ave 1, Singapore

Abstract–Safety is important in a lithium-ion battery power system. It is necessary to adopt an effective fault diagnosis method to keep the battery power system in the good working status. In this paper, Genetic Algorithm (GA) is integrated to build a single hidden layer Back-Propagation Neural Network (BPNN) for fault diagnosis. In the process of training the neural network, GA is used to initialize and optimize the connection weights and thresholds of the neural network. Several faults are detected by the proposed GA optimized fault diagnosis scheme. Simulation results show that the proposed fault diagnosis scheme provides satisfactory results.

Keywords–Back-propagation neural network, fault diagnosis, genetic algorithm, lithium-ion battery.

I. INTRODUCTION

Compared with other commonly used batteries (lead acid battery, NiMH battery, NiCd battery, etc), lithium-ion batteries are featured by high energy density, high power density, long life and environmental friendliness and thus have found wide applications in the area of consumer electronics and automotive applications. However, lithium-ion batteries must operate within the safe and reliable operating area that is restricted by temperature and voltage windows. Exceeding the restrictions of these windows will lead to rapid attenuation of battery performance and even result in safety problems [1] [6].

To predict and solve the safety problems, people would like to develop new robust battery power system with fault diagnosis that works under severe situations [10]. As presented in [2], the ability to detect and diagnose faults is essential for the safe and reliable control of mechanical and electrical systems. In all systems, every fault has effects on measurements, but the relationships between the faults and the measurements are not very obvious. Through the measurement data, different approaches for fault detection and diagnosis have been developed in recent years. Riascos developed a fault diagnosis system using Bayesian networks for proton exchange membrane fuel cells [4]. An artificial neural network ensemble method was developed for fault diagnosis in proton exchange membrane fuel cell system [8]. A method based on entropy for fault detection of the connection of lithium-ion power batteries was proposed [9]. An improved BP Neural Network for transformer fault diagnosis [5] was presented. Moreover, a model-based fault detection of a battery system for a hybrid electric vehicle [3] was proposed.

But in most of modern lithium battery power systems, there are just simple protection circuits to protect the batteries when the critical faults developed (e.g. low battery voltage). There are few fault diagnosis and prediction features in those battery power systems. In this paper, an on-line fault detection and diagnosis system is proposed for lithium-ion battery power system by applying genetic algorithm based back-propagation neural network approach. It aims to detect and diagnosis critical faults in the battery power system through on-line monitoring of the measurable parameters (e.g. voltage, current, and temperature) [7].

In summary, the proposed approach provides a feasible and effective method to detect and diagnosis critical faults in the battery power system. This paper is organized as follows. The modeling of the proposed approach is addressed in Section 2. In Section 3, the experiment data are collected from the battery testing platform, and used to train and evaluate the models which have been built in section 2. Lastly, the conclusions together with future works are drawn in Section 4.

II. MODELING OF BPNN AND GA-BPNN

1. BPNN Based Fault Detection Model

In the BPNN model, learning takes place during the propagation of input patterns from the input nodes to the output nodes. The outputs are compared with the desired target values and an error is produced. Then the weights are adapted to minimize the error.

The relation of output $Q_i^{(l)}$ and input $Q_j^{(l-1)}$ of layer j is defined as [4]:

$$Q_i^{(l)} = fs[I_i^{(l)}] \tag{1}$$
$$I_i^{(l)} = \sum w_{ij}^{(l)} Q_j^{(l-1)} \tag{2}$$

Equation (1) can be transformed into:

$$F_s(I) = 1/(1 + exp(-I)) \tag{3}$$

The initial values of weights are assumed to be zero, and the weight between the jth neuron of the $(k-1)^{th}$ layer and the ith neuron of the kth layer is defined as $w_{ij,k}$. The weight adaptation equation is given by the following.

$$W_{ij,k}(t_n) = W_{ij,k}(t_{n-1}) - \frac{\alpha E(t_n)}{W_{ij,k}(t_{n-1})} \Delta W_{ij,k}(t_{n-1}) \tag{4}$$

where $0<\alpha<1$, and $E = 1/2 \sum(y_i - b_i)^2$. $I = 1...n$, y_i is i^{th} actual output, b_i is i^{th} simulation output. The structure of BPNN is shown in Figure 1.

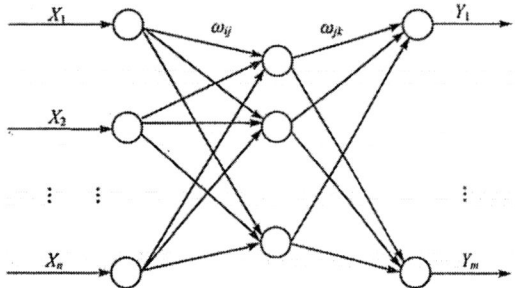

Fig. 1: Topology structure of BPNN model

In the paper, the BP neural network model is built and carried out in the MTALAB platform. The neural network includes three layers: input layer, hidden layer and output layer. Each layer consists of a number of fundamental processing units called neurons. Function 'newff' is called to create a BP neural network. Transfer function for the hidden layer and the output layer neuron is set to sigmoid-type function of 'tansig' and 'logsig' respectively. 'trainscg' is chosen as the training function, and the learning rate, training times and other relevant parameters are set.

The BPNN is built and simulated as per the following sequence:

Step 1: Experiment data acquisition and importing;

Step 2: Data selection and normalization (simple pre-processing to the training data and testing data);

Step 3: BPNN structure and parameters initialization (Max training epoch *net.trainParam.epochs* is set to 200. Min training error *net.trainParam.goal* is set to 0.001. And the learning rate *net.trainParam.lr* is set to 0.1.);

Step 4: BPNN neural network training using the training data, update the network parameters (e.g. $W_{ij,k}(t_n)$) to the model when the training process finished.

Step 5: Detect the faults by inputting the testing data to the model which has been trained above;

Step 6: Compare the simulation result and actual faults.

2. GA-BPNN Based Fault Detection Model

In this paper, a Genetic Algorithm based Back Propagation Neural Network (GA-BPNN) approach is proposed for the fault detection and diagnosis in the lithium-ion battery power system. GA is used to optimize the weights and thresholds of BP neural network to get the best initial weights and thresholds (individual). The flow chart of GA-BPNN model is shown in Figure 2.

2.1 Initialize the GA parameters

In population initialization, each individual is an encoded vector which consists of the weights of input layer - hidden layer and hidden layer - output layer, the thresholds of hidden layer and output layer. So the number of elements in each individual can be calculated as (5),

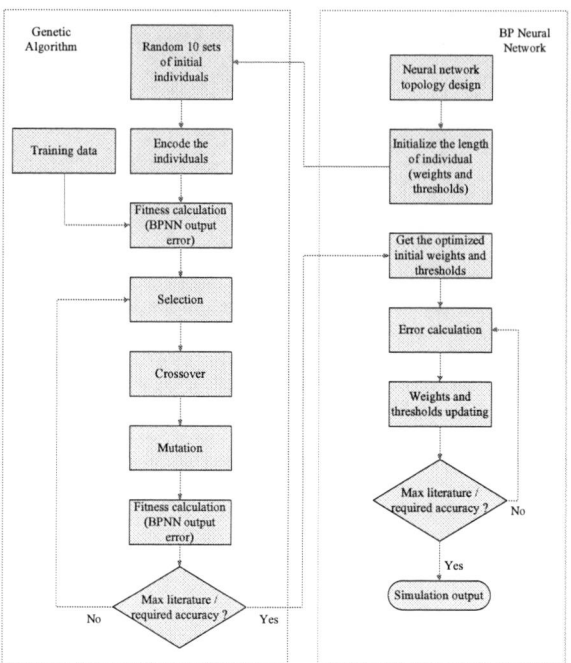

Fig. 2: Flow chart of GA-BPNN model

$$N_{sum} = N_{input} \times N_{hidden} + N_{hidden}$$
$$+N_{hidden} \times N_{output} + N_{output} \qquad (5)$$

where N_{sum} is the number of elements in each individual, N_{input} is the number of elements in input layer of the BPNN, N_{hidden} is the number of elements in hidden layer of the BPNN, N_{output} is the number of elements in output layer of the BPNN.

The size of the population affects the end result of genetic optimization and the execution efficiency of genetic algorithm, the common population size is 10 to 160. In this paper, the population size is set to 10.

2.2 Fitness function

Fitness function is used in GA during each iteration of the algorithm to evaluate the quality of all the proposed individuals in the current population. The fitness function evaluates how good an individual in a population. In this paper, the fitness function is defined as the sum of the absolute error values between simulation output and actual output of the training data sets.

$$F = k \left(\sum_{i=1}^{n} abs(y_i - o_i) \right) \qquad (6)$$

where n is the number of neurons in output layer of the GA-BPNN model, y_i is the actual output of neuron i, o_i is the simulation output of neuron i, k is coefficient.

2.3 The selection function

The selection function chooses parents for the next generation based on their scaled values from the fitness scaling function. An individual can be selected more than once as a parent, in which case it contributes its genes to more than one child. In this paper, fitness proportionate selection or roulette-wheel selection is applied for the selected function, with this method, the probability p_i of each individual i can be described as follows,

$$f_i = {^k/_{F_i}} \quad (7)$$

$$p_i = \frac{f_i}{\Sigma_{j=1}^{N} f_j} \quad (8)$$

where F_i is the fitness value of individual i, N is the number of individuals in the population, k is coefficient.

2.4 Crossover and Mutation

GA uses the individuals in the current generation to create the children that make up the next generation. Besides elite children, that corresponds to the individuals in the current generation with the best fitness values.

Crossover involves in selecting vector entries, or generations, from a pair of individuals in the current generation and combines them to form a child. In this paper, a real number based crossover method is applied. The individual k and individual l performing crossover operation at element j can be described.

$$a_{kj} = a_{kj}(1-b) + a_{lj}b \quad (9)$$

$$a_{lj} = a_{lj}(1-b) + a_{kj}b \quad (10)$$

Mutation operation is to apply random changes to a single individual in the current generation to create a child. Both processes are essential to the genetic algorithm. Crossover enables the algorithm to extract the best generations from different individuals and recombine them into potentially better children. Mutation adds to the diversity of a population and thereby increases the likelihood that the algorithm will generate individuals with better fitness values.

III. RESULT AND DISCUSSION

1. Training Data and Testing Data Collection

In the proposed fault detection and diagnosis system, four types of faults (voltage sensing fault, temperature sensing fault, battery cell fault and external short circuit fault) are considered. The definitions of system input/output are shown in Table 1, 18 groups of experiment data captured from the battery testing system for the neural network training is shown in Table 2, and 9 groups of the experiment data which are different from the training data are used to test the generalization capability of the trained neural network model, as shown in table 3. Battery testing system is shown in Figure 3.

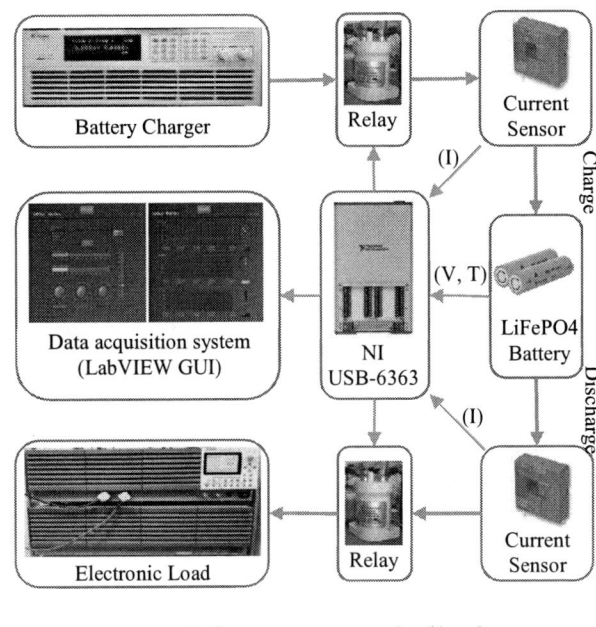

Fig. 3: Battery testing system

Table 1: Input/output definition of the fault diagnosis system

Signal type	Symbol	Description
Inputs	X1 (V)	Battery cell voltage
	X2 (I)	Battery cell current
	X3 (°C)	Battery cell temperature
	X4 (0/1)	System on/off status
	X5 (V)	voltage after the system output relay
Output	F1 (0/1)	Battery cell voltage sensing fail
	F2 (0/1)	Battery cell temperature sensing fail
	F3 (0/1)	Battery cell fail
	F4 (0/1)	External short circuit

Table 2: Experiment data for neural network training

X1	X2	X3	X4	X5	F1	F2	F3	F5
0.01	31.5	5.15	1	3.28	1	0	0	0
0.01	29.5	4.55	1	3.15	1	0	0	0
0.02	24.6	0	0	0.01	1	0	0	0
0.01	24.5	0	0	0.01	1	0	0	0
3.3	11	5.5	1	3.3	0	1	0	0
3.25	11.1	4.56	1	3.25	0	1	0	0
3.1	11	0	0	0.01	0	1	0	0
3.05	11	0	0	0.01	0	1	0	0
0.03	24.7	0.02	1	0.03	0	0	1	0
0.02	24.5	0.03	1	0.02	0	0	1	0
0.55	24.6	0	0	0.01	0	0	1	0
0.65	24.5	0	0	0.01	0	0	1	0
0.01	50.5	60.5	1	0.01	0	0	0	1
0.01	55.8	59.8	1	0.01	0	0	0	1
3.32	28.5	5.1	1	3.32	0	0	0	0
3.2	29.5	5.2	1	3.2	0	0	0	0
3.15	24.8	0	0	0.01	0	0	0	0
3.1	24.5	0	0	0.02	0	0	0	0

978-1-5090-0064-7/15 $31.00 © 2015 IEEE

Table 3: Experiment data for neural network testing

X1	X2	X3	X4	X5
0.02	30.5	4.95	1	3.2
0.03	24.7	0	0	0.01
3.05	11	5.05	1	3.05
3.2	11.2	0	0	0.01
0.03	24.7	0.02	1	0.03
0.56	24.7	0	0	0.01
0.02	51	55.5	1	0.02
3.28	30.5	5	1	3.28
3.25	24.2	0	0	0.01

2. Comparison of Simulation Results

In this paper, the BPNN networks with GA optimization (GA-BPNN) and without GA optimization (BPNN) were built and trained respectively to compare the performance. After training the BPNN model built in section II using the training data from Table 2, the testing data from Table 3 were imported to the model to obtain the simulation result. The comparison between actual faults and simulation results for BPNN model was shown in Table 4.

Table 4: Comparison between estimated faults and actual faults (BPNN)

No	Actual Faults				Simulation Result			
1	1	0	0	0	1.079 *	0.1246	-0.0088	-0.0367
2	1	0	0	0	0.8478	0.2689	0.2243	-0.1634
3	0	1	0	0	0.0725	1.181	-0.1931	0.019
4	0	1	0	0	-0.3275	0.5859	0.3755	-0.296
5	0	0	1	0	1.322	2.2882	0.4631	-1.2196
6	0	0	1	0	0.5111	0.2776	0.5211	-0.3131
7	0	0	0	1	0.379	0.2314	-0.1843	0.6254
8	0	0	0	0	-0.0257	-0.121	0.0762	0.0599
9	0	0	0	0	-0.1736	0.0396	0.216	-0.1016

* The items with underline will be recognized as a fault.

From Table 4, one of the faults (No. 5) was not detected correctly based on the testing data. Correct value was around 88.89%. The comparison between estimated faults and actual faults in BPNN model was shown in Figure 4 (A).

Similar with BPNN, the training data from Table 2 were imported to the GA-BPNN model to train the network, and the testing data from Table 3 were imported to the GA-BPNN model to evaluate the model performance.

In the paper, we initialized the parameters for GA-BPNN model as follows: population size: 10 individuals; number of generations: 20; mutation rate: 10% of individuals in each generation; crossover rate: 20% of individuals in each generation; weight and threshold values: initial values for the weights and the thresholds are random draw from [-3, 3].

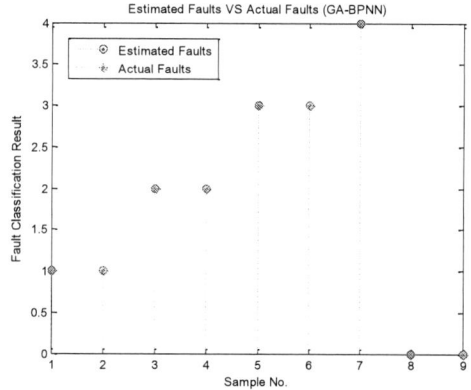

(B) Comparison between estimated faults and actual faults in GA-BPNN

Fig. 4: Comparison between estimated faults and actual faults

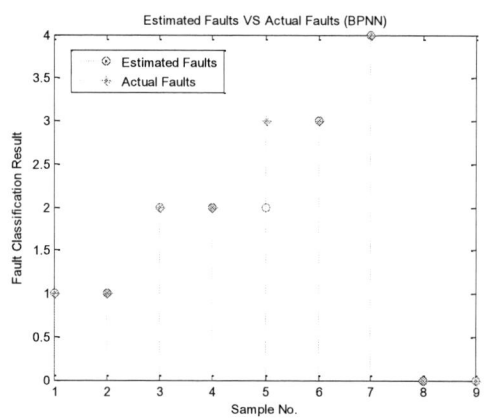

(A) Comparison between estimated faults and actual faults in BPNN

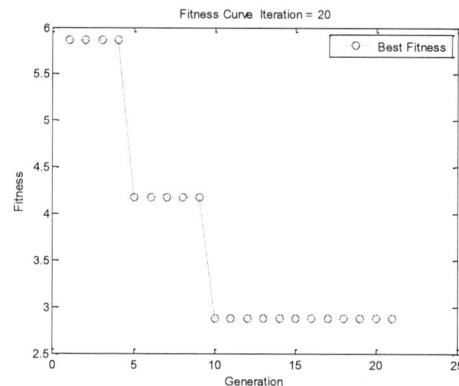

Fig. 5: Fitness curve in GA-BPNN model

Following the flow chart shown in Figure 2, the individuals would be selected according to their fitness values and finished the crossover and mutation process every iteration, the fitness curve was shown in Figure 5. When the iterations came to an end, the final optimized weights and thresholds from Genetic Algorithm were used to update the GA-BPNN model. The same training data from Table 2 were used to train the model and the same testing data from Table 3 were used to evaluate the performance of the model. The comparison between simulation results and actual faults for GA-BPNN model was shown in Table 5.

Table 5: Comparison between estimated faults and actual faults (GA-BPNN)

No	Actual Faults				Simulation Result			
1	1	0	0	0	1.0782*	-0.0644	0.1194	0.0532
2	1	0	0	0	0.9837	-0.0008	0.0314	-0.0166
3	0	1	0	0	0.172	1.1544	-0.0512	-0.07
4	0	1	0	0	0.0136	0.9702	-0.0192	0.0277
5	0	0	1	0	0.0008	-0.0014	0.997	0.0007
6	0	0	1	0	0.1431	0.0393	0.9089	-0.063
7	0	0	0	1	0.3733	0.4036	-0.5066	0.7523
8	0	0	0	0	-0.0262	-0.0665	-0.0627	0.0449
9	0	0	0	0	0.0615	0.1484	-0.0363	-0.0272

* The items with underline will be recognized as a fault.

From Table 5, all faults can be detected correctly based on the testing data. Correct value is 100%. The comparison between estimated faults and actual faults in GA-BPNN model is shown in Figure 4 (B).

The error comparison in BPNN / GA-BPNN model is shown in the Figure 6. It can be observed that the fluctuation range of the output error in BPNN model varies from -0.7 to +1.8. Comparing with BPNN model, the output error in GA-BPNN model is smaller and more stable. Therefore, it provides better fault diagnosis performance in a battery power system when the GA-BPNN model is applied.

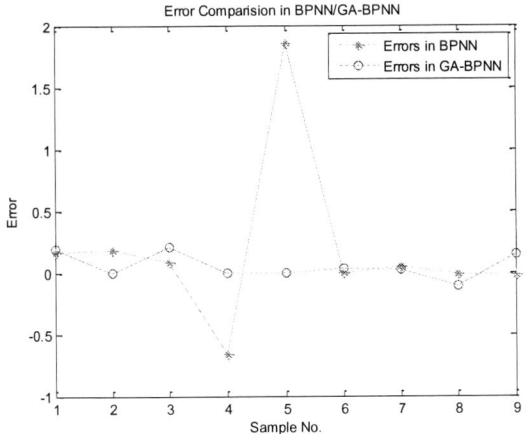

Fig. 6: Error curve between estimated faults and actual faults in BPNN/GA-BPNN models

IV. CONCLUSION

In this paper, a genetic algorithm based back-propagation neural network approach for fault diagnosis in Lithium-ion battery system is proposed. Models of BPNN and GA-BPNN based fault diagnosis systems are built and the experiment data are imported to both models to evaluate the performance. From the comparisons, it shows the proposed GA-BPNN model in the fault diagnosis system performs higher value and lower error range. Therefore, compared with the conventional BP neural network, the GA-BPNN model provides better fault diagnosis performance in a battery power system.

In the future, the GA-BPNN fault diagnosis model will be integrated into the lithium-ion battery model, the real-time battery information will be shared among different modules, to achieve the on-line diagnosis and control to the battery power system.

ACKNOWLEDGMENT

The author would like to thank the master students from Southwest Jiaotong University, SMI (SMI-2013-MA-05), Newcastle University and Temasek Polytechnic for their research support during the project.

REFERENCES

[1] S.A. Gadsden, K. R. McCullough, S.R. Habibi, Fault detection and diagnosis of an electro hydrostatic actuator using a novel interacting multiple model approach, in: American Control Conference (ACC), San Francisco, California, 2011.

[2] S. Andrew Gadsden and Saeid R. Habibi, Model-Based Fault Detection of a Battery System in a Hybrid Electric Vehicle, Journal of Energy and Power Engineering 7 (2013) 1344-1351.

[3] Luis Alberto M. Riascos, Marcelo G. Simoes, Paulo E. Miyagi, On-line fault diagnostic system for proton exchange membrane fuel cells, Journal of Power Sources 175 (2008) 419–429.

[4] K. Mohammadi, A.R. Mohseni Monfared and A.-Molaei Nejad, Fault Diagnosis of Analog Circuits with Tolerances By Using RBF and BP Neural Networks, 2002 Student Conference on Research and Development Proceedings, Shah Alam, Malaysia

[5] Jian-Da Wu, Peng-Hsin Chiang, Yo-Wei Chang, Yao-jung Shiao, An expert system for fault diagnosis in internal combustion engines using probability neural network, Expert Systems with Applications 34 (2008) 2704–2713.

[6] Chin CS. Model-Based Simulation of an Intelligent Microprocessor-Based Standalone Solar Tracking System.

In: Katsikis, V, ed. MATLAB - A Fundamental Tool for Scientific Computing and Engineering Applications. Rijeka, Croatia: InTech, 2012, pp.251-278.

[7] Chin C, Mesbahi E. Problem-Based Learning Approach for Martronics. In: IEEE International Conference on Teaching, Assessment and Learning for Engineering. 2012. The Hong Kong Polytechnic University, Hong Kong: IEEE.

[8] Meng Shao, Xin-Jian Zhu, Hong-Fei Cao, Hai-Feng Shen. An artificial neural network ensemble method for fault diagnosis of proton exchange membrane fuel cell system. Energy 67 (2014) 268 – 275.

[9] Lei Yao, Zhenpo Wang, Jun Ma. Fault detection of the connection of lithium-ion power batteries based on entropy for electric vehicles. Journal of Power Sources 293 (2015) 548 – 561.

[10] Chin C , Lau M, Low E, Seet G. Robust and decoupled cascaded control system of underwater robotic vehicle for stabilization and pipeline tracking. Proceedings of the Institution of Mechanical Engineers, Part I: Journal of Systems and Control Engineering 2008, 222(4), 261-278.

BIOGRAPHIES

Zuchang Gao was born in China in April 1983. He received MEng degree in Electrical Engineering from the Southwest Jiaotong University, China. Since Sep 2009, his research activities are concerning fuel cells, power electronics, and energy management. Since Oct 2013, he is pursuing his PhD degree concerning smart battery power system in Newcastle University, UK

Cheng S. Chin (M'01-SM'09). He received his Ph.D in Applied Control Engineering from Nanyang Technological University (NTU) in 2008 and M.Sc. in Advanced Control and Systems Engineering from The University of Manchester in 2001. He also graduated with a B.Eng in Mechanical and Aerospace Engineering from NTU in 2000. He published around 50 journals, conferences and book chapters. He had authored a book titled "Applied Control Engineering using MATLAB and Simulink" published by CRC Press in 2012. He is a Fellow of IMarEST, member of IET, FHEA, C.Eng and EURING. His research interests are in applied control systems design of mechatronics systems and noise control.

Wai L. Woo (M'08-SM'12). He is currently Director of Singapore Operations. His major research is in the mathematical theory and algorithms for nonlinear signal and image processing. This includes areas of machine learning, blind source separa tion, multidimensional signal processing, signal/image deconvolution and restoration. He has an extensive portfolio of relevant research supported by a variety of funding agencies. He has published over 200 papers on these topics on various journals an d international confer ence proceedings. He serves as Associate Editor to several international signal processing journals including IEEE Transactions on Neural Networks, IET Signal Processing, ISRN Machine Vision, and Journal of Computers. Dr Woo is a Member of the Institution Engineering Technology (IET) and Senior Member of Institution of Electrical an d Electronic Engineering (IEEE).

Junbo Jia received his B.Eng. degree in 1988, the M.Eng. degree in 1996, both from electrical engineering in SouthWest Jiaotong University, China, and Ph.D. degree in 2011 from Nanyang Technological University, Singapore. Junbo developed the electric railway over-head line testing train for Zhengzhou, China Railway Bureau; developed the testing systems for mobile

substation testing coaches which belong to Fuzhou, Chengdu and Kunming Electric Railway Bureau and Electrical Bureau of Ministry of Railway, China, individually. The fund was around US$3m. In 1996, he was a lecturer, and in 1998, he became an Associate Professor in SouthWest Jiaotong University, Chin a. In 2001, he joined Temasek Polytechnic Engineering School as a lecturer, and in 2010, he became a senior lecturer; his research fields are pow er electronic, fuel cell modelling and advanced control as well as energy management for electric vehicle. Junbo published more than 30 journal papers and international conference papers on the program of railway station flexible chains prefabricating and drawing, the implementation of the electrical testing vehicles kinetic state software based on dataflow, the port communication technology, the embedded system in the programmable solar panel simulator, and the research of the transient behavior of PEM Fuel Cell.

Wei-Da Toh received the B.Eng. and M.Eng. degrees in electrical engineering from the National University of Singapore in 2007 and 2010, respectively. In 2007, he was with National University of Singapore, as a research engineer. From 2010 to 2013, he was with Institute of Microelectronics (IME), A*STAR, Singapore, as a research engineer, worked on digital integrated circuits design for biomedical applications. In 2013, he joined DSO National Laboratories, Singapore, as a member of technical staff, involved in FPGA based digital controller design. Since 20 14, he is with Temasek Polytechnic, as a project engineer, working on battery management system. His current research interests include algorithm opti mization for digital circuits and low-power digital design.

978-1-5090-0064-7/15 $31.00 © 2015 IEEE

Step-Less Voltage Regulation on Radial Feeder with OLTC Transformer-DVR Hybrid

Vishal Verma[1]   Ritika Gour[2]

[1,2]Electrical Engineering Department, Delhi Technological University

[1] vishalverma@dce.ac.in, [2] riti113@gmail.com

Abstract—On Load Tap Changing (OLTC) transformer acts to regulate the load voltage on load by increasing or decreasing the input power flow whereas, the capacity supported DVR regulates the voltage with reactive power transfer. As the OLTC operates in discrete steps, it brings in quantum jump in both the phase and magnitude of voltage at PCC, whereas sensitive load requires step-less operation. For step-less voltage regulation over extended range a hybrid combination of DVR with OLTC control of transformer is envisaged. The paper presents a simple and fast control strategy for hybrid control of both OLTC and DVR to desire stepless voltage regulation of the radial feeder under load perturbation and source side disturbances. A simple control algorithm is proposed which control the DVR and OLTC transformer to regulate the voltage in the radial feeder. Simulated results under MATLAB environment are presented to validate the effectiveness of control scheme.

Keywords—Capacity limited voltage source converters, DVR, OLTC transformer, voltage regulation.

I. INTRODUCTION

Voltage disturbance is one of the severe most power quality problems which may lead consumer to substantial loss and damage [1]. These disturbances are caused due to system fault, load switching (sudden increase or decrease of large load), extensive use of non-linear load in the distribution system are integration of distributed generation sources to name a few [1]. Therefore, voltage in the distribution system is required to be kept regulated within prescribed limits. Some devices used in distribution system for regulating and balancing the voltage are tap changing transformers, UPS (Uninterrupted Power Supply), and custom power devices like DVR (Dynamic Voltage Restorer), UPQC (Unified Power Quality Conditioner), D-STATCOM (Distribution STATCOM) etc. The OLTC transformers are used for regulating the voltage by controlling in majority with the real power flow. By varying the number of turns, which in turn forms the discrete steps eventually lead to transient in the system each time the tapping of the transformer is changed [2]. The rapid changes caused by intermitted loading cannot be corrected by the OLTC, as it will not able to follow the voltage build-up process, eventually leading to voltage collapse [2]. Hence, OLTC transformer does not operate independently but are reported for operation in combination with dynamic regulating device such as reactive power control provided in shunt [3-4], [14-16].

DVR, a custom power device that injects voltage at an angle with line voltage such that voltage at load side is balanced and regulated [5-6]. Different topologies for different applications are reported literature [7]-[12]. In addition to voltage disturbances compensation, DVR in addition is reported to effectively compensate harmonics, reduce transients in voltage, and effectively limits the fault currents [13]. The maximum injection capability of DVR which is restricted by the energy storage on DC bus and transformer ratio is required to be extended for capacitor supported DVR to figure out a viable series compensation solution. Moreover, due to limited capacity of capacitor supported DVR, the compensation with greater depth is not achievable, leaving behind OLTC as only other alternative to achieve the compensation upto desired level. The research is therefore done to extend the range of DVR and operate OLTC steplessly to arrive at viable solution for voltage regulation and compensation of other power quality problems.

The paper proposes a new approach of using OLTC transformer along with the DVR to regulate the voltage for an industrial plant supplied by a radial feeder. This approach aims at increasing the response time of OLTC transformer, linearizes, the steps involved in regulating the voltage by providing coarse compensation and during intermediate requirements between the steps, employing DVR for voltage regulation, and thus decreases the rating of the DVR for higher depth of compensation. The computer simulation is performed using MATLAB/SIMULINK software. The simulation results are presented to validate the effectiveness of hybrid system in compensating the voltage disturbances with greater depth through a viable stepless cost effective solution.

II. SYSTEM CONFIGURATION

Fig. 1 shows the block diagram of proposed OLTC transformer and DVR connected on a radial distribution feeder. The considered system for study comprises of a source, an OLTC transformer, a DVR and a three phase load. V_{Sabc} depicts a three phase voltage source supplying 415 V (line-to-line) and the feeder drop is realized as Z_S. The OLTC transformer is rated for 15kVA as distribution transformer within a range of ± 0.05 pu. DVR is shown to be comprised of injection transformer T_r, ripple filter, voltage source converter and capacitor on the DC side. The DVR injects voltage into the feeder through the injection transformer T_r, which is a star-connected 5.5 kVA low leakage impedance coupling transformer. The injected voltage of the DVR is kept in quadrature with the feeder current for ascertaining the avoidance of involvement of real power in compensation realized through IGBT bridge acting as voltage source converter (VSC), with a DC link capacitor LC ripple filters (L_f, C_f) for three phase VSC is connected to reduce the ripple voltage and provide a requisite voltage across secondary transformer to enable injection across the primary side of transformer and assist in self-support of DC bus. The capacity of VSC used to realize DVR is limited to 5kVA, against fact of coupling transformer (5.5 kVA) to avoid saturation. The DVR is inserted into the feeder by providing a near short circuit across the secondary terminals of transformer by operating

978-1-5090-0064-7/15 $31.00 © 2015 IEEE

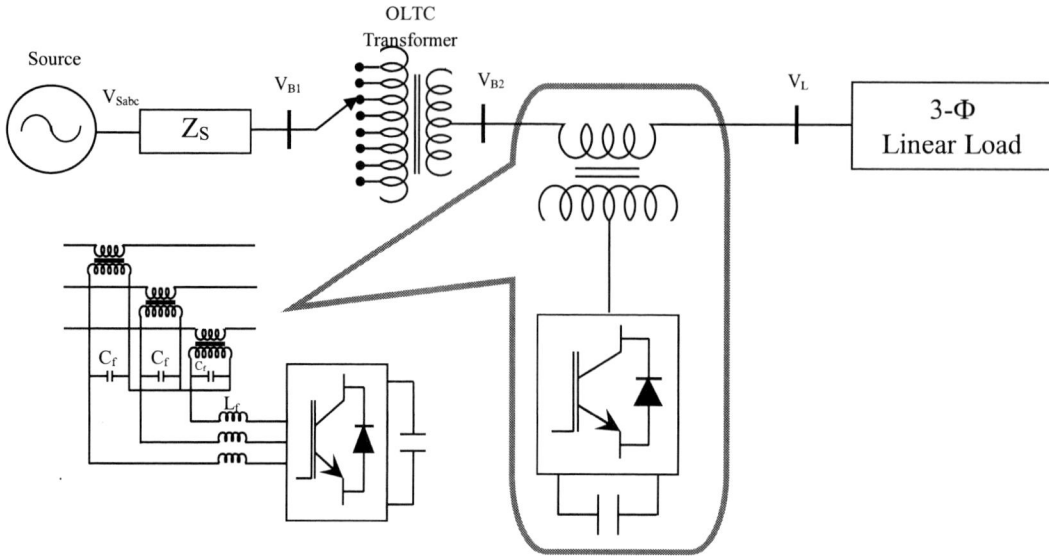

Fig. 1: System Configuration

the bottom switches of the VSC, and thereon involved into the circuit by applying the PWM switching. A linear 3-phase RL load consuming both real and reactive power (Z_L) is considered connected at the receiving end of the radial feeder. The connected load has the capacity of 10.8kVA.

III. CONTROL SCHEME

The proposed control of OLTC in tandem with DVR is implemented by estimating the linked capacity of DVR and up on its exhaustion, change the tap of the OLTC transformer by utilizing the synchronous reference frame (SRF) theory to avoid switching transients. As in radial feeder the power flows in one direction from source to load, so both OLTC transformer and DVR have to compensate for voltage disturbance downstream in the feeder so that the voltage at the load terminal experiences the minimum disturbances. Fig. 2(a) shows the basic block diagram of control scheme adopted for controlling the tap position of OLTC transformer and Fig. 2(b) shows the block diagram of control scheme adopted for the control of DVR module.

The tap position of the OLTC transformer depends upon two parameters: (i) Magnitude of Source voltage magnitude V_S, and, (ii) Capacity of the DVR. The source voltage is sensed, and its magnitude is calculated. The OLTC transformer changes its tap position only when the voltage to be fed by the DVR goes past its maximum capacity. This is done by sensing the difference between the voltage prior to DVR; V_{B2} as shown in Fig. 1 and the reference rated load voltage $V_L{}^*$. If this voltage difference is less than the DVR maximum capacity V_{DVRmax} then OLTC transformer will stick to the current tap position, and, DVR alone will regulate the voltage to rated value. When this difference is increases beyond the DVR maximum capacity, the OLTC transformer switches to higher tap position to alter the ratio of transformer and provides a voltage regulation within ± 0.05 pu which is further topped up to rated value by the DVR.

DVR unit is required to consume a small real power to offset the losses incurred in VSC, whereas, capacity dependent partial/full voltage regulation is obtained by injection of voltage in phase quadrature with line current. The voltage towards the load side is sensed for generating

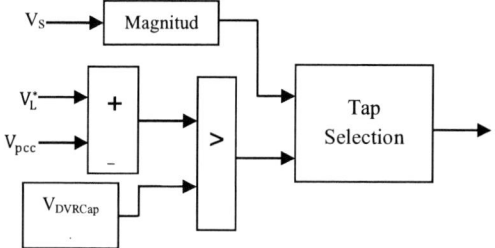

Fig. 2(a): OLTC Transformer tap selecting control

the gating signal for the DVR. The deviation of the load side voltage from the reference value keeping the tap at fixed value will decide the depth of compensation provided by DVR. Whereas, the voltage on the DC bus of the VSC is kept constant by creating a drop of voltage in phase with current to input real power equivalent to meeting the losses in DVR. The estimation of losses in VSC is computed by a PI controller which forms the d-axis component of voltage.

$$V_d(n) = V_d(n\text{-}1) + K_P(V_{de}(n) + V_{de}(n\text{-}1)) + K_I V_{de}(n) \quad (1)$$

where,

$$V_{de} = V_{DC}{}^* - V_{DC}; \quad (2)$$

and, K_P and K_I are the proportional and integral gain of the PI controller.

The magnitude of the load side voltage is regulated by compensation, the reference of which is estimated using another PI block. The q-axis component so estimated pertains to the voltage in quadrature with current and referred as, reference DVR voltage.

$$V_q(n) = V_q(n-1) + K_P (V_{de}(n) + V_{de}(n-1)) + K_I V_{Le}(n) \quad (3)$$

where,

$$V_{Le} = V_L{}^* - V_L \quad (4)$$

and, K_P and K_I are the proportional and integral gain of the other PI controller.

The reference estimates of injection voltage on d-q axis so obtained are reverse Park transformed to abc reference with the help of synchronizing signal, $\sin\theta$ and $\cos\theta$; shown in Fig. 2(b) synchronous with the line current, depicted in eqn. (5).

$$\begin{bmatrix} V_{DVRa}{}^* \\ V_{DVRb}{}^* \\ V_{DVRc}{}^* \end{bmatrix} = \begin{bmatrix} \cos\theta & \sin\theta & 1 \\ \cos\left(\theta - \dfrac{2\pi}{3}\right) & \sin\left(\theta - \dfrac{2\pi}{3}\right) & 1 \\ \cos\left(\theta + \dfrac{2\pi}{3}\right) & \sin\left(\theta + \dfrac{2\pi}{3}\right) & 1 \end{bmatrix} \begin{bmatrix} V_d{}^* \\ V_q{}^* \\ V_0{}^* \end{bmatrix} \quad (5)$$

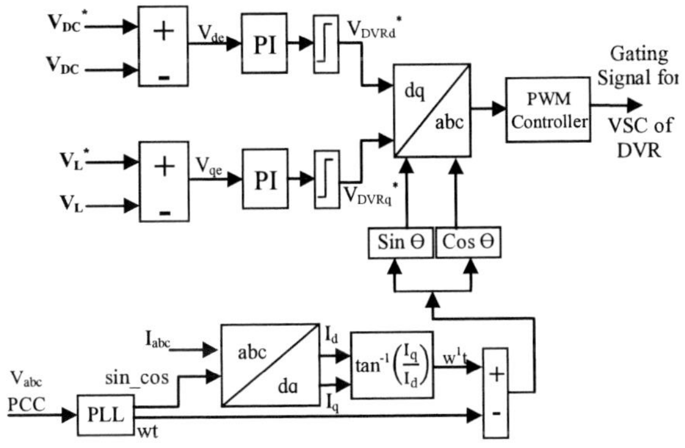

Fig. 2(b) Block Diagram of Control Scheme

These synchronizing signals are obtained from PLL working primarily on source side voltages (PCC) (sensed just before the coupling transformer) and phase modified by coupling the phase delay of line current. The computation block diagram is shown in Fig. 2 (b).

IV. MATLAB SIMULATION

A three phase radial feeder system is simulated under MATLAB/SIMULINK environment using SimPowerSystems toolbox. The system consists of a three phase source, series RL branch depicting the feeder drop, OLTC transformer, DVR module having injection transformer, VSC with its control scheme, and a series RL, 3-Φ, balanced linear load of 10.8 kVA rating. The DVR module is placed downstream after OLTC transformer in the feeder. The DVR compensate for the drop in the feeder, voltage difference caused due changes in tap positions of the OLTC transformer and the reactive power demand of the load. The considered system is shown in the Fig. 3. The OLTC transformer and the DVR module are

controlled in such a way, that the voltage at the load end is kept regulated at rated value. The parameters of the simulation model are given in the Table-I.

TABLE 1: PARAMETERS OF CONSIDERED SYSTEM

Parameter	Values
Feeder Impedance	L_s = 10mH, R_s = 1
Load	10.8kVA, 3-Φ, balanced load
Ripple filter	C_r = 10μF, L_r = 1mH
DC bus voltage	V_{DC} = 300V
DC bus capacitance	C_{DC} = 1000μF
AC line voltage (Load Side)	V_{LL} = 415V, 50Hz.
AC line voltage (Source Side)	V_{LL} = 11e3V, 50Hz.
PWM switching frequency	10 kHz
OLTC Transformer	15 kVA
Injection Transformer	5kVA, 100V / 500V

V. RESULTS

Performance of the OLTC transformer in tandem with the DVR for regulation of voltage on radial feeder evaluated through simulations is under MATLAB/SIMULINK environment using and SimPowerSystems toolbox and is evaluated through results shown in Fig. 4(a-e) under source side disturbance and in Fig. 5(a-e) during load perturbations. From t = 0.05 sec to t = 0.45 sec, under voltage situation is observed on the source side with supply side voltage lowers around 0.76pu as depicted in Fig. 4(a). The OLTC transformer tries to maintain the load by changing the OLTC tap position to provide voltage regulation within ±0.05 pu of the rated value as seen in Fig. 4(b). At t = 0.15 sec when the DVR unit is switched into the circuit to provide the requisite compensation for the feeder drop and to compensate the difference created in voltage due to OLTC transformer tap position, the voltage at load terminals is well regulated to the rated value as the same is shown in Fig 4(c). At t = 0.45 sec and up to t = 0.7 sec, the source side voltage is returned to 1 pu as shown in Fig. 4(a). The OLTC transformer changes its tap position and the DVR regulates the load side voltage to the rated value by providing compensating voltage injected into feeder and the mitigation drop across transformer. The delay observed in the settling of the load voltage is due to the difference of the dynamics of OLTC transformer and that of DVR. At t = 0.7 sec when the source side voltage rise to 1.36 pu depicting an overvoltage condition as shown in Fig 4(a). The hybrid system comes into action again since, the change in voltage is more than maximum capacity of the DVR, the OLTC transformer changes its tap position and reduces real power flow into the circuit this provides load side voltage regulation within ± 0.5pu, thereby leaving the regulation job to the DVR to bring back the load side voltage to 1 pu as shown in Fig. 4(c). Fig. 5 shows the performance analysis of OLTC transformer and the DVR under load perturbation. The load side voltage is kept regulated at 1 pu to evaluate the performance of hybrid construction of OLTC transformer in tandem with DVR under load perturbation. As seen from Fig. 5(a), from t = 0.15 DVR when is switched into the feeder the load side voltage is regulated to 1 pu by injection of voltage in quadrature with the line current as shown in Fig. 5 (d-f). The kVA load remain constant till t = 0.4 sec. At t = 0.4 sec when the kVA load demand is increased by 50% which in term increases the feeder current as observed in Fig. 5(d), it may be observed that the voltage injected by the DVR is sufficient to regulate the load side voltage to 1 pu as shown in Fig. 5(c). Further at t = 0.7 sec when the load demand is increased to 250%

of initial load demanded, the increased demand of the load is beyond the capacity of DVR therefore tap of the OLTC transformer is changed. The OLTC transformer changes the tap as per the requisite voltage range. And thus the load bus voltage is regulated to the rated value by the combined operation of OLTC transformer and DVR as shown in Fig. 5(c)

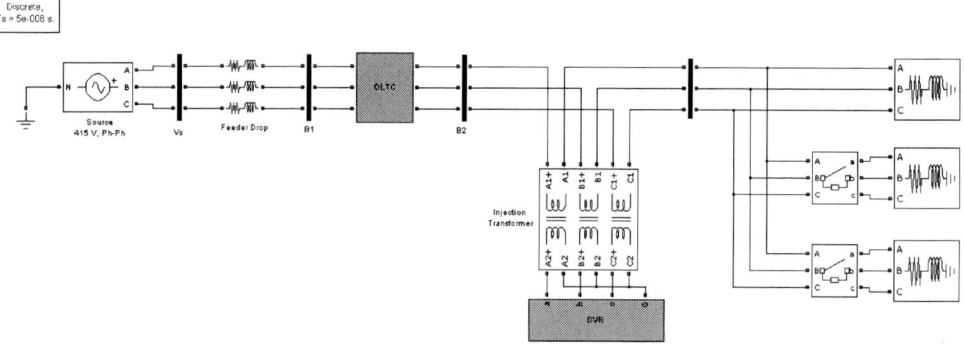

Fig. 3: Simulation Diagram of OLTC Transformer

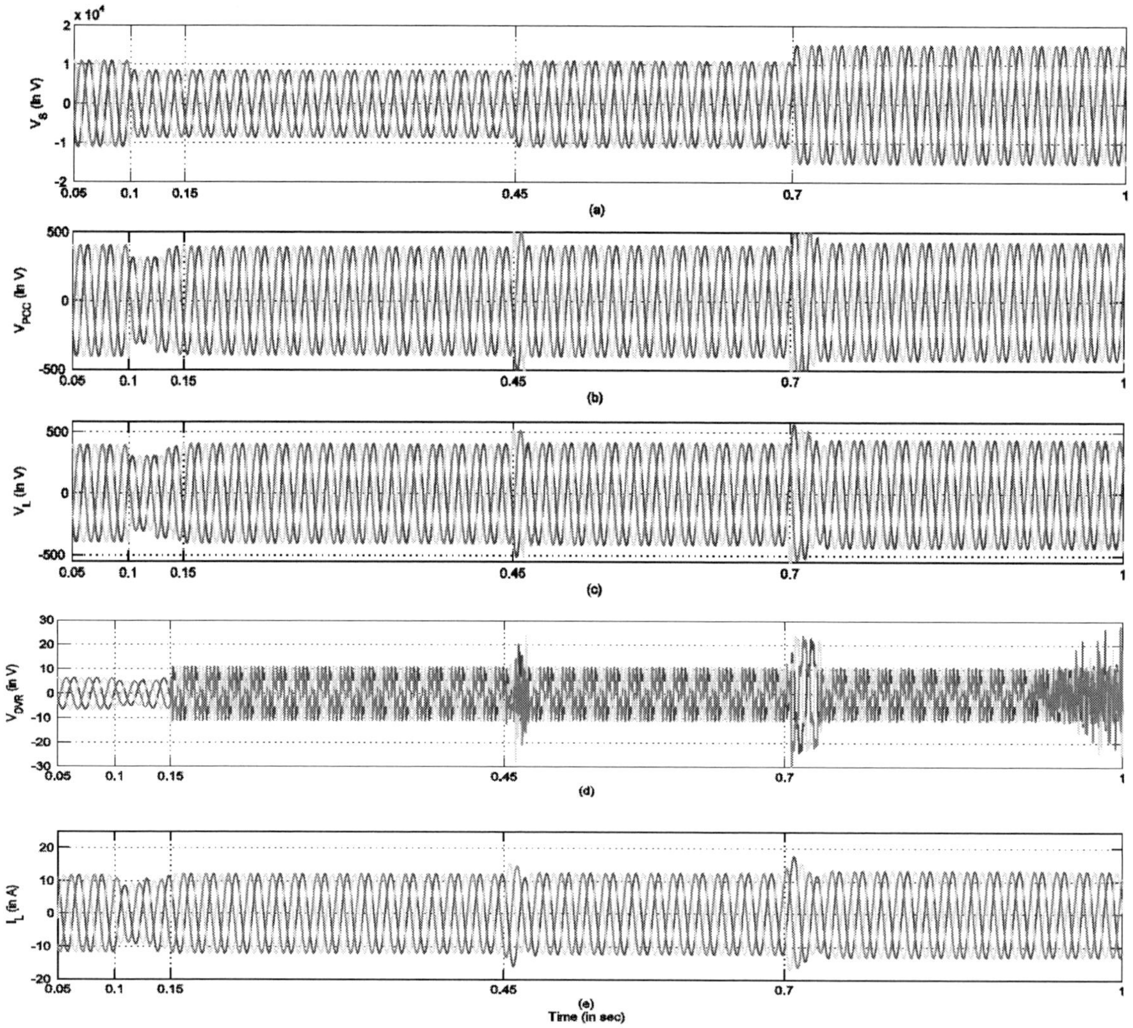

Fig.4: Simulation results for the operation of OLTC Transformer and DVR under source side disturbance

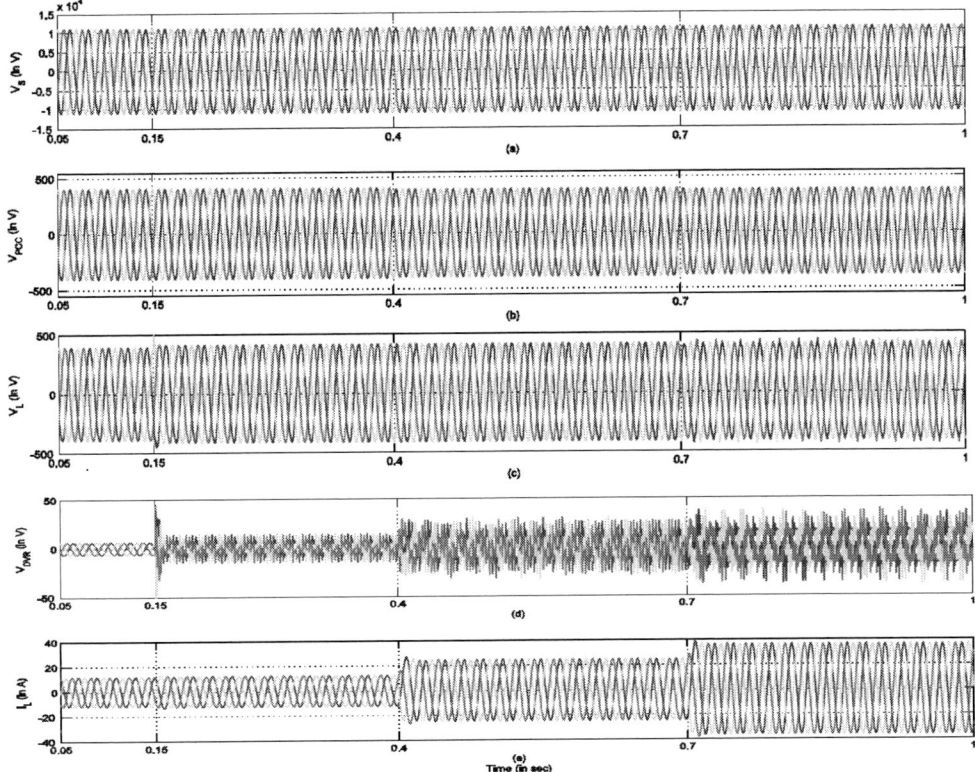

Fig. 5: Simulation results for the operation of OLTC Transformer and DVR under Load Perturbations

VI. CONCLUSION

The combined operation for OLTC transformer and DVR for regulating the voltage in the radial feeder has been demonstrated using MATLAB simulation results. The voltage regulation for source side disturbances and load perturbations on radial feeder is obtained and is close to the present value. DVR is controlled to regulate the voltage to rated value in way to inject a voltage in phase quadrature with the line current and thereby dealing with reactive power only. The OLTC transformer changes its tap position only when the voltage difference is beyond the injection capability of the DVR. The combination of dynamic device DVR with the OLTC transformer makes the regulation faster, near step-less, and the transients are suppressed to a great extent and persist only for some time.

REFERENCES

[1] T. Davis, G.E. Berm and C.J. Melhom, "Voltage Sags: Their impact on the utility and industrial customers," *IEEE Tran. Industry Applications,* vol.3, pp.549-558, May June 1998.

[2] T. X. Zhu, S. K. Tso, and K. L. Lo, "An Investigation into the OLTC Effects on Voltage Collapse", *IEEE Trans. Power System,* vol. 15, pp. 515-521, Aug 2002.

[3] C. Gao, M. A. Redfern, "Advanced Voltage Control Strategy for On-Load Tap-Changer Transformer with Distributed Generations.", *in Proc. Sep. 2011 Universities' power Engineering Conference*, pp. 1-6.

[4] WEI Bo, ZHANG Yong-jun, "Evaluation Model and Regulation Effect Index of OLTC in Power Systems.", *in Proc. Power Engineering and Automation Conf.,* pp.-86-89.

[5] A. Ghosh and G. Ledwich,"Power Quality Enhancement Using Custom Power Devices." Kluwer Academic Publishers, 2002, pp. 333-378.

[6] A Ghosh and G. Ledwich, "Compensation of distribution system voltage using DVR", *IEEETrans. Power Delivery,* vol. 17, no. 4, pp. 1030 – 1036, Oct. 2002

[7] Bhim Singh, P. Jayaprakash, D. P. Kothari, Ambrish Chandra, Kamal-Al-Haddad, "Indirect Control of Capacitor Supported DVR for Power Quality Improvement in Distribution System", *in Proc. 2008, IEEE Power and Energy Society Conf.,* pp. 1-7.

[8] Bhim Singh, P. Jayaprakash, D. P. Kothari, Ambrish Chandra and Kamal-Al-Haddad, "Control of Reduced Rating Dynamic Voltage Restorer with Battery Energy Storage System", *in Proc. 2008 IEEE Power India Conf. and Power System Technology,* pp. 1-8

[9] ArindamGhosh, Amit Kumar Jindaland Avinash Joshi, "Design of a Capacitor-Supported Dynamic Voltage Restorer (DVR) for Unbalanced and Distorted Loads" *IEEE Trans Power Delivery,* vol. 19, pp. 405-413, *2004.*

[10] Andres E. Leon, Marcelo F. Farias, Pedro E. Battaiotto, Jorge A. SolsonaandMaría Inés Valla, "Control Strategy of a DVR to Improve Stability in Wind Farms Using Squirrel-Cage Induction Generators", *IEEE Trans. Power System,* vol. 26, pp. 1609-1617,Aug 2011.

[11] ParagKanjiya, VinodKhadkikar, H. H. Zeineldin, Member and BhimSingh, "Reactive Power Estimation based control of self Supported Dynamic Voltage Restorer (DVR)" *in Proc. 2012 IEEE Harmonics and Quality of Power Conf.,* pp. 785-790.

[12] R. Strzelecki and G. Benysek, "Control Strategies and Comparison of the Dynamic Voltage Restorer", *in Proc.*

2008 IEEE Power Quality and Supply Reliability Conf., pp. 79-82.

[13] S.H. Hosseini, M. Abapour, M. Sabahi, "A Novel Improved Combined Dynamic Voltage Restorer (DVR) Using Fault Current Limiter (FCL) Structure" *in Proc. of ICEM, IEEE Oct. 2007*,pp. 98-101.

[14] B.B.Zad, J. Lobry, F. Vallee, O. Dureix, "Improvement of on-load tap changer performance in voltage regulation of MV distribution systems with DG units using D-STATCOM ", *in Proc. of*International Conference and Exhibition on Electricity Distribution, IEEE, June 2013, pp.1-4.

[15] Yasir Muhammad, Muhammad NaeemArbab, "Optimization in Tap Changer Operation of Power Transformer using Reactive Power Compensation by FACT devices", *in Proc. of* Control and System Graduate Research Colloquium, IEEE, Aug 2013, pp. 27-31.

[16] *K. S. Jeong, Y. S. Baek, J. S. Yoon, B. H Chang, H. C. Lee, G. J. Lee*, "EMTP Simulation of a STATCOM-Shunts-OLTC Coordination for Local Voltage Control", *in Proc. of* Transmission & Distribution Conference & Exposition: Asia and Pacific, IEEE Oct, 2012, pp. 1-4.

Some Considerations of Arc Protection and Breaker Design for High Frequency AC Power Distribution Systems

Junfeng Liu.[1], C. D. Xu[1], Jun Zeng[2], K. W. Eric Cheng[1]

[1] Department of Electrical Engineering, The Hong Kong Polytechnic University, Hong Kong
[2] New Energy Center, South China University of Technology
E-mail: jf.liu@connect.polyu.hk

Abstract–high frequency power distribution already has a numerous applications in electric vehicle, renewable energy microgird, and telecommunication system. In order to propel high frequency power distribution into larger power grade, an effective ac arc protection scheme is significant for safety and reliability. However, almost all of the studies of arc fault are focused on dc or low frequency ac (LFAC). This paper presents protection scheme and examines electrical characteristics for high frequency arc fault. Meanwhile, high frequency breaker is examined with ZCS feature, and residual current devices (RCD) design is explored for high frequency power distribution as well. First, the location of arc fault detector (AFD) and interrupting devices (ID) is studied from system level; and the single phase arc grounding fault is analyzed for high frequency sinusoidal waveforms. Second, zero-current-switching (ZCS) feature of high frequency breaker is discussed from monitor circuit and breaker circuit. Third, trigger deficiency of high frequency RCD is demonstrated and corresponding solution is examined. Lastly, the simulation results further testify the effectiveness of theoretical analysis.

Keywords–Arc fault, high frequency, protection scheme.

I. INTRODUCTION

In view of the component saving and efficiency improvement, high frequency ac (HFAC) power distribution has already been adopted in many regions, such as electric vehicle, renewable energy microgrid, as well as telecommunication and computer system [1-3]. The most of the applications have a smaller power grade and limited distribution range. When the power transmission capacity is increased, high frequency distribution has a wider range of applications with the outstanding merits. However, the use of high frequency distribution complicates the system protection as some traditional protection devices used for LFAC or dc cannot be adopted directly in HFAC distribution systems. Traditionally, a high-speed CB (circuit breaker) can interrupt the fault current within less than 0.01s[4]. However, the interrupt time is still too long for high frequency sinusoidal waveforms. Therefore, some known issues are associated with fuses and CBs in high frequency power distribution, such as different arc characteristic and long interrupt time. In order to decrease the operation time and improve the operation speed, some scholars utilize power-electronic switches, such as thyristors and gate-turn-off thyristor(GTO)[4]. However, the power losses of power-electronic components are much higher than the mechanical switch. Therefore, a CB implemented by the combination of a mechanical switch and a power-electronic switch has been discussed in [5]. Considering the quick dynamic characteristic, a comprehensive

consideration of protection system is required for the design of high frequency power distribution.

In addition to high speed requirement, the zero-current-switching (ZCS) feature is also significant for high frequency breaker, especially for the scenario of high voltage and large current. Therefore, a novel breaker design is required for high frequency system with ZCS feature. Meanwhile, RCD of 50-60 Hz utility system has improper operation under higher frequencies scenario [12]. A new design method is needful for high frequency RCD.

In this paper, (1) practical locations of protection devices are discussed for HFAC power distribution. (2) Since the arc grounding fault is a main issue in medium voltage power distribution networks and it is difficult to detect the high frequency arc fault by the existing CB, the corresponding electrical characteristics of arc fault are examined for high frequency sinusoidal waveforms. (3) A breaker design is proposed with ZCS feature; and two sides are considered for practical implementation. (4) The deficiency of RCD is discussed for high frequency system; and possible design method is introduced for high frequency RCD as well. (5) The simulation accomplished by the Matlab and PSIM further testifies the previous analysis.

II. ARC IN HIGH FREQUENCY AC DISTRIBUTION

Arc fault is an unexpected sparking in an power circuit, which is the cause of electrical fire accidents An arc protection scheme requires the installation of arc fault circuit interrupter (AFCI) or arc fault detector (AFD) with interrupting device (ID) in high frequency power source side, transmission track and load side.

As shown in Fig.1, the arc fault can be categorized into three types: series, parallel and ground. Since the fault current is much larger than the normal current, the parallel and ground arc are easy to be detected by the conventional CB. However, the series arc current is difficult for detecting as the fault current is close to the normal current of CB. Several detection methods of series arc fault have been examined from time domain to frequency domain. Considering the random noise characteristics in arcing current and voltage, the most popular detection method is the frequency spectrum based analysis. The arc characteristics need be discussed in depth before the design of arc fault detection. Therefore, the corresponding discussion of arc fault is examined for high frequency power distribution in next section.

978-1-5090-0064-7/15 $31.00 © 2015 IEEE

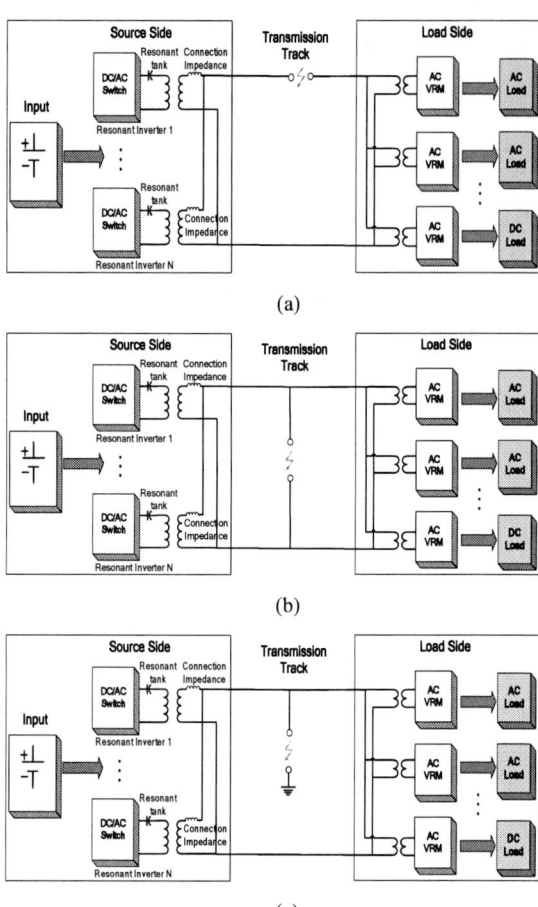

(a)

(b)

(c)

Fig.1: Arc fault in high frequency power distribution. (a). series. (b). parallel. (c). ground

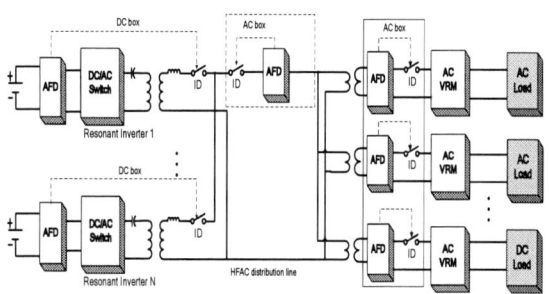

Fig. 2: A simple protection scheme for high frequency power distribution system

Beside the analysis of electrical characteristics, the locations of the AFDs and IDs are significant for the safety and reliability of the entire system. The proper location of arc fault detection can minimize the effect of the arc fault by de-energizing the faulted branch. Considering the trade-off of performance and system cost, a simple protection scheme as shown in Fig. 2 is proposed for high frequency power distribution system. AFD is installed to detect arc fault on the power unit, distribution line, and load side. When an arc fault occurs in the power unit, the power unit is islanded from the transmission line as the ID is turned off by AFD. If an ac arc fault occurs in a load, AFD of local load unit can disable the load branch by turning off the ID. Since the branch with arc fault is completely disabled, the entire distribution system can continue the

operation with flexibility and reliability. If an ac arc fault happens in the transmission line, AFD of transmission line will cut down the line using the ID. Therefore, a comprehensive protection can also be accomplished by the proposed scheme.

If the dc source is obtained by photovoltaic (PV) or battery, arc fault solution need be carefully designed[6]. If an arc fault happens in PV panels or line connection of PV panels, the module type PV generation can improve the protection and ease maintenance. On the other hand, the cooperation of protection devices can effectively improve the performance by splitting the entire system and locating the arc fault[7].

III. ANALYSIS OF HIGH FREQUENCY ARC FAULT

In order to design the effective arc detection, the electrical characteristics of arc fault are desired for high frequency sinusoidal waveform. Traditionally, the ac current has a zero-crossing every half of operational period. Since the fast change of high frequency dynamics, it is difficult for arc fault to find the equilibrium state. Conventionally, the fault current should be close to sinusoidal waveform with fast change; while the fault voltage should have a large voltage distortion. The fault current should be in phase with the fault voltage caused by the physical characteristic of ac arc. The typical waveforms of HFAC arc is demonstrated in Fig. 3 including fault current and fault voltage. It can be found the fault voltage has saddle-shaped waveform. The voltage of leading point is burning voltage; while the voltage of lagging point is quenching voltage.

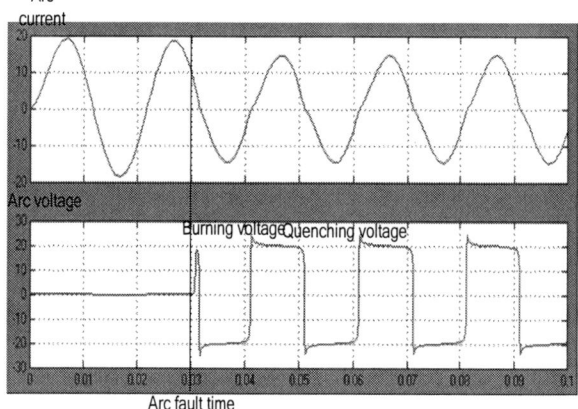

Fig. 3: The general waveforms of ac arcs fault[8]

Furthermore, it can be found that high frequency noise is obvious in voltage and current waveforms. The arcing current eliminates before zero crossing point of normal current and reignites after zero crossing point of normal current. The region of zero current is flat in each half cycle.

It is concluded from previous analysis that ac arc has a nonlinear load characteristic. When the fault current is close to zero-crossing point, arc resistance is almost infinity caused by the large fault voltage. The root cause is that the arc path becomes gaseous insulator from conductor. When the fault current is large enough, the arc resistance is small due to the good conductivity. The arc

978-1-5090-0064-7/15 $31.00 © 2015 IEEE   186

can be viewed as a metallic conductor with the large fault current.

The arc fault can be further denoted by dynamic elements. The two-port electrical network as shown in Fig. 4a is used to demonstrate nonlinear characteristic of arc fault.

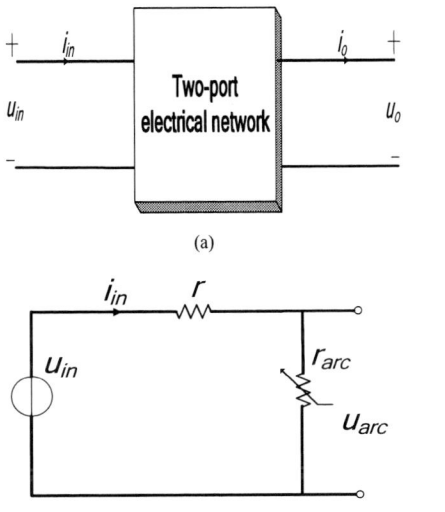

(a)

(b)

Fig. 4: (a) the equivalent circuit of two-port network. (a). two-port network. (b). simplified circuit of CB

The functional relation of two-port network is

$$u_i = ri_{in} + r_{arc}i_{in} \qquad (1)$$

where r is the parasitic resistance of CB, and r_{arc} is the arc resistance that is a nonlinear variable. u_{in} and i_{in} are the input voltage and input current of CB respectively. Hence, the simulation model of CB can be further demonstrated by Fig. 4b. The method of arc fault detection is frequency domain analysis that is derived by DFT(Discrete Fourier transform).

$$X(k) = \sum_{i=1}^{N} x(i)e^{(-2\pi j)(i-1)(k-1)/N} \qquad (2)$$

where N is data length. A simple method of frequency analysis is demonstrated in Fig. 5 with arc fault of transmission track and loads.

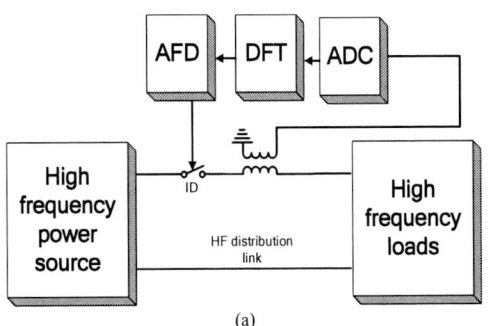

(a)

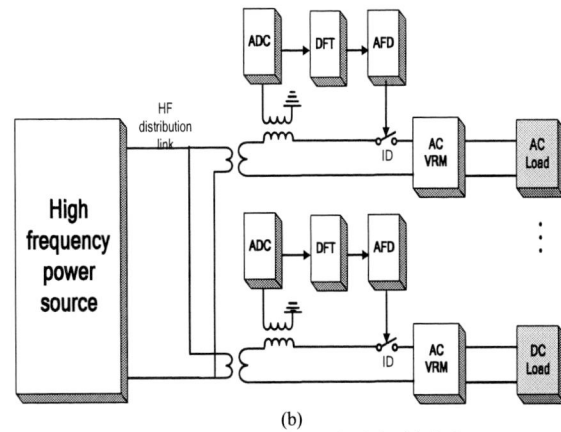

(b)

Fig. 5: frequency based analysis method for high frequency ac arc. (a). arc in distribution link. (b). arc in load side

The current signal is measured by a transformer in both arc fault of distribution link and load side. A high speed Analog to Digital Converter(ADC) is the second step to accomplish the conversion of analog signals. The frequency based analysis is accomplished by DFT that is implemented by digital signal processor. Lastly, AFD is the last operation to discriminate the arc signal from the normal signals considering the high frequency characteristics. Hence, the fault analysis based on frequency spectrum is also effective for power distribution with almost 20 kHz distribution frequency.

IV. FURTHER CONSIDERATION OF ZCS(ZERO CURRENT SWITCHING) BREAKER FOR HIGH FREQUENCY ARC FAULT

A desirable characteristic of a circuit breaker is arc-less of current interruption. Considering the characteristics of high frequency arc, the new design method is required for circuit breaker in high frequency power distribution. Since the high frequency ac passes through the zero point with quick dynamics, the effective arc quenching is required to open the switching device with the zero arcs current. The original idea is similar with the circuit breaker in LFAC; however, the design and implementation become more complicated for high frequency zero-crossing characteristics. The design of circuit breaker largely contains the monitoring circuit, control circuit and breaker circuit. The typical diagram of circuit breaker is illustrated in Fig. 6.

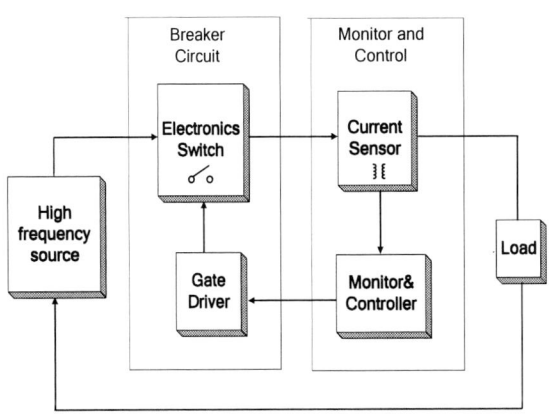

Fig. 6: the diagram of circuit breaker

978-1-5090-0064-7/15 $31.00 © 2015 IEEE

1. Enhancement of measurement and control for ZCS feature

It can be found that the circuit breaker can be accomplished with ZCS or ZVS feature from two sides. The first method is to consider the quick dynamic response. A fast monitoring and control circuit is required due to the high frequency zero-crossing characteristics. The traditional methods are not enough for the overload and short-circuit of HFAC. Some quick monitor algorithms are necessary for high frequency breaker circuit; meanwhile, the advanced microprocessor unit is indispensible for zero-crossing monitoring of digital implementation. A control system of automatic switch is demonstrated in Fig. 7 with high frequency dynamic characteristics.

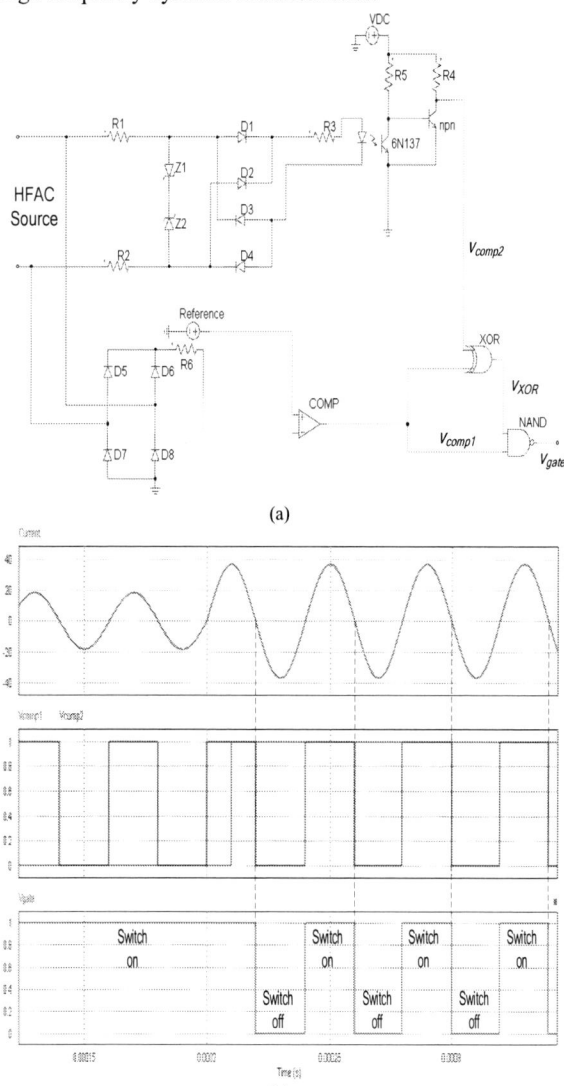

(a)

(b)

Fig. 7: Control diagram of high frequency system. (a) Block diagram of control system. (b) Simulation results of control system logic

The current signal collected from power circuit is modeled as an AC voltage source, and the voltage signal is converted into a DC signal by full-bridge rectifier. Then this signal viewed as an input signal is compared with a reference voltage. If the input signal is smaller than the reference voltage, the output v_{comp1} is logic low; if the input signal is larger than the reference voltage, the output

v_{comp1} is logic high. Meanwhile, the output v_{comp2} is generated by the zero-crossing comparing circuit that is formed by the optocoupler 6N136, transistor, and four diodes. v_{XOR} is derived by XOR operation of v_{comp1} and v_{comp2}. Lastly, the gate signal of breaker switch v_{gate} is obtained by the NAND logic between v_{comp1} and v_{XOR}. The logic output is simulated as shown in Fig. 7b. The required waveforms are accomplished by the proposed logic circuit. It can be found that the breaker circuit tries to turn off when the detected current is larger than the normal value. The turning off point of breaker is the falling edge of zero-crossing current. Hence, ZCS feature of the breaker can be accomplished by the enhancement of monitor and control circuit. Since the monitor and control circuit is implemented by discrete analogue circuit, a quick dynamic response becomes possible for high frequency power distribution.

2. Enhancement of breaker circuit for ZCS feature

The second method is to consider the breaker circuit, and the design requirement is rapid operations of breaker circuit. This feature enables electronics breaker to limit current and the let current significantly below the short-circuit current in comparison with high speed electro-mechanical circuit breakers. The existing breaker based on power electronics is implemented by the thyristor and GTO, both of which are not good choice for high frequency system. The low voltage protections can be accomplished by IGBT with the reducing response time. A typical topology of electronic circuit breaker is shown in Fig. 6a that is formed by one IGBT and four diodes[10].

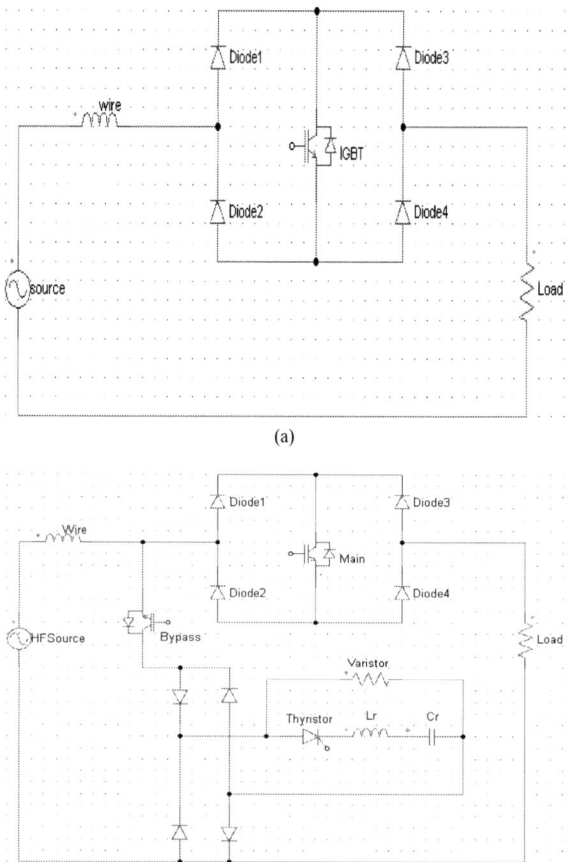

978-1-5090-0064-7/15 $31.00 © 2015 IEEE   188

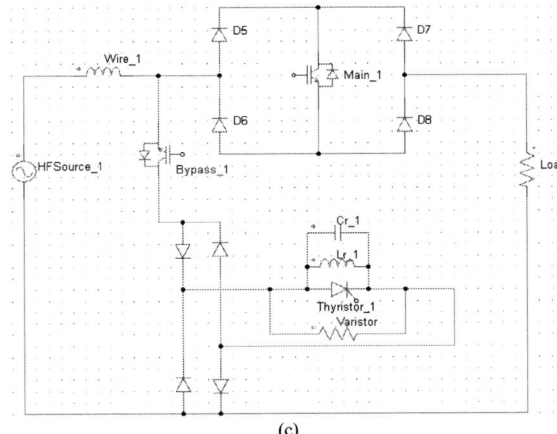

(c)

Fig. 8: Topology of electronic circuit breaker. (a). IGBT based circuit breaker. (b). resonant topology A of ZCS breaker. (c). resonant topology B of ZCS breaker

The positive current flows through diode1, IGBT and diode4; while the negative current flows through diode2, IGBT, and diode3. Therefore, the current passing through IGBT is a DC current, which makes the control simpler. Furthermore, the ZCS breaker can be accomplished by rectifier, thyristor and the addition of inductors and capacitors. Firstly, the rectifier converts the high frequency AC to DC; secondly, the resonance of inductor and capacitor makes the current with zero-crossing point; lastly, thyristor accomplishes the tuning off with ZCS feature. The possible topologies are demonstrated in Fig. 6b&c both of which have similar structure. The difference between Fig. 6b&c is that the location of resonant tank that is used for zero-crossing control. The resonant tank is used to change the current frequency, so that a slow response is achieved due to the circuit resonance. The suitable selection of resonant components is significant for current suppression and period matching of resonant loop and input voltage.

V. LEAKAGE PROTECTION IN HIGH FREQUENCY SYSTEM

In low voltage electrical installations, residual current devices(RCDs) are considered for protection against electric shock and a fire caused by leakage currents. $I_{\Delta n}$ is residual operating current. When residual current is equal to or less than $0.5 \times I_{\Delta n}$, residual current devices do not operate for a sinusoidal waveform. In utility system, the residual operating current shall not exceed 30 mA. Several issues have to be confronted by leakage protection of high frequency power distribution, including (1) the definition of the residual operating current $I_{\Delta n}$ for the higher operating frequency; and (2) RCD design for high frequency system.

1. Residual operating current of high frequency distribution system

Table 1 shows the relative effect of various frequencies on human cells and tissues. Frequencies above 10 kHz are safer than dc voltage, while high frequency systems cause a lower perception and paralysis effect than 60 Hz utility system.

Table 1: Relative effects of electric current on human cells and tissues[11]

Frequency, Hz	dc	10	60	400	1k	10k
Perception and paralyses, relative	0.2	0.9	1	0.7	0.6	0.2

Furthermore, the relation curve of current influence to frequency is illustrated in Fig. 8. The logarithmic coordinates is adopted for X axis, and linear relation can be obtained.

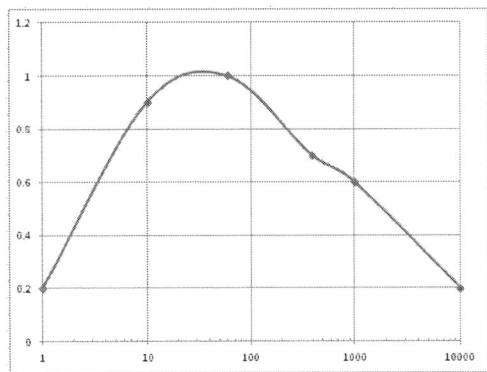

Fig. 9: the relation of electric current on human cells and tissues to operating frequencies.

It can be observed from Fig. 9 that the electric current on human cells and tissues is linearly declining when the operating frequency is larger than 100 Hz. It can be estimated that the residual operating current should be significantly larger than 30mA for high frequency system over 20 kHz distribution frequencies.

2. RCD design of high frequency system

Residual current device designed for 50-60 Hz utility system has improper operation under higher frequencies scenario. Residual current frequency exceeding 50/60 Hz raises the tripping current of residual current devices[12]. When the operating frequency is above 400-500 Hz, these traditional residual devices do not trip even with residual current equaling to a few amperes. Hence, it is not acceptable for high frequency protection device against electric shock and a fire. The reason is that a higher value of earth fault current is required for higher frequency to achieve the same level of magnetic flux density in the current transformer core. Meanwhile, the impedance of secondary circuit winding and the electromechanical relay increases along with the higher residual current frequency. The secondary current is too small to make the electromechanical relay trip in higher frequency residual current. Therefore, a new RCD design need be considered for the elimination of tripping current rising in high frequency system. A possible solution is to change AC voltage with DC voltage using rectifier circuit[12]. The rectifier can avoid the increasing impedance along with the rising frequency. The current amplitude flowing in the electromechanical relay is almost the same regardless of frequency variation. A structure diagram of RCD is illustrated in Fig. 8 for high frequency system.

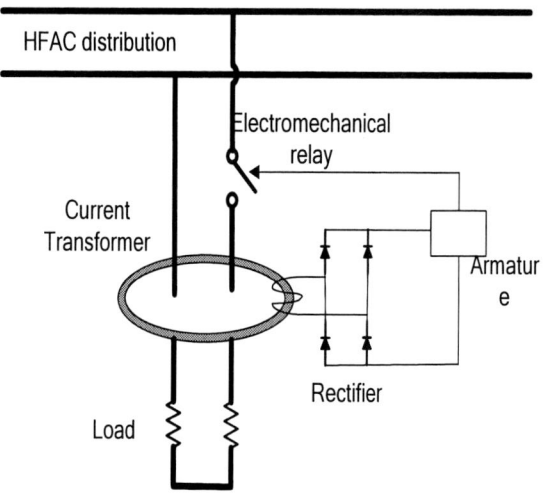

Fig. 10: the structure diagram of high frequency RCD

High frequency current difference is collected by current transformer, and secondary voltage is generated by the leakage current. The rectifier accomplishes the conversion from high frequency AC to DC voltage. DC voltage is used for the on/off control of electromechanical relay. In order to operate under high frequency, current transformer needs optimizing magnetic properties and electromechanical relay has fast response[13]. An RCD simulation of high frequency is demonstrated in Part VI. The currents flowing in the electromechanical relay are compared for the scenarios with different frequencies.

VI. SIMULATION VERIFICATION

Based on the abovementioned analysis, the simulation verification can be divided into three parts. One is the arc fault simulation of high frequency system; the second one the simulation of high frequency ZCS breaker; the last one is the simulation of high frequency RCD. The simulation results will be demonstrated as below.

1. Arc simulation results of high frequency distribution
It is expensive to simulate arc fault in an actual distribution system. Therefore, an effective simulation platform is significant for arc fault analysis. A simple simulation circuit is established as show in Fig. 9. An CB is used to turn on or turn off the short-circuit current. CB adopts Cassie model that is implemented by Simulink AMB(Arc model blockset)[9]. The series RL branch denotes the impedance of transmission line and load. The arc voltage and current are derived by voltage and current measurement. The scopes are used to display the waveforms of arc voltage, arc current and nonlinear resistance.

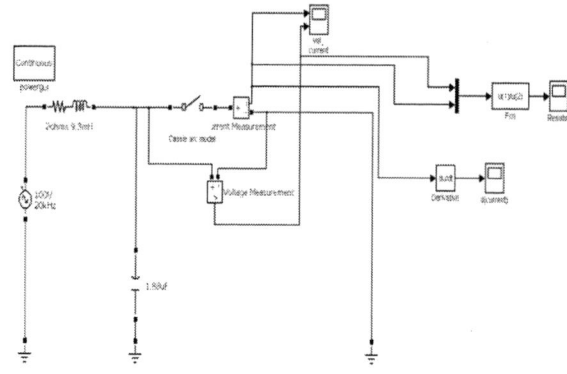

Fig. 11: the simulation model of arc fault circuit

The simulation parameters are demonstrated as below:

input voltage: 100V
input frequency: 20 kHz
line resistor: 2Ω
line impedance: 9.3mH
ground capacitor: 1.88μF

The simulation time is 0.001s, and the arc fault happens at 0.0003s.

The simulation waveforms of arc current and arc voltage are shown in Fig. 12 respectively. Although the waveforms of arc currents and voltages vary significantly comparing LFAC with HFAC, there are some basic characteristics that are same to low frequency and high frequency system. The arc voltage has a large distortion as a saddle-shaped waveform; and the fault current is in phase with the fault voltage.

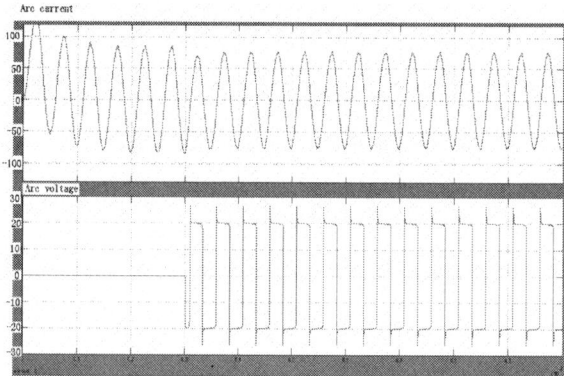

Fig. 12 Simulation waveforms of arc current and arc voltage

The current shoulders at zero-crossing points become small in high frequency system. Meanwhile, it can be found that the arc voltage waveform has the larger distortion comparing with LFAC system. The peak value of arc voltage is larger than that in LFAC system. The resistance waveforms of HFAC arc fault are shown in Fig. 12.

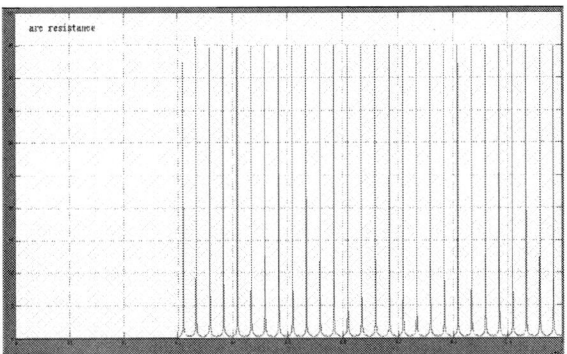

Fig. 13: simulation waveform of arc resistance

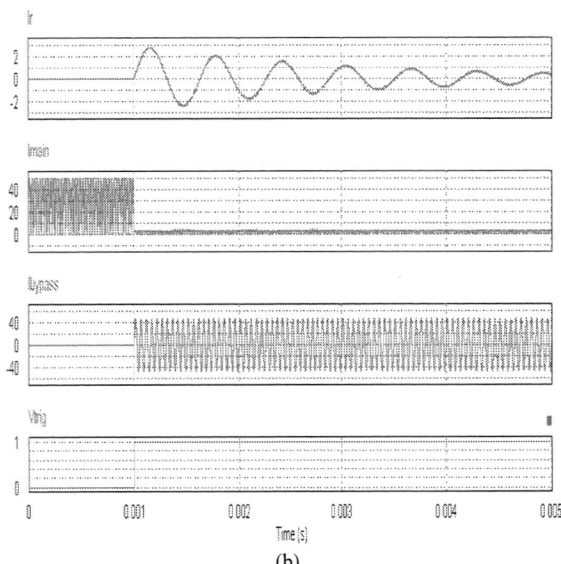

(b)

Fig. 14: simulation of ZCS breaker. (a). simulation model. (b). simulation waveforms.

It can be found from Fig. 13 that arc resistance is almost infinity when the fault current is close to zero-crossing point. Hence, the nonlinear load characteristic can also be observed in high frequency power distribution. The simulation results agree well with the theoretical analysis.

1. Simulation results of high frequency breaker with ZCS
In order to verify the ZCS feature of the proposed breaker circuit, an PSIM simulation is conducted based on the Fig. 14a. The simulation results are demonstrated in Fig. 12b. The input of simulation model is a high frequency current source with peak value of 50A, and output frequency of 20 kHz. When the fault is detected, the bypass switch is conducted with bidirectional capability. After the bypass current is rectified by full-bridge rectifier D_5-D_8, the resonance is generated by the L_r and C_r. The thyristor can cut down the current at the zero-crossing point of resonant current due to low resonant frequency. Varistor is the dissipation resistor. i_{in} is the input current, i_{main} is the current of main switch, i_{bypass} is the current of bypass branch, and i_r is the current of resonant tank.

Resonant inductor L_r is 1mH; and resonant capacitor C_r is 10μF. It can be found from Fig. 14b that the fault is detected at the time of 0.001s. The current of main loop is reduced to zero; and input current passes through the bypass branch. The resonant current i_r is the low frequency sinusoidal waveform that is determined by the resonant parameters. Consequently, the traditional utility breaker can be used as main switch with low frequency dynamics.

A. Simulation results of high frequency RCD
Different distribution frequencies are adopted for the performance verification of high frequency RCD with rectifier operations. The currents flowing in the electromechanical relay are compared for the different frequencies scenarios. RCD waveforms of 100 Hz distribution system are demonstrated in Fig. 15a; and RCD waveforms of 20 kHz distribution system are demonstrated in Fig. 15b.

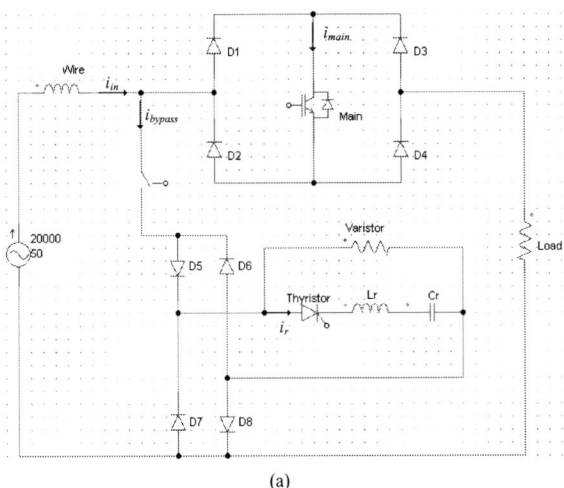

(a)

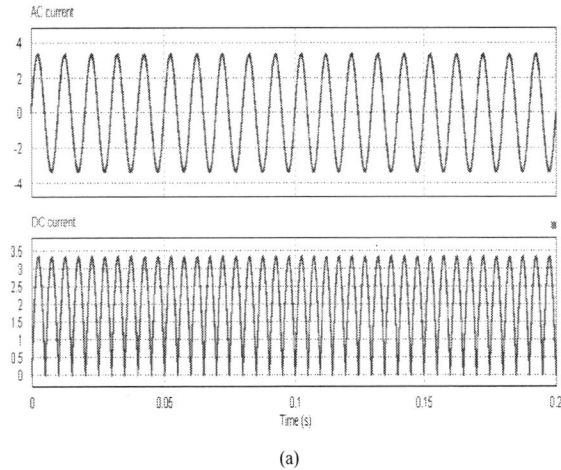

(a)

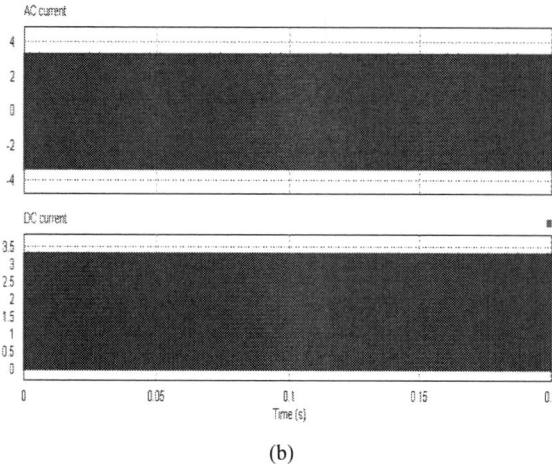

(b)

Fig. 15: RCD waveforms. (a). in 100 Hz distribution system. (b) in 20 kHz distribution system.

AC current is the current in front of rectifier operation; while DC current is the current waveform behind of the rectifier operation. It can be found that the impedance of secondary circuit winding and the electromechanical relay cannot change the RMS(Root Mean Square) of DC current that is used for the control of the electromechanical relay. Consequently, a fixed tripping current to high frequency RCD can be achieved by rectifier operation.

VII. CONCLUSION

In view of component saving and efficiency improvement, HFAC system is viewed as an effective distribution approach for some applications. Comparing with LFAC system, arc fault become more complicate for HFAC power distribution. In order to guarantee safety and reliability, a simple protection scheme is presented with isolated AFDs and IDs. The novel scheme improves the protection performance by splitting the entire system. In addition to systematic study, the protection devices and possible issues are examined for high frequency distribution system separately.

Firstly, the electrical characteristic of high frequency arc fault is analyzed, and the traditional detection method of frequency spectrum is introduced into high frequency arc fault as well. Secondly, high frequency breaker is examined with ZCS feature, and the possible enhancements can be conducted from the monitor/control circuit and breaker circuit. Thirdly, the leakage protection scheme is studied for high frequency system. The relations of residual operating current with operational frequency are examined. The possible issues of traditional RCD are reviewed. The new design method is discussed based on rectifier operations. Finally, the simulation results obtained by Simulink and PSIM further verify the theoretical analysis.

ACKNOWLEDGMENT

The authors gratefully acknowledge the financial support of the Research Grants Council (RGC) of The Hong Kong SAR under the project reference PolyU 5133/10E.

REFERENCES

[1] Junfeng Liu; Cheng, K.W.E.; Yuanmao Ye, "A Cascaded Multilevel Inverter Based on Switched-Capacitor for High-Frequency AC Power Distribution System," *IEEE Transactions on Power Electronics*, vol.29, no.8, pp.4219,4230, Aug. 2014

[2] Bose, B.K.; Min-Huei Kin; Kankam, M.D., "High frequency AC vs. DC distribution system for next generation hybrid electric vehicle," *International Conference on Industrial Electronics, Control, and Instrumentation, IEEE IECON*, vol.2, pp.706-712, 5-10 Aug 1996.

[3] Junfeng Liu; Cheng, K.W.E.; Jun Zeng, "A Unified Phase-Shift Modulation for Optimized Synchronization of Parallel Resonant Inverters in High Frequency Power System," *IEEE Transactions on Industrial Electronics*, vol.61, no.7, pp.3232,3247, July 2014

[4] Salomonsson, D.; Soder, L.; Sannino, A., "Protection of Low-Voltage DC Microgrids," *IEEE Transactions on Power Delivery*, vol.24, no.3, pp.1045,1053, July 2009

[5] Meyer, J.-M.; Rufer, A., "A DC hybrid circuit breaker with ultra-fast contact opening and integrated gate-commutated thyristors (IGCTs)," *IEEE Transactions on Power Delivery*, vol.21, no.2, pp.646,651, April 2006

[6] Gab-Su Seo; Hyunsu Bae; Bo-Hyung Cho; Kyu-Chan Lee, "Arc protection scheme for DC distribution systems with photovoltaic generation," *Renewable Energy Research and Applications (ICRERA), 2012 International Conference on*, vol., no., pp.1,5, 11-14 Nov. 2012

[7] HyungSeok Kim; SeongWoo Kim; GiPoong Gwon; DongRyul Lee; SeungWoo Seo, "Multi-Level Arc Fault Circuit Interrupter with Collaborative Communications for Smart Grid," *2010 IEEE International Conference on Communications Workshops (ICC)*, vol., no., pp.1,5, 23-27 May 2010

[8] Gregory, G.D.; Kon Wong; Dvorak, R., "More about arc-fault circuit interrupters," *38th IAS Annual Meeting. Conference Record of the Industry Applications Conference, 2003*, vol.2, no., pp.1306,1313 vol.2, 12-16 Oct. 2003

[9] Ran Yu; Zhouxing Fu; Qingliang Wang; Shangbin Sun; Haidong Chen; Qi Xu; Xuejie Chen, "Modeling and simulation analysis of single phase arc grounding fault based on MATLAB,", *2011 International Conference on Electronic and Mechanical Engineering and Information Technology (EMEIT)*, vol.9, no., pp.4607,4610, 12-14 Aug. 2011

[10] Jicheng Yu; Yingying Tang; Karady, G.G., "Design of an IGBT based electronic circuit breaker," *North American Power Symposium (NAPS), 2010*, vol., no., pp.1,5, 26-28 Sept. 2010

[11] M. R. Patel, Spacecraft Power Systems. Boca Raton, FL: CRC Press, 2005.

[12] Czapp, S., "The impact of higher-order harmonics on tripping of residual current devices," *Power Electronics and Motion Control Conference, 2008. EPE-PEMC 2008. 13th*, vol., no., pp.2059,2065, 1-3 Sept. 2008

[13] Czapp S. "Elimination of the Negative Effect of Earth Fault Current Higher Frequency on Tripping of Residual Current Devices" *Electronics and Electrical Engineering. – Kaunas: Technologija*, 2009. – No. 3(91). – P. 85–88.

A novel initial rotor position detection method for PM synchronous motors

Zou Xunhao, Shenghua Huang, Qin Zhuqian, Xiaodong Hu, Kan Guangqiang

State Key Laboratory of Advanced Electromagnetic Engineering and Technology (Huazhong University of Science and Technology), China
E-mail: zouxh1992@qq.com

Abstract —This paper presents a novel initial rotor position estimation method for PM synchronous motors in EV application based on high frequency voltage injection. Mathematical model of PMSM are presented. Instantaneous current value rather than the amplitude of the corresponding current is utilized to calculate the rotor position. The filter to get the amplitude of the current is not needed and the accuracy of the result is irrelevant with the amplitude and frequency of the injected voltage.From the results of the experiment, the mentioned technique proves to be appropriate for initial rotor position detection of PMSM.

Keywords–PMSM, high frequency signal injection, neural network, rotor position

I. Introduction

Sensorless control techniques for PMSM have gained many researchers attention in recent years. These techniques can be classified into back EMF detection[1-2] or high frequency signal injection[3-5]. The initial rotor position detection is a special case of sensorless control. The accuracy of the initial rotor position detection is of great influence of the start-up of the motor. Inaccuracy of the initial rotor position may cause less start-up torque or momentarily reversing rotation, which is not allowed in some area such as the EV application.

The rotor position detection method based on back electromotive force uses voltage models[6-7], observers in synchronous or stationary frame[8], or Kalman filters[9]. But the amplitude of back EMF is proportional to the rotor speed, it's not applicable at zero speed.

For the sensorless rotor position detection method based on high frequency voltage injection, we can identify the rotor position from the magnitude of the resulted current signals which are affected by rotor saliency[10]. A filter is needed to get the amplitude of the currents. Both the voltage amplitude injected and the filter designed will have a great influence on the accuracy of the current amplitude detection, and further they will affect the accuracy of the rotor position detection. In this paper, a detection method without any filter is proposed. The basic idea is to fit the current equations with a few sample data, and the rotor position can be calculated with the current equation parameters. Experimentations have been carried out to verify the effectiveness of the proposed method.

II. Techniques based on injection of a high-frequency voltage vector

The equations of the PMSM in αβ reference frame are derived as (1) and(2).

$$u_\alpha = Ri_\alpha + D\psi_\alpha - \omega\psi_\beta$$
$$u_\beta = Ri_\beta + D\psi_\beta + \omega\psi_\alpha \tag{1}$$

$$\psi_\alpha = [L_{s0} + L_{s2}\cos(2\theta)]i_\alpha + L_{s2}\sin(2\theta)i_\beta + \psi_f\cos\theta \tag{2}$$
$$\psi_\beta = L_{s2}\sin(2\theta)i_\alpha + [L_{s0} - L_{s2}\cos(2\theta)]i_\beta + \psi_f\sin\theta$$

where

$$L_{s0} = (L_{sd} + L_{sq})/2$$
$$L_{s2} = (L_{sd} - L_{sq})/2 \tag{3}$$

The voltage rotating at a constant frequency ω_h injected is given by

$$\begin{bmatrix} u_\alpha \\ u_\beta \end{bmatrix} = U\begin{bmatrix} \cos(\omega_h t) \\ \sin(\omega_h t) \end{bmatrix} \tag{4}$$

Neglecting the stator resistance, the corresponding high-frequency current derived from (1),(2) and (4) when the rotor speed is zero is given by

$$\begin{bmatrix} i_\alpha \\ i_\beta \end{bmatrix} = k\begin{bmatrix} (L_{s0} - L_{s2}\cos(2\theta))\sin(\omega_h t) + L_{s2}\sin(2\theta)\cos(\omega_h t) \\ -L_{s2}\sin(2\theta)\sin(\omega_h t) - (L_{s0} + L_{s2}\cos(2\theta))\cos(\omega_h t) \end{bmatrix} \tag{5}$$

where

$$k = \frac{U}{\omega_h(L_{s0}^2 - L_{s2}^2)} > 0$$

If it's denoted that

$$a = k \bullet (L_{s0} - L_{s2}\cos(2\theta))$$
$$b = k \bullet L_{s2}\sin(2\theta)$$
$$c = -k \bullet L_{s2}\sin(2\theta)$$
$$d = -k \bullet (L_{s0} + L_{s2}\cos(2\theta))$$

It can be easily seen that the angle between rotor flux and a-axis can be calculated as

$$\theta = \frac{\arctan\dfrac{c-b}{a+d}}{2} \tag{6}$$

As can be seen from(6), although a,b,c,d are proportional to k, theta calculation is irrelevant with k, which means that the amplitude and frequency of the voltage injected have little influence on the accuracy of theta estimation. This is the main advantage of the proposed method compared with the traditional ones.

Since the neural network technology is widely used in parameter estimation[11], the Adaline algorithm[12] is used to calculate the parameters a,b,c,d, which are derived from(5) and respectively shown in

$$a(k+1) = a(\text{k}) + \eta \sum_{j=1}^{n} \sin(\omega_h t_j)(a(k)\sin(\omega_h t_j) + b(k)\cos(\omega_h t_j) - i_{\alpha j})$$

$$b(k+1) = b(\text{k}) + \eta \sum_{j=1}^{n} \cos(\omega_h t_j)(a(k)\sin(\omega_h t_j) + b(k)\cos(\omega_h t_j) - i_{\alpha j})$$

$$c(k+1) = c(\text{k}) + \eta \sum_{j=1}^{n} \sin(\omega_h t_j)(c(k)\sin(\omega_h t_j) + d(k)\cos(\omega_h t_j) - i_{\beta j})$$

$$d(k+1) = d(\text{k}) + \eta \sum_{j=1}^{n} \cos(\omega_h t_j)(c(k)\sin(\omega_h t_j) + d(k)\cos(\omega_h t_j) - i_{\beta j})$$

(7)

where $i_{\alpha j}, i_{\beta j}$ are the *jth* sample data of currents in αβ reference frame and n is the overall sample data number. With no more than 50 times iteration shown in(7), the estimated a,b,c,d will converge to the right point. With the help of sin/cos table, only addition and multiplication are needed to calculate parameters a,b,c,d, so it's easy to implement with DSP.

It should be noted that the theta angle we calculated is within 0-π, which means that θ may be aligned with N pole or S pole. And additional technique should be utilized to distinguish the polarity. There is always slight magnetic saturation when a PMSM is made. So the voltage aligned with d-axis will cause supersaturation of the magnet path as shown in Fig.1(a), and the voltage aligned with negative d-axis will cause desaturation as shown in Fig.1(b). The d-axis inductance L_{sd} will decrease with supersaturation of the magnet path and increase with the desaturation of the magnet path.

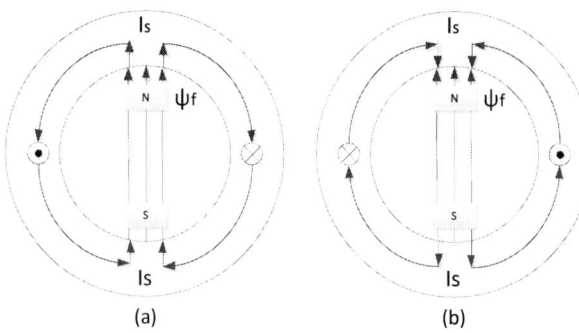

(a)   (b)

Fig. 1. Influence of stator magnetomotive force to d-axis flux

From(3) and(5), we can get d-axis current shown as(8). It's obvious that i_d will increase with the decrease of L_{sd} and vice versa. By applying voltage with the same amplitude and period aligned with θ and $(\theta+\pi)$, we can get the amplitude of the corresponding currents denoted by $|I_1|$ and $|I_2|$. The rotor angle should be θ if $|I_1| > |I_2|$, and $(\theta+\pi)$ if $|I_1| < |I_2|$.

$$i_d = k(L_{s0} - L_{s2})\sin(\omega_h t) = \frac{U}{L_{Sd}\omega_h}\sin(\omega_h t) \quad (8)$$

III. EXPERIMENTAL VERIFICATION

To validate the performance of the proposed method, a typical IPMSM is employed for testing. The parameters of the machine are shown in Table I. The motor is controlled by a Texas Instruments 28335DSP. A high-frequency voltage was injected with amplitude Vh=10V and frequency fc=200Hz.

TABLE I
PMSM parameters

Parameter	Unit	Value
Pole pairs		4
Permanent flux ψ_f	Wb	0.236
Nominal phase resistance	Ω	0.0212
d axis reluctance	mH	0.5307
q axis reluctance	mH	1.3386
Rated speed	Rpm	1028
Rated power	KW	70

The three phase currents at θ =0° are shown in Fig.2. Currents i_α, i_β shown in Fig.3 are displayed through DAC7565 chip. The AD sample frequency is 5k, and 25 sample data are collected for parameters a,b,c,d calculation. And the parameters' value will soon converge to the right point as shown in Fig.4. In fact, more data can improve the accuracy of the rotor position estimation, but will increase the calculation burden. The rotor position detected by position transducer (resolver in this experiment) and the estimated position are listed in Tables II. It can be observed that the largest error of estimated position is about 5.7°(electric angle), and that is small enough for the right start-up of the motor.

TABLE II
Rotor position(rad)

Position detected by resolver	Position estimated	error
0	0.02	0.02
0.76	0.71	-0.05
1.35	1.27	-0.08
1.89	1.90	0.01
2.33	2.28	-0.05
3.28	3.22	-0.06
3.70	3.66	-0.04
4.30	4.24	-0.06
4.91	4.81	-0.10
5.78	5.80	0.02

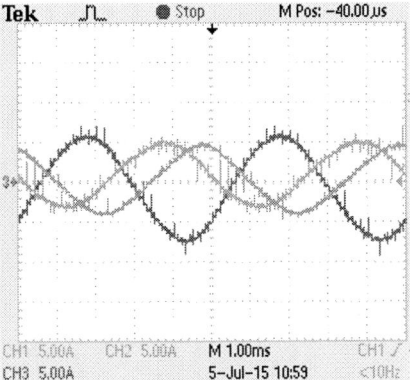

Fig. 2. Three phase current waveforms

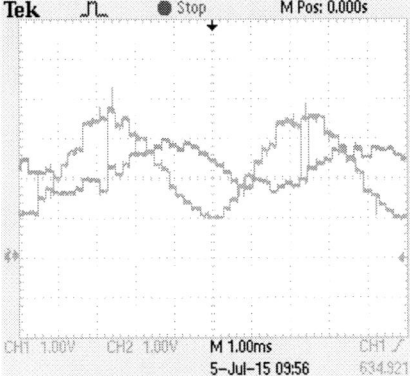

Fig. 3. i_α, i_β waveforms

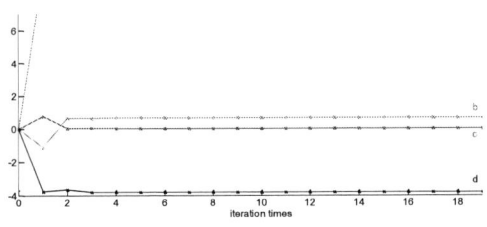

Fig.4. parameters calculation

IV. CONCLUSION

A novel initial rotor position detection method for IPM synchronous motors has been proposed in this paper. This method is based on the traditional high frequency voltage injection method, but the instantaneous current value rather than the amplitude of the corresponding currents i_α, i_β are utilized to calculate the rotor position, thus the filter to obtain the current amplitude can be removed and algorithm can be implemented with DSP in a very short time. Moreover, the amplitude and frequency of the voltage injected have little influence on the accuracy of theta estimation. The experiment has verified the proposed method.

ACKNOWLEDGMENT

This work is supported by the National Natural Science Foundation of China (Grant No. 51277053) and Infineon HUST student innovation project.

REFERENCES

[1] S. Morimoto, K. Kawamoto, M. Sanada, and Y. Takeda, "Sensorless control strategy for salient-pole PMSM based on extended EMF in rotating reference frame," IEEE Trans. Ind. Appl., vol. 38, no. 4, pp.1054–1061, Jul./Aug. 2002.

[2] Genduso F , Miceli R , Rando C , et al . Back EMF sensorless-control algorithm for high-dynamic performance PMSM[J] . IEEE Trans . on Industrial Electronics , 2010 57(6) : 2092-2100 .

[3] Nicola Bianchi, Silverio Bolognani, Ji-Hoon Jang and Seung-Ki Sul, "Comparison of PM Motor Structures and Sensorless Control Techniques for Zero-Speed Rotor Position Detection" IEEE Trans. Power Electronics, vol. 22, pp. 2466–2475, Nov. 2007.

[4] Jang, J.H., Sul, S.K., Ha, J.I. et al.: 'Sensorless drive of surface-mounted permanent-magnet motor by high-frequency signal injection based on magnetic saliency', IEEE Trans. Ind. Appl., 2003, 39, (4), pp. 1031–1039

[5] Nakashima, S., Inagaki, Y., and Miki, I.: 'Sensorless initial rotor position estimation of surface permanent-magnet synchronous motor', IEEE Trans. Ind. Appl., 2000, 36, (6), pp. 1598–1603

[6] Genduso F , Miceli R , Rando C , et al . Back EMF sensorless-control algorithm for high-dynamic performance PMSM[J] . IEEE Trans . on Industrial Electronics , 2010 57(6) : 2092-2100 .

[7] K. D. Hurst, T. G. Habetler, G. Griva, and F. Profumo, "Zero-speed tacholess IM torque control: simply a matter of stator voltage integration," IEEE Trans. Ind. Applicat., vol. 34, pp. 790–795, July/Aug. 1998.

[8] A. B. Kulkarni and M. Ehsani, "A novel position sensor elimination technique for interior permanent-magnet synchronous drive," IEEE Trans. Ind. Applicat., vol. 28, pp. 144–150, Jan./Feb. 1992.

[9] J. S. Kim and S. K. Sul, "High performance PMSM drives without rotational position sensors using reduced order observer," in Conf. Rec. IEEE-IAS Annu. Meeting, 1995, pp. 75–82.

[10] Consoli A , Scarcella G , Testa A . Industry application of zero-speed sensorless control techniques for PM synchronous motors[J] . IEEE Transactions on Industry Applications , 2001 , 37(2) : 513-521 .

[11] K. Liu, Q. Zhang, J. T. Chen, Z. Q. Zhu, J. Zhang, and A.W. Shen, "Online multiparameter estimation of non-salient pole PM synchronous machines with temperature variation tracking," IEEE Trans. Ind. Electron., vol. 58,no. 5, pp. 1776–1788, May 2011.

[12] B. Widrow and M. A. Lehr, "30 years of adaptive neural networks: Perceptron, madaline, and backpropagation," Proc. IEEE, vol. 78, no. 9, pp. 1415–1442, Sep. 1990.

BIOGRAPHIES

Zou Xunhao obtained his BSc degrees in the college of electrical engineering from Zhejiang University in 2013. He is now working towards his master degree in Huazhong University of Science and Technology. His main research interests are control of PM synchrounous motors and generators.

Shenghua Huang is now the professor of Huazhong University of Science and Technology. His main research interests include new special motor and its control system, power electronic devices and systems and the applications of power electronics in Power system.

Qin Zhuqian was born in China in 1989.He obtained his B.E. degree of electrical engineering from China University of Mining and Technology. He is now studying the master degree in HuaZhong University of Science and Technology, and his research direction is power electronics and power drives.

Xiaodong Hu, Master Degree, dedicated to optimiz ation on structure of PM machine and electromagne tic actuators with permanent magnetic.

Kan Guangqiang obtained his BSc degrees in the college of electrical engineering from southwest jiaotong University in 2014. He is now working towards his master degree in Huazhong University of Science and Technology. His main research interests are control of PM synchrounous motors and power electronics.

Cascaded High Gain Micro-Converter for Storage-less PV Fed Rural Telecom Systems

Vishal Verma   VandanaArora

Department of Electrical Engineering,Delhi Technological University, INDIA

E-mail: vishalverma@dce.ac.in

Abstract–For ubiquitous wireless mobile network, micro towers are becoming necessity and feeding them in remote locations, unapproachable to grid, has become even more daunting task. A cost-effective, sustainable, low maintenance and environment-friendly electricity generation is highly solicited, and these can be offered alternatively through renewable energy sources. PV fed battery-less system typically suits such application which can cater to rural intra-village telecom and information system, which are solicited majority during day times. This paper proposes a cascaded high gain single stage micro-converter which does the MPPT and boosts the voltage upto usable level. The scheme evacuates the maximum power amidst insolation changes in a limited range, whereas, the load controller provides the matching of power for maintenance of voltage. The performance of the system is evaluated under Matlab/Simulink environment. Presented simulation results show close conformity with design and validates the effectiveness of the control proposed scheme employed.

Keywords–Micro converter, MPPT, Load Control, Single stage conversion.

I. INTRODUCTION

Sustainable supply of power plays a vital role in working of telecom base station. Still some remote rural areas are devoid of grid availability. The Base Transceiver Stations (BTS) which are generally not very heavily loaded and demand the usage majority in day time, are supplied through diesel generator set, which is expensive and demands recurring fuel and maintenance cost with heavy pilferage [1].

Current research shows that energy can suitably be harnessed through the Photovoltaic (PV) sources for small power applications. The ample availability of solar energy, which in India is less intermittent, makes it favourableas compared to the other renewable energy sources, when small power generation is considered. Often intermittency is overcome through multiple sources with multi-input converters to supply to a single output [1].

Since the output power from PV panel vary largely with insolation variation, the maximum power point (MPP) tracking (MPPT) becomes important to harness the power with maximum efficiency [1]. Many MPPT techniques discussed in literature varies on the basis of complexity, number of sensors required, convergence speed, effectiveness, and popularity and on other aspects. Due to low power storage-less application under varying insolation level fast tracking becomes necessity and therefore Incremental Conductance (InC) method is employed for fast dynamic response and higher accuracy [2].

Because of PV Panel having low output voltage and exhibit higher current characteristics, a dc-dc boost converter is often required to be integrated with panel to realize usable voltage level. Multiple - input single output converters (MISO) provides a cost-effective and flexible

way to interface two or more sources to produce a usable voltage. Reported literature [3]-[7] reveals variety of isolated and non-isolated topologies of Single Input Single Output (SISO) and MISO. Limited voltage gain boost converter pass stresses on power switches, diodes are affected by equivalent series resistance (ESR) of inductors and capacitors, and also suffers from serious reverse recovery problem when operated at high duty ratio[3]. The isolated topologies are constructed around multi-winding transformer based on half-bridge or full bridge topologies, which makes the converter complex and costly. In [4], a multi-input boost type converter is reported. In [5],a family of multi-port dc-dc converters based on the combination of DC link voltages are discussed, whereas, multi-input buck boost topology is reported in [6] for hybrid renewable sources. A multiple-input high step-up boost converter is introduced for PV applications but, the topologies employ complex control as reported in [7].

In this paper, analysis and operation control of a new MISO DC-DC high gain boost converter (HGBC) for telecom power applications is presented. This converter system is a voltage cascaded connection of two DC-DC HGBC converter units which operates on two isolated PV sources to produce single output voltage. As solar energy is intermittent on the count of variation in insolation and temperature, oversized panels with a load controller is used to curb the intermittency in output. The proposed cascaded high gain micro-converter (CHGMC) system integrating two isolated PV panels is operated with MPPT method utilizing the InC. Due to availability of higher gain, the converters are intentionally operated with lower duty cycle to avert the stresses on power devices and high frequency for reduction in ripples. The results are presented under MATLAB/Simulink environment amidst both insolation variation and load perturbations.

II. SYSTEM CONFIGURATION

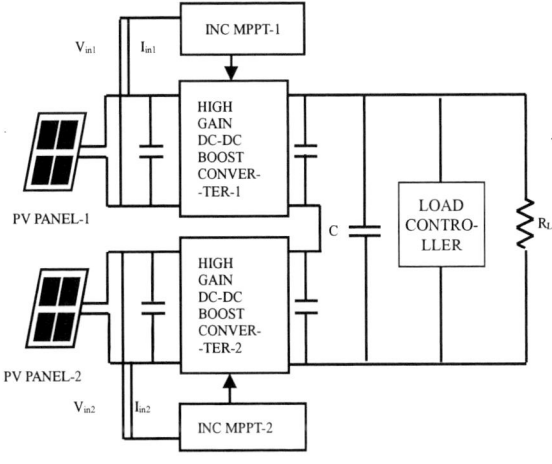

Fig. 1: Schematic Diagram of Proposed System

978-1-5090-0064-7/15 $31.00 © 2015 IEEE

Fig.1 shows schematic diagram and modes of operation of the Multi-input CHGMC for telecom applications using two PV panels and a load controller. Each HGBC is controlled for MPPT by using InC method which senses current and voltage from PV panel and appropriately produces duty cycle to ensure maximum power evacuation from PV panel. A load controller is used at the output side to match the load as per requirement.

III. MODELLING OF PV MODULE

Solar cell is the basic element of each Solar PV System which converts solar energy directly into electrical energy. PV panel hosts series and parallel connected PV cells to produce standardized voltage and current outputs. The equivalent circuit of practical PV cell is shown in Fig. 2, and the relevant characteristic equation shows that current of a PV cell is a function of voltage:

$$I = I_{PV} - I_o \left[e^{\left(\frac{V + R_S I}{V_t a} \right)} - 1 \right] - \frac{V + R_S I}{R_P} \qquad (1)$$

Where I is the total output current from Solar cell, I_{PV} is the current generated due to insolation, I_d is the Diode current, I_o is the reverse saturation or leakage current of the diode, V_t is the thermal voltage of PV Module. V_t for N_s number of PV cells connected in series and as given as = $(N_s KT/Q)$, where, K is the Boltzmann Constant $(=1.380503 \times 10^{-23}$ J/°K), Q is the Electron Charge $(=1.60217646 \times 10^{-19}$ C), T is the Temperature (°Kelvin) and a = Diode ideality factor $(1 < a < 1.5)$.

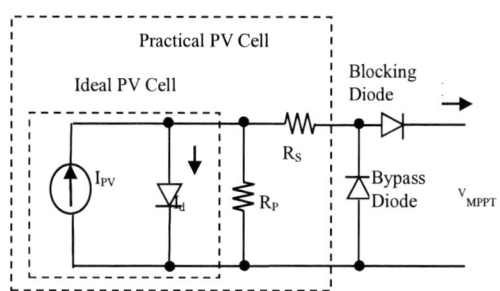

Fig.2: Equivalent Circuit of PV Cell

For simulation purpose, we have used manufacturer's datasheet presented in Table 1, for open-circuit voltage, short circuit current, the voltage at the maximum power point(MPP), the current at the MPP, the open circuit voltage/temperature coefficient(K_v), the short-circuit current/temperature coefficient(K_I), and the maximum experimental peak output power ($P_{max,e}$).This information is provided at standard test conditions i.e. at 1000 W/m² irradiation and 25°C. The other parameters like the light generated current, diode saturation current, diode ideality constant, series and parallel resistance etc. can be computed.
For model verifications, the parameters given in Table 1 of KC200GT Kyocera panel areconsidered. Both Fig. 3 (a) or(b) shows the simulated I-V Curves and P-V curves respectively at different insolation levels. The simulated result matches closely with datasheet, thus validating the model developed.

Table 1: Parameters of the model of KC200GT solar array

Parameters	Values
I_{mp}	7.61 A
V_{mp}	26.3 V
$P_{max,m}$	200.143 W
I_{sc}	8.21 A
V_{oc}	32.9 V
$I_{0,n}$	9.825 X 10^{-8} A
I_{pv}	8.214 A
A	1.3
R_p	415.405 Ω
R_s	0.221 Ω

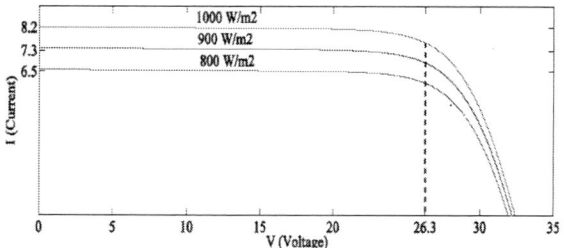

Fig. 3: (a) I-V Characteristics of PV Array

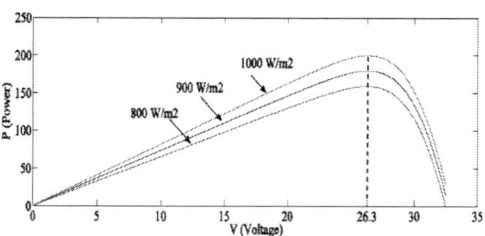

Fig. 3: (b) P-V Characteristics of PV Array

IV. INCREMENTAL CONDUCTANCE (InC) BASED MPPT

The method is based on the fact that the slope of the power curve of the PV panel at the MPP is zero, positive on the left, and negative on the right of the MPP, and the same is shown in Fig.4(a) for providing more clarity. The same may be expressed mathematically as:

$$\begin{cases} dP/dV = 0, & \text{at MPP} \\ dP/dV > 0 & \text{left of MPP} \\ dP/dV < 0 & \text{right of MPP} \end{cases} \qquad (2)$$

Since,

$$\frac{dP}{dV} = \frac{d(IV)}{dV} = I + V\frac{dI}{dV} \cong I + V\frac{\Delta I}{\Delta V} \qquad (3)$$

Therefore,

$$\begin{cases} \Delta I/\Delta V = -I/V, & \text{at MPP} \\ \Delta I/\Delta V > -I/V & \text{left of MPP} \\ \Delta I/\Delta V < -I/V & \text{right of MPP} \end{cases} \qquad (4)$$

The MPP can thus be tracked by comparing the

instantaneous conductance (I/V) to the incremental conductance (ΔI/ΔV) as shown in the flowchart in Fig.4(b). Thus in this method input impedance of switching converter is adjusted to a value that matches optimum impedance of PV panel.

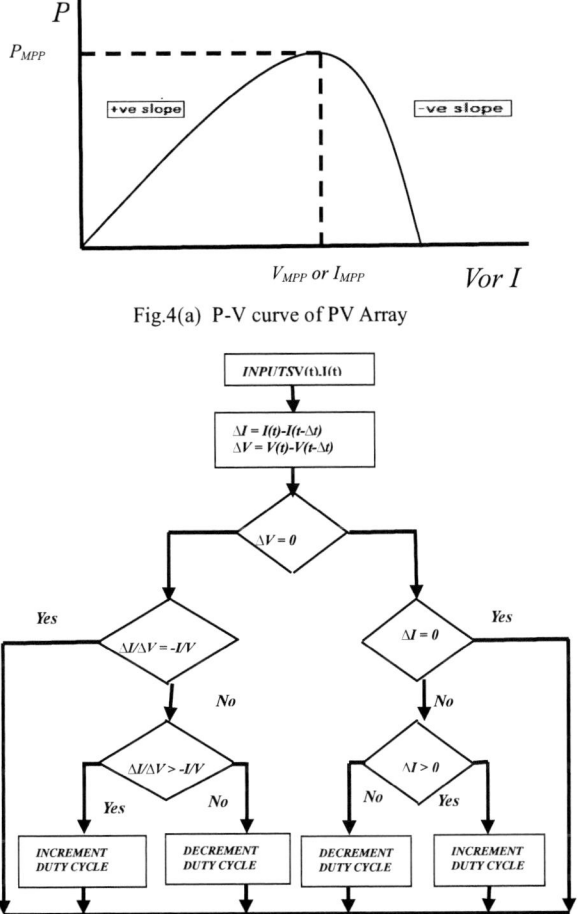

Fig.4(a) P-V curve of PV Array

Fig.4(b) Flow Chart of InC MPPT Method

V. ANALYSIS OF THE PROPOSED CHGMC

Converter employs four active switches (S_1, S_2, S_3 and S_4) and four inductors (L_1, L_2, L_3 and L_4) having same value, two diodes, one output capacitor and load resistance. For steady state analysis some assumptions are made:- 1) All components are ideal, i.e., the ON-state resistance $R_{DS(ON)}$ of the active switches, the forward voltage drop of the diodes, and ESRs of inductors and capacitors are ignored and 2) all capacitors are of sufficiently large value such that the voltages across capacitors remains constant. Since CHGMC is a cascaded structure formed using two HGBCs working on different duty cycles D_1 and D_2,

respectively, thus steady state analysis is carried out for three operating modes under the assumption that $D_1 < D_2$; Mode 1 where all the switches are closed; Mode 2 where one pair of switches are opened and other pair is still closed and Mode 3 where all the switches are open. The circuit representations of Mode 1, Mode 2 and Mode 3 is depicted in Fig. 5(a), Fig. 5(b) and Fig. 5(c) respectively. Assuming duty cycle for Mode 1 to be D' as $D_1 < D_2$, so D_1 is the period for which all the switches are closed, accordingly $D' = D_1$. The duration for Mode 2 is assumed as D"T(where 'T' corresponds to time period for switching frequency considered). Since this represents the period between D_1T and $D_2 T$, thus $D" = D_2 - D_1$. Mode 3 is assumed to be carried on for time D'''T which represents the period for which all the switches are open, thus $D''' = 1 - D_2$.

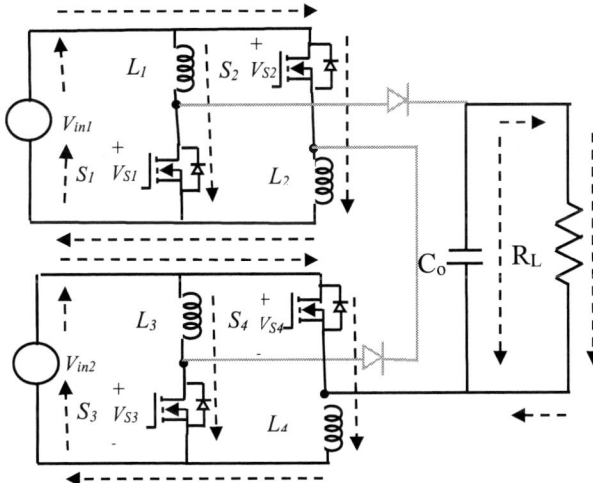

Fig.5 (a)Circuit Configuration with depiction of operation(Mode-1)

Mode 1-When all the Switches of both HGBCs areClosed
For HGBC-1 :When switches are closed the current flows into the inductors making diodes ineffective for transfer any power to the load. Thus voltage equation around the path containing source - 1, inductor L_1 and L_2, and switch $S_1$and S_2 will be represented as:

$$V_{L1;L2} = V_{in1} = L_{1;2} \frac{di_{L1;L2}}{dt} \tag{5}$$

$$\frac{di_{L1;L2}}{dt} = \frac{V_{in1}}{L_{1;2}} \tag{6}$$

For HGBC - 2 : Similarly writing mesh equations using KVL around the path containing source - 2, inductor L_3 and L_4, and switch S_3 and S_4will be expressed as:

$$V_{L3;L4} = V_{in2} = L_{3;4} \frac{di_{L3;L4}}{dt} \tag{7}$$

$$\frac{di_{L3;L4}}{dt} = \frac{V_{in2}}{L_{3;4}} \tag{8}$$

The rate of change of current depends upon the input voltage which is assumed maintaining current, so the current increases linearly while all the switches are closed. Thus rate of change in inductor currents can be expressed as:

$$\frac{di_{L_1,L_2;L_3,L_4}}{dt} = \frac{\Delta i_{L_1,L_2;L_3,L_4}}{D'T} = \frac{V_{in1;in2}}{L_1} \tag{9}$$

978-1-5090-0064-7/15 $31.00 © 2015 IEEE   199

Therefore, expression for Δi_{L_1} switches closed is expressed as:

$$\left(\Delta i_{L_1,L_2;L_3,L_4}\right)_{for\ D'=D_1} = \frac{V_{in1;in2}D'T}{L_{1,2;3,4}} \qquad (10)$$

where $D' = D_1$.

Mode 2- When one pair of switches are opened and other pair of switches are closed ($D''T = (D_2 - D_1)T$)

For HGBC – 1: When switches S_1 and S_2 are opened, inductor voltage polarity gets reversed and instead it

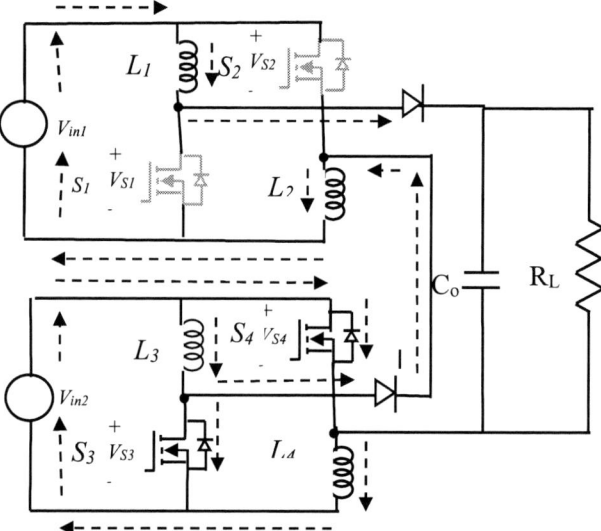

Fig. 5:(b)Circuit Configuration with depiction of operation(Mode-2)

releases energy and boosts the voltage and net power is transferred to the load through forward biased diode. Thus voltage equation for the loop so formed can be expressed as:

$$V_{L1;L2} = \frac{(V_{in1} - V_{o1})}{2} = L_{1;2}\frac{di_{L1;L2}}{dt} \qquad (11)$$

The rate of change current in inductors $L_1;L_2$rises linearly while switch is open.

$$\frac{di_{L1;L2}}{dt} = \frac{(V_{in1} - V_{o1})}{2L_{1;2}} \qquad (12)$$

Thus, solving for$\Delta i_{L1;L2}$,

$$\left(\Delta i_{L1;L2}\right)_{for\ D''=D_2 - D_1} = \frac{(V_{in1} - V_{o1})\ D''T}{2L} \qquad (13)$$

For HGBC – 2: During this time the switches S_3 and S_4remains closed, as a result two current start flowing in the inductor, one is because of cascaded structure which is flowing through load and the other component is drawn from source-2. Since the inductor is having large value there will be no spurt change in current thus resulting in decrease in drawl of current from source-2 to maintain the same current in the inductors, thus rate of change in inductor currents shall continue to be same. The voltage across inductors will accordingly be expressed as:

$$V_{L3;L4} = V_{in2} = L_{3;4}\frac{di_{L3;L4}}{dt} \qquad (14)$$

Solving for $\Delta i_{L3;L4}$, for $D'' = D_2 - D_1$

$$\left(\Delta i_{L3;L4}\right)_{for\ D''=D_2 - D_1} = \frac{V_{in2}D''T}{L_{3;4}} \qquad (15)$$

Mode 3- When all the Switches of both the HGBCs are Opened ($D''' = 1 - D_2$)

For HGBC-1: During this interval the status is maintained same as the switches $S_1$and S_2remained opened, as per previous conditions. Thus the power shall be continued to be supplied with boost in voltage.

$$v_{L_1,;L_2} = \frac{V_{in1} - V_{o1}}{2} = L_{1;2}\frac{di_{L1;L2}}{dt} \qquad (16)$$

The rate of change of current remains constant, and the same is expressed as:

$$\frac{di_{L_1,L_2}}{dt} = \frac{\Delta i_{L_1,L_2}}{D'''T} = \frac{V_{in_1} - V_{o_1}}{2L_{1,2}} \qquad (17)$$

For HGBC-2: When switches S_3 and S_4 are opened, polarity of the voltage across inductor gets reversed and inductor starts acting like a source to supply the power, and provide further boost in voltage. Thus inductor voltages may be expressed as:

$$v_{L_3,;L_4} = \frac{V_{in2} - V_{o2}}{2} = L_{3;4}\frac{di_{L3;L4}}{dt} \qquad (18)$$

The rate of change of current is constant, and the same may be expressed as:

$$\frac{di_{L_3,L_4}}{dt} = \frac{\Delta i_{L_3,L_4}}{D'''T} = \frac{V_{in_2} - V_{o_2}}{2L_{3,4}} \qquad (19)$$

Solving for (Δi_L) for $D''' = 1 - D_2$

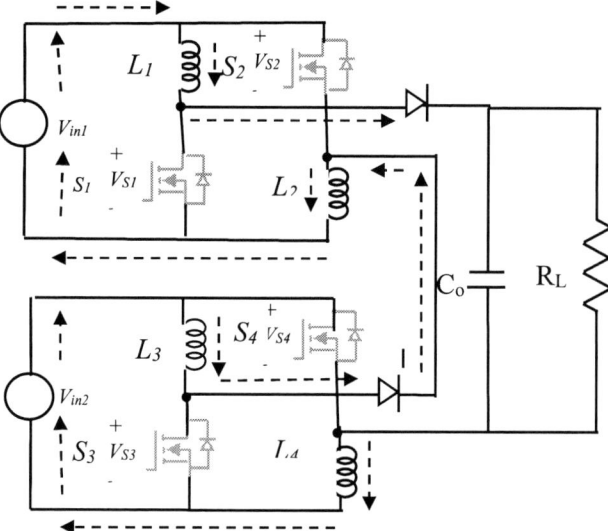

Fig. 5:(c)System Configuration with depiction of operation(Mode-3)

$$\left(\Delta i_{L_1,L_2;L_3,L_4}\right)_{D'''=1-D2} = \frac{(V_{in_1;in_2} - V_{o_1;o_2})D'''T}{2L_{1,2;3,4}} \qquad (20)$$

For steady state operation, net change in inductor current for each HGBC must be zero hence:

For HGBC-1:

$$\left(\Delta i_{L_1,L_2;L_3,L_4}\right)_{D'} + \left(\Delta i_{L_1,L_2;L_3,L_4}\right)_{D''} + \left(\Delta i_{L_1,L_2;L_3,L_4}\right)_{D'''} = 0 \quad (21)$$

$$\frac{V_{s1}D'}{L} + \frac{(V_{in1}-V_{o1})D''}{2L} + \frac{(V_{in1}-V_{o1})D'''}{2L} = 0 \quad (22)$$

$$\frac{V_{s1}D_1}{L} + \frac{(V_{in1}-V_{o1})(D_2-D_1)}{2L} + \frac{(V_{in1}-V_{o1})(1-D_2)}{2L} = 0 \quad (23)$$

On solving the above expression

$$\frac{V_{o1}}{V_{in_1}} = \frac{1+D1}{1-D1} \quad (24)$$

For HGBC-2:

$$\left(\Delta i_{L_1,L_2;L_3,L_4}\right)_{D\prime} + \left(\Delta i_{L_1,L_2;L_3,L_4}\right)_{D\prime\prime} + \left(\Delta i_{L_1,L_2;L_3,L_4}\right)_{D\prime\prime\prime} = 0 \quad (25)$$

$$\frac{V_{in2}D'}{L} + \frac{V_{in2}D''}{L} + \frac{(V_{in2}-V_{o2})D'''}{2L} = 0 \quad (26)$$

$$\frac{V_{in2}D_1}{L} + \frac{V_{in2}(D_2-D_1)}{L} + \frac{(V_{in2}-V_{o2})(1-D_2)}{2L} = 0 \quad (27)$$

On solving the above expression

$$\frac{V_{o2}}{V_{in2}} = \frac{1+D2}{1-D2} \quad (28)$$

Thus, Overall Voltage gain, $M = \frac{V_o}{V_{in}} = \frac{V_{o1}+V_{o2}}{V_{in1}+V_{in2}} \quad (29)$

Voltage Stress on Switches are computed for S_1, S_2, S_3, S_4 and diode D_{o1} and D_{o2} are derived as

$$\begin{cases} V_{S1} = V_{S2} = \frac{V_{o1}+V_{in1}}{2} \\ V_{D_{o1}} = V_{o1} + V_{in1} \end{cases} \quad (30)$$

$$\begin{cases} V_{S3} = V_{S4} = \frac{V_{o2}+V_{in2}}{2} \\ V_{D_{o2}} = V_{o2} + V_{in2} \end{cases} \quad (31)$$

VI. CONTROL THEORY

The control of power and maintenance of voltage of CHGMC is employed in two stages in the first stage MPPT of each PV panel is done, which computes the duty cycle for operation of HGBCs in cascade structure even with changes in insolation. The duty cycle is kept less than 0.6 to minimize the voltage stresses in switch and diodes. The cascade connection substantiates the voltage gain of each panel, due to the circuit current of CHGMC being equal. The control scheme dealing with MPPT is shown in Fig. 5.

In the second stage of control the power equalization is done with the help of load controller. The output power of CHGMC is matched to telecom load power by absorbing the excess power from the DC bus into the load controller. The power matching control is obtained by estimating the current requisite to draw into the load by keeping the DC bus voltage constant. The reference current is thus estimated as:

$$i_{ref}^* = \left(k_p + \frac{k_i}{s}\right)(V_{dc}^* - V_{dc}) \quad (32)$$

The switching signal for the load controller is derived through hysteresis current based control.

VII. PERFORMANCE EVALUATION

Performance of the system is evaluated considering variation in insolation keeping temperature constant. Only resistive load is considered to imitate power supply in the circuit. Simulated results are studied to gauge the performance of the CHGMC is evaluated in two stages,

akin to control strategy. The variation of voltage, current and power due to variation of the insolation are scaled and are presented in the Fig. 7, Fig.8 and Fig.9.The current is shown un-scaled, whereas, voltage is scaled by 1/10 times, power by 1/100 times and insolation level as 1/1000 times. The difficult operating conditions for CHBMC are considered, where one panel is put to changing insolation levels while the other is kept constant. Since the current in CHGMC have to be same the two HGBCs will operate at different duty cycles to enable evacuation of generated power producing different output voltages under different insolation conditions, enacting global MPPT of the string. Fig.3and Fig.4shows the insolation inputs(G), output

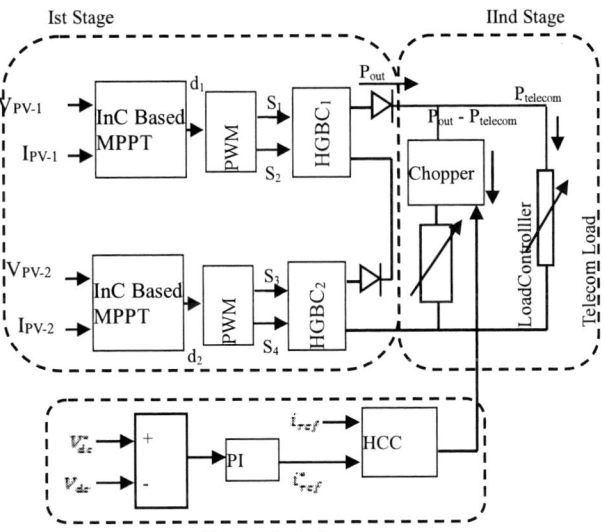

Fig. 6: Block diagram depicting Control scheme for operation of CHGMC

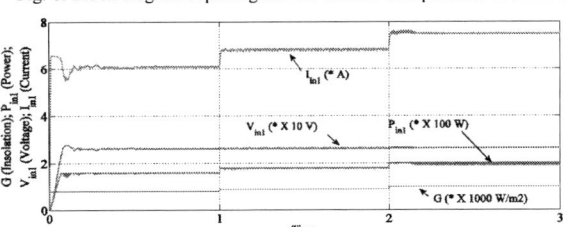

Fig.7: Dynamics of PV panel-1 under MPPT with insolation variation

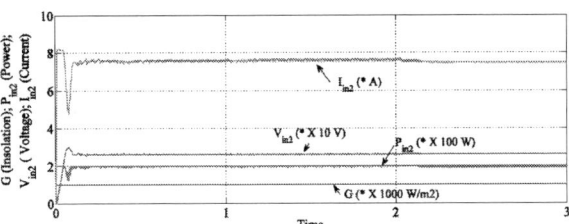

Fig.8: Dynamics of PV panel-2 under MPPT with constant insolation

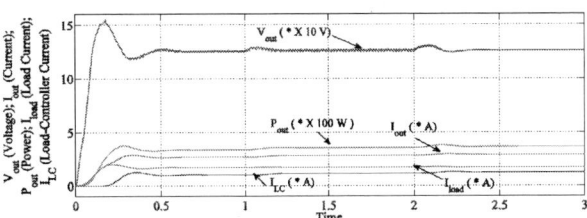

Fig.9: Dynamics of CHGMCunder insolation variation with load control

voltages(V_{in}), currents(I_{in}) and power (P_{in})evacuated from

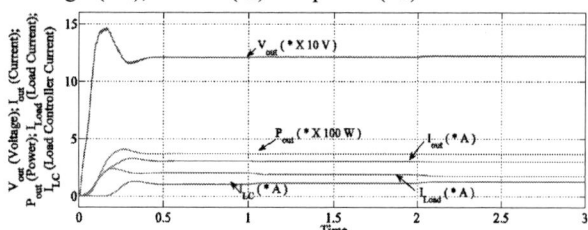

Fig..10: Dynamics of load controller under load perturbations

PV panel-1 and2 respectively, where insolation variation is considered at different times, viz, from t=0 to t=3s @800W/m², t=1 to t=2s @900 W/m² and from t=2 to t=3s @1000 W/m², whereas, panel-2 is kept at an insolation of 1000 W/m²throughout. MPPT-1 for panel-1 ensures that output power for 800W/m² is evacuated @160W; for 900W/m² is evacuated @180W and for 1000W/m² is evacuated @200W (Fig.6), andpanel-2 at insolation of 1000 W/m² evacuates output of 200W (Fig.7).

The tracking power equivalent to global MPP may be observed in Fig.8, where CHGMC maintains an output of 125.7 V except small transients during step change in the insolation as aforesaid, where output power and current varies as per insolation variations respectively as, 336W& 2.69A; 354W & 2.81A; and, 369W & 2.91A, establishing a conversion efficiency of 92%. It may also be observed that power is efficiently equalized by load controller keeping load current constant throughout. The voltage stresses are low during aforesaid insolation variations. as the voltage across $S_{1,2}$varies from 43V to 46V, whereas, for $S_{3,4}$varies between 50V to 46V.

The effectiveness of the load controller is also gauged under perturbing load varied between 180-220W keeping constant insolation incidents on each panel of 1000W/m². It may be observed from Fig.9 that power output is rapidly matched by load controller maintaining the output voltage constant and evacuation of generated power.

VIII. CONCLUSION

A cascaded high gain boost micro converter configuration has been proposed for storage less PV fed rural telecom systems. The observed performance of the proposed micro converter in isolated mode has demonstrated the ability of the proposed structure and control to effectively boost voltage at required level and supply power to the rural telecom system under limited changes of insolation levels. The cascaded structure and control is capable of achieving global MPPT automatically, maximizing the output. The result demonstrates that the scheme effectively maintain the output voltage constant with the help of load controller, and extract the maximum available power, thus always operating at global MPPT, under perturbation of insolation levels and load. The dynamics of the system is quite fast and overall conversion efficiency is high baring power decoupling through load controller and the steady state gain obtained closely matches with that obtained for analysis. The scheme has an advantage of simplicity, and also offers modularity thus enabling easy maintenance. The scheme with such modular microconverter may be

extended to address issues of partial shading effectively as is evident in string PV system. In a nutshell it is perceived that such modular, rugged and low maintenance approach shall pave the way for better and effective utilization of PV for off-grid and grid connected application in future.

REFERENCES

[1] Mangu, B.; Fernandes, B.G.; "Multi-input Transformer Coupled DC-DC Converter for PV-Wind based Stand-Alone Single-Phase Power Generating System", Proc.of IEEE ECCE'14, pp.5288-5295.

[2] T. Esram, P. L. Chapman, "Comparison of Photovoltaic Array Maximum Power Point Tracking Techniques", IEEE Trans. on energy conversion, vol. 22, No.2, pp. 439-449, June 2007.

[3] D. C. Lu, K. W. Cheng and Y. S. Lee, " A single-switch continous-conduction-mode boost converter with reduced reverse-recovery and switching losses, IEEE Trans. Ind. Electron., vol. 50, no. 4, pp. 767-776, Aug. 2003.

[4] Di Napoli, A., Crescimbini, F., Solero, L., Caricchi, F., Capponii, F. G., "Multi-input DC-DC power converter for power flow management in hybrid vehicles", Proc. IEEE Ind. Appl. Conf., 2002, pp. 1578-1585.

[5] Tao, H., Kotsopoulos, A., Duarte, J.L., Hendrix, M.A.M. "Family of multiport bidirectional DC–DC converters", Proc. IEEE Electron. Power, 2008, pp. 451–458.

[6] Dobbs, B.G., Chapman, P.L., "A multiple-input DC–DC converter topology", IEEE Power Electron. Lett., 2009, 1, (1), pp. 6–9.

[7] Zhou, L.-W., Zhu, B.-X., Luo, Q.-M.,"High step-up converter with capacity of multiple input",IET Power Elect.,2011,(5), pp.524–531.

Static Model of a 2x25kV AC Traction System

Mariia Plakhova, Bassam Mohamed and Pablo Arboleya
Department of Electrical Engineering
University of Oviedo
Gijón, Spain
Email: mariiaplakhova@gmail.com, engbassam@gmail.com, arboleyapablo@uniovi.es

Abstract—**Transport system, and especially railways as a part of it, has a highly significant place in the modern world. Railway system has a lot of advantages, comparing to the other type of transport, such as: comfort, economy, better control on the travel time and schedule and less risk factors. Development of high performance computers, growing complexity of traction drives and power supplies made possible and, moreover, necessary to model railway systems in order to provide its efficient planning, design and maintenance. This paper introduces a basis of the traction system modelling for high speed railway systems. First, a brief review of existing traction systems is given. Second, a 2x25kV AC bivoltage network is explained in details. Finally, a mathematical model of the static AC system has been proposed.**

Keywords—*Power system, power system modelling, railway system.*

I. INTRODUCTION

According to the research, people have known about the railway system from early 6th century BC [1]. Nevertheless, it took a long time for the railway system to refine and get the familiar modern form. Creation of the steam engine and introduction of the first steam locomotive, based upon it, showed up the new stage for the transport system. Further advancement in railway and locomotive technologies caused the development of the first electrified railways, which was performed by Siemens in 1879 [2]. Following technical revolution brought out the power supply systems and traction motors and made them a key parts of the modern electrified transport systems. DC motors supplied from a low-voltage DC (1.5kV from the beginning of 20th century and 3kV from 1930s) have been used in the early traction systems due to its simplicity and ease to control. There are two main directions in railway systems: low-voltage DC transmission networks for drives with DC traction motors, and low-frequency (16.7Hz in Central Europe and 25Hz in the United States) high voltage AC transmission networks that were updated up to the industrial values (50 and 60Hz respectively) with the establishment of high-voltage electrification [3]. Today, the standard range of voltages is defined by standards EN 50163 [4] and IEC 60850 [5]. The most common railway systems are following:

- DC high voltage systems: 3kV with the distance between substations 15-30 km.
- DC medium voltage systems: 1.5kV with the distance between substations 15-30 km.
- DC low voltage systems: 0.6 - 1.4kV with the distance between substations 1-6 km.
- AC single phase systems: 15kV/16.7Hz, 25kV/50Hz.
- AC three-phase systems: 25kV/50Hz.

The most common voltage value for the majority of tram and metro systems is 750V DC. However, there are also some non-standard values of voltage, e.g. metro-transit systems of London and Milan use 630V DC.

In [6], [7] a detailed explanation of all different feeding topologies for AC and DC railways is provided, however a summary of the basic structure of both, DC and AC, traction power systems is illustrated on Figure 1.

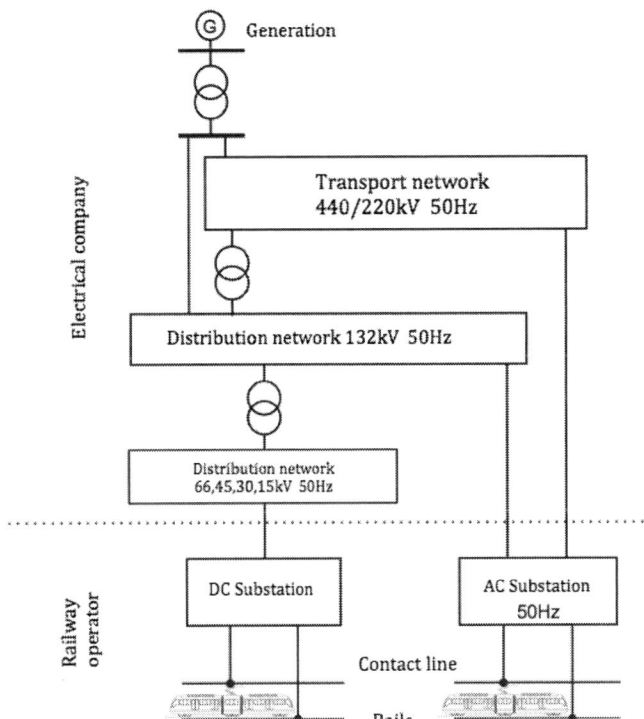

Fig. 1: Basic Structure of DC and AC Power Systems [6]

Over the last 20 years the railway traction systems went trough a lot of changes as a result of a significant progress in power electronics and microprocessor fields. Nevertheless, DC power supply is still the most common and economic for the public transportation, whereas AC at industrial frequency is more useful and efficient for the long-distance lines. Nowadays, there are no alternatives to the fast travel on the land. Thus, the work on improvement of AC traction systems is substantial.

978-1-5090-0064-7/15 $31.00 © 2015 IEEE

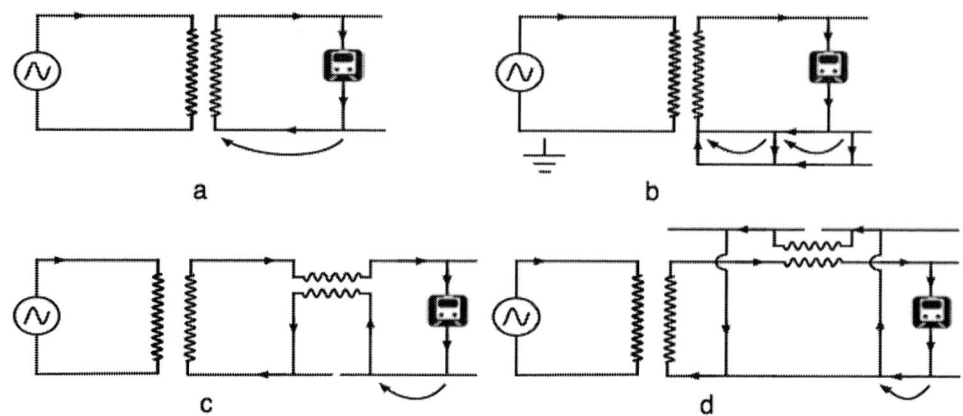

Fig. 2: AC railway feeding systems: (a) Direct feed; (b) Direct feed with return conductor; (c) Booster transformer feed; (d) Booster transformer with return conductor

II. AC TRACTION POWER SYSTEM

An electrified line is similar to a typical power transmission and distribution system. Main difference is that trains move and change operation modes constantly, thus varying power consumption over a wide range. The number of other factors, such as train speed, track layouts, traffic demand, and drivers' behaviour, can also affect power demand [8].

AC power-supply systems are widely used in Europe and allover the world (Great Britain, Spain, Portugal, Italy, Taiwan, Hong Kong etc.). The development of the commercial high-speed railways in 1980s has expanded the use of AC power-supply systems, involving larger power flows compare to DC systems. The most commonly adopted AC traction system (1x25kV or 2x25kV that can be called dual [9] or bivoltage at 50Hz) was designed for the lines with high power requirements.

In general any railway system can be divided into a number of electrically isolated sectors. The single phase or three-phase network feeds each sector through the traction sub-stations, which are modified to guarantee the operation in case of failure. The direct connection of the feed transformer is common in AC railroads. Booster transformers (BTs) and autotransformers (ATs), set into a feeding section, are widely used in AC system to improve transmission efficiency and system regulation, decrease rail-to-earth voltage and prevent electromagnetic interference to the telecommunication circuits, located nearby.

Simple/direct feeding. This is the simplest and the cheapest option of feeding power. It achieves by the direct connection of the traction feed transformer to the catenary and rails of each substation (Figure 2a). However, there are next disadvantages: high feeding impedance with large losses, high rail-to-earth voltage (safety issues) and the earth currents as the side-product that can cause interference in the telecommunication circuits, that are located nearby. In order to reduce the leakage current a return conductor can be added to the system. In this case current is forced to flow rather in a conductor, than in a rail, thus the impedance traction current return path is reduced (Figure 2b).

Booster transformers (BTs). First BT, rated at about 150 kVA has been used in Japan in 1964 and it could improve the characteristics of the feeding circuit. BTs are usually located along the catenary at distance 3-4 km. The primary and the secondary windings are connected across a gap of the contact wire and across the insulated rail section respectively (Figure 2c). The purpose of a return conductor is the same as in case of direct feeding. It is preferable to incorporate a conductor in parallel with the rails for the return current (see Figure 2d).

Autotransformer (AT) power feeding. The first AT was presented in Philadelphia in early of XX century [10], after it was installed in Japan in 1972, and starting from 1981 more and more countries included ATs in their railway systems, developing new standards for the AC electrification system. AT feeding is shown on Figure 3. It combines the advantage of higher-voltage power transmission with the benefit of using standard 25 kV/50 kV equipment. The AT winding is connected between the catenary and an auxiliary feeder, with the rails tied to an intermediate point. The principle of AT operation is following: the train draws current from the two ATs, located nearby, the supply current from each of AT depends on the location of the train. Rail currents flow through the AT windings as illustrated in order to maintain Ampere-turn balance in cores. The AT system operates by balancing voltage [10]. This is its main advantage over BT feeding systems. As for another benefits of the AT system, it is easier to maintain, because this topology allows to separate a lot of substations.

III. A 2x25kV AC BIVOLTAGE TRACTION SYSTEM

Currently, high speed railways are widely implemented and used in the whole world. This type of railways demands higher values of power, and for this purpose the feeding voltage should be also increased. Thru this fact, nowadays, new lines are often electrified in AC and existing AC lines are converted into bivoltage (e.g. 2x25kV, 50Hz), allowing to get higher feeding voltage. In this configuration power is transferred using high voltage, which then is reduced by autotransformers to the suitable for the trains or distribution network level. This configuration has been described in [9], [11], [12], [13].

Figure 3 represents a 2x25kV 50 Hz bivoltage traction system.

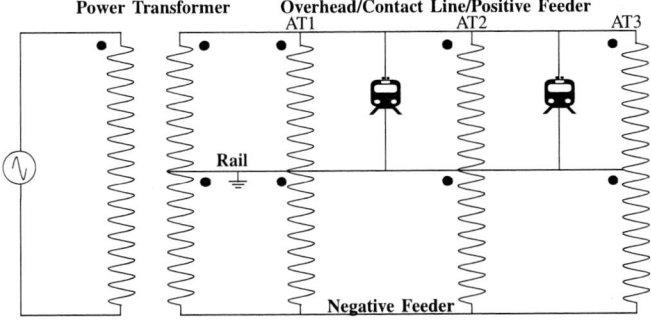

Fig. 3: AC bivoltage traction system

As it is showed on Figure 3 the secondary windings of the power transformers are placed at ESSs and have a central tap connected to the rails (ground), while the poles are connected to the contact/overhead line (or positive feeder) and the negative feeder respectively. From the design and maintenance point of view, it is common to set up the same voltage for the both, positive and negative feeder. Nevertheless, the positive voltage can be defined by standards, while the negative one can be chosen freely. ATs with a unity turn ratio are usually used in this configuration, and this is a specific characteristic for such schemes. ATs are connected to contact line, rail and negative feeder at certain intervals, and break the electrical section fed by one substation into different compartments. These compartments are called cells and in general the length of each cell can be from 10 km up to 50 km.

The Figure 4 shows an example of the current distribution in bivoltage configuration. There are two cells in the scheme, represented in the Figure 4: the first one is between the substation and the first AT and the second one is between the first and the second ATs. The most important part to study is placed in the second cell, where the train is located. The train is consuming a current $I = 400A$, which consists of two parts: the first part is current aI comes from the first AT, while the second part $(1 - a)I$ comes from the second AT. The value of current, supplying by each AT, depends on the factor a, which defines the location of the train. In this example $a = 0.75$ has been assumed, thus the values of current, supplying by the first and the second ATs, are $300A$ and $100A$ respectively. The return current flows through the rails to the closest ATs, then it splits into two, so the current flowing through the rails of the first cell is zero. Also, the current flowing from the second cell to the power transformer through the positive and negative feeders is $200A$, half of the train current. The empty cell (without train) has the feeding and return system of 50kV, so it is possible to increase the transport capacity of the line together with the distance between the nearby stations. There are five different loops in this configuration: two in the first cell (without train) with the current $I/2$ and three in the second cell (with train), and two of them have $(1 - a)I$ current and one loop has aI current. Thus, the current distribution of this scheme can be represented by the current flow in these five loops.

The decrease of current though the rails is one of the benefits of the bivoltage configuration. It makes the rails potentials and the losses also to be reduced. Among the other advantages, the bivoltage design for high speed railways makes the mainte-

nance procedures simpler and reduces the level of interference with the telecommunication systems.

It is also possible to implement bivoltage configuration in DC traction systems, using DC/DC controlled converters instead of the ATs.

IV. STATIC MATHEMATICAL MODEL

ATs can break any electrical section fed by one substation (one power transformer) into different cells. Cells are usually assumed to be independent in order to avoid the problem of unbalanced network and do not complicate the solving procedure. It allows to make estimation of voltage, current and load flow faster and easier.

Figure 5 illustrates a bivoltage AC system with one cell and one train in it. The network can be divided into two parts: the high voltage feeding network, which is connected to the AC traction network through the primary side of the power transformer, and the AC traction network by itself, that includes the secondary side of the power transformer, ATs and trains. The source voltage, primary side current and impedance are represented by V_{src}, I_p and Z_p respectively, while all values, that describe the AC traction network, can be separated into different groups, such as node voltages, line currents (current trough the positive feeder I_{PF}, current trough the rails I_{GND} and current through the negative feeder I_{NF}) and line impedances $(Z_{PF}, Z_{GND}$ and $Z_{NF})$, also taking into account currents and voltages of the secondary side of the power transformer and ATs $(I_1, I_2, V_1, V_2, I_{ATw1}, I_{ATw2}, V_{ATw1}, V_{ATw2})$ together with the characteristics of the train $(V_t$ and $I_t)$.

System will be called static, if it does not consist trains. The mathematical model of the static system is represented by a system of equations that describes the network. This set of equations includes Kirchhoff Voltage and Current Laws (KVL and KCL) for all branches and nodes, power transformer and ATs' equations. The general form of the equations, which describe the power transformer and ATs', is given below.

Two of the power transformer equations describe the relation between the primary side and each group of the windings on the secondary side:

$$V_p = \frac{N_p}{N_{p1}} V_1 \tag{1}$$

$$V_p = \frac{N_p}{N_{p2}} V_2 \tag{2}$$

where N_p is the number of turns of the power transformer primary winding, while N_{p1} and N_{p2} are the number of turns of the power transformer secondary winding first and second groups respectively (here and later the winding that connects the positive feeder and the rails will be considered as the first winding group, and the winding that connects the rails and the negative feeder will be the second winding group. This will also work for the ATs windings). From the Figure 5 V_p, V_1 and V_2 are the voltage of the primary side of the power transformer and voltages of the first and second winding groups of the secondary side of the power transformer.

978-1-5090-0064-7/15 $31.00 © 2015 IEEE

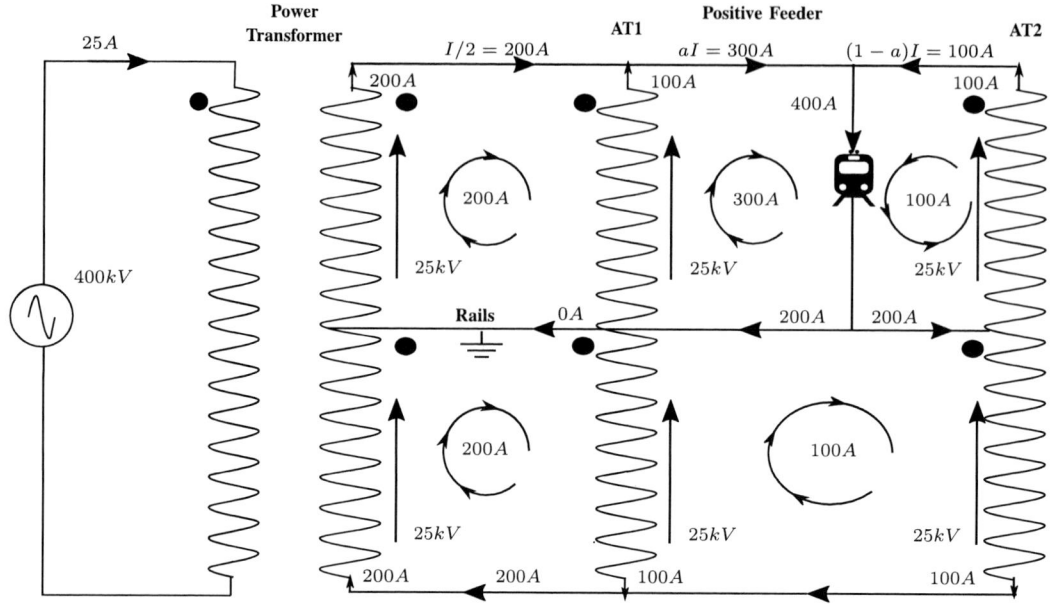

Fig. 4: Current distribution in an AC bivoltage traction system

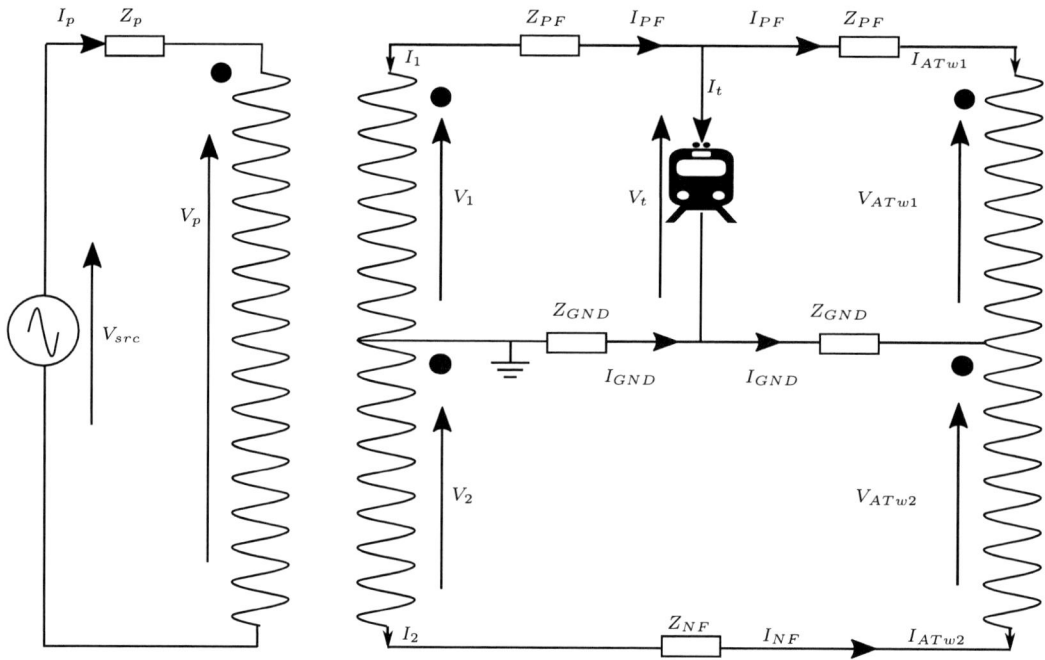

Fig. 5: AC bivoltage system with one cell and one train

Power equation of the power transformer will be also included into the set of equations, which describe the network:

$$I_p V_p = I_1 V_1 + I_2 V_2 \qquad (3)$$

I_1 and I_2 are the currents through the power transformer secondary windings.

Each AT will add two equations to the system (one that describes the voltage relation between first and second winding groups, and the power equation):

$$V_{ATw1} = \frac{N_2}{N_1} V_{ATw2} \qquad (4)$$

$$I_{ATw1} V_{ATw1} + I_{ATw2} V_{ATw2} = 0 \qquad (5)$$

In order to obtain the network, i.e. obtain all currents and voltages in the system, the next data is required:

- Cell Data that consists all information about all cells: number of cells in the network, length of each cell, turn ratio of ATs and resistance and impedance values for the contact line, rails and the negative feeder.

- Train Data that includes all information about trains: number of trains, number of cells, in which each train is located, together with distance and active and reactive power values. Active and reactive power of the train can be obtained, knowing the power consumption of the train and its power factor.
- Value of the source voltage.
- Number of turns for the primary and secondary side of the power transformer.

Based on the described mathematical model, a power flow simulator for the traction system can be built. Any developed and properly working simulator is a tool that can be successfully used by any railway company. In order to use this tool the company should provide necessary information, required to solve the system.

V. CONCLUSIONS

A rapid increase of the usage of high speed railways requires to be able to create a proper models and simulators for the traction systems in order to provide its efficient design, operation and maintenance. For this purpose it is important to understand the structure and operation principles of the existing traction systems, and especially a bivoltage configuration due to its wide implementation in the current high speed railroads. This paper has provided a review of the present railways and has established a mathematical model for the bivoltage network. In further works, the proposed model will be used to create a simulator to analyse the power flow in a 2x25kV AC bivoltage traction systems.

REFERENCES

[1] M. Lewis, "Railways in the greek and roman world," http://www.sciencenews.gr/docs/diolkos.pdf.

[2] "Siemens history site: In focus," http://www.siemens.com/history/en/news/electric_railway.htm.

[3] R. Hill, "Electric railway traction. part 1: Electric traction and dc traction motor drives," *Power Engineering Journal*, Feb. 1994.

[4] "Railway application. supply voltages for traction systems," 2005.

[5] "Railway applications - supply voltages of traction systems. ed.4.0," 2014.

[6] B. Kiessling, S. A. Puschmann, R., and T. Vega, *Lineas de contacto para ferrocarriles electrificados.* Siemens-Aktiengesellschaft, 2008.

[7] B. Kiessling, S. A. Puschmann, R., and E. Schneider, *Contact Lines for Electric Railways; Planning, Design, Implementation, Maintenance.* Siemens-Publicis-Publishing, 2012.

[8] T. Ho, Y. Chi, J. Wang, and K. Leung, "Load flow in electrified railway," *2nd IEE International Conference in Power Electronics, Machines and Drives*, 2004.

[9] E. Pilo, L. Rouco, and A. Fernandez, "A reduced representation of 2x25kv electrical systems for high-speed railways," *IEEE/ASME Joint Rail Conference*, 2003.

[10] R. Hill, "Electric railway traction. part 3: Traction power supplies," *Power Engineering Journal*, Dec. 1994.

[11] S. Raygani, A. Tahavorgar, S. Fazel, and B. Moaveny, "Load flow analysis and future development study for an ac electric railway," *IET Electrical System in Transportation*, 2012.

[12] E. Pilo, *Power Supply, Energy Management and Catenary Problems.* WIT Press, 2010.

[13] V. Zakarukin and A. Krukov, "Methods of joint simulation for external power supplies and ac traction systems (written in russian)," Ph.D. dissertation, State Railway University, Irkutsk, 2011.

BIOGRAPHIES

Mariia Plakhova received the B.Sc in Industrial Electronics and Automation and the M.Sc in Energy Management from the National Aviation University (Ukraine) in 2012 and 2014. She also received the M.Sc degree in Sustainable Transportation and Electrical Poweer Systems from the University of Oviedo, Gijon, Spain, in 2015. Her master thesis was focused on the development of an AC High speed traction system simulator. Now, she is working on modelling and simulation of AC railway traction networks.

Bassam Mohamed received the M.Sc degree from the University of Oviedo, Gijon, Spain, in 2014. He is now pursuing his Ph.D studies in the Department of Electrical Engineering at the University of Oviedo. His master thesis was focused on implementing power flow and optimal power solver for transmission networks. Now, he is working on modelling and simulation of AC and DC micro-grid and railway traction networks.

Pablo Arboleya (SM' 13) Received the M.Eng. and Ph.D. (with distinction) degrees from the University of Oviedo, Gijón, Spain, in 2002 and 2005, respectively, both in electrical engineering. He is Senior member of the IEEE Power and Energy Society since 2013 and was a recipient of the University of Oviedo Outstanding Ph.D. Thesis Award in 2008. Nowadays, he works as an Associate Professor in the Department of Electrical Engineering at the University of Oviedo (with tenure since 2010).

978-1-5090-0064-7/15 $31.00 © 2015 IEEE

Graph Theory Approach for a 2x25kV AC Bivoltage Traction System

Mariia Plakhova, Bassam Mohamed and Pablo Arboleya
Department of Electrical Engineering
University of Oviedo
Gijón, Spain
Email: mariiaplakhova@gmail.com, engbassam@gmail.com, arboleyapablo@uniovi.es

Abstract—**Electrical power system is one of the largest and essential nowadays engineering systems that is used in every single field of the modern life. Over the years, the power system advanced significantly, developing from local isolated networks with small-scale load and generation to large networks, such as power plants, railways etc. With increasing the size of the systems, power flow management has become an important task that requires an application of proper methods and techniques to solve power flow problem. In this paper, the authors have applied a graph theory approach as an effective method for solving the load flow in a 2x25kV AC bivoltage traction systems. Obtaining a static system, proposed method is able to overcome numerous problems in the system topology and dimensions.**

Index Terms—**Graph theory, power system, railway system, power system modelling.**

I. INTRODUCTION

First application of the graph theory to solve different mathematical problems was around 300 years ago and started from Leonardo Euler and his try to walk linking seven bridges in Königsberg [1]. From this time forward, graph theory, the study of networks in form of the graphs, which are symbolic representation of systems and connections within these systems, has developed from its recreational purposes into an independent and useful study. The beginning of the XX century brought Poincare theory, which described the principles of algebraic topology with help of graph theory and using incidence matrix. Based on this theory, in 1916, the Kirchhoff's laws were formulated, applying the graph theory [2]. It started a new stage in power system analysis and modelling, following by numerous improvements and innovations in graph theory. Currently, the graph theory is widely used to solve different problems. As for power system analysis and modelling, there are three main graph theory applications [3]:

- Nonlinear networks analysis and modelling. Based on the graph theory, nonlinear systems can be described and solved, applying e.g. "two-graph modified nodal analysis" for the sensitivity analysis of periodically switched linear networks or switching signal flow graph method for large signal analysis in super-lift converters.

- Observability analysis in power systems, which allows to estimate the internal states of the system based on the knowledge about its external states.
- Power flow techniques. Applying graph theory, large-dimension power flow problem with a numerous of constraints can be modified into a set of medium-dimension problems, thus the network can be obtained easier and faster.

Railways and its power systems are significant parts of the traction systems, and require proper planning, effective operation and well-timed service. Thus, it is important to find an accurate, simple and robust power flow method, and it has been done in [4], [5], [6], [7]. The scope of this paper is a power flow problem. The graph theory based method will be used for the network description. The paper is built as follows: the first section provides a brief theoretical background, while next section gives an information about a 2x25kV AC bivoltage systems and defines nodes and lines numeration procedure, and matrices construction for the static (without trains) AC system, and the conclusions are given in the last section.

II. BRIEF INTRODUCTION TO GRAPH THEORY

Graph theory describes properties of networks using graphs, which are symbolic representations of systems and connections within these systems. Graph consists vertices or nodes (substations or trains in case of railways), and lines or edges, which join nodes. Any graph in graph theory can have the next properties:

- Graph G will be an oriented/directed graph (or digraph), when its edges are identified with ordered pairs of nodes. Otherwise graph G will be undirected/non- oriented graph.
- Graph G is a simple graph, when there is no self-loops or multi-edges, i.e. there is no edges starting and ending in the same node.
- Graph G is called dense, if the number of edges is twice higher than the order of the graph[1]. Otherwise graph is called sparse.

[1]The order of the graph is defined by the number of nodes

978-1-5090-0064-7/15 $31.00 © 2015 IEEE

Figure 1 shows a simple graph G, that includes two finite sets N and E (nodes and edges accordingly). Vertices are represented as 1, 2..n, while edges are represented as $e_1, e_2..e_k$. Each edge is defined by a pair of nodes (i, j), where i is a source node/tail, and j is a destination node/head.

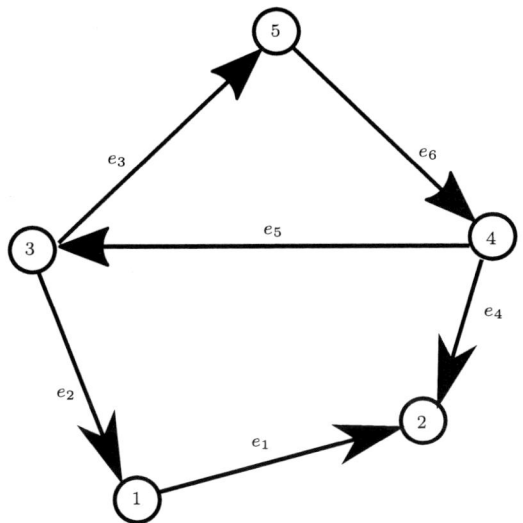

Fig. 1: Simple graph G

Thus, simple graph G (Figure 1) consists:

- Nodes $N = \{1, 2, 3, 4, 5\}$;
- Edges $E = \{e_1, e_2, e_3, e_4, e_5, e_6\}$, where:
$$e_1 = (1, 2);$$
$$e_2 = (3, 1);$$
$$e_3 = (3, 5);$$
$$e_4 = (4, 2);$$
$$e_5 = (4, 3);$$
$$e_6 = (5, 4).$$

The direction of the edges will define a positive reference for currents in a given system.

It is possible to divide graph G into subgraphs, in which nodes and edges are subsets of a present simple graph G. Figure 2 shows one of the subgraphs from a graph G: a subgraph H consists nodes $N_h = \{3, 4, 5\}$ and edges $E_h = \{e_3, e_5, e_6\}$.

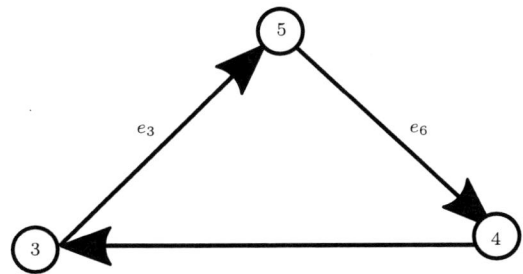

Fig. 2: Subgraph H of a simple graph G

Numeration of the nodes and edges in any graph can be done in any order, depending on the designer of the system. Nevertheless, enumeration of the network elements (nodes/lines) plays a significant role in the description of the railway systems, especially when it comes to the solving of the system with the large amount of trains. Irrelevant numeration can cause difficulties during the solving procedure.

A pair of nodes are adjacent if there is an edge, that connects them (e.g. node 1 and node 2 in a graph G are connected by an edge e_1). The number of edges that connects a node with another vertices verifies the degree of a node $(d(i))$. Node with a degree one is called a leaf or pendant node. Depending on the direction of the edge, whether it goes in or go out from the node, a vertices can have the following types of degree: in-degree (number of incoming edges) and out-degree (number of outgoing edges).

For a graph G with n nodes the matrix Λ with dimensions (n, n) can be built. This matrix Λ relates the connection between nodes.

$$\Lambda_{i,j} = \begin{cases} \text{number of edges between } i \text{ and } j & i \neq j \\ \text{number of self-loops in } i & i = j \end{cases}$$

However, in most of studied cases in order to simplify the solution procedure, all graphs are usually simple digraphs. Thus:

$$\Lambda_{i,j} = \begin{cases} 1 \text{ if exists adjacency between i and j and } (1 > j) \\ 0 \text{ other cases} \end{cases}$$

Adjacency matrix Λ can be represented as following:

$$\Lambda_{i,j} = \begin{pmatrix} 0 & 1 & 1 & 0 & 0 \\ 0 & 0 & 0 & 1 & 0 \\ 0 & 0 & 0 & 1 & 1 \\ 0 & 0 & 0 & 0 & 1 \\ 0 & 0 & 0 & 0 & 0 \end{pmatrix} \tag{1}$$

Usually, adjacency matrices are defined as symmetric. However, to get the greater efficiency obtaining the sparse matrices, it is better to define them as an upper triangular, thus avoiding the redundant information storage.

Furthermore, the graph G with n nodes and k edges can be fully defined by the k, n dimension incidence matrix Γ:

$$\Gamma_{k,n} = \begin{cases} 1 & \text{, the tail/source of the edge } k \text{ is the node } n \\ -1 & \text{, the head/destination of the edge } k \text{ is} \\ & \text{the node } n \\ 0 & \text{, other cases} \end{cases}$$

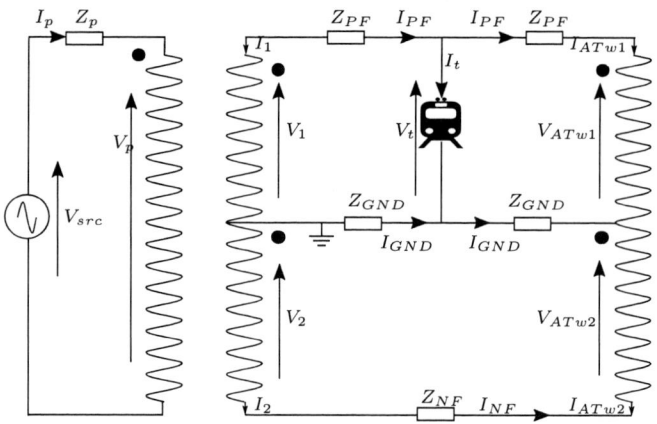

Fig. 3: AC bivoltage system with one cell and one train

$$
\Gamma_{k,n} = \begin{pmatrix}
1 & -1 & 0 & 0 & 0 \\
-1 & 0 & 1 & 0 & 0 \\
0 & 0 & 1 & 0 & -1 \\
0 & -1 & 0 & 1 & 0 \\
0 & 0 & -1 & 1 & 0 \\
0 & 0 & 0 & -1 & 1
\end{pmatrix}
\quad (2)
$$

Applying graph theory to solve railway systems trains and substations will be consider as nodes, and the lines, connecting them, as edges. The network by itself can be considered as a fixed network, i.e. the number of trains and topology can be fixed, even if the number of trains changes. Thus, the number of trains during the simulation will be consider constant, even if some of trains will be out of stage.

III. STATIC SYSTEM FORMULATION BASED ON GRAPH THEORY APPROACH

A. 2x25kV AC Bivoltage Traction System

With the growing tendency towards the high speed railways, the power demands and, thus, the feeding voltages are expected to be increased. For this purpose, most of the commonly adopted traction systems are AC systems and, moreover, designed as 1x25kV or 2x25kV (so-called dual [8] or bivoltage network). Such configuration allows to increase the feeding voltage in order to transfer required power. The voltage can be later reduced by autotransformers (ATs) to the suitable for the trains and distribution network level. This scheme has been described in [8], [9], [10], [11].

A typical 2x25kV 50 Hz bivoltage traction system is shown on Figure 3.

Figure 3 illustrates a bivoltage AC system with one cell and one train in it. The compartments of the electrical section, located between each pair of ATs, is called a cell, and its length is usually can be from 10 km up to 50 km. In this specific configuration, the turn ratio of the ATs is generally equal to one. The network, represented on Figure 3, can be divided into two parts: the high voltage

feeding network, which is connected to the AC traction network through the primary side of the power transformer, and the AC traction network by itself, that includes the secondary side of the power transformer, ATs and trains. The source voltage, primary side current and impedance are represented by V_{src}, I_p and Z_p respectively, while all values, that describe the AC traction network, can be separated into different groups, such as node voltages, line currents (current trough the positive feeder I_{PF}, current trough the rails I_{GND} and current through the negative feeder I_{NF}) and line impedances (Z_{PF}, Z_{GND} and Z_{NF}), also taking into account currents and voltages of the secondary side of the power transformer and ATs ($I_1, I_2, V_1, V_2, I_{ATw1}, I_{ATw2}, V_{ATw1}, V_{ATw2}$) together with the characteristics of the train (V_t and I_t).

Bivoltage configuration has numerous advantages: current through the rails can be decreased, thus, losses can be reduced, simple and easy maintenance, the level of the electromagnetic interferences with the adjacent telecommunication systems can also be reduced.

B. Numeration Procedure and Matrix Formulation

As it was mentioned above, using the graph theory approach for the railways it is important to choose a proper way to numerate network elements, so the further solving of equations that describe the network will not be complicated. Before start the numeration procedure, it is critically to point out that all changes will occur in the AC traction system (right side of the power transformer on the Figure 3) and not in the high voltage feeding network. Due to this fact the whole high voltage system can be described just by one equation of KVL, which will remain constant, and the numeration procedure will be applied to the elements of the AC traction system.

The first node is a central tap in the secondary side of the power transformer, so the nodes that belongs to the ground line will be numerated first, then the nodes of the positive feeder and the nodes of the negative feeder will be numerated the last. However, in case of line numeration, the procedure will start from the negative feeder, following by the ground lines and only then the lines of the positive feeder. This way of numeration has been chosen to simplify the further verification of the results for the dynamic system (trains are connected only to the rails and positive feeder, since the values of the negative feeder are not going to be affected by the allocation of the trains). The graphic representation of the numeration procedure is shown on the Figure 4.

N in this figure represents the number of cells. The index of the node depends on the number of the cells and the last node of the static system has 3(N + 1) index.

Figure 5 illustrates the static AC network with two cells. Node and line numeration have been done following the procedure, described previously.

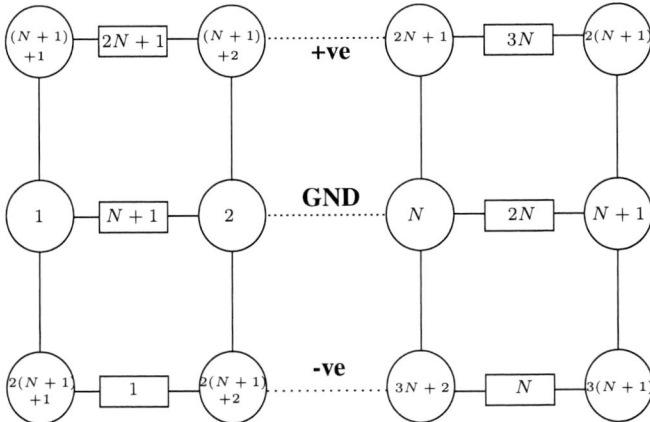

Fig. 4: The graphic representation of the numeration procedure

In case of the static system three node incidence matrices are presented: the first one Γ_{line} represents the real connections between ATs and between the secondary side of the power transformer and ATs, while the second and the third ones (Γ_{w1} and Γ_{w2}) define the connections within winding groups of the ATs and the secondary side of the power transformer. The number of columns in these matrices is equal to the number of nodes in the network, while the number of rows is different: for line incidence matrix Γ_{line} it is equal to the number of lines, and for the transformer incidence matrices Γ_{w1} and Γ_{w2} it is equal to the number of transformers (it includes ATs and the windings of the secondary side of the power transformer). Each line is defined by two nodes, the first one is the source node ("from" node) and the second one is the destination node ("to" node). These names refer to the direction of the currents. Thus, following the procedure of building incidence matrix, the next result for the configuration, shown on the Figure 5, is achieved.

$$\Gamma_{line} = \begin{bmatrix} 0 & 0 & 0 & 0 & 0 & 0 & 1 & -1 & 0 \\ 0 & 0 & 0 & 0 & 0 & 0 & 0 & 1 & -1 \\ 1 & -1 & 0 & 0 & 0 & 0 & 0 & 0 & 0 \\ 0 & 1 & -1 & 0 & 0 & 0 & 0 & 0 & 0 \\ 0 & 0 & 0 & 1 & -1 & 0 & 0 & 0 & 0 \\ 0 & 0 & 0 & 0 & 1 & -1 & 0 & 0 & 0 \end{bmatrix} \quad (3)$$

$$\Gamma_{w1} = \begin{bmatrix} -1 & 0 & 0 & 1 & 0 & 0 & 0 & 0 & 0 \\ 0 & -1 & 0 & 0 & 1 & 0 & 0 & 0 & 0 \\ 0 & 0 & -1 & 0 & 0 & 1 & 0 & 0 & 0 \end{bmatrix} \quad (4)$$

$$\Gamma_{w2} = \begin{bmatrix} 1 & 0 & 0 & 0 & 0 & 0 & -1 & 0 & 0 \\ 0 & 1 & 0 & 0 & 0 & 0 & 0 & -1 & 0 \\ 0 & 0 & 1 & 0 & 0 & 0 & 0 & 0 & -1 \end{bmatrix} \quad (5)$$

The mathematical model of the static system is represented by a system of equations that describes the network. This set of equations includes Kirchhoff Voltage and Current Laws (KVL

and KCL) for all branches and nodes, power transformer and ATs' equations.

Using defined node incidence matrices all the KVL's and KCL's can be written in a general form.

$$V_{src} = I_p * Z_p + V_p \quad (6)$$

$$\Gamma_{line} * V_N = Z * I_{line} \quad (7)$$

$$(\Gamma_{line})^T * I_{line} + (\Gamma_{w1})^T * I1_{trans} + (\Gamma_{w2})^T * I2_{trans} \quad (8)$$

Equations (6) and (7) represent the KVL for the feeding system and for the lines of the AC traction system, while equation (8) interprets the KCL in the traction network.

The vector V_N includes the voltages in the traction network nodes (nodes 1 to 9) and the vector I_{line} contains line currents (lines 1 to 6), whereas vectors $I1_{trans}$ and $I2_{trans}$ hold the values of the transformers currents for the first and the second winding groups respectively. Z is the impedance vector that includes the impedances of the lines (first impedance of the lines of the negative feeder, then rails and afterwards the contact line).The impedances of the lines that belong to negative feeder are static and indicate the real topology of the network. They are not going to be changed with the introduction of the trains into the system.

For the final description of the static system, equations of the power transformer and ATs should be included. Its general forms are given below. Equations (9)-(11) describe the power transformer, while equations (12) and (13) represent the ATs.

$$V_p = \frac{N_p}{N_{p1}} V_1 \quad (9)$$

$$V_p = \frac{N_p}{N_{p2}} V_2 \quad (10)$$

where N_p is the number of turns of the power transformer primary winding, while N_{p1} and N_{p2} are the number of turns of the power transformer secondary winding first and second groups respectively (here and later the winding that connects the positive feeder and the rails will be considered as the first winding group, and the winding that connects the rails and the negative feeder will be the second winding group. This will also work for the ATs windings). From the Figure 3 V_p, V_1 and V_2 are the voltage of the primary side of the power transformer and voltages of the first and second winding groups of the secondary side of the power transformer.

$$I_p V_p = I_1 V_1 + I_2 V_2 \quad (11)$$

I_1 and I_2 are the currents through the power transformer secondary windings.

$$V_{ATw1} = \frac{N_2}{N_1} V_{ATw2} \quad (12)$$

$$I_{ATw1} V_{ATw1} + I_{ATw2} V_{ATw2} = 0 \quad (13)$$

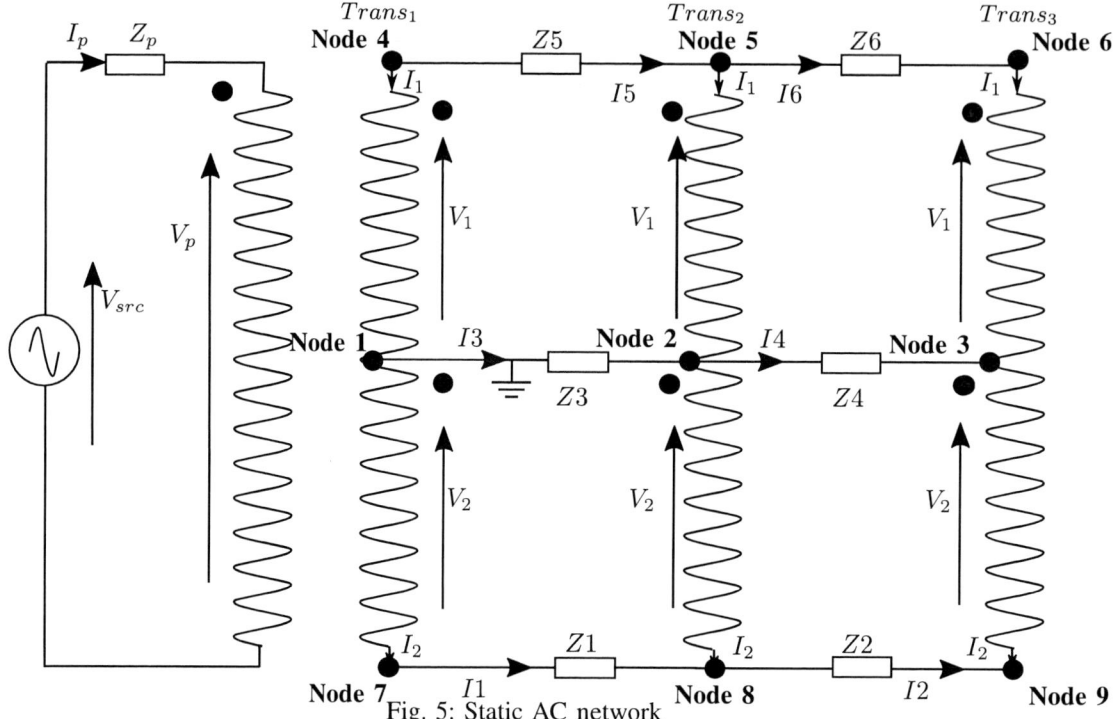

Fig. 5: Static AC network

IV. CONCLUSIONS

In this paper, graph theory approach has been implemented in order to describe a static bivoltage traction system. Applying proposed method, the Kirchhoff current and Kirchhoff voltage laws can be written in an easy compact matrix form. Next step to develop the proposed technique for solving AC traction systems will be to implement suggested method for the dynamic system, i.e. for the system that includes trains.

REFERENCES

[1] N. L. Biggs, K. E. Lloyd, and R. J. Wilson, *Graph Theory 1736 - 1936.* Oxford University Press, 1986.

[2] I. Cederbaum, "Some applications of graph theory to network analysis and synthesis," *Circuits Syst., IEEE Trans.*, vol. 31, no. 1, pp. 127–139, 1984.

[3] M. Coto, P. Arboleya, and C. Gonzales-Moran, "On the use of graph theory for railway power supply systems characterization," *Intelligent Industrial Systems*, vol. 1, no. 2, pp. 64–68, 2015.

[4] P. Arboleya, G. Diaz, and G. M., "Unified ac/dc power flow for traction systems: A new concept," *IEE Transaction on Vehicular Technology*, 2012.

[5] L. Abrahamsson and L. Soder, "Fast estimation of relations between aggregated train power system data and traffic performance," *IEEE Trans. Veh. Technol.*, vol. 60, no. 1, pp. 16–29, 2011.

[6] C. Pires, S. Nabeta, and J. Cardoso, "Dc traction load flow including ac distribution network," *IET Elect. Power Appl.*, vol. 3, no. 4, pp. 289–297, 2009.

[7] T. Smed, G. Andersson, G. Sheble, and L. Grigsby, "A new approach to ac/dc power flow," *IEEE Trans. Power Syst.*, vol. 6, no. 3, pp. 1238–1244, 1991.

[8] E. Pilo, L. Rouco, and A. Fernandez, "A reduced representation of 2x25kv electrical systems for high-speed railways," *IEEE/ASME Joint Rail Conference*, 2003.

[9] S. Raygani, A. Tahavorgar, S. Fazel, and B. Moaveny, "Load flow analysis and future development study for an ac electric railway," *IET Electrical System in Transportation*, 2012.

[10] E. Pilo, *Power Supply, Energy Management and Catenary Problems.* WIT Press, 2010.

[11] V. Zakarukin and A. Krukov, "Methods of joint simulation for external power supplies and ac traction systems (written in russian)," Ph.D. dissertation, State Railway University, Irkutsk, 2011.

BIOGRAPHIES

Mariia Plakhova received the B.Sc in Industrial Electronics and Automation and the M.Sc in Energy Management from the National Aviation University (Ukraine) in 2012 and 2014. She also received the M.Sc degree in Sustainable Transportation and Electrical Poweer Systems from the University of Oviedo, Gijon, Spain, in 2015. Her master thesis was focused on the development of an AC High speed traction system simulator. Now, she is working on modelling and simulation of AC railway traction networks.

Bassam Mohamed received the M.Sc degree from the University of Oviedo, Gijon, Spain, in 2014. He is now pursuing his Ph.D studies in the Department of Electrical Engineering at the University of Oviedo. His master thesis was focused on implementing power flow and optimal power solver for transmission networks. Now, he is working on modelling and simulation of AC and DC micro-grid and railway traction networks.

Pablo Arboleya (SM' 13) Received the M.Eng. and Ph.D. (with distinction) degrees from the University of Oviedo, Gijón, Spain, in 2002 and 2005, respectively, both in electrical engineering. He is Senior member of the IEEE Power and Energy Society since 2013 and was a recipient of the University of Oviedo Outstanding Ph.D. Thesis Award in 2008. Nowadays, he works as an Associate Professor in the Department of Electrical Engineering at the University of Oviedo (with tenure since 2010).

978-1-5090-0064-7/15 $31.00 © 2015 IEEE

A Hybrid Open MP/MPI Parallel Computing Model Design on the SMP Cluster

Ying Xu[1, 2]*   Tie Zhang[2]

[1] Department of Mechatronics and Control Engineering, Shenzhen University, PRChina
* E-mail: yxu@szu.edu.cn
[2] Department of Mechanical and Automotive Engineering, South China University of Technology, PRChina
E-mail: merobot@scut.edu.cn

Abstract–In the integrated power system, the traditional parallel algorithm is difficult to achieve system electromagnetic transient simulation, so a hybrid parallel computing model was proposed. Two kinds of parallel programming models (OpenMP shared-memory and MPI message-passing) are widely used on the SMP (Symmetric Multi-Processors) cluster. By comparing different programming paradigms for parallelization on the nodes of a SMP cluster, we obtain a parallel computing model that shows high performance. When parallel communication is low, using MPI programming purely will be more efficient. Otherwise, the hybrid MPI/OpenMP parallel programming model can be used to achieve better results. The main characteristic of the hybrid programming model is that the whole project is divided into segments according to the hierarchical structure of the task being executed, and this division is based on the principle of system load balance. Task level process adopts MPI for inter-SMP nodes communication, and uses OpenMP for intra-SMP node parallelization inside each processor. Experimental results show that this hybrid parallel computing model yields high performance and works very effectively.

Keywords–MPI, OpenMP, HPC, SMP cluster, hybrid mpi/openmp programming, parallel computing

I. INTRODUCTION

Parallel computing has been widely used in electric power system simulation and analysis. The basic idea of parallel computing is to decompose the large scale of the original problem into several smaller sub-problems, and each sub problem can be calculated by coordinating mutual information. With the fast development of computer processor techniques and their related hardware facilities, high performance parallel computing becomes more and more simple. Although great progress has been achieved in hardware development, little improvement has been made to the parallel computing in software. Generally, parallel computing mainly includes two models: the Message-Passing (MPI) model [1] and the Shared- Memory (OpenMP) model [2]. Both of them have advantages and disadvantages. Programming by using the application interface in Message- Passing model has high performance in parallel computing, but it is difficult to make a detailed analysis of the relationship in data, and the cost in data communication is always high. OpenMP represents the mainstream in Shared-Memory model, with its good performance of simple programming and high parallel performance. However, it required that all the OpenMP threads derived from the same process be based on Shared-Memory model.

Currently, Symmetric Multi-Processors (SMP) cluster has been widely used in the field of avigation, petroleum prospecting, weather forecasting, medical treatment and so on. Therefore, under the current parallel programming model, it is vital to enhance furthest the high capacity of parallel computing of SMP cluster [3]. In this paper, we make full use of the advantages of the two kinds of the models by using them on SMP cluster. On one hand, when the communication among processes is heavy, we use hybrid hierarchical programming model, while the Message-Passing model is used for programming among the nodes and the Shared-Memory model is applied for programming in the nodes, in this way, advantages of both these two parallel programming methods are furthest embodied. On the other hand, when the communication among processes is lower, higher efficiency can be obtained by purely using Message-Passing model. This programming approach plays a very important role in the high performance parallel computing on clusters.

II. CHARACTERISTICS OF DIFFERENT PROGRAMMING MODELS

The advantages and disadvantages of different parallel programming models on a SMP cluster will be discussed as follows [4,5].

1. MPI mode

MPI belongs to the distributed memory programming mode. It provides a complete set of API standard interfaces [6,7]. In the Message-Passing model, an application consists of many parallel tasks, and every task has its data that are need to be deal with. The data exchange among tasks is achieved by the apparent Message-Passing messages. MPI can realize complete asynchronous communication efficiently, it is independent on hardware facility and has well transplantable characteristic. Moreover, it can provide all kinds of flexible parallel ports of C and Fortran for users.

MPI can provide us a well solution to parallel programming in cluster, however, it also has quite a few drawbacks. For example, long delay time in MPI communication causes a problem, although data transformation among processors that have Shared-Memory in the same node is equal to data transformation happens in different nodes, this problem still exists. Furthermore, in order to achieve high parallel computing performance, we should divide the core logic of the program into several pieces properly, and project them to distributed process set, it is really a big challenge.

978-1-5090-0064-7/15 $31.00 © 2015 IEEE   213

2. OpenMp mode

A Shared-Memory model consists of several parallel threads that are derived from one single process, and data exchange can be realized by using hidden shared data. OpenMP is a popular Shared-Memory model lately [8], because it can define a series of development interfaces that are transplantable and extensible. Therefore, OpenMP has advantage of easy to program, high efficiency, good transplantable characteristic, and convenience of being independent on the parallel communication.

OpenMP can be very efficient when it is applied in a single multiprocessor computer, but the most widely used structure of cluster lately is based on a combination of several multiprocessor computers. In this case, OpenMP will show its limits in application, generally speaking, we can use DSM (distributed shared memory) to realize virtual Shared-Memory on clusters in large scale, however, it will cause problems of resources wasting and parallel performance reduction .

3. Mixed mode MPI/OpenMP programming

The approach of hybrid MPI/OpenMP programming can make full use of the advantages of the former two programming models, the whole project can be divided into process level and thread level according to the hierarchical structure of the task to be accomplish. Process level consists of several MPI task processes, and it works in the Message-Passing mode. While in thread level, many OpenMP sub-task threads are derived from the inside of every MPI task, all these threads works in the Shared-Memory mode, Figure 1 shows the structure of hybrid MPI/OpenMP programming. For further information see [9,10].

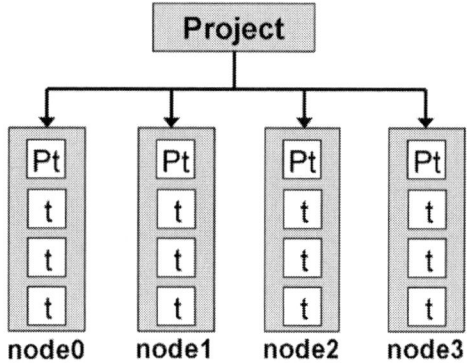

Fig. 1: The structure of hybrid MPI/OpenMP programming

This programming model has the following advantages:
 Make programming process very simple. By dividing the whole problem into several task blocks according to the hierachical structure of the project, the division of data structure becomes relative simpler, therefore, users don't need to consider the problem of data relativity. Every task is accomplished by a Message-Passing process, then thread level division can be done inside of the process, and use several OpenMP threads to accomplish the parts that are able to in be parallel in the process.

Decrease of the parallel communication capacity. In the Mixed mode MPI/OpenMP programming, Message-Passing model works only in task level and the minute division of data is not necessarily required, most data transformation takes place inside the tasks, therefore, comparing with pure MPI programming, the data communication capacity among of tasks is much less, while the speed of parallel computing is faster.

Avoid the problem of replicated data. Because Shared-Memory model is adopted inside the mixed model MPI/OpenMP programming, OpenMP threads inside the same node doesn't need to recopy and transfer the same data, therefore, this task can be done efficiently by data shared. However, in the pure MPI programming, each independent equal data should be reserved in memories, although it is different process inside one single node. In addition, data communication is time consuming among these processes. In this case, the mixed model MPI/OpenMP programming not only saves the EMS memory resource, but also decreases the time of message transforming.

Better system load balance. The problem of system load balance is considered at task level division period. Those parts that cost more time is carried out by multi-task so that every task consumes similar execute time. In this way, delays of other process, that results from the overload of the MPI program at a certain interface, can be avoided, which leads to the improvement of parallel computing efficiency.

Easy for debugging and realization. Because of using hierarchical structure, mixed mode MPI/OpenMP programming uses Message-Passing model in task level processes, and adopts Shared-Memory model inside the process. Therefore, program debugging can be done in different blocks, and it's relatively simple to realize.

4. The characteristics of hybrid programming

The hybrid programming model is composed of the MPI program in the upper level and the OpenMP program in the lower level. When writing MPI program, one should avoid the deadlock phenomena between sending and receiving functions, and be clear about the data flow between MPI programs. Writing OpenMP programs, one has to take into account the data variations caused by multiple iteration of multi-thread in nested loops. Usually, private variables and critical sections have to be created . In the test program below, critical sections are employed in order to get the numerical value of pi, since different iterations are data related, in each iteration, function (1) should be recalculated according to different values of x. The Conjugate Gradient Method Program employs private variables in each for iterations.

III. TESTING METHODS AND RESULTS

Two representative examples of multi- programming programs are employed to run comparisons between pure MPI programming and hybrid MPI/OpenMP programming.

1. Hardware environment for Testing

The environment is composed of a lenovo ShenTeng 6800, four Itanium2 processors (64bit EPIC architecture) at each node, and 3G of DDR memory with ECC check facility. The high performance of parallel computing is achieved over the Qsnet of the Quadrics Ltd. Britain with the Elan3 protocol. The Linpack performance testing shows that the peak speed of the system reaches to 192Gflops.

2. Testing Methods

2.1 The calculation of the π value

A calculus can be made over equation (1),

$$f(x) = \frac{4}{1+x^2} \tag{1}$$

to get the π value as shown below by equation (2).

$$\int_0^1 f(x)dx \approx \sum_{i=1}^n f\left(\frac{2 \times i - 1}{2 \times N}\right) \times \frac{1}{N}$$
$$= \frac{1}{N} \times \sum_{i=1}^n f\left(\frac{i - 0.5}{N}\right) = \pi \tag{2}$$

Note that N is the number of sub-intervals divided from the interval (0,1) with i as the index for each sub-interval. Equation (2) shows that the integral can be conducted for each sub-interval through the parallel running of multiple processes and threads, and then the π value can be achieved out of the sums of those integrals.

2.2 The Conjugate Gradient Method

The conjugate gradient method is a iterative method. In order to solve the equation $Ax=b$, a quadratic function is formed:

$$q(x) = \frac{1}{2} x^T Ax - x^T b \tag{3}$$

When the minimum of the quadratic equation $q(x)$ is achieved, x is exactly the solution to the equation.

The iterative process starts with an initial value x_0, and goes according to the equation (4).

$$x(k) = x(k-1) + a(k)d(k) \tag{4}$$

Note that current x value is obtained from x value in the last iteration; $a(k)$ and $d(k)$ are the step for iteration and the direction vector, respectively, both of which can be calculated with corresponding equations.

The key to this method is the calculation of the inner products of large scale matrices, since $a(k)$, $d(k)$, and $x(k)$ can got from inner product calculations. In high performance computers, a matrix can be divided into sub-matrices whose inner product calculation can be distributed over multiple processes and threads which will in turn be passed to the major process to produce the final result. We can also activate multiple parallel process to compute $a(k)$, $d(k)$, and $x(k)$ simultaneously.

3. The testing results and analysis

Four nodes of high performance computers (16 processors altogether) are employed to run the test. The method for testing the pure MPI program is demonstrated by figure 2. P stands for a processor within a same node, while the

number on the processor stands for the identifiers of processes. For example, in figure 3, the y-coordinate value corresponds to x-coordinate value of 10 implies the result produced by 10 MPI processes. Those ten processes distribute in the node machine as shown by figure 2 where one can find the node number and processor number corresponding to processes 1 to 10.

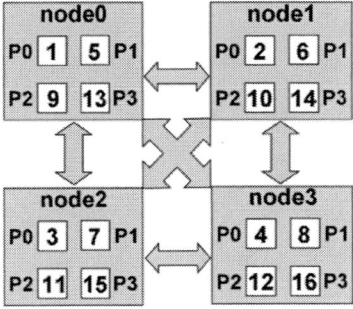

Fig. 2: The structure of pure MPI program running on the SMP cluster

The hybrid programming mode works in two ways: one is by running two parallel OpenMP threads inherited from a MPI process; the other one is by activating four threads. In figure 3, the testing method applied with respect to x-coordinate value of 6 for the MPI+OpenMP(2) is that one process is run at each of the three node machines, and each process is equipped with 2 OpenMP threads. MPI+Open(4) can be understood similarly.

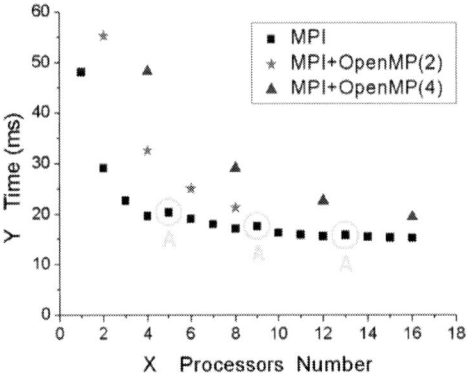

Fig. 3: Comparisons between pure MPI programming and hybrid MPI/OpenMP programming on the calculation of the π value

One can see from figure 3 that the pure MPI programming enjoys the highest efficiency, while the hybrid programming is slower than the pure MPI programming. This is because the cost for communication is small in the determination of the π value, and the communication is mainly for the computing of calculus of small regions. On the other hand, the hybrid programming introduces more cost for the OpenMP communications. Therefore, the more the OpenMP threads are, the longer the program runs. We can see that the time for MPI drops dramatically at the first four sample points while it stables at the following ones. It's because in the first four sample points, each process monopolizes a node, which maximizes the bandwidth between CPU and memory. A trace on the points marked

978-1-5090-0064-7/15 $31.00 © 2015 IEEE

by "A" shows that the program slows down when the number of MPI processes running on a node increases, which brings down the hit rate of memory and increases the communication cost.

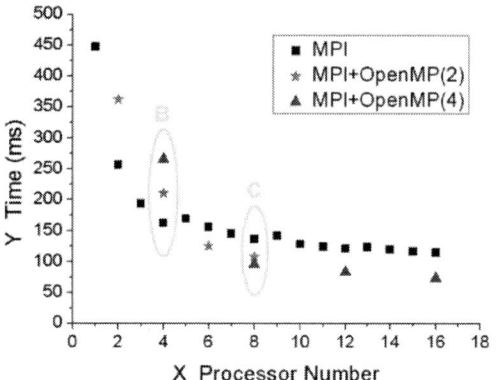

Fig. 4: Comparisons between pure MPI programming and hybrid MPI/OpenMP programming on the calculation of the conjugate gradient

In the test of getting the Conjugate Gradient solution to an equation, advantages of hybrid programming at the presence of large amounts of communication. Figure 4 demonstrates the similar MPI characteristics to that by figure 3. From the points marked by "B" in figure 4 we can see that when four processors are employed, a MPI program monopolizes a node; MPI+OpenMP(2) takes up two nodes with a MPI processes and two threads on each node; MPI+OpenMP(4) runs only on one node with one process and four threads. At this time, the speed of the hybrid programming is overwhelmed by that of the pure MPI programming, since multiple OpenMP threads run on different CPUs of a node, which brings down the CPU's storing bandwidth and the hit rate of the memory. A trace of the points marked by "C" shows that the speed advantage of the pure MPI programming is compromised with the increase of the amount of communication caused by the increase of the number of MPI processes. Testing shows that the hybrid programming mode with four MPI processes, each with four inherited threads, enjoys a 36% time-cut compared to the pure MPI programming mode with 16 processes.

VI. CONCLUSION

As a conclusion, the pure MPI programming bares relatively higher efficiency in parallel programs involving low communication cost, while the hybrid programming implies easier coding. When it comes to large scale parallel programs requiring complex communication, it's appropriate to divide the program into levels, such as the process-level (MPI) and the thread-level (OpenMP), and load balance among processes should be taken care of. Tasks requiring long execution time need to be further divided so that it can be finished with multiple processes. The MPI programming is applied on the process-level. It's better to let a node monopolized by one process in order to maximize the bandwidth between CPUs and memory. When the nodes are not enough, multiple processes can also share a node but monopolizing different processors, which brings down the efficiency a little. The OpenMP programming mode is applied on the thread level parallel computing. OpenMP threads of a same process must run on the same node. The hybrid programming mode absorbs the advantages of the two parallel programming modes to achieve higher performance-cost-ratio. Of course, with the rapid development of hardware, much more effective software programming modes are to be explored in the future.

ACKNOWLEDGMENT

This paper is financially supported by National Natural Science Foundation of China (No. 61403259) and Science and Technology Research and Development Foundation of Shenzhen (No. JCYJ20140418182819128)

REFERENCES

[1] MPI, MPI: A Message-Passing Interface standard. Message Passing Interface Forum, 1995, June, http://www.mpiforum.org/.

[2] OpenMP: Simple, Portable, Scalable SMP Programming. http://www.openmp.org/.

[3] Zhou Hongwei, Deng Rangyu, Dai Zefu, Yan Xiaobo, Zhang Ying, The virtual open page buffer for multi-core and multi-thread processors, Proceedings - 16th IEEE International Conference on High Performance Computing and Communications, HPCC 2014, P290-297

[4] Miachael J. Quinn, Parallel Programming in C with MPI and OpenMP ,Mc-Graw-Hill Education Press, USA, 2004, P323-347.

[5] L. Smith and M. Bull, Development of hybrid mode MPI/OpenMP applications, Scientific Programming, 2001, Vol. 9, No 2-3, P83-98..

[6] Si Min, Peña Antonio J., Balaji Pavan, Takagi Masamichi, Ishikawa Yutaka, MT-MPI: Multithreaded MPI for many-core environments[C], Proceedings of the International Conference on Supercomputing, 2014, P125-134

[7] Saillard Emmanuelle, Carribault Patrick,Barthou, Static/dynamic validation of MPI collective communications in multi-threaded[C]. context,Proceedings of the ACM SIGPLAN Symposium on Principles and Practice of Parallel Programming, PPOPP,2015-January, P279-280

[8] Dong Feng, Xu Xiatian, Zhang Xu, Parallel contingency analysis solution based on OpenMP[C], 2014 North American Power Symposium, NAPS 2014, November, P978-981.

[9] D. Pekurovsky, T. Kaiser and L. Nett, OpenMP and Hybrid-Model Performance Issues, presented at SciComp 2000,August, http://www.spscicomp.org/2000/.

[10] Mangual Osvaldo, Teixeira Marvi, Lopez-Roig Reynaldo, Hybrid MPI-OpenMP versus MPI implementations: A case study[C], ASEE Annual Conference and Exposition, Conference Proceedings, 2014, June.

BIOGRAPHIES

Ying Xu received the B.S. degree in control engineering from the Jilin University of Technology in 1997, and M.E. and Ph.D. degrees in electrical and computer engineering from the Jilin University in 2005 and 2009, respectively. From 1997 to 2013, she was an Assistant Professor with the State Key Laboratory of Automotive Simulation and Control, Jilin University, Changchun, China. Since 2013, she has been teaching with the College of Mechatronics and Control Engineering, Shenzhen University, Shenzhen, China. Her current research interests include information fusion, computer simulation, vehicle dynamic modeling, and automotive electrical and electronic control technology.

 **Tie Zhang** received the B.S. and Ph.D. degrees in mechanical design and theory from the South China University of Technology, Guangzhou, China, in 1996 and 2001, respectively.

Since 2001, he has been a Professor at the College of Mechanical and Automotive Engineering, South China University of Technology, Guangzhou, China. He current research interests include industrial robots technologies, service mobile robot, robotized equipment, Mechatronics integration equipment.

Comparison and Analysis of DBD-Ozonizers Powered by Current- and Voltage-Mode Power Supplies

Le Wang[1]   Xiongmin Tang[2]   Shuaijie Luan[3]   Sizhe Chen[4]

[1,2,3,4] School of Automation, Guangdong University of Technology
[1] E-mail: wanglelynn@163.com
[2] E-mail: tangxiongmin@126.com
[3] E-mail: luanshuaijie@163.com
[4] E-mail: cszscut@126.com

Abstract–Equivalent parameters and power of dielectric barrier discharge (DBD) ozonier powered by both current-and voltage-mode power supplies are researched in the paper. The relationship among equivalent parameters, discharge power and the voltage(or current) power supply are achieved by QV Lissajous graphics .Compared to DBD type ozonier powered by voltage-mode power supply, DBD-ozonizer powered by current-mode power supply has smaller change range of the equivalent parameters but the higher discharge power under the same conditions. So the current- mode power supply is more suitable for DBD type ozonier.

Keywords–Current–mode power supply, dielectric barrier discharge, load characteristics, voltage-type power supply

I. INTRODUCTION

Dielectric barrier discharge (DBD) is a gas-discharge form widely used in the large-scale industrial [1,2]. It can generate low temperature plasma that has high energy at atmospheric pressure and widely used in surface modification of materials, excimer light sources, environmental protection and ozone synthesis, so it is significant to study the load characteristics and the equivalent parameters of DBD [3,5].

Due to the complexity of the DBD discharge mechanism, discharge power of dielectric barrier discharge type ozonizer is researched with different power supply topologies [6,7,8]. In power supply, voltage-mode inverter is researched by the majority of researchers and manufacturers because of its simple structure and control. In the Literature [9,10], the influence of voltage and frequency on the DBD type load equivalent parameters is analyzed under the voltage-mode series resonant inverter power supply. In contrast, the current-mode power supply is used gradually by researchers because of light current impact, load adaptability and other advantages. Investigations in application of DBD type load in the conditions of current-mode power supply are conducted but only the design of power supply is analyzed. The effects of current-mode and voltage-mode power supply on load characteristics of DBD are not studied. So it is necessary to analyze load characteristics of the DBD type load with current-mode and voltage-mode power supply. Because DBD type ozonizer is a typical DBD load, the DBD type ozonizer is adopted in the experiment system.

II. EXPERIMENTAL SYSTEM

The structures of the experimental system are shown in Fig.

1. The physical and structure of ozone generator is given in Fig.2, the value of the test capacitor is 2.2nF. The High frequency step-up transformer T is made of Mn-Zn ferrite core, and the ratio is N=1:17. The ratio of current transformer CT is 1:20. A622 TEK current probe, HVP-15HF high voltage probe and TDS2012B TEK scope is used.

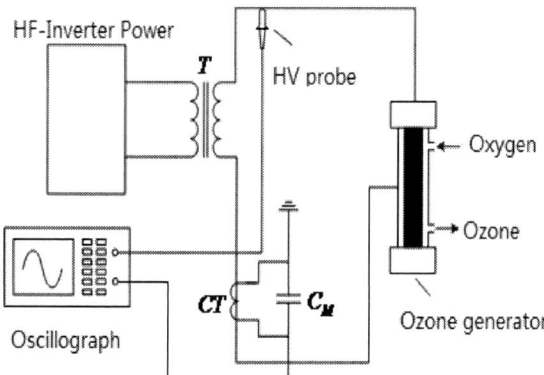

Fig .1: The structures of experimental system

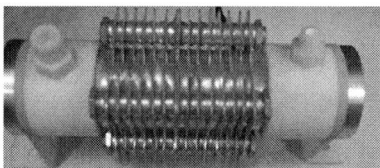

Fig. 2: The DBD ozone generator

The conditions for experiment are as follows. The power supply switching frequency is 10 kHz, the concentration quotient of oxygen is 99.2%, the gas volume flow is 1 L/min, the pressure is 101kPa and the ambient temperature is 20 ± 3℃.

III. TOPOLOGICAL STRUCTURE AND WORKING WAVE FORM OF POWER SUPPLY

The topological diagram of voltage-type inverter power supply is presented in Fig. 3 (a). The U_{1d} is the input DC voltage and the $S_{11} \sim S_{14}$ IGBT form a full bridge inverter circuit. The topological diagram of current-mode inverter is shown in Fig. 3 (b), The U2d is input DC voltage, L is a inductor and it's valued is 4mH, Each bridge arm is composed of fast recovery diode $D_{21} \sim D_{24}$ and $S_{21} \sim S_{24}$ MOSFET, the output current of the inverter is square wave current, which is continuous adjustable. The inverter

978-1-5090-0064-7/15 $31.00 © 2015 IEEE

controls the voltage across the DBD type ozonizer by DC adjusting DC voltage and fixed value of operating frequency are 10 kHz. The waveforms of inventers are shown in Fig. 4.

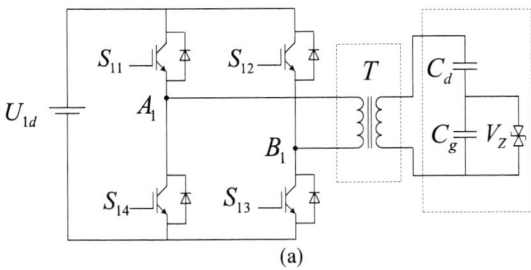

(a)

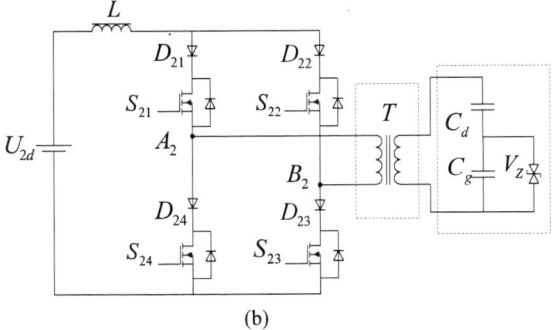

(b)

Fig. 3: Topological structure of power supply,(a) voltage- mode power supply, (b) current-mode power supply

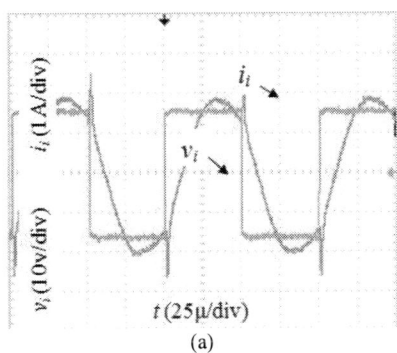

(a)

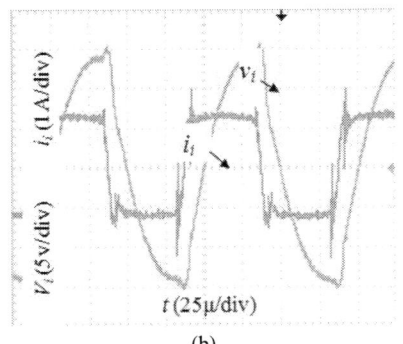

(b)

Fig. 4: waveforms of power supply, (a) voltage- mode power supply, (b) current-mode power supply

IV. PARAMETER MEASUREMENT METHOD OF DBD TYPE LOAD

The typical Lissajous figure for DBD type ozone generator is shown in Fig. 5, u_{cm} is the Terminal voltage of test capacitor, u_c is the voltage across the ozone tube. The discharge stage is AB and CD and the non discharge stage corresponds to AD and BC, respectively. The equivalent parameters are:

$$C_d = K_{AB}C_M$$

$$C_p = K_{DA}C_M$$

$$C_g = \frac{C_d C_p}{C_d - C_p}$$

According to the Literature [4], in each work cycle, the discharge power is Q-V loop area, i.e.

$$P = fSC_M$$

where C_M denotes the test capacitor, f represents the power operating frequency, and S is the Lissajous graphics area.

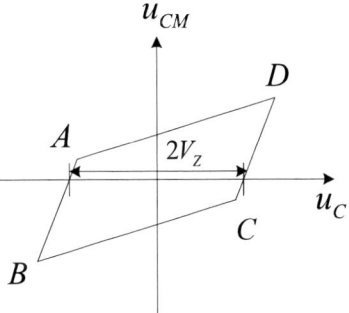

Fig. 5: Lissajous graphics for DBD type ozone generator

V. LISSAJOUS GRAPHICS UNDER DIFFERENT CONDITIONS

In the experiment, the voltage regulation of the discharg tu be is realized by adjusting the DC input voltage of the inve rter and adjustable rang is the normal working voltage for discharge tube V_{pp}=3~7.5kV. In order to reduce the test error caused by accidental factors, the test data in the test are stable running 10min after the test results. The Lissajous graphics in the condition of two styles of power supplies are shown in Fig. 6 and Fig. 7, respectively.

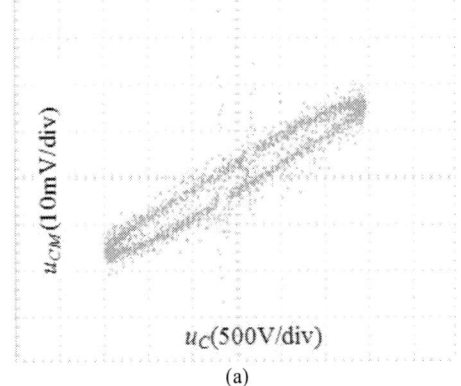

(a)

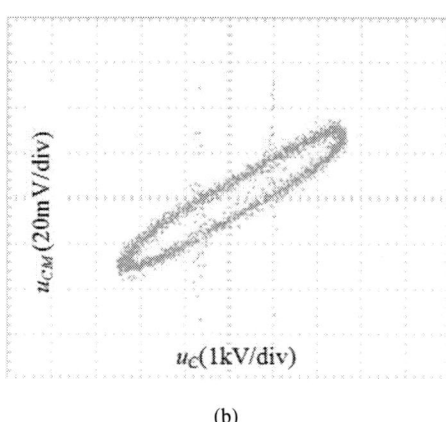

(b)

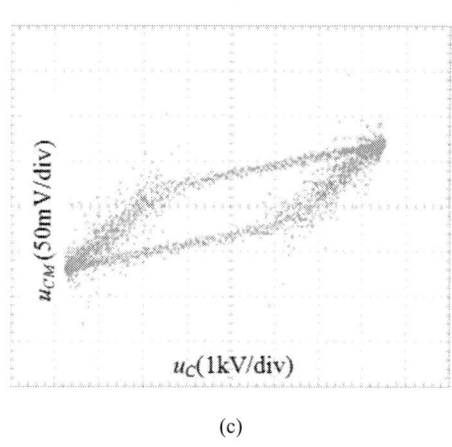

(c)

Fig. 6: Q-V Lissajous graphics with the Voltage-type power supply,(a) V_{pp} =3kV, (b) V_{pp} =5kV,(c)=7kV

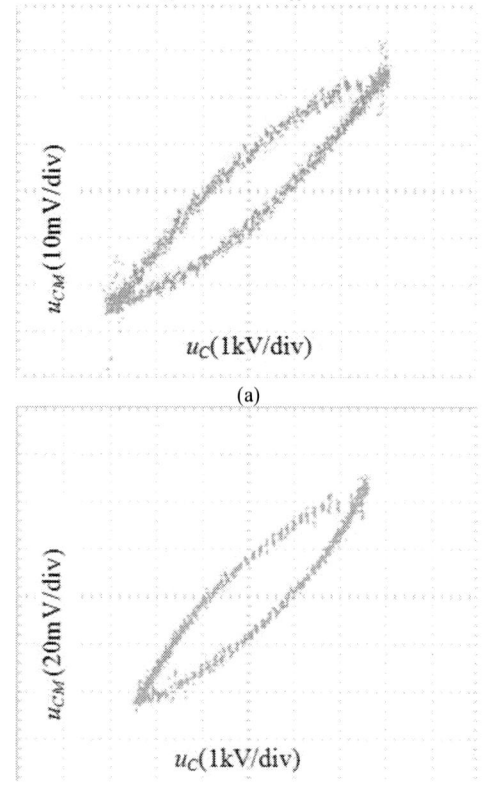

(a)

(b)

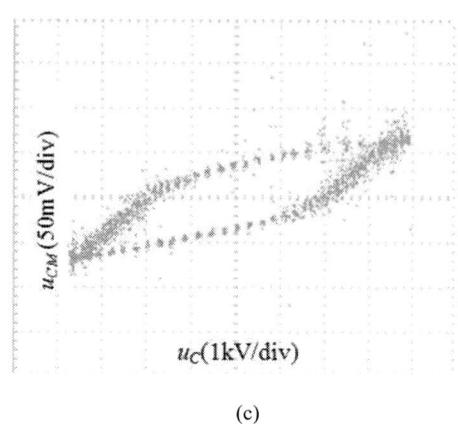

(c)

Fig. 7: Q-V Lissajous graphics with the Current-mode power supply, (a) V_{pp} =3kV, (b) V_{pp} =5kV, (c) V_{pp} =7kV

VI. EXPERIMENTAL RESULT ANALYSIS

With the adjustment of V_{pp} , the curves of equivalent C_g , C_d , C_p and V_z are made in Fig. 8-Fig.11.

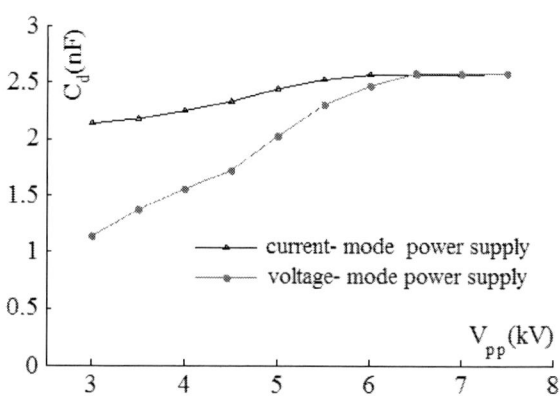

Fig. 8: the curve of between C_d and V_{pp}

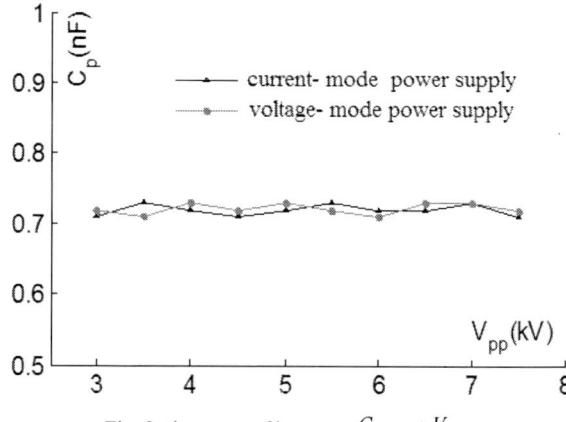

Fig. 9: the curve of between C_p and V_{pp}

978-1-5090-0064-7/15 $31.00 © 2015 IEEE   220

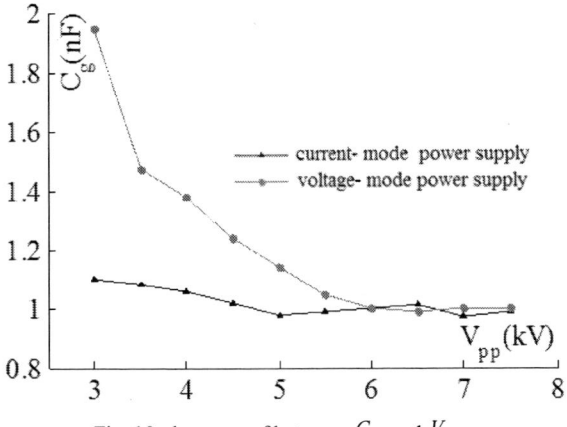

Fig. 10: the curve of between C_g and V_{pp}

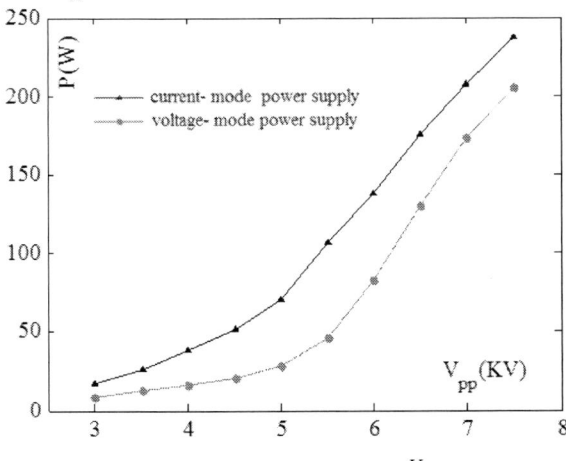

Fig. 11: the curve of between V_z and V_{pp}

The relationship between the discharge power P and the applied voltage V_{pp} are shown in Fig. 12. With the increasing of V_{pp}, the discharge gap voltage increases, the number of the discharge channels increases, the charge ability is enhanced, the discharge power P is increased. In a certain range, the discharge power P of both voltage- and current-mode power supply increases linearly along with the increase in applied voltage. However, the discharge power is higher when the current-mode power supply is used with the same V_{pp}.

Fig. 12: the curve of between P and V_{pp}

It is shown that the discharge channel is easy to be established under current-mode power supply, and the influence of the applied voltage across DBD type ozonizer is smaller as shown in Fig.12. So current-mode power supply is recommended for DBD-ozonizer.

VII. CONCLUSIONS

Comparison and analysis of DBD type ozonizer powered by current- mode and voltage- mode power supply are made in the paper. The experimental results show that the discharge power of current-mode source is higher than that of voltage-mode source at the same input voltage. It means that the current-mode power supply is more suitable for DBD type load

ACKNOWLEDGMENT

The authors grateful acknowledge the financial support from National Natural Science Foundation of China under the project numbers 51207026, 51307025, and Natural Science Foundation of Guangdong Province under the project number 2015A030313487.

REFERENCES

[1] Jodzis S, Smoliński T, Sówka P. Ozone Synthesis Under SurfaceDischarges in Oxygen Application of a Concentric Actuator[J]. IEEE Transactions on Plasma Science, 2011, 39(4): 1055-1060.

[2] J Marcos Alonso, Carlos Ordiz, Dalla Costa, et al. High-voltage Power Supply for Ozone Generation Based on Piezoelectric Transformer[J]. IEEE Transactions on Industry Application, 2009, 45(4): 1513-1523 .

[3] Amjad M, Salam Z . Analysis Design and Implementation of Multiple Parallel Ozone Chambers for High Flow Rate [J]. IEEE Transactions on Industrial Electronics , 2014, 61(2): 753-756

[4] Amjad M, Salam Z, Facta M, et al. A Simple and Effective Method to Estimate the Model Parameters of Dielectric Barrier Discharge Ozone Chamber. IEEE Transactions on Instrumentation and Measurement, 2012, 61(6): 1676-1683.

[5] Alonso JM, Rico Secades M. Low Power High-voltage High-frequency Power Supply for Ozone Generation[C]. Industry Applications Conference, 2002,13:(18).

[6] J Marcos Alonso, Jorge Garcia, Antonio, et al. Analysis Design and Experiment of a High Voltage Power Supply for Ozone Generation Based on Current fed Parallel Resonant Push-pull Inverter[J]. IEEE Transactions on Industry Applications 2005, 41(5): 1364-1371.

[7] Ordiz C, Alonso JM, Dalla, et al. Development of a High-voltage Closed Loop Power Supply for Ozone Generation, Applied Power Electronics Conference and Exposition, 2008, 24(28): 1861-1867.

[8] Vijit Kinnares, Prasopchok Hothongkham. Circuit Analysis and Modeling of a Phase-shifted Pulse-width Modulation Full-bridge Inverter-fed Ozone Generator With Constant Applied Electrode Voltage[J]. IEEE Transactions on Power Electronics, 2010, 25(7): 1739-1751.

[9] Shengpei Wang, Nakaoka M, Konishi Y. PDM and PWM Hybrid Power Control of a Voltage-source Type High-frequency Inverter for Ozonizer Applications. Power Electronics and Letter Symbols for Quantities, ANSI Standard Y10.5-1968

[10] Koudriavtsev O, Konishi Y, Nakaoka M. A Novel Pulse-density Modulated High Frequency Inverter for Silent Discharg Type Ozonizer. IEEE Transactions on Industry Applications, 2002, 38(2): 369-378.

BIOGRAPHIES

Le Wang is a graduate student at Guangdong University of Technology, the main research direction is the power electronics.

Xiongmin Tang received the B.Sc., M.Sc., and Ph.D. degrees all from the Hunan University, Changsha, China, in 1999, 2004, and 2007, respectively. Dr. Tang currently is an Associate Professor in the School of Automation, Guangdong University of Technology, Guangzhou, China. His main research interests include power electronics technology, special power supply for plasma generation and its application.

Shuaijie.Luan is a graduate student at guangdong Univerisity of Technology, the main research direction is the power electronics.

Sizhe Chen received the B.Sc. and Ph.D. degrees from the South China University of Technology, Guangzhou, China, in 2005 and 2010, respectively. Dr. Chen currently is an Associate Professor in the School of Automation, Guangdong University of Technology, Guangzhou, China. His general research interests include the control and power electronics technology in renewable energy.

Research of Phase-Shift Series-Load Resonant Power Supply for DBD-Ozone Generator

Shuaijie Luan[1]   Xiongmin Tang[2]   Le Wang[3]   Sizhe Chen[4]

[1,2,3,4] School of Automation, Guangdong University of Technology
[1]E-mail: luanshuaijie@163.com
[2]E-mail: tangxiongmin@126.com
[3]E-mail: wangleynn@163.com
[4]E-mail: cszscut@126.com
[4]E-mail: cszscut@126.com

Abstract–In this paper, the power supply of a full bridge series-load resonant DBD type ozone generator is analyzed in detail under phase-shift control. The working modes of the circuit in two main states is obtained by combining the on-off state of power switches and the equivalent circuit of DBD type ozone generator. The experimental results indicate that: the series-load resonant power supply of ozone generator not only can realize the soft switching, but also can adjust the discharge power of ozone generator by changing the phase-shift angle under phase-shift control strategy. The series-load resonant power supply is therefore very suitable for ozone generator.

Keywords–DBD, ozone generator, phase-shift control, power supply, series resonant

I. INTRODUCTION

Dzone is an allotrope of oxygen，it is a kind of Green Oxidant with strong oxidizability and bactericidal performance. So it can be used in the sewage treatment, air purification, medical pharmaceutical, food processing, textile bleaching, and many other areas. At present, there are many methods to produce ozone, and the method of Dielectric Barrier Discharge (DBD) has an absolute advantage over other methods, so it is the main method of producing ozone in industry. Because the DBD type ozone generator contains the insulating medium and discharge gap, the discharge phenomenon is different from the ordinary discharge. It is generally considered that when the DBD circuit is stable, it contains two different working modes in a full cycle, which are called non discharge and discharge. The study shows that the discharge phenomenon of the gas in the discharge tube is related to the applied voltage and field strength of air gap at medium and low frequency (50Hz~10 kHz) condition. When the circuit is in the non-discharge stage, the electric field intensity in the air gap is not enough to make the gas breakdown, so the ozone generator can be equivalent to the circuit which dielectric capacitor(C_d) is in series with the air gap capacitor(C_g). With the increase of the applied voltage, field strength of air gap is increasing too, when it increases to a certain value, air gap capacitance will be breakdown, and at the same time the voltage between air gap is essentially unchanged, the voltage also known as discharge sustain voltage (V_z), and its equivalent circuit can be replaced by a voltage source in the opposite direction of input voltage or a zener diode in the state of reverse breakdown.

Power supply is the main component of DBD ozone generator, and the voltage, frequency and waveform of the power supply have great influence on the discharge efficiency of the ozone generator. The performance and quality of the power supply system will be the key to the ozone yield when the structure, gas source, gas flow rate and cooling system of the generator are determined. Voltage mode and current mode load resonant inverter power supply are two kinds of power supply modes that most widely used and researched in ozone generator field. These two topologies of power supply are mostly transplanted from induction heating circuit, and the capacitive characteristic of the load is fully utilized during the design, through connecting with compensation inductance, so that the ozone generator load works in the resonant or quasi resonant state, which is conducive to reduce the harmonic pollution and improve the efficiency of ozone generator.

II. TOPOLOGY STRUCTURE AND PHASE-SHIFT CONTROL STRATEGY OF SERIES RESONANT POWER SUPPLY

Fig. 1 shows the topology of series resonant power supply. U_d is a DC input power supply; T is as a step-up transformer; L_s is a series compensation inductors (including the leakage inductance of the boost transformer);S_1~S_4 are as power switches (IGBT or MOSFET), and D_1~D_4 are the corresponding anti-parallel diode; R_e and C_e are the equivalent resistance and equivalent capacitance of the load of ozone generator; U_{ab} is the inverter output voltage, I_o is the inverter output current.

In the industrial production, it is often needed to adjust the ozone concentration according to the field condition. That is, the power of the ozone generator system can be adjusted. When the ozone generator system uses a series resonant inverter power supply, compared with other power adjusting strategy, the phase-shift control strategy has incomparable advantages, and the control strategy is used in power supply for inductor heating.

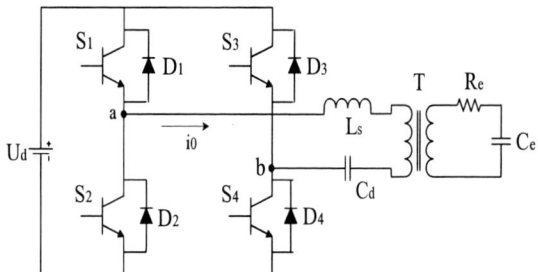

Fig. 1: Series resonant inverter for ozone generator.

Phase-Shift Control (PSC) is a kind of PWM control strategy. In the control strategy, the driving signal of diagonal pairs switches in the inverter bridge legs (S_1, S_4) and (S_2, S_3) should have a stagger angle β, which called phase- shift angle, and the driving signal of the two switches (S_1, S_2) and (S_3, S_4) in the same bridge is complementary. So a zero voltage interval is inserted between the positive and negative edges of the inverter output voltage. The phase shifting angle β is adjustable in the range of $0^0 \sim 180^0$, and the duty ratio of the inverter output voltage can be adjusted by adjusting the β, then the effective value of the output voltage can be changed, and power regulation is achieved ultimately. The principle of the phase-shift control strategy is shown in Fig 2, where $U_{gs1} \sim U_{gs4}$ are the driving signals of the switches $S_1 \sim S_4$, and U_o is the inverter output voltage.

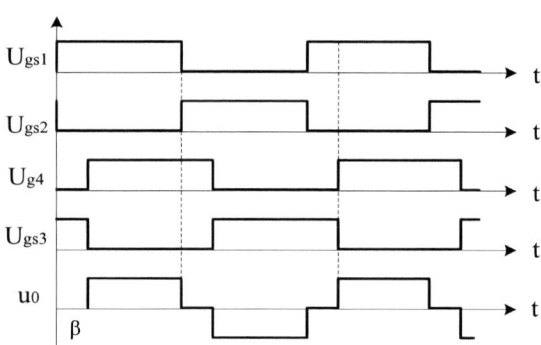

Fig. 2: Principle of phase-shift PWM modulation

III. THE WORKING STATES ANALYSIS WITH PHASE SHIFTING CONTROL

When the series resonant power supply for ozone generator is under phase-shift control, there are three kinds of working modes, the different working states are shown in Fig 3 (a) ~(c) respectively. In mode A, all the switches are zero- voltage turn-on; In mode B, switches of the leading leg and the lagging leg are zero-voltage turn-on and zero-current turn-off respectively; in mode C, although all of the switches are zero-current turn-off, the current must be in a non zero state at the time of turn-on, and the reverse recovery of the diode is inevitable. In order to reduce the switching losses and improve power efficiency, the series resonant power supply for ozone generator usually works in mode A or mode B.

In order to analyze the various working modes of the state A and state B, the equivalent circuit of the load of ozone generator is converted to the primary side of the transformer, then $R=n^2 R_e$, $C=C_e/n^2$. During the analysis, the following conditions should be satisfied:

- The power supply of ozone generator has been working in the stable state.
- The switching devices are ideal.
- The impact of the blocking capacitance C_d is ignored.
- The impact of the dead-time is ignored.
- The positive direction of the inverter output current is $U_d^+ \rightarrow S_1 \rightarrow S_4 \rightarrow U_d^-$.

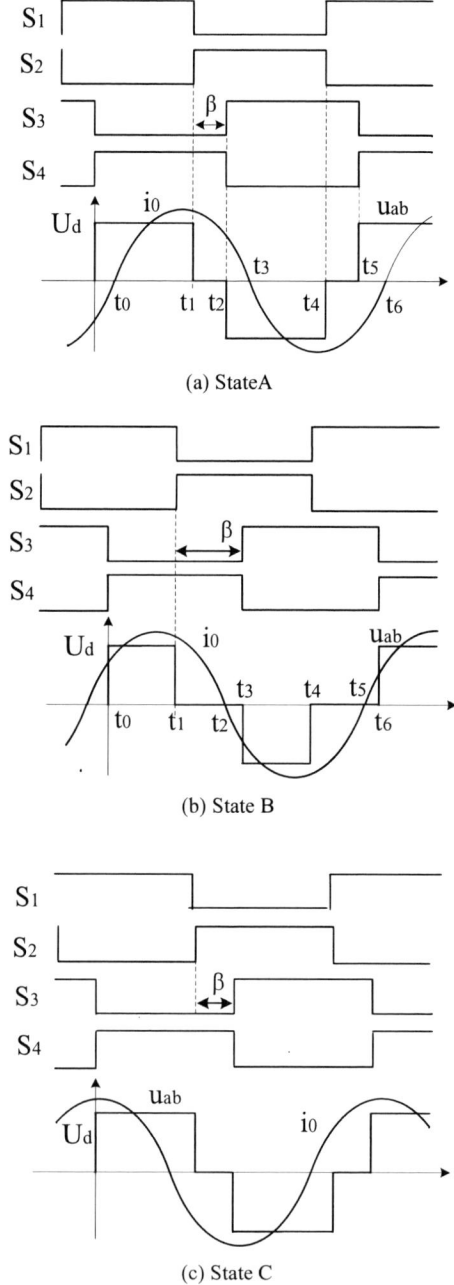

Fig. 3: Working states under phase-shift control strategy

1. Working modes in state A

Mode 1 (t_0~t_1): Before t_0 , the inverter output current i_0 is negative and flows along D_1 and D_4 . The voltage across the switch tube S_1 and S_4 is clamped to zero. At t_0 , the direction of i_0 is changed, and because the positive trigger pulse is applied to S_1 and S_4 , they switch on ZVS. The equivalent circuit of the power supply for ozone generator in the mode is shown in Fig 4 (a) with bolt lines.

Mode 2 (t_1~t_2): At t_1, S_3 switches but S_1 turns off, and S_4 maintains the conducting state. Because the inverter output current is positive, the inductor current cannot be mutated and D_2 starts to flow through, then i_0 continues to decrease. The equivalent circuit of the power supply for ozone generator in the mode is shown in Fig 4 (b) with bolt lines.

Mode 3 (t_2~t_3): At t_2 , S_4 turns off and S_1 is applied to the positive trigger pulse. Because the direction of i_0 maintains positive and S_3 does not satisfy the condition of switching, the inverter output current flow through the loop of $U_d^- \rightarrow D_2 \rightarrow D_3 \rightarrow U_d^+$ in the mode. The equivalent circuit of the power supply for ozone generator in the mode is shown in Fig 4 (c) with bolt lines.

Mode 4 (t_3~t_4): At t_3, i_0 drops to zero, because the positive trigger pulse is applied to S_2 and S_3, they switch on ZVS. After t_3 , the DC input power supply starts supplying the load through S_2 and S_3, and i_0 is changed from zero to negative according to the law of resonance. The equivalent circuit of the power supply for ozone generator in the mode is shown in Fig 4(d) with bolt lines.

Mode 5 (t_4~t_5): At t_4, S_2 turns off and the positive trigger pulse is applied to S_1, and S_4 maintains the conducting state. Because the inverter output current is negative, the inductor current cannot be mutated and D_1 starts to flow through. The equivalent circuit of the power supply for ozone generator in the mode is shown in Fig 4(e) with bolt lines.

Mode 6 (t_5~t_6): At t_5, S_3 turns off and S_4 is applied to the positive trigger pulse, though S_1 have long been applied to the positive trigger pulse, i_0 maintains negative that moment, then S_1 does not satisfy the switching condition, so i_0 flows through the loop of $U_d^- \rightarrow D_4 \rightarrow D_1 \rightarrow U_d^+$ in the mode until i_0 turns to zero again at t_6. The equivalent circuit of the power supply for ozone generator in the mode is shown in Fig 4 (f) with bolt lines.

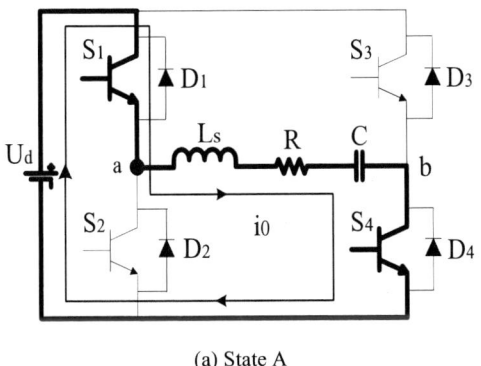

(a) State A

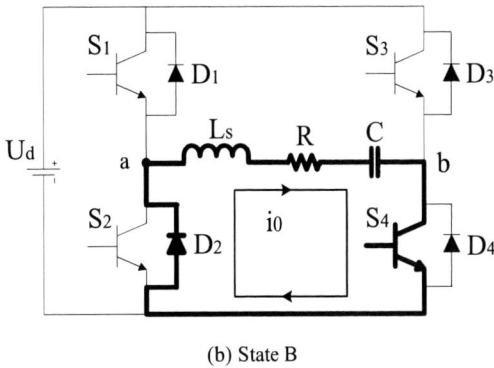

(b) State B

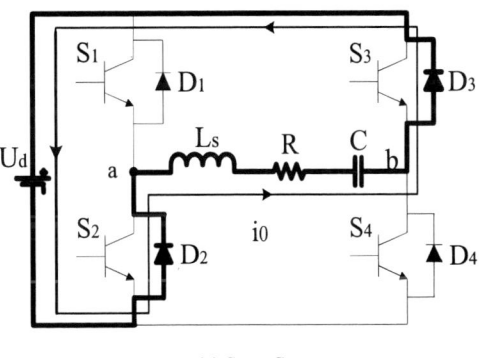

(c) State C

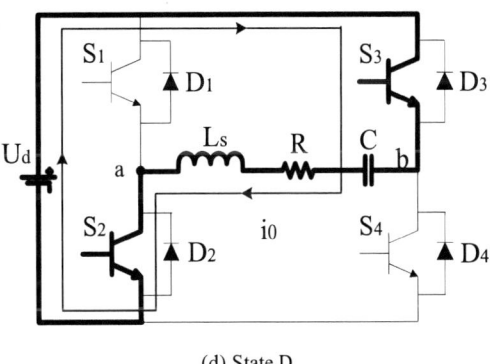

(d) State D

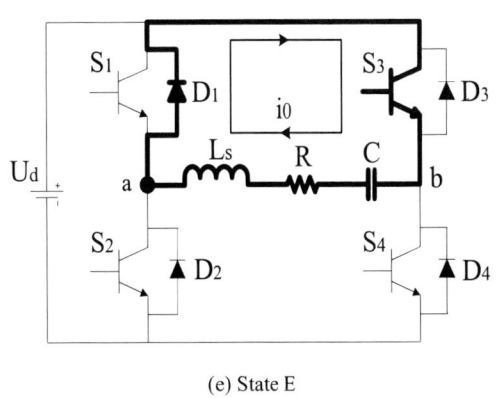

(e) State E

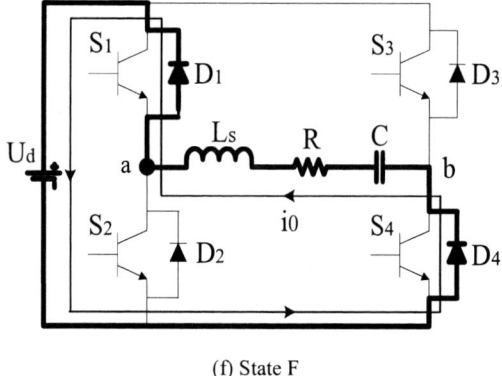

(f) State F

Fig. 4: Equivalent circuits of each mode in state A

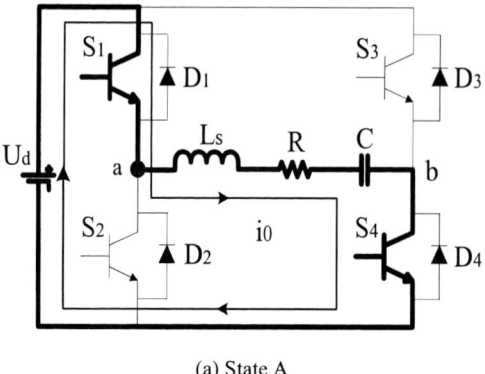

(a) State A

2. Working modes in state B

Mode 1 ($t_0 \sim t_1$): S_1 was triggered before t_0 and S_4 switches at t_0. Because the direction of i_0 is positive before t_0, S_4 switches in large current. After t_0, the DC input power supply starts supplying the load along the circuit of $U_d^+ \to S_1 \to S_4 \to U_d^-$ until S_1 turns off at t_1. The equivalent circuit of the power supply for ozone generator in the mode is shown in Fig 5 (a) with bolt lines.

Mode 2 ($t_1 \sim t_2$): At t_1, S_1 turns off and S_2 is applied to the trigger pulse, Because i_0 maintains positive that moment, S_2 does not satisfy the switching condition, then D_2 starts to flow through and i_0 decreases constantly. The equivalent circuit of the power supply for ozone generator in the mode is shown in Fig 5(b) with bolt lines.

Mode 3 ($t_2 \sim t_3$): At t_2, i_0 drops to zero and S_2 switches on ZVS. D_4 starts to flow through. The equivalent circuit of the power supply for ozone generator in the mode is shown in Fig 5(c) with bolt lines.

Mode 4 ($t_3 \sim t_4$): At t_3, S_4 turns off and S_3 is applied to the positive trigger pulse. Because the direction of i_0 is already negative, S_3 satisfy the switching conditions. So the DC input power supply starts supplying the load along the circuit of $U_d^+ \to S_3 \to S_2 \to U_d^-$. The equivalent circuit of the power supply for ozone generator in the mode is shown in Fig 5 (d) with bolt lines.

Mode 5 ($t_4 \sim t_5$): At t_4, S_2 turns off and S_1 is applied to the positive trigger pulse. Though the positive trigger pulse is applied to S_4, it does not satisfy the switching conditions because the direction of i_0 maintains negative.Then i_0 goes through D_1. The equivalent circuit of the power supply for ozone generator in the mode is shown in Fig 5(e) with bolt lines.

Mode 6 ($t_5 \sim t_6$): At t_5, i_0 changes from negative to positive and S_1 satisfy the switching conditions. Then S_1 switches on ZVS. The direction of i_0 turns positive after t_5, and then D_3 starts to flow through. The equivalent circuit of the power supply for ozone generator in the mode is shown in Fig 5 (f) with bolt lines.

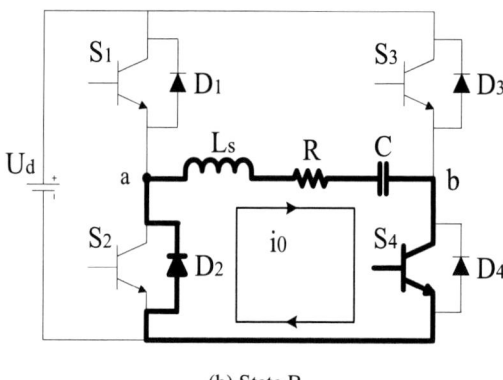

(b) State B

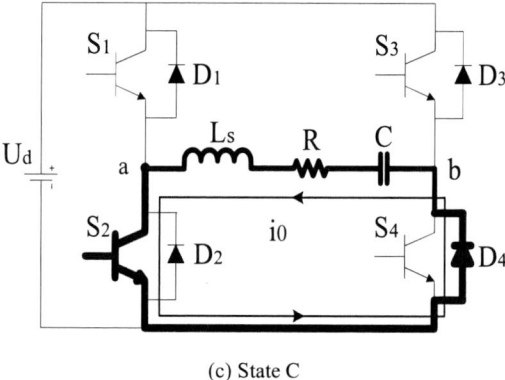

(c) State C

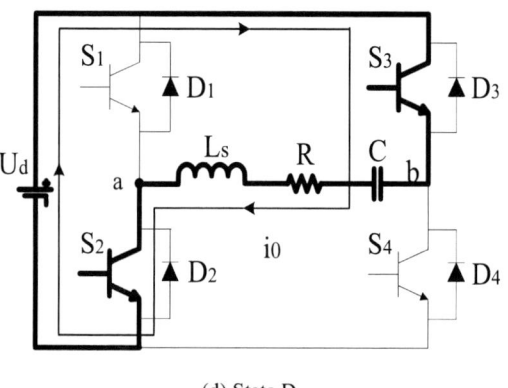

(d) State D

978-1-5090-0064-7/15 $31.00 © 2015 IEEE

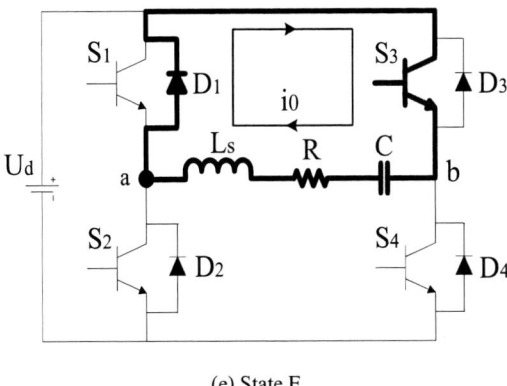

(e) State E

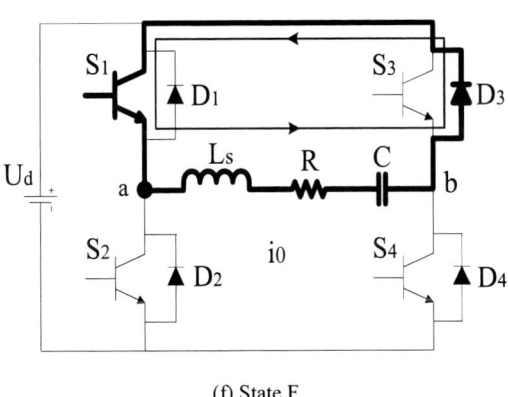

(f) State F

Fig. 5: Equivalent circuits of each mode in state B

IV. EXPERIMENTAL RESULTS

Fig. 6 shows the driving signals of the switches with phase shifting control. The dead time is set to 20us.

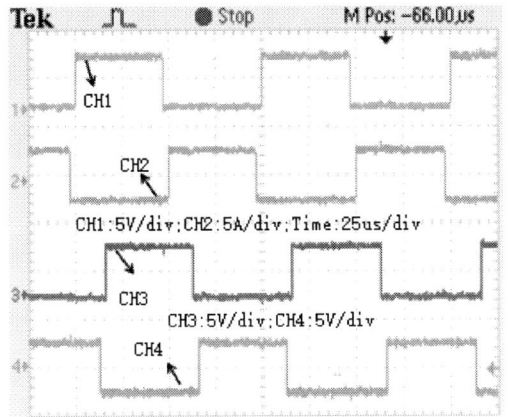

Fig. 6: Driving signal of all the switches

The equivalent resistance of the ozone generator (R_e) is about 15 KΩ, the equivalent capacitance is about 0.5nF, and the leakage inductance is about 1.48mH. And the resonant frequency of the load is about 10 kHz. The output voltage and output current waveform of the inverter when phase shift angle is zero are shown in Fig 7(a), and the corresponding current and voltage waveforms of the ozone generator are shown in Fig 7(b).

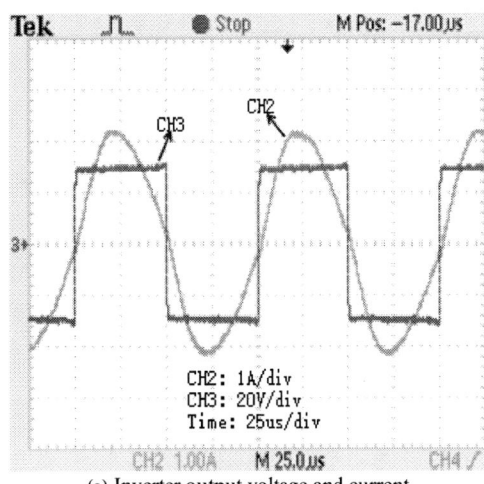

(a) Inverter output voltage and current

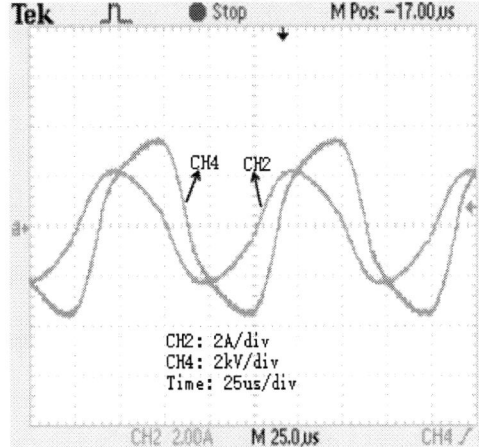

(b) voltage and current of the Ozone generator

Fig. 7: Related electrical waveforms at the resonant frequency

Because the load of the DBD type ozone is capacitive, in order to make the switches of inverter work in the ZVS state, the power supply should works in the inductor state. So the switching frequency is slightly greater than the resonance frequency. The waveforms of the voltage across the ozone generator, the output voltage and current of inverter and the current through the ozone generator are shown in Figs. 8~10, respectively, in different angle β. It is indicated that all the switches switch on ZVS.

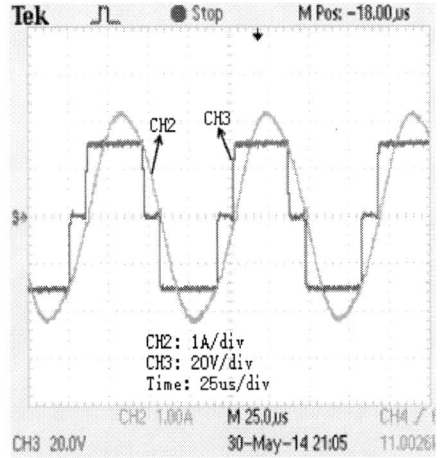

(a)Inverter output voltage and current

978-1-5090-0064-7/15 $31.00 © 2015 IEEE

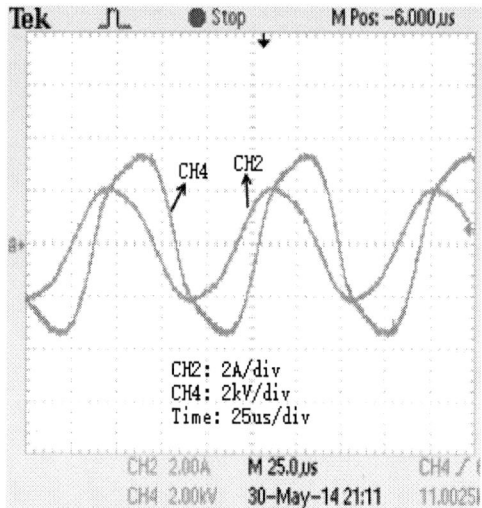

(b) voltage and current of Ozone generator

Fig. 8: Related electrical waveforms at 11KHz and $\beta=0^0$

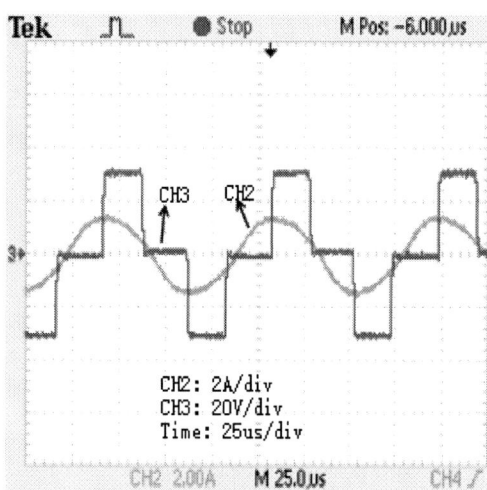

(a) Inverter output voltage and current

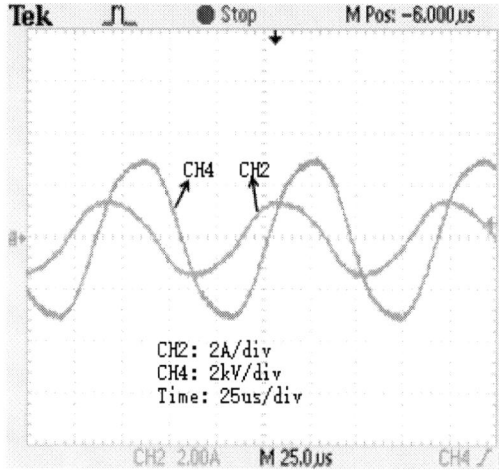

(b) voltage and current of Ozone generator

Fig. 9: Related electrical waveforms at 11KHz and $\beta=90^0$

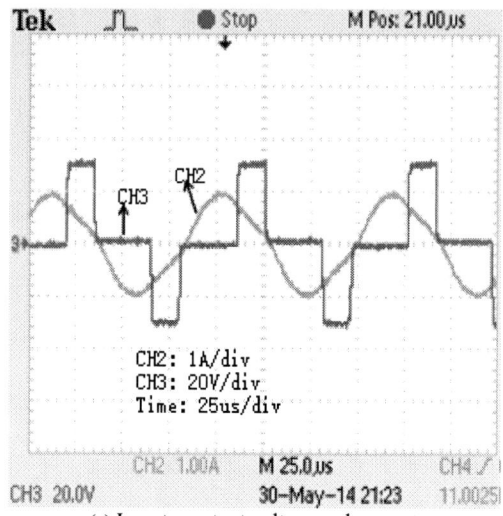

(a) Inverter output voltage and current

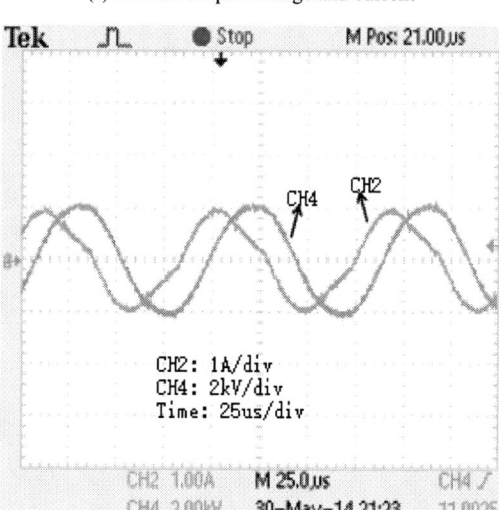

(b) voltage and current of Ozone generator

Fig. 10: Related electrical waveforms at 11KHz and $\beta=120^0$

V. CONCLUSION

With the equivalent resistance-capacitance (RC) model of the DBD type ozone generator, all working modes of the series-load resonant power supply for ozone generator are analyzed in detail and verified by experiments. The experimental results indicate that: the series-load resonant power supply of ozone generator not only can realize the soft switching, but also can adjust the discharge power by changing the phase-shift angle under phase-shift control strategy.

ACKNOWLEDGMENT

The authors grateful acknowledge the financial support from National Natural Science Foundation of China under the project numbers 51207026, 51307025, and Natural Science Foundation of Guangdong Province under the project number 2015A030313487.

REFERENCES

[1] Jodzis S, Smoliński T, Sówka P. Ozone Synthesis Under Surface Discharges in Oxygen Application of a Concentric

Actuator[J]. IEEE Transactions on Plasma Science, 2011, 39(4): 1055-1060.

[2] J Marcos Alonso, Carlos Ordiz, Dalla Costa, et al. High-voltage Power Supply for Ozone Generation Based on Piezoelectric Transformer[J]. IEEE Transactions on Industry Application, 2009, 45(4): 1513-1523.

[3] Zhen Liu, Wen Jiang, Wang Hai Cheng, et al. A Novel ZVS Double Switch Flyback Inverter and Pulse Controlled Dimming Methods for Flat DBD Lamp[J]. IEEE Transactions on Consumer Electronics, 2011, 57(3): 995 -1002.

[4] J Marcos Alonso, Jorge Garcia, Antonio, et al. Analysis Design and Experiment of a High Voltage Power Supply for Ozone Generation Based on Current fed Parallel Resonant Push-pull Inverter[J]. IEEE Transactions on Industry Applications 2005, 41(5): 1364-1371.

[5] Ordiz C, Alonso JM, Dalla, et al. Development of a High-voltage Closed Loop Power Supply for Ozone Generation, Applied Power Electronics Conference and Exposition, 2008, 24(28): 1861-1867.

[6] Vijit Kinnares, Prasopchok Hothongkham. Circuit Analysis and Modeling of a Phase-shifted Pulse-width Modulation Full-bridge Inverter-fed Ozone Generator With Constant Applied Electrode Voltage[J]. IEEE Transactions on Power Electronics, 2010, 25(7): 1739-1751.

[7] Amjad M, Salam Z. Analysis Design and Implementation of Multiple Parallel Ozone Chambers for High Flow Rate[J]. IEEE Transactions on Industrial Electronics，2014, 61(2): 753-756.

[8] Amjad M, Salam Z, Facta M, et al. A Simple and Effective Method to Estimate the Model Parameters of Dielectric Barrier Discharge Ozone Chamber. IEEE Transactions on Instrumentation and Measurement, 2012, 61(6): 1676-1683.

[9] Alonso JM, Rico Secades M. Low Power High-voltage High-frequency Power Supply for Ozone Generation[C]. Industry Applications Conference, 2002, 13(18):

[10] Koudriavtsev O, Konishi Y, Nakaoka M. A Novel Pulse-density Modulated High Frequency Inverter for Silent Discharg Type Ozonizer. IEEE Transactions on Industry Applications, 2002, 38(2): 369-378.

[11] Shengpei Wang, Nakaoka M, Konishi Y. PDM and PWM Hybrid Power Control of a Voltage-source Type High-frequency Inverter for Ozonizer Applications. Power Electronics and Variable Speed Drives, 1998, 40(45): 21-23 .

BIOGRAPHIES

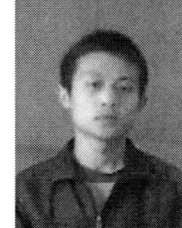

Shuaijie Luan is a graduate student at Guangdong University of Technology, and the main research field is power electronics.

Xiongmin Tang received the B.Sc., M.Sc., and Ph.D. degrees all from the Hunan University, Changsha, China, in 1999, 2004, and 2007, respectively. Dr. Tang currently is an Associate Professor in the School of Automation, Guangdong University of Technology, Guangzhou, China. His main research interests include power electronics technology, special power supply for plasma generation and its application.

Le Wang is a graduate student at Guangdong University of Technology, and the main research field is power electronics.

Sizhe Chen received the B.Sc. and Ph.D. degrees from the South China University of Technology, Guangzhou, China, in 2005 and 2010, respectively. Dr. Chen currently is an Associate Professor in the School of Automation, Guangdong University of Technology, Guangzhou, China. His general research interests include the control and power electronics technology in renewable energy.

A Control Method for Permanent-Magnet Synchronous Motor with Unbalanced Cable Resistor

Zou Xunhao, Shenghua Huang, Qin Zhuqian, Xiaodong Hu, Kan Guangqiang

State Key Laboratory of Advanced Electromagnetic Engineering and Technology (Huazhong University of Science and Technology), China
E-mail: zouxh1992@qq.com

Abstract — The three phase cable connecting the inverter and PMSM may be unbalanced in some situation such as the deep drilling platform. This paper presents the mathematical model of PMSM with unbalanced cable resistance. The traditional PI current regulator cannot work well within the situation. The hysteresis band controller can operate well because it's independent of load parameters but the switching frequency may be very high to get good performance, which is not suitable. So a novel proportional integral resonant controller is proposed to eliminate the harmonic current caused by the asymmetry of the stator resistance. From the results of the experiment, the mentioned control strategy proves to be appropriate for PMSM with unbalanced cable resistance.

Keywords–PMSM, unbalanced resistor, resonant controller

I. INTRODUCTION

The resistance of the cable connecting an inverter and the load is not taken into account in most occasions. However, in some special occasions such as the deep drilling platform, the wire connecting the PMSM and inverter can reach as long as several kilometers. As shown in Fig.1, one phase of the PMSM is connected to the inverter through the pipe wall and the other phases are connected with copper wire. So the resistance of the load which includes both the motor and cable could be unbalanced among the three phases.

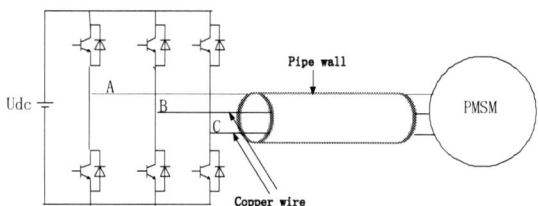

Fig.1 deep drilling platform schematic

This paper presents the mathematical model of the PMSM with unbalanced stator resistor is proposed since the cable resistor unbalance is equivalent to the stator resistance unbalance. The side effect caused by the asymmetry is analyzed and two methods are proposed in the following section. A hysteresis band controller can perform well because it's independent with load parameters[1], but the switching frequency and the current harmonics is a paradox. Another method is to add a resonant term in parallel with the traditional PI controller. Resonant term can help eliminate sinusoidal disturbance and has been applied in many area such as grid connected converter and motor control[2-8]. It can be found that the PIR controller can reach good performance in a constant acceptable switching frequency.

II. THE MATHEMATICAL MODEL AND CONTROL STRATEGY

The equations of the PMSM in abc reference frame are derived as (1).

$$
\begin{bmatrix} u_a \\ u_b \\ u_c \end{bmatrix} = p\begin{bmatrix} \psi_a \\ \psi_b \\ \psi_c \end{bmatrix} + \begin{bmatrix} R_a & 0 & 0 \\ 0 & R_b & 0 \\ 0 & 0 & R_c \end{bmatrix}\begin{bmatrix} i_a \\ i_b \\ i_c \end{bmatrix} \quad (1)
$$

It should be noted that the resistor coefficient is a matrix now because the three phases' resistance are no longer the same. By applying abc to dq transformation on(1), the mathematical model of PMSM in dq reference frame can be obtained as (2).

$$
\begin{aligned}
u_d &= \frac{R_d + R_q}{2}i_d + \frac{R_d - R_q}{2}\cos 2\theta \cdot i_d + R_{dq}\sin 2\theta \cdot i_d + R_{dq}\cos 2\theta \cdot i_q \\
&\quad + \frac{R_q - R_d}{2}\sin 2\theta \cdot i_q + L_d\frac{di_d}{dt} - \omega_r L_q i_q \\
u_q &= \frac{R_q - R_d}{2}\sin 2\theta \cdot i_d + R_{dq}\cos 2\theta \cdot i_d + \frac{R_d + R_q}{2}i_q + \frac{R_q - R_d}{2}\cos 2\theta \cdot i_q \\
&\quad - R_{dq}\sin 2\theta \cdot i_q + L_q\frac{di_q}{dt} + \omega_r(L_d i_d + \psi_f)
\end{aligned}
$$
$$(2)$$

where

$$
R_d = \frac{4R_a + R_b + R_c}{6}, \quad R_q = \frac{R_b + R_c}{2}, \quad R_{dq} = \frac{R_c - R_b}{2\sqrt{3}}
$$

As we can see from (2), the relationship between stator voltages and currents are no longer linear since the existence of time variant coefficients containing sin2θ and cos2θ term resulted from the asymmetry of the stator (cable) resistance. With the traditional PI current regulator, it will inevitably cause three phase current imbalance and torque ripple, and will ultimately lead to speed ripple. The terms involving sin2θ and cos2θ can be considered as disturbances which have twice the frequency of the rotor's electrical frequency since id and iq should be DC value. According to internal model principle, to completely suppress a periodic disturbance, an internal model of the periodic disturbance must be established in the controller [9]. In this paper, the PIR current controller is constructed by connecting a resonant term in parallel with a PI controller. The transfer function of a resonant term is

$$
G(s) = \frac{k_r s}{s^2 + \omega_0^2}
$$

978-1-5090-0064-7/15 $31.00 © 2015 IEEE

The PIR control diagram of the system is presented as Fig.2.

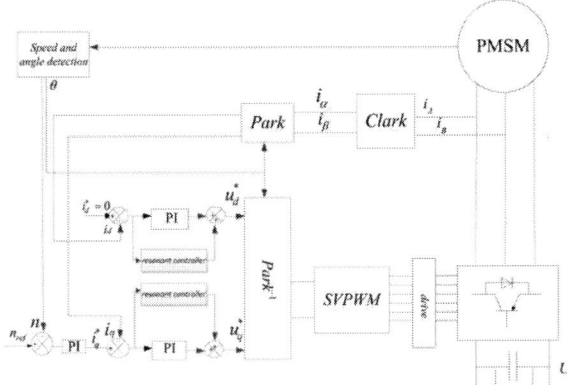

Fig. 2.   Control diagram of PMSM with unbalanced stator resistance

III.   SIMULATION AND EXPERIMENTAL VERIFICATION

1. Simulation Results

First, the proposed PIR controller and the hysteresis band controller are simulated respectively in matlab/Simulink model. A 3Ω resistor is connected in phase a between the PMSM and inverter to represent the cable resistor. The parameters of the system are listed in TABLE I. The switching frequency is 5k with a PIR controller. Speed reference is 900rpm and id reference is kept -3A during the test.

TABLE I
system parameters

Parameter	Unit	Value
Pole pairs		3
stator resistance	Ω	1.6
Cable a resistance	Ω	3
Cable b resistance	Ω	0
Cable c resistance	Ω	0
Rated power	KW	3
Load	Nm	5

The three phase current waveforms are shown in Fig.3. Fig.4 shows the dq-axis current and Fig.5 shows the fft analysis of d-axis current((a) is the result of PIR controller and (b) is the result of the hysteresis band controller). It can be seen that both methods work well in simulation. But the switching frequency when hysteresis band controller adopted can reach as high as 50k, which is not acceptable.

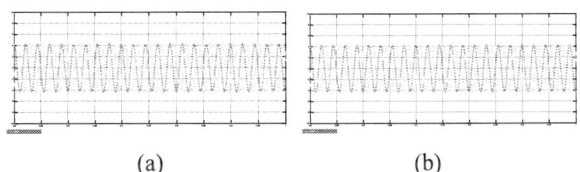

(a)   (b)

Fig.3 Three phase currents waveforms

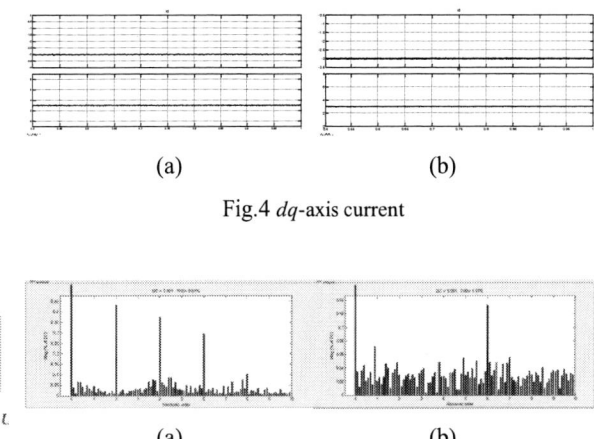

(a)   (b)

Fig.4 dq-axis current

Fig.5 fft analysis of d-axis current

2. Experimental Results

Experimental verification is implemented with a typical PMSM. The parameters of the motor and the system are the same as TABLE I. The motor is controlled by a Texas Instruments 28335DSP. Speed reference is 300rpm and id reference is kept -3A during the test. Three phase currents at steady state are displayed in Fig.6 with traditional PI current controller and Fig.7 with PIR current controller. The id,iq current waveforms and corresponding fft analysis are displayed in Fig.8 with traditional PI current controller and Fig.9 with PIR current controller. All these figures are plotted with data acquired by the scope.

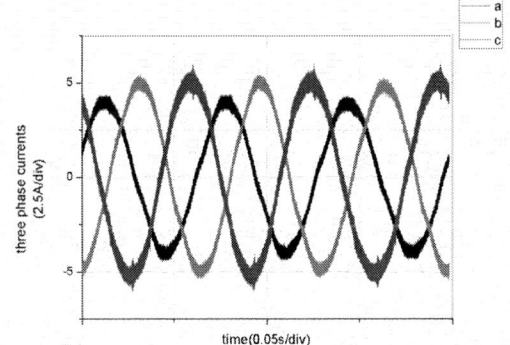

Fig. 6.   Three phase currents waveforms(PI)

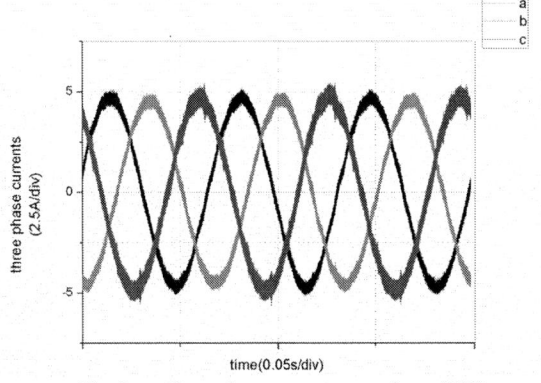

Fig. 7.   Three phase currents waveforms(PIR)

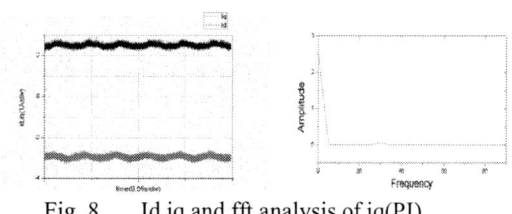

Fig. 8. Id,iq and fft analysis of iq(PI)

Fig. 9. Id,iq and fft analysis of iq(PIR)

We can see that the 2nd harmonic order(30Hz) of the current in *dq* reference frame is the main harmonic caused by the asymmetry of cable resistor and can be eliminated through PIR regulator. It demonstrates the effectiveness of the proposed control strategy.

IV. CONCLUSION

The cable resistance may vary among phases in some special occasions such as in the deep drilling platform. Based on the mathematical model of PMSM with unbalanced stator resistor in dq reference frame, a resonant controller in parallel with the PI controller is proposed in this paper to replace the traditional current controller. The method is proved effective by experiments. But with the speed of the motor varying, the resonant term has to change accordingly, and further study is needed to deal with the situation.

ACKNOWLEDGMENT

This work is supported by the National Natural Science Foundation of China (Grant No. 51277053) and Infineon HUST student innovation project.

REFERENCES

[1] M. P. Kazmierkowski and L. Malesani, "Current control techniques for three-phase voltage-source PWM converters: A survey," IEEE Trans. Ind. Electron., vol. 45, no. 5, pp. 691–703, Oct. 1998.

[2] D. Basic, V. S. Ramsden, and P. K. Muttik, "Harmonic filtering of high power 12-pulse rectifier loads with a selective hybrid filter system," IEEETrans. Ind. Electron., vol. 48, no. 6, pp. 1118–1127, Dec. 2001.

[3] S. Fukuda and T. Yoda, "A novel current-tracking method for active filters based on asinusoidal internal model," IEEE Trans. Ind. Appl., vol. 37, no. 3, pp. 888–895, May/Jun. 2001.

[4] R. Bojoi, L. Limongi, D. Roiu, and A. Tenconi, "Frequency-domain analysis of resonant current controllers for active power conditioners," in Proc. 34th Annu. IEEE IECON, Nov. 2008, pp. 3141–3148.

[5] C. Xia, B. Ji, and Y. Yan, "Smooth speed control for low speed high torque permanent magnet synchronous motor using proportional integral resonant controller," IEEE Trans. Ind. Electron., vol. 62, no. 4, pp. 2123–2134, Apr. 2015.

[6] D. N. Zmood and D. G. Holmes, "Stationary frame current regulation of PWM inverters with zero steady-state error," IEEE Trans. Power Electron., vol. 18, no. 3, pp. 814–822, May 2003.

[7] D. N. Zmood, D. G. Holmes, and G. H. Bode, "Frequency-domain analysis of three-phase linear current regulators," IEEE Trans. Ind. Appl., vol. 37, no. 2, pp. 601–610, Mar./Apr. 2001.

[8] Y.-R. Mohamed and E. El-Saadany, "A control scheme for PWM voltage source distributed-generation inverters for fast load-voltage regulation and effective mitigation of unbalanced voltage disturbances," IEEE Trans. Ind. Electron., vol. 55, no. 5, pp. 2072–2084, May 2008.

[9] Y. Sato, T. Ishizuka, K. Nezu, and T. Kataoka, "A new control strategy for voltage-type PWM rectifiers to realize zero steady-state control error in input current," IEEE Trans. Ind. Appl, vol. 34, pp. 480–486, May/June 1998.

BIOGRAPHIES

Zou Xunhao obtained his BSc degrees in the college of electrical engineering from Zhejiang University in 2013. He is now working towards his master degree in Huazhong University of Science and Technology. His main research interests is control of PM synchrounous motors and generators.

Shenghua Huang is now the professor of Huazhong University of Science and Technology. His main research interests include new special motor and its control system, power electronic devices and systems and the applications of power electronics in Power system.

Qin Zhuqian was born in China in 1989.He obtained his B.E. degree of electrical engineering from China University of Mining and Technology. He is now studying the master degree in HuaZhong University of Science and Technology, and his research direction is power electronics and power drives.

Xiaodong Hu, Master Degree, dedicated to optimiz ation on structure of PM machine and electromagne tic actuators with permanent magnetic.

Kan Guangqiang obtained his BSc degrees in the college of electrical engineering from southwest jiaotong University in 2014. He is now working towards his master degree in Huazhong University of Science and Technology. His main research interests are control of PM synchronous motors and power electronics.

Design of a Wireless Charging System Composited of Antenna and Circuits in HF Band

Chi-Fang Huang, Yu-Wei Weng

Graduate Institute of Communication of Engineering, Tatung University Taipei, TAIWAN

ras@ttu.edu.tw

Abstract–This work present a systematic design procedure of wireless charging system operating at the frequency 13.56MHz different from the ones used in the current standards. Such a mechanism of power transfer is based on the magnetic induction coupling in terms of electromagnetics. What more important, this work describes both of the design details of loop antenna for magnetic coupling and the circuit of charging. Measurement data are provided in this report.

Keywords–Magnetic induction coupling; wireless charging; RFID; loop antenna.

I. INTRODUCTION

Since the commercial success of personal consumer portable products, say, MP3 players, digital cameras, Pads, and mobile phones, etc., the vast supply of battery charging lines has been made from factories. Unfortunately, usually the charging lines supplied by different vendors are not replaceable by each other's. Moreover, the products supplied even by one certain company, the charging lines may evolve somehow later, and the old ones are abandoned after all. This is a serious environmental issue in past decades, namely, electronics pollution.

Based on a concept of "without cables", wireless charging had been proposed for years, yet the progress is quite slow. The main reason is commercial profit of cable/line industry. On the other hand, the standard of power transmission is also a question. Currently, there have been some transmission standards of wireless charging proposed, for example, A4WP [1] and Qi [2] standards. Both of them are based on the magnetic induction coupling concept [3] in the near field for the energy transfer. However, the operating frequencies used for them are quite different. A4WP uses the ISM band 6.78MHz which is the exact half of the 13.56MHz of the RFID technology [4]. The frequency used for Qi chargers is located between about 110 and 205 kHz for the low power Qi chargers up to 5 watts and 80-300 kHz for the medium power Qi chargers. To the latter, for an efficient coupling of magnetic field, usually a ferrite material [3] included in a coil loop is used.

As a proprietary technique, this work proposed a wireless charging system, see Fig. 1, based on the same structure of loop antenna [5] operating at 13.56MHz of RFID.

This adaptation of higher frequency is supposed to gain higher power transfer efficiency, especially, no ferrite material is used to reduce the system weight. It is worth to mention that, in the present work, a charging circuit is also designed instead that only the demonstration of power transfer is given. The components of the whole system are: charging loop antenna (power supplying end), receiving loop antenna, matching and rectifying circuit, and charging

circuit-module for battery. Details of the system are to be described in the following part.

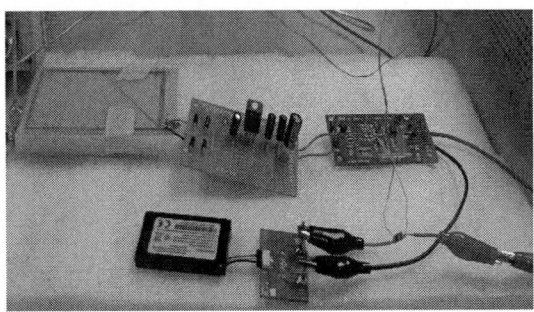

Fig. 1: The Proposed System of Wireless Charging System

II. STRUCTURE OF PROPOSED SYSTEM

Referring to the Fig. 2, the test bed designed in work is composed with the following parts.

Transmitter (power supplier): Signal generator + Amplifier + Antenna (planar loop)

Receiver (charging hardware): Antenna (planar loop) + Rectifier circuit + charging module + Battery (for test)

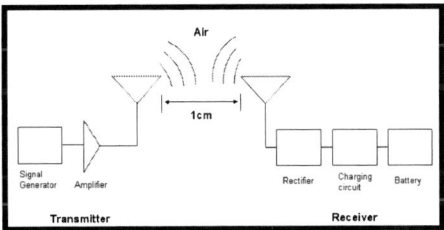

Fig. 2: The Diagram of the Proposed System

The Lab equipment of signal generator and amplifier are to simulate the power supplied with an assigned operating frequency. The planar loop antennas for both ends are PCB based as shown in Fig. 3(a). The design of antenna is carried out by the tool of full-wave electromagnetic simulation package CST [6]. The Fig. 3(b) shows the simulated spatial distribution of magnetic field excited by the loop. On the other hand, the input impedance of this planar antenna can be found by a simulation of this loop structure. As shown in Fig. 3(c), it is:

$$Z_{in} = 15.18 + j117.07 \quad \Omega \qquad @13.56\text{MHz}$$

As what expected, in addition to the resistance of loop, the input impedance Z_{in} is obviously inductive as well. The final measured Z_{in} is,

$$Z_{in} = 2.47 + j115.45 \quad \Omega \qquad @13.56\text{MHz}$$

This measurement was carried out on a test fixture (model: 3680) provided by Anritsu as shown in Fig. 4.

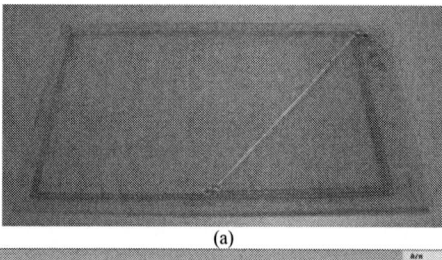

(a)

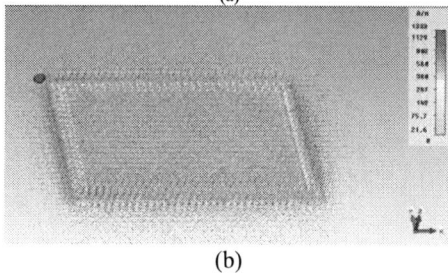

(b)

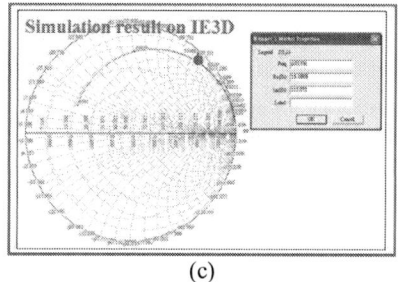

(c)

Fig. 3: A planar loop antenna

However, at the power-supplying end, this differential-input antenna with inductive impedance must be connected with a *50 Ω* coaxial cable leading to the amplifier. Therefore, a Balun should be designed and offered. In the present work, some passive components are used for such a circuit. It is shown in Fig. 5. They are one capacitor and one inductor to form the Balun which is connected with the following loop antenna mentioned above. To verify the matching effectiveness, the whole circuit is measured as for its return lose as shown in Fig, 6. The measured return loss is -22.02dB at the operating frequency 13.56MHz. Namely, it is well matched with the signal generator based on the *50 Ω* line impedance.

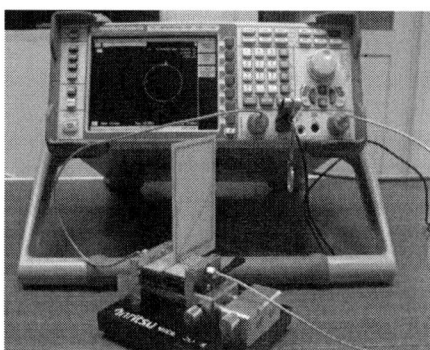

Fig. 4: Test fixture (model: 3680) of Anritsu

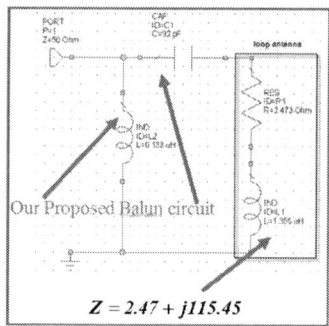

Fig. 5: The proposed Balun

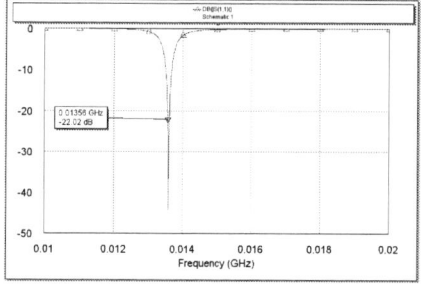

Fig. 6: Measured return loss of the Balun plus loop antenna

III. RECEIVING END

On the other hand, at the receiving end, the planar loop antenna has to be complex-conjugated match [3] with the following rectifier circuit. The capacitor C1 shown in the Fig. 7 is for matching the inductive loop antenna's impedance as for that purpose. The resistor R1 is the assumed load of charging from the rectifier. In this our proposed circuit, 4 Schottky diodes (Fairchild 1N8519) are used to form a full wave rectifier that is with higher efficiency of AC-DC conversion than half-wave rectifier.

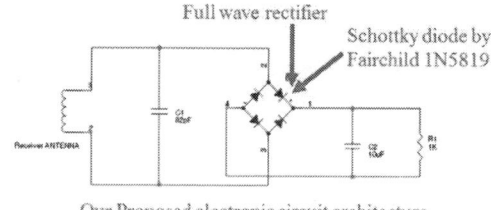

Fig. 7: Receiving antenna, matching circuit and rectifier

IV. TEST

By the system described in the previous section, firstly, if R1 = *1k Ω*, the efficiency of such a wireless charging is tested. Given the conditions that, input power = 2.11W and the separation distance between the two planar loop antennas is 1cm (see Fig. 1), the resultant voltage on R1 is 21.8V, so the received power is 0.4754W and the efficiency is 22.523%.

978-1-5090-0064-7/15 $31.00 © 2015 IEEE

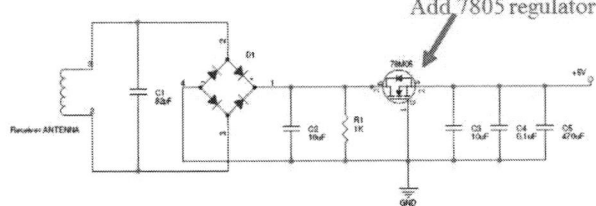

Fig. 8: A regulator is added

However, for functioning of charging to a real device, a regulator is further necessary. In the present test bed, as shown in Fig. 8, a regulator is added to supply a stable 5V voltage. What more is, usually a type of battery is associated with a charging circuit provided by the industrial provider as shown in Fig. 9. Yet, for the purpose of evaluating the charging efficiency, an electronic load is to replace the battery. A power amplifier is following the signal generator, and the output power is set to be 5.14W. Referring to the Fig. 10, by adjusting the load, the max. efficiency is 12.96% happening when the charging is with 4.15V/160.4mA.

By applying the charging system to a real mobile phone, the phone can be demonstrated to work well as shown in Fig. 11.

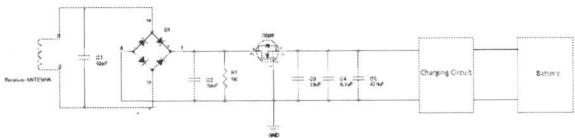

Fig. 9: The complete charging circuit

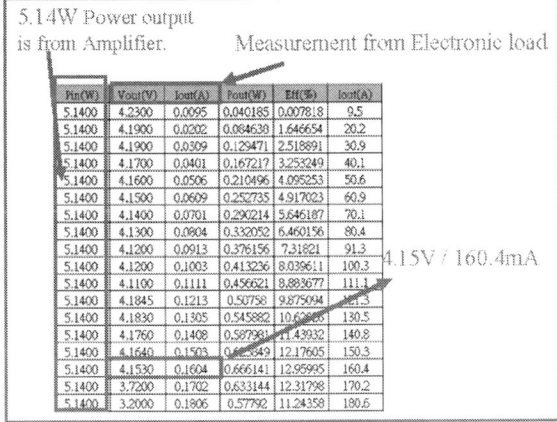

Fig. 10: Efficiency Evaluation

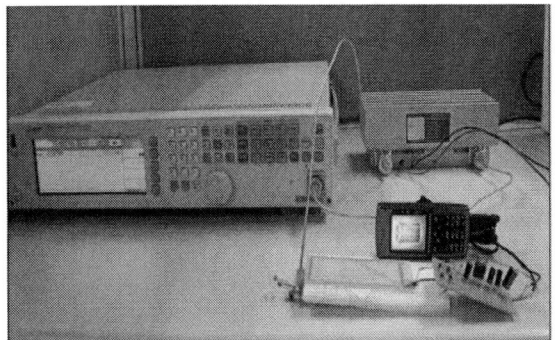

Fig. 11: Real wireless charging into a mobile phone by the present system

V. CONCLUSION

This paper presents a design procedure for a wireless charging system, which is operating at the frequency 13.56GHz, same as that used by RFID. The power transfer is based on the mechanism of magnetic induction coupling. Especially, for antenna, balun, matching circuit, rectifier and charging module, the design techniques of them are all explained. Such a procedure is very easy to follow for engineers, and further hardware optimization or charging efficiency in the future can be achieved by this way.

REFERENCES

[1] R. Tseng, B. von Novak, S. Shevde and K. A. Grajski, "Introduction to the Alliance for Wireless Power Loosely-Coupled Wireless Power Transfer System Specification Version 1.0," IEEE Wireless Power Transfer Conference 2013, Technologies, Systems and Applications, May 15-16, 2013 Perugia

[2] http://www.radio-electronics.com/info/power-management/wireless-inductive-battery-charging/qi-wireless-charging-standard.php, accessed July 3, 2015

[3] David K. Cheng, Field and Wave Electromagnetics, Addison-Wesley, 1989.

[4] C.-F. Huang, "Low-Cost Solution for RFID Tags in Terms of Design and Manufacture," Chapter in Current Trends and Challenges in RFID, ISBN 978-953-307-356-9, Edited by Cristina Turcu, InTech, July 2011.

[5] A. Boswell, A. J. Tyler, and A. White, "Performance of a small loop antenna in the 3-10 MHz band," IEEE Antennas and Propagation Magazine, pp. 51-56, Vol. 47, Issue: 2, April 2005

[6] https://www.cst.com/ accessed July 4, 2015

Energy Storage Variation Ratio and Power Efficiency of Cuk converter in continuous mode

Wenzheng Xu[1]   K.W.E.Cheng[2]

[1,2] Department of Electrical Engineering, The Hong Kong Polytechnic University, Hong Kong
[1]E-mail: xuwenzheng2012@gmail.com
[2] E-mail: eeecheng@polyu.edu.hk

Abstract–The amount of storage energy in comparison to the output energy for basic topologies deserves deep research which was proposed over ten years ago. It has been analyzed and demonstrated that the larger is the energy variation ratio of power devices including inductors and capacitors, the larger is power loss, thus lower efficiency. This paper extends this theory to Cuk converter and derived the energy storage variation ratio and its relationship with power loss. Simulation in conducted in PSIM platform and verified the energy ratio's impact on efficiency via detailed calculation. Finally a sample Cuk converter is built, and experiment results further demonstrated the energy ratio's relationship with power loss and efficiency from one perspective, which could help us understand and design a converter better.

Keywords–Cuk converter, energy ratio, efficiency, power loss, switched-mode power converters.

I. INTRODUCTION

The amount of energy variation in power devices like inductors and capacitors for basic topologies deserves detailed study. It has been analyzed and demonstrated that the larger is the energy variation ratio of power devices including inductors and capacitors, the larger is power loss, thus lower efficiency [1]. In paper [1], basic switched-mode converters including buck, boost, and buck-boost converter are analyzed especially their energy storage variation ratio in both continuous and dis-continuous mode. The energy-storage factor is also applied on isolated power convertors using integrated magnetics including flyback and forward converters [2], which shows similar result as expected.

This paper extends this theory to Cuk converter in continuous mode and derived the energy ratio and its relationship with power loss, since Cuk converter has low switching losses and the highest efficiency among non-isolated dc–dc converters and deserves more research [3]. The calculation result agrees with the energy ratio theory and matches the principle that Cuk converter can be regarded as combination of boost and buck converter. Simulation in conducted in PSIM platform and verified the energy ratio's impact on efficiency. Hardware of Cuk converter is built and result of both simulation and experiment demonstrated the energy ratio's relationship with power loss, which could help us understand and design a converter better.

II. CALCULATION OF THE ENERGY RATIO

The variation of energy stored in an inductor and capacitor during continuous mode is represented by S_L and S_C. The corresponding energy ratio is represented by R_{SL} and R_{SC}.

According to paper [1], basic equations of the storage energy theory in continuous mode are listed firstly for convenience:

$$S_L = \frac{1}{2}L(i_{L\max}{}^2 - i_{L\min}{}^2) = L\overline{i_L}\Delta i_L \tag{1}$$

$$R_{SL} = \frac{S_L}{V_o I_o T_s} = \frac{L\overline{i_L}\Delta i_L}{V_o I_o T_s} \cdot \tag{2}$$

$$S_C = C\overline{V_C}\Delta V_C \tag{3}$$

$$R_{SC} = \frac{S_C}{V_o I_o T_s} \tag{4}$$

Cuk converter can be regarded as a boost converter plus an inversed buck converter. Its typical topology is shown in figure 1. In the following discussion, we assume that C_1 is large enough that V_{C1} is constant.

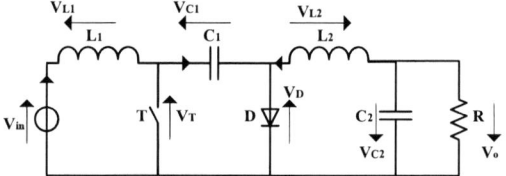

Fig.1: Circuit diagram of a Cuk converter

Considering the volt-second balancing of the two inductors, we can get:

$$V_{in}DT_s + (V_{in} - V_{C1})(1-D)T_s = 0 \Rightarrow V_{C1} = \frac{1}{1-D}V_{in} \tag{5}$$

$$(V_{C1} - V_o)DT_s - V_o(1-D)T_s = 0 \Rightarrow V_{C1} = \frac{1}{D}V_o \tag{6}$$

Solving the above equations we can get:

$$\frac{V_o}{V_{in}} = \frac{D}{1-D}, V_{C1} = V_{in} + V_o \tag{7}$$

The voltages across the two inductors during the ON and OFF states are:

$$v_{L1\_on} = V_{in} \tag{8}$$

$$v_{L1\_off} = V_{in} - V_{C1} = -V_o \tag{9}$$

$$v_{L2\_off} = -V_o \tag{10}$$

$$v_{L2\_on} = V_{C1} - V_o = V_{in} \tag{11}$$

The changes of L_1 and L_2's currents are:

978-1-5090-0064-7/15 $31.00 © 2015 IEEE

$$\Delta I_{L1} = \frac{V_{in}DT_s}{L_1} = \frac{V_o(1-D)T_s}{L_1} \quad (12)$$

$$\Delta I_{L2} = \frac{V_{in}DT_s}{L_1} = \frac{V_o(1-D)T_s}{L_2} \quad (13)$$

The variations of the energy stored in the inductors are:

$$S_{L1} = L_1 \cdot \overline{i_{L1}} \cdot \Delta i_{L1} = L_1 \cdot I_{in} \cdot \frac{V_{in}DT_s}{L_1} = V_o I_o DTs \quad (14)$$

$$S_{L2} = L_2 \cdot \overline{i_{L2}} \cdot \Delta i_{L2} = L_2 \cdot I_o \cdot \frac{V_o(1-D)T_s}{L_2} = V_o I_o(1-D)Ts \quad (15)$$

Obviously,

$$R_{SL1} = D \quad (16)$$

$$R_{SL2} = 1-D \quad (17)$$

$$R_{SL(1+2)} = R_{SL1} + R_{SL2} = 1 \quad (18)$$

It is just as expected, because the Cuk converter can be regarded as combination of boost converter and buck converter, thus for the R_{SL} of inductors, it should be equal to D just as boost converter (L_1 plays key role) plus $1-D$ just as buck converter (L_2 plays key role). And R_{SL} is the same as buck-boost converter.

For capacitor C_1, the voltage ripple can be calculated as figure 2 represents.

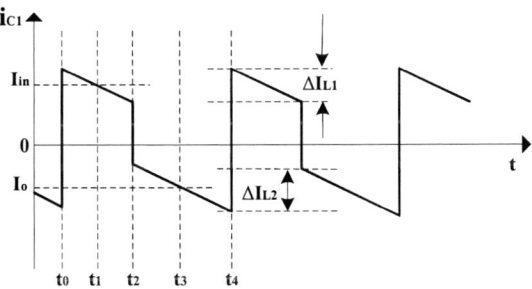

Fig. 2: Ideal waveform of capacitor C1's current

$$\Delta V_{C1} = \int \frac{i_{C1}dt}{C_1}$$
$$= \frac{1}{C_1} \cdot \frac{1}{2} \cdot (1-D)T_s \cdot [(I_{in} - \frac{1}{2}\Delta I_{L1}) + (I_{in} + \frac{1}{2}\Delta I_{L1})] = \frac{(1-D)T_s I_{in}}{C_1} \quad (19)$$

$$S_{C1} = C_1 \overline{V_{C1}} \Delta V_{C1} = C_1 \cdot (V_{in} + V_o) \cdot \frac{(1-D)T_s I_{in}}{C_1}$$
$$= (1-D)(V_{in} + V_o) \cdot T_s I_{in}$$
$$= (1-D)[\frac{(1-D)}{D} \cdot V_o + V_o] \cdot T_s \cdot \frac{D}{(1-D)} I_o = V_o I_o T_s$$
$$(20)$$

$$R_{SC1} = 1 \quad (21)$$

This is because the Capacitor C_1 needs to handle all the

energy just the same as the inductor in buck-boost converter. During OFF state, the capacitor is charged from V_{in} and the inductor L_1, and during ON state the capacitor will discharge to transmit energy to the load side.

For capacitor C_2, the voltage ripple can be calculated by figure 3.

$$\Delta V_{C2} = \int \frac{i_{C2}dt}{C_2} = \frac{1}{C_2} \cdot \frac{1}{2} \cdot \frac{T_s}{2} \cdot \frac{\Delta I_{L2}}{2} = \frac{(1-D)V_o \cdot T_s^2}{8L_2 C_2} \quad (22)$$

$$S_{C2} = C_2 \overline{V_{C2}} \Delta V_{C2} = C_2 \cdot V_o \cdot \frac{(1-D)T_s^2 V_o}{8L_2 C_2} = \frac{(1-D)V_o^2 T_s^2}{8L_2} \quad (23)$$

$$R_{SC2} = \frac{S_{C2}}{V_o I_o T_s} = \frac{(1-D)T_s R}{8L_2} = \frac{1-D}{4K_2}, K_2 = \frac{2L_2}{RT_s} \quad (24)$$

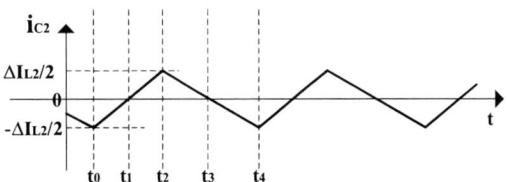

Fig. 3: Ideal waveform of capacitor C2's current

It is the same as buck converter, because the role capacitor C_2 plays is exactly the same as buck converter. In this way, we can derive the overall energy ratio of a Cuk converter:

$$MSEF = R_{SL1} + R_{SL2} + R_{SC1} + R_{SC2} = 2 + (1-D)/4K_2 \quad (25)$$

Summarize the above result with overall energy ratio of other topologies, we can derive table 1.

Table 1: Comparison energy factor of four topologies under continuous mode

Converter	R_{SL}	R_{SC}	MSEF
Buck	1-D	(1-D)/4K	(1-D)(1+1/4K)
Boost	D	D	2D
Buck-boost	1	D	1+D
Cuk	1	$1+(1-D)/4K_2$	$2+(1-D)/4K_2$

So we conclude that the energy ratio of Cuk converter is basically higher than buck, boost, and buck-boost converter. This is because more components are involved, and the capacitor C_2 as well as the two inductors would handle all the energy variation during a cycle. All the calculation results are reasonable and easy to understand.

It is also obvious that the energy ratio is only affected by K_2 while K_1 has nothing to do with the result. Figure 4 shows the value of energy ratio MSEF of Cuk converter as duty cycle D varies when K_2 equals to 1 and 2 respectively. Besides the principle that larger duty cycle proves less power loss thus higher efficiency, we can also observe that in continuous mode, when the duty cycle is large, the difference of power loss of different K_2 is small. When the duty cycle is small, K_2 has greater influence on the efficiency.

978-1-5090-0064-7/15 $31.00 © 2015 IEEE

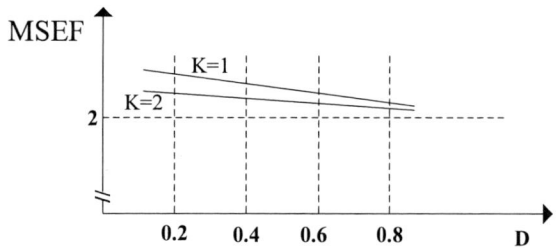

Fig. 4: MSEF vs D for Cuk converter in continuous mode

So when designing a Cuk converter, the rated duty cycle is expected to be a bit larger in order to gain high efficiency. Simulation and experiment proof is given below.

III. SIMULATION OF CUK CONVERTER

1. Cuk converter's modeling in PSIM

Simulation of Cuk converter's operation in continuous mode is conducted on PSIM platform. For each inductor and capacitor, a resistor is set in series with them to represent the ESR of them. We will conduct simulation of Cuk converter and Buck converter firstly with the same ratings. The overall diagram of Cuk converter is shown below.

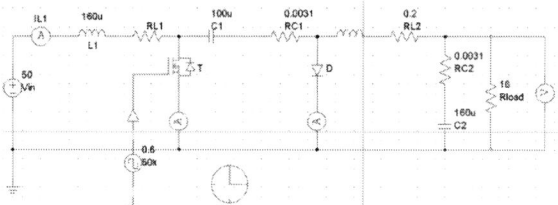

Fig.5: Cuk converter's diagram in PSIM

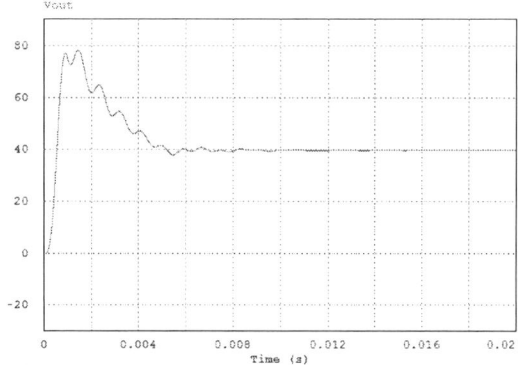

Fig.6: Output voltage waveform of Cuk converter in PSIM simulation

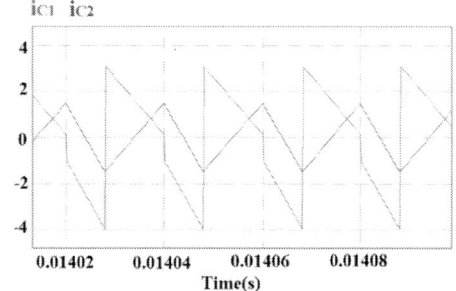

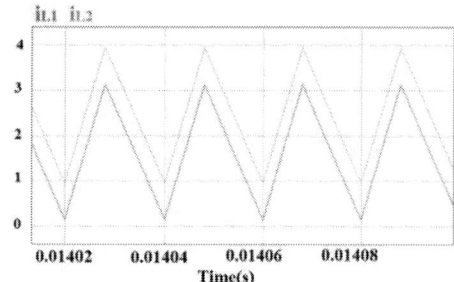

Fig.7: Waveform of, I_{C1}, I_{C2}, I_{L1} and I_{L2} respectively

The parameters of each device have been marked in the diagram. In this case, $K_1=K_2=2L/RT_s=1$. When $D=0.5$, i.e. voltage ratio is 1, the output power is 100W. Similarly, when $V_{in}=60V$ and $D=0.4$, voltage ratio is 0.667, the output voltage and power is the same. All relevant diagrams when $D=0.4$ are shown below. $V_{in}=60V$, $D=0.4$, $L_1=L_2=160\mu H$, $C_1=C_2=160\mu F$, and $R_{load}=16\Omega$.

Because $L_1=L_2$ and $C_1=C_2$, all the waveforms are exactly the same as expected.

2. Inductor and capacitor's power loss modeling

The power loss of an inductor mainly comes from the core loss, ac copper loss and dc copper loss, which can be described in the following formula:

$$P_{loss} = P_{core} + P_{ac} + P_{dc}$$
$$= Kf^x \Delta B^y V_e + I_{rms-ripple}^2 R_{ac} + I_{rms-dc}^2 R_{dc} \tag{26}$$

In the above equation, K represents the core's material constant, f represents frequency in kHz, x represents frequency factor and its value normally ranges from 1 to 3, ΔB represents the variation of flux density, y represents flux density factor and its value normally ranges from 2 to 3, V_e represents effective volume of the core, and R_{ac} represents the equivalent ac resistance of the inductor. Meanwhile, the core loss consists of hysteresis loss, eddy current loss and residual loss [3]. Since DC loss is much lower than AC and core loss, we neglect it in our following discussion.

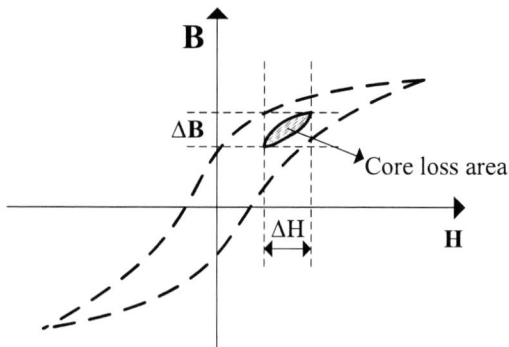

Fig.8: B-H diagram showing core loss area

It is generally difficult to calculate the power loss of an inductor accurately, especially the core loss because it is affected by many factors. So we can only give an estimated calculation based on simulation result and experience values. Figure 8 represents the relationship between core loss and ΔB of inductors in Cuk converter's continuous operation, and the special phenomena of remaining inductor current existing in complex topologies have been explored and explained in paper [4]. Based on the above equation, we assume y=2.5 which well reflects the ratio of core loss and ΔB. There is no doubt that the core loss increases greatly as the frequency increases, but since we are going to analyze the relationship between efficiency and MSEF, the frequency is constant at 50kHz in our following simulation and analysis.

To set up the model in PSIM, we selected a real inductor whose datasheet is listed below. The detailed methodology of estimating the inductor L_2's total power loss is described in table 3 below. In simulation, D is varied to give different MSEFs for the simulation, while the output power lever is about 100W, V_{in} is adjusted then. This is because if power level varied too much, it's not easy to compare the efficiency fairly only via MSEF.

Now we calculate the power loss of inductor L_2. According to parameters of the inductor's datasheet and experience values, we assume: $x=2$, $V_e=8.95\times10^{-6}m^3$, $f=50$kHz, $y=2.5$, $K=1.2\times10^{-3}$ which is described in formula (26), and assume the design of the inductor is good that eddy loss is constrained. Paper [5] gives detailed methodology of estimating power loss of inductor.

The calculation result of inductor L_2 makes sense and agrees with the R_{SL2}, i.e. when D increases, R_{SL2} decreases thus the power loss of L_2 decreases, too. Detailed data is shown in table 3. We also learned that the core loss is much larger than ac loss.

Table 3: The inductor L2's detailed power loss calculation

D		0.4	0.45	0.5	0.55	0.6
V_{in}	V	60	50	41	34	29
V_o	V	39.21	40.06	39.99	40.29	40.35
R_{SL2}		0.6	0.55	0.5	0.45	0.4
$\overline{I_{L2}}$	A	2.449	2.506	2.500	2.540	2.600
Δi_{L2}	A	2.951	2.790	2.385	2.285	2.100
$\overline{I_{L2}}\cdot\Delta I_{L2}$	A^2	7.227	6.992	5.962	5.803	5.460
$I_{ac\,max}$	V	1.476	1.395	1.193	1.142	1.050
$I_{ac\,ripple\,rms}$	A	0.852	0.805	0.689	0.659	0.606
R_{ac}	Ω	0.04	0.04	0.04	0.04	0.04
AC loss	W	0.029	0.026	0.019	0.017	0.014
ΔB	Gs	6530	6173	5277	5056	4646
$(\Delta B)^{2.5}$	T$^{2.5}$	0.345	0.300	0.202	0.182	0.142
Core loss	W	9.252	8.041	5.433	4.881	3.952
Total loss	W	9.281	8.067	5.452	4.898	3.967

We can see from the data that, as duty ratio D increases, the power loss of the inductor L_2 decreases, the same way as its energy variation ratio R_{SL2}. This result agrees with the proposed energy ratio theory.

3. Capacitor's power loss modeling

The power loss of a capacitor consists of many factors, and capacitors (including super-capacitors) deviate from the ideal capacitor equation in a number of ways [6]. Some of them, such as leakage current and parasitic effects are linear, or can be assumed to be linear, can be dealt with by adding virtual components to the equivalent circuit of the capacitor [7]. The usual methods of network analysis can then be applied in different converter models. In other cases, such as with breakdown voltage, the effect is non-linear and normal (i.e., linear) network analysis cannot be used, the effect must be dealt with separately [8] [9]. There is yet another group, which may be linear but invalidate the assumption in the analysis that capacitance is a constant. Such an example is temperature dependence. Finally, combined parasitic effects such as inherent inductance, resistance, or dielectric losses can exhibit non-uniform behavior at variable frequencies of operation.

All dielectrics (except vacuum) have two types of losses. One is a conduction loss, representing the flow of actual charge through the dielectric. The other is a dielectric loss due to movement or rotation of the atoms or molecules in an alternating electric field.

Dielectric loss tangent of ceramic capacitors is dependent upon specific characteristics of the dielectric formulation, level of impurities, as well as microstructural factors such as grain size, morphology, and porosity [10]. The calculation or modeling of its power loss is generally hard, but can be estimated since the dielectric loss can be derived.

According to the book [11], we can use an estimation of all kinds of loss as the following equation shows.

$$P = I_{rms}^{\ 2}ESR + V^2\omega C(DF) \qquad (28)$$

Where V represents the rms value of voltage applied on the capacitor, and DF represents the dissipation factor whose experience values can be found in book [11]. We also selected a 160μF polypropylene capacitor and adopt its data to our simulation. It's ESR is 3.1mΩ and in this case, DF=0.0005.

Table 4: Power loss of capacitor C2 of the Cuk converter

D		0.4	0.45	0.5	0.55	0.6
V_{in}	V	60	0.45	41	34	29
R_{SC2}		0.6	50	0.5	0.45	0.4
$\overline{U_{C2}}(V_o)$	V	39.20		39.99	40.29	40.35
Δu_{C2-pp}	mV	47.04	40.06	39.87	36.18	32.21
$I_{C2rp-rms}$	A	0.862	43.96	0.731	0.634	0.590
$R_{ac(ESR)}$	mΩ	3.1	0.8059	3.1	3.1	3.1

Table 4 summarizes the calculation of power loss of capacitor C_2 based on simulation results. The result shows that power loss of capacitor C_2 decreases as the duty cycle increases. The other clue is the voltage variation cross C_2, which has key relationship with the capacitor loss, also decreases as R_{SC2} decreases. Similar to the result of inductor L_2, the outcome also agrees with the energy ratio theory.

Since we are focusing on the power loss of inductors and capacitors, we assume that power loss of the switch and diode is much smaller thus we neglect it.

For inductor L_2 and capacitor C_2, their power loss variation as D varies matches their energy ratio, i.e. R_{SL2} and R_{SC2} respectively. But the estimation methodology of power loss of the devices might be inaccurate, all the above data in simulation is estimated, and we made some assumptions and adopted some experience values. Hardware is essential besides theoretical calculation to demonstrate the relation between energy ratio and overall efficiency.

IV. HARDWARE EXPERIMENT DEMONSTRATION

To further demonstrate the relationship between energy ratio and power efficiency, a Cuk converter is built as figure 9 shows. The output load resistance is 22Ω, and we use ir2111 as the MOSFET driver with a frequency of 60 kHz. Figure 9 shows the photo of the hardware platform, and figure 10 shows the waveform of MOSFET's driving signal and v_{DS} respectively.

Firstly an experiment was conducted while the input voltage is kept constant at 20V and duty cycle is varied accordingly. Output voltage of the load is measured by the multi-meter. Experiment data and relevant calculation result is listed in table 5.

Fig.9: Hardware platform of a Cuk converter

It is clearly that overall efficiency increases as the duty cycle increases. It agrees with the energy ratio theory because according to equation (25), the larger is the duty cycle, the smaller is the total energy variation, and thus the larger is the efficiency. So this experiment result supports the energy ratio theory and demonstrates it well.
In the above experiment, the input voltage is kept constant thus the overall power level varies with the duty cycle. This may have influence on the power loss and efficiency because of higher current and higher heat loss. Thus another experiment is conducted where the output power level is kept constant and the input voltage varies together with duty cycle. In this way, we get groups of data as

shown in table 6. Similarly, the result also agrees with the expectation based on the energy variation theory and demonstrates that power loss of Cuk converter has negative correlation with duty cycle in continuous mode. Some of the data does not follow the rule, but it is probably caused by measuring error.

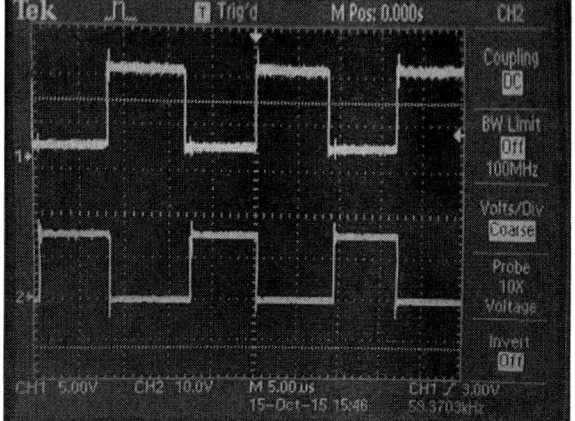

Fig.10: Waveform of driving signal and V_{DS} of the MOSFET

Table 5: Experiment result with constant input voltage

V_{in}	I_{in}	V_{out}	P_{out}	P_{in}	Efficiency
20	0.53	14.81	9.9698	10.60	94.05%
20	0.62	16.03	11.6800	12.40	94.19%
20	0.70	17.08	13.2603	14.00	94.72%
20	0.87	19.01	16.4264	17.40	94.40%
20	1.14	21.84	21.6812	22.80	95.09%
20	1.27	23.10	24.2550	25.40	95.49%
20	1.45	24.67	27.6640	29.00	95.39%
20	1.65	26.42	31.7280	33.00	96.15%

Table 6: Experiment result with constant output power

V_{in}	I_{in}	V_{out}	P_{out}	P_{in}	Efficiency
14.1	1.39	20.10	18.36	19.60	93.70%
16.7	1.16	20.00	18.18	19.37	93.86%
19.4	1.02	20.20	18.55	19.79	93.73%
20.5	0.94	20.00	18.18	19.27	94.35%
23.4	0.82	20.00	18.18	19.19	94.76%
26.4	0.73	20.00	18.18	19.27	94.34%

The power level is improved to 300W finally, and the result of efficiency versus duty cycle is the same. Whereas, the efficiency of a Cuk converter depends on many factors, and there is no definite relationship between efficiency and duty cycle. However, what we've done above offers a perspective to look through the philosophy and principle of power loss of a converter and its energy variation ratio. Further research will be conducted in Cuk converter of discontinuous mode, and other isolated DC-DC converters as well.

978-1-5090-0064-7/15 $31.00 © 2015 IEEE

V. Conclusion

The energy storage theory is extended to Cuk converter in continuous mode. The MSEF value which represents the total energy storage variation in all power devices during the operation of the converter is derived and the result is reasonable according to the principle of Cuk converter's operation. Simulation is conducted in PSIM platform and calculation based on the simulation result data proved the effectiveness of the MSEF model, i.e. power loss of the device increases as the energy ratio increases. The more energy variation takes place in certain device, the more power loss would be produced.

A Cuk converter has been built and two experiments also demonstrated the energy ratio theory. The efficiency ranges from 94% to 96% as the duty cycle increases from 0.4 to 0.6. Whereas the efficiency depends on many complicated factors rather than duty cycle only, however this research provides a perspective to look through the philosophy and principle of power loss of a converter and its energy variation ratio.

References

[1] Cheng K W E. Storage energy for classical switched mode power converters. IEE Proceedings-Electric Power Applications, 2003, 150(4): 439-446.

[2] Cheng K W E, Lu Y. Formulation of the energy-storage factor for isolated power convertors using integrated magnetics. IEE Proceedings-Electric Power Applications, 2005, 152(4): 837-844.

[3] Safari A, Mekhilef S. Simulation and hardware implementation of incremental conductance MPPT with direct control method using cuk converter, IEEE Transactions on Industrial Electronics, 2011, 58(4): 1154-1161.

[4] Zhu M, Luo F L, He Y. Remaining Inductor Current Phenomena of Complex DC–DC Converters in Discontinuous Conduction Mode: General Concepts and Case Study. IEEE Transactions on Power Electronics, 2008, 23(2): 1014-1019.

[5] Boglietti A, Cavagnino A, Lazzari M, et al. Predicting iron losses in soft magnetic materials with arbitrary voltage supply: an engineering approach, IEEE Transactions on Magnetics, 2003, 39(2): 981-989.

[6] Leuchter J, Bauer P, Zobaa A F. Power electronics and energy management of hybrid power sources with supercapacitors, Power Electronics, Machines and Drives (PEMD 2010), 5th IET International Conference on. IET, 2010: 1-6.

[7] Luo F L, Ye H. Small signal analysis of energy factor and mathematical modeling for power dc–dc converters, IEEE Transactions on Power Electronics, 2007, 22(1): 69-79.

[8] Luo F L, Ye H. Energy factor and mathematical modelling for power DC/DC converters, IEE Proceedings-Electric Power Applications, 2005, 152(2): 191-198.

[9] Wei H, Batarseh I. Comparison of basic converter topologies for power factor correction, Southeastcon'98. Proceedings. IEEE, 1998: 348-353.

[10] 10 Fanglin L, Hong Y. Investigation of DC modulated single-stage power factor correction AC/AC converters. Transactions of China Electrotechnical Society, 2007, 22(5): 92-103.

[11] Gray L. Johnson, "Solid State Tesla Coil", available at http://www.g3ynh.info/zdocs/refs/Tesla/Johnson2001_ssTeslacoil.pdf, cited in 2015

Biographies

Wenzheng Xu received his B.E.E. degree from the Department of Electrical Engineering , Beijing Jiaotong University, Beijing, China, in 2012, and received the M.Sc. degree from Department of Electrical and Electronic Engineering, The University of Hong Kong, Hong Kong, in 2013. He is currently working toward the Ph.D. degree in the Department of Electrical Engineering, the Hong Kong Polytechnic University.

From April to September in 2013, he was a part-time research assistant in Department of Electrical and Electronic Engineering in the University of Hong Kong, where he was involved in researches about smart grid development in China. From September 2013 to June 2015, he was a full-time research associate in Department of Electrical Engineering in the Hong Kong Polytechnic University, where he was the team leader for a silicon-carbide power devices based dc-dc converter project. His research interest includes power electronics topologies and control for switch mode converters.

K.W.E.Cheng obtained his BSc and PhD degrees both from the University of Bath in 1987 and 1990 respectively. Before he joined the Hong Kong Polytechnic University in 1997, he was with Lucas Aerospace, United Kingdom as a Principal Engineer.

He received the IEE Sebastian Z De Ferranti Premium Award (1995), outstanding consultancy award (2000), Faculty Merit award for best teaching (2003) from the University, Brussels Innova Energy Gold medal with Mention (2007), Consumer Product Design Award (2008), Electric vehicle team merit award of the Faculty (2009). Special Prize and Silver Medal of Geneva's Invention Expo (2011) and Eco Star award (2012) He has published over 250 papers and 7 books. He is now the professor and director of Power Electronics Research Centre of the university. His research interests are all aspects of power electronics, electromagnetics, motor drives, EMI and energy saving.

978-1-5090-0064-7/15 $31.00 © 2015 IEEE

Study of AC-DC Converters with Isolated Transformers Applied to High Frequency AC-DC Power Conversion

X. D. Xue, J. Mei, Raghu Raman S, J. F. Liu, and K. W. E. Cheng

Department of Electrical Engineering, The Hong Kong Polytechnic University, Hong Kong
E-mail: xd.xue@polyu.edu.hk, jie.mei@polyu.edu.hk, raghu.s.raman88@gmail.com, jf.liu@connect.polyu.hk, eeecheng@polyu.edu.hk

Abstract–At high frequency, such as 50 kHz, the behaviors of isolated transformers and diodes in rectifiers are considerably different from ones at low frequency, such as 50 Hz. The analysis and the experiment in this paper show that it suffers from poor power factor and low power output if conventional AC-DC converters with isolated transformer are directly applied to high frequency AC-DC power conversion. Thus, three topologies of AC-DC converters applied to high frequency AC-DC power conversion are proposed, studied and evaluated, to convert 50kHz/36V AC voltage to DC voltage of 12 V. The analysis and experimental results show that the AC-DC converters with an isolated transformer, a resonant capacitor and a class-E rectifier and with an isolated transformer, a resonant capacitor and a full-bridge rectifier are suitable for 50 kHz AC-DC power conversion. Therefore, this study provides the valuable and reasonable solutions to high frequency AC-DC power conversion.

Keywords–Converter, high frequency, rectifier, transformer.

I. INTRODUCTION

High frequency AC distribution operating at 20-50 kHz is a candidate for some applications, such as the space station, the space platform, the electric vehicles, the renewable energy micro-grid, as well as the telecommunication and computer system [1]-[4]. The potential merits of high frequency AC system are compact high frequency transformers, considerable reduction in amount and volume of electrical components, improvement in dynamic response, degradation or elimination of acoustic noise, and improvement in safety.

Among the many different types of loads connected to the high frequency AC bus are dc loads such as lighting lamps in vehicles and other electronic instrumentation in vehicles. However, few reported studies deal with the AC-DC converters applied to high frequency AC-DC power conversion. This study is focused on this issue.

The analysis in this paper will show that they suffer from poor power factor and low power output if conventional AC-DC converters with isolated transformer and full-bridge rectifier and with isolated center-tapped transformer and full-wave rectifier are directly applied to high frequency AC-DC power conversion. Three new schemes will be proposed for applying to high frequency AC-DC power conversion in this study. The proposed high frequency AC-DC converters includes the AC-DC converter with an isolated transformer, a resonant capacitor and a class-E rectifier, the AC-DC converter with an isolated center-tapped transformer, two resonant capacitors and two class-E rectifiers, and the AC-DC converter with an isolated transformer, a resonant capacitor and a full-bridge rectifier.

II. CONVENTIONAL AC-DC CONVERTERS APPLIED TO HIGH FREQUENCY AC POWER CONVERSION

1. Conventional AC-DC converter with isolated transformer and full-bridge rectifier

Fig. 1 illustrates the typical schematic diagram of a conventional AC-DC converter with an isolated transformer and a full-bridge rectifier. It is popular in application of AC-DC power conversion at low frequency. Such a converter is applied to 50 kHz AC-DC power conversion. In this case, the RMS voltage of 50 kHz AC supply is 36 V, the capacity of the 50 kHz transformer is 133 VA, the primary voltage is 36 V and the secondary voltage is 9.3V. The full-bridge rectifier consists of four ultrafast diodes. The design objective of the converter is that the rated DC output voltage is 12 V and the rated DC output power is 100W.

The experimental results are illustrated in Fig. 2. It can be seen that the secondary voltage decreases considerably, the peak forward voltage increases significantly and the DC output voltage reduces significantly if the load current increases. Furthermore, the DC output voltage and the DC output power are significantly smaller than the expected values, respectively.

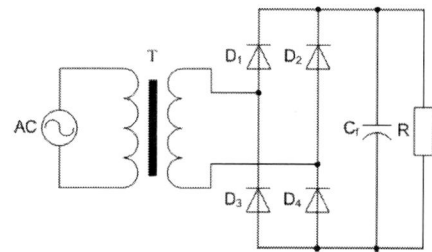

Fig. 1: Topology of conventional AC-DC converter with an isolated transformer and a full-bridge rectifier

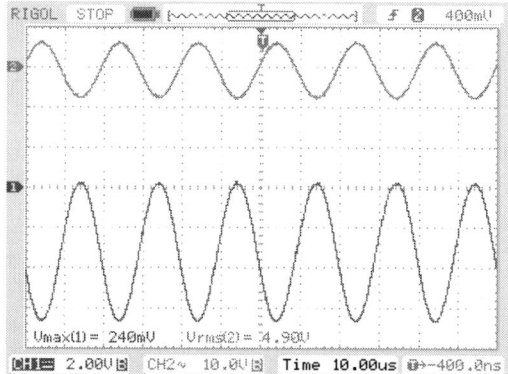

(a) No-load (Ch1: forward voltage of diode, 4V/Div; Ch2: secondary voltage, 20V/Div)

978-1-5090-0064-7/15 $31.00 © 2015 IEEE

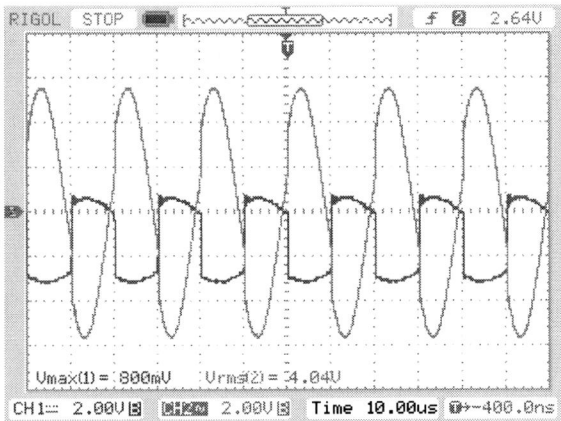

(b) R = 1 ohm (Ch1: forward voltage of diode, 4V/Div; Ch2: secondary voltage, 4V/Div)

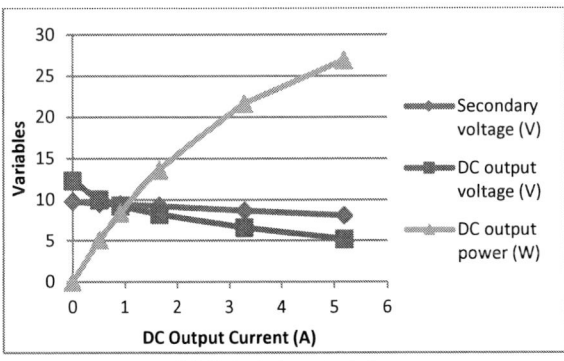

(c) Output characteristics

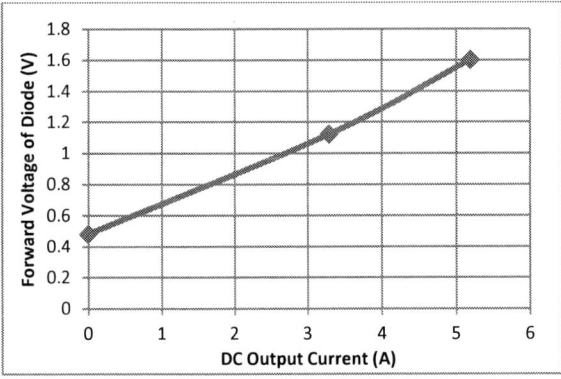

(d) Peak forward voltage of a diode

Fig. 2: Experimental results of conventional AC-DC converter with an isolated transformer and a full-bridge rectifier

2. Conventional AC-DC converter with center-tapped isolated transformer and full-wave rectifier

The typical schematic diagram of a conventional AC-DC converter with a center-tapped isolated transformer and a full-wave rectifier is depicted in Fig. 3. The main parameters of the center-tapped transformer are described as: capacity = 110 VA, frequency = 50 kHz, primary voltage = 36 V, and rated secondary voltage = 11 V.

The experimental results are shown in Fig. 4. It can be observed that the secondary voltage is non-sinusoidal waveform and decreases, the peak forward voltage increases significantly and the DC output voltage reduces considerably if the load current increases. Furthermore, the

DC output voltage and the DC output power are considerably smaller than the expected values, respectively.

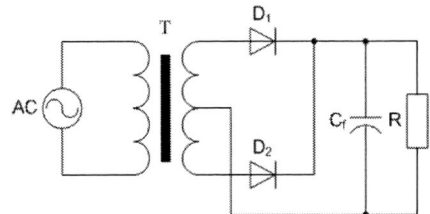

Fig. 3: Topology of conventional full-wave AC-DC converter with a center-tapped isolated transformer

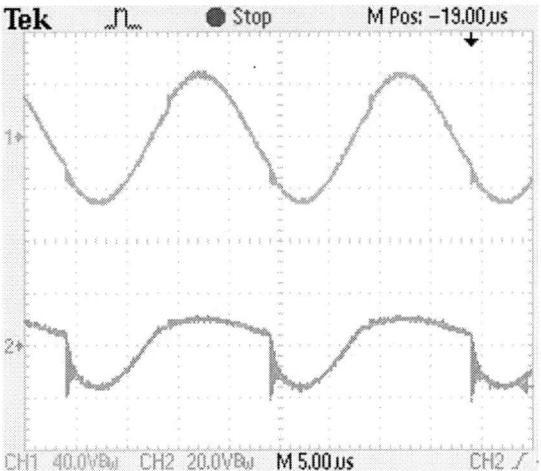

(a) R = 1 ohm (Ch1: primary voltage, 40V/Div; Ch2: secondary voltage, 20V/Div)

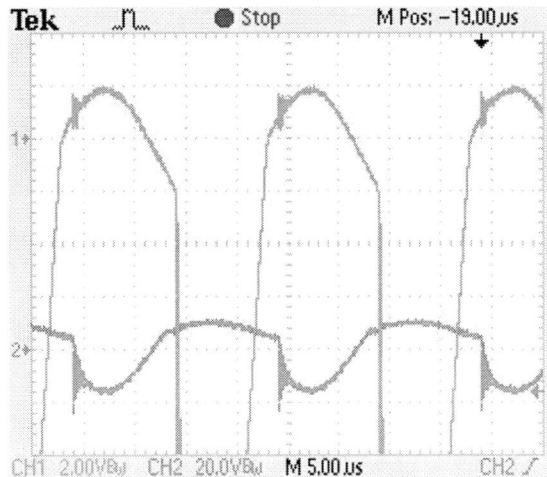

(b) R = 1 ohm (Ch1: forward voltage of a diode, 2V/Div; Ch2: secondary voltage, 20V/Div)

It can be seen from the converter circuits shown in Fig. 1 and Fig. 3 that the DC output voltage depends on the secondary voltage and the forward voltage. Due to the high AC frequency, the secondary leakage reactance and the resistance of the secondary winding become large. It should be pointed out that the secondary leakage reactance is much more than that at low frequency, respectively. Thus, such a large secondary leakage reactance results in considerable reduction in the secondary voltage if the

978-1-5090-0064-7/15 $31.00 © 2015 IEEE   243

current increases. It is obvious that the above analysis is agreement with the experimental results.

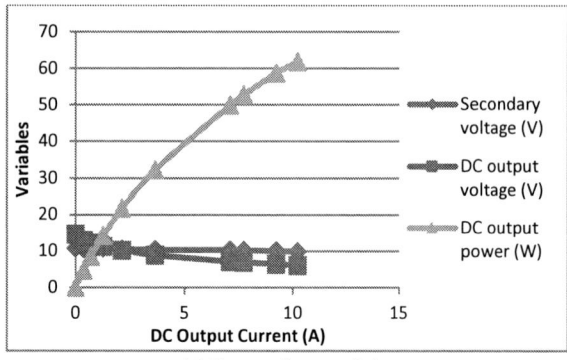

(c) Output characteristics

(d) Peak forward voltage of a diode

Fig. 4: Experimental results of conventional AC-DC converter with center-tapped isolated transformer and full-wave rectifier

In addition, the forward voltage of diodes increases with increment in DC output current. Consequently, the DC output voltage decreases significantly if the DC output current increases. In summary, the experimental results and the theoretical analysis show that both the conventional AC-DC convert with an isolated transformer and a full-bridge rectifier and the conventional AC-DC converter with a center-tapped isolated transformer and a full-wave rectifier are not suitable for high frequency AC-DC power conversion.

III. PROPOSED AC-DC CONVERTERS APPLIED TO HIGH FREQUENCY AC-DC POWER CONVERSION

To obtain the expected DC output voltage and DC output power, three schemes are proposed in this paper, in which a capacitor is connected with the secondary winding in parallel and operates under resonance to maximize the DC output power. Such a capacitor is named the resonant capacitor. Three proposed schemes have different topologies of the converters and will be analyzed and studied in this section.

1. AC-DC converter with isolated transformer, resonant capacitor and class-E rectifier
Based on the class-E rectifier presented in [5], the AC-DC converter with an isolated transformer, a resonant capacitor and a class-E rectifier is proposed as shown in Fig. 5, where r_1 and r_2 represent the resistances of the

primary and secondary windings, respectively, L_1 and L_2 the leakage inductances of the primary and secondary windings, L_m the magnetizing inductance (core loss is neglected), N_1 and N_2 the numbers of the primary and secondary coil turns, C_r the resonant capacitance, C_f the filter capacitance, and R the resistive load. The main parameters of the isolated transformer are as follows: capacity = 110 VA, frequency = 50 kHz, primary voltage = 36 V and secondary voltage = 11V.

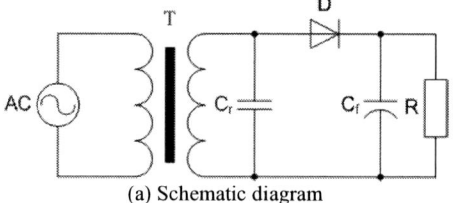

(a) Schematic diagram

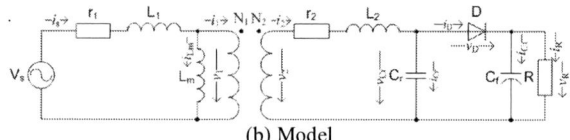

(b) Model

Fig. 5: AC-DC converter proposed in scheme 1

Referring to Fig. 5, the model of the proposed converter can be expressed as

$$v_1 = V_s - i_s r_1 - L_1 \frac{di_s}{dt} = L_m \frac{di_m}{dt} \quad (1)$$

$$i_s = i_{Lm} + i_1 \quad (2)$$

$$v_2 = v_1 \left(\frac{N_2}{N_1}\right) \quad (3)$$

$$i_2 = i_1 \left(\frac{N_1}{N_2}\right) \quad (4)$$

$$v_{Cr} = v_2 - i_2 r_2 - L_2 \frac{di_2}{dt} \quad (5)$$

$$i_{Cr} = C_r \frac{dv_{Cr}}{dt} \quad (6)$$

$$i_{Cf} = C_f \frac{dv_R}{dt} \quad (7)$$

$$i_R = \frac{v_R}{R} \quad (8)$$

$$v_{Cr} = v_D + v_R \quad (9)$$

If $v_{Cr} > v_R$, one has

$$i_2 = i_{cr} + i_D \quad (10)$$

$$i_D = i_{Cf} + i_R \quad (11)$$

If $v_{Cr} \leq v_R$, one has

$$i_2 = i_{Cr} \quad (12)$$

$$i_D = i_{Cf} + i_R = 0 \quad (13)$$

The experimental results of the AC-DC converter with an isolated transformer and a class-E rectifier are depicted in Fig. 6.

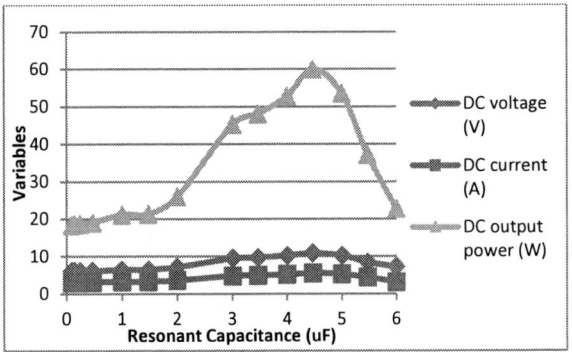

(a) Effect of resonant capacitance (V_s = 14.2 V and R = 2.0 ohm)

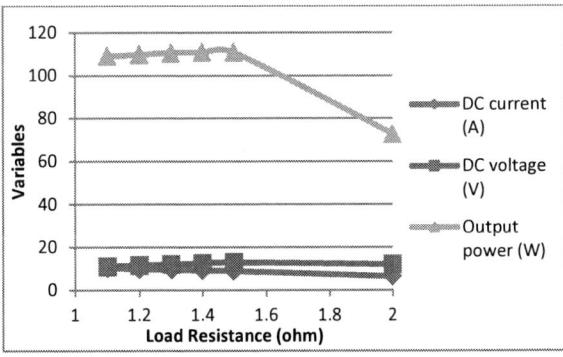

(b) Output characteristics (C_r = 4.47 µF)

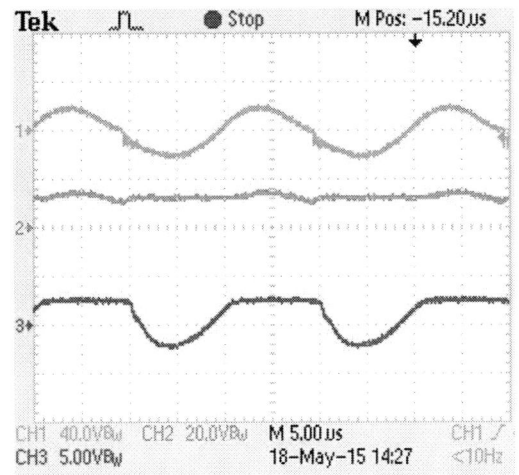

(c) R = 1.506 ohm and C_r = 4.2 µF (Ch1: secondary voltage, 40V/Div; Ch2: DC voltage, 20V/Div; Ch3: secondary current, 50A/Div)

It can be observed from Fig. 6 that (a) the resonant capacitance has the considerable effect on the DC output and the optimal resonant capacitance can be found to maximize the DC output power, (b) the DC output voltage and the DC output power can reach to the expected values, respectively if the resonant capacitance is equal to the optimal value, (c) the primary voltage, current and secondary voltage are sinusoidal, and the secondary current is non-sinusoidal, (d) there are the large secondary current, the large resonant current and the large forward current due to the topology of the class-E rectifier, and (e) the DC output voltage is considerably flat.

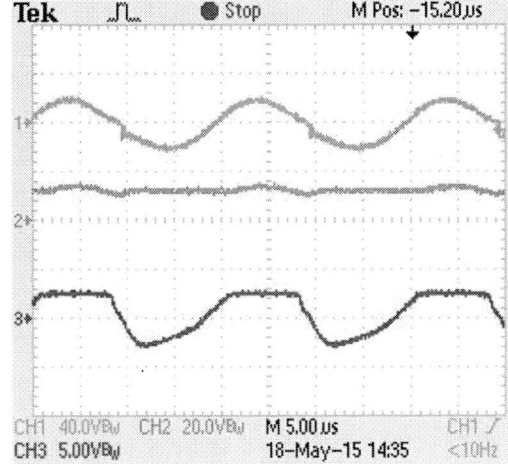

(d) R = 1.506 ohm and C_r = 4.2 µF (Ch1: secondary voltage, 40V/Div; Ch2: DC voltage, 20V/Div; Ch3: resonant current, 50A/Div)

Fig. 6: Experimental results of AC-DC converter with an isolated transformer and a class-E rectifier

2. AC-DC converter with isolated center-tapped transformer, resonant capacitors and class-E rectifiers

The proposed AC-DC converter with a center-tapped transformer, two resonant capacitors and two class-E rectifiers is shown in Fig. 7, where the secondary winding consists of two same secondary coils via the center-tapped approach and the number of each coil turns is equal to that of the secondary winding in Fig. 5. The main parameters of the isolated center-tapped transformer are as follows: capacity = 110 VA, frequency = 50 kHz, primary voltage = 36 V, and secondary voltage = 11V.

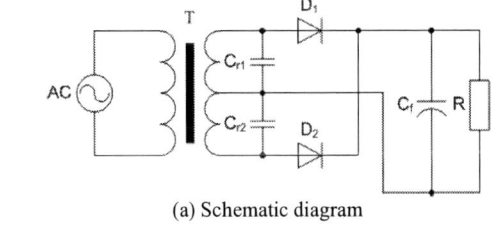

(a) Schematic diagram

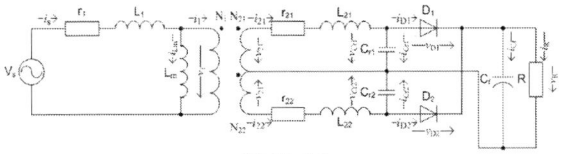

(b) Model

Fig. 7: AC-DC converter proposed in scheme 2

From Fig. 7, the dynamic equations of the model can be given as

$$v_1 = V_s - i_s r_1 - L_1 \frac{di_s}{dt} = L_m \frac{di_m}{dt} \tag{14}$$

$$i_s = i_{Lm} + i_1 \tag{15}$$

$$v_{21} = v_1 \left(\frac{N_{21}}{N_1} \right) \tag{16}$$

978-1-5090-0064-7/15 $31.00 © 2015 IEEE   245

$$i_{21} = i_1 \left(\frac{N_1}{N_{21}} \right) \tag{17}$$

$$v_{Cr1} = v_{21} - i_{21} r_{21} - L_{21} \frac{di_{21}}{dt} \tag{18}$$

$$i_{Cr1} = C_{r1} \frac{dv_{Cr1}}{dt} \tag{19}$$

$$v_{22} = -v_1 \left(\frac{N_{22}}{N_1} \right) \tag{20}$$

$$i_{22} = -i_1 \left(\frac{N_1}{N_{22}} \right) \tag{21}$$

$$v_{Cr2} = v_{22} - i_{22} r_{22} - L_{22} \frac{di_{22}}{dt} \tag{22}$$

$$i_{Cr2} = C_{r2} \frac{dv_{Cr2}}{dt} \tag{23}$$

$$i_{Cf} = C_f \frac{dv_R}{dt} \tag{24}$$

$$i_R = \frac{v_R}{R} \tag{25}$$

$$v_{Cr1} = v_{D1} + v_R \tag{26}$$

$$v_{Cr2} = v_{D2} + v_R \tag{27}$$

If $v_{Cr1} > v_R$, one has

$$i_{21} = i_{Cr1} + i_{D1} \tag{28}$$

$$i_{D1} = i_{Cf} + i_R \tag{29}$$

If $v_{Cr1} \leq v_R$, one has

$$i_{21} = i_{Cr1} \tag{30}$$

$$i_{D1} = i_{Cf} + i_R = 0 \tag{31}$$

If $v_{Cr2} > v_R$, one has

$$i_{22} = i_{Cr2} + i_{D2} \tag{32}$$

$$i_{D2} = i_{Cf} + i_R \tag{33}$$

If $v_{Cr2} \leq v_R$, one has

$$i_{22} = i_{Cr2} \tag{34}$$

$$i_{D2} = i_{Cf} + i_R = 0 \tag{35}$$

Fig. 8 illustrates the experimental results of the AC-DC converter with an isolated center-tapped transformer and two class-E rectifiers.

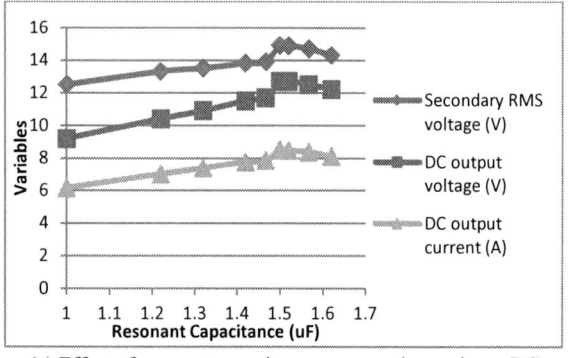

(a) Effect of resonant capacitance on secondary voltage, DC output voltage and DC output current

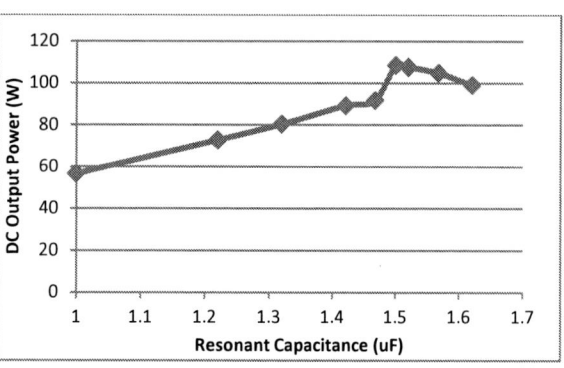

(b) Effect of resonant capacitance on DC output power

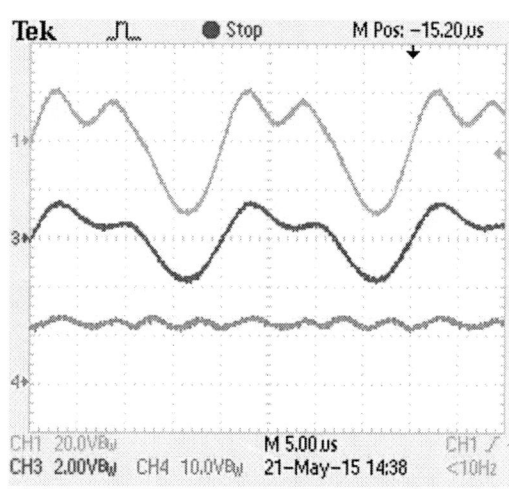

(c) $C_r = 1.5$ μF (Ch1: secondary voltage, 40 V/Div; Ch3: secondary current, 10A/Div; Ch4: DC voltage, 10V/Div)

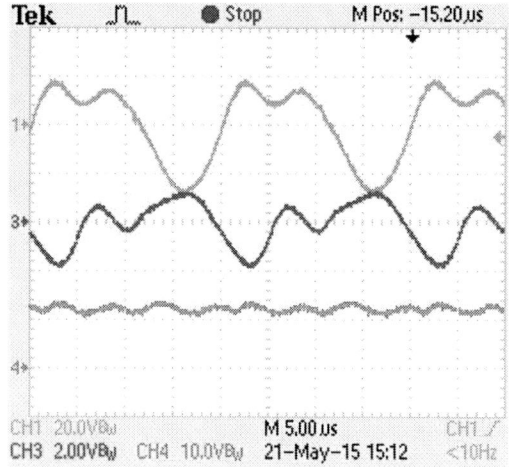

(d) $C_r = 1.5$ μF (Ch1: secondary voltage 1, 20 V/Div; Ch3: resonant current 1, 20A/Div; Ch4: DC voltage, 10V/Div)
Fig. 8: Experimental results of AC-DC converter with an isolated center-tapped transformer and two class-E rectifiers (R=1.507 ohm)

It can be observed from Fig. 8 that (a) the resonant capacitance has the significant effect on the DC output, and the optimal resonant capacitance can be found to maximize the DC output power, (b) the DC output voltage and the DC output power can reach to the expected values, respectively, (c) the primary current, and the secondary voltage and current are non-sinusoidal and have large harmonics, (d) in comparison with the scheme 1, there are the small secondary current, the small resonant current, and the small forward current due to the topology of two class-E rectifiers, and (e) the DC output voltage has the ripple.

3. AC-DC converter with isolated transformer, resonant capacitor and full-bridge rectifier

Fig. 9 illustrates the schematic and model of the proposed AC-DC converter with an isolated transformer, a resonant capacitor and a full-bridge rectifier. The transformer is the same with that in the scheme 1.

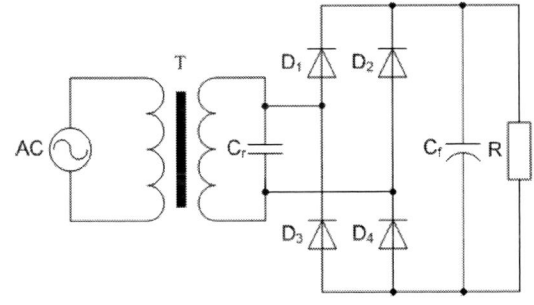

(a) Schematic diagram

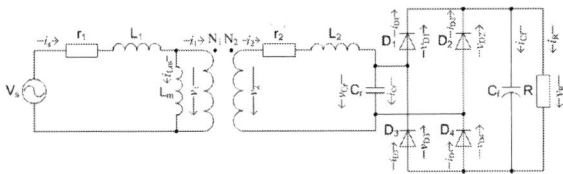

(b) Model

Fig. 9: AC-DC converter proposed in scheme 3

Referring to Fig. 9, the dynamic equations of the model can be expressed as

$$v_1 = V_s - i_s r_1 - L_1 \frac{di_s}{dt} = L_m \frac{di_m}{dt} \tag{36}$$

$$i_s = i_{Lm} + i_1 \tag{37}$$

$$v_2 = v_1 \left(\frac{N_2}{N_1} \right) \tag{38}$$

$$i_2 = i_1 \left(\frac{N_1}{N_2} \right) \tag{39}$$

$$v_{Cr} = v_2 - i_2 r_2 - L_2 \frac{di_2}{dt} \tag{40}$$

$$i_{Cr} = C_r \frac{dv_{Cr}}{dt} \tag{41}$$

$$i_{Cf} = C_f \frac{dv_R}{dt} \tag{42}$$

$$i_R = \frac{v_R}{R} \tag{43}$$

If $v_{Cr} > v_R$, one has

$$i_2 = i_{Cr} + i_{D1} \tag{44}$$

$$i_{D1} = i_{D4} = i_{Cf} + i_R \tag{45}$$

$$v_{Cr} = v_{D1} + v_{D4} + v_R \tag{46}$$

If $-v_{Cr} > v_R$, one has

$$-i_2 = -i_{Cr} + i_{D2} \tag{47}$$

$$i_{D2} = i_{D3} = i_{Cf} + i_R \tag{48}$$

$$-v_{Cr} = v_{D2} + v_{D3} + v_R \tag{49}$$

The experimental results of the AC-DC converter proposed in scheme 3 are shown in Fig.10. It can be seen that (a) the resonant capacitance has the significant effect on the DC output, and the optimal resonant capacitance can be found to maximize the DC output power, (b) the DC output voltage and the DC output power can reach to the expected values, respectively, (c) the primary voltage and current, and the secondary voltage and current are sinusoidal, (d) there is the small forward current due to the topology of the full-bridge rectifier, and (e) the DC output voltage is flat.

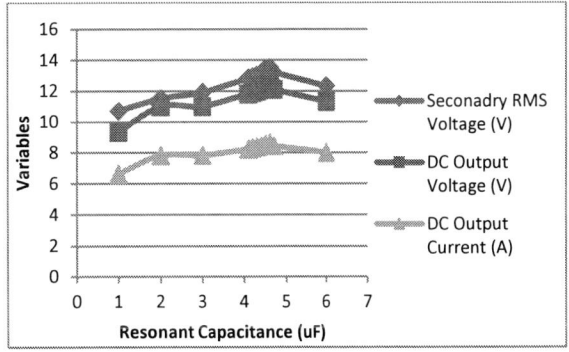

(a) Effect of resonant capacitance on secondary voltage, DC output voltage and DC output current

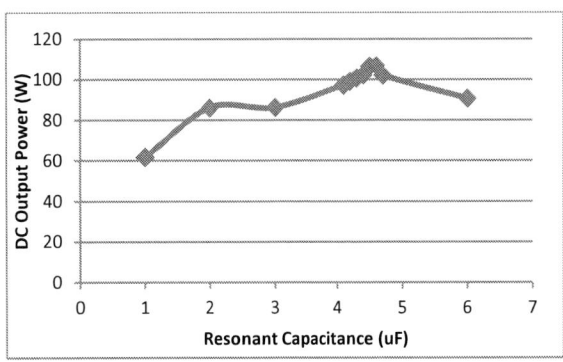

(b) Effect of resonant capacitance on DC output power

978-1-5090-0064-7/15 $31.00 © 2015 IEEE

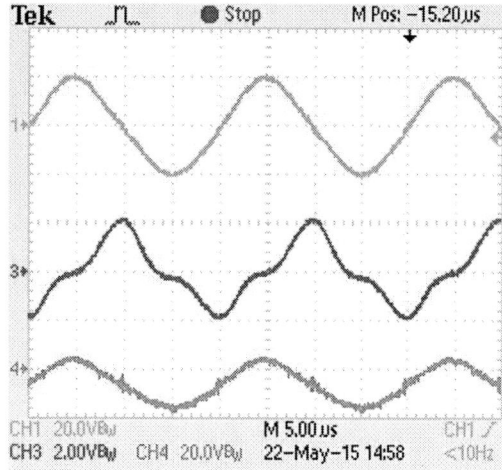

(c) $C_r = 4.6$ μF (Ch1: secondary voltage, 20V/Div; Ch3: AC current flowing to full-bridge rectifier, 20A/Div; Ch4: forward voltage of diode 1, 20V/Div)

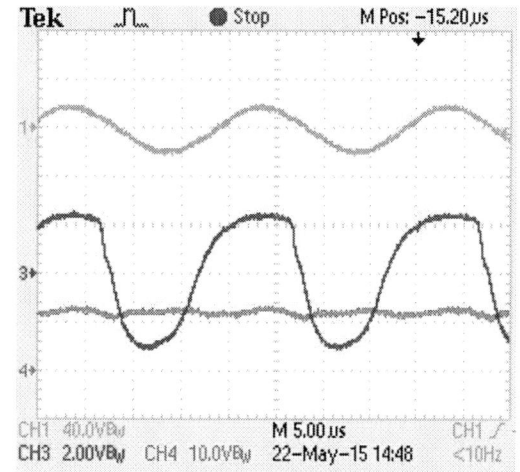

(d) $C_r = 4.6$ μF (Ch1: secondary voltage, 40V/Div; Ch3: current in resonant capacitor, 20A/Div; Ch4: DC output voltage, 10V/Div)

Fig. 10: Experimental results of AC-DC converter with isolated transformer, resonant capacitor and full-bridge rectifier (R = 1.507 ohm)

4. Comparison and evaluation

Based on the analysis of three proposed converters and the experimental results, the comprehensive comparison between three discussed converters is described in the table 1. Three marks are used to evaluate each performance index, which are good, medium and bad.

Taking into account all the performance indexes in the table 1, it can be see that the scheme 1 and the scheme 3 are better than the scheme 2. Furthermore, the scheme 3, i.e., the AC-DC converter with an isolated transformer, a resonant capacitor and a full-bridge rectifier, is preferred among three schemes, if the harmonics of the secondary current and the resonant current, and the forward current of the diode are the prior evaluation indexes.

Using the AC-DC converter in the scheme 3, the experimental results of the output performance are

illustrated in Fig. 11. It can be seen that the output performance is satisfactory. Thus, the analysis and the experimental results verify that the AC-DC converter prior proposed in this study is effective and reasonable, and can be applied to real high frequency AC-DC power conversion.

Table 1: Comparison between three schemes

Performance index	Scheme1	Scheme2	Scheme3
Complicacy of transformer	Good	Medium	Good
Complicacy of rectifier	Good	Medium	Medium
Harmonics of primary voltage	Good	Medium	Good
Harmonics of primary current	Good	Bad	Good
Harmonics of secondary voltage	Good	Bad	Good
Harmonics of secondary current	Medium	Bad	Good
Harmonics of resonant current	Medium	Bad	Good
Magnitude of resonant current	Medium	Good	Medium
Capacitance of resonant capacitor	Medium	Good	Medium
Magnitude of forward current	Bad	Good	Good
Magnitude of forward voltage	Medium	Bad	Good
DC output power	Good	Good	Good
DC output voltage	Good	Good	Good
Ripple of DC output voltage	Good	Bad	Good

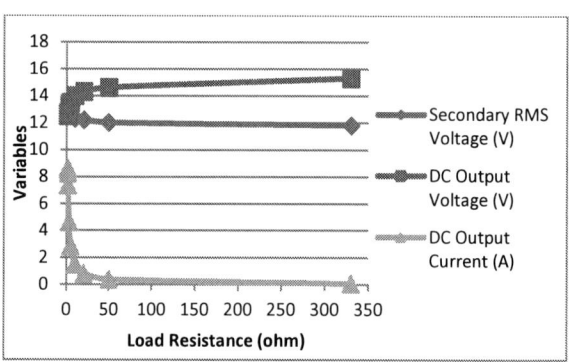

(a) Effect of resistive load on secondary RMS voltage, average DC voltage, average DC current

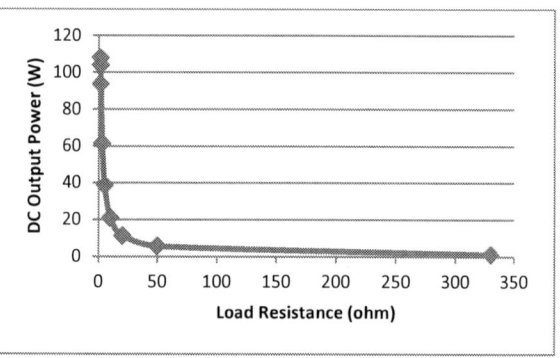

(b) Effect of resistive load on DC output power

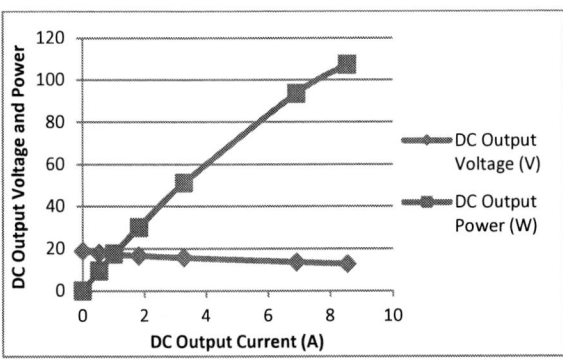

(c) Effect of DC output current on DC output voltage and DC output power

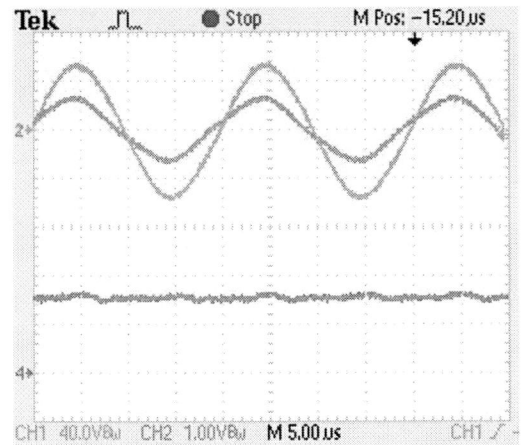

(d) V_R = 15.53 V and I_R = 3.29 A (Ch1: primary voltage, 40V/div; Ch2: primary current, 10A/div; Ch4: DC output voltage, 10V/div)

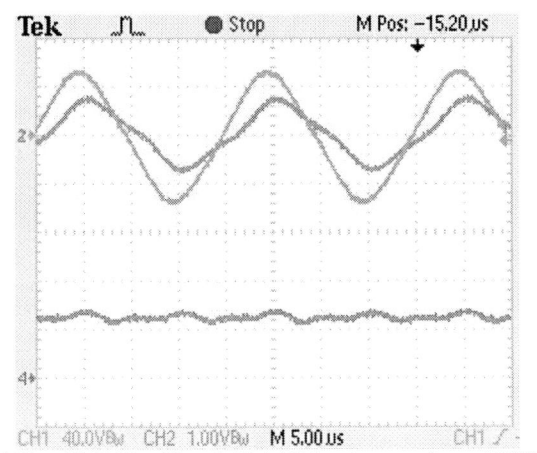

(e) V_R = 12.54 V and I_R = 8.55 A (Ch1: primary voltage, 40V/div; Ch2: primary current, 10A/div; Ch4: DC output voltage, 10V/div)
Fig. 11: Experimental results of AC-DC converter with an isolated transformer, a resonant capacitor and a full-bridge rectifier

IV. CONCLUSION

The study in this paper has shown that conventional AC-DC converters with isolated transformers and rectifiers are not suitable for high frequency AC-DC power conversion. Three topologies of AC-DC converters have been studied for applications to high frequency AC-DC power conversion, which are the AC-DC converter with an isolated transformer, a resonant capacitor and a class-E

rectifier, the AC-DC converter with an isolated center-tapped transformer, two resonant capacitors and two class-E rectifiers, and the AC-DC converter with an isolated transformer, a resonant capacitor and a full-bridge rectifier. Three prototypes are developed to examine three proposed schemes. The experimental results and the analysis have demonstrated that three proposed AC-DC converters are effective and able to meet the design objective. The comparative study of three schemes has indicated that the AC-DC converter with an isolated transformer, a resonant capacitor and a full-bridge rectifier and the AC-DC converter with an isolated transformer, a resonant capacitor and a class-E rectifier are better than the one with an isolated center-tapped transformer, two resonant capacitors and two class-E rectifiers. Furthermore, the AC-DC converter with an isolated transformer, a resonant capacitor and a full-bridge rectifier is preferred in this paper. This study provides the reasonable and valuable solutions to AC-DC converters applied to high frequency AC-DC power conversion.

ACKNOWLEDGMENT

The authors would like to thank to the part support from the Innovation and Technology Fund of Hong Kong under Project ITS/036/14.

REFERENCES

[1] Junfeng Liu., C. D. Xu, K. W. Eric Cheng, K.W.E. Cheng, "Some Considerations of Arc Protection and Breaker Design for High Frequency AC Power Distribution Systems", *5th IEEE International Conference on Power Electronics Systems and Applications (PESA2013)*, pp. 1-8.

[2] B.K. Bose, Min-Huei Kin, and M.D. Kankam, "High frequency AC vs. DC distribution system for next generation hybrid electric vehicle," *International Conference on Industrial Electronics, Control, and Instrumentation, IEEE IECON*, vol.2, 1996, pp.706-712.

[3] Vatche Vorperian, and Raymond B. Ridley, "A Simple Scheme for Unity Power-Factor Rectification for High Frequency AC Buses", *IEEE Transactions on Power Electronics*, vol. 5, no. I , January 1990, pp. 77-87

[4] Junfeng Liu; Cheng, K.W.E.; Jun Zeng, "A Unified Phase-Shift Modulation for Optimized Synchronization of Parallel Resonant Inverters in High Frequency Power System", *IEEE Transactions on Industrial Electronics*, vol.61, no.7, pp.3232,3247, July 2014

[5] Gurhan Alper Kendir, Wentai Liu, Guoxing Wang, Mohanasankar Sivaprakasam, Rizwan Bashirullah, Mark S. Humayun, and James D. Weiland, "An Optimal Design Methodology for Inductive Power Link With Class-E Amplifier", *IEEE Transactions on Circuits and Systems,* vol. 52, no. 5, May 2005, pp. 857-866.

978-1-5090-0064-7/15 $31.00 © 2015 IEEE

Integration Development for Supercapacitor Controlled Distributed Generation System

Xiaolin Wang[1]   K.W.Eric Cheng[2]   Yongquan Nie[2]

[1,2] Department of Electrical Engineering, The Hong Kong Polytechnic University, Hong Kong
[1] E-mail: xiaolinee.wang@connect.polyu.hk

Abstract–With the construction of large-scale distributed power stations, the distributed generation power imbalance becomes a serious concern as many distributed generation systems and large power grids cannot be consistent with users' load. Wind power, which is used worldwide, is regarded as the typical example of the distributed generation power. In the wind power plants, though reactive power can be easily compensated locally by adding static var devices, the compensation of active power can be difficult. In order to solve the problem, this paper conducts research on the supercapacitor (SC) energy storage system controlled wind power integration, striving to find ways to make the entire output smooth and stable. This method utilizes the characteristics of SC to compensate the intermittent wind power. Case studies on a test system are presented using wind data to substantiate the problem.

Keywords– Supercapacitor, wind power generation; energy storage system, droop control method, active power compensation.

I. INTRODUCTION

Nowadays wind farm is one of the fastest developing renew energy plants. Wind energy growing in Denmark is fast, which constitutes more than 33% of national electricity consumption. Denmark plans to be 50% in the 2020 and 100% fossil fuel free country by 2050 [1]. The clean energy will reduce the CO_2 emission over 10 billion ton worldwide. However, solar energy, wind energy, tidal energy or some regenerative braking energy released within a short time can impact power grids, or even cause a large scale of electrical incidents. With the approaching of the energy crisis, wind power, which is renewable and cannot be consumed up, has been gradually taken as the substitution [2]-[5].

However, as an intermittent energy source, wind power cannot be predicted: it peaks for the moment but may fade away the next minute. This nature makes its output fluctuate so greatly that the wind power cannot be constantly accepted by the grid.In the common wind power plants, though reactive power can be easily compensated by static var devices or capacitance array [6], the compensation of active power will be difficult. The active power is compensated by fixed speed generator wind turbine and doubly-fed induction generator wind turbine. Due to the wear and tear on the equipment, the total efficiency is low. But for the energy storage system integration with wind farm could easily solve this problem. It can compensate the reactive power balancing the voltage level of the main bus, meanwhile modifying the active power in a wide range.

The energy storage system includes battery, flywheel, superconducting energy storage and supercapacitor. [7]-[9]

show that low frequency (0.01Hz-1Hz) imbalance contributes the main part of the instability problem in the power grids.

Supercapacitor (SC) is called electrochemical double layer capacitor (EDLC) [10]. Though it is electrochemical devices, there is no chemical reaction involved. Using SC is suitable for balancing this part in seconds. The integration of the supercapacitor system and wind plants to reduce the uncertainty of wind power can enhance the reliability and security of the grid.

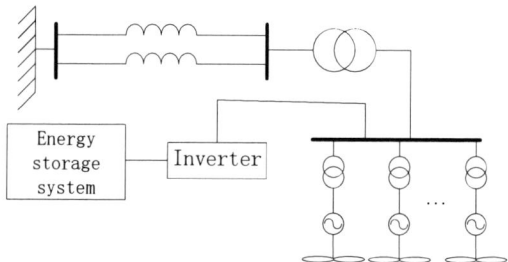

Fig.1: The integration of the energy storage system and wind plants

First of all, this paper will introduce the supercapacitor energy storage systems. Afterwards, this paper proposed two control strategies (decoupled P-Q control and droop control) for paralleled supercapacitor energy storage system and made a simulation on PSCAD to test their performance respectively. By using droop control method, this paper will simulate wind power integration possible with the control of supercapacitor energy storage system and analyzed its performance on the output of active power, reactive power and voltage. Simultaneously, a case study is made to obtain the transient response of the whole system when coming across instant three-phase grounding fault, further demonstrating its stability.

II. MATHEMATICAL MODELLING

A. Wind power model

Dynamic system in the wind turbine includes rotor, gear box, wind blades, generator, and controller. The mathematic model of the wind turbine shows below [2].

$$T_{turbine} = \frac{1}{2} \pi \rho \frac{C_p}{\lambda} R^3 v_{wind}^2 \times \frac{\Omega_N}{P_N} \times 10^{-3} \qquad (1)$$

where, $T_{turbine}$ is the torque of the blades. R is the radius of the blade. ρ is density of the air. v_{wind} is the speed of the wind . λ is tip speed ratio $\lambda = \dfrac{\varpi R}{v_{wind}}$. P_N is the rated output power of the wind turbine. Ω_N is the mechanical angular speed of the blades. C_p is the power factor of the wind turbine.

S_Σ is the summation of total equivalent rated power. S_i is the rated power of the ith wind generation system.

$$S_\Sigma = \sum_{i=1}^{n} S_i \qquad (2)$$

B. SC electrical circuit modelling
Due to the fact that SC is electrochemical devices, it is necessary to know its unique features during the application. Two different electrical modelling of supercapacitor have been proposed.

RC model is the simplest model of the SC as shown in Fig.2. It only has one branch that consists a resister to simulate the equivalent series resistor (ESR) and a capacitor to model the SC's capacitor during charging and discharging period.

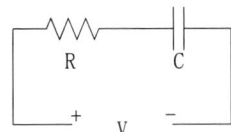

Fig.2: Equivalent electrical circuit for RC model

RC parallel branch model is more accurate than the previous one based on the three different branches (fast term branch, medium term branch and long term branch). Three branches have different time constant from seconds to minutes. It is common and wildly used in simulation of SC [7].

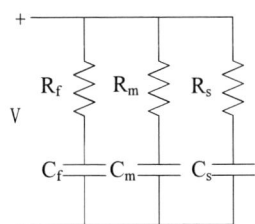

Fig.3: RC parallel branch model

III. PROBLEM FORMULATION

A. Problem of integrated modelling
Derived from (1), the output torque of the wind turbine is proportional to the square of the wind speed v_{wind}. The $T_{turbine}$ varies significantly as the wind speed changes quickly. Installation of the SC energy storage system can solve this problem. As shown in Fig. wind turbine and SC modules are integrated in the same distributed generation system. SC energy storage system could balance the active

and reactive power of the intermittent wind power with proper control scheme.

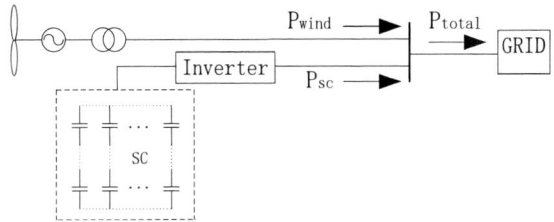

Fig.4: Integration of SC and wind power generation system

B. Droop control
The SC system and wind turbine integration system can be seen as a microgrid system. In microgrid, distributed sources are located in multiple places and it is hard to have a uniform control for the inverters installed at SC system which is in parallel with the wind power generator. To guarantee the stability and power quality of wind power generation, it is advised to use the well-known P/Q droop control method [11]-[15].

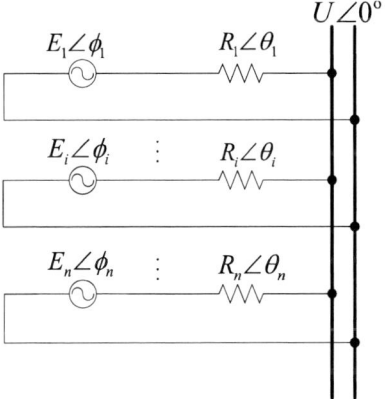

Fig.5: The paralleled multi-inverter system

The paralleled multi-converter system is shown in Fig.5. $R_i \angle \theta_i$ is the output impedance of the ith inverter. E_i is the output voltage and ϕ_i is the angle between output voltage and bus voltage. U is the bus voltage. Deriving from fundamental circuit principles, the output current of the ith inverter is shown in (3).

$$I_i = \frac{E_i \angle \phi_i - U \angle 0}{R_i \angle \theta_i} \qquad (3)$$

The output apparent power is:

$$S_i = E_i \angle \theta_i \cdot I_i^* = P_i + jQ_i \qquad (4)$$

Therefore, the output active power P_i and reactive power Q_i of the ith inverter are calculated to be (5) and (6), respectively.

$$P_i = \frac{1}{R_{zi}}\left[\left(E_iU\cos\phi_i - U^2\right)\cos\theta_i + E_iU\sin\phi_i\sin\theta_i\right] \tag{5}$$

$$Q_i = \frac{1}{R_{zi}}\left[\left(E_iU\cos\phi_i - U^2\right)\sin\theta_i - E_iU\sin\phi_i\cos\theta_i\right] \tag{6}$$

As in the high voltage networks, the line impedance can be regarded as inductive reactance and the output impedance of the inverter should also be inductive reactance. Therefore, the angle θ_i of the impedance approximates 90°. （7）and （8）are simplified as below accordingly:

$$P_i = \frac{E_iU}{X_i}\sin\phi_i \tag{7}$$

$$Q_i = \frac{E_iU\cos\phi_i - U^2}{X_i} \tag{8}$$

When ϕ_i is sufficiently small, （9）and （10）can be further simplified as below:

$$P_n = \frac{E_iU}{X_i}\sin\phi_i \approx \frac{E_iU}{X_i}\cdot\phi_i \tag{9}$$

$$Q_i = \frac{E_iU\cos\phi_i - U^2}{X_i} \approx \frac{E_iU - U^2}{X_i} \tag{10}$$

After differentiating both sides of (9) and (10) simultaneously, the dynamics of system output power are obtained.

$$\Delta P_i \approx \frac{E_iU}{X_i}\cdot\Delta\phi_i \tag{11}$$

$$\Delta Q_i \approx \frac{U}{X_i}\cdot\Delta E_i \tag{12}$$

(11) and (12) are the principles of droop control. Derived from the above two equations, the dynamics of active power output are determined by the variation of power angle ϕ_i while the dynamics of reactive power output are mainly determined by the changes in inverter's output voltage E_i. Therefore, the desired active and reactive power output can be controlled separately and efficiently. The flow chart of the droop control method is depicted in Fig.6. In practical operation, the bus voltage, active and reactive power outputs of wind generation are continuously monitored to provide up-to-date information to the control center. The control center then calculates the optimal PWM control for the inverter.

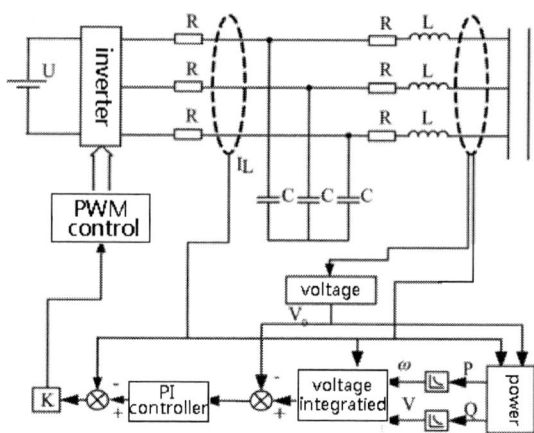

Fig.6: Flow chart of the droop control

IV. A CASE STUDY

A. Basic system configuration

Based on the consumption of the electricity, this paper will set up a wind plant with 100MVA. Meanwhile, the SC system should be at least 100MVA to compensate the wind plant. The rated volume of the system is set to be 200MVA.

In order to examine the stability of the whole system, there is no control device installed on the wind turbine itself. In the case study, noise is added to the wind speed signal and the whole system is tested under the wind speed variation shown in Fig.7.

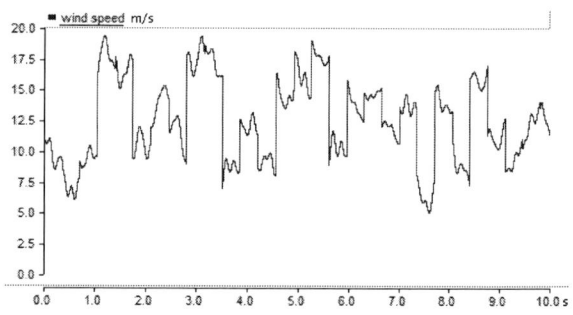

Fig.7: model of wind speed variation

B. Simulation results under normal conditions

In general, the goal of the whole system is to make the total active/reactive power output as smooth as possible even if the wind speed changes quickly.

When the desired active power output of the whole system is set to be $P_{ref} = 0.7$, the simulation results are shown in Fig.7 , It can be seen that the output voltage, active and reactive power are all settled down after around 1.5s with the control scheme. The stabilized value of the active power output perfectly matches P_{ref}. The value of the reactive power output can be easily adjusted with local static var compensation.

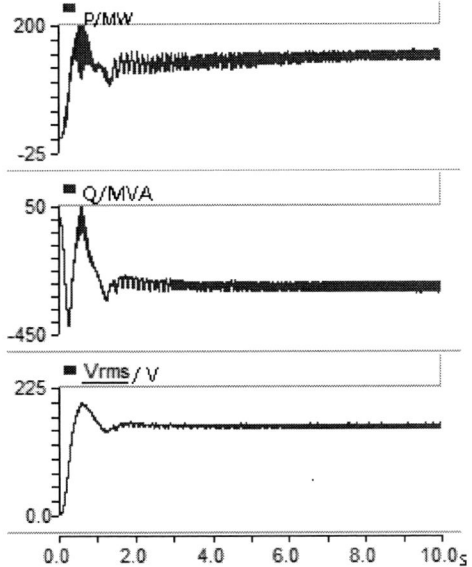

Fig.7: Simulation results under normal conditions

C. Simulation results with short circuit

To assess the reliability of the whole system, a short time three phase grounding fault is added to the common node of the system at t=4.0s and removed after 0.05s. The desired active power output is still assumed to be $P_{ref} = 0.7$.

The simulation results are shown in Fig.8.

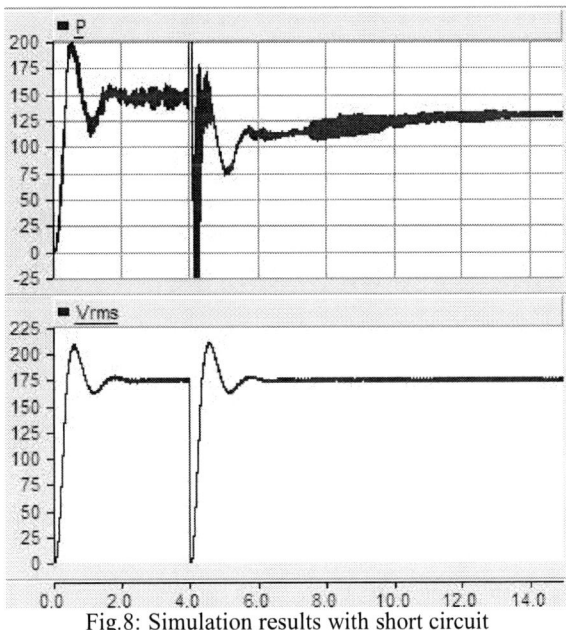

Fig.8: Simulation results with short circuit

Generally speaking, instantaneous three phase grounding fault may trigger the protection devices or even cause blackout if it is too serious. The proposed control system, however, can restore the system within 1s. Although the stabilized active power output is less than the desired P_{ref}, it is still within the acceptable margin.

V. CONCLUSION

Based on the energy storage characteristics of the SC, the common wind power is integrated with SC storage system. Using droop control method, this paper made wind power integration possible with the control of SC energy storage system to smooth the intermittent output of active power, reactive power and voltage. The simulation results agreed with the theoretical prediction and verified the theory proposed of improving the stability and reliability of the grid.

Future work includes the applications of SC power balancing system to other renewable distributed generation system such as solar, tide and hydro power.

REFERENCES

[1] Energy Policy in Denmark, Danish Energy Agency, Dec. 2012.

[2] C. Luo, H. Banakar, "Strategies to smooth wind power fluctuations of wind turbine generator." *IEEE Trans on Energy Conversion*, 2007, 22 (2): 64-69.

[3] S. Lamichhane, N. Mithulananthan, "Influence of Wind Energy Integration on Low Frequency Oscillatory Instability of Power System", *Australasian Universities Power Engineering Conference, AUPEC 2014*, Sep. 2014.

[4] P. A. Coppin, "Using intelligent storage to smooth wind energy generation," *Conf. Electr. Energy Storage Appl. Technol.*, San Francisco, CA, Sep. 2007.

[5] S. Teleke, M. E. Baran, A. Q. Huang, "Control Strategies for Battery Energy Storage for Wind Farm Dispatching", *IEEE Trans. on Energy Conversion*, vol. 24, no. 3, Sep. 2009.

[6] L. Zubieta, and R. Boner, "Characterization of double-layer capacitors for power electronics applications," *IEEE Trans. on Industry Applications*, vol. 36, no. 1, pp. 199-205, Feb. 2000.

[7] A. Tuohy and M. O'Malley, "Impact of Pumped Storage on Power Systems with Increasing Wind Penetration", *IEEE Trans. on Energy Conversion*, Jul. 2009

[8] P. Hu, R. Karki, R. Billinton, "Reliability evaluation of generating systems containing wind power and energy storage", *IET Generation Transmission Distribution*, vol. 3, iss. 8, pp. 783–791

[9] R. H. Lasseter, "Microgrids," Power Engineering Society Winter Meeting, IEEE, 2002.

[10] R. Majumder, and B. Chaudhuri, "Improvement of Stability and Load Sharing in an Autonomous Microgrid Using Supplementary Droop Control Loop," *IEEE Trans. on Power Systems*, vol. 25, no. 2, pp. 796-808, Apr. 2010.

[11] L. Xiaonan, and S. Kai, "State-of-Charge Balance Using Adaptive Droop Control for Distributed Energy Storage Systems in DC Microgrid Applications," *IEEE Trans on Industrial Electronics*, vol. 61, no. 6, pp. 2804-2815, Dec. 2013.

[12] K. Rouzbehi, and A. Miranian, "A Generalized Voltage Droop Strategy for Control of Multiterminal DC Grids," *IEEE Trans on Industry Applications*, vol. 51, no. 1, pp. 607-618, Jan. 2015.

[13] K. Jaehong, and J. M. Guerrero, "Mode Adaptive Droop Control With Virtual Output Impedances for an Inverter-Based Flexible AC Microgrid," *IEEE Trans on Power Electronics*, vol. 26, no. 3, pp. 689-701, Mar. 2011.

[14] H. Bevrani, and S. Shokoohi, "An Intelligent Droop Control for Simultaneous Voltage and Frequency Regulation in Islanded Microgrids," *IEEE Trans. on Smart Grid*, vol. 4, no. 3, pp. 1505-1513, Aug. 2013.

[15] J. M. Guerrero, and J. C. Vasquez, "Hierarchical Control of Droop-Controlled AC and DC Microgrids—A General Approach Toward Standardization," *IEEE Trans. on Industrial Electronics*, vol. 58, no. 1, pp. 158-172, Jan. 2011.

BIOGRAPHIES

Xiaolin Wang obtained her bachelor degree in Electrical Engineering and Automation from Three Gorges University, Yi Chang, China, in 2012. She received her master degree in Electrical Engineering from the Hong Kong Polytechnic University, Hong Kong, in 2013. Currently, she is a Ph.D. student in the department of Electrical Engineering at the Hong Kong Polytechnic University, Hong Kong.

K.W.E.Cheng obtained his BSc and PhD degrees both from the University of Bath in 1987 and 1990 respectively. Before he joined the Hong Kong Polytechnic University in 1997, he was with Lucas Aerospace, United Kingdom as a Principal Engineer.

He received the IEE Sebastian Z De Ferranti Premium Award (1995), outstanding consultancy award (2000), Faculty Merit award for best teaching (2003) from the University and Silver award of the 16th National Exhibition of Inventions, Faculty Engineering Industrial and Engineering Services Grant Achievement Award (2006), Brussels Innova Energy Gold medal with Mention (2007), Consumer Product Design Award (2008), Electric vehicle team merit award of the Faculty (2009), Special Prize and Silver Medal of Geneva's Invention Expo (2011) and Eco Star award (2012). He has published over 250 papers and 7 books. He is now the professor and director of Power Electronics Research Centre.

Yongquan Nie received his B.Eng. degree in Electrical Engineering and Automation from Huazhong University of Science and Technology, Wuhan, China, in 2012. Currently, he is pursuing the Ph.D. degree in the Department of Electrical Engineering, The Hong Kong Polytechnic University, Hong Kong.

His research interests include EV load forecasting, power system planning and state estimation.

High Frequency AC Auxiliary Power Source for Future Vehicles

Raghu Raman S[1], Junfeng Liu, X. D Xue, K W Eric Cheng

Department of Electrical Engineering, The Hong Kong Polytechnic University, Hong Kong
[1]Email: raghu.raman.1990@ieee.org

Abstract – High frequency AC power distribution system have been discussed for a few decades now. It finds applications in computer power supplies, aerospace power, lighting systems, micro grid and telecommunication equipment. This paper applies the high frequency AC power distribution system into a commercial vehicle. Prototype of a high frequency AC power source, rated at 500 VA with distribution frequency of 50 kHz, to be employed in a future vehicle for auxiliary power supply is designed and presented in the paper. DC voltage to the HF inverter is fed from a front-end hard switched boost converter. The converter-inverter system's performance is evaluated from the obtained simulation and preliminary experimental results.

Keywords – DC to DC converter, high frequency AC power distribution, inverter.

I. INTRODUCTION

High frequency alternating current (HFAC) power distribution systems (PDS) offer plethora of advantages over classical direct current (DC) PDS and low frequency alternating current (LFAC) PDS. LFAC system refers to voltage buses with frequency of 50 Hz or 60 Hz, i.e. utility systems. PDS with the transmission frequency greater than utility frequency is commonly referred to as HFAC PDS. For e.g., in 1980s, NASA proposed a 20 kHz, 440 V_{rms}, sinusoidal PDS on its space station [1] & [2].

Principle advantage of HFAC PDS over DC PDS is that it eliminates the necessity for two power conversion stages in the power distribution framework as emphasized in Fig. 1. Reduction in the power conversion stages, which in turn means fewer components, would translate to improved efficiency, lowered cost and higher reliability. Additionally, DC PDSs exhibit poorer transient response due to large filters. A low frequency AC PDS has inferior power density and slower dynamic performance when compared to HFAC PDS. To summarize, HFAC PDS offer the following advantages –

- Significant cost reduction due to fewer power conversion stages and lower component count when compared to DC PDS.
- Improved system efficiency and system integration due to fewer power conversion operations.
- A high-voltage low-current (HVLC) PDS can easily be designed using compact high frequency transformers. HVLC PDS reduces copper loss, eliminates the necessity for larger heat sinks and other auxiliary cooling arrangements. In contrast, DC PDSs, especially in computer power supply architectures, aim at low voltage high current architectures which demand complicated interleaving converters in order to divert current among different phases.
- Higher reliability due to fewer component count and better heat dissipation.

- HF transformer not only offers the flexibility to meet loads at different voltage levels but also provides effective galvanic isolation for the electrical system and a possibility to employ wireless power.
- Higher power density is obtained at high frequency operation due to appreciable reduction in the size of passive components.
- HFAC PDSs offer improved dynamic response due to smaller filters.
- Frequencies greater than 10 kHz are safer to human beings than DC. 60 Hz and 400 Hz systems cause a higher perception and paralysis effect than DC [3].

In spite of several exceptional advantages, there are a few drawbacks to the system as listed out below –

- At high frequencies, skin and proximity effects magnify. This leads to higher ohmic losses and demand improved cable design [1].
- High frequency power distribution amplifies impedance in the transmission line. This increases the reactive power in the system and leads to poorer transmission efficiency.
- Connecting high frequency inverters in parallel is challenging owing to difference in phases of voltage where series connection is limited due to differences in voltage magnitudes.
- Higher Electromagnetic Interference (EMI) effect limits HFAC applications.

HFAC PDS, as elaborated in [4], find application in telecommunication, aerospace, computer power supply, auxiliary power supply units for automotive, micro-grid and lighting systems.

At present, vehicles employ 12 V DC, for power distribution, from a battery pack and alternator unit to cater several electrical units like motors, power converters, actuators, lighting, computer control and entertainment systems. If such a 12 V DC fed auxiliary power supply unit is rated at 1 kW, it would mean that the distribution current at full load is close to 90 A. Such high currents in the system often result in higher copper and semiconductor conduction losses, larger component size and distribution cable diameter, larger heat sinks and other cooling requirements. References [5] and [6] present a detailed study of a HFAC PDS for vehicles, suggests merits of 3-φ ac distribution in vehicles operating at 400 Hz or 800 Hz, and possible usage of induction motors replacing their DC counterpart.

Traditional auxiliary PDS use battery as the lone power source. In a DC PDS, the DC voltage is inverted, stepped up and rectified before being filtered to obtain good quality DC for power transmission. On the load side, inverter is utilized to step down the voltage to obtain the

978-1-5090-0064-7/15 $31.00 © 2015 IEEE

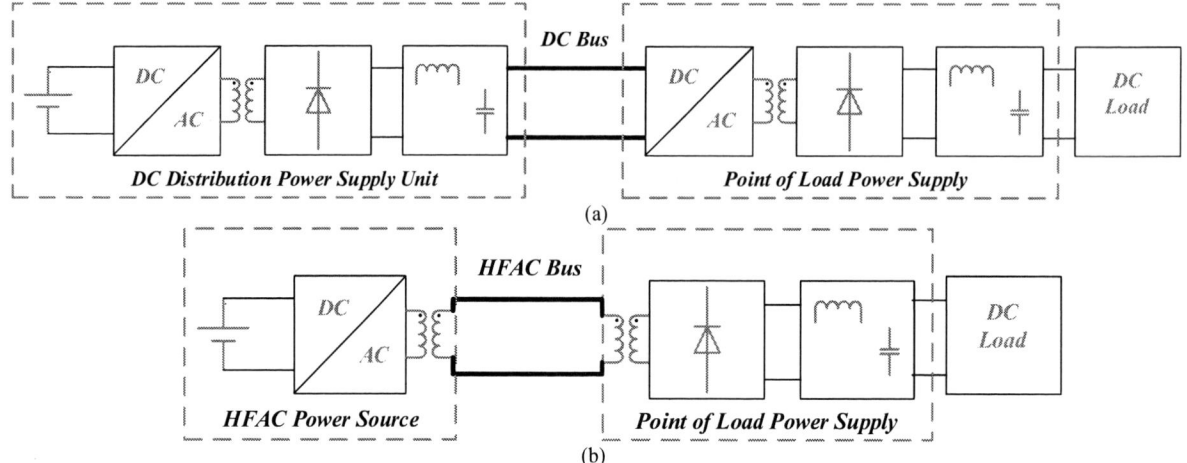

Fig. 1: Highlighting the differences in power distribution architecture (a) DC PDS (b) HFAC PDS

required magnitude DC for the connected loads. In HFAC systems, the need for the rectifier and filter on either ends are eliminated. Additionally, the inverter on the load side isn't required. This paper presents a simple boost converter front end resonant inverter as a HFAC power source. Basic analysis, simulation and experimental results of a 500 VA prototype are included. A description of DC distribution system presently being used in vehicles is briefed in section II. Section III covers the design specifications and topology analysis. Simulation and experimental results are discussed in section IV.

II. EXISTING DC DISTRIBUTION IN VEHICLES

Auxiliary electrical power system unit in a modern day's vehicle is depicted in Fig. 2. The primary power source is a 12 V lead acid battery. A 3-φ alternator is employed to generate power to cater loads such as ignition, fuel-injection, electronic control units, etc. Design of the system should ensure that battery capacity and alternator's output power should account for the on-board load demand. Alternator supplies most of the electrical energy to the auxiliary electrical units on-board a vehicle except during starting. The alternator is excited independently during starting via the pre-excitation circuit powered by 12 V battery. This pre-excitation circuit induces a voltage in the stator that drives excitation current.

Mechanical input to the alternator is from car engine; therefore the output voltage and frequency varies with the engine speed. A voltage regulator is installed in order to obtain constant frequency and voltage to ensure stable DC output to the connected loads and battery. Regulated AC voltage is rectified, filtered and distributed via DC bus while a part of the power generated is used to charge the battery.

The proposed HFAC PDS would utilize the DC output from the alternator and the battery to generate high frequency AC power that can be distributed at a suitable voltage, preferably high voltage to reduce current levels, via a HF transformer. The secondary of HF transformer, nearer to the load side, would either step up or step down AC voltage to feed point of load AC – DC converters.

Auxiliary electronic loads in a modern car (or vehicle), to name a few, are as follows. They include Anti-lock braking system (ABS), headlamp adjustment systems, sensing and trigger systems for airbags, parking aid assistant, power window and sun roof, air-conditioning and heating, central locking, wash-wipe control, seat adjustment, electronic transmission, entertainment systems, Global positioning systems (GPS) and Navigation, car phone etc. With so many individual electronic loads, it can be visualized that a modern car will need hundreds of connectors and several hundred meters of cable length. By employing HFAC distribution, it would be possible to realize wireless power transmission thereby reducing the number of connectors.

III. TOPOLOGY ANALYSIS

Circuit diagram of the boost converter front end based HFAC power source is shown in Fig. 3. Table 1 details design specifications for the prototype converter – inverter system.

Table 1: Design requirements of HFAC power source

Max. output power	500 VA
R.M.S output voltage	36 V
Distribution frequency	50 kHz
Input voltage range	$26 - 30$ V$_{DC}$
Input V$_{DC}$ to Inverter	60V

1. Boost Converter Design

To design a CCM boost converter, parameters to be considered include inductor current ripple (ΔI_L), output voltage ripple (ΔV_{out}) and duty cycle (D). These parameters depend on inductance, capacitance and output voltage respectively. These can be determined from (1) to (3). ΔI_L is chosen to be 20% to 40% of I_{out}.

$$\Delta I_L = \frac{V_{IN(min)}D}{f_s L} \tag{1}$$

$$\Delta V_{out} = \frac{I_{out(max)}D}{f_s C} \tag{2}$$

$$D = \frac{V_{out} - (V_{IN(min)}\eta)}{V_{out}} \tag{3}$$

978-1-5090-0064-7/15 $31.00 © 2015 IEEE

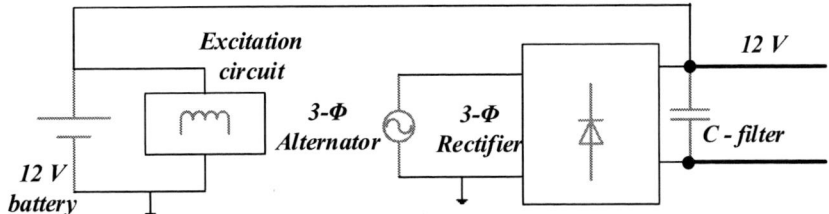

Fig. 2 Existing DC distribution system in a vehicle

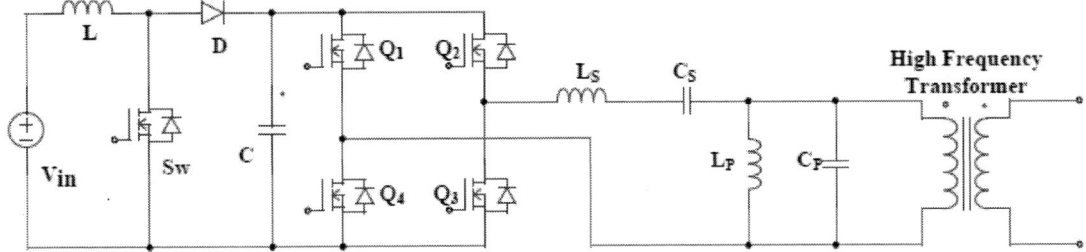

Fig. 3 Topology of the converter – inverter system

where,

$V_{IN(min)}$ is the minimum input voltage
$I_{out(max)}$ is the maximum output current
V_{out} is the desired output voltage
η is the converter efficiency
L is the inductance
C is the capacitance
f_S is the switching frequency

Values of L, C, f_S, and D are chosen as per required specifications. With increase in f_S, the size of passive elements decreases. It is important to determine maximum current flowing through the switches. This helps to select correct components. Maximum current through the MOSFET, $I_{SW(max)}$, is determined using (4). For a boost converter, maximum voltage that appears across its drain and source is equal to V_{out}.

$$I_{SW(max)} = \frac{I_{out(max)}}{1-D} + \frac{\Delta I_L}{2} \qquad (4)$$

$I_{SW(max)}$ is also the peak current through the inductor. This value is essential in choosing the right copper wire diameter and in calculating inductor design parameters. Forward current rating of the diode, usually Schottky, is equal to the peak output current. The MOSFET and diode must be able to handle power dissipation given by (5) and (6) respectively.

$$P_{MOS} = I^2{}_{MOS(rms)} R_{ds(on)} \qquad (5)$$

$$P_D = I_{D(avg)} V_f \qquad (6)$$

$$I_{MOS(rms)} = \sqrt{D \left(\frac{\Delta I^2{}_L}{12} + I^2{}_{in} \right)} \qquad (7)$$

$$I_{D(avg)} = I_{in}(1-D) - I_{out} \qquad (8)$$

where,

P_{MOS} is the power dissipation rating of MOSFET

P_D is the power dissipation rating of power diode
$I_{MOS(rms)}$ is R.M.S current through the MOSFET
I_{in} and I_{out} are avg. i/p & o/p current respectively
$R_{ds(on)}$ is MOSFET on-state resistance
V_f is the diode forward voltage drop
$I_{D(avg)}$ is the average diode current

Output capacitor with low equivalent series resistance (ESR) is selected to ensure efficient operation. The ESR also contributes to the output ripple (ΔV_{ESR}), given by:

$$\Delta V_{ESR} = ESR \left(\frac{I_{out(max)}}{1-D} + \frac{\Delta I_L}{2} \right) \qquad (9)$$

2. High Frequency Inverter Design

Full bridge voltage source inverter is fed by the front end boost converter. LCLC tank is used to realize a resonant inverter and obtain sinusoidal output voltage and current with frequency of 50 kHz. The inverter switches are turned on and off alternatively. Phase shift control strategy is employed to maintain constant output voltage. Design includes calculating the LCLC parameters and selection of inverter switches. The full bridge inverter switches have to block a maximum voltage of $V_{in(DC)}$. Switches must be able to accommodate peak value of load current at full load. Additional compensation in the current rating is required due to resonant circuitry. Design of LCLC resonant tank is based on [7]. The design equations for fully resonant inverter include:

$$L_{S1} = \frac{V_0 \sqrt{\varepsilon}}{I_0 \omega_s} \qquad (10)$$

$$C_s = \frac{I_0}{\omega_s V_0 \sqrt{\epsilon}} \qquad (11)$$

$$L_{p1} = \frac{V_0}{I_0 \omega_s \sqrt{\varepsilon}} \qquad (12)$$

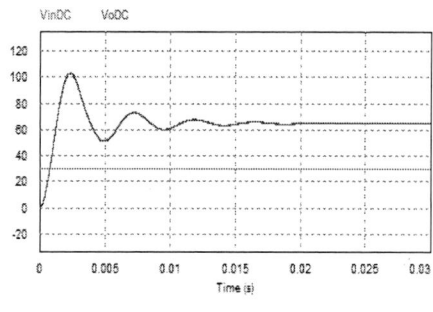

(a)

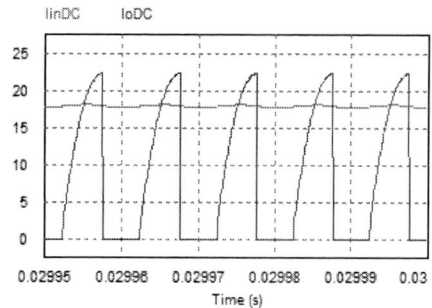

(b)

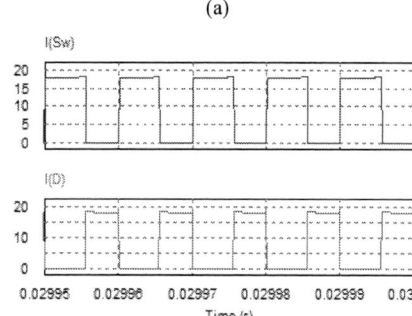

(c)

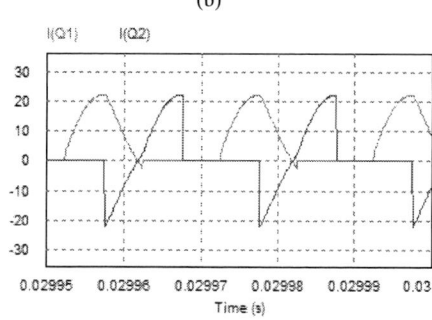

(d)

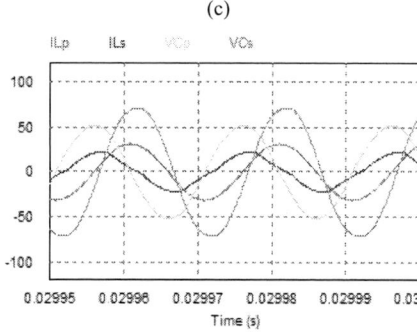

(e)

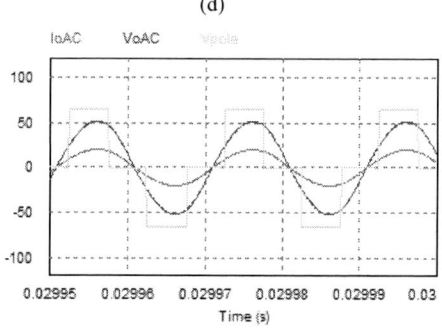

(f)

Fig. 4 500 VA simulation results (a) Boost converter's i/p and o/p voltage (b) Boost converter's i/p and o/p current (c) Converter's MOSFET and diode current (d) Inverter switch current (e) Current flowing through and voltage across resonant inductors and resonant capacitors respectively (f) AC o/p current, o/p voltage and pole voltage

$$C_s = \frac{I_0 \sqrt{\epsilon}}{\omega_s V_0} \qquad (13)$$

$$L_{p2} = \frac{k_2 R_0}{\omega_s} \qquad (14)$$

$$\varepsilon = \omega^2 L_s C_p \qquad (15)$$

The design requirement is that ε should be around 1.7 to have lower THD.

$$L_{s2} = k_1 L_{p2} \qquad (16)$$

$$L_s = L_{s1} + L_{s2} \qquad (17)$$

$$L_p = \frac{L_{p1} L_{p2}}{L_{p1} + L_{p2}} \qquad (18)$$

$$\omega_s = 2\pi f_s \qquad (19)$$

where,

V_0 is the AC output voltage
I_0 is the AC output current
ω_s is the angular frequency

f_s is the switching frequency = 50 kHz
R_0 is the equivalent output resistance
L_s is the series resonant inductor
C_s is the series resonant capacitor
C_p is the parallel resonant capacitor
L_p is the parallel resonant inductor

The constants $k_1 = 0.1$ and $k_2 = 4$ to ensure better voltage regulation and lower current stress respectively. V_o is chosen to be 36 V_{rms} as it falls under the category of safer voltages. Voltages beyond 48 V_{rms} are considered high and require other improved protection mechanisms as per international standards which make it more expensive. The input V_{dc} to the inverter is decided based on convenience of designing the front end converter with 12 V battery as the source. A two stage front end DC –DC converter (from 12V – 70 V) is considered for the final design.

978-1-5090-0064-7/15 $31.00 © 2015 IEEE   258

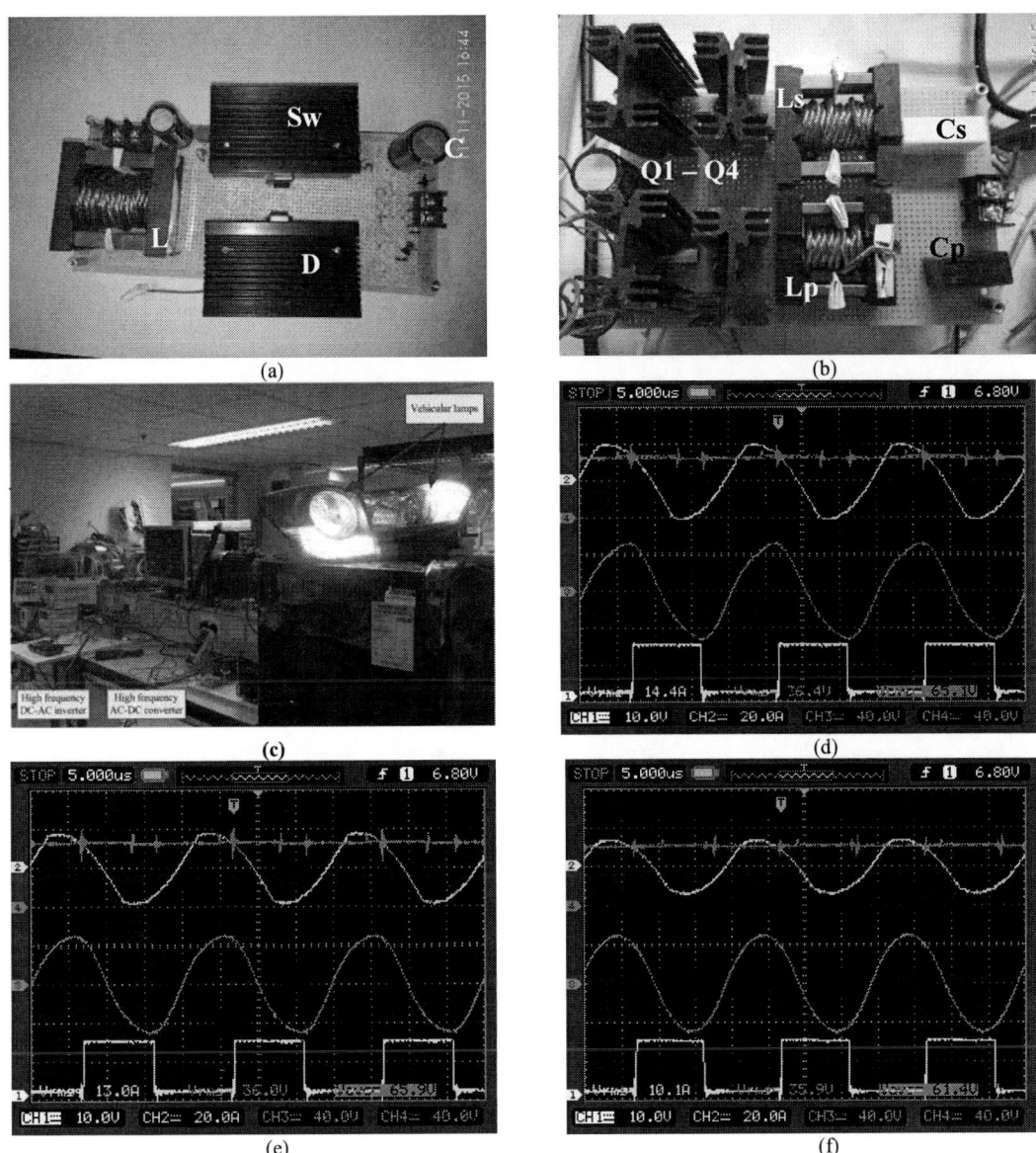

Fig. 5 (a) Boost converter hardware (b) Resonant inverter hardware (c) HFAC system under test (d) (e) (f) experimental waveforms under different loads. Ch.1 Reference signal. Ch.2 o/p current. Ch.3 o/p voltage. Ch. 4 DC Input to Inverter

Table 2: Components of the HFAC system

Boost inductor filter	300μH
Boost output capacitor filter	330μF
Series resonant inductor	14μH
Series resonant capacitor	1μF
Parallel resonant inductor	5.3μH
Parallel resonant capacitor	1.5μF
MOSFETs	IRFP4568PbF
Diode	40EPF02
Boost PWM chip	TL494
Boost MOSFET driver	ICL7667
Inverter control chip	UC3875
Inverter MOSFET driver	IR2110

IV. RESULTS AND DISCUSSION

Simulation results are shown in Fig. 4. They help in choosing right components for the application and also verify design equations. Preliminary experimental

outcomes shown in Fig. 5 validate the simulation results and theoretical calculations. Future work include upgrading HFAC system to higher power levels, detailed analysis on EMI effects, THD analysis, introducing zero voltage switching in the inverters and develop a unified controller for the system.

V. CONCLUSION

This paper discusses the design of a 500 VA HFAC power distribution system. HFAC PDS offers several advantages over DC PDS and LFAC PDS clearly elaborated. The existing DC PDS in a vehicle is explained and the merits of replacing it with HFAC PDS are highlighted. Design of the front end boost converter and resonant inverter is discussed in detail. Design equations required for determining converter and inverter system's passive elements have been presented. Simulation results are used to choose the system's MOSFETs and diode. Preliminary experimental results match simulation and theoretical computations.

ACKNOWLEDGMENT

Authors would like to thank the Innovation and Technology Fund of Hong Kong for their support (Project ITS/036/14).

REFERENCES

[1] Status of 20 kHz Space Station Power Distribution Technology. NASA Publication, TM 100781.

[2] Renz, David D., et al. Design considerations for large space electric power systems. No. N-8324552; NASA-TM-83064; E-1535. National Aeronautics and Space Administration, Cleveland, OH (USA). Lewis Research Center, 1983.

[3] Patel, Mukund R. Spacecraft power systems. CRC press, 2004, ch. 22, sec. 22.7, pp. 539-543.

[4] Jain, P.; Pahlevaninezhad, M.; Pan, S.; Drobnik, J., "A Review of High-Frequency Power Distribution Systems: For Space, Telecommunication, and Computer Applications," in Power Electronics, IEEE Transactions on , vol.29, no.8, pp.3852-3863, Aug. 2014

[5] Masrur, M.A.; Sitar, D.S.; Sankaran, V.A., "Can an AC (alternating current) electrical system replace the present DC system in the automobile? An investigative feasibility study. I. System architecture," in Vehicular Technology, IEEE Transactions on , vol.47, no.3, pp.1072-1080, Aug 1998

[6] Masrur, M.A.; Sitar, D.S.; Sankaran, V.A., "Can an AC (alternating current) electrical system replace the present DC system in the automobile? An investigative feasibility study. II. Comparison and tradeoffs," in Vehicular Technology, IEEE Transactions on , vol.47, no.3, pp.1081-1086, Aug 1998

[7] Sun Dejun, "Research on High Frequency Output Soft-Switching Inverter" MS. thesis, EE Dept., Nanjing university of aeronautics and astronautics, March 2007

Measurement and Analysis of EMI Radiated by a High-Frequency AC Distribution System in Vehicles

Z. Y. Jiang[1], X. D. Xue[2], J. Mei[3], Eric K. W. Cheng[4]

Department of Electrical Engineering, The Hong Kong Polytechnic University, Hong Kong
[1] E-mail: zy.jiang@polyu.edu.hk
[2] E-mail: eexdxue@polyu.edu.hk
[3] E-mail: jie.mei@polyu.edu.hk
[4] E-mail: eeecheng@polyu.edu.hk

Abstract – This paper presents a literature review on the high-frequency AC distribution system and an experimental relationship for the radiated electromagnetic noise level on the 50kHz AC distribution system in the vehicle system, with regards to the required safety level for human exposures, based on the ICNIRP guidelines. The Electro-Magnetic Interference (EMI) level of the 500VA/50kHz AC distribution system is measured and the experimental results are analyzed, complying with the standard CISPR16/EN55016. The experimental results verify the viability and safety of the developed high frequency AC distribution system consisting of the inverter, transmission and converter. Furthermore, the effectiveness of different topologies of transmission cables is also examined.

Keywords–AC distribution, EMI, high frequency, vehicles

I. INTRODUCTION

When compared with the commonly used DC distribution circuits in the present vehicle system, the high-frequency AC distribution system being analyzed in this paper enjoys the advantages of smaller size and lower transmission loss. With more extended-range electric vehicles being put into the market nowadays, there is an increasing need for a larger space in car, to install more batteries. Meanwhile, this need correspondingly leads to the size minimization of other vehicle systems with an unchanged total car body, but not to sacrifice their performances at the same time. This 500VA 50kHz AC distribution circuit demonstrates a good scale-down proposal for this trend, to use a 36 volts AC distribution design instead of a normal 12 volts DC system. It includes three main components, the DC-DC converter together with the DC-AC inverter serving the power supply, the one-meter long transmission cables for connection, as well as AC loads and AC-DC converters for DC loads. However, AC operation will induce the electromagnetic noise, which can be further categorized into the conducted type and the radiated type. The conducted EMI from the AC power source will create the common-mode and differential-mode noise, and thus can degrade the overall system performance, which has already been fully researched in [1]-[3]. While for the other radiated type EMI in high frequency, it may result in harm to human beings directly, on which there are few literature reviews or research reports.

This paper presents the experimental results of a proposed distribution circuits for different power levels and testing positions within an EMI-shielded chamber.

With reference to the International Commission on Non-Ionizing Radiation Protection (ICNIRP) Guidelines [4] as well as the BS EN55016 [5] and IEEE C95.1 [6], the results have verified that this AC distribution circuit achieves the required safety standard and does not harm the human beings, under its normal operation. Additionally, the experiment demonstrates a significant radiated EMI decrease for the two transmission cables when a twisted-pair topology is used, other than a normal parallel distribution wiring.

II. HIGH FREQUENCY AC DISTRIBUTION SYSTEM

A typical DC distribution system in the vehicle is rated 12 volts and serves the power supply for a number of electrical units like electrical motors, power converters, generators, lightings, entertainment systems, control systems for computer and actuators in the vehicle. The major disadvantages in this system are the low voltage level and its corresponding slow respondent time. In this circumstance, the required current in the distribution circuit will be large, which gives rise to a poor performance in system efficiency, a high battery requirement and the big equipment size.

In order to overcome the above shortcomings, a high frequency AC distribution system is proposed with 50kHz, as Figure 1 shows below,

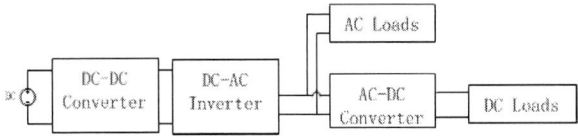

Fig. 1: 50kHz AC Distribution Circuit

For the power supply, it first steps up the supply voltage from dc 12 volts to dc 70 volts through a boost dc-dc converter, and then the circuit utilizes a dc-ac inverter to transfer the dc input to a 36 volts ac output with a 50kHz frequency. These designs in voltage level and frequency cater for the considerations of both the favor in large power transfer and the must in safe human operation. After the transmission cables up to one-meter long, different distributed loadings (AC type or DC type) are connected to the circuit. In the test, the output of the inverter ran from 250VA to 550VA continuously, with a fixed 36-volt voltage level.

978-1-5090-0064-7/15 $31.00 © 2015 IEEE

III. EMI MEASUREMENT AND ANALYSIS

To test the radiated electromagnetic interference (both the magnetic and electric field), the measurement procedures that defined in the standard CISPR16/EN55016 [5] need to be followed.

1. Test Setup

The test apparatus (shown in Figure 2a, 2b) include the near-field probes for the electrical and magnetic field detection, an EMI-shielded chamber for the placement of the equipment under test (EUT), to set them free from the other electromagnetic interference and an EMI test receiver. In the test, the 6-cm loop probe as well as the 3.6-cm ball probe (Model 7405 from ETS-LINDGREN) were connected to the RF input of the EMI test receiver through a 50-ohm termination, for detecting the magnetic and electrical field separately on different probe.

The test receiver (R&S ESR) was set to display the real-time EMI values (up-peak, down-peak and r.m.s.) in the maximum-hold mode with a 10-dB attenuation, since it is important to record the highest possible EMI that may incur to the human beings. Also, the EMI frequency scan function was used to determine the actual frequency of the EMI generated by different sources.

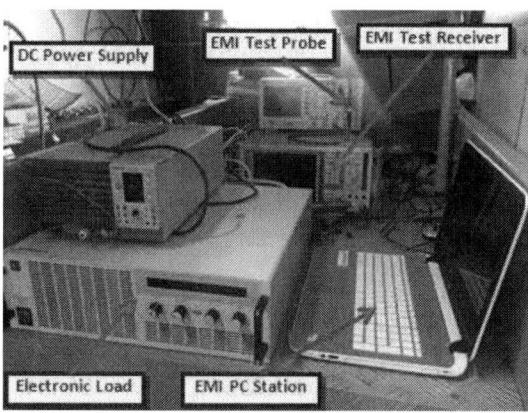

Fig.2a: EMI test apparatus

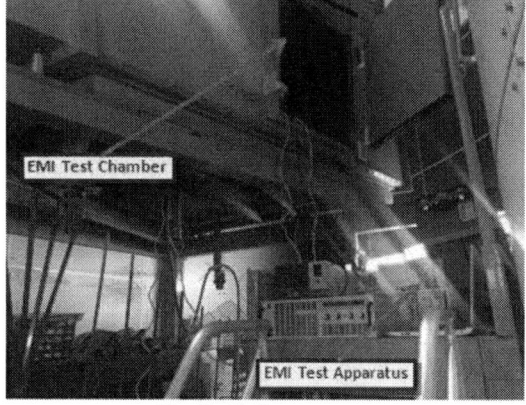

Fig. 2b: EMI test setup

2. Measured EMI Results and Analysis

The radiated electromagnetic interference strength is primarily governed by the power level of the EUT as well as the distance that from the surface of the measured device to the actual position, when a fixed system working frequency 50kHz is maintained. This can be expressed alternatively as the electrical and magnetic field strength (dBμV/m) being determined by different electrical loadings, and the increasing vertical locations of the near-filed probes.

The measurement results for the DC-AC inverter in the power supply side of the circuit are shown in the below Figure 3a, 3b.

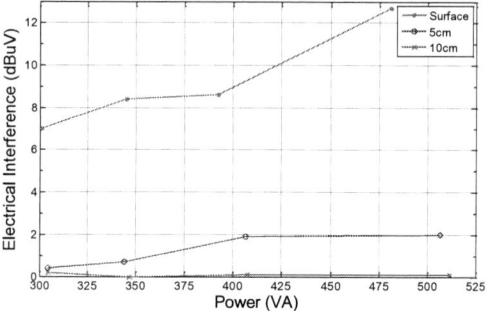

(a). Radiated Electrical Interference on Inverter

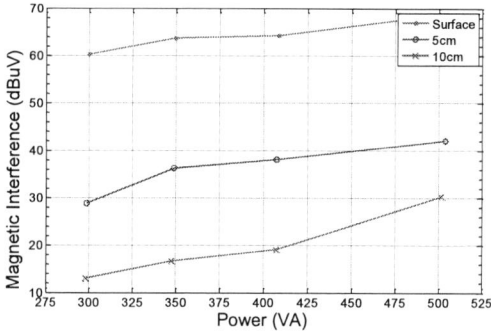

(b). Radiated Magnetic Interference on Inverter
Fig. 3 Measured EMI results of inverter

The Figure 3a reveals that for the DC-AC inverter, the electrical interference strength only goes up to 12dBμV/m (0.0133pT) in the surface, with the full-load power of 500VA. Also, it can be seen that the electrical interference strength decreases sharply to almost zero when the testing surface is only raised up to 10 centimeters, indicating that there is little EMI influence there. The above two relationships indicate that there is little influence for the electrical interference, which agrees with the conduction nature of the electrical interference field. Consequently, little impact will be coming from the electrical interference when we concern the radiated electromagnetic interference, if no conduction medium actually exists. The Figure 3b illustrates the measured magnetic interference of the DC-AC inverter. As expected, it displays much clearer relationships between the magnetic interference strength and either the power level or the test surface distance. With the test surface being raised, the magnetic interference strength from the inverter under the full-load condition, 520VA, decreases from the maximum 68dBμV/m (surface-test) to 42dBμV/m (5cm-test), and then to 31dBμV/m (10cm-test). A similar magnetic interference strength drop occurred when the inverter power level decreased from 500VA to 300VA. Besides the impact of different test levels, the magnetic interference strength on this inverter displayed a characteristic of having larger increases in both the starting-up region (250VA-350VA) and the full-load to over-load region (450VA-550VA) than the normal working area (350VA-450VA), suggesting a nonlinear increase. Compared with the electrical interference, the findings for the magnetic interference strength also matched with its radiation nature of field propagation.

Similar findings for the electrical and magnetic interference are displayed in the below graphs: the EMI results for the transmission cables with twisted pairing topology are displayed in the Figure 4a, 4b, in addition, the EMI measurement data for the AC-DC converter are displayed in the Figure 5a, 5b.

It can be summarized from the comparisons between the three tests of all the three components, that both the magnetic interference strength and the electrical interference strength generally demonstrate a smaller fluctuation under the surface test than the other two tests. This can be understood as the surface-test can display more convincing results than the 5cm-test or the 10-cm test, since the interference strength itself is strong enough to cover the influence of other unwanted sources. In other words, besides the electrical interference, the 10-cm magnetic interference results on or under the level of 300VA-operation reveal less accurate information than the other situations.

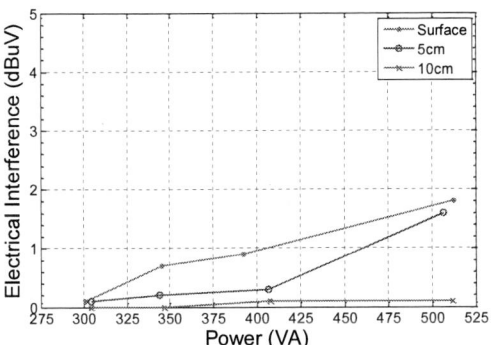

Figure 4a. Radiated Electrical Interference on Transmission Cables with Twisted Pairing Topology

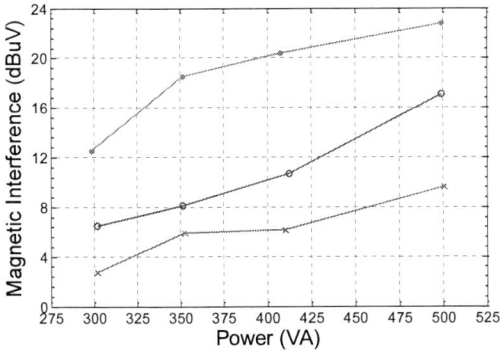

Figure 4b. Radiated Magnetic Interference on Transmission Cables with Twisted Pairing Topology

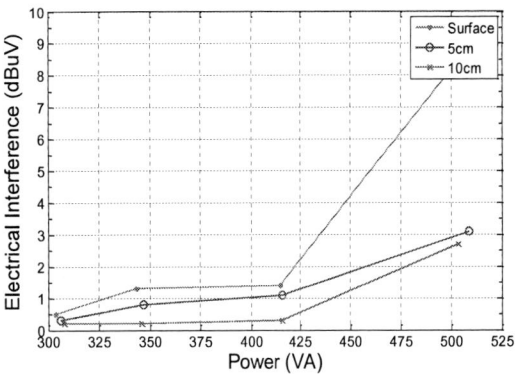

Figure 5a. Radiated Electrical Interference on AC-DC Converter

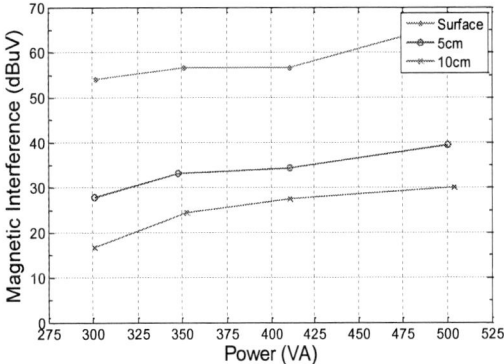

Figure 5b. Radiated Magnetic Interference on AC-DC Converter

It can be summarized that for the radiated electromagnetic interference, the magnetic interference has the dominant impact. To reduce the electromagnetic interference of the high-frequency AC distribution system, it is suggested to operate the system within the normal working range and to run it less in the full load or overload condition.

3. Improvement in EMI of Transmission Lines

In the hardware prototype of the 500VA high frequency AC distribution circuit, the transmission lines are selected to be one pair of one-meter long cables without EMI-shielding. The AC voltage level of the transmission lines is 36 volts and the working frequency is 50 kHz. Two methodologies for the cable wirings are proposed and tested on EMI performance. The first one is to use two parallel cables, since in reality for protection the cables themselves may be too large and too tough to bend. The second proposal is to use smaller-size cables in twisted pairing but to include more separate pairs, which would be definitely generating more cost, compared with the first topology. For the EMI performance on the two topologies of transmission cables, the similar experiments on different test locations were carried out, under the conditions that the power supply was providing a total power of 420VA and the near-field probe was placing in the horizontal center of the cables. The compared results for both the electrical and magnetic interferences are shown in the below Figure 6.

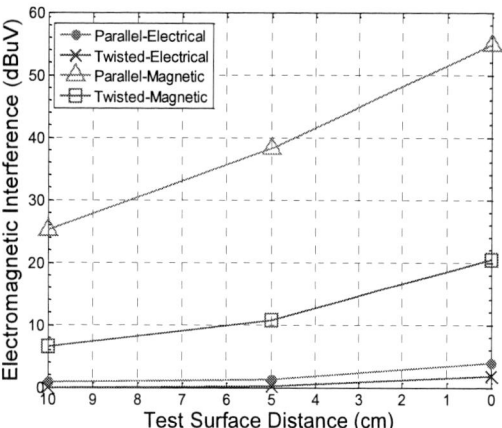

Figure 6. Radiated Electromagnetic Interference on Transmission Cables with Twisted Pairing Topology versus Paralleling Topology

Like the previous analysis, Figure 6 illustrates a sharp descending relationship between the electromagnetic interference and the surface distance. Moreover, given a fixed power level and test location, it can be observed that the transmission cables with twisted pairing topology enjoys a 36dBμV/m magnetic interference strength decrease (electrical interference is very small and neglected in this case), when compared with that with the paralleling topology, in other words, 64% decaying. It is to say, the radiated EMI can be weakened efficiently with these twisted pairs, and nevertheless the tradeoffs are the corresponding extra cable materials and the labor costs to manipulate the topology. Consequently, the final decision can only be achieved after a thorough consideration on the both sides, especially in reality the actual power that needed to be distributed within a vehicle, exceeds the simulated 500VA case a lot.

4. EMI Frequency

The electromagnetic interference frequency depends on the source frequency of the EUT, i.e., 50kHz of the high-frequency AC distribution system. The frequency scan function in the EMI test receiver enables the user to scan a wide range of frequency (30kHz-70kHz) around the probe and to determine the dominant one from the range, as the peak marker shown in the Figure 7a-c. The three waveforms from the receiver reveal the frequency scan results on the surfaces of the inverter, the twisted-pair transmission cables and the converter. With the 10dB attenuation on the data, it marks out the below dominant frequencies in the scan range, with 47.04dBμV in 50.00kHz for inverter, 8.9dBμV in 50.08kHz for transmission cables as well as 49.06dBμV in 50.08kHz for the converter.

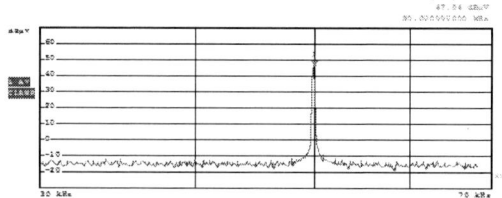

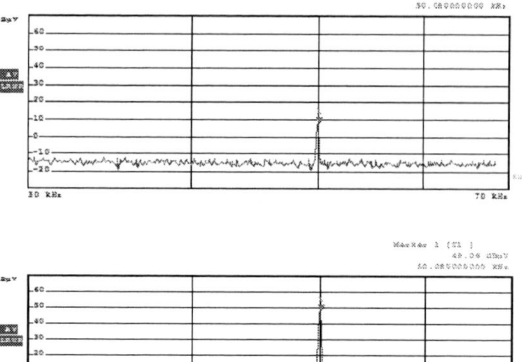

Figure 7a-c (from up to down). EMI test receiver frequency scan on the surface for inverter, transmission cable and converter under 420VA operation
(x-axis: 30kHz to 70kHz; y-axis: -30dBμV to 70dBμV)

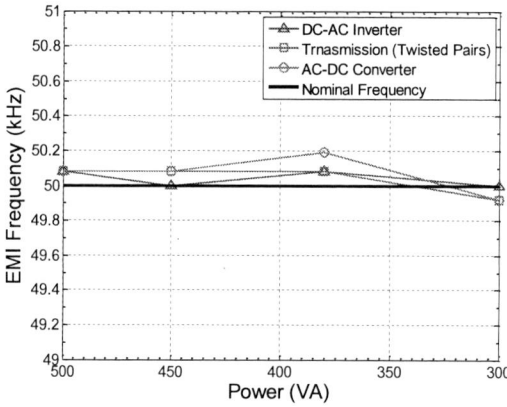

Figure 8. EMI frequency for inverter, transmission and converter

In addition, the above Figure 8 shows all the summarized results of the scans on the surface of EUTs that were under 420VA operation specifically. It demonstrates that the phenomenon that the EMI frequencies for the three main components of this 50kHz AC distribution circuit follow the nominal working frequency of the system closely, with a variance of +/-0.3%, and this correlation grows a bit looser when the power goes down, according with the fact that the EMI influence of the EUTs on the surrounding environment reduces.

IV. MEETING EMI EXPOSURE GUIDELINES

The EMI level from this proposed high-frequency AC distribution system in the vehicle should comply with the guidelines from the International Commission on Non-Ionizing Radiation Protection (ICNIRP), with an exposure characteristics subgroup in the general public exposure. This international series of guideline was created for limiting human beings exposure to the time-varying electromagnetic field, so as to reduce the potential harm to health. Inside the guideline, the reference levels are taken as the average value for the whole body, and are obtained from the fundamental limitations by mathematics modeling and by extrapolations from the laboratory findings in the pre-defined frequencies. Meanwhile, there

978-1-5090-0064-7/15 $31.00 © 2015 IEEE

are another two guidelines from IEEE [6] and ARPANSA [7], which are both developed from the previous international standard. They are to give more explanations and suggestions on the EMI standards, like separate levels on different parts of human body, or the spot value restriction versus the average whole value.

To look into the IEEE standard C95.1, it shows that within the frequency range of 0.003-0.1MHz under both controlled and uncontrolled environments, the maximum Permissible Exposure is restricted under a r.m.s. value of 614V/m (175.7dBμV/m), while that of the magnetic interference field strength is 163A/m (215.7dBμV/m). Both of the two reference values are much higher than the test results displayed in this paper, indicating a very safe environment on EMI for the inverter, transmission and converter. Also, it is undoubted that with larger power consumption in the real application, the EMI problem in the system is still controllable, not only because the large gaps between the reference values and the test values (usually in 10cm distance case), but also because that we can install the extra EMI-shielded shells to hold the inverters and converters inside, as another effective way to reduce the EMI in the worst case.

V. CONCLUSION

With a working frequency of 50kHz, the AC distribution circuit in this paper handles a power range from 250VA to 500VA well. To ensure its radiated EMI safety, a series of EMI tests on each separate component were performed with regards to different power levels as well as different surface levels. The results show different decaying patterns on the magnetic interference strength with power level dropping and the testing surface rising, and the results also reveal that the electrical interference is small enough to be neglected. EMI improvements of the transmission cables are further analysed between two wiring topologies including paralleling and twisted-pairing. Besides, the frequency scan function of the receiver proves that the EMI frequency closely follows the frequency of the source that generates it. Finally, the international guideline is listed to qualify this high-frequency AC distribution system on the safe human exposure level for electromagnetic interference.

ACKNOWLEDGMENT

The authors would like to thank to the part support from the Innovation and Technology Fund of Hong Kong under Project ITS/036/14.

REFERENCES

[1] E. Zhong, S. Chen, and T. A. Lipo: "Improvement in EMI performance of inverter-fed motor drives," in APEC 94 Conf. Rec., pp. 608-614, Vol. 2, 1994.

[2] S. Ogasawara, H. Ayano, and H. Akagi, "Measurement and Reduction of EMI Radiated by a PWM Inverter-Fed AC Motor Drive System", IEEE/IAS Annual Meeting, pp. 1072-1079, 1996.

[3] A. Hesener, "Electromagnetic Interference (EMI) in Power Supplies", Fairchild Semiconductor Power Seminar, pp.1-16, 2010-2011

[4] "Guidelines For Limiting Exposure to Time-Varying Electric, Magnetic, and Electromagnetic Fields (Up to 300GHz)", ICNIRP Guidelines – Health Physics, 494-512, Vol. 74, No. 4, April 1998.

[5] "Specification for Radio Disturbance and Immunity Measuring Apparatus and Methods. Radio Disturbance and Immunity Measuring Apparatus. Coupling Devices for Conducted Disturbance Measurements", BS EN 55016-1-2, British Standards Institution, July, 2014

[6] "IEEE Standard for Safety Levels with Respect to Human Exposure to Radio Frequency Electromagnetic Fields, 3kHz to 300GHz", IEEE Standards Coordinating Committee 28 on Non-Ionizing Radiation Hazards, IEEE C95.1-1991, September, 1991

[7] "Maximum Exposure Levels to Radiofrequency Fields – 3 kHz to 300 GHz", ARPNSA – Radiation Protection Standard, Series.3, May 2002

Characterization and modeling of copper foil conductor for high frequency power distribution

C. D. Xu, K. W. E. Cheng, Raghu Raman S, J. F. Liu

[1] Department of Electrical Engineering, The Hong Kong Polytechnic University, Hong Kong
E-mail: 10902353r@connect.polyu.hk

Abstract–This paper investigates characteristics of copper foil conductor for 50 kHz high frequency AC power distribution. Characterization includes parameter estimation, power loss analysis and electro-magnetic field simulation using FEM software. Experiments have been conducted to obtain parameters for the double layer copper foil transmission line. Further, a model for the high frequency AC transmission line is proposed based on the obtained experimental results

Keywords– Copper foil. high frequency power distribution power transmission,

I. INTRODUCTION

High frequency power distribution has been proposed for decades [1]. It has been verified to be more reliable, efficient and environmental friendly. It has brought out several interesting topologies of high frequency power source and point of load converters [2-4]. However, the transmission high frequency line still remains a challenge. Many results published in the area of ac current transmission, in YMCO [5-6], are not economical and applicable for now. In this paper, Three configuration of the power cable and 50Hz, 50 kHz of the source frequency are proposed to investigate the characteristic for high frequency distribution, which indicates that the copper foil conductor is suitable for high frequency power distribution..

II. OHMIC LOSS

Ohmic loss is simulated in Ansoft; eddy current solution is given by

$$P = \int_{Vol} \frac{J \cdot J^*}{2\sigma} dVol \qquad (1\text{-}1)$$

where,

J is the current density;

J* is the complex conjugate of the current density;

σ is the conductivity in siemens/meter.

Ohmic loss is used to compute the power loss in a structure. For the impedance boundaries, ohmic loss is given by

$$P = \sqrt{\frac{\int \omega \mu_0 \mu_r}{8\sigma}} \int_{Sur} H_t \cdot H_t^* ds \quad (\text{Watts}) \qquad (1\text{-}2)$$

where,

ω is the angular frequency, which is 2πf,

f is the frequency. The source currents and voltages oscillate at the frequency during the solution.

S is the conductor's conductivity in siemens/meter;
μ_r is the conductor's relative permeability;

μ_0 is free space's permeability, which is 4π×10-7 H/m.

H_t is the tangential component of H on the impedance boundary;

H_t^* is the complex conjugate tangential component of H on the impedance boundary.

The ohmic loss density (P_0) is given by the following:

$$P_0 = \frac{1}{2}\text{Re}(E \cdot J_c^*) \qquad (1\text{-}3)$$

This loss is calculated in the softwar of Ansoft and from current density for the conduction copper foil.

III. TWO DIMENSIONAL FIELD ANALYSIS

The analysis of the electromagnetic field can be derived according to the Maxwell's equations.

$$\nabla \times \mathbf{H} = \mathbf{J} + \frac{\partial \mathbf{D}}{\partial t} \qquad (1\text{-}4)$$

$$\nabla \times \mathbf{E} = -\frac{\partial B}{\partial t} \qquad (1\text{-}5)$$

By Maxwell equation (1-5) and

$$\mathbf{B} = \nabla \times \mathbf{A} \qquad (1\text{-}6)$$

$$\nabla \times \mathbf{E} = -\nabla \times \frac{\partial \mathbf{A}}{\partial t} \qquad (1\text{-}7)$$

where A is the magnetic vector potential.
For linear isotropic medium, the following expressions are valid –

$$\mathbf{D} = \varepsilon \varepsilon_0 \mathbf{E} ;$$
$$\mathbf{B} = \mu_r \mu_0 \mathbf{H} ;$$
$$\mathbf{J} = \sigma \mathbf{E} ;$$

The,

$$\frac{1}{\mu_r \mu_0}(\nabla \times \nabla \times \mathbf{A}) = \mathbf{J}_s - \sigma \frac{\partial \mathbf{A}}{\partial t} , \qquad (1\text{-}8)$$

where, J_s is the excitation current density in the transmission line.

IV. SIMULATION

In order to figure out the suitable transmission line for high frequency, three configurations of the transmission line are chosen to investigate the characteristics of the line. The ohmic loss and magnetic field of the various configurations of the line is studied by throughput AC current of 10A 50Hz and 50kHz. The parameters of the wire are illustrated in Table 1. Fig. 1 shows the ohmic loss in the line. Fig. 1 (a) shows a traditional round wire of the transmission line. The ohmic loss of the line in 50 kHz is obviously larger than the 50 Hz. Also there is stronger skin effect and higher proximity loss in the transmission line. In Fig. 1(b), ohmic loss is moderate in the all five small

978-1-5090-0064-7/15 $31.00 © 2015 IEEE

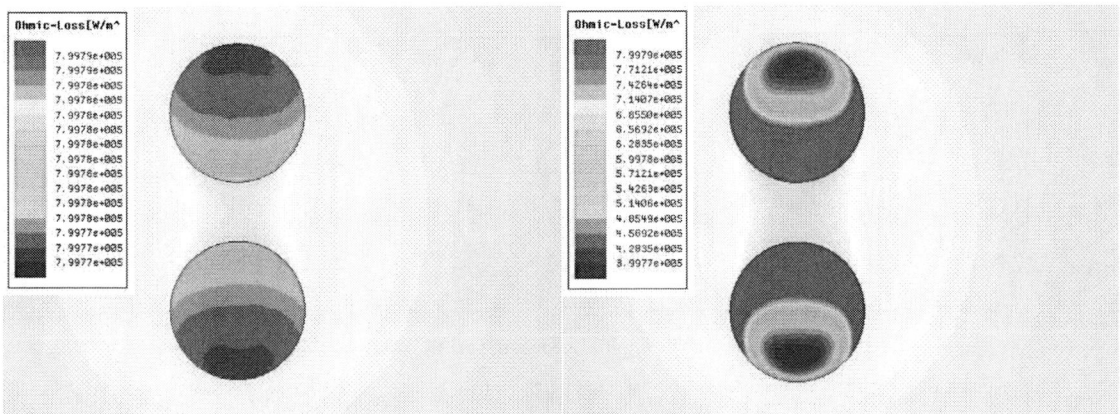

(a) Ohmic loss of two AWG17 operating on 50Hz(left) and 50 kHz(right)

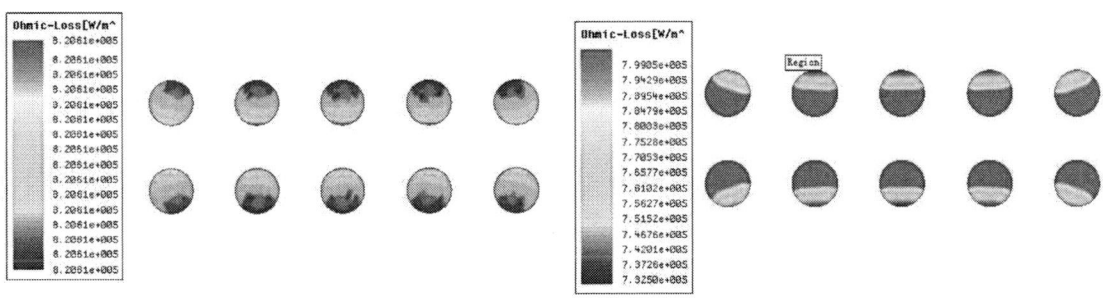

(b) Ohmic loss of five AWG34 operating on 50Hz(left) and 50k Hz(right)

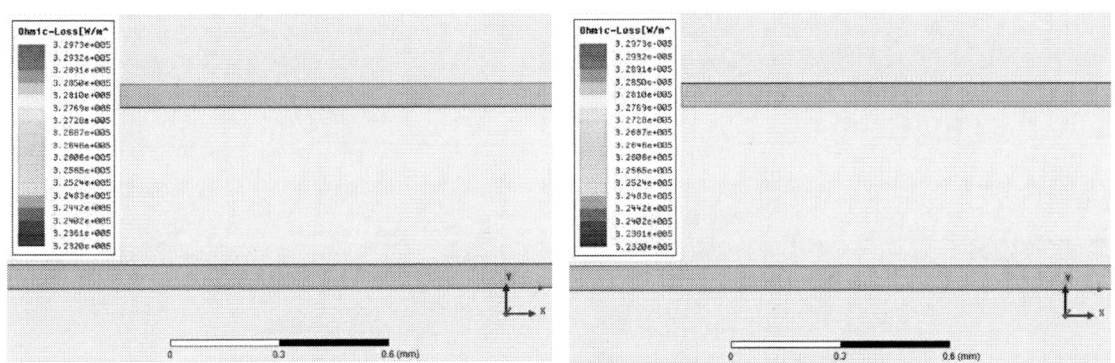

(c) Ohmic loss of two copper foil operating on 50Hz(left) and 50k Hz(right)

Fig. 1 The power loss of different transmission line and different frequency.

wires. But skin effect still is still significant. However, as shown in Fig. 1 (c), the power loss in foil transmission line when the distributed line is operated on 50 kHz is much smaller than the above two configurations. And there is no skin effect in the transmission line. We can also see that The Ohmic loss is not very different between 50 Hz and 50 kHz, Which means when the transmission line is designed as double layer copper foil structure, the Ohmic loss doesn't affect much by the frequency. That is, the structure are suitable for high frequency distribution.

Fig. 2 shows magnetic field of the line. As shown in Fig 2. (a) Magnetic interference of a given line is higher when it is closer to other transmission lines and smaller when it is farther away other transmission lines. It is larger when operating on 50 kHz. In Fig. 2 (b) it is similar with Fig. 2(a), except the magnetic field is smaller when it is at thesame distance from the transmission line, because the

skin effect is much smaller when there are five slim wire than using one wire. As shown in Fig. 2(c), the magnetic field is much stronger when it is between the two copper foil lines, and it is very small outside the copper foil, therefore, there will be very small electromagnetic interference nearby, which is environmental friendly. The value of the difference magnetic field between the 50Hz and 50kHz is not larger than the above two configurations.

TABLE 1 PARAMETER OF THE TRANSMISSION LINE

Configuration	Total Area (mm²⁾	Spacing(mm)
AWG17	1.04	0.0245
5 pieces of AWG34	5×0.0201	0.0245
Copper foil	25×0.065	0.0245

978-1-5090-0064-7/15 $31.00 © 2015 IEEE

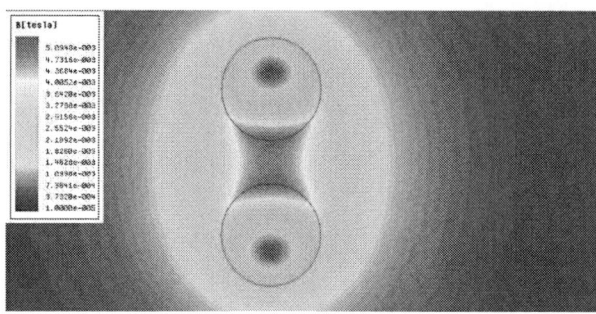 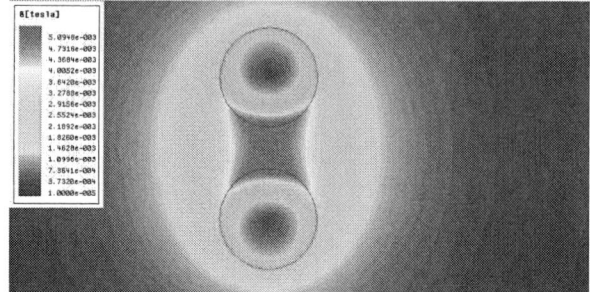

(a) Magnetic field of two AWG17 operating on 50Hz(left) and 50k Hz(right)

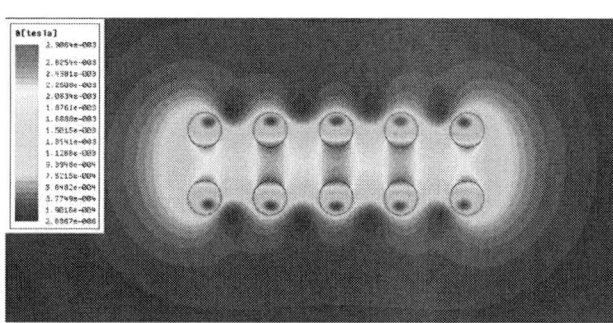 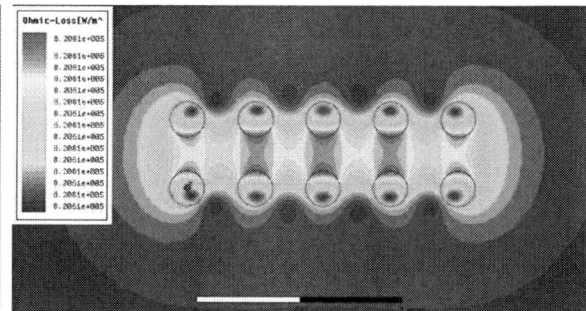

(b) Magnetic field of five AWG34 operating on 50Hz(left) and 50k Hz(right)

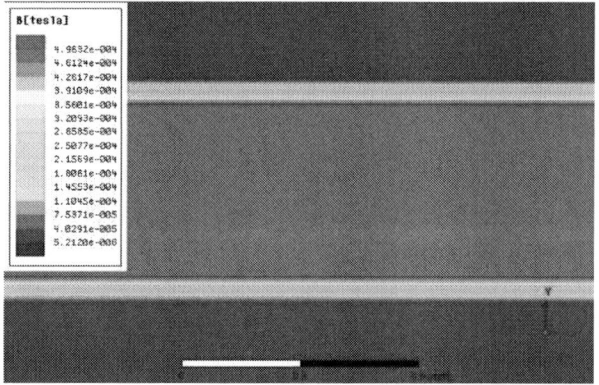 

(c) Magnetic field of two copper foil operating on 50Hz(left) and 50k Hz(right)

Fig. 2 The magnetic field of different transmission line and different frequency.

V. MODELLING BASED ON EXPERIMENTAL RESULT

Double layer copper foil transmission line has proved that it has the advantages of low power loss and no electromagnetic emission from the impedance characteristics analysis of the difference type power lines. The experiment has been conducted to obtain the two port equivalent model of the proposed distribution line which will provide the guide for the further study.

Fig. 3 shows the proposed double layer distribution line, the basic parameters of it are illustrated in Table 2. The data has been obtained by spectrum analyzer and has been shown in Fig. 4 and Fig. 5. Compared with Fig. 4 and 5, the resistor in series is almost constant under the frequency of 10M Hz but the resistor in parallel is getting high from 100k to 10 M. Therefore, the suitable frequency can be up higher to 100k for the proposed distribution line. Based on the experimental result, the two port equivalent circuit can

be obtained as shown in Fig. 6, which will be guide to the further analysis and study.

TABLE 2. **Parameter of the double layer copper foil**

Configuration	Total Area (mm2)
Foil Thickness	0.035mm
Length	230cm
Total Thickness	0.076mm
Width	25mm

978-1-5090-0064-7/15 $31.00 © 2015 IEEE

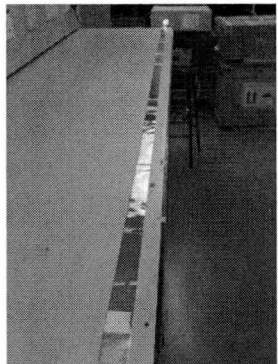

FIG. 3 The Characteristic of the proposed distribution line

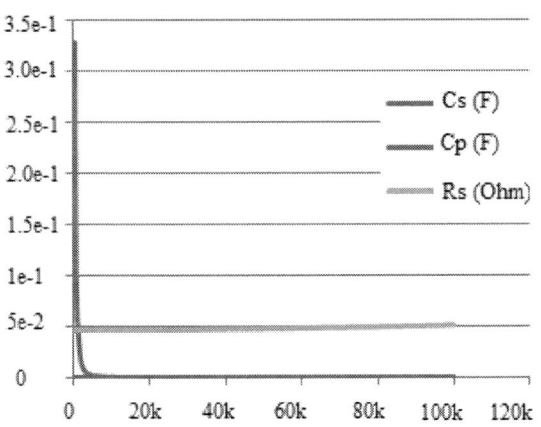

Fig. 4 The parameter of the test double layer copper foil
transmission line

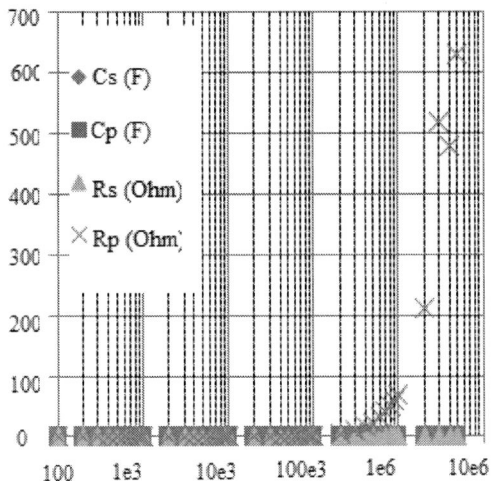

Fig. 5 The Characteristic of the proposed distribution line
Modelling of the proposed power distribution line for high
frequency

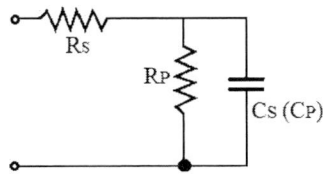

Fig. 6 Equivalent model of the proposed power distribution line
for high frequency

VI. CONCLUSION

From the ohmic loss and magnetic field point, the copper foil is more suitable for the high frequency distribution line. The experimental results have been used to model the proposed distribution line, which also shows that the frequency can be up higher till 100 kHz. Characteristics and configurations have to be investigated to figure out the suitable transmission line for high frequency and will be reviewed in further study.

ACKNOWLEDGMENT

Authors would like to thank the Innovation and Technology Fund of Hong Kong for their support (Project H-ZDAN).

REFERENCES

[1] I. G. Hansen, G. R. Sundburg, " space station 20 kHz power management and distribution system", IEEE PESC'86, pp. 676-683.

[2] P.K. Jain; M. Tanju, A 20 kHz Hybrid Resonant Power Source for the Space Station, IEEE Trans on Aerospace and Electronics Systems, vol. 25, No. 4, July 1989, 491-496.

[3] Patrick C. K. Luk and Andy S. Y. Ng; "High Frequency AC Power Distribution Platforms" Power Electronics in Smart Electrical Energy Networks Power Systems 2008, pp 175-201

[4] Tanju, M.C.; Jain P.K.;A High Performance AC/DC Converter for High frequency Power Distribution System: Analysis, Design and Experimental Results, IEEE Trans on Power Electronic, 9:3, pp. 275-280, May 1994.

[5] E. Demencik, P. Usak, M. Polak, G. A.Levin and P.N. Barnes, "Hall probe based system for study of AC transport current distribution in YBCO coated conductors at frequencies up to 700Hz" IEEE Transactions on Applied Superconductivity, Vol. 17 , No. pp. 3175-3178, Jun 2007

[6] Min Zhang, J Kvitkovic, S V Pamidi, T A Coombs, "Experimental and numerical study of a YBCO pancake coil with a magnetic substrate" , Superconductor Science and technology,Vol. 25, No. 12, pp. 1-6, 2012

978-1-5090-0064-7/15 $31.00 © 2015 IEEE

Energy Factor Analysis for Switched Reluctance Motors Under Various Conditions

Jingwei Zhu[1]   K.W.E Cheng[2]

[1][2] Department of Electrical Engineering, The Hong Kong Polytechnic University, Hong Kong
[1]E-mail: 15901658r@connect.polyu.hk
[2]E-mail: eeecheng@polyu.edu.hk

Abstract-The switched reluctance motor (SRM) shows great advantages of structural simplicity, high reliability, wide speed range with high efficiency, making it attractive to study its stored energy further. This paper accomplishes the modeling of the SRM and its control system on the basis of the software Matlab/Simulink. Furthermore, it explores the laws of energy factor [1] variation under various speeds and conducting angles, including switch-on and switch-off angle with the same software and confirms them in theoretical analysis. Besides, it gives three control methods for comparison about energy factor under different rotating speed, which is essential to the determination of the optimal one for the least energy loss.

Keywords-SRM, Matlab/Simulink, modeling, energy factor, control

I. INTRODUCTION

The SRM is a typical research area based on modern power electronics and computer-aided controllers. This kind of machine has various advantages, such as a simple structure, good high-speed performance, high stability of power converter circuits and low cost. Besides, it has a feature of high start torque and low current within a wide range of operation speed and power, which is suitable for frequent start & stop and positive & negative motoring. Hence, it also has a widespread application on speed regulating and high efficiency occasions.

An important factor for electrical drive is high efficiency and low loss. In [1], the concept "energy factor", which means the storage energy of the reactive components [2] as compared with the output energy, is proposed. According to [1], with the energy factor increasing, the experimental efficiency descends. Therefore, in order to gain high efficiency for the SRM, low energy factor is necessary. Some papers has considered energy optimized control strategy for the SRM. A control approach to minimize the power consumption of a voltage controlled SRM by on-line adjustment of the turn-on angle when the motor runs at steady-state has been introduced in [3]. Document [4] gives a novel generating control strategy of SRM to optimize motor efficiency over a wide speed range by means of storing the electric energy in terms of kinetic energy in a flywheel. Literature [5] develops an adaptive iterative learning control strategy to optimize the exciting current profile by iteratively tuning the control parameters of θ_{on}, θ_{off} and duty ratio of the applied PWM voltage waveform to achieve the purposes of minimization electromagnetic energy conversion loss. Also, in [6], a new improved C-dump power converter is presented to allow the reduction of the switch losses, thus enhancing the whole efficiency.

This paper conducts research, simulation and mathematical

proof on the energy factor of the SRM with the variation of rotating speed, turn-on and turn-off angle. And then it discusses the impact of different parameters on the energy factor. Furthermore, it provides three different control methods and compares their energy factor under various speeds to gain a compound control method for the SRM to lessen loss and enhance efficiency.

II. MODELLING OF SWITCHED RELUCTANCE MOTORS AND DRIVES

The SRM model in the Simulink is composed of two parts: one is linked with the power converter circuits and the other is the calculation of the motor parameters according to the DC voltage. The function block of the latter is as follows in Figure 1:

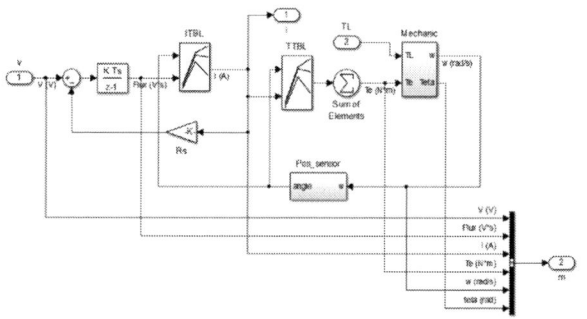

Figure 1. the model of the SRM

On the basis of voltage equation,

$$U = IR + \frac{d\varphi}{dt} \tag{1}$$

equation (2) can be obtained,

$$\varphi = \int (U - IR)\, dt \tag{2}$$

Then from the $\varphi - I$ curve and rotor position, winding current I can be determined via two-dimensional table. Furthermore, current I and rotor position in another 2D table ensure the electromagnetic torque for the SRM. Eventually, according to the expressions,

$$T_e - T_l - D\omega_r = J\frac{d\omega_r}{dt} \tag{3}$$

$$\theta = \int \omega_r dt \tag{4}$$

the rotating speed of the motor ω_r and rotor position can be obtained. The theory of the whole SRM system is shown below in Figure 2:

978-1-5090-0064-7/15 $31.00 © 2015 IEEE

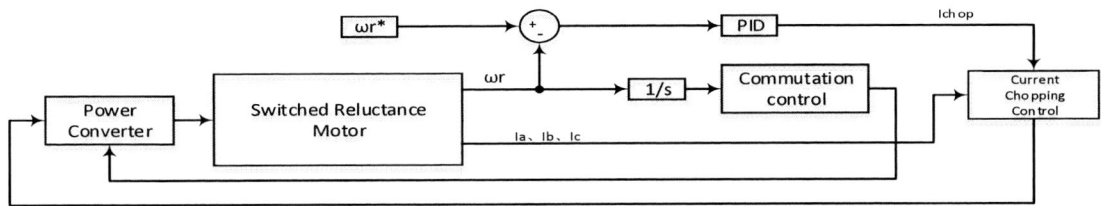

Figure 2. SRM system with current chopping control method

This machine is a 6/4 three-phase SRM. The specific data of it are in the following Table 1:

Table 1: Motor Data

Stator pole	6	Friction(Nms)	0.02
Rotor pole	4	Load(Nm)	1
Stator resistance (Ohm)	0.01	DC Bus Capacitor(μF)	1000
Inertia(kg**m²**)	0.05	DC Bus Voltage (V)	120

The aligned position (center line of the stator tooth coinciding with that of the rotor tooth) is set as the zero position as shown below in Figure 3:

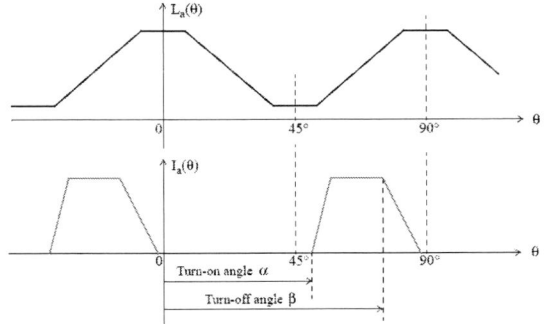

Figure 3. inductance & current variation rules based on rotor position

Current Chopping Control (CCC) is a common control method for SRM. When a SRM is running at a low speed, it has a small rotating EMF and a high value of di/dt. Also, because of long period for inductance increasing, we usually utilize CCC to prevent overcurrent in each phase. It regulates the upper and lower limit of the permitted current and keep $\theta_{on}, \theta_{off}$ constant, in order to limit the value of current within an expected range. Its principle is in the following shown in Figure 4:

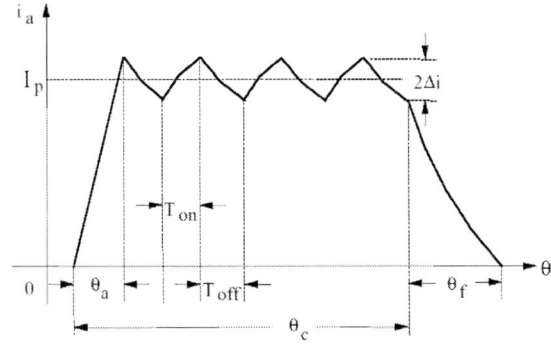

Figure 4. I_p is the chopping current of the winding with the upper limitation $I_p + \Delta i$ and the lower limitation $I_p - \Delta i$

Since the topology of the power converters is conventional, both freewheeling and regenerating process can be achieved during the chopping period. For regenerating, both IGBT switches are turned off and i_a, φ_a quickly decay under negative voltage $-U_{dc}$, which is named hard chopping [8]. For freewheeling, one switch is on while the other is off, to make i_a, φ_a attenuate slowly in a zero-voltage condition, which is called soft chopping [8]. The current curves based on Matlab/Simulink of both methods with regenerating commutation are as follows in Figure 5 and 6:

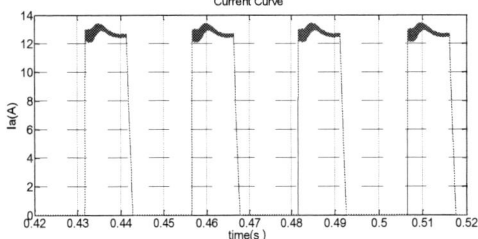

Figure 5. i_a under a hard chopping condition

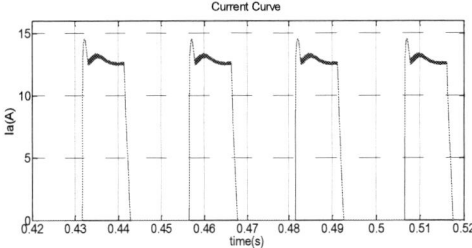

Figure 6. i_a under a soft chopping condition

From the simulation result, soft chopping reduces the times of chopping because it needs more time decay current during the freewheeling process, thus decline the switch loss, torque ripple and current ripple which is essential to weaken the copper loss. Furthermore, since the regenerating process transfers the energy back to the battery, energy of this part isn't utilized to drive the machine instantly and high frequency recharging is detrimental to the battery, while the freewheeling process use the energy to power the motor.

For commutation control, a constant turn-on and turn-off angle is valued to keep the main switches on when the rotor angle is between them and off otherwise. In this model, the switch-on angle is 40°and the switch-off angle is set at 75°.

III. ENERGY ANALYSIS FOR SPEED VARIATION UNDER CCC CONDITION

Theory analysis: from the torque formula, we know that

$$T_e - T_l - D\omega_r = J\frac{d\omega_r}{dt} \qquad (5)$$

978-1-5090-0064-7/15 $31.00 © 2015 IEEE   271

So, when the load is constant, set as T_0, and speed is stable ($\frac{d\omega_r}{dt} = 0$),

$$T_e = T_0 + D\omega_r \qquad (6)$$

Based on $T_e = C_T\Phi I$, $\varphi = N\Phi$, $W_f = \int I d\varphi - T_l\omega_r - W_{other}$, where W_f is the energy storage in the motor winding, W_f is a linear function of the rotating speed ω_r . Furthermore,

$$R_{sl} = \frac{3 \cdot 4 W_f}{P T_s} = \frac{3 \cdot 4 W_f}{T_l \cdot \omega_r \cdot T_s} = \frac{3 \cdot 4 W_f}{T_l \cdot 2\pi} \qquad (7)$$

Hence, $R_{sl} \propto W_f$ and the energy factor R_{sl} is a linear function of the speed ω_r .Simulation on the basis of Matlab/Simulink can be made to confirm it and the form of data is as follows in Table 2:

Table 2: Energy factor under various rotational speed

n(r/min)	R_{sl}	W_f(J)
50	1.010	0.53
100	1.108	0.58
200	1.242	0.65
300	1.413	0.74
400	1.566	0.82
500	1.738	0.91
600	1.872	0.98
700	2.044	1.07
800	2.196	1.15
900	2.349	1.23
1000	2.464	1.29

According to the data above, a curve of the energy factor R_{sl} is gained in Figure 7:

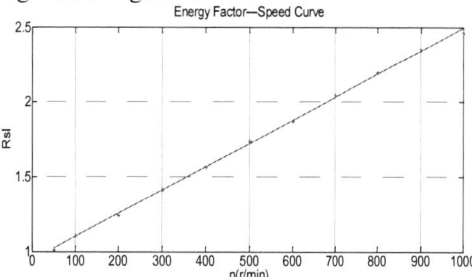

Figure 7. Energy Factor under different speeds

The simulation confirms the theoretical analysis that the energy factor R_{sl} is a linear function of the speed ω_r.

IV. ENERGY ANALYSIS FOR CONDUCTING ANGLE VARIATION

The value of conducting angle determines the time of current's effect on the motor. There are two modes of changing it: one is that switch-on angle varies, the other is the switch-off one altering.

(i) To start with, the switch-off angle is fixed and switch-on angle varies. The data and curves of simulation are as follows in Table 3 and Figure 8:

On the basis of the information in the table, a curve of the energy factor R_{sl} is obtained in Figure 8:

Table 3: Stored Energy & Energy Factor with Various Turn-on Angle

θ_{on}(°)	R_{sl}	W_f(J)
35	2.599	1.36
38	2.082	1.09
40	1.872	0.98
42	1.872	0.98
43	1.910	1.00
44	1.891	0.99
45	1.872	1.00
46	1.891	0.99
47	1.872	0.98
48	1.891	0.99
50	1.929	1.01
53	2.044	1.07
55	2.254	1.08

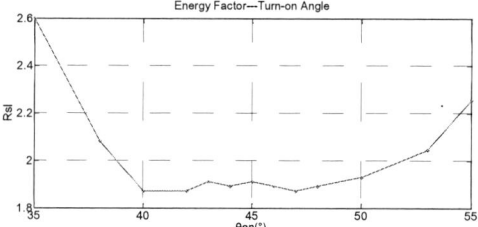

Figure 8. Energy Factor variations with changing turn-on angle and constant turn-off angle at 75° at the rotating speed 600r/min

From the simulation result, we can observe that the energy factor descends until it reaches 40° , then keeps constant till about 47° and finally it goes up again.

The reason for the curve variation trend can be explained in theory. The voltage equation has two forms:

$$U_s = iR + \frac{d\varphi}{dt} \qquad (8)$$

$$0 = iR + \frac{d\varphi}{dt} \qquad (9)$$

The equation (8) and (9) are for the conducting and freewheeling situation respectively. For equation (8), neglecting the component iR,

$$
\begin{aligned}
U_s &= \frac{d(Li)}{dt} \\
&= L\frac{di}{dt} + i\frac{dL}{d\theta}\frac{d\theta}{dt} \\
&= L\frac{di}{dt} + i\omega_r\frac{dL}{d\theta} \qquad (10)
\end{aligned}
$$

According to the laws for the inductance development with the rotor position variation, we can presume that during the conducting angle, the inductance is a linear function of the rotor position. Thus, equation (10) can be transformed to

$$U_s = L\frac{di}{dt} + i\omega_r k \ (k > 0) \qquad (11)$$

So through conversion,

$$
\begin{aligned}
I &= \frac{U_s - U_s e^{-\frac{k\omega_r t}{L}}}{k\omega_r} \\
&= \frac{U_s - U_s e^{\frac{k\theta}{L}}}{k\omega_r} \qquad (12)
\end{aligned}
$$

Also, on the basis of equation (8),

$$U_s = \frac{d\varphi}{d\theta} \cdot \frac{d\theta}{dt}$$
$$= \omega_r \frac{d\varphi}{d\theta}$$

Therefore,

$$d\varphi = \frac{U_s}{\omega_r} d\theta \qquad (13)$$

Because of

$$W_f = \int_{\theta_{on}}^{\theta_{off}} I d\varphi - W_0 \qquad (14)$$

W_0 containing output energy and other loss, substitute (12) and (13) into (14) and simplify:

$$W_f = \frac{U_s^2}{k\omega_r^2} \left(\theta_{off} + \frac{L}{k} e^{-\frac{k\theta_{off}}{L}} - \theta_{on} - \frac{L}{k} e^{-\frac{k\theta_{on}}{L}} \right) - W_0 \qquad (15)$$

Then, W_f differentiates θ_{on} and we can acquire,

$$W_f'(\theta_{on}) = \frac{U_s^2}{k\omega_r^2} \left(e^{-\frac{k\theta_{on}}{L}} - 1 \right) < 0 \qquad (16)$$

So, W_f is a decreasing function of θ_{on} and the first drop can be explained.

Analogically, for equation (9),

$$I = I_0 e^{-\frac{k\theta}{L}} \qquad (17)$$

Since both I and φ are decreasing functions of θ_{on}, W_f declines with θ_{on} augments.

For the following energy factor stable and rising period, with θ_{on} increasing, the conducting angle decreases. The conducting angle can be too small to keep the current and torque continuous as shown in Figure 9. Therefore, when the torque becomes zero, the voltage source will not provide energy for motoring, but store it in the coil, thus boosting the energy factor.

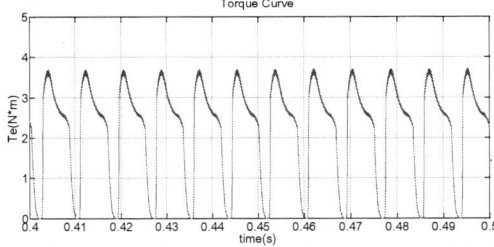

Figure 9. Electromagnetic torque curve when turn-on angle is 53°

(ii) Then, the switch-on angle is fixed and switch-off angle varies. The data and curves of simulation are as follows in Table 4 and Figure 10:

On the basis of the information ahead, a curve of the energy factor R_{sl} is obtained in Figure 10:

Table 4: Energy & Energy Factor with Various Turn-off Angles

$\theta_{off}(°)$	R_{sl}	$W_f(J)$
65	1.711	0.896
66	1.690	0.885
67	1.736	0.909
68	1.751	0.917
69	1.792	0.938
70	1.812	0.949
71	1.830	0.958
72	1.845	0.966
73	1.889	0.989
75	1.872	0.980
76	1.928	1.010
77	1.961	1.027
78	2.019	1.057
79	2.074	1.086
80	2.097	1.098
83	2.116	1.108
84	2.166	1.134
85	2.143	1.122
86	2.160	1.131
87	2.156	1.129

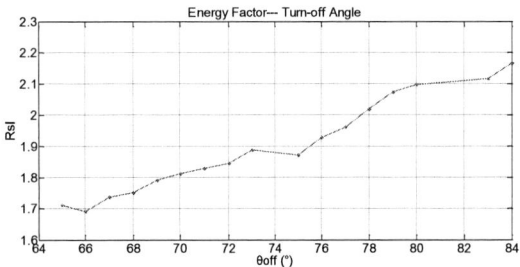

Figure 10. Energy Factor variations with changing turn-off angle and constant turn-on angle at 40° at the rotating speed 600r/min

From the simulation result, we can see that the energy factor rises with the turn-off angle θ_{off} increasing. This can also be explained by theoretical analysis similar to that of turn-on angle changing. According to the equation (12) (13) (14) and (17), we can calculate and gain that as θ_{off} expands, the energy factor raises. The variation of the turn-off angle θ_{off} doesn't produce any braking torque because the maximum of the angle is 90° and the inductance hasn't reached the dropping region. This should be interpreted by the expression (18)[7]

$$T_e = \frac{1}{2} i^2 \frac{dL}{d\theta} \qquad (18)$$

V. COMPARISON OF VARIOUS CONTROL METHODS IN ENERGY STORAGE

(i) CCC Method

This control method has been discussed in detail in the above paragraph and the current curve of it is shown in Figure 11.

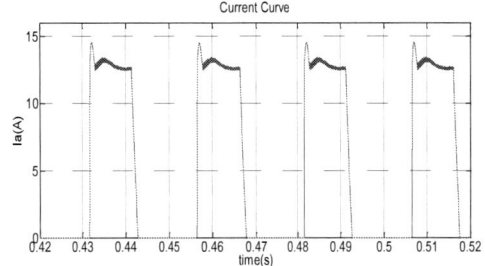

Figure 11. i_a gained by CCC method

(ii) Voltage PWM Control

This voltage PWM control utilizes soft chopping as well. It is a control method of taking advantages of the length of effective time provided by the DC source U_s on the conducting phase to change the average value of exciting voltage. Therefore, the electromagnetic torque can be controlled. For instance, one switch keeps conducting while the other is turned on and off with a duty of $D = \frac{t_{on}}{T_{chop}}$. And the schematic diagram of it is as follows in figure 12:

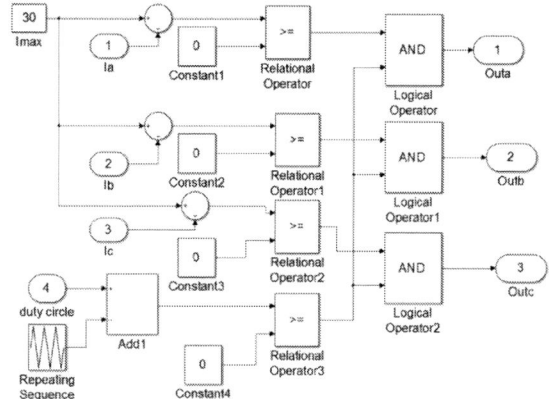

Figure 12. Schematic diagram of voltage PWM control

In this method, each phase current has an upper limit of 30 ampere. The modulating wave derives from speed difference between the standard one and actual one and then pass a PID regulator. The carrier wave has an amplitude of 1V and a frequency of 2 kHz. Through the comparison of the modulating and carrier waves, PWM signal is produced which is utilized to control the main switches. The simulation result of the phase current with the software Matlab/Simulink is in Figure 13:

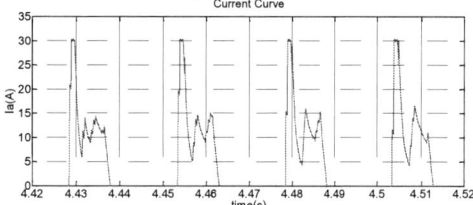

Figure 13. i_a obtained by voltage PWM method

(iii) Angle Position Control (APC)

When the rotating speed is high enough for the SRM to reach its maximal electromagnetic power, the torque will begin to drop as a result of the successive speed raising. To slower the torque reduction and keep power constant, APC is a suitable method. It regulates the speed via adjusting the switch-on or switch-off angle. The on-off of the main switch is only once during a rotor cycle. In this paper, the value of the turn-on angle is controlled to be compensated for the torque decrease. And the control principle is in Figure 14 below.

About this control method, the commuting angle is got from a PID controller passed by the difference between the standard speed and actual one. Furthermore, the upper limit and lower limit of the saturation part is 70 ° and 35 ° respectively. And the switch-off angle is fixed at

75°. The simulated current curve of APC is shown in Figure 15:

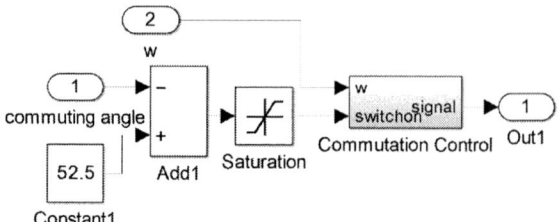

Figure 14. APC principle

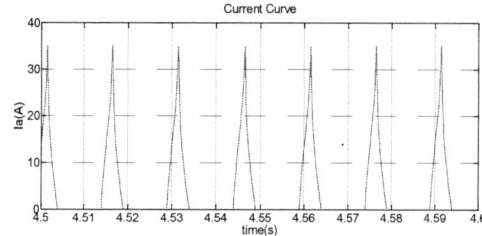

Figure 15. Current variation of APC at the speed 600r/min

The image shows that the peak value is higher than the two methods ahead, but the switching times and conducting time each phase is much less than the former two. It has a widespread application in high speed conditions to produce sufficient torque for the SRM.

For low-speed motoring, CCC and voltage PWM control is better. The data of energy factor of these two methods under different speed conditions can be acquired in Table 5.

Table 5: Energy Factor Comparison Between CCC and Voltage PWM Control Under Various Speeds

$n(r/min)$	$R_{sl}(CCC)$	$R_{sl}(PWM)$
50	1.010	1.102
100	1.108	1.031
200	1.242	0.949
300	1.413	1.276
400	1.566	1.322
500	1.738	1.409
600	1.872	1.751
700	2.044	1.979
800	2.196	2.284
900	2.349	2.527
1000	2.464	2.699

The energy factor curves of both methods is displayed below in Figure 16:

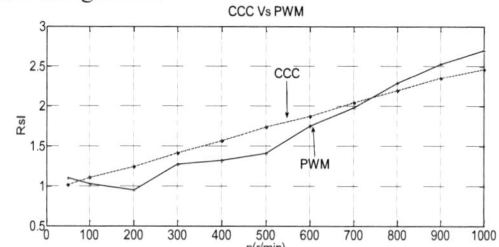

Figure 16. Energy factor comparison of CCC and voltage PWM control methods

From Figure 16, we can get the information that for lower energy factor, when the rotating speed is between 80r/min and 740r/min, voltage PWM control is a better choice and when speed is lower than 80r/min or higher than 740r/min,

CCC can be more suitable.

VI. CONCLUSION

This paper completes the simulation and confirmation of energy factor under various speeds, turn-on and turn-off angles respectively, which is conducive to ensuring the optimal speed, switch-on and switch-off angle for the SRM performance. About this simulation result, the energy factor increases as the rotating speed raises. It has some reference value for speed limit if there is loss requirement. Besides, the optimal turn-on angle is 40° and turn-off angle is 75°, where enough electromagnetic torque should also be taken into consideration. From the comparison of three control methods, when speed is higher than the fundamental one, APC is better to provide a greater torque; when speed is lower than the fundamental one, it's necessary to compare the energy factor to choose a better control method. In this simulation experiment, when speed is lower than 80r/min or higher than 740r/min, CCC will cause less energy loss and when it is between the mentioned two speed values, voltage PWM control method is a better choice.

In sum, the proposed methodology offers some reference values in choosing the optimal speed, turn-on and turn-off angles, and control methods.

REFERENCE

[1] K.W.E Cheng, "Storage energy for classical switched mode power converters", IEE Proc.-Electr. Power Appl. Vol. 150. No.4, July 2003

[2] Wong, R.C., Owen, H.A., and Wilson, T.G.: "An efficient algorithm for the time-domain simulation of regulated energy-storage dc-to-dc converters", IEEE Trans. Power Electron., 1987, 2, pp. 154-168

[3] Philip Came Kjaer, Peter Nielsen, Lars Andersen, Frede Blaabjerg, " A New Energy Optimizing Control Strategy for Switched Reluctance Motors", IEEE Transaction on industrial applications, Vol. 31, No. 5, September/October, 1995

[4] Jianbo Sun, Zhe Kuang, Shuanghong Wang, and Yi Chen, "Efficiency Optimal Control of Switched Reluctance Machine over Wide Speed Range Applied to Flywheel Energy Storage System", Electromagnetic Launch Technology (EML), 2012 16th International Symposium on.

[5] Shun-Chung Wang, Yi-Hua Liu, Shun-Jih Wang, Yih-Chien Chen, Shou-Zhuang Lin, "Adaptive Iterative Learning Control of Switched Reluctance Motors for Minimizing Energy Conversion Loss and Torque Ripple", Power Electronics Specialists Conference, 2007. PESC 2007. IEEE

[6] Krzysztof Tomczewski, Krzysztof Wrobel, "Improved C-dump converter for switched reluctance motor drives", Power Electronics, IET, 2014.

[7] R.Krishnan, "Switched Reluctance Motor Drives Modeling, Simulation, Analysis, Design, and Applications", Industrial Electronics Series, 2001.

[8] Miller T J E. "Switched Reluctance Motors and Their Control" [M]. Oxford: Magna Physics Publishing and Clarendon Press, 1993.

BIOGRAPHIES

Jingwei Zhu received his B.S. degree in 2015 in Electrical Engineering & Automation from Zhejiang University, Hangzhou, China.
He joined the department of Electrical Engineering, The Hong Kong Polytechnic University as a postgraduate. His main research interests are in the field of power electronics and motor drives.

K.W.E.Cheng obtained his BSc and PhD degrees both from the University of Bath in 1987 and 1990 respectively. Before he joined the Hong Kong Polytechnic University in 1997, he was with Lucas Aerospace, United Kingdom as a Principal Engineer.
He received the IEE Sebastian Z De Ferranti Premium Award (1995), outstanding consultancy award (2000), Faculty Merit award for best teaching (2003) from the University and Silver award of the 16th National Exhibition of Inventions, Faculty Engineering Industrial and Engineering Services Grant Achievement Award (2006), Brussels Innova Energy Gold medal with Mention (2007), Consumer Product Design Award (2008), Electric vehicle team merit award of the Faculty (2009), Special Prize and Silver Medal of Geneva's Invention Expo (2011) and Eco Star award (2012). He has published over 250 papers and 7 books. He is now the professor and director of Power Electronics Research Centre.

Luminous Flux Meter Failures and Potential Solution

William F. Chen[1]   K.W.E. Cheng [2]

[1,2] Department of Electrical Engineering, The Hong Kong Polytechnic University, Hong Kong
E-mail: William.f.chen@connect.polyu.hk
[2] E-mail: Eric-cheng.cheng@polyu.edu.hk

Abstract-This paper discusses the current problems that lie with using luminous flux meters as a method of data measurement and why there must be a need for a replacement method. The meters are a common product on the market but they are surprising inefficient when in use. Currently, their only useful for finding the luminous flux of an area but it is impossible for the flux of an individual light to be found using the meter. With the presence of LED lights and drivers now it may be possible to find the flux of an individual piece of lighting using a new method. The new method involves the use of a potential correlation between heat, efficiency and lumen output within LED drivers. This journal acknowledges the current situation of lighting, including common lighting information, testing and testing equipment. Then an in-depth look will be provided about the current problems along with the potential solution. The solution will include a look into the theory, initial or possible equations and an experiment to test and prove the theory.

Keywords-Heat-Efficiency Correlation, LED, LED Driver, Luminous Flux Meter

I. INTRODUCTION

Lighting is an integral part of daily life and it is so common that it can often be overlooked. It can also be easily purchase in most stores for household use and there is a wide range of manufactures available for larger lighting. However, it is still important to understand the products being purchase and what the information that is provided by manufactures really means. This way everyone can make a proper purchase for each surrounding.

Luminous flux or illuminance is used as a common indicator of wherever a lighting product is of high or low quality. Obviously, a higher number would indicate a brighter light and a low number would mean a poor quality product. Most of the time lighting companies will provide the data for their products when they are release. Some of these data will include the Colour rending index, luminous flux, power factor, beam angles and color temperature.

1. Colour Rendering Index (CRI)

The colour rendering index or CRI is an indication of how well the lighting is able to reveal the colour of an area. The range of the CRI is measured on a scale of 0 to 100%. A higher number will indicate that the product does a really good job of revealing colour while a low number will indicate a poor ability in revealing colour. The test that is conducted for CRI can be conducted very easily. The test starts with a set of blocks or samples with base colours,

the amount will depend on the test but there are commonly at east eight. Then the light will be shined upon the colours individually to see how well the light does in revealing the original colour. Lights that does a poor job in illuminating the colour or changes the colour in anyway stands out and will score very low. When the colours are revealed properly and have the correct shine and brightness on it, the CRI score will be high. [5] Generally, products on the market will have a CRI percentage that is over 80.

2. Luminous Flux

One of the most important numbers that consumers will inspect is the Luminous Flux. The luminous flux is represented by the unit lumens and in the most basic terms, the luminous flux is a representation of energy. It refers specifically to the energy that is produce by the lighting product when on. However, most of the time the luminous flux is judge by the brightness level. Although, technically brightness and energy is two different topics in this scenario they are related. When it comes to lighting, a product that can produce a higher level of lumens will also produce a brighter light source. Since, there is a correlation between flux and brightness, the luminous flux data becomes very important to consumers because they want to purchase the correct brightness of lighting for their uses.

3. Power Factor and Beam Angle

Power factor is another number that manufacturs will provide to consumers so they can better understand the product. The power factor is a number that usually from zero to one and it gives a clear indication of how well the product is at using the supplied power. By definition the power factor is the ratio of real power to apparent power. The real power is the power or electricity that is actually used by the lighting to light up and the apparent power is the amount of electricity that is actually taken in by the lighting. Small lighting systems generally do not need to store electricity to function, so it is ideal for the light to use as much of the power they intake as they can. This would mean that an ideal power factor number for a lighting product is as close to one as possible. This would mean a minimum loss or waste of power. The beam angle is a simple number that gives the consumers an idea on how much area the light can cover when lit. When the angle is large the light is more evenly disbruted and a small angle would mean the area of lighting is very narrow.

4. Colour Temperature

The colour temperature will simply give the colour of the lighting, usually with the unit K or Kelvin. Low temperatures are usually used by households, i.e. 2500-3500K, since the color is a warm yellow. When the temperature increases the light will become brighter, with the colour changing to white around 6000K and bright blue when over. All of the data provided gives consumers a general sense of how the product will function in ideal operating conditions and if they fit the consumers needs.

II. EQUIPMENT

All lighitng manufacturers have a job of providing the previously mentioned data to the public and their consumers. Although the results may seem short and simple, the task of actually producing this data is expensive and time consuming. Most of the cost for producing this data is the cost of equipment. The equipment that is required to produce data includes items like power suppliers, power meters, spectrometers and an intregating sphere. Power suppliers and meters are common items that will be found in any lab or factory. They have the simple job of allowing the manufactures to decide how much power to supply to their products. Through this method the manufacturs can decide the ideal operating settings for their products. The meters will also be used when the lighting products are going through the final quality test. Almost all the products will be check to ensure they are operating at the correct power and malfunctioning products will be quickly found.

1. Spectrometer

Another piece of important equipment is the spectrometer, which is used to find the intensity of lighting products. When consumers closely examine the specifications of their lighting products they will come across many graphs and most of the data from the graphs are provided by a sprectometer. The spectrometer has the ability to relate intensity in a few different ways including wavelengths, frequency, momentum and energy. The way a spectrometer works is by attaching the lighting product to the socket and then directing the light to the centre of the spectrometer. The centre of the spectrometer is usually occupied by a crystal or a form of liquid. The purpose of this is that when the light hits it the crystal will be able to separate all the different colours of the light. At different angles the light will be able to give off a different colour of the light. Then the light separated by the crystal will be directed to receivers that will intake the given data. The data will most commonly be provided in the form of wavelengths and frequency. Even though the data may seem confusing and intimidating initially, it can be quite simple. The graphs will basically notify you of how intense the lighting can be when operating and what is their ideal intensity. [1]

This data is very useful for manufactures because they will know if the colour or radiation from their light is dangerous and if adjustments are needed. This is important information because depending on the location of use, it is not always ideal to have a high intensity. For example, when the lighting is used in a small home high intensity can be brighter but it can cause your home to become hot and high intensity for long periods is a potential safety hazard to your body and eyes. Figure 1 is a picture of a small scale spectrometer with crystal attached.

Fig. 1: A common spectrometer on market [4]

2. Integrating Sphere

The most important and expensive piece of equipment for manufacturs is the integrating sphere. The sphere is usually used to provide most of the information that the manufacturers give to the consumers. The appearance of the integrating sphere is, as the name implies, in the form of a sphere. The sphere can be opened up to install the light for testing and closed up for testing. The benefit of this is function is that during testing the lighting product will be undisturbed and unaffected by the surrounding. Meaning the data received from this equipment will be as accurate as possible for each piece of lighting product. [2] Figure 2 is an image that shows a common integrating sphere and some of its functions.

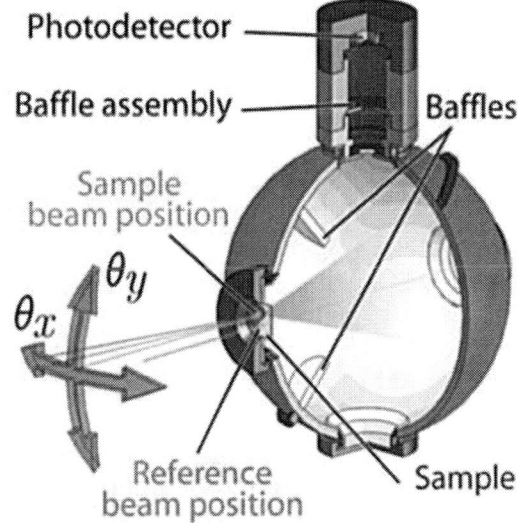

Fig. 2: A common integrating sphere [3]

When manufacturers complete all these task and tests, their goal is to show to consumers that their product is safe and superior to other companies. This is great for consumers because it gives them great detail about the product they will be paying for. Most of the data provided by companies are correct and they remain consistent throughout their lifetime. The only piece of data that changes over time and is almost impossible to prove for consumers is the luminous flux.

978-1-5090-0064-7/15 $31.00 © 2015 IEEE   277

III. THE ISSUE

This is true because for consumers there is actually no real way to get an accurate reading of luminous flux of a lighting product in a living environment. Of course, that is unless consumers have an integrating sphere in their homes. The only piece of equipment regular consumers can use for detecting lumens at home is a lumen reader.

Fig. 3: A common Luminous Flux meter available on market in use

Figure 3 showcases a luminous flux meter, available for sale to consumers, which is being used to check the flux of an area. The general instructions for using the meter is to stand directly under the light being tested, hold the meter at waist height and check the results produce from the meter. The issue that arises from this method is that lumen readers are inefficient. Good lumen readers will intake the lumen detected around an area and provide the data. The problem is that there is no way for the data to be correct because there will always be interference from the surrounding. Most lighting now have a beam angle that is over 180 degrees, so even if a room only has one piece of lighting it will still be spread out all over the room. If there is more than one piece of lighting in the area then all data will be of all the lighting products in the room. Even if there is only one piece of lighting in the room, the data will still be affected by daylight or any other lighting that can enter the room. Also, when using the flux meter it is extremely hard to record consistent data because any movement of the meter will cause a change in the data. And it will be very hard to avoid movement because the meter is handheld, which is hard to stabilize over a period of time.

Obviously most households have no need to check their lights for flux and data is of little concern. However, there are instances where luminous flux data is a very important piece of data. A prime example would be the data for street lighting. Street lighting is crucial to driving safety so

the lumen level for such a product is very important data. Street lighting must maintain a certain level of lumen and once it falls under that level, new lights must be installed. Without sophisticated equipment it is impossible for a normal maintenance team to collect the lumen levels of one lighting. The best they can do is record the results of a whole area and if that area is under illuminated then they have to change all the lights in that area. This method is highly inefficient and outdated. It is very likely that only one or two lights in an area are causing the levels to drop but they have to change all the lights in that area. Also the data is not 100% accurate because they have to test when the surrounding is not completely dark, meaning that sunlight also factors into the data.

The truth is that a change has to be made now and a new method has to be developed. Technology is now improving rapidly and the next move in lighting will the implementation of LED lighting worldwide, so the focus will be developing a method of calculating luminous flux for LED lighting.

IV. POTENTIAL SOLUTION

LED lighting, like other forms of traditional lighting, will produce heat. For traditional lighting the heat is a production of the infrared radiation rays produce from the light being on. LED lighting however does not produce any infrared radiation and the lighting heat is coming from the backside of the LED chips. This heat is natural from being on and having current passing through it. However, LED lights have another area that produces heat, which is the LED driver. LED lights need a driver to function because they have to convert the input electricity and control the amount of current that is passed to the LED chips. If the current is allowed to flow freely, the LED chips will be damaged easily if too much current flows through. This heat produce from the drivers will be the important factor that will spur on this new method of calculating luminous flux. If a LED driver is able to pass all the voltage it intakes to the LED chips then there will be no heat produce at all. At the moment there are no drivers that function at 100% efficiency because usually LED chips function at a very low voltage while standard voltage is much higher. Therefore, there is always an excess amount of voltage flowing in to the driver. To rid themselves of the heat, the driver must convert it to heat so it can be dissipated.

This means that there is a direct relationship between the efficiency and the heat produce by the driver. If temperature is low then the driver is working at a high efficiency and the LED lights will be working at full capacity and at max luminous flux level. If the driver is giving off a tremendous amount of heat then efficiency is very low and the luminous flux will be low also. Therefore, by using this relationship, it can be stated that there is a direct correlation between luminous flux and the heat produce by the LED drivers.

V. EXPERIMENT AND EQUATION

1. Experiment

978-1-5090-0064-7/15 $31.00 © 2015 IEEE   278

The next step to this process is to put this theory into actual testing. By matching LED lights with the most compatible LED drivers, then testing can be done. The test will involve using many different LED lighting products and recording their heat levels over time. If a correlation can be found then a new equation can be develop.

The testing will be done with multiple forms of lighting, including light bulbs, spot lights and large street lighting. The ideal subject for this experiment would be the street light because the size of the lighting is perfect and this type of lighting is the target audience. The most important factor of the experiment will still be decided by the drivers that is used. The lighting during the experiment will be fed many different amounts of voltage for long periods of time to check the response of the drivers. The LED drivers and lighting products will all be purchase from the market because the results are only valuable if it can be applied to products that already on the market. The only thing that will be controlled during the experiment is the voltage input. The experiment will start by feeding the lighting the exact amount of voltage it needs to operate. For example, if the street lighting is 180W than that is the amount that will be provided to it. In this scenario, for the theory to hold true, there should be no heat or very little heat emitted from the LED drivers. The temperature for this case will have to be very close to zero.

As the experiment progresses the input voltage will become higher and higher, leading to more current flowing in to the lighting. When the voltage increases going into the LED driver then the driver will have to do more work to keep the current flowing to the lighting constant. The purpose of this process will be to record data that can be plotted on a graph and then a correlation may be found visually. If the theory holds true then an equation can be proven for a correlation between heat from LED drivers and driver efficiency. The test then would have to be redone by including the luminous flux and lumen outputs. This part of the testing will be conducted with an integrating sphere and the length of the testing period will also be longer. With the use of an integrating sphere, accurate lumen data can be obtained and used to observe another layer of the lighting. With lumen data observations can be made on how much the effcency levels affect the lumen output. The lumen data will also give a clear indication of what the range of proper operation is for the piece of lighting. Also by completing the experiment a second time, it can help verify that the intial is complete and reliable. If there are large disrepences then parts of the experiment must be redone.

2. Equations

At this point without conducting the experiment it is hard to formulate a proper set of equations. Even if equations are formulated it can not be verified until testing is completed. Therefore, this journal is only introducing the problem and a theory for a proper solution. However, at this moment the initial part of the final equation is common knowledge to most people. The initial part is refering to the efficiency of LED drivers.

$$\eta = \frac{E_{used}}{E_{Supplied}}$$

where,

η = Efficiency
E_{used} = Energy Used by Lighting
$E_{Supplied}$ = Total Energy Supplied to Driver

This is the common efficiency equation for power usage. this equation is the start of the whole relationship and by deriving this equation it can be stated that:

$$E_{Supplied} - E_{used} = E_{Heat}$$

E_{Heat}, in this case would be used to represent the Energy that is being converted to heat so it can be dissipated. For the final equation to be correct there also needs to be confirmation on how the excess energy is converted to heat. At this moment it can be potential energy or thermal energy. Depending on the energy type, the final equation will be very different. By completing the experiment and analyzing the data, the final equation and correlation will be found.

VI. CONCLUSION

In the future LED drivers that will be used for large projects, like street or highway lighting, will have an innate ability to measure heat and wireless transfer that data to a computer that can calculate the luminous flux with that data. By pursuing this concept that involve a simple test and idea, a real change can be made in the field of LED lighting. Then items like lumen readers will be unnecessary and consumers will finally be able to have an accurate reading of their lighting output quickly and efficiently.

REFERENCES

[1] Walker, Kris. "Spectrometer Technology and Applications." AZO Materials (2014). Web. 1 Nov. 2015.
[2] Poikonen, Tuomas, Pasi Manninen, Petri Karha, and Erkki Ikonen. "Multifunctional Integrating Sphere Setup for Luminous Flux Measurements of Light Emitting Diodes." AIP (2010).
[3] "Integrating Sphere." Light-Material Interactions. California Institute of Technology, 2015.
[4] "Spectrometer Goniometer." Holmarc Opto-Mechanics PVT Ltd. 2014. Web. 1 Nov. 2015.
[5] Crist, Ry. "Shining a Light on High-CRI LEDs." CNET. CNET, 1 May 2014. Web. 1 Nov. 2015.

An Adaptive Modulation Scheme for Fundamental Frequency Switched Multilevel Inverter with Unbalanced and Varying Voltage Sources

Y. C. Fong[1]   K. W. Eric Cheng[2]

Department of Electrical Engineering, The Hong Kong Polytechnic University, Hong Kong
[1] E-mail: yc-chi.fong@connect.polyu.hk
[2] E-mail: eeecheng@polyu.edu.hk

Abstract–Stepped modulation operates at fundamental frequency allows the use of low speed switching devices and mitigates the switching loss and EMI concerns suffered by many other hard-switched inverters adopting carrier-based pulse-width modulation (PWM) such as space vector PWM and sinusoidal PWM. Half-height (HH) method is one of the simplest ways to compute switching angles of stepped sinusoidal output. Similar approaches have been utilized in digital-analog signal conversion and synthesis. However, there is limited research on HH-based modulation for power inverters. In this paper, a computational modulation scheme based on HH method was studied. The proposed modulation scheme is capable for real-time computation with most present microprocessors. This paper will demonstrate the implementation of the proposed modulation scheme at balanced voltage and the techniques of adaption to varying and unbalanced sources. Capability of having minimum total harmonic distortion (THD) under equal voltage sources of HH method was verified. The performance of the proposed modulation scheme was examined using numeric calculation and simulation with a 7-level cascade H-bridge inverter and the results are presented.

Keywords–multilevel VSI, Staircase modulation, THD

I. INTRODUCTION

Multilevel level inverters have drawn increased attention from the fields of electrical and power engineering due to their potential of modularization as well as applications in high-voltage direct current (HVDC) and high power electric propulsion system. Fundamental frequency switched multilevel inverters allow the possibility of low distortion under fundamental frequency operation. This eliminates the switching loss, EMI issues and the need of bulky filter which exist in many hard-switched carrier based inverters. A number of studies have been done on investigation of different techniques of step modulation [1-5] yet many of them on the basis of balanced and constant voltage. Due to the unitized features of energy storage elements (ESE) including electrochemical batteries and super-capacitors, it is possible to power energy storage system (ESS) based multilevel inverters using multiple isolated or series-connected voltage sources rather than a single voltage source. Unlike capacitive voltage divider, capacitor clamped or charge-pump based multilevel inverters [5-10], direct connection to multiple voltage sources dismisses the benefit of cycle voltage balancing among capacitors by adding specified switching states; on the other hand, this eliminates extra energy process by temporary storage elements, thus higher energy efficiency

is expected [11]. As terminal voltage of ESE is a function of the ESE state-of-charge (SoC), variation and unbalance of source voltage is principally inevitable during operating because of the capacity deviation and inequality of load allocation among storage units. Output voltage of stepped output multilevel inverters can be easily distorted by the voltage variation and imbalance.

In this study, a computational program based feed-forward scheme cooperating with modified half-height (HH) method is proposed for real-time modified sine wave output synthesis with stepped multilevel inverters. The modulation and control program is aimed to be realized with most of the available microprocessors in present. Therefore, solving complicated non-linear system is avoided and curve fitting approach was employed in the program instead. Source voltage variation and unbalance are also considered by the proposed modulation scheme so that the modulation index and firing angles are adjusted according to the states of ESS. Analysis of the gain error and THD in the stepped sinusoidal output of multilevel voltage source inverters employing simple HH method under balanced voltage will be discussed in the next section. Then, working principle and performance of the proposed feed-forward control scheme and the modification and adjustment of HH method will be illustrated

II. MULTILEVEL INVERTERS WITH STEPPED OUTPUT

Multilevel inverters combine a number of DC voltage sources to synthesize AC output and aim for high power and high voltage applications in the industries. There is a wide range of multilevel VSI topologies. The most well-known examples include cascaded H-bridge converter (CHBI), diode-clamped inverter (DCLI), and modular multilevel converter (M2C). Fig. 1 shows the topology of a single phase 7-level CHBI and Fig. 2 illustrates the synthesis of 7-level stepped sine wave. A 7-level CHBI contains three isolated voltage sources; each of the voltage sources is conducted by an H-bridge alternatively at specified firing angles so as to generate the required sinusoidal voltage at the output. Corresponding to different arrangements of firing angles, the output fundamental amplitude and THD content can be adjusted with the same sources. Modulation index, m, of a stepped modulation based inverter could be defined as the ratio of fundamental output voltage to the maximum attainable output voltage:

$$m = \frac{\pi}{4} \cdot \frac{V_1}{V_{tol}} \qquad (1)$$

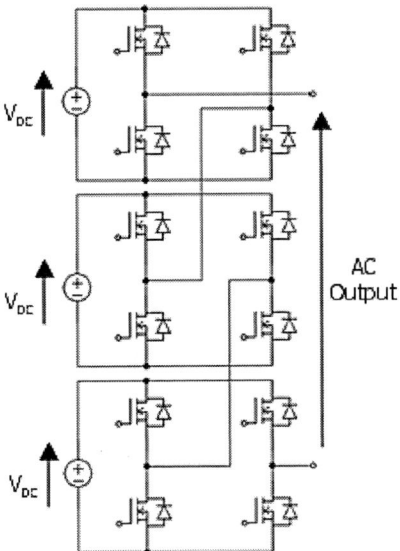

Fig. 1: Topology of a 7-level CHBI

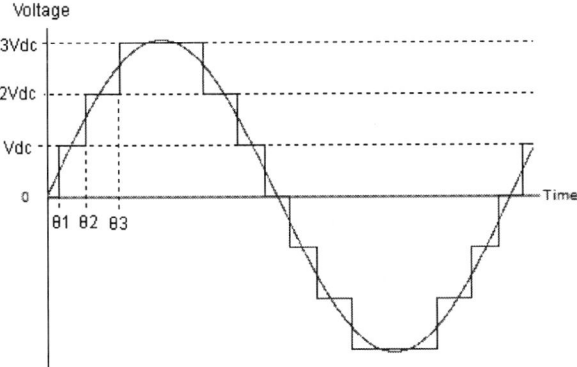

Fig. 2: The staircase modified sine wave synthesized by a 7-level CHBI

where V_{tol} is the total source voltage and the THD is defined as

$$THD = \frac{\sqrt{\sum_{n=2}^{\infty} V_n{}^2}}{V_1} \qquad (2)$$

and it can be simplified to

$$THD = \frac{\sqrt{V_{rms}{}^2 - V_{1rms}{}^2}}{V_{1rms}} \qquad (3)$$

There are a number of techniques to calculate the required switching angles such as geometric methods that compare the shape [1] or area [3] between the stepped output and referencing waveform; and numeric methods to eliminate the specific harmonics [4-5] or to minimize the total THD [2]. In general, geometric methods are usually simpler than the latter; therefore, they are more suitable for real-time applications.

III. THE HALF-HEIGHT (HH) METHOD

Luo [1] has carried out an investigation on the THD of stepped sine wave synthesized with several geometric switching angle computation methods including Equal-phase (EP), Half-equal-phase (HEP), Half-height (HH) and Feed-forward (FF) methods. The results suggested that HH method attains the lowest THD among the four simple methods and the resultant values were close to the minimum attainable values. Unfortunately, no information about the fundamental amplitude of output voltage was included and analysis of the concerned calculation techniques under different modulation indices was not in the scope of that study. However, the ability of adjusting or regulating output voltage could be very important in many applications. For example, the controller of electric motor drive adjusts the output amplitude and frequency of inverters based on torque and speed requirement and modulator should be able to actuate the switches to synthesize the output corresponding to the voltage set-point. Especially that the sources are not precisely tuned in ESS powered inverters, switching angles have to be adjusted according to the actual source voltage level. In view of this, further investigation on the gain error and THD of simple HH method are presented in this paper, then, the corresponding improvement on this method is suggested.

HH method synthesizes the desired sinusoidal output by setting the switching angle at the time that the value of the referencing sine function is equal to the half-height of source levels. In other words, the ith switching angle is determined by the following:

$$\theta_i = sin^{-1}\left(\frac{\pi}{4} \cdot \frac{\frac{V_{dci}}{2} + \sum_{k=1}^{i-1} V_{dck}}{mV_{tol}}\right) \qquad (4)$$

For balanced voltage source, the switching angles can be simplified as

$$\theta_i = sin^{-1}\left(\frac{\pi}{4} \cdot \frac{i - \frac{1}{2}}{ms}\right) \qquad (5)$$

and the resultant fundamental amplitude of the n-step output, generated by a (2s+1) levels VSI is

$$V_1 = \frac{4}{\pi} \cdot \sum_{i=1}^{s} V_{dci} cos(\theta_i) \qquad (6)$$

Recalling the definition of THD in (3), the overall root-mean-square (RMS) value of the s-step waveform and its fundamental value can be calculated as

$$V_{rms} = \sqrt{\frac{2}{\pi} \cdot \left[\sum_{i=1}^{s-1}(\theta_{i+1} - \theta_i)\left(\sum_{k=1}^{i} V_k\right)^2 + \left(\frac{\pi}{2} - \theta_s\right)\left(\sum_{j=1}^{s} V_j\right)^2\right]} \qquad (7)$$

and

$$V_{1rms} = \frac{V_1}{\sqrt{2}} \qquad (8)$$

Table 1: Comparison of THD values of HH method at different modulation index and the best firing angle in [1]

Number of Levels	Best Firing Angle	HH (@m = 0.785)	HH (@THD$_{min}$)	
5	16.42%	17.61%	16.42%	(@m = 0.88)
7	11.53%	12.23%	11.53%	(@m = 0.85)
9	8.9%	9.37%	8.9%	(@m = 0.83)
15	5.31%	9.37%	5.31%	(@m = 0.81)
25	3.18%	3.27%	3.18%	(@m = 0.80)

IV. ANALYSIS OF HH METHOD AT DIFFERENT MODULATION INDEX

According to the definition in (1) and considering a multilevel inverter which is supplied by balanced voltage sources, if the modulation index setting, m of the referencing sinusoidal signal in (5) varies from 0.6 to 1.3, the output index of HH method would equivalent from around 0.5 to 0.95 (Fig. 3). The corresponding THD values were obtained using (1) and (6) to (8), parts of the results are plotted in Fig. 4. In Fig. 3, a generally attenuated output is observed; the repetitive pattern in the input-output functions is reasoned by the fact that a high level VSI is actually employs lower number of levels when the m input is low. By observing the THD values plotted in Fig. 4, the lowest attainable THD value by HH method is not occurred at $m = \frac{\pi}{4} \approx 0.785$, but at a slightly higher m setting; and the same THD values as "best firing angles" were observed at these points (Table 1). Liu et al. [2] have developed a mathematically proved optimization algorithm to obtain the switching angles for stepped sine output with minimum THD:

$$\theta_i = sin^{-1}(\frac{i-\frac{1}{2}}{s-\frac{1}{2}} \cdot \rho) \qquad (9)\ [2]$$

in which was actually solving a non-linear optimization equation:

$$s \cdot m = \sum_{i=1}^{s} \sqrt{1 - \left[\frac{i-\frac{1}{2}}{s-\frac{1}{2}} \cdot \frac{V_{dc}(s-\frac{1}{2})}{\frac{1}{V_1}\sum_{j=1,3...}^{\infty} V_j^2 + \frac{\pi\lambda V_1^2}{8nV_{dc}}}\right]^2} \qquad (10)\ [2]$$

and substitute $\rho = \frac{V_{dc}(s-\frac{1}{2})}{\frac{1}{V_1}\sum_{j=1,3...}^{\infty} V_j^2 + \frac{\pi\lambda V_1^2}{8nV_{dc}}}$, which could be solved by iterative methods like Newton-Raphson. But if we compare equation (5) and (9), it can be observed that optimal THD could also be obtained with HH method by substituting $\frac{\pi}{4mn} = \frac{\rho}{n-\frac{1}{2}}$ as a result of the fact that every different m setting result unique m outputs. The output THD of a 7-level stepped VSI with different switching angles evaluation techniques, including real-time optimal THD in [2], voltage-second equalization in [3], selective elimination of 5th and 7th harmonics, as well as HH method were compared. This comparison verified that the output THD of HH method lies on the optimal curve (Fig. 5). This implies that minimum THD could be attained with simple geometric methods like HH based on the premise that the mismatch of m between output and input is corrected.

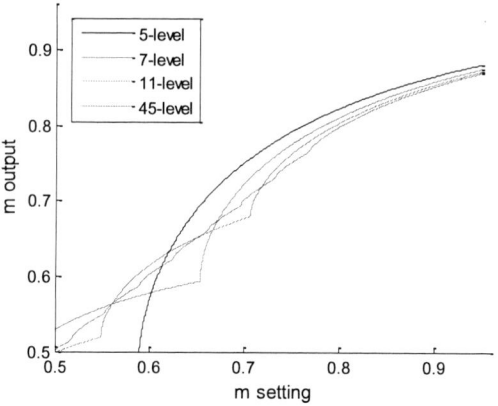

Fig. 3: Input-output characteristic of simple HH method under balanced voltage sources

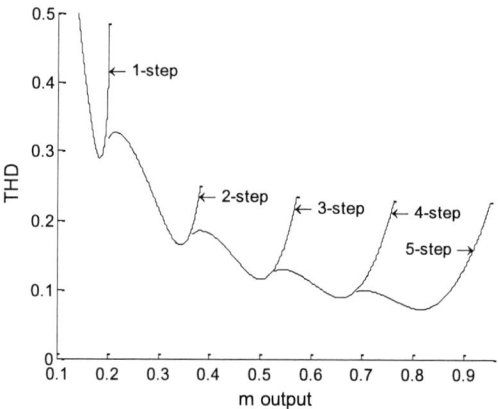

Fig. 4: The output THD of an 11-level VSI using different numbers of switches

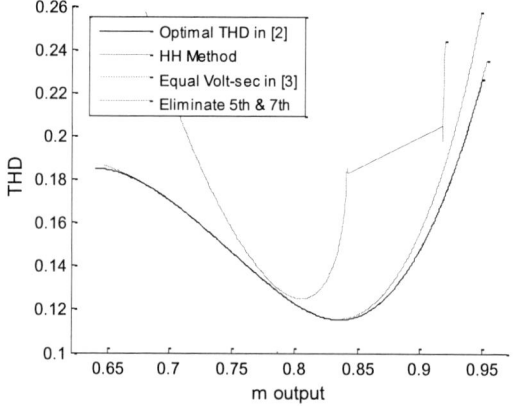

Fig. 5: Output THD of 7-level VSI employing different firing angle evaluation methods

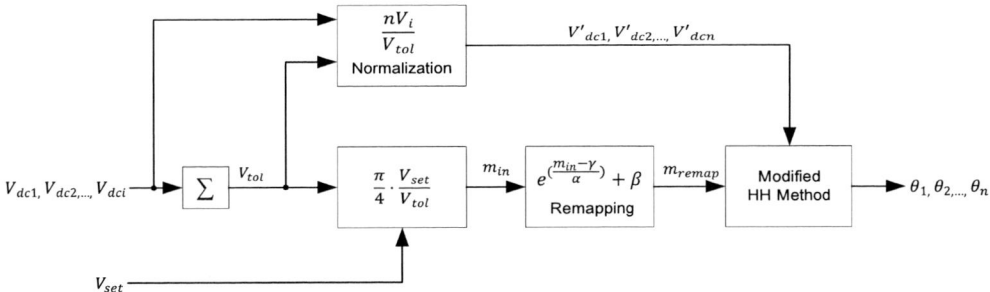

Fig. 6: Simplified functional block diagram of the proposed modulation scheme

V. PROPOSED MODULATION SCHEME

The proposed modulation scheme is fundamentally based on an open-loop and feed-forward system (Fig. 6) that input parameters are the voltage set-point and source voltage levels measured by the ESS management or monitoring units whereas the output is a group of switching angles. Overall voltage, V_{tol}, is the summation of source voltage values while normalized voltage is calculated based on average source voltage which indicates the voltage imbalance. Set-point modulation index, m_{in}, is obtained with (1), which is then remapped to m_{remap} using (13). With m_{remap} and voltage imbalance factors, switching angles, $\theta_1, \theta_2, ..., \theta_i$, are computed by an improved HH method which includes a level allocation scheme with corresponding HH models and can adjust the firing angles in response to voltage imbalance.

Disregarding the instrumental errors in V_{dci}, gain error and output THD defined in (11) and (3) were main performance indicators in this study. These two factors are primarily governed by the accuracy of the remapping algorithm and modified HH method.

$$Gain\ error\% = \frac{m_o - m_{in}}{m_{in}} \times 100\% \qquad (11)$$

1. Modulation Remapping of HH Method

HH method is a simple geometric way to approximate the referencing sinusoidal signal and the resultant fundamental amplitude is slightly different from the set-point which can be observed in Fig. 3. Since the input-output function of modulation index of HH-based modulation is non-linear, it would be demanding for the processor to solve this function online. Instead of iteratively solving the complicated function in (9)-(10), one direct way to deal with this problem would be using results computed offline to perform curve fitting and construct a remapping function to match the set-point and output indices. In this study, staircase modulated VSIs of up to 101-level with input modulation index m_{in} ranging from 0.5 to 1.3 were considered. In the aforementioned range, the relationship between input modulation index, m_{in}, output modulation index, m_o, and number of VSI levels, n, could be fitted with the following function:

$$m_o = \alpha\ ln(m_{in} - \beta) + \gamma \qquad (12)$$

this suggests that the input modulation index to HH function can be remapped by the following:

$$m_{remap} = e^{(\frac{m_{in} - \gamma}{\alpha})} + \beta \qquad (13)$$

where

$$\alpha = 0.1062 \left(\frac{n-1}{2}\right)^{-1.6} + 0.07497$$
$$\beta = \frac{n}{n-1} \cdot \frac{\pi}{4} - \frac{1}{243.5(\frac{n-1}{2})^{-1.6} + 13.72.} \qquad , and$$
$$\gamma = 0.02704 \left(\frac{n-1}{2}\right)^{-1.6} + 0.9762$$

The input modulation index of the proposed modified HH method is remapped via (13) before passing to HH evaluation function. Since α, β and γ are only dependent on the number of levels n, which is a finite number. These three parameters can either be calculated online or offline and stored in a $3 \times \frac{n-1}{2}$ array. The modulation index input-output characteristic of the remapped HH method in a range of $0.5 < m_{in} \leq 0.95$ was examined; the results of 5, 7, 11 and 45-level VSIs are plotted in Fig. 7. The gain errors of original HH method (Fig. 8) and proposed modified HH method (Fig. 9), in the range of $0.5 < m_{in} \leq 0.95$, for 5 to 101 level VSIs were compared. The results show a significant reduction in input-output deviation when the proposed remapping function is adopted.

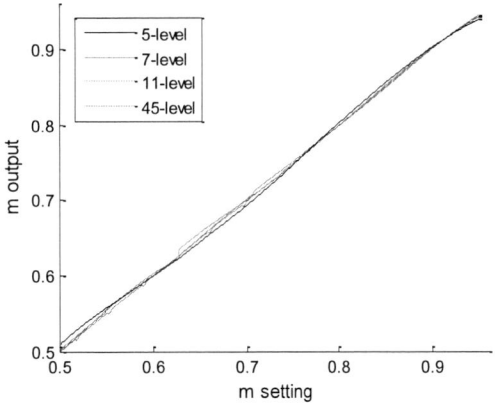

Fig. 7: Input-output characteristic of remapped HH method under balanced voltage sources

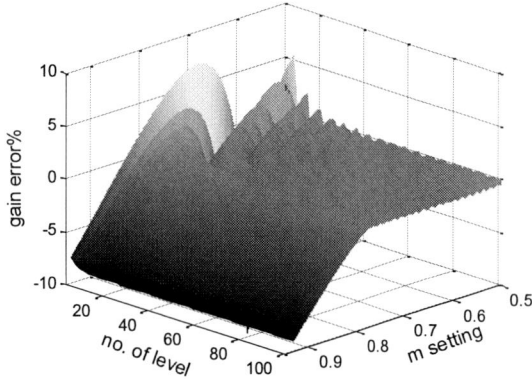

Fig. 8: Gain error of simple HH method in the range of study

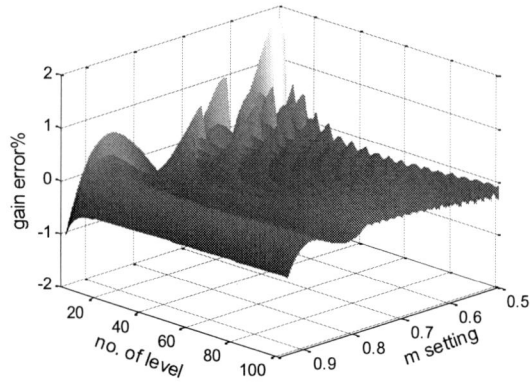

Fig. 9: Gain error of remapped HH method in the range of study

2. Active Number of Levels in Staircase Modulation

The staircase sinusoidal output generated by an n-level HH based VSI contains less number of steps when the required voltage amplitude is significantly lower than the total source voltage. Although different numbers of active sources can attain identical fundamental amplitude by selecting different equivalent input modulation indexes m_{in}, the analysis based on (3), (7)-(8) shows that allocating a higher number of active levels results less THD in output (Fig. 4). Therefore, one criteria to determine the number of active sources, a, is when $m_{HHa} \geq \frac{\pi}{4} \cdot \frac{2a-1}{n}$. This can be evaluated with the following curved fitted formula:

$$m_{amin} = -\frac{\sqrt{2}}{a^{0.189}(n-1)} + 0.7829 \qquad (14)$$

for $a \geq 2$ and $n \geq 5$

and m_{in} sending to the remapping function is then scaled by $\frac{n-1}{2a}$.

If $m_{in} < \frac{\pi\sqrt{5}}{3(n-1)}$, only one voltage source is deployed ($a = 1$) to generate a simple 3-level modified sine wave, and the firing angle can be directly computed using:

$$\theta_1 = cos^{-1} \frac{m_{in}(n-1)}{2} \qquad (15)$$

3. Response to Voltage Variation

In this study, variation of the voltage sources is categorized into two types:

    1) Overall voltage variation

    2) Voltage imbalance

In an ESS powered multilevel inverter, overall source voltage variation is mainly due to the increase or decrease in total charge of ESS during charging or discharging operation whereas capacity inequality among ESEs and importer duty allocation can lead to voltage imbalance. Overall voltage variation can be corrected by tuning modulation index setting m_{in} in the controller but this is not applicable for voltage imbalance. In this study, two techniques of handling voltage imbalance were examined.

3.1. Angle Compensation Technique

Angle compensation method bases on normalization of source voltage values and evaluates the switching angles according to the fundamental amplitude calculated in (6). The resultant switching angles are adjusted with (16) to compensate the effect of voltage variation to the output fundamental amplitude.

$$\theta_i = cos^{-1} \left(\frac{\cos \theta_{HHi}}{V'_i} \right) \qquad (16)$$

The main shortcomings are the additional distortion on the output voltage since this compensation does not conform to the sinusoidal geometry and θ_i cannot be adjusted beyond θ_{i-1} or zero. As a result, this is only capable for small voltage variation.

3.2. Unbalanced HH Technique

HH method is capable of handling unbalanced voltage sources following (4). Unlike the aforementioned technique that angle compensation is evaluated only with source voltage variation and fundamental amplitude. Unbalanced HH method conforms to sinusoidal characteristic. Although this may not be able to fully compensate for the fundamental amplitude change. The following simulation result shows the effectiveness of this technique.

VII. SIMULATION RESULT

In order to examine the performance of the proposed modulation scheme as well as to compare the effectiveness of different techniques in response to voltage fluctuation, the output voltage waveform of a fundamental frequency switched 7-level inverter with stationary and varying voltage sources was evaluated in MATLAB. The voltage set-point changed with time; the corresponding gain error and THD were analyzed and compared with a referencing condition of balanced 100V constant voltage sources. (Fig. 10-12) The simulation result shows that angle compensation method has limited effect in correcting the gain error caused by voltage imbalance. The THD grew significantly when the voltage imbalance rose beyond 15%. In contrast, the resultant gain error and THD value of output waveform produced with unbalanced HH method were close to that of the referencing condition.

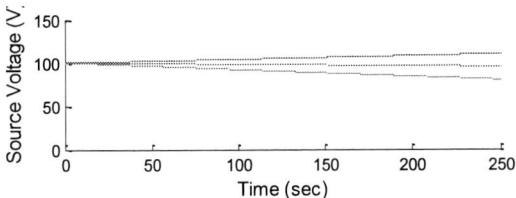

Fig. 10: The three voltage sources of the 7-level multilevel VSI varied from 100 to 110, 95 and 80 respective in simulation

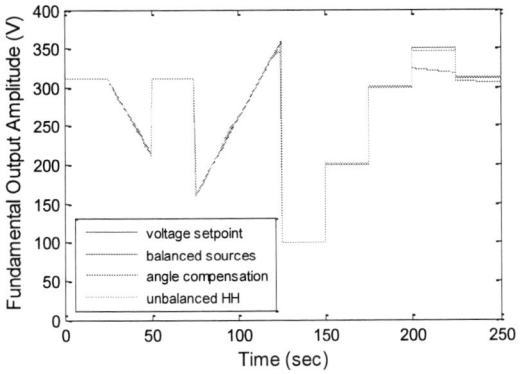

Fig. 11: Fundamental amplitudes of the outputs and compared with balanced sources and set-point voltage as reference

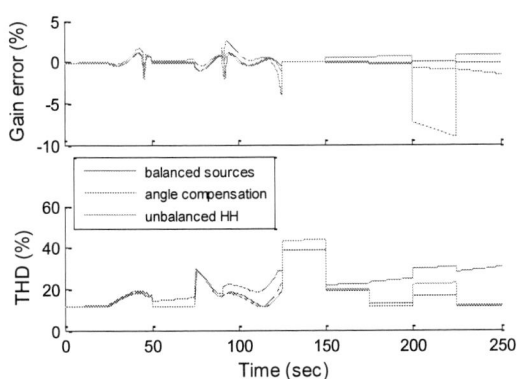

Fig. 12: The gain error and THD of the outputs

VIII. CONCLUSION

HH method has been employed in many single processing and synthesizing applications due to its simplicity and accuracy. This study analyzes the input-output characteristic and output THD of simple HH method comprehensively and has found that by adopting a remap function for the amplitude of referencing sine function, HH method is found to be an effective way to modulate the staircase sine wave of multilevel VSI and attain minimum output THD. With the nominalized voltage levels, unbalanced HH method can adjust the switching angles in respond to voltage variations and imbalance. Simulation result shows that the proposed modulation scheme is adaptive to varying voltage sources such as batteries and super-capacitors in ESS powered multilevel VSI and is a promising alternative for fundamental frequency switched VSI having accuracy and less distorted output.

REFERENCES

[1] F. L. Luo, "Investigation on best switching angles to obtain lowest THD for multilevel DC/AC inverters," *IEEE Conference on Industrial Electronics and Applications (ICIEA)*, 8th, vol., no., pp.1814-1818, 19-21 June 2013

[2] Y. Liu; H. Hong; Huang, A.Q., "Real-Time Calculation of Switching Angles Minimizing THD for Multilevel Inverters With Step Modulation," *IEEE Trans. Ind. Electron.*, vol.56, no.2, pp.285-293, Feb. 2009

[3] D. W. Kang, H. C. Kim, T. J. Kim, D. S. Hyun, "A simple method for acquiring the conducting angle in a multilevel cascaded inverter using step pulse waves," *IEE Proceedings in Electric Power Applications*, vol. 152, issue:1, pp.103-111, Jan. 2005

[4] Chiasson, J.N.; Tolbert, L.M.; McKenzie, K.J.; Zhong Du, "A unified approach to solving the harmonic elimination equations in multilevel converters," *IEEE Trans. Power. Electron.*, vol.19, no.2, pp.478-490, Mar. 2004

[5] Ozpineci, B.; Tolbert, L.M.; Chiasson, J.N., "Harmonic optimization of multilevel converters using genetic algorithms," *IEEE Power. Electron. Letters*, vol.3, no.3, pp.92-95, Sept. 2005

[6] Peng, F.Z., "A generalized multilevel inverter topology with self voltage balancing," *IEEE Trans. Ind. Applicat.*, vol.37, no.2, pp.611-618, Mar./Apr. 2001

[7] Gui-Jia Su, "Multilevel DC-link inverter," *IEEE Trans. Ind. Applicat.*, vol.41, no.3, pp.848-854, May-June 2005

[8] Ceglia, G.; Guzman, V.; Sanchez, C.; Ibanez, F.; Walter, J.; Gimenez, M.I., "A New Simplified Multilevel Inverter Topology for DC/AC Conversion," *IEEE Trans. Power. Electron.*, vol.21, no.5, pp.1311-1319, Sept. 2006

[9] Y. Ye; Cheng, K.W.E.; J. Liu; K. Ding, "A Step-Up Switched-Capacitor Multilevel Inverter With Self-Voltage Balancing," *IEEE Trans. Ind. Electron.*, vol.61, no.12, pp.6672-6680, Dec. 2014

[10] J. Liu; Cheng, K.W.E.; Y. Ye, "A Cascaded Multilevel Inverter Based on Switched-Capacitor for High-Frequency AC Power Distribution System," *IEEE Trans. Power. Electron.*, vol.29, no.8, pp.4219-4230, Aug. 2014

[11] K. W. E. Cheng, "Storage energy for classical switched mode power converters," *IEEE Proc. in Electric Power Applica.*, vol.150, issue:8, pp.439-446, July 2003

Large Signal Stability Analysis of Aircraft Electric Power System Based on Averaged-Value Model

Yanbo Che[1]   Xiaokun Liu[1]   Zhangang Yang[2]

[1] Key Laboratory of Smart Grid of Ministry of Education, Tianjin University, Tianjin, China
E-mail: ybche@tju.edu.cn
[2] Department of Aviation automation, Civil Aviation University of China, Tianjin, China
E-mail: zgyang@cauc.edu.cn

Abstract–More electric aircraft (MEA) is using more electrical power on board and the stability analysis of the aircraft electric power system (AEPS) becomes more important. This paper presents a modeling study and stability analysis of the EPS for MEA. Averaging technique is used to model the aircraft EPS, which includes a generator, an autotransformer rectifier and an inverter-motor load. The proposed model is accurate and simple for the large signal stability analysis of MEA power system. Brayton-Moser mixed potential and Lyapunov stability theorems are employed to estimate the large-signal stability region of the system. Simulation results are used to verify the effectiveness and feasibility of the proposed analytical method.

Keywords–Brayton-Moser mixed-potential theory, large signal stability region, nonlinear averaged-value model

I. INTRODUCTION

Nowadays, all countries around the world is facing great challenges of fuel consumption and pollution, and the more electric aircraft concept is considered to be an attractive solution for this problem among international aerospace industry[1-5]. Many subsystems that previously ran using hydraulic, mechanical, and pneumatic systems will be fully or partially replaced with electrical systems in the more electric aircraft. The increase in the requirement for electrical energy on the more electric aircraft brings great challenges to the design and analysis of the aircraft power system.

Aircraft electrical power system can be divided into generating system, distribution system and various onboard loads. The increase use of electrical energy has led to a demand for rapid technology development, particularly in electronics. Power electronics become one of the most important enabling technologies for the more electric aircraft. For example, AC/DC converter is used in Transformer rectifier unit (TRU) to provide a High DC voltage and a lot of electrical loads are based on power electronic converters and drive systems [7-12].

Due to the rapid demand of electrical energy onboard, the AEPS is becoming more complex. Additionally, power electronic driven loads often behave as constant power loads having negative impedance that can significantly affect the power system stability. Therefore, stability analysis of AEPS is of great importance for the development of the more electric aircraft concept. Stability analysis of AEPS mainly includes electrical power system modeling and simulation, small signal and large signal stability analysis etc [13,14].

Several methods are commonly used for the modeling of aircraft power system. The first is the state-averaging modeling method, which is used for analysis of many power converters, particularly for converters in dc distribution system and single-phase ac systems [15-17]. However, this approach can be very complicated when used for large scale aircraft power system modeling [18]. The second is the dq-transformation theory, which is widely used in three-phase ac systems. For aircraft power system modeling, dq-transformation theory can be used to model ac/dc converter and dc/ac inverter, which will be models as time-varying transformers. The dq-transformation theory is suitable for functional modeling and simulation for AEPS [19-20]. Moreover, linearization method can be used with dq-model to investigate the small-signal stability of the aircraft power system [21-23]. But the dq-model is still complicated for the large signal stability analysis of AEPS. The third method is the averaged-value modeling method, which has been used for six and twelve pulses rectifier. With the average-value method, the ac power source together with the rectifier can be modeled as dc voltage source with good accuracy. The averaged-value model is suitable for not only fast simulation but also large signal stability analysis [24].

Most of previous papers focus on the small signal stability analysis with the linearized model around the equilibrium point and the classical eigenvalues or frequency domain approaches. Although small signal stability analysis can be useful to understand the behavior around the equilibrium point, the large signal asymptotic region of the system can not be obtained and the system's behavior under large disturbances will not be understand. Moreover, despite the behavior of the system under large disturbances can be obtained by solving the system's differential equations for some particular conditions, the computational burden and the time cost can be very high. And it can not be able to help us to get insight into the influence of different parameters on the system's behaviors. Thus, large signal stability analysis with Lyapunov stability theorems is necessary.

In this paper, averaging technique is used to model the auto-transformer rectifier unit (ATRU). Based on the averaged-value ATRU model, a model suitable for large signal stability of aircraft power system is build. And Brayton-Moser mixed potential and Lyapunov stability theorems are employed to estimate the large-signal stability region of the system. In the last section of this paper, simulation results are given to verify the effectiveness and feasibility of the proposed analytical method.

978-1-5090-0064-7/15 $31.00 © 2015 IEEE

II. AIRCRAFT POWER SYSTEM MODEL

The large aircraft power system considered in this paper is shown in Fig.1. There is one 250kVA synchronous generator, which is connected directly to the engine shaft. Then, the output of the generator will have a variable frequency related to the speed of the turbine. In addition, the generator is operated with a generator control unit (GCU) to control the main high-voltage AC bus. Large loads on the AC bus include the wing ice protection system (WIPS), the flight control system represented in this figure by two electromechanical actuators (EMA), and the autotransformer rectifier unit which feed the DC high-

voltage bus. The most significant loads on the dc buses are the environmental control system (ECS) which maintains the temperature and pressure of the passenger cabin of the aircraft. Additionally, a 28V DC bus fed from a buck-boost converter unit is connected on the dc high voltage bus.

It can be seen in Fig.1 that there are many electrical loads based on electronic converters (AC/DC, DC/DC, DC/AC converters) driving various loads. The influence of them on the large signal stability of the system will be investigated in this paper.

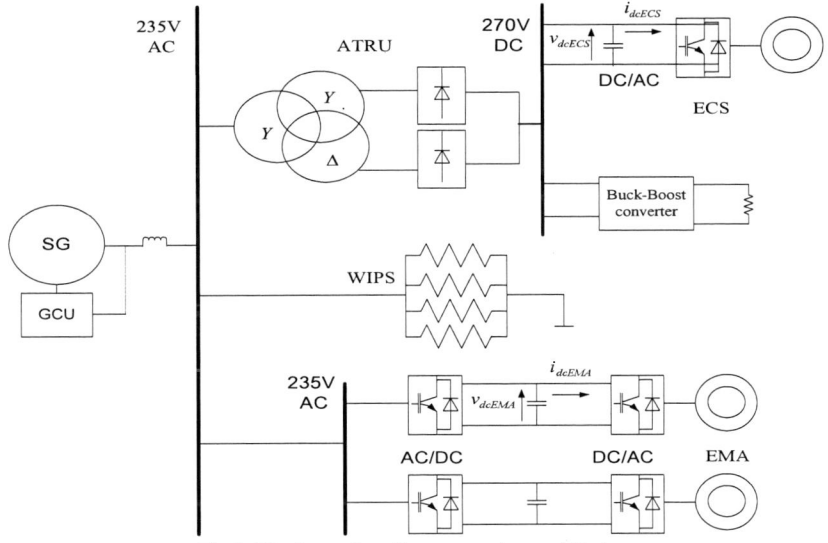

Fig.1: The large aircraft power system architecture

Fig.2 gives a typical simplified aircraft electric power system. As shown in the figure, the system is comprised of a generator, an autotransformer rectifier unit and an inverter-motor load system.

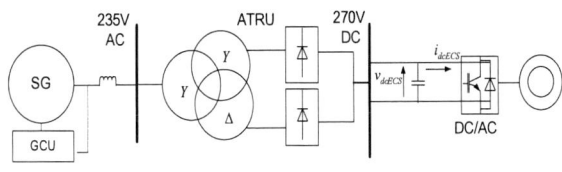

Fig.2: A typical simplified aircraft electric power system

Due to the tight regulation of the generator control unit, the generator can be model as an ideal voltage source. And like in many other papers, the inverter-motor system will be model as a constant power load. As for the ATRU, this paper proposes an averaged-value model [24], which is simple but accurate enough for large-signal stability analysis. Fig.3 shows the DC side equivalent circuit for aircraft power system based on the averaged-value modeling technique.

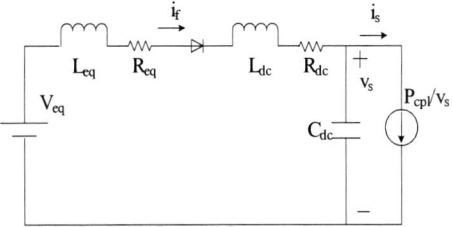

Fig.3: Equivalent circuit for aircraft power system

where, $V_{eq} = \dfrac{12\sqrt{3}}{\pi}\sin(\dfrac{\pi}{12})V_m$, V_m is the AC voltage amplitude. $L_{eq} = 2L_p + L_s$, L_p and L_s represent the primary and secondary leakage inductances of autotransformer, respectively. $R_{eq} = \dfrac{3w}{2\pi}[2(1-\dfrac{3}{n})L_p + L_s]$ is due to overlap effects of the rectifier, $n = 6.464$. L_{dc}, R_{dc}, C_{dc} are the parameters of the filter. p_{cpl} represents the constant power load.

III. BRAYTON-MOSER THEORY

Brayton-Moser mixed-potential theorem is used to study the large signal stability of the aircraft power system. The method can derive a scalar function, known as mixed potential function, which can be used to build Lyapunov function of the system for large signal stability analysis.

The general Brayton-Moser mixed potential function is the following:

$$P(i,v) = -A(i) + B(v) + D(i,v) \tag{1}$$

where $A(i)$ and $B(v)$ are the nonlinear function of the system's currents and voltages, respectively. $D(i,v) = i^T \cdot \gamma \cdot v$, γ is a constant matrix depending on the topology.

And the Lyapunov energy function can be obtained by the equation:

$$P^*(i,v) = \frac{u_1 - u_2}{2} \cdot P(i,v) + \frac{1}{2} P_i^T \cdot L^{-1} \cdot P_i + \frac{1}{2} P_v^T \cdot C^{-1} \cdot P_v \tag{2}$$

where

$$\begin{aligned} u_1 &= \min\left[\lambda(L^{-1/2} \cdot A_{ii} \cdot L^{-1/2})\right] \\ u_2 &= \min\left[\lambda(C^{-1/2} \cdot B_{vv} \cdot C^{-1/2})\right] \end{aligned} \tag{3}$$

$$A_{ii} = \frac{\partial^2 A(i)}{\partial i^2}, \quad B_{vv} = \frac{\partial^2 B(v)}{\partial v^2}$$

$$P_i = \frac{\partial P(i,v)}{\partial i}, \quad P_v = \frac{\partial P(i,v)}{\partial v}$$

According to the Brayton-Moser Theory, if

$$u_1 + u_2 > 0 \tag{4}$$

then, for all state variables i,v among certain domain D

$$P^*(i,v) \to \infty \tag{5}$$

And the domain D is the asymptotic stability region.

IV. Large Signal Stability Analysis

Based on the equivalent circuit in Fig.3, the equations for the system can be written as follows:

$$L_f \frac{di_f}{dt} = -R_f \cdot i_f - v_s + V_{eq} \tag{6}$$

$$C \frac{dv_s}{dt} = i_f - \frac{P_{cpl}}{v_s} \tag{7}$$

where，$L_f = L_{eq} + L_{dc}$, $R_f = R_{eq} + R_{dc}$.

And the mixed potential function for the system can be written:

$$\begin{aligned} P(i_f, v_s) &= -\frac{1}{2} R_f \cdot i_f^2 + \int_0^{v_s} \frac{p_{cpl}}{v_s} dv_s + i_f \cdot (V_{eq} - v_s) \\ u_1 &= \min\left[\lambda(L^{-1/2} \cdot A_{ii} \cdot L^{-1/2})\right] = \frac{R_f}{L_f} \end{aligned} \tag{8}$$

$$u_2 = \min\left[\lambda(C^{-1/2} \cdot B_{vv} \cdot C^{-1/2})\right] = -\frac{P_{cpl}}{Cv_s^2}$$

And the Lyapunov energy function of the system is as follows:

$$P^*(i,v) = \frac{u_1 - u_2}{2} \cdot P(i,v) + \frac{1}{2} \cdot \frac{1}{L_f}(V_{eq} - R_f \cdot i_f - v_s)$$

$$+ \frac{1}{2} \cdot \frac{1}{C}\left(\frac{P_{cpl}}{v_s} - i_f\right) \tag{9}$$

According to the Brayton-Moser stability theorem, the critical condition will be given by:

$$u_1 + u_2 = 0 \tag{10}$$

$$u_1 = -u_2 = \frac{R_f}{L_f} = \frac{P_{cpl}}{Cv_s^2} \tag{11}$$

$$\rightarrow \quad v_{s\min} = \sqrt{\frac{P_{cpl} \cdot L_f}{C \cdot R_f}} \tag{12}$$

And the critical energy for the system will be:

$$P^*(i,v) = \min P^*(i, v_{s\min}) \tag{13}$$

According to the Lyapunov stability theory, the asymptotic stability region can be obtained when the energy is less than the critical energy.

V. Simulation Validation and Analysis

In order to verify the effectiveness of the stability region obtained by Brayton-Moser Theorem, this section presents the simulation results of the system of Fig. 3. Table 1 gives the parameters for the system.

Table 1: Parameters for simulation and analysis

Parameter	Value	Parameter	Value
V_m	162V	L_{dc}	50uH
f	400Hz	C_{dc}	20uF
L_p	30uH	R_{dc}	0.5Ω
L_s	164uH	P	0.5/1.5/2.5kW

Fig.4 shows the comparison between the simulation results of the system of Fig.3 for different initial points (i_s, v_s) and estimated stability region obtained by Brayton-Moser Method. Note that the result has been shifted to the origin. As shown in the figure, the stability region is smaller than the real one, which is because of the conservation characteristic of the Lyapunov method. However, the region can still cover most of large disturbances of the aircraft power system.

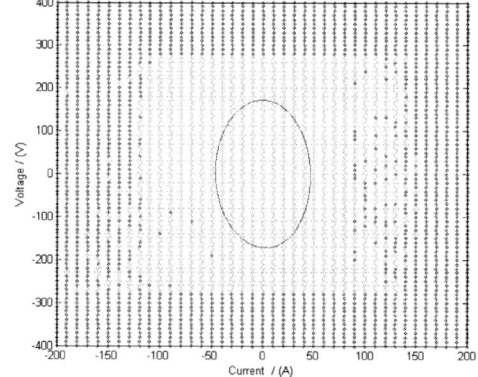

Fig.4: Stability region obtained by simulation and Brayton-Moser Method

Fig.5 gives the three-dimensional and contour plots of the Lyapunov energy function of the system. As can be seen from the figure, the minimum energy is located at the equilibrium point, which has been shifted to the origin point. And this confirms the existence of a stability region around the equilibrium point.

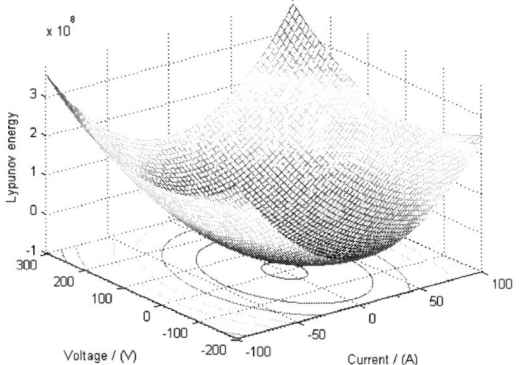

Fig. 5: Three-dimensional and contour plots of the Lyapunov energy function of the system.

Fig.6 gives the phase portrait and the waveforms of the system's voltage v_s and current i_s of an initial point on the stability region boundary obtained by the Brayton-Moser Method. The system will finally converge to the equilibrium point.

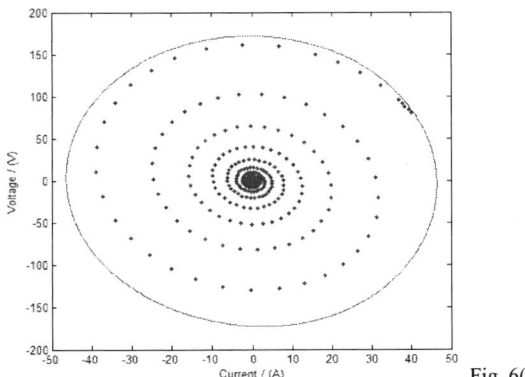

Fig. 6(a): The phase portrait of an initial point on the stability region boundary

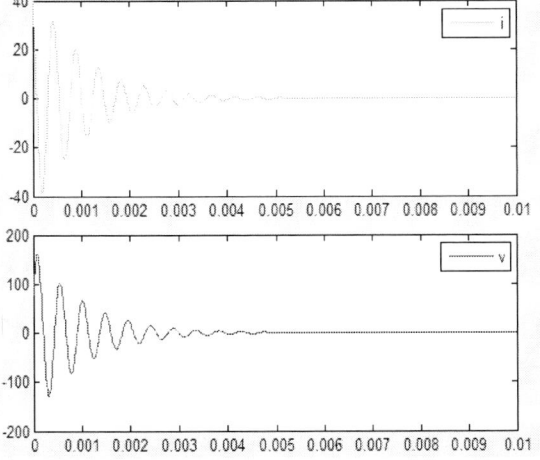

Fig. 6(b): The waveforms of system's current and voltage

Fig. 7 shows the influence of parameter changes on the stability region. As can be seen from the Fig. 7, the value of constant power p_{cpl} will affected both the voltage v_s and current i_s greatly, but the value of capacitor C_{dc} will have a greater impact on the current i_s.

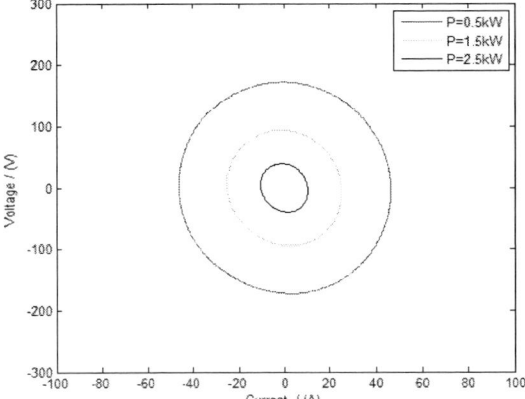

Fig. 7(a): The influence of Parameter of constant power

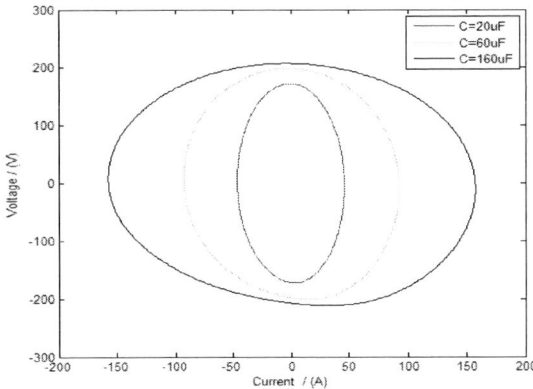

Fig. 7(b): The influence of Parameter of capacitor C_{dc}

VI. CONCLUSION

This paper proposes an averaged-value method to model the aircraft electric power system. The proposed model is simple but accurate for the large signal stability analysis of MEA power system. Based on the model of the aircraft power system, Brayton-Moser mixed potential and Lyapunov stability theorems are employed to estimate the large signal stability region of the system. It is shown that the load on the aircraft can significantly affect the stability region of the system. The simulation and analytical results are given to illustrate related parameters' influence on the systems stability margins. This paper provides not only a useful tool for understanding the behavior of the system under large signal disturbances, but also will help the designer to build criteria for system components sizing.

ACKNOWLEDGMENT

The authors would like to express thanks to support from National Natural Science Foundation of China (No. 51407185).

978-1-5090-0064-7/15 $31.00 © 2015 IEEE

REFERENCES

[1] Pat Wheeler, Sergei Bozhko, "The more electric aircraft: Technology and challenges", IEEE Electrification Magazine, Vol. 2, No. 4, 2014, pp. 6-12.

[2] Evgeni Ganev, "Selecting the best electric machines for electrical power-generation systems: High-performance solutions for aerospace more electric architectures", IEEE Electrification Magazine, Vol. 2, No. 4, 2014, pp.13-22.

[3] Benjamin B. Chol, "Propulsion power-train simulator: Future turboelectric distributed propulsion aircraft", IEEE Electrification Magazine, Vol.2, No. 4, 2014, pp. 23-24.

[4] Kaushik Rajashekara, "Power Conversion Technologies for Automotive and Aircraft Systems", IEEE Electrification Magazine, Vol. 2, No. 2, 2014, pp. 50-60.

[5] J.C. Shaw, S.D.A. Fletcher, P.J. Norman, "More Electric Power Systems for an Environmentally Responsible Aircraft", 47th International Universities Power Engineering Conference, 2012, pp. 1-6.

[6] Bulent Sarlioglu, "Advances in AC-DC Power Conversion Topologies for More Electric Aircraft", IEEE Transportation Electrification Conference and Expo, 2012, pp. 1-6.

[7] A. Garcia, J. Cusido, J.A. Rosero, "Reliable electro-mechanical actuators in aircraft", IEEE Aerospace and Electronic Systems Magazine, Vol.23, No.8, 2008, pp. 19-15.

[8] J.W. Bennett, B.C. Mecrow, D.J. Atkinson, "Safety-critical design of electromechanical actuation systems in commercial aircraft", IET Electric Power Applications, Vol.5, No.1, 2011, pp. 37-47.

[9] O. Bennouna, N. Langlois, "Development of a Fault Tolerant Control for Aircraft Electromechanical Actuators, Electrical Systems for Aircraft Railway and Ship Propulsion", 2012, pp. 1-5.

[10] O. Bennouna, N. Langlois, "Modeling and Simulation of Electromechanical Actuators for Aircraft Nacelles, Proceedings of the 9th Symposium on Mechatronics and its Applications", Amman, Jordan, 2013, pp. 1-5.

[11] M. Rottach, C. Gerada, P.W. Wheeler, "Design optimization of a fault-tolerant PM motor drive for an aerospace actuation application", 7th IET International Conference on Power Electronics, Machines and Drives, 2014, pp. 1-6.

[12] L. Castellini, M. D. Andrea, N. Borgarelli, "Analysis and design of a Linear Electro-Mechanical Actuator for a High Lift System", International Symposium on Power Electronics, Electrical Drives, Automation and Motion, 2014, pp. 243-247.

[13] Y. Ji, M.R. Kuhn, "Modeling and Simulation of Large Scale Power Systems in More Electric Aircraft", IEEE 14th Workshop on Control and Modeling for Power Electronics, 2013, pp.1-6.

[14] Ashraf Tantawy, Xenofon Koutsoukos, Gautam Biswas, "Aircraft Power Generators: Hybrid Modeling and Simulation for Fault Detection", IEEE Transactions on Aerospace and Electronic Systems, Vol. 48, No.1, pp. 2012, 552-571.

[15] Liqiu Han, Jiabin Wang, David Howe, "State-space average modeling of 6- and 12-pulse diode rectifiers", 2007 European Conference on Power Electronics and Applications, 2007, pp. 1-10.

[16] Emadi Ali, "Modeling and Analysis of Multiconverter DC Power Electronics Systems Using the Generalized State-Space Averaging Method", IEEE Transactions on Industrial Electronics, Vol. 51, No. 3, 2004, pp. 661-668.

[17] Tao Yang, Serhiy Bozhko, Jean-Mark, "Dynamic Phasor Modeling of Multi-Generator Variable Frequency Electrical Power Systems", IEEE Transactions on Power Systems, PP(99), 2015, pp. 1-9.

[18] A. Griffio, Jiabin Wang, "Large Signal Stability Analysis of 'More Electric' Aircraft Power Systems with Constant Power Loads", IEEE Transactions on Aerospace and Electronic Systems, Vol. 48, No.1, 2012, pp. 477-489.

[19] K-N. Areerak, T. Wu, S.V. Bozhko, "Aircraft Power System Stability Study Including Effect of Voltage Control and Actuators Dynamic", IEEE Transactions on Aerospace and Electronic Systems, Vol. 47, No. 4, 2011, pp. 2574-2589.

[20] K-N. Areerak, S.V. Bozhko, G.M. Asher, "Stability Analysis and Modeling of AC-DC Systems with Mixed Load Using DQ-Transformation Method", IEEE International Symposium on Industrial Electronics, 2008, pp. 19-24.

[21] C.I. Hill, K. Areerak, Tao Yang, "Automated Stability Assessment of More Electric Aircraft Electrical Power Systems", International Conference on Electrical Systems for Aircraft, Railway, Ship Propulsion and Road Vehicles, 2015, pp. 1-6.

[22] T. Wu, S.V. Bozhko, G.M. Asher, "High speed modeling approach of aircraft electrical power systems under both normal and abnormal scenarios", IEEE International Symposium on Industrial Electronics, 2010, pp. 870-877.

[23] Tao Yang, Serhiyi Bozhko, Greg Asher, "Functional Modeling of Symmetrical Multipulse Autotransformer Rectifier Units for Aerospace Applications", IEEE Transactions on Power Electronics, Vol. 30, No. 9, 2015, pp. 4704-4713.

[24] A.Baghramian, A.J. Forsyth, "Averaged-Value Models of Twelve-Pulse Rectifiers for Aerospace Applications", Second International Conference on Power Electronics, Machines and Drives, 2004, pp. 220-225.

BIOGRAPHIES

Yanbo Che was born in Shandong, China. He received his B.S. degree from Zhejiang University, Hangzhou, China, in 1993. He received his M.S. and Ph.D. degrees from Tianjin University, Tianjin, China, in 1996 and 2002, respectively. Since 1996, he has been engaged in teaching and scientific research of power electronic technology and power systems. He is presently an Associate Professor in the School of Electrical Engineering and Automation at TianjinUniversity. His current research interests include power electronics, new energy and micro-grids.

Xiaokun Liu received his B.S. degree from Hebei University of Technology, Tianjin, China, in 2014. Since September 2014, he has been working towords his M.S. degrees at Tianjin University, Tianjin, China. His current research interests include power electronics, electric vehicles.

Zhangang Yang received his M.S. and Ph.D degrees from Tianjin University, Tianjin, China, in 2007 and 2011, respectively. Since Aprial 2011, he has been working in Department of Aviation automation, Civil Aviation University of China, Tianjin, China. His current research interests include more electric Aircraft power system, power systems and micro-grids.

Scenario Method in Economical Dispatch of Microgrid Connected to Distribution System

QIN Chuan, DONG Ping, LIN Yun, FENG Yongqin

School of Electric Power, South China University of Technology, Guangzhou, China

Abstract–The intermittent power has unpredictable property, specific to the situation that microgrid containing intermittent power operates by connected to distribution system, this paper build a dynamic economical dispatch model with the objective of minimizing the cost of conventional generator. Considering intermittent power's uncertainty, this paper applies the scenario method, by means of Latin hypercube sampling, Cholesky Decomposition and Scenarios Elimination to create the error scenarios, and solves the dynamic economical dispatch model above-mentioned by using CONOPT that is under in General Algebraic Modeling Software(GAMS). Numerical examples verify the correctness of the proposed model. The method's effectiveness and practicability also can be proved.

Keywords–Intermittent power, scenario method, dynamic economical dispatch, microgrid operation with distribution system.

NOMENCLATURE

Sets

t	Set of time segment.
g	Set of diesel generator.
b	Set of battery.
i	Set of nodes in microgrid.
s	Set of scenarios.
k	Set of nodes in distribution system.

Parameters

P_g^{min}	Minimum allowed output of diesel generator.
P_g^{max}	Maximum allowed output of diesel generator.
R_g^{down}	Maximum decrease of diesel generator output between two contiguous period of time.
R_g^{up}	Maximum increase of diesel generator output between two contiguous period of time.
$P^{max}_{b,t}$	Maximum allowed power of battery's charging and discharging.
S_{b0}	Initial electric power of battery.
S_{bmax}	Maximum allowed electric power of battery.
ξ	Efficiency coefficient of battery's charging.
V_k^{min}	Minimum allowed voltage magnitude.
V_k^{max}	Maximum allowed voltage magnitude.

Variables

$P_{g,t}$	Output of diesel generator in t.
$P'_{b,t}$	Discharging power of battery.
$P''_{b,t}$	Charging power of battery.
$P_{is,t}$	Active power at node i.
$P_{ks,t}$	Active power at node k.
$V_{i,t}$	Voltage magnitude at node i.
$V_{k,t}$	Voltage magnitude at node k.

I. INTRODUCTION

In today's world, we are facing shortage of energy resources and serious situation of environmental pollution, the Distributed Generation(DG) is a new way of power generation. With advocacy of insisting the sustainable development strategy in modern social, distributed generation technologies containing new energy will become the development trend of power industry. The microgrid, which consists of Distributed Generation, energy storage devices, energy conversion devices, load and relay protection equipment, is a small power generation system and distribution system. It can either operate by connected to the distribution system, and operate in isolation. Existing research and practice show, to structure microgrid with Distributed Generation, is a effective way to play DG's efficiency, then microgrid is connected to large power grid, also microgrid and large power grid support each other, this situation has great social and economic importance. This can be seen, microgrid technology is also key to the development trend of smart distribution network in the future. Literature [1] presents a dispatch model of microgrid containing various DG, use the Particle Swarm Optimization to solve the problem that microgrid operate with distribution system and in isolation. Literature[2] does the research on microgrid economical operation problem under operating with distribution system and in isolation respectively. Literature[3] analyzes the influences of different dispatching modes of microgrid operating to economical dispatch strategy of microgrid. Specific to the problem that DG's low permeability in distribution system, literature[4] connects DG to IEEE30-bus standard system, proposes coordinated control strategy and optimization of microgrid.

The DG in microgrid, such as wind power and photovoltaic power, impact the stable operation of microgrid. To deal with the problem that intermittent power outputs have uncertainty, there are some researches about scenario method[5][6]. According to some known data, scenario method imagines a variety of possible future environment, combines various kinds of uncertainty that affects optimization results, that each kind of uncertainty corresponds to a scenario, in order to achieve approximation between uncertain future scenarios and certain scenarios. Now, we solve the economical dispatch problem with considering the uncertainty of intermittent power outputs by the means of scenario method, build a dynamic economical dispatch model with the objective of minimizing the cost of conventional generator, use the General Algebraic Modeling Software(GAMS) to solve the model above-mentioned. Numerical examples verify the correctness of the proposed model. The method's effectiveness and practicability also can be proved.

II. SYSTEM MODEL

A. Mathematical Model of Microgrid's Structure and Its

Elements

Figure 1 shows a connection diagram of a real islanded microgrid. Generating equipment of intermittent power, includes wind driven generator and solar photovoltaic panels, there are lead-acid batteries and diesel generator in this microgrid. When the battery energy storage and intermittent power outputs cannot satisfy the load demand, the diesel generator, what also is conventional generator, contributes to providing the necessary support to load demand.

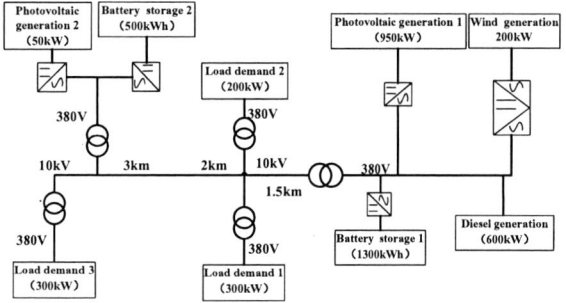

Fig. 1 Connection diagram of a real islanded microgrid

The simplified connection diagram of the islanded microgrid showed in Figure 2. The line parameters: $R=1.2\Omega/km$, $X=0.33\Omega/km$. Power standard is taken as 1MW, $S_B=1MW$. The voltage reference of high voltage side and low voltage side are taken as 10kV and 0.38kV respectively.

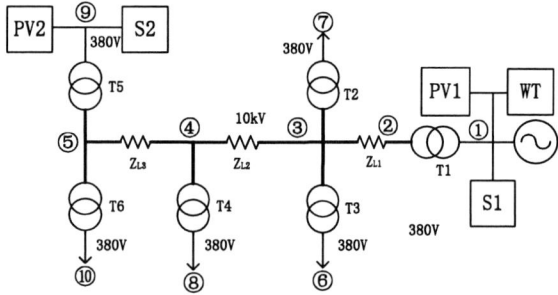

Fig. 2 Simplified connection diagram of the islanded microgrid

According to Figure 2, we build the mathematical model of lead-acid batteries, diesel generator and microgrid's structure:

$$P_g^{min} \le P_{g,t} \le P_g^{max} \quad \forall g \in G, \forall t \in T \quad (1)$$

$$-R_g^{down} \le P_{g,t} - P_{g,t-1} \le R_g^{up} \quad \forall g \in G, \forall t \in T \quad (2)$$

$$-\sum_{t=1}^{T} P_{b,t}' + \xi \sum_{t=1}^{T} P_{b,t}'' = 0 \quad \forall b \in N_b, \forall t \in T \quad (3)$$

$$P_{b,t}' \cdot P_{b,t}'' = 0 \quad \forall b \in N_b, \forall t \in T \quad (4)$$

$$0 \le P_{b,t}', P_{b,t}'' \le P_{b,t}^{max} \quad \forall b \in N_b, \forall t \in T \quad (5)$$

$$0 \le S_{b0} - \sum_{t=1}^{T'} P_{b,t}'' + \xi \sum_{t=1}^{T'} P_{b,t}'' \le S_{b\,max} \quad \forall b \in N_b, \forall T' \in T \quad (6)$$

$$P_{is,t} - V_{i,t} \sum_{j \in i} V_{j,t} \left(G_{ij} \cos\theta_{ij,t} + B_{ij} \sin\theta_{ij,t} \right) = 0$$
$$\forall i \in N, \forall t \in T \quad (7)$$

In the above formula, T which is the total numbers of

dispatch period, is generally divided into 24 segments, each segment is 1 hour, $\forall t \in T$. (1) is the constraint of upper and lower bounds of diesel generator, (2) is the climbing constraint of diesel generator. (3) is storage balance of lead-acid batteries in the total numbers of dispatch period. (4) is the status of lead-acid batteries. (5) is the capacity constraint of lead-acid batteries. (6) is the constraint of lead-acid batteries' state of charge(SOC). (7) represents the power flow constraints of each node in microgrid.

B. Mathematical Model of Microgrid Connected to Distribution System

What is used as distribution system in this paper, is a PG&E69 69-bus distribution system in America[7]. The schematic diagram of a PG&E69 69-bus distribution system is shown in Figure 3.

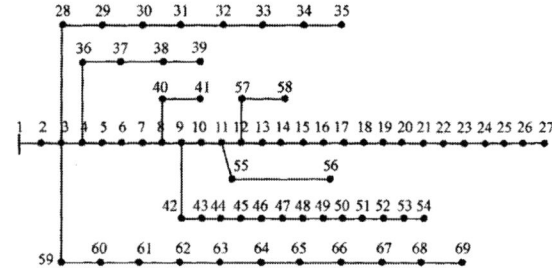

Fig.3 Schematic diagram of a PG&E69 69-bus distribution system

The mathematical model of this system is as follows:

$$P_{ks,t} - V_{k,t} \sum_{l \in k} V_{l,t} (G_{kl} \cos\theta_{kl,t} + B_{kl} \sin\theta_{kl,t}) = 0 \quad (8)$$

$$V_k^{min} \le V_{k,t} \le V_k^{max} \quad (9)$$

In the above formula, (8) is the power flow constraints of each node in this system. (9) is the constraint of upper and lower bounds of node voltage.

The simplified connection diagram of whole system is as follows. we choose node 50 as connection point of microgrid and distribution system, and choose node 1 as connection point of distribution system and large power grid.

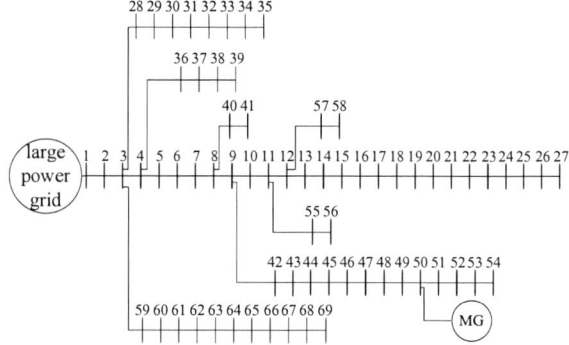

Fig. 4 Simplified connection diagram of microgrid connected to 69-bus distribution system

We consider that minimizing the cost of conventional generator is objective function, and build the dynamic economical dispatch model.

$$\min \ \sum_{t=1}^{T}\sum_{g\in G} F(P_{g,t}) \qquad (10)$$

Formula(10) is the objective function of this problem. The cost-function of conventional generator is as follow.

$$F(P_{g,t}) = aP_{g,t}^2 + bP_{g,t} + c \qquad (11)$$

III. SCENARIO METHOD

The intermittent power have unpredictable property, while the existing predictive technique cannot guarantee that the predictied value is completely accuracy, we need to use scenario method to considering the uncertainty of intermittent power outputs, in order to simulate the actual situation of intermittent power. By means of Latin hypercube sampling, Cholesky Decomposition and Scenarios Elimination[8][9], we create the error scenarios on the basic of predicted value of intermittent power outputs. We choose different standard deviation, to represent the different uncertain degrees of wind power and photovoltaic power respectively.

Using scenario method, the original model needed to be corrected[10]. Uncertain economic dispatch model's constraints consist of error scenario constraints and predicted scenario constraints. The predicted scenario constraints are the same as those in the original model. The error scenario constraints are as follow:

$$P_g^{s,\min} \le P_{g,t}^s \le P_g^{s,\max} \qquad \forall g\in G, \forall t\in T \qquad (12)$$

$$-R_g^{down} \le P_{g,t}^s - P_{g,t-1}^s \le R_g^{up} \qquad \forall g\in G, \forall t\in T \qquad (13)$$

$$-\sum_{t=1}^{T} P_{b,t}^{',s} + \xi \sum_{t=1}^{T} P_{b,t}^{'',s} = 0 \quad \forall b\in N_b, \forall t\in T \qquad (14)$$

$$P_{b,t}^{',s} \cdot P_{b,t}^{\bullet,s} = 0 \quad \forall b\in N_b, \forall t\in T \qquad (15)$$

$$0 \le P_{b,t}^{',s}, P_{b,t}^{\bullet,s} \le P_{b,t}^{\max} \quad \forall b\in N_b, \forall t\in T \qquad (16)$$

$$0 \le S_{b0} - \sum_{t=1}^{T'} P_{b,t}^{'',s} + \xi \sum_{t=1}^{T'} P_{b,t}^{'',s} \le S_{b\max} \quad \forall b\in N_b, \forall T'\in T \ (17)$$

$$P_{is,t}^s - V_{i,t}^s \sum_{j\in i} V_{j,t}^s \left(G_{ij}\cos\theta_{ij,t}^s + B_{ij}\sin\theta_{ij,t}^s \right) = 0$$
$$\forall i\in N, \forall t\in T \qquad (18)$$

In the above formula, $s = 1, 2, \ldots, S$, S is the number of error scenarios.

In error scenario constraints, all variables of the original model have its corresponding variable, the variable that provides output to load demand in microgrid, should be used to form a transfer error constraint with its corresponding variable. The transfer error constraint is the key part of scenario method, it represent that the variable transform one error scenario from another error scenario, it considers all error scenarios' influence to objective function synthetically, by solving the problem, the optimal results can finally be achieved.

$$\left| P_{g,t} - P_{g,t}^s \right| \le \Delta_g \qquad \forall g\in G, \forall t\in T \qquad (19)$$

IV. CALCULATION ANALYSIS

In microgrid, the capacity of diesel generator is 600KW, the capacity of wind generation is 200KW, and the capacity of photovoltaic generation 1 and photovoltaic generation 2 are 950KW and 50KW respectively. The location of photovoltaic generation 1 is node 1, and the location of photovoltaic generation 2 is node 9.

According to the historical data of wind power plant and photovoltaic power station, the predicted value of intermittent power outputs is shown in Figure 4.

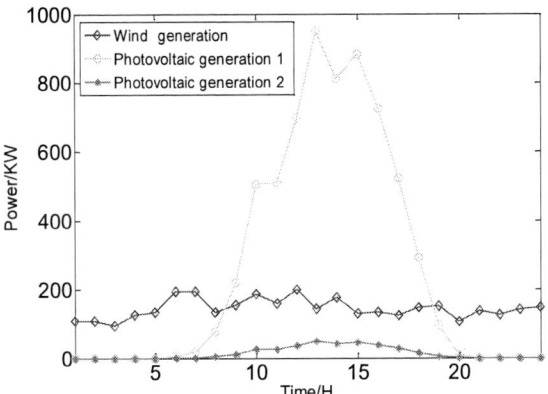

Fig. 5 Predicted power outputs of wind power and PV power

As we can see from Figure 5, the output of PV power is related to light intensity, the numerical value of PV power output is large from 12th period of time to 16th period of time, among a day. By using GAMS to solve the problem that is proposed in this paper, and creating 10, 20 and 50 error scenarios respectively, we can get these results as follow.

Tab 1 Calculation

numbers of error scenarios	10	20	50
object function value($)	599.6	597.28	595.17
calculating time(min)	16	31	73

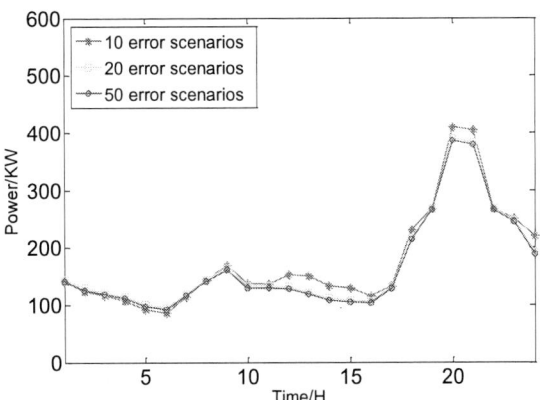

Fig. 6 Outputs of conventional generator in different numbers of error scenarios

As can be seen in Figure 6, there is a little distinction

among the results in different numbers of error scenarios. Obviously, the more error scenarios are created, the more approximate between simulation and reality. Meanwhile, more error scenarios will cost more calculating time of solving the problem, which can be seen in Table.1. Compared to Figure 5, we can see, when the intermittent power provides the main output to load demand, the conventional generator provides less output, therefore the cost of conventional generator can be reduced. And when the intermittent power provide cannot provide enough output to load demand, the conventional generator can provide the support timely. The result shows that the conventional generator and intermittent power can support each other to satisfy the load demand, they also contribute to maintaining stable operation of microgrid together.

V. CONCLUSION

Intermittent power outputs have uncertainty, this situation brings some influences to solving deterministic optimization problem. This paper improves the previous economical dispatch model, and builds a dynamic economical dispatch model with the objective of minimizing the cost of conventional generator, considers the uncertainty of intermittent power outputs by the means of scenario method, and uses GAMS to solve the problem that is proposed. Calculation verifies the correctness of the proposed model. The method's effectiveness and practicability also can be proved.

REFERENCES

[1] HAN Hongfei. The Distributed Generation Coordinated Optimal Scheduling[D]. North China Electric Power University,2014.

[2] CHEN Jie, YANG Xiu, ZHU Lan, ZHANG Meixia. Genetic algorithm based economic operation optimization of a combined heat and power microgrid [J]. Power System Protection and Control, 2013,41(8):7-15.

[3] CHEN Jie, YANG Xiu, ZHU Lan, ZHANG Meixia. Comparison of microgrid economic operation among different dispatch modes [J]. Electric Power Automation Equipment, 2013, 33(8):106-113.

[4] WANG Ruiqi. Research on Multi-objective Optimization Design and Coordinated Control of Distributed Generation and Microgrid[D].Shan Dong University, 2013.

[5] YE Rong, CHEN Haoyong, WANG Gang, CHEN Pan. A Mixed Integer Programming Method for Security-constrained Unit Commitment with Multiple Wind Farms [J]. Automation of Electric Power Systems, 2010, 34(5): 29-33.

[6] ZHAO Wenmeng, LIU Mingbo. A Dynamic Reduction Based Multi-cut Method for Solving Stochastic Unit Commitment with Wind Farm Integration [J]. Automation of Electric Power Systems, 2014, 38(9): 26-33.

[7] ZENG Xueqiang. Studies on reactive power optimization and algorithm of distribution network containing wind power generators[D]. Southwest Jiaotong University, 2011.

[8] LOH W L. On Latin hypercube sampling[J]. The Annals of Statistics, 1996, 24(5): 2058-2080.

[9] HEITSCH H, ROMISCH W. Scenario reduction algorithms in stochastic programming[J]. Computational Optimization and Applications, 2003, 24(2/3): 187-206.

[10] FU Yimu, LIU Mingbo. Scenario decomposition method for multi-objective stochastic dynamic economical dispatch problem[J]. Automation of Electric Power Systems, 2014, 38(9): 34-40.

Shorting-time Transmission Line Maintenance Scheduling Method Based on Credibility Theory

LIN Yun, DONG Ping, QIN Chuan, FENG Yongqin

School of Electric Power, South China University of Technology, Guangzhou, China

Abstract-Scheduling transmission line maintenance is important to the operation of power system. Traditionally, transmission line maintenance scheduling formulated as a stochastic programming problem, which cannot express operation risk accurately. In terms of Short-term transmission line maintenance scheduling, line forced outage rate is based on $(100km \cdot year)^{-1}$. In other words, in the same region, the longer the transmission line with the same voltage level is, the higher the forced outage rate is. However, this is not conform to the actual. The power system contains twofold uncertainty, consisting of random and fuzzy. Therefore, we need to evaluate random and fuzzy of the maintenance plan comprehensively. Credibility Theory provides a comprehensive assessment for the fuzziness and randomness, which is the key to solve the problem that traditional line maintenance scheduling model is not applicable to practice. In this paper, combined with the IEEE-RBTS system, the model and algorithm of short-term transmission line maintenance scheduling based on credibility theory is presented.

Keywords–Transmission line maintenance; Short-term scheduling; Credibility theory; Genetic algorithm.

I. INTRODUCTION

Scheduling transmission line maintenance is one of research topics in the operation risk. In order to ensure electric power will be transmitted normally, traditional research is to schedule transmission line maintenance periodically and organize repair work in time after failure. Therefore, with the constraints of power grid operation and manpower, transmission line maintenance schedule is to realize the goal that the sum of maintenance cost and power loss cost will be the minimum.

According to schedule period, maintenance schedule can be divided into daily plan, weekly plan, monthly plan, seasonal plan and annual plan. Weekly plan should schedule the start maintenance time of equipment in detail. According to the load and operation mode of the power grid, dispatch department makes coordinate correction based on the weekly plan, which is called daily plan. Short-term schedule is a general designation of weekly plan and daily plan.

With the rapid development of modern industry and science technology, there is a growing tendency of the grid size and transmission line structure.,which brings about increasing influence factors and potential possibility of failure. However, when the line maintained, it ought be out of service, which will give rise to reducing power supply reliability. Nowadays, Grid Corporations schedule transmission line maintenance just by simple computer programs. Therefore, it's time to get the issue how to schedule transmission line reasonable onto their agenda.

The power system contains twofold uncertainty, consisting of random and fuzzy. The model which takes random into account simply isn't applicable, because it's unable to measure the operation risk in practice. In terms of short-term maintenance schedule, the line forced outage rate is given by the unit of $(100km \cdot year)^{-1}$. In other words, in the same region, the longer the transmission line with the same voltage level is, the higher the forced outage rate is. Obviously, it doesn't conform to actual. Therefore, we should evaluate random and fuzzy of the maintenance plan comprehensively. Credibility Theory provides a comprehensive assessment for the fuzziness and randomness, which is the key to solve the problem that traditional line maintenance scheduling model is not applicable to practice.

In this paper, the model of short-term transmission line schedule based on credibility theory is put forward and solved by an efficient genetic algorithm. Taking both randomness and fuzziness into consideration, the maintenance schedule will be more practical and effective.

In the field of system operation risk, research on generation maintenance schedule has made great achievements[1]–[5] while transmission line maintenance schedule is in its infancy. At present, a majority of research formulate transmission line maintenance schedule as a stochastic programming problem, modeling by reliability theory.

In the paper[6], transmission line starting repair time is introduced as a control variable. The goal is to minimize the shortage of power supply and increased risk due to line maintenance in the whole planning period, solved by a relative heuristic algorithm based on tabu-research. In the paper[7], the generation and transmission line joint maintenance schedule under deregulation is decomposed into multi-objective integer programming problem and stochastic programming problem. However, modeling with the sum of maintenance costs, it doesn't balance the interest between the grid company and the power plant rationally.

Credibility Theory [8][9] provides a comprehensive assessment for the fuzziness and randomness. It lays a solid theoretical foundation for short-term transmission line maintenance schedule, which is a twofold uncertainty problem. The axiomatic system of Fuzzy Theory is accomplished in 2004 and has developed for just ten years. As a result, there is vacant in the research field of transmission line maintenance schedule, given the late start of its study.

The paper[10]~[12]come up with a algorithm to solve operation risk problem based on Credibility Theory in the power system at the first time. From then on, the research on uncertainty problem based on Credibility Theory in electric power field goes further and further.

The paper[2] is modeled by maintenance schedule for thermal power generating units and hydroelectric generating set with twofold uncertainty, which handles effectively the fuzziness of water volume in practice.

The paper[13] is modeled by short-term transmission line maintenance schedule based on Credibility Theory. The aim is to minimize the random fuzzy expected value of the sum of the maintenance cost and interruption loss. Introducing the model based on Credibility Theory will have great practical significance.

Genetic Algorithm operates parallelly from many points instead of one points.Therefore, not only does it reduce the possibility of acquiring local optimum solution, but also it improves efficiency by large-scale parallel operation calculation principle.Equipped with powerful function and uncomplicated calculation principle, Genetic Algorithm has been widely used in maintenance schedule optimization problem. Compared with Tabu search algorithm, Genetic Algorithm and Simulated annealing algorithm, the paper[14] finds out Genetic Algorithm can acquire a better effect in optimization problem. Genetic Algorithm is used in the paper[15] to solve the maintenance schedule of large thermal power generating units, considering the constraints that may occur in the maintenance and formatting genes to indicate different maintenance schedule. The paper[16] tries to combine network topology and Genetic Algorithm, finding that it will work out optimized maintenance schedule and cut down interruption loss.

II. CREDIBILITY THEORY AND ITS APPLICATION IN SHORT-TERM TRANSMISSION LINE MAINTENANCE SCHEDULING

The work of *Fuzzy Sets* by Zadeh in 1965 is the symbol of new-born fuzzy mathematics. Based on fuzzy sets, fuzzy mathematics sticks into inaccuracy and uncertainty realm, and provides insights for processing fuzzy information of real world. In 2004, Baoding Liu completed the axiomatization of fuzzy theory based on measure theory, and integrated probability theory and fuzzy theory to evaluate the randomness and fuzziness. This is the foundation of credibility theory.

Credibility theory as mathematical tool of fuzziness processing has been fruitfully applied into fields of engineering optimization and investment portfolio [2]. Literatures[9-11] initially proposed an evaluation algorithm of parallel risks in electrical system, and dived into a field of solving uncertain problems of electrical system based on credibility theory.

Credibility theory axiomatized fuzzy theory based on measure theory, and derived a comprehensive evaluation method for randomness and fuzziness. To make this paper self-contained, the basic definitions and concepts of credibility measure and its algorithms are briefly reviewed.

Assuming Θ is non-empty set, $P(\Theta)$ is the power set of Θ, and P_{os} satisfies the three axioms, we call P_{os} the

possibility measure, and $(\Theta, P(\Theta), P_{os})$ possibility space; if A is a set of $P(\Theta)$, A^{C} is complementary set of A, $N_{ec}\{A\} = 1 - P_{os}\{A^{C}\}$ is defined as necessity measure of A .while $C_{r}\{A\} = \dfrac{1}{2}(P_{os}\{A\} + N_{ec}\{A\})$ is defined as credibility measure of A.

Definition 1 (Expectation of Fuzzy Variable)

Assuming fuzzy variable ξ, and the expectation of ξ is:

$$E_{pro-fuz}[\xi] = \int_{0}^{+\infty} C_r\{E_{pro}[\xi(\theta)] \geq r\}d_r - \int_{-\infty}^{0} C_r\{E_{pro}[\xi(\theta)] \leq r\}d_r$$

(1)

According to Eqs.()and (), the model for the short-time transmission line maintenance scheduling considering both randomness and fuzziness can be formulated as follows:

$$Min \quad E_{pro-fuz}\left[\sum_{t}\sum_{k=1}^{N} C_{kt}(1-x_{kt}) + \sum_{t} \psi_{pro-fuz,t}\right] \quad (2)$$

s.t.

$$\sum_{t}\sum_{i} r_{kj}(1-x_{kt}) \leq \beta_{jt} \quad \text{for each material resource j} \atop \text{in period t} \quad (3)$$

$$\sum_{k \in N_r}(1-x_{kt}) \leq b_r \text{ for labour power in route r} \quad (4)$$

$$\begin{cases} x_{kt} = 1, & t < e_k \text{或} t > l_k + d_k \\ x_{kt} = 0, & S_k \leq t \leq S_k + d_k \\ x_{kt} \in \{0,1\}, & e_k \leq t \leq l_k \\ \sum_{t} x_{kt} = l_k - e_k - d_k, & e_k \leq t \leq l_k \end{cases}$$

(5)

$$E_{pro-fuz}[E_{ENS\,pro-fuz}(\xi_{FOR,i})] < \varepsilon_t \quad (6)$$

$$\sum_{i \in NG} P_i + \sum_{j \in NC} D_j = \sum_{j \in NC} C_j \quad (7)$$

$$P_{min} \leq P \leq P_{max} \quad (8)$$

$$T \leq T_{max} \quad (9)$$

$$0 \leq D \leq C \quad (10)$$

where,

C_{kt}	maintenance cost of line k in period t
$x_{kt}=0$	line k maintained in period t
$x_{kt}=1$	otherwise
$\psi_{pro-fuz,t}$	random fuzzy variable of interruption loss in period t
$E_{pro-fuz}$	expected value of random fuzzy variable
r_{kj}	material resource j needed to maintenance line k
β_{jt}	number of resource j available in period t
e_k	earliest available start time for maintenance of line k
l_k	latest available start time for maintenance of line k
d_k	duration of maintenance
S_k	beginning period of line k in period t

ε_t — the maximum acceptable expectation of energy not supply

P_{max} P_{min} — the maximum and the minimum of the generation output

T_{max} — the maximum of the power flow

In this model, the objective function of Eq(2) is decomposed into two parts:the maintenance cost and random fuzzy interruption loss cost over the operational period.

Eqs.(3)-(5) are the maintenance window, material resource and crew resource constraints,respectively. Eq(6) is the expectation of random fuzzy interruption loss constraint. Eq(7) is the constraint of power balance.

III. GENETIC ALGORITHMS OF TRANSMISSION LINE MAINTENANCE SCHEDULING

Genetic algorithm is an adaptive probability optimization algorithm based on biological heredity and evolution mechanism, stemming from computer simulation of biological system. In the iterated process, the superior items will be preserved while inferior items will be eliminated. In short-term transmission line maintenance scheduling, this process is that the scheduling will be improve toward optimal solutions under constrained conditions.

The conduction of Genetic Algorithm covers six necessary steps to achieve the optimal results:

● Chromosome Coding: considering the start maintenance time of each transmission line to be a control variable and using integer encoding. Assuming that there are three transmission lines to be maintained in a daily scheduling, every individual has three control variables and varies from 1 to 21 if d_k =4. for example, determining line 1,4 and 7 to be maintained, [10 9 13] indicates line 1 starts to be maintained at 10:00 while line 4 at 9:00 and line 7 at 13:00.

● Construction of Fitness Function: fitness is used in Genetic Algorithm to measure the superiority degree to approaching or obtaining optimal value of items in population of optimization.

After constructing a chromosome, objective function value of every individual should be calculated to assess their adaptability. The procedures are described as below:

1. According to individual's genes, maintenance schedule is formed.
2. Work out maintenance cost which is decomposed into fixed cost and adjustment cost.
3. Conduct "N-1" fault analysis in every hour when no transmission line maintained and calculate expected value of random fuzzy interruption loss. According to Credibility Theory, methods are as below:

a) Let $e = 1$;

b) Generate θ_k from Θ of forced outage rate so that

$$Pos\{\theta_k\} \geq \delta , k = 1, 2, \cdots, N$$

c) Calculate $E_{pro}[EENS_{pro}]$ by active power correction

as shown in <mark>Error! Reference source not found.</mark> .
Let

$$a = Min_{1 \leq k \leq N} E_{pro}[EENS_{pro}(\theta_k)]$$

$$b = Max_{1 \leq k \leq N} E_{pro}[EENS_{pro}(\theta_k)]$$,

d) Generate $r \in [a, b]$, $k = 1, 2, \cdots, N$

If $r \geq 0$, then $e = e + Cr\{\theta \in \Theta \mid E_{pro}[EENS_{pro}] \geq r\}$;

If $r < 0$, then $e = e + Cr\{\theta \in \Theta \mid E_{pro}[EENS_{pro}] \leq r\}$

$$E_{pro-fuz}[EENS_{pro-fuz}] = a \vee 0 + b \wedge 0 + e \cdot (b-a) / N$$

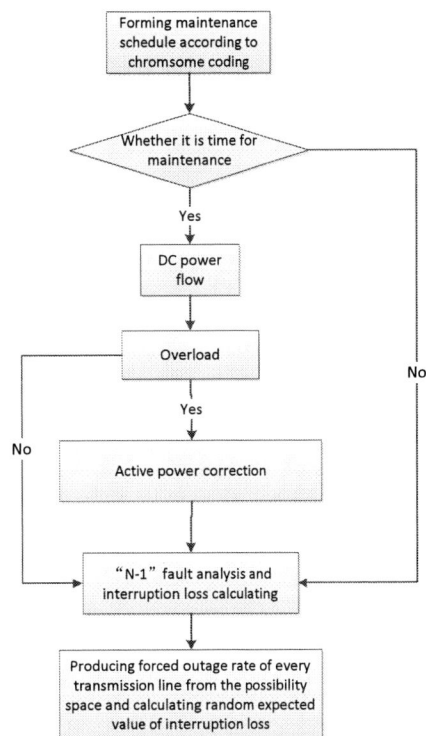

Fig 1 the flowchart of calculation for $E_{pro}[EENS_{pro}]$

III. TECHNICAL INFORMATION

4. Calculate expected value of random fuzzy interruption loss in maintenance time in sequence. "N-1" fault analysis is also taken into account. So expected value of random fuzzy interruption loss can be calculated.

5. The expected value operator is linear so that objective function value is easliy calculated by adding maintenance cost and interruption loss cost. Genetic Manipulation: the genetic manipulation is composed with three steps including selection, crossover and variation. Selection is choosing a certain superior items to increase the global convergence and computing efficiency based on some rules. random ergodic and sampling are used. Crossover is exchange parts of gene from pairs of items to form new items, preserving superior genes of last generations. In the paper, single-point crossover is used. Variation means a gene change without direction under very low probability, and form a new item. With function mutbga in GA

978-1-5090-0064-7/15 $31.00 © 2015 IEEE

toolbox, variation will be realized.

● Convergence Criteria: we define that if the iterated computation satisfies one of conditions below, the results are optimal: first, the minimal value of fitness function is less than a given value and is kept unchanged after several generations; second, the iterated generations exceed a given critical point. In the paper, iteration will be ended when iteration number comes to 100.

● Choice of Parameters: the parameters of Genetic Algorithm cover the amount of population, rate of intersection, rate of variation, and etc. Let the amount of population to be 50, rate of crossover 0.7 and rate of variation 1/3.

● Constrained Conditions: penalty function method is used in the paper. In other words, it decreases fitness values of a given item when it dissatisfies the constrained conditions, and reduce the probability of that item to be copied and pasted in next generation.

IV. NUMERICAL CASE STUDIES

Based on the described method, a Matlab progranm was developed. The IEEE-RBTS was employed to demonstrate the effectiveness of the presented method.

In the paper, the maximum acceptable expectation of energy not supply is assumed as 5% of hourly load while interruption loss 4RMB/kW·h.The fixed part of maintenance cost is assumed as 20000RMB/per line while feasible part is added 10RMB if maintenance schedule advanced or delayed for an hour.A daily maintenance schedule of IEEE-RBTS for Tuesday of the 52th week is studied.

1. The influence of the representation of random fuzzy variables

Based on Credibility Theory, different representations of the line forced outage rate are studied, such as triangular fuzzy number, rectangular fuzzy number and trapezoidal fuzzy number.The forced outage rates of nine transmission lines are divided into four levels, as shown in **Error! Reference source not found.**.

Case A1: line forced outage rate is represented by a triangular fuzzy number

Case A2: line forced outage rate is represented by a rectangular fuzzy number

Case A3: line forced outage rate is represented by a trapezoidal fuzzy number

Table 1 Forced outage rate of every transmission line in CaseA1-A3

Level	Line Number	Case A1	Case A2	Case A3
1st	4,5,8,9	(0.0010, 0.0020,0.0025)	(0.0010, 0.0025)	(0.0010,0.001, 0.0020,0.0025)
2nd	1,6	(0.0015, 0.0025,0.0030)	(0.0015, 0.0030)	(0.0015,0.002, 0.0025,0.0030)
3rd	2,7	(0.0050, 0.0060,0.0065)	(0.0050, 0.0065)	(0.0050,0.005, 0.0060,0.0065)
4th	3	(0.0040, 0.0050,0.0060)	(0.0040, 0.0060)	(0.0040,0.004, 0.0050,0.0060)

Table 2 shows the maintenance schedule results of Case A1-A3. Meanwhile, random fuzzy expected value of interruption loss, interruption loss cost and maintenance cost for a whole day are shown in Table 3.

Table 2 Maintenance schedule of CaseA1-A3

Line Number	The original plan	The optimization plan
1	10:00~13:00	10:00~13:00
2	10:00~13:00	10:00~13:00
3	10:00~13:00	14:00~17:00

Table 3 Random Fuzzy risk and economy result of CaseA1-A3

Index	Case A1	Case A2	Case A3
Interruption loss(kW·h)	7248.39	7076.83	7200.80
Interruption loss cost(RMB)	28993.56	28307.32	28803.20
Maintenance cost(RMB)	60100	60100	60100
Total cost(RMB)	89093.56	88407.32	88903.20

Observing the results of Case A1-A3, although membership function form of fuzzy variable representing transmission line forced outage is different, it does not change the maintenance schedule remarkably. However, Case A1-A3 have different operation risk and economy loss risk.

2. The influence of the maximum acceptable expectation of energy not supply

There are 10 cases in total designed to study the The influence of the maximum acceptable expectation of energy not supply:B1,B2,...,B10, representing 1%,2%,...,10% of the daily load respectively. The optimization schedule is shown in Table 4 while random fuzzy risk and economy result are shown in Table 5.

Table 4 Maintenance schedule of Case B1-B10

Line Number	The original plan	B1~B3	B4	B5~B10
1	10:00~13:00	10:00~13:00	10:00~13:00	10:00~13:00
2	10:00~13:00	10:00~13:00	10:00~13:00	10:00~13:00
3	10:00~13:00	3:00~6:00	5:00~8:00	14:00~17:00

Table 5 Random Fuzzy risk and economy result of Case B1-B10

Index	The original plan	B1~B3	B4	B5~B10
Interruption loss(kW·h)	31190.44	7125.51	7196.45	7248.39
Interruption loss cost(RMB)	124761.70	28502.05	28785.78	28965.93
Maintenance cost(RMB)	60000	60280	60150	60100
Total cost(RMB)	184761.73	88782.05	88935.78	89065.93

Assuming the maintenance cost difference between the original plan and the optimization schedule is $\Delta C_{ma\mathrm{int}}$, the interruption loss cost difference between the two is ΔC_{EENS}, we can see that with the increase of the maximum acceptable expectation of energy not supply, interruption loss cost and maintenance cost vary in two directions. $\Delta C_{ma\,int}$ decreases because the optimized schedule gets closer to the original plan. However, the decrease of $\Delta C_{ma\,int}$ is less than the increase of ΔC_{EENS} so that there is growing tendency in total cost. Compared with the 2nd list in Table 5, B1-B10 are more superior to the original plan in both reliability and economical efficiency.

3. The influence of the interruption loss cost unit and adjustment maintenance cost unit

In order to verifyrelativity between optimization maintenance schedule and the two cost units, Case C1-C10 are designed. On the basis of B1-B10, interruption loss cost unit changes into 4RMB/kW • h while the adjustment maintenance cost is 100RMB for every advance or delay. The optimization maintenance schedule and corresponding result are shown in Table 6 and **Error! Reference source not found.**.

Table 6 Maintenance schedule of Case C1-C10

Line Number	The original plan	C1~C3	C4	C5~C10
1	10:00~13:00	10:00~13:00	10:00~13:00	10:00~13:00
2	10:00~13:00	10:00~13:00	10:00~13:00	10:00~13:00
3	10:00~13:00	3:00~6:00	6:00~9:00	14:00~17:00

Table 7 Random Fuzzy risk and economy result of Case C1-C10

index	The original plan	C1~C3	C4	C5~C10
interruption loss(kW·h)	31190.44	7125.51	7031.38	7248.39
interruption loss cost(RMB)	12476.18	2850.21	2812.55	2896.60
maintenance cost(RMB)	60000	65600	62000	62000
Total cost(RMB)	72476.18	68450.21	64812.55	64896.60

Compared with Table 5 and **Error! Reference source not found.**, we can draw the conclusions below:
In terms of Case B1-B10, as the maximum acceptable expectation of energy not supply grows, the total cost indicates a trend of increase. However, when it turns to C1-C10, the total cost grows firstly and then decreases. So a conclusion can be reached that different adjustment maintenance cost unit and interruption loss cost unit exert different influence on the total cost.

Although the influence is different,optimization maintenance schedule cut down random fuzzy expected value of interruption loss and total cost to a great degree,

indicating that both power sysytem reliablity and economic benefit rise.

4. The optimization performance of genetic algorithm

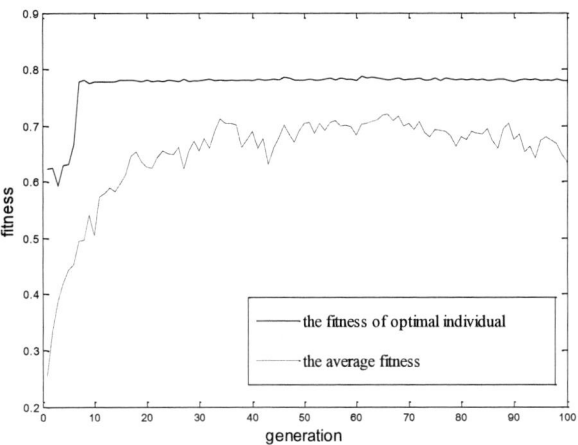

Fig 2 The optimization performanceof genetic algorithm

As shown in the Fig 2 , in the iterative process of genetic algorithm, a feasible solution can be found when seventh generation produced. The solid line represents the average fitness of the whole population while The dashed lines representsthe fitness of optimal individual. With the growth of the generations, the fitness of optimal individual and the average fitness increase rapidly at the beginning, suggesting that the solution is infeasible. Then the solution gets closer and closer to feasible solution. When it comes to some generation, the fitness of optimal individual shows slight fluctuation and the average fitness goes smoothly.

V. CONCLUSION

To schedule the short-time transmission line maintenance, the fuzziness and randomness of transmission lines' forced outage are facts in practice and impact to final maintenance decision. In this paper, based on Credibility Theory, a new model is established for maintenance scheduling considering the twofold random and fuzzy uncertainty. To solve the problem, a genetic algorithm applied to the model is proposed. Schedule period is divided into two parts:period when no line maintained and period when lines to be repaired. Simulation results of IEEE-RBTS proven the proposed model and approach are feasible to solve the short-time transmission line schedule considering the twofold random and fuzzy uncertainty.

REFERENCES

[1] Wang Jian,Wen Fushuan,Yang Rengang, "A Preliminary Investigation on the Regulation Mechanism for Maintenance Schedule of Generating Units in the Electricity Market Environment"Automation of Electric Power Systems,Vol.28, No.7,2004

[2] Feng Yongqing,Wu Wenchuan,Zhang Boming, "hydro-thermal Generateor Maintenance Scheduling Based on Credibility Theory" Proceedings of the CSEE,Vol.26,No.13,2006

[3] Ding Ming,Feng Yongqing, "Research on the Modeling and Algorithm to Global Generator and Transmission Maintenance Scheduling" Proceedings of the CSEE, Vol.24, No.5, 2004

[4] Feng Yongqing, Wu Wenchuan ,Zhang Boming, et al. "Power system operation risk evaluation based on credibility theory(2)"[J]. Automation of Electric Power Systems,Vol.30, No.2, 2006

[5] Feng Yongqing, Wu Wenchuan ,Zhang Boming, et al. "Power system operation risk evaluation based on credibility theory(3)"[J]. Automation of Electric Power Systems,Vol.30, No.2, 2006

[6] Feng Yongqing, Wu Wenchuan ,Zhang Boming, et al "Short-time transmission line maintenance scheduling based on Credibility Theory " Proceedings of the CSEE,Vol.27, No.4, 2007

[7] Feng Yongqing, Wu Wenchuan ,Zhang Boming, et al "Short-time transmission line maintenance scheduling based on Credibility Theory " Proceedings of the CSEE,Vol.27, No.4, 2007

[8] Huang Xianchao, "Research and Application of Distribution Maintenance Scheduling Methods"

[9] Chen Shaohua,Yang Peng, "Application of genetic algorithm to scheduling maintenance of thermal generating units" J. Wuhan Univ. of Hydr. & Elec. Eng. Vol.32, No.5, 1999

[10] Liu Yongmei,Sheng Wanxing, "Optimization Model for Distribution Equipment Maintenance Scheduling Based on Network Topology and Genetic Algorithm" Power System Technology, Vol.31 No.21,2007

[11] Li Dang, "Optimization of on Power Supplying Equipment Maintenance Scheduling based on Improved Genetic Algorithm" Practical management techniques,pp 33-37

[12] Schilling M.T., Leite da Silva A.M., Billinton R., et al. "Bibliography on power system probabilistic analysis". IEEE Transactions on Power Systems, 1990

[13] Allan R.N., Billinton R., Breipohl A.M. , et al. "Bibliography on the application of probability methods in power system reliability evaluation". IEEE Transactions on Power Systems, 1994

Model of Inverter in More Electric Aircraft Based on Generalized State Space Averaging Approach

Yanbo Che[1]   Guojian Liu[1]   Zhangang Yang[2]   Xiaokun Liu[1]

[1] Tianjin University Key Laboratory of Smart Grid of Ministry of Education Tianjin 300072 China
E-mail: lab538@163.com
[2] Civil Aviation University of China School of aviation automation Tianjin 300300 China
E-mail: zgyang@cauc.edu.cn

Abstract–The variable frequency power supply system has advantages of energy saving and light weighting. With its application in the new aircraft, DC distribution system is becoming important direction of more or all electric aircraft in the future. More constant frequency load will be supplied with DC main bus by inverter. Aircraft electric power system operation may suffer from large signal disturbances, which affect the inverter's performance greatly. To address the issue, this paper presents a 12-pulse inverter model for large signal disturbance based on generalized state space averaging method. By comparing generalized state space averaging model analysis result with state space averaging and time domain simulation, it is very verified the model is more accurate and effective for dc main bus power distribution system.

Keywords–12-pulse inverter, electric aircraft, electric power system, generalized state space averaging, large signal disturbance.

I. INTRODUCTION

In order to reduce petrol consumption and the emission of greenhouse gas, the more electric aircraft has been put forward[1]. Variable frequency (AF) power supply system, with respect to constant frequency (CF) power supply system, need not the constant speed drive unit and secondary conversion device. Because AF is easy to start, generate power and significantly reducing aircraft weight. It was adopted by the B787 and A380. As the change of power supply system form CF to AF, the future candidate aircraft distribution power system would be mainly distributed through a DC main bus[2-6].

As an important part of the aircraft power system, inverter have large capacity, high frequency, frequent switching, etc. More constant frequency load will be supplied with dc main bus by inverter in the DC distribution system. When the aircraft operation status changed, aircraft electric power system will generate large signal disturbance which affects the inverter circuit inevitably in the new distribution mode. It will be directly related to the safety of all kinds of constant frequency AC loads[7-10].

Currently the stability analysis of the aircraft power system inverter focused on the aspect of small signal stability analysis. Its analysis methods are time-domain simulation and mathematical modeling approach. Time-domain simulation technology was used to analysis transient behavior of inverse transform part in [11]. Multiple inverter performance was simulated in time domain for aircraft power system in [12]. Time-domain simulation needs long time and cannot reveal underlying

operating principle. It only applies to small-capacity simple system analysis, can not meet the needs of large and complex systems.

Mathematical modeling approach is to seek analytical expressions to characterize the inverter. One of the most representative is state space averaging method. State space averaging method was applied to establish the model of inverter small-signal for single-phase and three-phase[13-14].

It also was used to set up the state space averaging model of electromechanical actuators and rotating rectifier[15-16]. State space averaging method assume that state variables only have small change within the switching cycle in the derivation. It cannot meet the fast changing and dynamic analysis of large signal.

However, the large signal disturbance model of more electric aircraft inverter circuit is not taken seriously. We need a practical modeling method that can effectively analyze the inverter circuit of more electric aircraft large signal disturbance. Rapid and large signal dynamic characteristic cannot be analyzed by the state space averaging method.

The generalized state space averaging was proposed based on the state space average. It considers not only constant component, also high-order components. So it can be applied to analysis large-signal disturbance and has been used to analysis large-signal dynamic characteristic of transformer rectifier[18-19].

This paper applied the generalized state space method, the foundation and harmonics were taken into consideration, to build the 12-pulse inverter of aircraft. It is used to analysis dynamic characteristic of large signal disturbance.

II. GENERALIZED STATE SPACE AVERAGING METHOD

Generalized state space averaging method employs the Fourier series with time-dependent coefficients. On the interval [t-T, T], the waveform $x(t)$ can be approximated with a Fourier series representation of the form

$$x(\mathrm{t}) = \sum_{k=-n}^{n} \langle x \rangle_k (\mathrm{t}) e^{jk\omega t} \qquad (1)$$

where

$\omega = 2\pi / t$ is fundamental angular frequency, n depends on the required degree of accuracy and $\langle x \rangle_k$ is k the Fourier coefficients, which is defined as

$$\langle x \rangle_k (t) = \frac{1}{T} \int_{t-T}^{t} x(t) e^{-jk\omega t} dt \qquad (2)$$

So a signal can be calculated and given by

$$x(t) = \langle x \rangle_0 + 2 \sum_{k=1}^{\infty} \left\{ \text{Re} \langle x \rangle_k \cos(k\omega t) - \text{Im} \langle x \rangle_k \sin(k\omega t) \right\} \qquad (3)$$

The following calculation principles were used in the derivation process

$$\langle x(t) + y(t) \rangle_k = \langle x \rangle_k (t) + \langle y \rangle_k (t) \qquad (4)$$

$$\langle x(t) y(t) \rangle_k = \sum_{i=-\infty}^{\infty} \langle x(t) \rangle_{k-i} \langle y(t) \rangle_i \qquad (5)$$

$$\frac{d}{dt} \langle x(t) \rangle_k = \left\langle \frac{dx(t)}{dt} \right\rangle_k - jk\omega \langle x(t) \rangle_k \qquad (6)$$

III. MODELING AND ANALYSIS OF INVERTER

1. The inverter circuit in power electronic

The circuit of 12-pulse voltage source inverter in aircraft power system is shown in Fig.1.It is controlled by SPWM,.

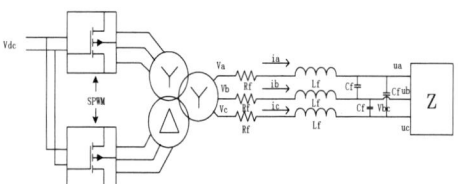

Fig.1: 12-pulse voltage source inverter

V_{dc} is the DC main bus voltage, which value is 270V. Taking into account neutral of the AC load side is not connected to neutral of the DC main bus, line voltage measurement and analysis were adopted when building the generalized state space averaging model.

2. Modeling of the inverter circuit

The switching function is 1 shows a closed 0 off. Using SPWM inverter technology, switching function can be replaced by Fourier series

$$s_1(t) = \sum_{n=1,\ odd}^{\infty} A_n \sin(n\omega t) \qquad (7)$$

The line voltage as follow:

$$v_{ab1}(\omega t) = v_{a1}(\omega t) - v_{b1}(\omega t) =$$
$$\frac{m\sqrt{3}V_{dc}}{2n} \sum_{n=1,\ odd}^{\infty} A_n \sin[n(\omega t + \varphi + \frac{\pi}{6})] \qquad (8)$$

where, m is amplitude modulation ratio, n is the transformer ratio, An is Fourier coefficients, φ is initial phase angle. Form symmetrical relationship we can obtain

$$\begin{cases} v_{bc1} = v_{ab1}(\omega t - 2\pi/3) \\ v_{ca} = v_{ab1}(\omega t + 2\pi/3) \\ v_{ab2} = v_{ab1}(\omega t + \pi/6) \\ v_{bc2} = v_{ab1}(\omega t - \pi/2) \\ v_{ca2} = v_{ab1}(\omega t + 5\pi/6) \end{cases} \qquad (9)$$

where v_{ab2}、 v_{bc2}、 v_{ca2} are voltage of secondary side of inverter.

The loads, whose value is R, were considered symmetrical loads in delta connection. Define virtual line current i_{ab}、 i_{bc}、 i_{ca}. Because the loads, phase voltage and current are symmetrical so

$$i_{ab} = \frac{1}{3}(i_a - i_b); i_{bc} = \frac{1}{3}(i_b - i_c); i_{ca} = \frac{1}{3}(i_c - i_a) \quad (10)$$

And

$$\begin{bmatrix} v_{ab} \\ v_{bc} \\ v_{ca} \end{bmatrix} = \begin{bmatrix} v_a - v_b \\ v_b - v_c \\ v_c - v_a \end{bmatrix} = \begin{bmatrix} L_f \frac{di_a}{dt} - L_f \frac{di_b}{dt} \\ L_f \frac{di_b}{dt} - L_f \frac{di_c}{dt} \\ L_f \frac{di_c}{dt} - L_f \frac{di_a}{dt} \end{bmatrix} +$$
$$R_f \begin{bmatrix} i_a - i_b \\ i_b - i_c \\ i_c - i_a \end{bmatrix} + \begin{bmatrix} u_{ab} \\ u_{bc} \\ u_{ca} \end{bmatrix} \qquad (11)$$

$$\begin{bmatrix} i_{ab} \\ i_{bc} \\ i_{ca} \end{bmatrix} = C_f \frac{d}{dt} \begin{bmatrix} u_{ab} \\ u_{bc} \\ u_{ca} \end{bmatrix} + \frac{1}{Z_L} \begin{bmatrix} u_{ab} \\ u_{bc} \\ u_{ca} \end{bmatrix} \qquad (12)$$

R_f、 L_f、 C_f are resistance, inductance, capacitance value of low pass filter.

The state equations of inverter current were constructed

$$\frac{d}{dt} \begin{bmatrix} i_{ab} \\ i_{bc} \\ i_{ca} \end{bmatrix} = -\frac{R_f}{L_f} \begin{bmatrix} i_{ab} \\ i_{bc} \\ i_{ca} \end{bmatrix} - \frac{1}{3L_f} \begin{bmatrix} u_{ab} \\ u_{bc} \\ u_{ca} \end{bmatrix} +$$
$$\frac{\sqrt{3}mV_{dc}}{6nL_f} \begin{bmatrix} \sin(\omega t + \varphi + \frac{\pi}{6}) + \sin(\omega t + \varphi + \frac{\pi}{3}) \\ \sin(\omega t + \varphi + \frac{5\pi}{6}) + \sin(\omega t + \varphi + \pi) \\ \sin(\omega t + \varphi - \frac{\pi}{2}) + \sin(\omega t + \varphi - \frac{\pi}{3}) \end{bmatrix} \quad (13)$$

$$\begin{bmatrix} i_{ab} \\ i_{bc} \\ i_{ca} \end{bmatrix} = C_f \frac{d}{dt} \begin{bmatrix} u_{ab} \\ u_{bc} \\ u_{ca} \end{bmatrix} + \frac{1}{Z_L} \begin{bmatrix} u_{ab} \\ u_{bc} \\ u_{ca} \end{bmatrix} \tag{14}$$

3. Generalized state space averaging modeling

Assuming generalized state space averaging variable as follows:

$$\begin{cases} \langle u_{ab} \rangle_0 = x_1, \langle u_{bc} \rangle_0 = x_2, \langle u_{ca} \rangle_0 = x_3; \\ \langle i_{ab} \rangle_0 = x_4, \langle i_{bc} \rangle_0 = x_5, \langle i_{ca} \rangle_0 = x_6; \\ \langle u_{ab} \rangle_1 = x_7 + jx_8, \langle u_{bc} \rangle_1 = x_9 + jx_{10}, \\ \langle u_{ca} \rangle_1 = x_{11} + j x_{12}; \\ \langle i_{ab} \rangle_1 = x_{13} + jx_{14}, \langle i_{bc} \rangle_1 = x_{15} + jx_{16}, \\ \langle i_{ca} \rangle_1 = x_{17} + jx_{18}; \end{cases} \tag{15}$$

Because the DC main bus has less harmonic components, it is considered contains only the DC component. Combining with the generalized state space algorithms, the final generalized state space averaging modeling is determined by (16) and (17)

$$\begin{bmatrix} \dot{x}_1 \\ \dot{x}_2 \\ \dot{x}_3 \\ \dot{x}_4 \\ \dot{x}_5 \\ \dot{x}_6 \end{bmatrix} = \begin{bmatrix} -\frac{1}{RC} & 0 & 0 & \frac{1}{C} & 0 & 0 \\ 0 & -\frac{1}{RC} & 0 & 0 & \frac{1}{C} & 0 \\ 0 & 0 & -\frac{1}{RC} & 0 & 0 & \frac{1}{C} \\ -\frac{1}{3L} & 0 & 0 & -\frac{R_f}{L_f} & 0 & 0 \\ 0 & -\frac{1}{3L} & 0 & 0 & -\frac{R_f}{L_f} & 0 \\ 0 & 0 & -\frac{1}{3L} & 0 & 0 & -\frac{R_f}{L_f} \end{bmatrix} \begin{bmatrix} x_1 \\ x_2 \\ x_3 \\ x_4 \\ x_5 \\ x_6 \end{bmatrix} \tag{16}$$

$$\begin{bmatrix} \dot{x}_7 \\ \dot{x}_8 \\ \dot{x}_9 \\ \dot{x}_{10} \\ \dot{x}_{11} \\ \dot{x}_{12} \\ \dot{x}_{13} \\ \dot{x}_{14} \\ \dot{x}_{15} \\ \dot{x}_{16} \\ \dot{x}_{17} \\ \dot{x}_{18} \end{bmatrix} = \begin{bmatrix} -\frac{1}{RC} & \omega & 0 & 0 & 0 & 0 & \frac{1}{C} & 0 & 0 & 0 & 0 & 0 \\ -\omega & -\frac{1}{RC} & 0 & 0 & 0 & 0 & 0 & \frac{1}{C} & 0 & 0 & 0 & 0 \\ 0 & 0 & -\frac{1}{RC} & \omega & 0 & 0 & 0 & 0 & \frac{1}{C} & 0 & 0 & 0 \\ 0 & 0 & -\omega & -\frac{1}{RC} & 0 & 0 & 0 & 0 & 0 & \frac{1}{C} & 0 & 0 \\ 0 & 0 & 0 & 0 & -\frac{1}{RC} & \omega & 0 & 0 & 0 & 0 & \frac{1}{C} & 0 \\ 0 & 0 & 0 & 0 & -\omega & -\frac{1}{RC} & 0 & 0 & 0 & 0 & 0 & \frac{1}{C} \\ -\frac{1}{3L} & 0 & 0 & 0 & 0 & 0 & -\frac{R_L}{L_f} & \omega & 0 & 0 & 0 & 0 \\ 0 & -\frac{1}{3L} & 0 & 0 & 0 & 0 & -\omega & -\frac{R_L}{L_f} & 0 & 0 & 0 & 0 \\ 0 & 0 & -\frac{1}{3L} & 0 & 0 & 0 & 0 & 0 & -\frac{R_L}{L_f} & \omega & 0 & 0 \\ 0 & 0 & 0 & -\frac{1}{3L} & 0 & 0 & 0 & 0 & -\omega & -\frac{R_L}{L_f} & 0 & 0 \\ 0 & 0 & 0 & 0 & -\frac{1}{3L} & 0 & 0 & 0 & 0 & 0 & -\frac{R_L}{L_f} & \omega \\ 0 & 0 & 0 & 0 & 0 & -\frac{1}{3L} & 0 & 0 & 0 & 0 & -\omega & -\frac{R_L}{L_f} \end{bmatrix} \begin{bmatrix} x_7 \\ x_8 \\ x_9 \\ x_{10} \\ x_{11} \\ x_{12} \\ x_{13} \\ x_{14} \\ x_{15} \\ x_{16} \\ x_{17} \\ x_{18} \end{bmatrix} + \frac{\sqrt{3}mV_{dc}}{12nR_f} \begin{bmatrix} 0 \\ 0 \\ 0 \\ 0 \\ 0 \\ 0 \\ \sin(\varphi+\frac{\pi}{6})+\sin(\varphi+\frac{\pi}{3}) \\ -\cos(\varphi+\frac{\pi}{6})-\cos(\varphi+\frac{\pi}{3}) \\ \sin(\varphi-\frac{\pi}{2})+\sin(\varphi-\frac{\pi}{3}) \\ -\cos(\varphi-\frac{\pi}{2})-\cos(\varphi-\frac{\pi}{3}) \\ \sin(\varphi+\frac{5\pi}{6})+\sin(\varphi+\pi) \\ -\cos(\varphi+\frac{5\pi}{6})-\cos(\varphi+\pi) \end{bmatrix} \tag{17}$$

The line voltage and virtual line current of inverter circuit can be expressed by generalized state variables as (18).

$$\begin{cases} u_{ab} = x_1 + 2x_7 \cos\omega t - 2x_8 \sin\omega t \\ u_{bc} = x_2 + 2x_9 \cos\omega t - 2x_{10} \sin\omega t \\ u_{ca} = x_3 + 2x_{11} \cos\omega t - 2x_{12} \sin\omega t \\ i_{ab} = x_4 + 2x_{13} \cos\omega t - 2x_{14} \sin\omega t \\ i_{bc} = x_5 + 2x_{15} \cos\omega t - 2x_{16} \sin\omega t \\ i_{ca} = x_6 + 2x_{17} \cos\omega t - 2x_{18} \sin\omega t \end{cases} \tag{18}$$

IV. EXPERIMENTAL RESULTS

Taking the line voltage V_{ab} and virtual line current I_{ab} as the analysis objects. Time domain device, state space averaging and generalized state space averaging simulation models were built in MATLAB/Simulink. The parameters of inverter circuit are listed in the follow table 1, when analysis the 12-pulse inverter in aircraft power system.

When a large disturbance occurs at 10ms, the load from 80Ω changes to 40Ω. The large signal disturbance will occur and affect the inverter.

The signal and secondary side of the transformer line voltage waveform are expressed in Fig.2

Table 1: the parameters of inverter circuit

Description	Symbol	Value	Unit
Main bus voltage	V_{dc}	270	V
Amplitude modulation ratio	m	0.85	—
Fundamental frequency	f	400	Hz
Switching frequency	fv	10800	Hz
Initial phase angle	φ	0	Rad/s
Filtering resistor	R_f	0.1	Ω
Filtering inductor	L_f	4.3	mH
Filtering capacitor	C_f	3.2	μF
Transformer ratio	n	2	—

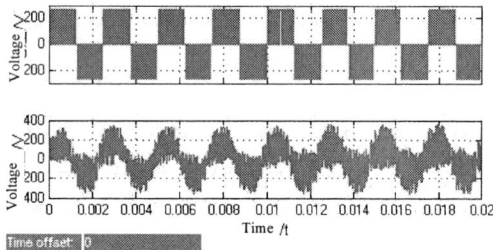

Fig.2: signal of SPWM and AC voltage waveform

The waveform output line voltage V_{ab} and virtual current I_{ab} are expressed in Fig.3 and Fig.4.

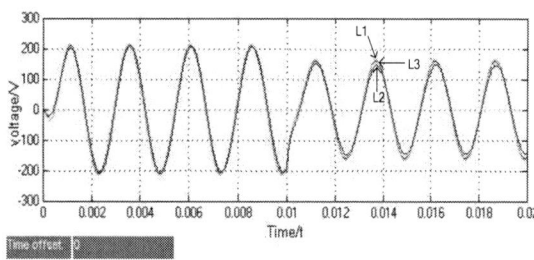

Fig. 3: waveform of V_{ab} voltage at load port

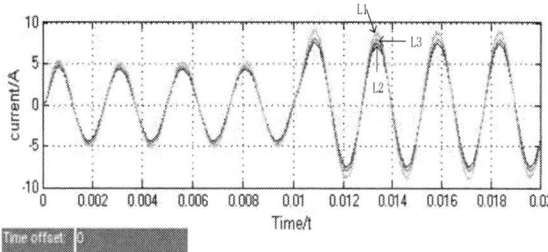

Fig. 4: waveform of I_{ab} at AC side

The L_1, L_2 and L_3 represent waveform of time domain device, state space averaging and generalized state space averaging respectly. By comparing the generalized state space averaging and state space averaging model analysis result with the time domain device simulation, the deviation of state space averaging model is about 6% when large signal disturbance occurred. The deviation of generalized state space averaging model is just about 3% .The accuracy of generalized state space averaging model is upgraded nearly doubled.

V. CONCLUSION

In this paper the generalized state space averaging approach have been utilized to build large-signal model. Experimental results were compared to time domain device model and traditional state space averaging model. When the inverter was attacked by large signal disturbance ,the generalized state space averaging model of inverter has better accuracy for large signal disturbance analysis.

In generalized state space model, n is the value of harmonic. The value of n depends on the wanted degree of accuracy. The more terms we consider, the more accuracy we have. If n approaches infinity, the approximation error is zero. The model can used to analysis large-signal disturbance.

ACKNOWLEDGMENT

The paper is funded by National Natural Science Foundation of China (No.51407185).

REFERENCES

[1] Quigley Jr R E, "More electric aircraft", Applied Power Electronics Conference and Exposition, APEC'93, Conference Proceedings, 7-11 Mar 1993,pp. 906-911.

[2] Rosero J A,Ortega J A,Aldabas E,et al, "Moving towards a more electric aircraft", Aerospace and Electronic Systems Magazine, Vol. 22, No. 3, 2007, pp.3-9.

[3] Bulent Sarlioglu,Morris C T, "More electric aircraft -review, challenges and opportunity for commercial transport aircraft", IEE Transportations on Transportation Electrification, Vol. 1, No. 1, 2015, pp.54-64.

[4] Roboam X, "New trends and challenges of electrical networks embedded in 'more electrical aircraft'" , Industrial Electronics (ISIE), 27-30 June 2011, pp.26-31.

[5] Cao W, Mecrow B C, Atkinson G J, et al, " Overview of electric motor technologies used for more electric aircraft (MEA)", IEEE Transactions on Industrial Electronics, Vol. 59, No. 9,2012, pp.3523-3531.

[6] Brombach J, Lucken A, Nya B, et al, "Comparison of different electrical HVDC-architectures for aircraft application", Electrical Systems for Aircraft, Railway and Ship Propulsion (ESARS),16-18 Oct 2012,pp. 1-6.

[7] Izquierdo D, Azcona R, Del Cerro F J L, et al, "Electrical power distribution system (HV270DC), for application in more electric aircraft", Applied Power Electronics Conference and Exposition (APEC),21-25 Feb 2010, pp.1300-1305.

[8] Avery C R, Burrow S G, Mellor P H, "Electrical generation and distribution for the more electric aircraft", Universities Power Engineering Conference (UPEC), 4-6 Sept 2007, pp.1007-1012.

[9] Zhang H, Mollet F, Saudemont C, et al, "Experimental validation of energy storage system management strategies for a local dc distribution system of more electric aircraft", IEEE Transactions on Industrial Electronics, Vol.57, No.12,

2010, pp. 3905-3916.

[10] T Wu, S V Bozhko, G M Asher, "High speed modeling approach of aircraft Electronical power systems under both normal and abnormal scenarios", International Symposium Industrial Electronics(ISIE), 4-7 July 2010, pp.870-877.

[11] Montealegre Lobo, Dufour C, Mahseredjian J, "Real-time simulation of More Electric Aircraft power systems", 2013 15th European conference on Power Electronics and Applications, 2-6 Sept 2013, pp.1-10.

[12] Mathew S R, Sai Kiran P V R, Anand M, et al, "Design and implementation of a three level diode clamped inverter for more electric aircraft applications using hardware in the loop simulator", International Conf. on Advances in Electronics, Computers and Computers Communications(ICAECC), 10-11 Oct 2014, pp.1-6.

[13] Hyun D Y, Lim C S, Kim R Y, et al, "Averaged modeling and control of a single-phase grid-connected two-stage inverter for battery application", Industrial Electronics Society, IECON, 10-13 Nov 2013, pp.489-494.

[14] Ahmed S, Shen Z, Mattavelli P, et al, "Small-signal model of a Voltage Source Inverter (VSI) Considering the Dead-Time Effect and Space Vector Modulation Types", Applied Power Electronics Conference and Exposition (APEC), 6-11 March 2011, pp.685-690.

[15] Wu T, Bozhko S, Asher G, et al, "Fast Reduced Functional Models of Electromechanical Actuators for More-Electric Aircraft Power System Study", SAE Technical Paper, 2008..

[16] Wu T, Bozhko S V, Asher G M, et al, "A Fast Dynamic Phasor Model of Autotransformer Rectifier Unit for More Electric Aircraft", 35th Annual Conference of IEEE Industrial Electronics, 3-5 Nov. 2009, pp.2531-2536.

[17] Sanders S R, Noworolski J M, Liu X Z, et al, "Generalized averaging method for power conversion circuits", IEEE Transactions on power Electronics, Vol. 6, No.2, 1991,pp.251-259.

[18] Darkhaneh H E, Gatabi J R, El-Kishky H, "A novel GSSA method for modeling of controllers in the multi-converter system of an Advanced Aircraft Electric Power System(AAEPS)", International Power Modulatoe and High Voltage Conference (IPMHVC), 3-7 June 2012, pp.795-798.

[19] Salem M, Jusoh A, Idris N R N, et al, "Modeling and simulation of generalized state space averaging for series resonant converter", Power Engineering Conference (AUPEC), Sept. 28 2014-Oct. 1 2014, pp.1-5.

BIOGRAPHIES

Yanbo Che was born in Shandong, Chian. He received his B.S. degree from Zhejiang University, Hangzhou, China, in 1993. He received his M.S. and Ph.D. degrees fromTianjin University, Tianjin, China, in 1996 and 2002, respectively. Since 1996, he has been engaged in teaching and scientific research of power electronic technology and power systems. He is presently an Associate Professor in the School of Electrical Engineering and Automation at TianjinUniversity. His current research interests include power electronics, new energy and micro-grids.

Guojian Liu was born in Hebei province, China. He obtained his BSc from Taiyuan University Of Technology. He is studying for a master's degree in Tianjin University and engaging in power electronics and aircraft power systems.

Zhangang Yang received his M.S. and Ph.D degrees from Tianjin University, Tianjin, China, in 2007 and 2011, respectively. Since Aprial 2011, he has been working in Department of Aviation automation, Civil Aviation University of China, Tianjin, China. His current research interests include more electric Aircraft power system, power systems and micro-grids.

High Power Capacitive Power Transfer for Electric Vhicle Charging Applications

Chris Mi[1]

[1]San Diego State University San Diego

Abstract—This paper focuses on high power capacitive power transfer (CPT) for electric vehicle charging applications. The CPT system can achieve several kW power transfer through an air-gap distance of 150 mm with the dc-dc efficiency higher than 90%. Four pieces of metal plates are used as the capacitive coupler to build up electric fields and transfer power. A double-sided LCLC compensation circuit topology is proposed for the coupler to boost the voltage on the plates. The plates can be arranged vertically, instead of horizontally, to maintain the system power at rotatory misalignment conditions. The CPT system can also be combined with the inductive power transfer (IPT) system, in which the capacitors between the plates resonate with the inductors of the coils. Three prototypes have been constructed to validate the high power CPT system, which demonstrate that the CPT technology is suitable for electric vehicle charging applications.

Keywords—capacitive power transfer, electric fields, high frequency resonant circuit, and electric vehicle charging

I. INTRODUCTION

Inductive power transfer (IPT) technology has been widely used in electrical vehicle charging applications. It utilizes Litz-wires to construct coils and generate magnetic fields to transfer power. The system output power has reached 7 kW with an efficiency of 96%, which is already comparable to the wire-connected vehicle charger [1]. However, the IPT system is sensitive to metal material, and the magnetic fields can generate eddy current losses in the metal.

Compare to the IPT technology, the capacitive power transfer (CPT) system utilizes electric fields to transfer power [2], which has two advantages. First, the electric fields can pass through metal material without generating significant power losses. Second, the CPT system adopts metal plates, instead of Litz-wires, to generate electric fields, so the system cost can be dramatically reduced.

The challenge of high power CPT technology comes from the small coupling capacitance between the plates. When the air-gap distance is around 150 mm, which is the ground clearance of the vehicle, the capacitance is in the range of several 10's of pF. It requires either too high switching frequency or too large resonant inductance for the capacitive coupler, which is difficult to realize in practice. The recent published CPT systems focus on low power and short distance applications, such as 1 mm.

In this paper, an LCLC compensation circuit topology is proposed for high power CPT application. The voltages on the metal plates are boosted to several kV level, so the transfer distance is significantly increased. A compact vertical plate structure is proposed to improve the rotatory misalignment ability. The CPT system can also work with the IPT system, which induces a novel IPT and CPT combined system.

II. A DOUBLE-SIDED LCLC COMPENSATED CPT SYSTEM

An LCLC compensation network is shown in Fig. 1 [3]. It is used to resonate with the capacitive coupler.

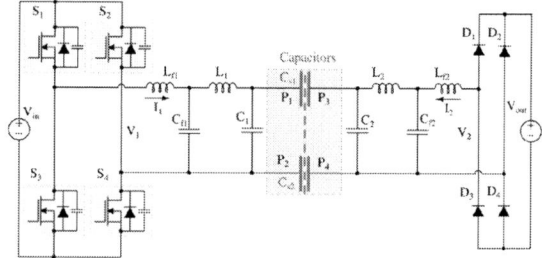

Fig. 1: A double-sided LCLC compensated CPT system

The coupler structure is shown in Fig. 2. P_1 and P_2 are horizontally arranged at the primary side as the transmitter. P_3 and P_4 are placed at the secondary side as the receiver.

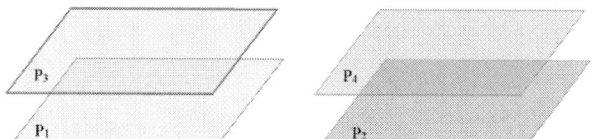

Fig. 2: Horizontal plates structure

The fundamental harmonics approximation (FHA) method is used and the equivalent circuit is shown in Fig. 3 (a).

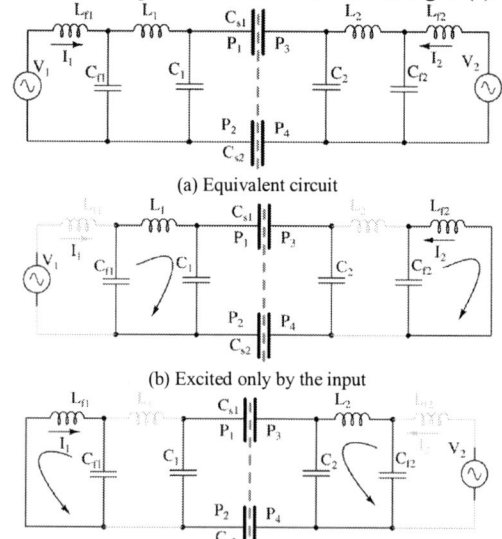

(a) Equivalent circuit

(b) Excited only by the input

(c) Excited only by the output

Fig. 3: Fundamental harmonics approximation of the circuit working principle

978-1-5090-0064-7/15 $31.00 © 2015 IEEE

The superposition theorem is used to analyze the two voltage sources separately, as shown in Fig. 3 (b) and (c). The relationship between the parameters is shown as,

$$\begin{cases} C_s = C_{s1}C_{s2}/(C_{s1}+C_{s2}) \\ C_{p1} = C_1 + C_2C_s/(C_2+C_s), L_{f1} = 1/(\omega_s^2 C_{f1}) \\ C_{p2} = C_2 + C_1C_s/(C_1+C_s), L_{f2} = 1/(\omega_s^2 C_{f2}) \\ L_1 = 1/(\omega_s^2 C_{p1}) + L_{f1}, L_2 = 1/(\omega_s^2 C_{p2}) + L_{f2} \end{cases} \quad (1)$$

The system output power is expressed as in (2) and Table I shows the parameters in a 2.4 kW prototype .

$$P_{out} = \omega_s C_s \frac{C_{f1}C_{f2}}{C_1C_2 + C_1C_s + C_2C_s} \cdot |V_1| \cdot |V_2| \quad (2)$$

TABLE 1: SYSTEM PARAMETER VALUES OF LCLC COMPENSATED SYSTEM

V_{in}	V_{out}	f_{sw}	L_{f1} (L_{f2})	C_{f1} (C_{f2})	C_1 (C_2)	L_1	L_2
265 V	280 V	1MHz	11.6μH	2.18nF	100pF	231μH	242μH

The prototype is constructed as shown in Fig. 4. Aluminum plates are used to make the capacitive coupler.

Fig. 4: Prototype of double-sided LCLC compensated CPT system

The experimental waveforms are shown in Fig. 5. The maximum output power is 2.4 kW with an efficiency of 90.8%.

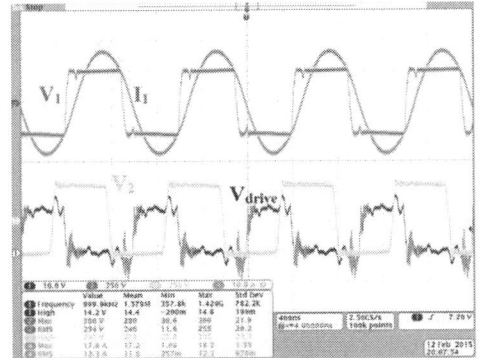

Fig. 5: Experimental waveforms of the LCLC compensated system

The misalignment test in Fig. 6 shows the output power is 2.1 kW with an efficiency of 90.7% at 300 mm misalignment.

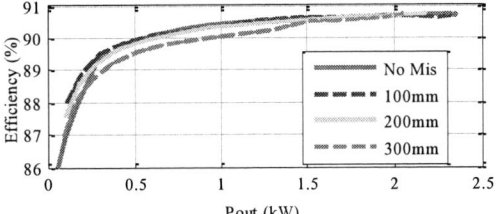

Fig. 6: Output power and efficiency of LCLC compensated system

III. A VERTICAL OVERLAPPING CAPACITIVE COUPLER

A vertically arranged coupler structure is shown in Fig. 7. The outer plates P_1 and P_3 are larger than inner plates P_2 and P_4. This structure can maintain couplings when there is rotatory misalignment.

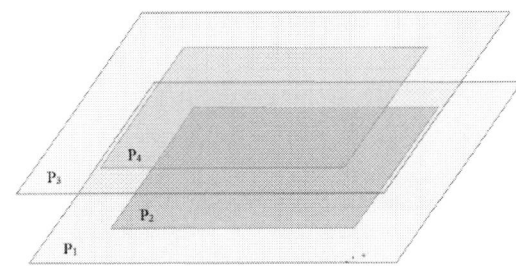

Fig. 7: A vertically arranged overlapping coupler structure

There are coupling capacitors between each pair of plates as shown in Fig. 8. The cross coupling capacitors C_{14} and C_{23} cannot be neglected.

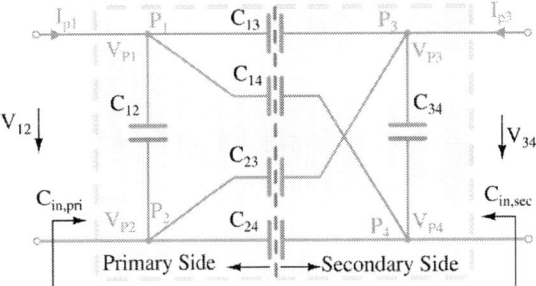

Fig. 8: Capacitive couplings between each pair of plates

The circuit model can be simplified using behavior sources as shown in Fig. 9. It is similar with the traditional behavior sources model of a transformers.

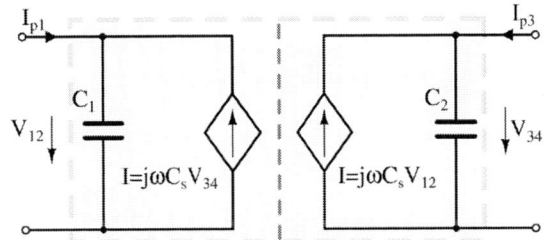

Fig. 9: Equivalent circuit model with behavior sources

978-1-5090-0064-7/15 $31.00 © 2015 IEEE

The Kirchhoff's current equations are used to derived the relationship between the parameters in Fig.8 and Fig. 9, which is shown as,

$$\begin{cases} C_1 = C_{12} + \dfrac{(C_{13} + C_{14}) \cdot (C_{23} + C_{24})}{C_{13} + C_{14} + C_{23} + C_{24}} \\ C_2 = C_{34} + \dfrac{(C_{13} + C_{23}) \cdot (C_{14} + C_{24})}{C_{13} + C_{14} + C_{23} + C_{24}} \\ C_s = \dfrac{C_{24}C_{13} - C_{14}C_{23}}{C_{13} + C_{14} + C_{23} + C_{24}} \end{cases} \quad (3)$$

It shows that there are two equivalent capacitors C_1 and C_2 at the input and output, respectively. They can be used to replace the capacitors in LCLC circuit, so the compensation circuit is simplified to LCL topology, as shown in Fig. 10.

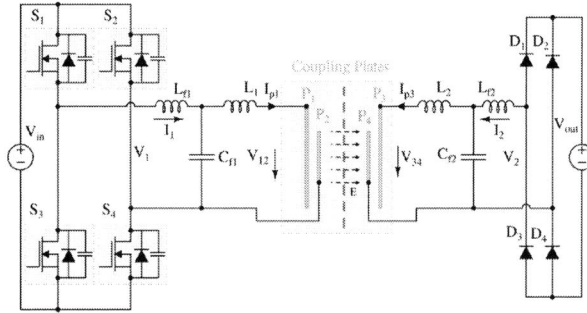

Fig. 10: A double-sided LCL compensated CPT system

Similar to the LCLC compensated system, the FHA method and superposition theorem are used to analyze the resonant circuit working principle. A 2.2 kW input power system is designed and the parameter values are shown in Table II.

TABLE 2: SYSTEM PARAMETER VALUES OF LCL COMPENSATED SYSTEM

V_{in}	V_{out}	f_{sw}	L_{f1} (L_{f2})	C_{f1} (C_{f2})	C_1 (C_2)	L_1	L_2
270V	270V	1MHz	2.90μH	8.73nF	381pF	69μH	70μH

The prototype is shown in Fig. 11. The four plates are vertically arranged and hold by PVC tubes.

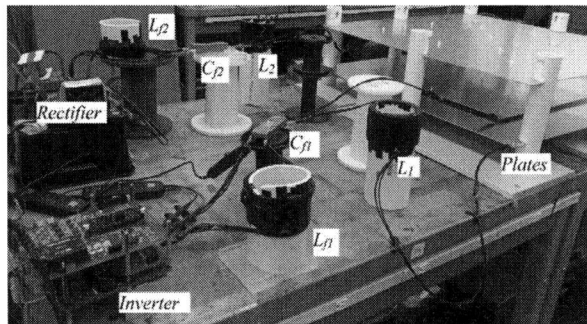

Fig. 11: Prototype of double-sided LCL compensated CPT system

The relationship between output power and efficiency is shown in Fig. 12. The system maximum output power is 1.88 kW with an efficiency of 85.87%. The efficiency can be improved through optimize the circuit parameters. In 360° rotatory misalignment test, the power ripple is within ±5.0%.

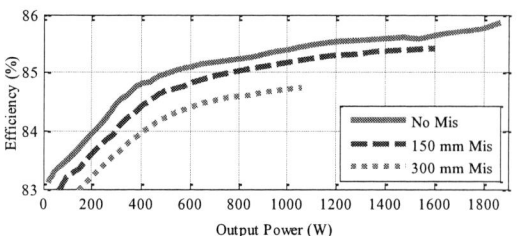

Fig. 12: Output power and efficiency of LCL compensated system

IV. IPT AND CPT COMBINED SYSTEM

The inductive and capacitive coupler can be combined together in a single system to transfer power through magnetic and electric fields simultaneously. The two couplers can resonate with each other. The circuit topology of the combined system is shown in Fig. 13.

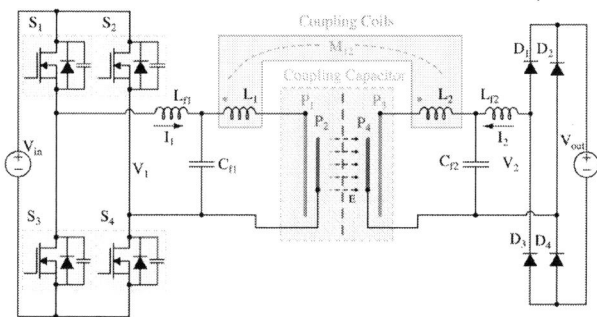

Fig. 13: A IPT and CPT combined system

The coils are inductively coupled, and the plates are capacitively coupled. The parameter values of a 3.0 kW input power system is shown in Table III.

TABLE 3: SYSTEM PARAMETERS OF IPT AND CPT COMBINED SYSTEM

V_{in}	V_{out}	f_{sw}	L_{f1} (L_{f2})	C_{f1} (C_{f2})	C_1 (C_2)	L_1	L_2
310 V	320 V	1MHz	14.2μH	1.78nF	96.1pF	256μH	264μH

The prototype is shown in Fig. 14. The two couplers are separated by 300 mm to reduce their interactions.

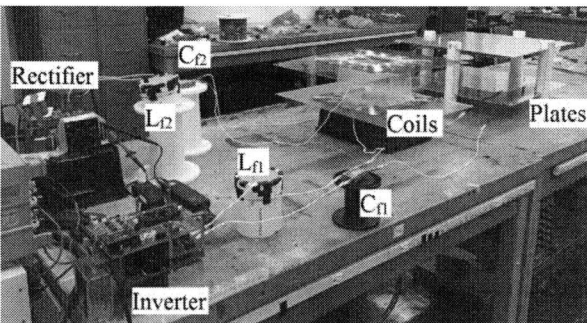

Fig. 14: Prototype of the IPT and CPT combined system

The power ratio between IPT and CPT system is 2.5. The power and efficiency is shown in Fig. 15. The maximum output power is 2.84 kW with an efficiency of 94.45%.

978-1-5090-0064-7/15 $31.00 © 2015 IEEE

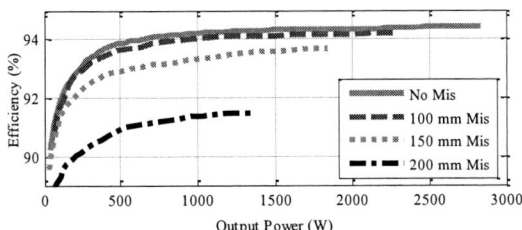

Fig. 15: Output power and efficiency of IPT and CPT combined system

REFERENCES

[1] S. Li, W. Li, J. Deng, T.D. Nguyen, and C. Mi, "A Double-Sided LCC Compensation Network and Its Tuning Method for Wireless Power Transfer," *IEEE Trans. Veh. Technol.*, vol.64, pp. 2261-2273, 2015.

[2] J. Dai, D. Ludois, "A Survey of Wireless Power Transfer and a Critical Comparison of Inductive and Capacitive Coupling for Small Gap Applications", *IEEE Trans. Power Electro.*, vol. 30, pp.6017-6029,2015.

[3] F. Lu, H. Zhang, H. Hofmann, and C. Mi, "A Double-sided LCLC-Compensated Capacitive Power Transfer System for Electric Vehicle Charging," *IEEE Trans. Power Electron.*, vol. 30, pp. 6011-4014, 2015.

The Design of an Electric Wing-in-Ground-Effect (WIG) Vehicle as Part of an Urban Air Transit System

Eric Bodlak[1]

[1] Department of Aerospace Engineering, University of Kansas, Lawrence, KS
E-mail: ebodlak@ku.edu

Abstract–An electric wing-in-ground-effect (WIG) vehicle was designed to supplement the existing public transportation network in densely populated coastal urban areas like Hong Kong, Incheon, the Persian Gulf, and (in particular) the San Francisco Bay Area. Routes and passenger volumes were modelled using circuit analysis, and the design was optimized to maximize the system's impact on traffic congestion and the resulting financial and environmental benefits. A concept of operations, including a battery-swapping procedure, was described, and aircraft performance was verified using a series of MATLAB simulations. It was determined that the transit system could reduce Bay Area greenhouse gas production by a total of 100 million kg annually and save commuters more than 10M commuting hours total and US$300 per person per year.

Keywords–Electric aircraft, ground effect vehicle, public transportation, San Francisco, traffic congestion

I. INTRODUCTION

As worldwide standards of living have improved over the past few centuries, global energy requirements have climbed to never-before-seen levels. This increased demand has brought with it concerns about climate change, environmental responsibility, and resource management. Now, more than ever, developed nations are working to cut emissions and improve efficiency, while exploring alternative means to power everything from communication devices to homes and automobiles. It is on this last item –the transportation sector – that this paper will be centered.

1. Motivation
In 2014, Americans spent nearly 7 billion extra hours waiting in delays due to traffic congestion [1]. While traveling an average of 22% slower than free-flow traffic, they burned an additional 12 billion liters of fuel, produced 28 billion kg of greenhouse gases, and wasted 160 billion U.S. dollars – a value larger than the respective GDPs of two-thirds of the world's nations [2]. Traffic

congestion is a function of several variables, including driving strategies, routing, and the vehicles being driven. Better traffic management and expanded infrastructure are certainly effective, but often neither are practical in established metropolitan areas.

Considering that much of the waste due to traffic congestion is tied to the combustion of excess fuel, it is apparent that many of symptoms of congestion could be alleviated by improving the average efficiency of the cars on the road. This would most effectively be done by replacing a portion of gas-powered commuter vehicles with electric cars, since these vehicles are well-suited for efficient stop-and-go driving – having no need to idle – and are often equipped with regenerative braking capabilities. Furthermore, it is typically more cost-effective for consumers to power electric vehicles, and average well-to-wheel emissions are lower than for their gas or diesel counterparts [3].

The widespread adoption of electric vehicles would not only reduce the excess of costs and emissions due to congestion – those reductions would apply to the whole commute, during which no fuel would need to be consumed. Currently, however, electric vehicles face several hurdles in the route to acceptance – specifically short ranges, lengthy charging times, lack of charging infrastructure, and high initial costs.

Since individuals are reluctant to purchase electric cars for themselves, perhaps it makes more sense to get electric vehicles on the road in some other way. Studies have shown that most commuters travel to work alone [4]. This means that far more vehicles occupy the roadways than are necessary to transport the commuter population. Increasing the average number of passengers per vehicle would be accompanied by a reduction of the total number of vehicles on the road. This is exactly what public transportation does, and it is for this reason that transit systems stand to play a major role in the reduction of traffic congestion in metro areas. Unfortunately, most systems (electric or not) are constrained by the same infrastructure and urban/geographic features that obstruct traffic in the first place. Aircraft, however, are not restricted by these features, and rely on established infrastructure only at the beginning and end of each route.

The intra-city nature of public routes limits their lengths considerably and suggests the interesting possibility of an air transit system designed to take advantage of the benefits of electric propulsion. These benefits include improved efficiency, a decreased need for cooling and maintenance, minimization of noise and emissions, and significant reduction in operating costs. None of the

978-1-5090-0064-7/15 $31.00 © 2015 IEEE

previously identified hurdles to electric vehicle adoption would be faced by transit passengers, as operations would be handled by the transit service itself.

In order to explore the feasibility of such a system, the San Francisco Bay Area was selected for further study. An electric WIG vehicle was proposed as the most effective aircraft for the design, since this choice reduces operating, infrastructure, and certification costs, and circumvents the issues that would arise with high volume operations in the controlled airspace surrounding the area's airports. Table 1 shows how San Francisco compares to the national average in terms of traffic congestion and its effects.

Table 1: Yearly Per-Commuter Waste Due to Traffic

	San Francisco	Nationwide Average
Time (h)	78	42
Fuel (L)	125	72
Cost (US$)	1675	960
Emissions (kg)	380	220

With its high congestion rate, coastal population, mild weather, and low-sea state, San Francisco presents itself as an ideal locale for a comparative electric WIG case study.

2. WIG Overview

Wing-in-ground-effect (WIG) craft are typically used in marine applications, where they operate just above the water in the "ground effect" region. There, the high pressure cushion formed between the underside of the wing and the ground plane allows them to support significantly greater loads than similarly-powered aircraft operating out of ground effect. They achieve speeds much higher than those reached by other types of marine vehicles, yet are regulated by marine organizations like the U.S. Coast Guard and IMO, rather than groups like the FAA and International Civil Aviation Organization [5][6]. This drastically reduces certification costs and nullifies concerns about airspace restrictions which, when coupled with the expected increase in efficiency, makes ground-effect vehicles an intriguing choice for certain applications. Additionally, their low operating heights and use over open water (rather than populated areas) make them appealing from a safety standpoint when compared to other airborne systems.

While the potential benefits of WIG craft have long been recognized, historically little research has been done with regards to their design, and a large market has yet to develop. Presently, the potential of ground-effect vehicles has been under-realized due to this lack of commercial and developmental experience.

II. DESIGN STRATEGY

The goal of this design project is to create a transit system that is feasible from a financial and technical standpoint, and that makes a measurable impact on the effects of traffic congestion in the Bay Area. Since the proposed design is limited to cross-bay travel, integration with other forms of public transportation is a critical step to achieving these goals.

1. Routing and Passenger Volume

Bay Area Rapid Transit (BART) is the most expansive and commonly-used public transportation service offered in the San Francisco Bay Area, with over 400,000 daily riders [7]. Three locations were identified to serve as BATWinG (Bay Area Transport, Wing-in-Ground-effect) hubs, as shown in Figure 1.

Figure 1: BATWinG Hubs and Existing Transit Lines

These locations allow for easy connection with the BART rail system and require a minimum infrastructural investment for the BATWinG service to begin operations. For the purposes of this paper, these operations are set to begin in 2020.

To estimate the number of passengers who would regularly use the BATWinG system, a set of 44 by 44 element matrices were constructed, containing the respective cost, ridership numbers, and travel time for each route in the BART network [7]. For comparison, similar tables were developed for the BATWinG design, based variables such as fleet size, speed of operation, and charged fare per passenger-km. Both sets of cost data accounted for the same dollar value of time as reported in Reference [1], scaled to 2020 dollars using the Consumer Price Index (CPI). BATWinG routes took advantage of BART connections and were evaluated with respect to each BART route, according to the following scenario:

• A passenger enters at one BART station and rides BART to the BATWinG station.
• The passenger then transfers to BATWinG and rides to a different BART station.
• From there, s/he rides BART to the final destination, where the BART/BATWinG system is exited.

The total time taken to complete this process, including transfers, was the value used for BATWinG travel times. The BATWinG route data were compared to the respective BART numbers, and the BATWinG system was assumed to take on the full number of passengers from the routes where customers were saved both time and money. Since it was more difficult to evaluate cost and time for specific car routes, BATWinG was assumed to serve the same percentage of U.S. 101/airport traffic as its BART share.

From there, passenger flow rates were calculated using circuit analysis, with passenger flow corresponding to current, and resistance representing travel time. Each of the BATWinG stops were modelled as sources with a potential of one nine-passenger "WIGful," to represent the goal of keeping no more passengers waiting than can be served by the next available aircraft. A combination of resistors in parallel, each representing a separate aircraft flow route, were used to match the expected passenger flow on each route as nearly as possible. The resistances and the number of parallel resistors on each route were varied to achieve this goal.

This was done for different combinations of aircraft cruise speeds and per-passenger-km fare rates, and tied into an aircraft sizing loop to yield both a routing strategy and an aircraft design that were optimized with respect to economic and environmental measures. The time to break even was chosen as a system indicator of economic feasibility, while the number of kilograms of greenhouse gas emissions prevented by the BATWinG network was selected as the environmental metric. System fares in US$/passenger-km were varied and the results of the model were normalized with respect to the minimum or maximum predicted value then plotted together in Figure 2.

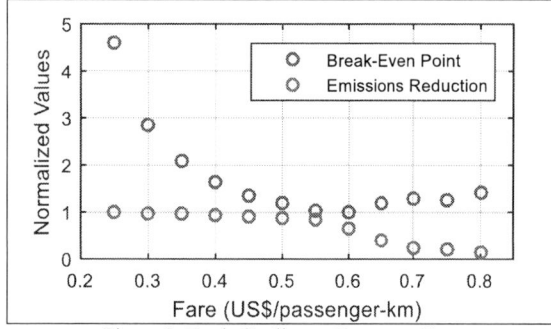

Figure 2: Trade Studies on Passenger Fares

From the figure it can be seen that the rate that yields the best combined results for the two measures is about US$0.55/passenger-km, where the time-to-break-even curve is at a near minimum and the reduction in emissions is near its maximum. According to the model, $0.20/passenger-km was the minimum economically feasible fare, while $1.90/passenger-km was the maximum fare that reduced emissions.

Matching the BART schedule and operating at US$0.55/passenger-km, BATWinG is predicted to serve 38,000 passengers per day. Figure 3 shows the BATWinG routes and their flow in WIGfuls per hour.

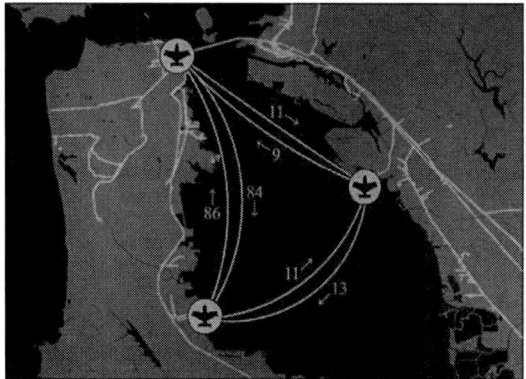

Figure 3: BATWinG Passenger Flow in WIGfuls/hour

2. Aircraft Sizing

The BATWinG was sized according to standard preliminary design methods [8], based on historical data and parametric studies which were subject to certain constraints. These constraints included a 150 m takeoff run and an overall span of 15 meters, which was intended to facilitate integration of the aircraft into existing ferry infrastructure in San Francisco Bay. Historical sizing data were limited, but a selection of relevant aircraft are shown on the sizing chart in Fig. 4, which plots the power and wing loading requirements versus takeoff lift coefficient ($C_{L,TO}$) for the 150 m takeoff constraint.

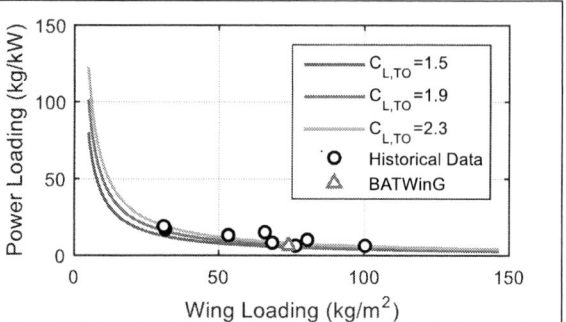

Fig. 4: Historical Sizing Chart for 150 m Takeoff Condition

To size the battery pack (intended to be swapped after each trip), a 0.39 payload-to-maximum-takeoff-weight ratio was enforced, based on historical values. The maximum route length (SFO ↔ Embarcadero) of 25 km was scaled by a factor of 1.5 to account for lifetime battery degradation to 80% of the initial capacity and variations in environmental conditions. Ragone plots were constructed for two battery chemistries, $LiCoO_2$ (high energy) and $LiFePO_4$ (high power), projected to 2020 values in Figure 5 [9],[10].

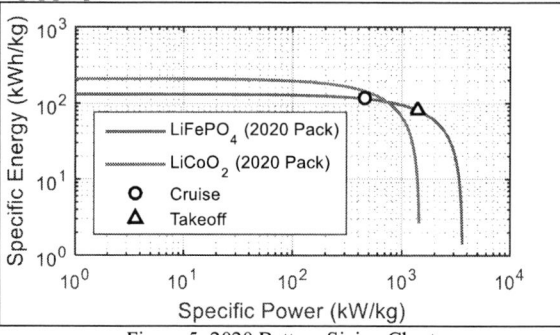

Figure 5: 2020 Battery Sizing Chart

A high specific power was required to meet the takeoff distance requirement, and so the $LiFePO_4$ chemistry was selected. $LiFePO_4$ is also the better choice from an environmental and safety standpoint, and can withstand many more discharge cycles than can $LiCoO_2$ batteries [9], which is critical if the BATWinG system is to be a financial success. Chemistries using lithium titanate in place of the conventional graphite anode combine high power with even better cycle life than the chosen G/LiFePO4, but sacrifice specific energy due to a lower cell voltage, and were rejected on this basis.

The routing/sizing model was varied iteratively to optimize the system with respect to the break-even point and total Bay Area emissions reduction, and the battery pack was sized to match the power and range requirements of the aircraft. Later state of charge simulations were used to verify the suitability of the battery pack for BATWinG operations. Following decisions regarding motor selection and passenger capacity, which are discussed in the next section, the aircraft was sized in accordance with Table 2, which summarizes the final aircraft specifications.

Table 2: Aircraft Specifications

Parameter	Value
Maximum Takeoff Weight	3780 kg
Empty Weight	2400 kg
Battery Pack Weight	460 kg
Battery Pack Capacity	156 Ah
Battery Pack Voltage	384 VDC
Peak Propulsive Power	570 kW
Wingspan	15 m
Wing Area	51 m^2
Cruise Speed	44 m/s
Cruise Range	25 km
Static Margin	4-18%
Cabin Capacity	9 passengers
Fleet Size	52 aircraft

III. CONFIGURATION SELECTION

1. Cabin and Cockpit

With the goal of rapidly exchanging both passengers and battery pack, a configuration was conceived with the pilot positioned above and behind the passenger cabin and the battery located in the nose. This frees up space for large gull wing doors that allow multiple rows of passengers to enter and exit the aircraft at the same time. This configuration also permits passengers to enter at floor level – making the aircraft much more accessible – and the front seats can be folded for wheelchair transportation. The desire to make this configuration work and to use an existing powerplant eventually led to the nine-seat configuration, shown in Figure 6.

Figure 6: Cockpit and Cabin Configuration

From a financial/logistics standpoint, more passengers per aircraft was potentially more profitable, but CG shift issues, motor sizing, and infrastructural considerations limited the cabin capacity of this design.

2. Propulsion System

Based on the historical power loading data and preliminary studies regarding power and passenger capacity, an estimate of the required motor output was made and compared to existing motor options. Since the takeoff power requirement is significantly higher than the power required during cruise, the decision was made to use two Tesla Motors 285 kW induction motors to propel the BATWinG during the takeoff phase, before switching off one of the motors once flight is achieved. Then, the motor can operate near its point of peak efficiency, which is expected to occur between 50-75% of the maximum motor load [11]. An advantage of choosing an induction motor for propulsion is that it is relatively efficient across a wide range of speeds, with a much flatter efficiency curve than that of a typical DC motor.

As shown in Figure 7, the propulsion system was conceived as a pair of motors attached to the same shaft by a set of reduction gears, each fitted with a freewheel hub to allow one or both of the motors to turn the shaft, depending on the scenario.

Figure 7: Dual Motor Concept

The motors would be installed within the centerbody of a 1.9 m ducted fan, where both would provide thrust during takeoff – when maximum power is required – and a single motor would operate during the cruise phase. In the event of a motor failure, the inactive motor could be switched on to complete the route.

The fan's pitchline blade angle was selected as 20 degrees, and a fixed gear ratio was chosen to place the fan's advance ratio at around 0.8 during cruise, ensuring efficient conversion of the motor's output to propulsive thrust [12].

Aiming to operate at about 65% of the motor's peak output of 285 kW, the final gear ratio of 4/7 was chosen to turn the fan at a cruise rotational rate of 1900 rpm. This gear ratio also places the propeller rpm at reasonably efficient values throughout the aircraft's expected speed range. Using momentum theory, the maximum static thrust was determined to be 11,750 N, and a maximum thrust curve was constructed for use in takeoff and state of charge simulations.

A ducted fan was incorporated into the design to reduce noise and intake diameter, and to provide a structure for mounting a shield as protection against waterfowl ingestion. Louvers were added to provide extra pitch authority and to create a more favorable pitching moment during the takeoff phase, increasing the aircraft's angle of attack and expediting the planing process. Small electric trolling motors were included in the wingtip floats for taxiing near BATWinG terminals.

3. Wing and Empennage Design

Preliminary analysis was performed on two potential wing planforms, both inverse delta wings. One was a low aspect ratio Lippisch-style wing, which is common for WIG designs, whereas the other was more modestly tapered, with a higher aspect ratio. Both wings used the Clark-Y airfoil, which is frequently used in WIG applications.

A mid-wing position with respect to the hull/fuselage structure was chosen to place the wing just above the water during the takeoff run. A slight anhedral angle and leading edge sweep were included for stability purposes, and winglets were added to take advantage of endplate effects and to increase the effective aspect ratio of the relatively low aspect ratio wing.

Aerodynamic modelling was done in XFLR5, a software that can apply lifting line, vortex lattice, and 3D panel methods to wing and airfoil analysis. The software has the ability to model ground effects and this feature was used to develop the design. Hybrid flaps/ailerons called flaperons were included to increase the lift generated by the wing during takeoff, and to aid in trimming the aircraft at low speeds.

Figure 8: BATWinG Wing and Empennage

An H-tail design was chosen for the BATWinG, since it offers a rudder area equivalent to a conventional tail, but the vertical tail remains much shorter than it would be in a conventional or T-tail. This is desirable from a structural, operational, and aesthetic viewpoint. Additionally, the vertical tail segments act as endplates that reduce the required area of the horizontal tail, which for a WIG is substantial. 40% of the horizontal and vertical surface areas were allotted to the elevators and rudders to ensure control authority in all flight phases. A summary of basic parameters is shown in Table 3.

Table 3: Wing and Empennage Sizing

Parameter	Value
Wing Airfoil	Clark Y
Wing Area	51 m^2
Wing Span	15 m
Mean Aerodynamic Chord	3.4 m
Wing Leading Edge Sweep	-10°
Dihedral Angle	-2.5°
Tail Airfoil	NACA 0015
Horizontal Tail Area	12 m^2
Horizontal Tail Span	7 m
Vertical Tail Area	6.5 m^2
Vertical Tail Height	2.3 m^2

4. Hull Design

The BATWinG was designed with a stepped hull to decrease resistance during the takeoff run. MATLAB simulations of the takeoff phase led to a reduction of the hull's beam width to 1.2 meters, in order to satisfy the takeoff constraint. Hull volume was selected in alignment with CFR Part 23 regulations for seaplane hull design that dictate that the total float volume should provide a buoyancy force 80% greater than the force required to support the aircraft's weight, and 10 percent of the total float volume was assigned to each of the tip floats in order to provide a righting moment in agreement with these regulations [13].

Figure 9: BATWinG Hull and Floats

5. Operations

Figure 10 shows a conceptual view of a BATWinG terminal. Each time a WIG docks, a ground crew member replaces the battery pack with a charged unit, while passengers exit the right side of the aircraft and load from the left. The dock features a U-shaped opening to accommodate the hull of the aircraft, which allows the majority of the aircraft weight to be supported by buoyant forces while the wings butt up against a pair of stops to provide a stable loading platform. The process is estimated to take up to five minutes, though future tweaks to the system could lead to an automated battery exchange procedure and shorter transfer times.

Figure 10: BATWinG Terminal Concept

6. Materials Selection

The BATWinG is structured like a typical aircraft with an aluminum interior frame. To reduce weight and corrosion concerns, a fiberglass skin is used in low impact areas of the exterior. As shown in Figure 11, aluminum or an aramid/fiberglass blend is used in areas that require greater damage tolerance. Aviation-grade acrylic glass was selected for its low weight and improved thermal properties.

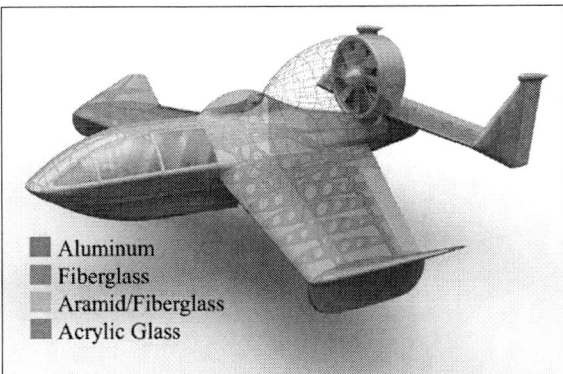

Figure 11: BATWinG Materials

7. Safety Features

To protect passengers from the risks associated with BATWinG travel, several safety features have been incorporated into the design. In the cabin, a firewall encases the battery pack and shields passengers from danger. The LiFePO$_4$ chemistry is a safety feature in its own right, as the fire risk associated with these batteries is low compared to that of other lithium-based cells. A fire extinguisher and escape hammer are located in the rear of the cabin and inflatable life vests are included under every seat. In the event of an emergency, a sea anchor can be deployed from the rear of the hull to stop the aircraft in a hurry. These features are shown in Figure 12.

Figure 12: Cabin Safety Features

Other safety/risk mitigation features include a protective cage around the fan to minimize the risk of foreign object ingestion, tail-mounted millimeter wave radar to see through fog, and an electronic assisted stability/collision avoidance system.

IV. PERFORMANCE VERIFICATION

1. Takeoff and Landing Performance

Figure 13: The BATWinG in Takeoff Mode

A 150 m constraint was placed on the takeoff run for the BATWinG design. This was done to reduce wake pollution in the Bay Area and to increase passenger comfort by rapidly transitioning to the smoother flight phase. The initial takeoff constraint for preliminary design was based on a modified FAR23 takeoff parameter (TOP) that was scaled to account for the greater resistance that accompanies a takeoff from the water. Later on, the takeoff run was modelled according to the methods in [14], which accounted for hull geometry and planing angle. A MATLAB simulation program was written to verify the results of the preliminary analysis. During this process, it was shown to be necessary to reduce the hull's width to 1.2 meters in order to meet the 150 m constraint. A breakdown of the forces involved in during takeoff can be seen in Figure 14.

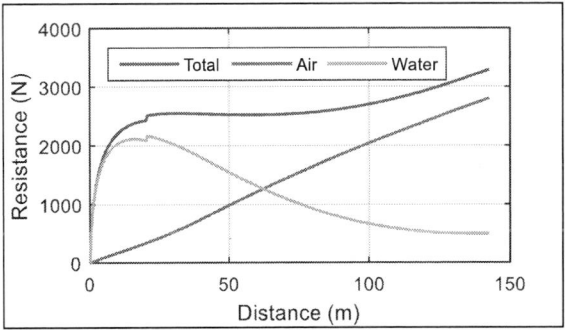

Figure 14: Results of Takeoff Run Simulation

According to the simulation, the BATWinG reaches its takeoff velocity of 25 m/s in 11 seconds for a no-wind condition. A 10 m/s tailwind reduces the velocity of the air flowing over the wings accordingly, and is expected to double the distance required for takeoff. On the other hand, the BATWinG could take off in a shorter distance, flying into a similar headwind.

A simulation was also run for the landing phase. Here, the stepped hull has a negative effect on stopping distance and the BATWinG takes approximately 400 m to stop in an unpowered state, as shown in Figure 15. This is still a relatively short distance and is probably acceptable.

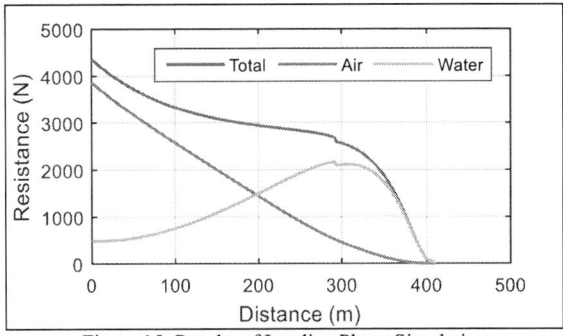

Figure 15: Results of Landing Phase Simulation

It should be noted that, if a shorter landing distance were deemed necessary, a motor could be reversed to apply a thrust that would decelerate the aircraft. Figure 16 shows the result of applying 70% of the cruise power to the task of braking the aircraft.

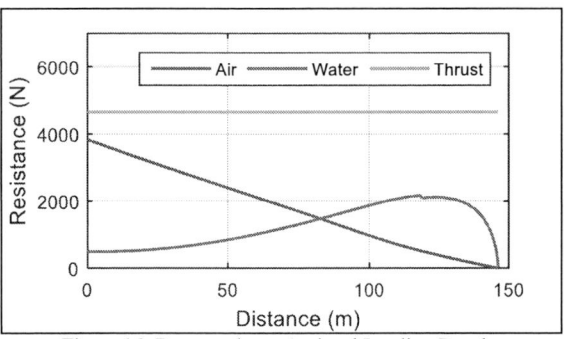

Figure 16: Reverse-thrust Assisted Landing Results

2. Range and State of Charge

The longest of the BATWinG routes – from Embarcadero to San Francisco International Airport – is 25 km. To verify the battery pack sized during the preliminary design process, a MATLAB program was written to simulate the state of charge (SOC) of the aircraft's battery along this route. From the resisting forces during takeoff/landing and the drag forces during flight, the following power profile was constructed for the battery pack as a function of route completion.

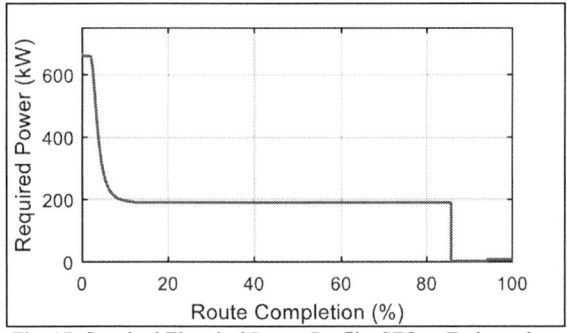

Fig. 17: Standard Electrical Power Profile: SFO to Embarcadero

Takeoff requires over three times the amount of power needed for cruising. For the purposes of this study, 88% of the energy from the battery was assumed to reach the motor, as reported by Tesla for its electric vehicles [15]. Additionally, a constant 6 kW draw was applied to represent auxiliary power usage, and taxiing with the wingtip-mounted trolling motors was estimated to require 5 kW of power [16].

Environmental conditions, especially variations in wind and wave measurements, can have a significant impact on the performance of an electric vehicle. To ensure that the aircraft can complete its routes under expected conditions, simulations included these effects, as shown in Figure 18. Temperature is also a factor, but San Francisco temperatures are mild and fairly consistent [17]. A thermal management system could be used to optimize battery pack performance if conditions demanded it.

It can be seen that headwinds significantly affect the final state of charge of the aircraft. If necessary, aircraft could use

newer batteries on windy days. This wouldn't be required often – based on a survey of the San Francisco weather from 2009-2014 [17], steady wind readings will reach 10 m/s less than one percent of the time, and should have minimal operational impact on the BATWinG transit system.

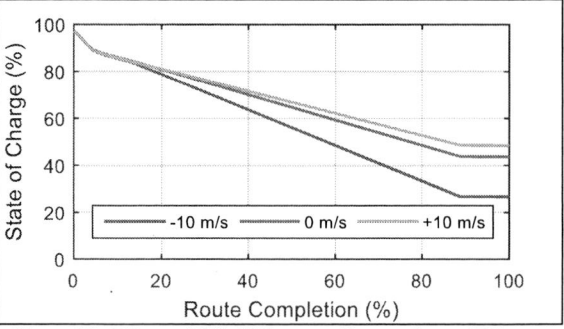

Figure 18: Effects of Wind/Waves on Battery State of Charge

Ground effect vehicles gain their efficiency by reducing the drag due to lift (induced drag), which is a function of an aircraft's span and height above the ground. Figure 19 shows aircraft state of charge performance at the BATWinG's minimum and maximum operating height, as well as out-of-ground-effect performance for reference.

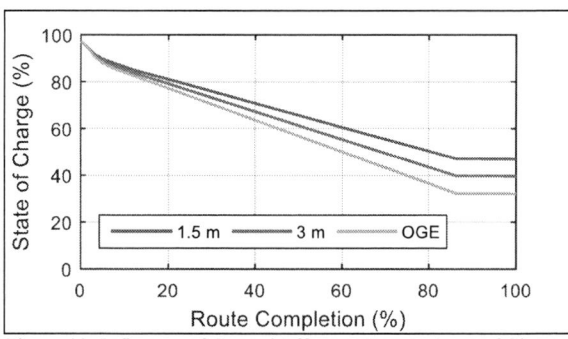

Figure 19: Influence of Ground Effect on Battery State of Charge

Ground-effect operations can be seen to preserve 10-20% of the battery's charge versus operations conducted outside the ground-effect region. To limit pilot-to-pilot variations, maximize efficiency, and retain its classification as a marine vehicle, the aircraft's altitude would be electronically controlled and limited.

BATWinG battery packs will be retired when batteries have degraded to 80% of their original capacity. The average discharge C-rate during BATWinG operations is expected to be between 3C-4C, while cycling to an average depth of 65%. Following the discharge portion of each cycle, high-power chargers will return packs to a fully charged state in under an hour. For this charge-discharge profile, each pack is expected to last about 7000 cycles [9],[18],[19] – approximately one year – before being replaced.

Figure 20 shows the changes in the BATWinG's state of charge profile over that time period.

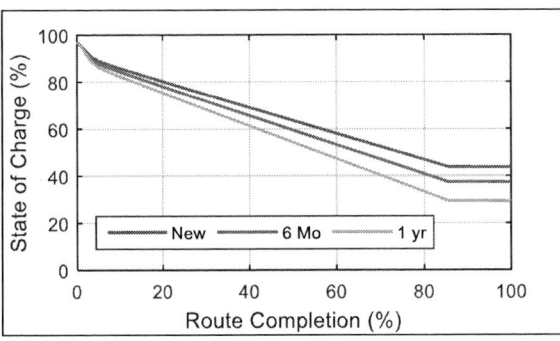

Figure 20: Influence of Battery Degradation on SOC Profile

3. Stability

Pitch stability has historically been a problem for wing-in-ground-effect vehicles. This is because WIG stability differs from the stability of other aircraft in that its neutral point and pitching moment vary with its height in the ground effect region. As the neutral point shifts rearward with a decrease in altitude, a downward pitching moment develops. For a craft operating near the ground, this can be disastrous.

As a response to this issue, the wings were given leading-edge sweep so that the wingtips generate lift first at a point ahead of the wing's aerodynamic center. This counteracts the downward moment that is so concerning. A WIG could also be stable when near the ground, but upon rising up, experience a destabilizing moment in the pitch-up direction – possibly one that cannot be controlled. Automatic stability control will be used to prevent both scenarios, trimming the aircraft according to its position in the ground-effect region. A MATLAB simulation was run to demonstrate the aircraft's trimmability throughout its speed range, in both the loaded and unloaded state. The results are presented in Fig. 21-Fig.23, where α is the aircraft's angle of attack, δ_E is the elevator deflection angle, and δ_F is the flaperon deflection angle.

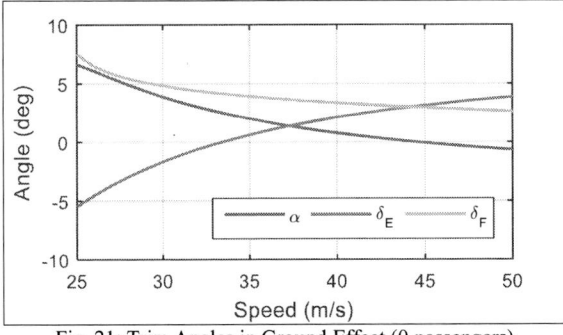

Fig. 21: Trim Angles in Ground Effect (0 passengers)

For reference, Fig.23 shows the trim state of the loaded aircraft out of ground effect, where a decreased need for upward elevator deflection can be seen.

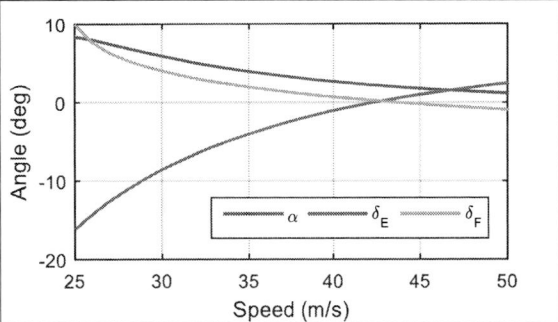

Fig.22: Trim Angles in Ground Effect (9 passengers)

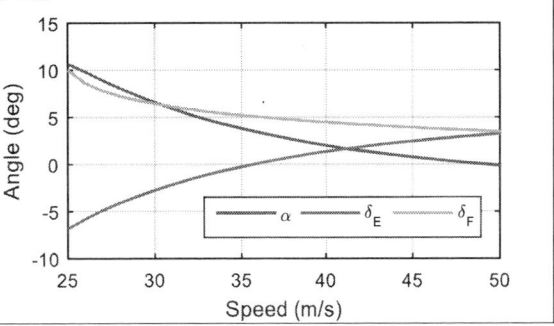

Fig.23: Trim Angles out of Ground Effect (9 passengers)

4. Cost/Revenue

Aircraft acquisition costs were estimated to be $US1.6 million, based on the methods found in Reference [14]. Fixed cost components included the costs of development, materials, manufacturing, testing, and certification. These costs were based on variables such as the weight of the airframe, cruise speed, and aircraft build number. Variable costs were calculated on a per aircraft per day basis. The most significant costs were for labor, taxes, electricity, maintenance, and battery replacements. Revenue was calculated at US$0.55/passenger-km, and did not include the effect of government incentives or the possible resale of battery packs for utilities purposes. Table 4 summarizes the expected cost and revenue in US$, along with some of the operational assumptions that led to these values.

5. Greenhouse Gases

The total reduction in greenhouse gas emissions was calculated by determining the fuel savings due to the BATWinG's effect on traffic congestion. ANL GREET software was used to compute the equivalent emissions output. The well-to-pump emissions associated with the BATWinG's consumption of California-generated electricity, plus the manufacture of 20 generations worth of aircraft batteries (priced at $30,000 apiece) were subtracted from the initial savings total. Emissions generated by manufacturing the vehicles themselves was not included in the analysis.

According to Reference [20], the correlation between traffic congestion and the number of cars on the road is not one-to-one. The model predicts that the BATWinG system will take about 9000 cars off of the San Francisco roadways during rush hour. Based on the effects of randomly removing cars from rush-hour traffic [20], and data from Reference [1], San Francisco's Travel Time Index of 1.41 could be reduced to 1.35, and its national ranking of third by this measure modified to sixth or seventh. After accounting for the rest of the estimated total of 26,000 cars removed from the highways daily, the final savings values are reported in Table 5.

Table 4: Cost and Revenue (Per Aircraft Values)

Parameter	Value
Number of Aircraft	52
Hours of Operation	4 am-midnight
Trips Per Day	81
Battery Pack Energy	60 kWh
Unit Cost of Electricity	US$0.12/kWh
Charging Efficiency	92%

Fixed Costs (US$)	
Acquisition Cost	$1,600,000
Infrastructure Costs	$300,000

Variable Costs and Revenue (US$)	
Battery Replacement Cost	$660/day
Cost of Electricity	$500/day
Cost of Maintenance	$120/day
Labor Costs	$1700/day
Other	$290/day
Taxes	$1,400/day
Revenue	$9,600/day
Time to Break Even	**< 2 years**

Table 5: BATWinG Effects on Greenhouse Gases

Source	Annual Value
Reduced Congestion	-90 million kg
Reduced Car Usage	-37 million kg
BATWinG Operations	+27 million kg
Total	-100 million kg

It can be seen that the BATWinG system would reduce San Francisco area greenhouse gas emissions by 100 million kg. The model also predicts commuter savings of $400 million dollars, or about $300 per commuter annually.

V. DISCUSSION AND FUTURE WORK

Because the BART portion of each route was generally the largest cost component, a wide range of BATWinG fares were possible. With more data on passenger acceptance of total route costs and/or collaboration with BART, rates could be optimized further.

Future studies would include experimental work to validate model assumptions, especially predictions regarding battery performance, and improvement to the model to better account for passenger volume fluctuations throughout the day. More work is needed to fully evaluate the logistics of such a system operating in the bay – including grid loading effects – and a battery recycling plan should be developed to maximize the BATWinG's financial and environmental benefits. A similar model could be applied world-wide wherever an electrified population is centered around a body of water with a low sea state. As an illustration of global potential, Figure 24 combines 90th percentile significant wave heights with a nighttime visualization of population density.

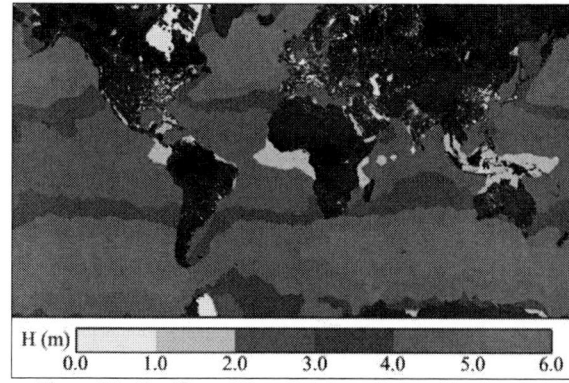

Figure 24: Significant Wave Height and Electrification [21]

Areas identified for future examination include Hong Kong, Incheon, and the Persian Gulf.

VI. CONCLUSION

The BATWinG air transit concept was explored via conceptual design and MATLAB modelling. It was found to perform well with respect to the stated goals of the study. The system was demonstrated to be profitable – an important factor in turning concept into reality – and was predicted to be effective in alleviating traffic congestion in the San Francisco Bay Area, reducing total greenhouse gas emissions, and saving commuters both time and money.

ACKNOWLEDGMENT

The author would like to thank Dr. Ron Barrett and Richard Bramlette for guidance on this project, Dhruv Chawla, Vidyasagar Jaju, Jeevan Teja Kolli, Ankur Patil, and Lauren Schumacher for their input to the design, and Cody Hill for the BATWinG acronym.

REFERENCES

[1] D. Schrank, B. Eisele, T. Lomax, J. Bak, "2015 Urban Mobility Scorecard", Texas A&M Transportation Institute, 2015.

[2] IMF World Economic Outlook Database. 2015. http://www.imf.org/external/pubs/ft/weo/2015/01/weodata/index.aspx

[3] Alternative Fuels Data Center, U.S. Dept. of Energy. 2015. http://www.afdc.energy.gov/vehicles/electric_emissions.php

[4] B. McKenzie, "Who Drives to Work? Commuting by Automobile in the United States: 2013", U.S. Census Bureau, 2015.

[5] United States Coast Guard. "Wing-In-Ground (WIG) Craft Interim Guidance." United States Coast Guard Navigation Center, 2001.

[6] International Maritime Organization, 2015. http://www.imo.org/en/OurWork/Safety/Regulations/Pages/WIG.aspx

[7] BART, 2015. http://www.bart.gov/about/reports/ridership

[8] D. Raymer, "Aircraft Design: A Conceptual Approach, 4th Edition", AIAA, 2012.

[9] J. Amirault, J. et al, "The Electric Vehicle Battery Landscape: Opportunities and Challenges," Center for Entrepreneurship & Technology, University of California, Berkeley, 2009.

[10] W. Bernhart, "The Li-Ion Battery Value Chain – Trends and Implications," Roland Berger Strategy Consultants, 2011.

[11] M. J. Melfi, "Motors, efficiency, and adjustable-speed drives," Power Electronics, 2011.

[12] B.W. McCormick, "Aerodynamics, Aeronautics, and Flight Mechanics", Wiley, New York, 1979.

[13] Code of Federal Regulations, "CFR 23.751 - Main Float Buoyancy", U.S. Government Publishing Office, 2011.

[14] S. Gudmundsson, "General Aviation Aircraft Design: Applied Methods and Procedures", AIAA, 2012.

[15] Tesla Motors, 2015. http://my.teslamotors.com/roadster/technology

[16] Torqeedo GmbH, 2015. http://www.torqeedo.com/en/products/outboards/cruise

[17] Weather Underground, 2015. http://www.wunderground.com/history

[18] J. Groot, M. Swierczynski, A.I. Stan, S.K. Kær, "On the Complex Ageing Characteristics of High-Power LiFePO4/Graphite Battery Cells Cycled with High Charge and Discharge Currents," J. Power Sources, 286, pp. 475-487, 2015.

[19] J. Wang, P. Liu, J. Hicks-Garner, E. Sherman, S. Soukiazian, M. Verbrugge, H. Tataria, J. Musser, P. Finamore, "Cycle-life Model for Graphite-LiFePO4 Cells", J. Power Sources 196, 3942-3948, 2011.

[20] P. Wang, T. Hunter, A. M. Bayen, K. Schechtner, M. Gonzalez, "Understanding Road Usage Patterns in Urban Areas", Scientific Reports, Article number: 1001, 2012.

[21] G. Holland, I. Young, "Atlas of the Oceans: Wind and wave Climate", Elsevier/Pergamon, 1996.

An ESS Charge Balancing Method Based on Current Allocation with Multi-source Power Converters for Electric Microcars

Y. C. Fong[1]   K. W. Eric Cheng[2]

[1] Department of Electrical Engineering, The Hong Kong Polytechnic University, Hong Kong
E-mail: yc-chi.fong@connect.polyu.hk
[2] Department of Electrical Engineering, The Hong Kong Polytechnic University, Hong Kong
E-mail: eeecheng@polyu.edu.hk

Abstract– Batteries are one of the most costly components in electric vehicles (EVs) yet they also limit the performance and lifespan of the vehicles. Voltage imbalance between series connected battery cells worsens the limitations but is inevitable due to the manufacturing tolerance of capacity and deviations of environmental and batteries' parameters. Industries have developed a wide range of solutions to keep cell voltage equalized including adding balancing circuits and specialized chargers to improve the battery capacity utilization and to extend the lifespan of whole battery pack. However, these methods are generally ineffective for the energy storage systems (ESSs) in EVs. Instead of developing circuitries for cell balancing, this study would like to investigate the effectiveness of battery charge balancing by load/charge allocation with multi-source power converters such as multilevel voltage source inverters (VSIs) that directly connect the batteries in ESSs as sources. A simulation model with an 11-level cascaded H-bridge inverter (CHBI) was employed to examine the proposed switching scheme. The simulation result is encouraging that voltage sources were balanced by utilizing a significant amount of the load/charging current as balancing current; the sources attained equal voltage within a short period in simulation environment.

Keywords–Charge equalization, multi-source power converter, multilevel VSI, Staircase modulation

I. INTRODUCTION

No matter in the types of pure electric vehicles or hybrid EVs (HEVs), most of the present electrified automobiles employ electrochemical energy storage elements as the main supply or energy buffer to provide electricity for their propulsion system. Since the electrochemical system voltage, for example, about 3.6V for lithium-based and 1.2V for nickel-based based batteries, is inadequate to supply the DC link voltage for propulsion applications, dozens or hundreds of battery cells are stacked together to form an ESS in the powertrain. Series connection of battery cells enables the capability of high voltage applications. However, this leads to charge imbalance along the series-connected battery cells in the stack which could bring about the issues of battery life degradation, reduction in available energy capacity as well as safety concerns. The lifespans of battery EVs (BEVs) are mainly limited by the costly batteries yet series connection has worsened the concern. Although electrochemical energy storage technology has been adopted in the industries for decades and battery performance is noticeably improved,

voltage imbalance between battery cells is inevitable. There is a wide range of solutions developed in industries to tackle the imbalance issues including online methods like adding electronics into the ESS for voltage equalization by the means of dissipation or charge exchange [1-3] and offline methods that require specified chargers or external voltage equalizing facilities. The effectiveness of these solutions is dissatisfying in the point of views of energy efficiency, balancing speed or cost. In an ESS, voltage unbalance between cells is principally a representation of unequal state-of-charge (SoC) ratio and, to some degree, variation of environmental and conditional parameters like temperature, duty and age between different cells. [4-5] Since the battery manufacturers allow few percent of capacity tolerance while most of the power drives and charger only utilize the most positive and negative terminals of the whole battery stack, the variation in capacity between cells becomes the main cause of voltage unbalance.

Unlike simple half-bridge, H-bridge inverters and the capacitive voltage divider, or charge-pump based multilevel inverters [6-7] that only cooperate with a single voltage source, power converters connecting multiple sources like particular multilevel VSIs can manipulate individual voltage source engaged in these topologies by allocating tailored loading cycles. By determining an appropriate loading allocation or priority of individual battery modules or cells, the multi-source converters permit battery voltage balancing through the discharging process when the EV is travelling. By adopting suitable switching schemes to allocate different average charging current between energy sources, the same concept can also be applied to the charging process of EVs.

This study mainly focuses on the charge/discharge allocation among individual energy storage elements in multilevel VSIs and investigates the relationship between the batteries' charge, switching states and firing angles, as well as the feasibility of using multilevel VSIs to perform equalization charging. Cascaded H-bridge multilevel inverter (CHBI) is one of the multilevel VSI topologies that have high flexibility and can provide abundant redundant switching states. This topology was selected as an example to illustrate the operation principle of the proposed current allocation method. A brief introduction to this topology and how the switching scheme utilizes the redundant switching states of CHBI are also included. An 11-level CHBI cooperating with current allocation was simulated to undergo charge/discharge processes with typical loading level of an electric microcar, the simulation result is presented in the last part of this paper.

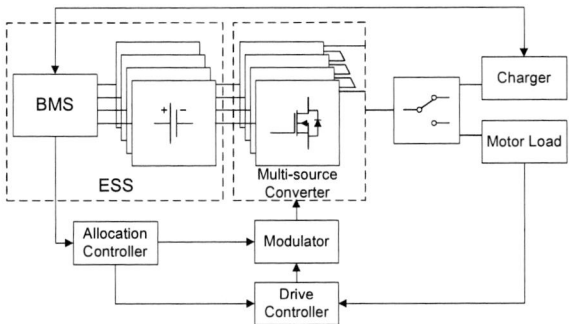

Fig. 1: Simplified functional block diagram for the load/charge allocation method

II. SWITCHING STATES OF CASCADED H-BRIDGE MULTILEVEL INVERTER (CHBI)

CHBI (Fig. 2) is one of the multilevel VSI topologies which provide highest number of redundant switching states. Although this topology requires a relative large number of active switches, which is *2(n-1)* for an n-level VSI, the voltage stress of each switch is limited to the corresponding single source level. CHBI is still one of favorable VSI topologies due its possibility of modularization with low voltage switches and comparably high fault tolerance [8]. A discrete power metal-oxide-semiconductor field-effect transistor (MOSFET) device with on-state resistance of less than 1 mini-ohm withstanding 60V for low voltage applications is presently available in the market. Unlike insulated-gate bipolar transistors (IGBTs), the on-state conduction loss of MOSFETs can be significantly reduced by parallel arrangement. Allowing the use of low voltage switching devices, such as MOSFETs, can potentially enable more efficient design in the future especially for low power vehicles such as electric microcars. At the same time, staircase modulation permits low switching frequency which reduces switching loss and electromagnetic interference while the output level and power quality is mainly dependent on the source voltage, number of output steps and the corresponding firing angles. This allows the controller to predict the VSI output parameters in real-time manner and manipulate the output and input accordingly.

Since CHBI cooperates with isolated voltage sources, it is possible to arrange series and anti-series connections between energy sources. For an n-level CHBI, the numbers of voltage sources, active switches and the total number of valid switching states are (1), (2) and (3) respectively.

$$n_l = \frac{n-1}{2} \tag{1}$$
$$n_{sw} = 2(n-1) \tag{2}$$
$$S = 2^{n-1} \tag{3}$$

Assuming the voltage sources of the n-level VSI are at balanced condition, the number of possible switching states for a k-level output can be represented as (4);

$$S_k = \sum_{i=k}^{\left\lceil \frac{n-1}{4} \right\rceil} \left(C_i^{\frac{n-1}{2}} \cdot C_{i-k}^{\frac{n-1}{2}-i} \cdot 2^{\frac{n-1}{2}-2i+k} \right) \tag{4}$$

neglecting the diversity of by-passing states and anti-series connection, CHBI can still provide (5) and (6) options for different source allocation respectively.

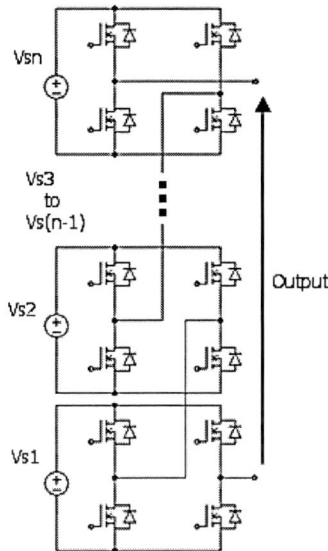

Fig. 2: Topology of a 2n+1-level CHBI

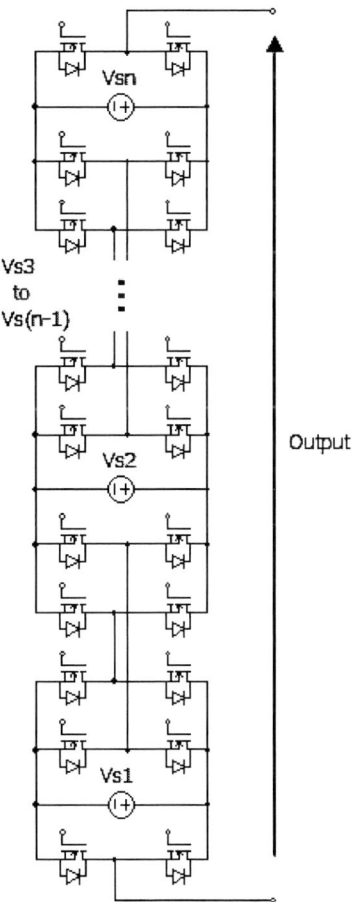

Fig. 3: Topology of a 2n+1-level M2SPC

$$S_k = \sum_{i=k}^{\left\lfloor \frac{n-1}{4} \right\rfloor} \left(C_i^{\frac{n-1}{2}} \cdot C_{i-k}^{\frac{n-1}{2}-i} \right) \tag{5}$$

$$S_k = C_k^{\frac{n-1}{2}} \tag{6}$$

This shows that anti-series connection of isolated voltage sources in CHBI permits not only simultaneous occurrence of charge and discharge among cells, but also higher flexibility in charge allocation by offering additional possible switching states.

Besides, there are some multilevel VSI topologies which allow more combinations of connection between voltage sources. For example, Goetz et al. [9] developed a novel modular multilevel series/parallel converter (M2SPC) (Fig. 3) which permits parallelization between adjacent sources by adding auxiliary bridges over CHBI. This design is found to have the capability of loss reduction by sharing current of the conducting batteries and MOSFETs and providing additional balancing current by direct interface between energy storage elements.

III. CURRENT ALLOCATION IN CHBI DURING DISCHARGING WITH STAIRCASE MODULATION

In view of the output current waveform of a CHBI with a fundamental amplitude I_a, phase angle φ and an angular frequency of ω; in order to simply the charge analysis, this study was based on an assumption that source level allocation is symmetrical along $\theta = \frac{\pi}{2}$; and switching pattern at positive and negative cycles are the same; the relationship between the amount of discharge in voltage sources and the corresponding firing angles θ_k during half a cycle is found to be

$$Q_{+ve,k} = I_a \int_{\frac{\theta_k}{\omega}}^{\frac{\pi-\theta_k}{\omega}} \sin\left(\omega t + \frac{\varphi}{\omega}\right) dt$$

$$Q_{+ve,k} = \frac{I_a}{\omega} [\cos(\theta_k - \varphi) + \cos(\theta_k + \varphi)] \tag{7}$$

while the average current is

$$I_{+ve,k} = \frac{I_a}{\pi} [\cos(\theta_k - \varphi) + \cos(\theta_k + \varphi)] \tag{8}$$

As $\theta_1, \theta_2, \theta_3, \ldots, \theta_n < \frac{\pi}{2}$ and $\theta_1 < \theta_2 < \theta_3 < \ldots < \theta_n$, whereas $\varphi < \frac{\pi}{2}$ under normal loading condition. It is reasonable to assume that $Q_{+ve,1} > Q_{+ve,2} > Q_{+ve,3} > \cdots > Q_{+ve,n}$ and $I_{+ve,1} > I_{+ve,2} > I_{+ve,3} > \cdots > I_{+ve,n}$.

Considering a loading power factor of 0.85 while $\theta_1 = 0.02\,\pi$ and $\theta_n = 0.45\,\pi$. Substituting these parameters into (8), the difference between $I_{+ve,1}$ and $I_{+ve,n}$ could be as large as $0.455 I_a$, which is equivalent to a balancing current of about $0.22 I_a$ sourcing from the 1st level and sinking to the nth level.

In addition, if anti-series switching states are employed, the amount of discharge from the anti-series connected sources can be diversified to (9) by allowing a switching pattern of reversely connecting the source k at θ_{k1}, then

by-passing at θ_{k2} and followed by forward connection at θ_{k3}

$$Q_{+ve,k} = I_a \left[\int_{\frac{\theta_{k3}}{\omega}}^{\frac{\pi-\theta_{k3}}{\omega}} \sin\left(\omega t + \frac{\varphi}{\omega}\right) dt \right. $$
$$\left. -2 \int_{\frac{\theta_{k1}}{\omega}}^{\frac{\theta_{k2}}{\omega}} \sin\left(\omega t + \frac{\varphi}{\omega}\right) dt \right]$$

$$Q_{+ve,k} = \frac{I_a}{\omega} [2\cos(\theta_{k2} - \varphi) - 2\cos(\theta_{k1} - \varphi)$$
$$+ \cos(\theta_{k3} - \varphi) + \cos(\theta_{k3} + \varphi)] \tag{9}$$

where θ_{k1}, θ_{k2} and $\theta_{k2} < \frac{\pi}{2}$. This implies that the controller can further increase the amount of discharge by setting $\theta_{k2} \leq \varphi$ or decrease the amount by setting $\theta_{k1} \geq \varphi$ to boost the equivalent balancing current.

IV. CURRENT ALLOCATION IN CHBI DURING CHARGING

CHBIs allow three modes of operation, the voltage sources can be connected to the charger forwardly, reversely or some of the sources are by-passed. With these possibilities, the most straight forward way to perform equalization charging is by-passing the charging current of sources with higher percentage SoC. The simulation model in this study employed "n-1" charging scheme that the maximum bus voltage of the charger is limited to the total voltage of four out of the five sources in the 11-level CHBI. This charge equalization was realized by by-passing the voltage source with the highest percentage SoC. The idea of source allocation can also be applied to dynamic number of by-passed sources as long as the charger allows constant current mode or the output inductive filter is large enough. In addition, pulse charging can be implemented with the active switches in multi-source converters to increase the charging speed and to extend batteries' lifespan [10-11]. The multilevel VSIs can be combined with constant current chargers and performs as current pulse generator so that pulse width or density is allocated according to the battery voltage or SoC. This primes an equivalent balancing current of up to half of the charging current between two sub-modules.

V. SIMULATION RESULT OF THE PROPOSED SWITCHING SCHEME

Since the main aim of this simulation was to examine the effectiveness of the proposed voltage balancing switching scheme, a simulation model with a simplified open-loop system regarding on the output amplitude and generating constant firing angles was implemented in MATLAB Simulink; the switching scheme only allocates the output priority of voltage sources according to their percentage SoC so that the voltage sources with higher SoC percent would undergo a large amount of discharge. An 11-level single phase CHBI with five lithium battery sources of nominal module voltage of about 37V having unequal capacity and initial SoC was employed.

1. Discharging Stage

During the discharging stage, an inductive load of about 2500W with a power factor of 0.92 was applied to the inverter; pre-calculated firing angles based on staircase modulation were used to synthesize a 50Hz, 11-level modified sinusoidal voltage output with modulation index 0.74. The initial SoC and capacities of the battery modules were defined as 80±4% and 200Ah ± 10% respectively. A firing angle arrangement logic based on sorting SoC was implemented for balancing discharge (Fig. 4). The SoC and voltage of each of the voltage sources during the discharge process of 2.5 hours were recorded. Fig. 5 and 6 show the output waveform of the CHBI and current waveforms of two unbalanced voltage sources. The percentage SoC and the corresponding terminal voltage values during discharge where depicted in Fig. 7. The simulation result indicates that the SoC of the five sources were basically equalized when they were discharged to around 70%, which implies the inverter utilized around 40% of average loading current as balancing current.

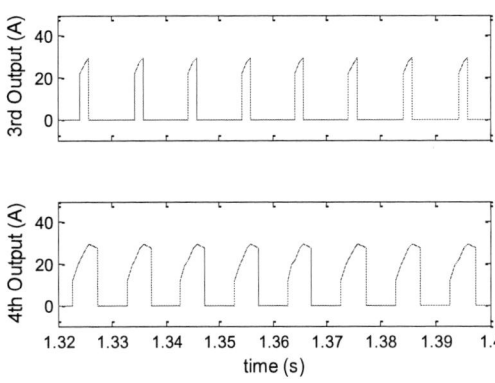

Fig. 6: The output current waveform of voltage sources having different SoC values (upper: the 3rd voltage source; lower: the 4th voltage source)

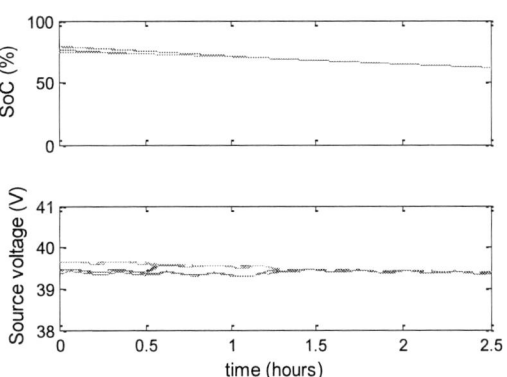

Fig. 7: The SoC and corresponding battery voltage during discharge

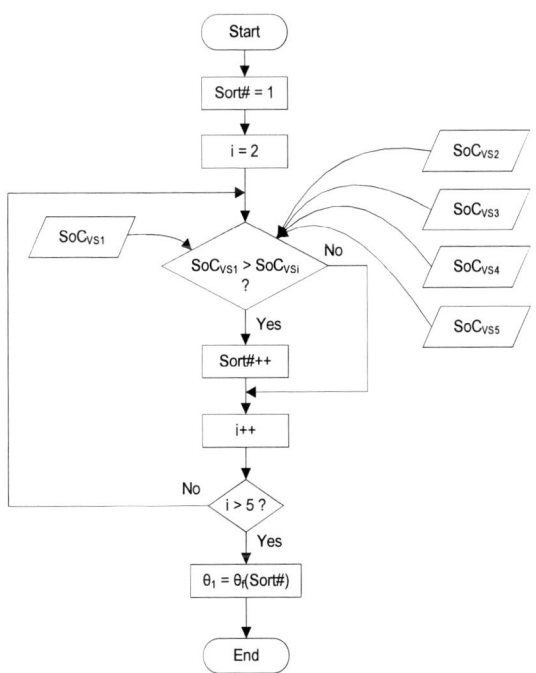

Fig. 4: The firing angle allocation logic for the H-bridge discharging the 1st voltage source in the simulated CHBI model

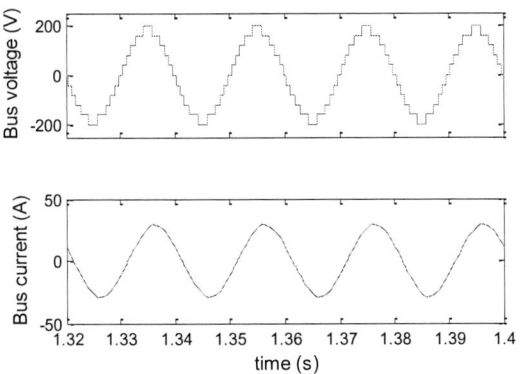

Fig. 5: The bus voltage and current waveform of the CHBI during discharge

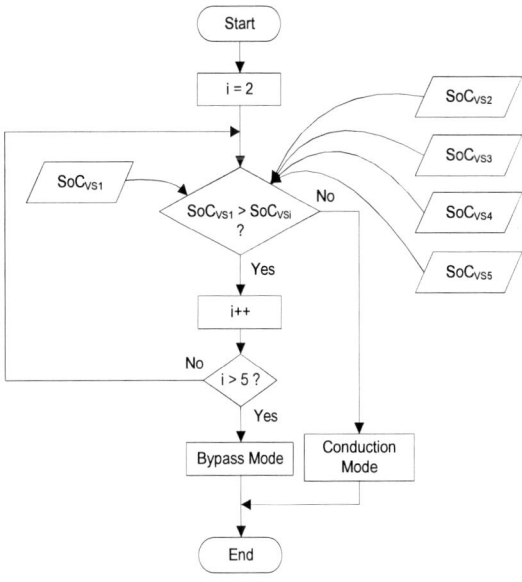

Fig. 8: The charging current allocation logic for the H-bridge charging the 1st voltage source in the simulated CHBI model

978-1-5090-0064-7/15 $31.00 © 2015 IEEE

Table 1: Comparison of different charge balancing techniques

Balancing Techniques	Balancing Current	Limitations
Current allocation	Eq. up to 50% of charging/loading current	- performs balancing only during load/charge condition
Onboard active balancers	< 5A	- high cost and complicated electronics
Onboard passive balancer	< 1A	- balances charge through dissipation
Offline balancing chargers	< 20A	- high cost and complicated charging facilities

2. Charging Stage

The same CHBI model was simulated to undergo a charging process. In this case, the initial SoC values were $10\pm4\%$ while the charger was set at constant current of 50A and constant voltage of 168V. At a sampling rate of 1sps and status updating rate of 1Hz, the voltage source module with highest SoC percentage was by-passed while the remainders were conducted (Fig. 8). The associated results are plotted in Fig. 9-11. The SoC of the voltage sources became balanced when they were charged to 25%, and then all battery modules underwent pulse charging after reaching equalization stage and charged to 100% at an equal rate.

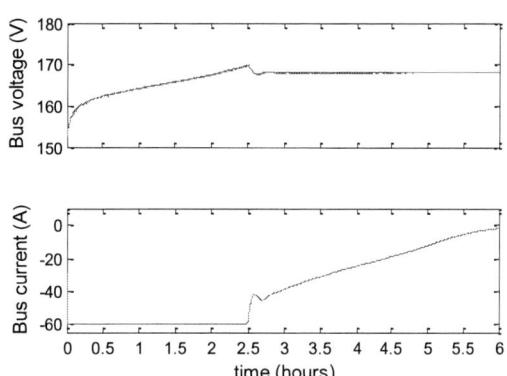

Fig. 9: The bus voltage and current waveform of the CHBI during charge

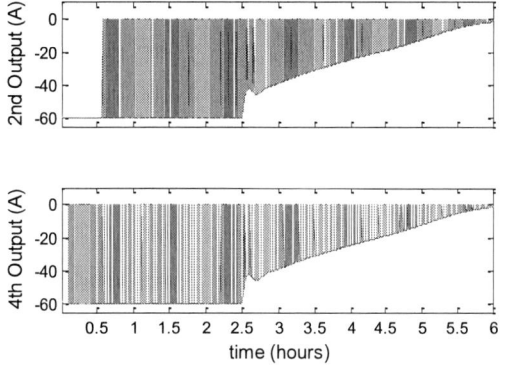

Fig. 10: The charging current of voltage sources having different initial SoC and capacity (upper: the 2nd voltage source; lower: the 4th voltage source)

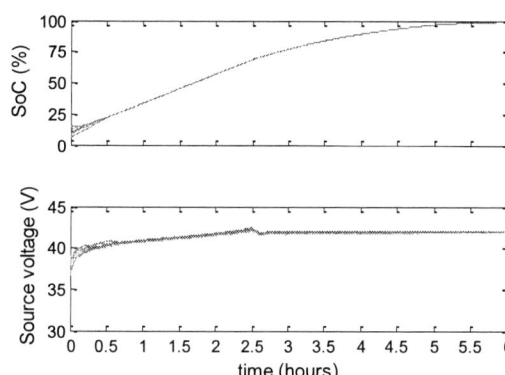

Fig. 11: The SoC and corresponding battery voltage during the charge period

VI. CONCLUSION

Cooperating with multi-source power converters like CHBIs, the proposed method facilitates charge equalization of EV's ESS by current allocation. In addition to staircase modulation, the same technique could also be extended to pulse-width modulation (PWM) or pulse-skipping modulation (PSM) given that the converter topologies permit diverse current allocation among voltage sources. The simulation result shows that the proposed equalization method allows high speed battery balancing and can keep batteries at balanced condition along the remaining charging and discharging cycle. Although 10-cell battery modules were employed to simplify the simulation model, this method can be extended to cell level to allow charge equalization between individual cells in the ESS or vice versa to larger scale for high power applications in which a rough equalization is tolerable. Along with the development of more compact, lightweight and more energy efficient semiconductor switches, multi-source power converters including multilevel VSIs could be implemented for smaller voltage levels to achieve lower output distortion and finer charge equalization between energy storage elements via suitable charge allocation schemes. Current allocation would be a promising method to enhance utilization factor of battery capacity as well as to extend the lifespan and reduce the maintenance cost of BEVs in the near future.

REFERENCES

[1] Ye Y.; Cheng, K.W.E.; Yeung, Y.P.B., "Zero-Current Switching Switched-Capacitor Zero-Voltage-Gap Automatic Equalization System for Series Battery String," in Power Electronics, *IEEE Trans. Power. Electron.*, vol.27, no.7, pp.3234-3242, July 2012

[2] Raman, S.R.; Xue, X.D.; Cheng, K.W.E., "Review of charge equalization schemes for Li-ion battery and super-capacitor energy storage systems," *International Conf. Advances in Electronics, Computers and Communication*, vol., no., pp.1-6, 10-11 Oct. 2014

[3] Ye, Y.; Cheng, K.W.E., "Modeling and Analysis of Series–Parallel Switched-Capacitor Voltage Equalizer for Battery/Supercapacitor Strings," *IEEE J. Emerg. Sel. Topics Power Electron.*, vol.3, no.4, pp.977-983, Dec. 2015

[4] Szumanowski, A.; Y. Chang, "Battery Management System Based on Battery Nonlinear Dynamics Modeling *IEEE Trans. Veh. Technol.*, vol.57, no.3, pp.1425-1432, May 2008

[5] Bhide, S.; Taehyun S., "Novel Predictive Electric Li-Ion Battery Model Incorporating Thermal and Rate Factor Effects," *IEEE Trans. Veh. Technol.*, vol.60, no.3, pp.819-829, March 2011

[6] Z. P. Fang, "A generalized multilevel inverter topology with self voltage balancing,", *IEEE Trans. Ind. Applicat.*, vol.37, no.2, pp.611-618, Mar./Apr. 2001

[7] Ceglia, G.; Guzman, V.; Sanchez, C.; Ibanez, F.; Walter, J.; Gimenez, M.I., "A New Simplified Multilevel Inverter Topology for DC/AC Conversion," *IEEE Trans. Power. Electron.*, vol.21, no.5, pp.1311-1319, Sept. 2006

[8] Rodriguez, J.; Hammond, P.W.; Pontt, J.; Musalem, R.; Lezana, P.; Escobar, M.J., "Operation of a Medium-Voltage Drive Under Faulty Conditions", *IEEE Trans. Power. Electron.*, vol.52, no.4, pp.1080-1085, Aug. 2005

[9] Goetz, S.M.; Peterchev, A.V.; Weyh, T., "Modular Multilevel Converter With Series and Parallel Module Connectivity: Topology and Control," *IEEE Trans. Power Electron.*, vol.30, no.1, pp.203-215, Jan. 2015

[10] L. R. Chen, "A Design of an Optimal Battery Pulse Charge System by Frequency-Varied Technique," *IEEE Trans. Power. Electron.*, vol.54, no.1, pp.398-405, Feb. 2007

[11] Savoye, F.; Venet, P.; Millet, M.; Groot, J., "Impact of Periodic Current Pulses on Li-Ion Battery Performance,", *IEEE Trans. Ind. Electron.*, vol.59, no.9, pp.3481-3488, Sept. 2012

Bootstrap Gate Driver and Output Filter of An SC-based Multilevel Inverter for Aircraft APU

Yuanmao YE ,K.W.Eric CHENG

Department of Electrical Engineering, The Hong Kong Polytechnic University, Hong Kong
yuanmao.ye@connect.polyu.hk eeecheng@polyu.edu.hk

Abstract–The objective of this paper is to propose a gate drive circuit and an output filter of a switched-capacitor–based multilevel inverter for aircraft APU. With the bootstrap methodology, only one voltage source is required to power the gate driver of all switches used in the multilevel inverter. With the LC filter, this inverter is capable of providing a pure sinusoidal output voltage waveform. Finally, the performance of the proposed multilevel inverter is evaluated with simulation results and experimental results of an eleven-level prototype inverter.

Keywords–Multilevel inverter, switched-capacitor, bootstrap capacitor driver, sinusoidal PWM.

I. INTRODUCTION

Aircrafts requires an auxiliary power unit (APU) to produce high frequency alternating current, usually 400 Hz. To obtain an output waveform as much as sinusoidal shape, multilevel inverter technique has been an alternative of conventional 2-level inverter. It is well known that the more the levels of an inverter, the more near sinusoidal its output voltage is. It also means the more power semiconductors and voltage sources or capacitors are required. Consequently, one of the key technologies for multilevel inverters is to use less components and simpler structures to obtain the more levels of output voltages.

The conventional multilevel inverters can be divided into three categories [1]: neutral-point-clamped [2], flying capacitors [3], and the H-bridge cascade [4]. One of their common drawbacks is that an excessive number of power semiconductor switches and capacitor sources employed that leads to the complex structure and higher power loss.

In the literature [5], a novel multilevel inverter is presented for high frequency applications. It is made up of a novel DC-DC multilevel converter and an H-bridge as shown in Fig.1. The key point of this inverter is the DC-DC conversion section which consists of multiple switched-capacitor (SC) cells. Each cell employs only one capacitor, one active switch and two diodes. The number of $n-1$ SC cells can compose an n-level DC-DC converter. They are connected to an H-bridge, a $(2n+1)$-level inverter can be easily derived. The structure is very simple and fewer components are required.

In order to promote this novel multilevel inverter for industrial applications, a simple gate drive circuit is developed by using bootstrap technique in this paper. It means that only one power supply is required to power the gate drive circuit for all switches employed in this inverter. This design philosophy contributes the small size and cost-effectiveness of the inverter.

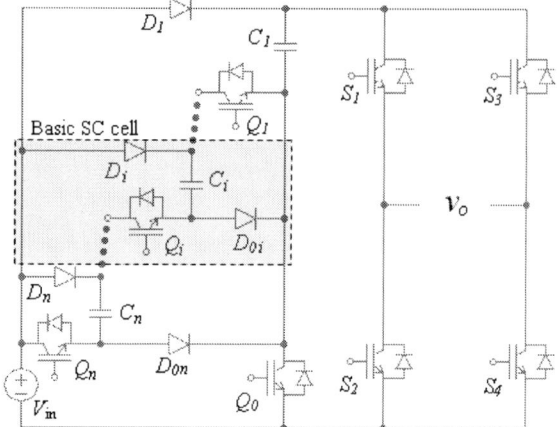

Fig.1: The multilevel inverter presented in [5]

To develop a pure sinusoidal output voltage waveform, an LC filter is added on the output terminal of this multilevel inverter in this paper.

Both simulation and experimental results of a seven-level inverter prototype are provided to evaluate the performance of the inverter.

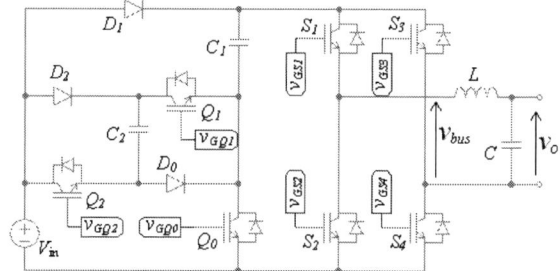

Fig.2: The seven-level topology of the multilevel inverter

II. CIRCUIT DESCRIPTION AND STATES ANALYSIS

1. Circuit Description

Fig.2 shows the topology of proposed inverter in seven levels. It is composed of a three-level DC-DC converter, a full bridge and an output low-pass filter. As mentioned before, the key point of the seven-level inverter is the section of DC-DC converter which consists of three active switches Q_0, Q_1 and Q_2, three diodes D_1, D_2 and D_{02}, and two capacitors C_1 and C_2. With different control strategies for the three active switches, the DC-DC conversion section is capable of converting the input voltage V_{in} in different levels, including $3V_{in}$, $2V_{in}$ and V_{in}. Like many other multilevel inverters aforementioned, the proposed inverter also includes an inverter bridge which employs four active switches S_1~S_4, and an output LC filter used for filtering higher harmonics.

2. States analysis

As mentioned before, with different control strategies, the circuit section of multilevel DC-DC converter of the proposed inverter is capable of converting the input voltage V_{in} in different levels. For the seven-level inverter as shown in Fig.2, there are three levels that can be produced by the multilevel converter section, including the levels of V_{in}, $2V_{in}$ and $3V_{in}$. With the combination of the operation of the inverter bridge, the inverter can provide seven levels of voltage: $3V_{in}$, $2V_{in}$, V_{in}, 0, -V_{in}, -$2V_{in}$ and -$3V_{in}$.

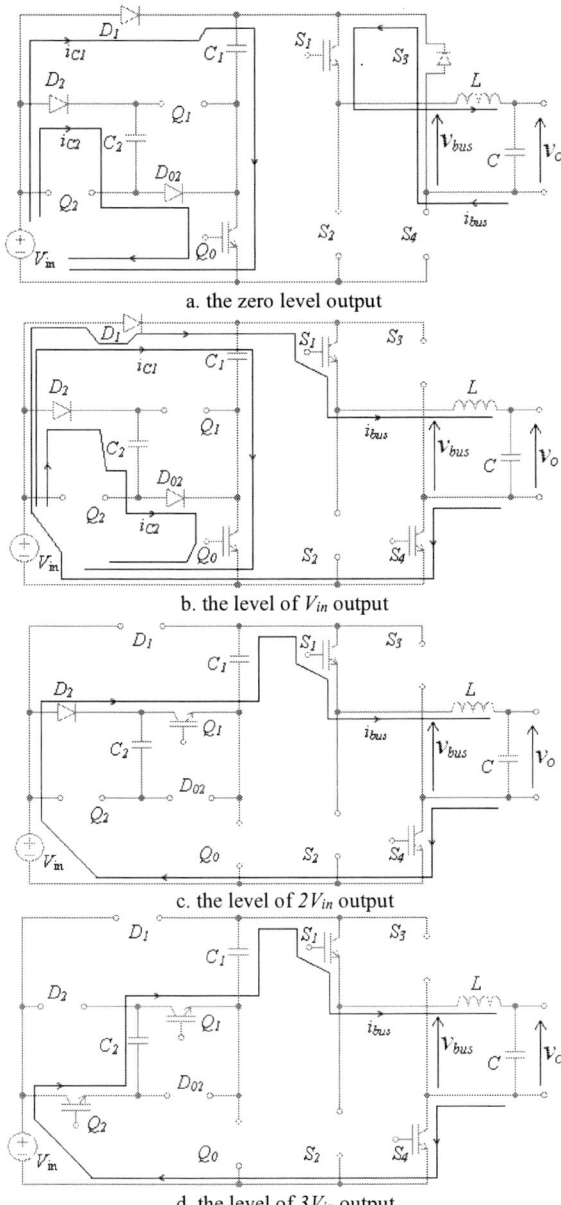

a. the zero level output

b. the level of V_{in} output

c. the level of $2V_{in}$ output

d. the level of $3V_{in}$ output

Fig.3: Working states for the proposed inverter

Specifically, when the switch Q_0 is turned ON and Q_1 and Q_2 being OFF, V_{in}, D_1, C_1 and Q_0 form a closed loop and C_1 is charged by input power V_{in}. And another closed loop is formed by V_{in}, D_2, C_2, D_{02} and Q_0, and C_2 is also

charged by V_{in}. In this case, when the switch S_1 is turned ON and other switches in the H-bridge maintains the OFF state as shown in Fig.3a, the output bus voltage v_{bus} is equal to 0. But if the switch S_4 is turned ON as well as shown in Fig.3b, the bus voltage will change to V_{in}. Of course, when the switch S_2 is turned ON, and S_1 and S_4 maintain OFF state, another 0 level and $-V_{in}$ could be produced by being OFF or ON of S_3.

In the DC-DC conversion section, when the switch Q_1 is turned ON and other switches are OFF, the capacitor C_1 is connected in series with input power V_{in} through Q_1 and D_2 and the DC-DC converter section output the voltage level of $V_{in}+V_{C1}$. Assuming the capacitance of C_1 is large enough, the $2V_{in}$ level can be produced by being ON of S_1, S_4 and OFF of S_2 and S_3 as shown in Fig.3c. Similarly, when S_1 and S_4 are turned OFF and S_2, S_3 are turned ON, the level of $-2V_{in}$ can be produced as the bus voltage v_{bus}.

When Q_1 and Q_2 are turned ON simultaneously while Q_0 being OFF, capacitors C_1, C_2 and input source V_{in} are connected in series by switches Q_1 and Q_2. Under the condition of the values of C_1 and C_2 are both large enough, the level of $3V_{in}$ can be output by the DC-DC converter section. In this case, if the switches S_1 and S_4 are turned ON and S_2 and S_3 being OFF, the bus voltage v_{bus} is equal to $3V_{in}$ as shown in Fig.3d. With similar method, the level of $-3V_{in}$ can be produced by turning switches S_1 and S_4 OFF, and S_2 and S_3 ON.

According the above analysis, the working states' combination of the seven-level version of the proposed inverter is concluded as shown in Tab.1. It can be seen from Tab.1 that there are eight working states for the inverter corresponding to seven voltage levels, including two zero level states. In each state, a maximum of only four switches are in conduction. And when the inverter operates alternatively in two adjacent states, there is only one or two switches' states needed to be changed.

TABLE I

Working states' combination of the seven-level inverter

No. of states	Bus voltage v_{bus}	Switching states						
		Q_0	Q_1	Q_2	S_1	S_2	S_3	S_4
1	$+3V_{in}$	0	1	1	1	0	0	1
2	$+2V_{in}$	0	1	0	1	0	0	1
3	$+V_{in}$	1	0	0	1	0	0	1
4	0	1	0	0	1	0	0	0
5		1	0	0	0	1	0	0
6	$-V_{in}$	1	0	0	0	1	1	0
7	$-2V_{in}$	0	1	0	0	1	1	0
8	$-3V_{in}$	0	1	1	0	1	1	0

3. Modulation Method

There are many modulation methods to control a multilevel inverter, such as classic carrier-based sinusoidal PWM (SPWM) method [6]. In this section, SPWM is also introduced to modulate the multilevel inverter, as follows. For the seven-level inverter, there are six carrier signals e_1~e_6 and a modulated sinusoidal signal e_S needed, as shown in Fig.4a which is the modulation logic circuit for the proposed seven-level inverter. Fig.4b shows the corresponding modulation waveforms, in which A_C is the amplitude of the carrier signals. If defining the symbol A_S

as the amplitude of the modulated sinusoidal signal, the modulation index M can be defined as

$$M = \frac{2A_S}{(N-1)A_C} \qquad (1)$$

where N is the number of the levels and it is odd. For the proposed seven-level inverter, $N=7$.

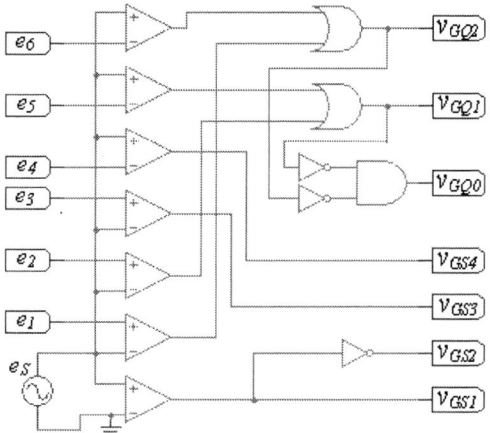

a. modulation logic circuit

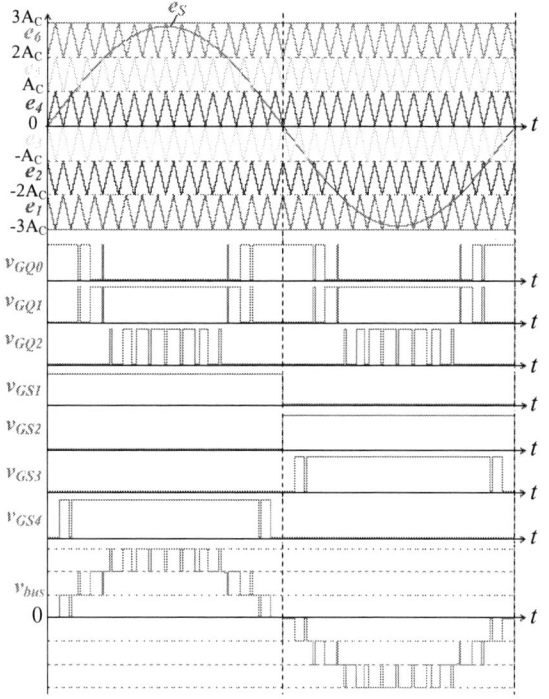

b. modulation waveforms

Fig.4: Modulation method for the proposed seven-level inverter

And the frequency modulation ratio can also be defined as

$$P = \omega_C / \omega_S \qquad (2)$$

where ω_C and ω_S are the angular frequencies of the carrier signal and modulated signal respectively. And the

desired output sinusoidal voltage therefore can be derived as (3).

$$v_O = \frac{N-1}{2} M V_{in} \sin \omega_S t \qquad (3)$$

III. GATE DRIVER AND OUTPUT FILTER

1. Gate driver for the proposed inverter

For multilevel inverters, a large number of active switching elements are required and the drive circuit is needed for each switch. In this respect, the cost and complexity of the gate driver depends on the number of active switches required for the multilevel inverter. For the proposed inverter, although the number of active switches employed is much lesser than conventional multilevel inverter, its gate drive circuit is still a very important issue. Bootstrap capacitor driver (*BSCD*) is a mature gate drive technique and has traditionally been applied in various bridge circuits [7]. Based on the special structure of the proposed inverter, *BSCD* technique is also introduced to drive the all active switches.

In the proposed topology which consists of a full bridge and a multilevel DC-DC converter, the gate drivers for the full bridge is very simple and is not elaborated in the text. For the DC-DC conversion section, active switches Q_1 to Q_n are actually connected in series with Q_0 though diodes D_1' to D_n' respectively. And Q_0 can be turned ON just after all other switches Q_1 to Q_n being OFF. The voltage for the gate driver of Q_0 therefore can be supplied directly by the signal power V_{gate}. And the voltage sources for the gate drivers of Q_1 to Q_n could be implemented by using a bootstrap capacitor for each switch as shown in Fig.5. Take the diver circuit of Q_1, BSCD-1 as an example, when switch Q_0 is turned ON while Q_1 being OFF, the capacitor C_{B1} is charged by the signal power V_{gate} though D_{11}, D_1' and Q_0, the energy is stored in C_{B1} and its voltage is eventually equal to V_{gate}. When Q_0 is turned OFF, switch Q_1 could be controlled by its trigger signal v_{GQ1} and voltage as well as power are supplied by the capacitor C_{B1}. For other switches Q_i (i=2, 3, ..., n), the gate drivers BSCD-i are totally the same as BSCD-1. The gate diver for the total inverter is therefore very simple and only one signal power V_{gate} is required.

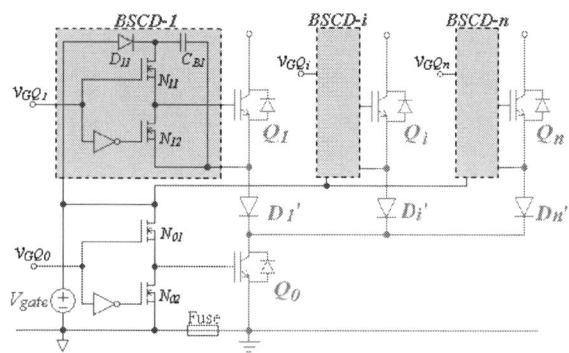

Fig.5: Gate driver for the proposed inverter

2. Output filter design for the proposed inverter

978-1-5090-0064-7/15 $31.00 © 2015 IEEE

Comparing with 2-level inverter, the output performance of the multilevel inverters is more satisfactory in the terms of harmonics. The output filters therefore are easier to be designed. Usually, the multilevel inverters only need to employ an *LC* low-pass output filter with reasonable parameters to provide satisfactory output sinusoidal voltage. The detailed design methods of *LC* filter for PWM inverters have been introduced in [8], [9] and the technique is very mature, that is mainly based the considerations of reactive power and output voltage harmonics. Simply, the design steps of a *LC* filter for PWM inverters can be summarized as follows.

1). to determine the filter cut-off frequency ω_f referring to the carrier signals frequency ω_C and the modulated signal frequency ω_S, i.e.

$$\omega_f = \frac{1}{\sqrt{LC}} \qquad (4)$$

and $\qquad \omega_S < \omega_f < \omega_C \qquad (5)$

2). to determine the inductance of the filter according to the principle of minimum reactive power. For the pure resistance load, the reactive power Q_{LC} caused by the *LC* filter could be approximately expressed as

$$Q_{LC} \approx \omega_S I_O{}^2 L + (\frac{\omega_S}{\omega_f{}^2} + \frac{\omega_S{}^3}{\omega_f{}^4}) U_O{}^2 \frac{1}{L} \qquad (6)$$

where U_O and I_O are the rms values of the output voltage and load current respectively. The minimum reactive power is obtained when $\partial Q_{LC}/\partial L = 0$. The value of the inductor L therefore can be calculated by

$$L = \frac{U_O}{I_O \omega_f} \sqrt{1 + (\frac{\omega_S}{\omega_f})^2} \qquad (7)$$

3). calculate the capacitance of the filter according the value of inductance *L* and the cut-off frequency ω_f, i.e.

$$C = \frac{1}{\omega_f{}^2 L} \qquad (8)$$

IV. SIMULATION EXPERIMENTAL VERIFICATION

1. Simulation Results

To verify the feasibility of the proposed gate driver and the LC output filter developed for the multilevel inverter of Fig.1, a simulation model is built based on the seven-level topology, the multicarrier SPWM technique and the gate driver structure as shown in Figs. 2, 4a and 5 respectively. Fig.6 shows the simulated results and the simulation parameters are chosen as following: the dc input voltage V_{in} is 24V; the capacitances of C_1 and C_2 both are 1000μF; the carrier signals frequency and the modulated signal frequency are 40kHz and 400Hz respectively; the modulation index *M* is 0.96 and the load resistance is 22Ω; the signal power V_{gate} is 15V and the bootstrap capacitor is 1μF; the output filter inductance *L* and capacitance *C* are 850μH and 2.2μF respectively.

Simulation results indicate that the multilevel inverter is capable of generating a pure sinusoidal output voltage waveform v_O and this is benefited from the bootstrap gate driver circuit and the LC filter developed in this paper.

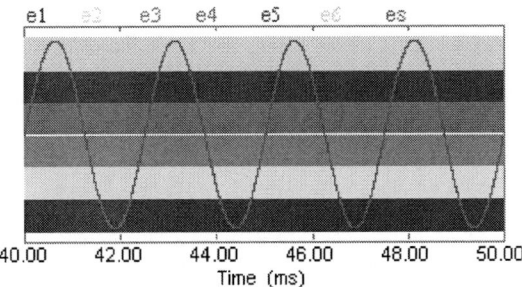

a. the carrier signals e_1~e_6 and the modulated signal e_s

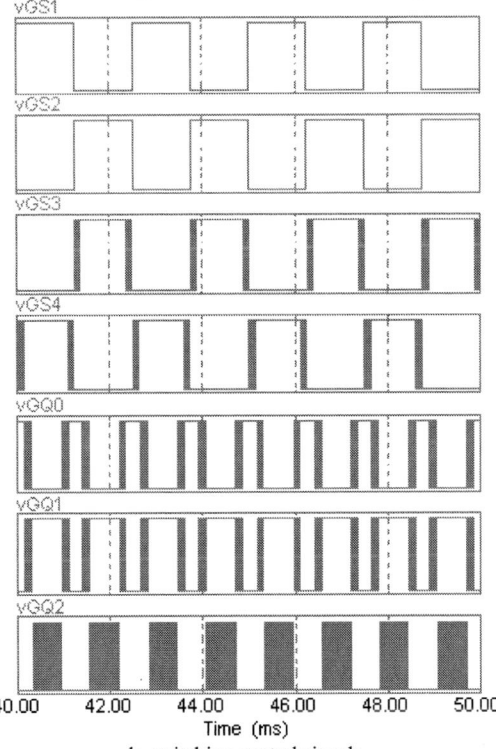

b. switching control signals

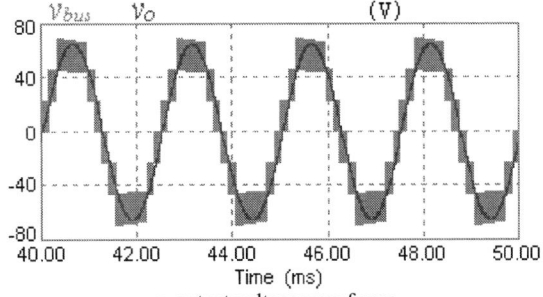

c. output voltage waveforms

Fig.6: Simulation results of the seven-level output

2. Experimental Results

A prototype of the seven-level version of the proposed inverter is developed to evaluate the performance of the proposed topology in the generation of a desired output voltage waveform. The basic parameters are the same as that used for simulation and the switches are selected as following: S_1~S_4 and Q_0 are MOSFETs IRFB4019PBF; Q_1 and Q_2 are MOSFETs IRFI540A; MBR10100 are used as the diodes D_1, D_2 and D_{02}. The modulation index M is still 0.96. The experimental results are shown in Fig.7. As shown in Fig.7a, the output voltage waveforms are basically the same as the simulation results aforementioned except for the amplitude, which is slightly lower than the theoretical value the simulation result because the voltage drops of the switching devices. Fig.7b shows the frequency spectrum of the bus voltage v_{bus}. It can be seen that the lower order harmonics are small but the higher harmonics cannot be neglected, especially those closed to the carrier signals frequency. This issue is easily solved by the output LC filter as shown in Figs.7c and 7d, which show the frequency spectrum of the output voltage v_O.

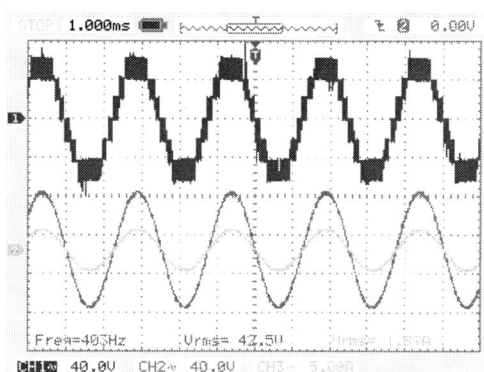

a. CH1: v_{bus}; CH2: v_O; CH3: i_O

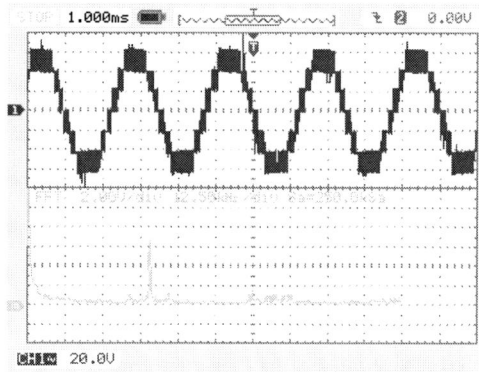

b. frequency spectrum of v_{bus} (0~125kHz)

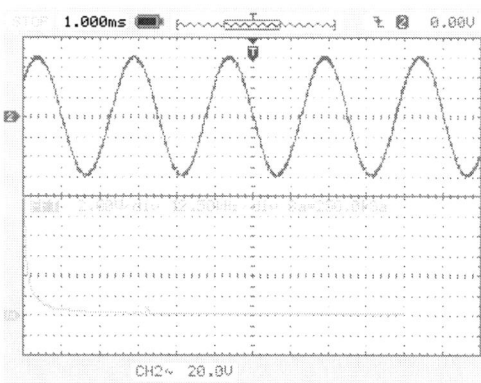

c. frequency spectrum of v_O (0~125kHz)

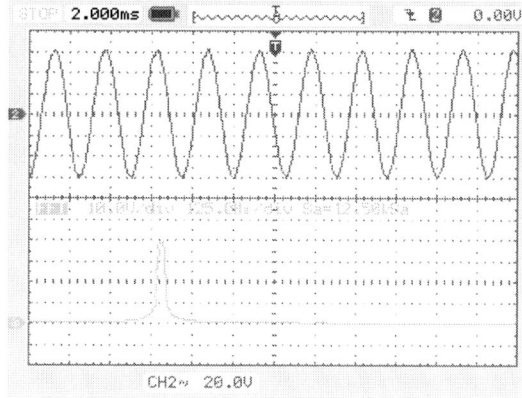

d. frequency spectrum of v_O (0~1.5kHz)

Fig.7: Experimental results of the seven-level inverter

IV. CONCLUSION

In this paper, a simple gate drive circuit and an output filter are developed for the multilevel inverter presented in [5]. With this bootstrap gate driver, only one voltage source is required to power the drive circuit of all switches employed in this inverter. This makes it has the advantages of small size and cost-effectiveness. With the LC filter, this inverter is capable of providing a pure sinusoidal output voltage waveform. And it is very suitable for aircraft APUs. Simulation and experimental results indicate the gate driver and the LC filter introduced in this paper provide a very well solution to promote the industrial applications of the multilevel inverter.

ACKNOWLEDGMENT

The authors grateful acknowledge the financial support of Innovation and Technology Fund of Hong Kong ITC and JL Ecopro Tech. Ltd under the project number UIM245.

REFERENCES

[1] J. Rodriguez, J. S. Lai, and F. Z. Peng, "Multilevel inverters: Asurvey of topologies, control, and applications", *IEEE Transaction on Industrial Electronics*, vol.49, no.4, pp.724–738, Dec. 2002.

[2] J. Rodriguez, S. Bernet, P.K. Steimer, and I.E. Lizama, "A survey on neutral-point-clamped inverters," *IEEE Transaction on Industrial Electronics*, vol.57, no.7, pp.2219–2230, Jul. 2010.

[3] M. F. Escalante, J. -C. Vannier and A. Arzande, "Flying capacitor multilevel inverters and DTC motor drive applications," *IEEE Trans. Ind. Electron.*, vol. 49, no. 4, pp. 809–815, Aug. 2002.

[4] M. Malinowski, K. Gopakumar, J. Rodriguez, M. A. Pérez, "A Survey on Cascaded Multilevel Inverters," *IEEE Trans. Ind. Electron.*, vol. 57, no. 7, pp. 2197–2206, Jul. 2010.

[5] Yuanmao Ye, K.W.E. Cheng, Junfeng Liu, and Kai Ding, "A Step-Up Switched-Capacitor Multilevel Inverter with Self Voltage Balancing," *IEEE Trans. Ind. Electron.*, vol. 61, no. 12, pp. 6672-6680, Dec. 2014.

[6] Mwinyiwiwa B., Wolanski Z., Yiqiang Chen, Boon-Teck Ooi, "Multimodular multilevel converters with input/output linearity," *IEEE Transactions on Industry Applications*, vol.33, no.5, pp.1214-1219, Sep./Oct. 1997.

[7] Graczkowski J.J., Neff K.L., Kou X., "A Low-Cost Gate Driver Design Using Bootstrap Capacitors for Multilevel MOSFET Inverters", *CES/IEEE 5th International Power Electronics and Motion Control Conference*, Aug. 2006.

[8] Soo-Hong Kim, Yoon-Ho Kim, Kang-Moon Seo, Sang-Seok Bang, Kwang-Seob Kim, "Harmonic analysis and output filter design of NPC multi-level inverters", *37th IEEE Power Electronics Specialists Conference*, June 2006.

[9] Hyosung Kim and Seung-Ki Sul, "A Novel Filter Design for Output LC Filters of PWM Inverters", *Journal of Power Electronics*, vol.11, no.1, pp.74-81, Jan. 2011.

New LED Lighting Design for Road Vehicles

William F. Chen[1]   K.W.E. Cheng [2]

[1] Department of Electrical Engineering, The Hong Kong Polytechnic University, Hong Kong
E-mail: William.f.chen@connect.polyu.hk
[2] Department of Electrical Engineering, The Hong Kong Polytechnic University, Hong Kong
E-mail: eric-cheng.cheng@polyu.edu.hk

Abstract-This journal will discuss the stagnant condition of current road vehicle lighting. This refers specifically to the design and technology of this field. The designs of vehicles have not been change for many decades and the technology have made strides recently with LED but it has failed to use all of it advantages. The focus will be on the versatility and controllability that is offered by LED lighting that is being ignored by the vehicle industry. This journal will look into these problems, provide a solution and promote a change in this field. The hope is that with the new design vehicles in the future will be even more efficient and safe for drivers.

Keywords-Vehicle Design, LED Lighting, Vehicle Lighting, Car Lighting

I. INTRODUCTION

In recent history great strides have been made in the field of electric vehicles. The new electric vehicles have completed the task of producing a product that is visually the same as the old gasoline power vehicles but at the same time completely different under the hood. In other words, the engines and hardware has broken off from tradition but the exterior has not been able to undergo a drastic change. A major part of the exterior image of the vehicle is the lighting, which has generally kept the same design. Of course there have been changes made in the past decade by replacing the tradition tungsten-halogen bulbs with LED lights. However, not enough has been done because the design is still mostly the same and the LED lights remain underutilized. If the LED lights used on vehicles can be used at its full capacity then vehicles today can truly undergo a transformation.

1. Light Emitting Diode (LED)

To fully understand the possible transformation, it is important to first understand what LED light is and what the capabilities are. LED or Light Emitting Diodes is a technology that has been available for several decades now. However, for most of the early years it remained an experimental item. The product truly expanded in the past decade when it became a household product. This was achieved by becoming an integral part of computer monitors, television sets and household lighting. When LED was made available for home use it was truly a breakthrough because it offered a replacement for the old fluorescent and energy-saving bulbs. The LED was literally better than the previous products in every way. LED lights are able to save more electricity, produce brighter lighting, have a longer lifespan, high efficiency and cause no harm to the environment. Therefore, it slowly expanded into many

different fields, going from household lighting to outdoor lighting to street lighting and now to vehicle lighting.

As much of the name implies, it is a diode that can come in many sizes and when a correct amount of voltage is applied, the diode will light up. Surprisingly, the specifications regarding LED lighting are fairly simple. When the LED is at its most basic set-up it only requires a voltage source, a resistor and a LED. Perhaps, the resistor is the biggest difference between LED and older lighting sources but it is a very important piece of the system. This is due to the fact that LEDs have a specific current range where it must operate in, once it is outside the range issues occur. Specifically, if the current is too low the LED will not light up and if it is too high it will cause harm to the LED and severely decrease its lifespan. That is why most current LED lighting products come with a pre-installed driver, which corrects the input before it reaches the LED. But besides the addition of a LED driver, most household lighting products also have the same image as traditional lighting. Meaning they suffer a similar fate as vehicle lighting where they have an advance technology but fail to reach their full capabilities. Now, common lighting products are limited due to the fact that they must fit in the original sockets and housings. However, LED vehicle lighting has no such limitations and improvements are still at a minimum. By following the designs of old, they miss out on LED lightings two greatest assets, versatility and controllability.

2. Versatilty of LED

LED chips have a tremendous amount of versatility because they can come in many different sizes, from small ones in the millimetres to large ones that can be several inches wide. They can also be arranged in many different ways, from the most basic forms of series and parallels to many different kinds of shapes and orientations. LED chips are also capable of lighting up in an assortment of colours versus the white or yellow of older lighting. At the moment, current LED lighting on vehicles are commonly seen only lined up in a series for the front lights and circular for the rear lights. As mentioned, this arrangement uses none of the versatility that is offered by LED lights.

3. Controllability of LED

The second asset, controllability, in many ways has been completely ignored. Controllability is a reference to LED chips ability to control brightness, flashing and personalization. First, brightness can easily be control in many different ways when it comes to LED chips. High quality LED has an innate ability of dimming therefore, the brightness level of the lighting can be controlled if there is a dimming option. As mentioned earlier, LED chips operate within a certain range and within that range the brightness of the lights will vary depending on the amount of current it receives. For example, if the range of a set of LED lights is 1-5 amperes then the light will be at its brightest when a full 5 amps of current. Of course the vice-versa is also true. This ability will be the base of a revolutionary idea that will be discuss later. A second ability is flashing. Flashing, currently on vehicles is used for signalling a lane change or for warning lights. Flashing displays was one of the first things that LED lights were used for; many of the common displayed outside businesses are the product of LED lights. Therefore, it is obvious that LED is a very suitable product to use this area. Current signalling lights can only be turned on and off and the flashing will always remain at one speed. LED allows for more control in this area and perhaps by changing to faster intervals of flashing, it will become more noticeable to other drivers leading to fewer accidents during lane changes. If the intervals can be control, then using warning signals will actually pass more information to drivers behind you. The last advantage of controllability is personalization. There is so much more that can done involving LED for a vehicle. Potentially the front lights can become larger, allowing for better lighting at night. The current spots currently allocated for license plates can be replace by a LED panel, eliminating the need for a physical license plate. An illuminated plate can be customized to the drivers preference and they can be better viewed by others and law enforcement officials at night. The possibilities are truly endless.

II. OBJECTIVE

However, these are just small suggestions that are part of a larger picture and a larger objective. The main objective is still to change the whole design of vehicle lighting design. The process will not be a simple one and many items will need to be change to have a complete system that takes advantage of LED lights.

III. DESIGN MODIFICATIONS

One proposition for change is to move forward by removing all controls for the main lighting and have the system be automated. This modification is a system change instead of design and the purpose is to make the drving experience easier. The fact is this is already an option on most vehicles, but only as a setting you must change to. By automating all the main lightings, drivers have one less thing to worry about. There are many instances and cases where accidents happen while driving at night simply because the light was not turned on. This change will also help drivers who forget to turn their light off, which results in a power loss and not having enough charge to start their cars the next day. Newer vehicles already have such a complex system within them that accurately predicting their surrounding and analysing if there is a need for lighting is very simple. Besides this system will only be an extension of the Daytime Running Light system or DRL, which is already in place in many countries around the world. DRL is basically a set of lights (usually LED) that cannot be turned off when driving so the vehicle is more noticeable on the road no matter the conditions or time.

1. Head Lamps

The next change will be on the current design of the lights. Traditional lighting, since the beginning of vehicles has always been the design four lights on the four corners or two head lights and two rear lights. There are actually a few problems with this design. One problem is, depending on the light that is used, the brightness level will not be enough. Especially with cheaper brands the middle section of the road will not be as illuminated as the sides of the road. Another problem is that if the head lights are too bright, they will actually overlap and leak into other lanes, which is hazardous in two way traffic because it can affect other drivers on the road. This means that a major problem with the current design is that the lights are not 100% illuminating the road in front of the driver; instead the brightest areas are the sides of the road. This problem can easily be corrected with tweaks in the design, specifically by changing the placement of the lights. The new lights would be installed in the middle section of the vehicle, where the grill plate usually lies. This design would actually only take up a small amount for space, using high brightness LED chips two parallel strips would be adequate. The top strip would be installed on the vehicle at an angle that is perpendicular to the ground, so it can fully illuminate the area in front of the car. The second strip would be lower and attached at an angle of 45-60 degrees in relation to the ground. The second strip can then focus purely on illuminating the ground in front of the vehicle and act as the new form of DRL. The DRL will automatically always remain on allowing them to be notice by other drivers when the car is in motion. By using this design many benefits arise. The main benefit is that it can solve the problems that were previously mentioned, especially helping the lights stay within one lane instead of crossing over to other lanes where the glare might become dangerous. Another benefit is that this design will free up a lot of space, allowing engineers to add or improve on the design. Figure 1 shows how the new design would look on a current vehicle.

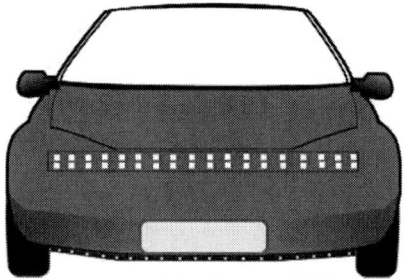

Fig. 1: Showcasing main head lamps and DRL on bottom

2. Signalling Lights

Obviously the new head light design leaves out an important part of the vehicle and driving, the signalling lights. They are instrumental to driving so they have not been forgotten only moved to a different area. The new placement of signalling lights would be on the side of the vehicle instead of the front. They will be located under the side fenders of the vehicle, using a similar two strip design as the head lights. One of the strips will be attached directly under the side fender of the car, so it is exactly parallel to the ground. This part will be set to produce a soft white lighting. The second set of strips to will be installed to the

978-1-5090-0064-7/15 $31.00 © 2015 IEEE   332

side fenders at an angle from the ground and this set will function as the main signal indicators. The colour will be same as the traditional signalling lights. The benefit of using this design is that when the light is activated the illumination is bright enough for everyone nearby to see and at the same time illuminate the lane to be cut into. The goal is to reduce the amount of accidents while drivers change lanes and with this design nearby drivers can fully observe if a lane is clear to move into. If there is a blinking yellow light on the ground of a lane then drivers know someone else will be moving into the lane. This light allows drivers to mark the area they will be cutting into and safety move into it. The white light is used as another safety precaution by acting as an extension of the DRL. The light is always on, allowing the perimeter of a vehicle to be illuminated on the ground. If visibility is poor, drivers further away have a hard time judging where cars are, the side DRL along with the head lights give a clear indication of the area that a vehicle is occupying. Figure 2 showcases this two strip system with DRL on bottom and signalling lights on the top.

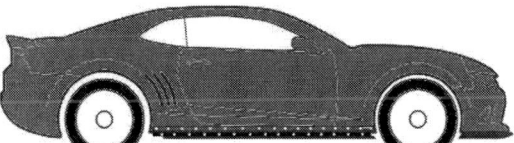

Fig. 2: Image of DRL and signalling lights design on vehicle

3. Rear Lighting

The rear lighting of a vehicle might be the most important part of a vehicle because it includes the brake lights. Obviously there is no proposal to remove the use of brake lights but again the infusion of design and LED technology can change it for the better drastically. The new concept for rear lighting is to; first, change the design to match the design of the front of the vehicle. This means that the lights will no longer be located on the edge of the rear; instead it will again all be concentrated towards the centre of the rear. Since this part of the vehicle is so important for safety, the lighting will be made bigger than the other parts. There will be three parts to the rear lighting. The first one will be the smallest; only one strip in length attached to the bottom of the rear would be used to complete the perimeter lights that are attached to the rest of the vehicle. This light will give a full view of the area the vehicle will occupy at night even from far away. The second set of lights will be located on the top section of the rear and it will be the main brake light. A third set of lights will be directly below the second set and these will act as the secondary brake light and reverse light. The secondary brake light will have the ability to function in two colours, white and red. When driving regularly at night the secondary brake lights will automatically be on always, but the light will also have the ability to turn white when the vehicle is put in reverse. This will send a clear signal to vehicles behind that it will be moving in reverse. These are standard functions and are very similar to traditional lighting. The biggest advancement will be displayed on the main brake lighting. The main braking light will use the previous mentioned technology of LED brightness control through current. This technique can be applied by using the braking pedal of a vehicle as control of the lighting. This means that there is a

direct relationship between how much the brake pedal is pushed and how much voltage is released to the main brake lights. By using high-end LED chips with a large range of operation, there will be a clear difference of brightness between a light brake and a hard brake. Of course this design will be more complicated than traditional brake lighting but the change is necessary. This design will have to be paired with a management system that will instantly calculate the force applied to the brakes and use this data to allocate the correct amount of current to the brake lights. Using this method provides a huge advantage in terms of safety. Now all the drivers behind a vehicle would be able to tell what type of breaking the vehicle in front is making. This allows drivers to instantly identify the breaking being made in front and make the same adjustments. Many times accidents occur because time is lost identifying or realizing that a hard brake is being applied. Figure 3 showcases a drawing of the mentioned design.

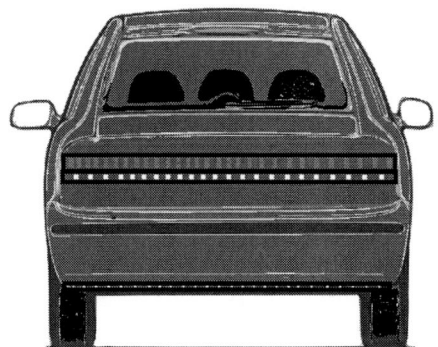

Fig. 3: A image showing design of triple layer system of main brake lights, secondary brake lights and DRL

IV. CONCLUSION

There was a time where technology could not support an inventor or designers ideas but that time has pass. Now the opposite might be true, where the technology has made huge strides and it is not being utilized. There are no reasons to not make changes besides the fact that it will be parting from tradition. The reason for changes might not necessary have to be due to design or aesthetics because arguments could be made that nothing is wrong in that department. But this change is not all about aesthetics, it is about safety first.

To be safe on the road when driving, the driver must have a clear sense or view of their surrounding and the road. Without these accidents are more likely to happen. The World Health Organization states that "Approximately 1.24 million deaths occurred on the world's roads in 2010"[1], which is a large amount of deaths and the rate has been consistent instead of dropping. The amount of non-fatal accidents annually is an even larger number. Although non-fatal accidents do not involve death it can still cause injuries or become an annoyance for drivers involved. Even in a small area like Hong Kong, the average number of accidents daily is 43.3 or almost 2 every hour of the day.[2] In the year 2014 there was over 6,600 cases of vehicle on vehicle accidents which resulted in 27 fatalities.[3] This amount of accidents in an area with only about 486,000 registered private vehicles[4] is cause for concern. Perhaps, this is not even the most concerning part because the truly frightening thing is in Hong Kong the number of accidents are actually increasing almost every year.[5]

This means that it is now time to revisited the current design of vehicles and review on ways to improve the safety. The first step has to be the lighting because the current lighting system and design will allow drivers to have a clear sense of the road and feel more secure. Also, the new DRL design will allow each vehicle to become more noticeable to other drivers and provide a zone for safe driving. This design will also finally use LED lighting to its full versatility and controllability.

REFERENCES

[1] "Number of Road Traffic Deaths." World Health Organization 2013, Global Health Observatory (GHO) Data sec.

[2] "Road Traffic Accidents by Hour of the Day and Day of the Week 2014." Hong Kong Transport Department (2015): Hong Kong.

[3] "Road Traffic Accidents by Type of Vehicle Collision and Severity (2004-2014)." Hong Kong Transport Department (2015): Hong Kong.

[4] "Motor Vehicle Involvements and Involvement Rates by Selected Class of Motor Vehicle." Hong Kong Transport Department (2015): Hong Kong.

[5] "Road Traffic Accidents by Type of Vehicle Collision (2004-2014)." Hong Kong Transport Department (2015): B7. Hong Kong.

Soft-switching Topologies for Switched Reluctance Motors in the Application of Electric Vehicles

Jingwei Zhu[1]   K.W.E Cheng[2]

[1] Department of Electrical Engineering, The Hong Kong Polytechnic University, Hong Kong
E-mail: 15901658r@connect.polyu.hk
[2] Department of Electrical Engineering, The Hong Kong Polytechnic University, Hong Kong
E-mail: eeecheng@polyu.edu.hk

Abstract- two topologies of soft-switching for the switched reluctance motor (SRM) are proposed in this paper to reduce the switching loss and EMI in the chopping and commuting periods. The design circuit of the topologies are shown and the feature of the current or voltage of resonant elements are analyzed in theory and confirmed by simulation of Matlab/Simulink. Besides, the energy loss of the soft-switching circuit for the second topology is calculated and compared with the conventional one. The loss is computed under different rotating speeds to simulate electrical vehicle (EV) running under various speeds and the advantage of the proposed circuit is proved.

Keywords-SRM, Matlab/Simulink, soft-switching, energy loss, topology

I. INTRODUCTION

With the increasingly serious environment pollution and energy crisis, EV has become a new trend for future industrial vehicle study due to its high efficiency and low emission. Motor drive is the central part of EV because of offering enough power for its performing. The SRM has the advantage over other kinds of motors as it has a simple structure, low cost, high efficiency, easy control, low starting current and high pull-up torque. Therefore, it has a potential prospect in the application of machine drives of EV. But in order to optimize a control strategy for SRM drive, high frequency is required for accurate current and flux linkage regulation, causing considerable switching losses and EMI. Soft switching technology is an efficient method in lowering these losses for EV in normal urban driving [1].

In [2], resonant DC link (RDCL) is utilized to achieve ZVS on and ZCS off, but the resonant inductor is in series with the inverter-bridge leading to unnecessary loss since both resonant and load current flow through this resonant inductor. Literature [3] introduces additional inductors and capacitors into the topology, however, there's a resonant inductor in the main loop as well. A novel topology is put forward in [4] with PWM method to reduce the switching losses, but the resonant inductor is still in the main circuit. The soft-switching topology in [5] can only attain ZVS on for phase switches. Document [6-8] solved the inverter problems above, but when there is no load, the formation of zero voltage between the DC bus encounters great difficulties.

This paper proposes two topologies to reduce the switching losses and settle the challenges above. The topology in section II utilizes only one auxiliary switch and remains the resonant inductor in the resonant circuit,

thus declining the current flowing through it greatly. Besides, the number of the auxiliary elements is reduced to a great extent. Section III gives an improved topology to achieve controlling the zero voltage duration freely and no current flowing through the resonant inductor during normal conducting time, although it adds two additonal auxiliary switches. Section IV makes a comparison of the energy losses between the improved topology in section III and the convential one; furthermore, some analysis is offered.

II. A SUITABLE SOFT-SWITCHING TOPOLOGY FOR THE SRM

The designed circuit is in the following Figure 1:

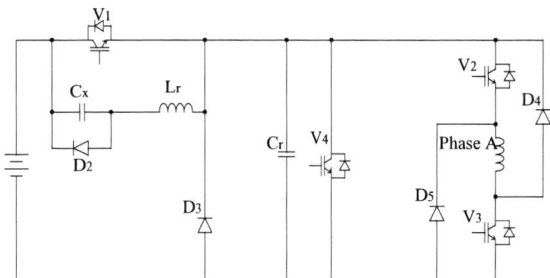

Figure 1. Schematic of the new soft-switching topology.

In this Figure 1, one phase of the SRM is simplified as a constant inductance during the soft-switching period since it is short enough. And the parameter of the elements in the circuit is in the following Table 1:

Table 1: Circuit Parameters for the Topology

E (V)	300	C_s (pF)	0.1
C_x (nF)	1000	L_r (μF)	60
C_r (nF)	30	L_A (mF)	10

V_1, V_2, V_3, V_4 of the switching components are all selected as IGBT. This proposed circuit can simplify the control of switching elements, reduce the number of switching and resonant components to lower the hardware cost. Besides, it has extra advantages of each switch turning on-off at zero voltage and the maximum enduring voltage is equal to that of the DC source. Meanwhile, because of no series voltage-divider capacitors between the DC bus, there's no potential variation of the neutral point [9].

The theoretical waveforms of the switching signal, voltage or current of the resonant components are shown as follows in Figure 2:

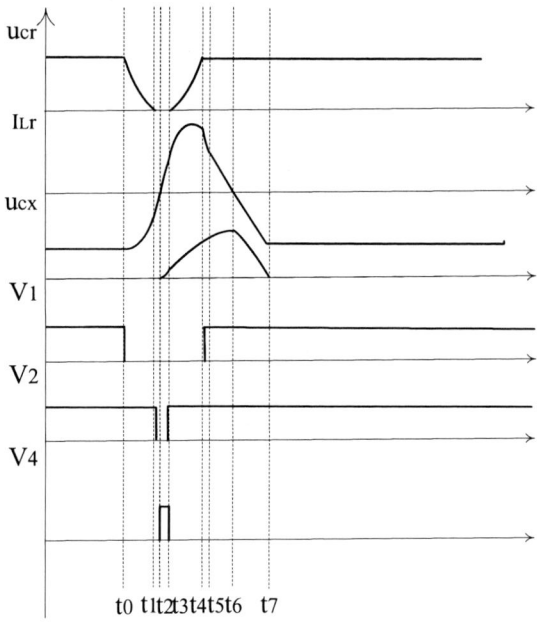

t0 t1t2t3t4t5t6 t7

Figure 2. theoretical waveform for the novel topology

This circuit can be divided into six working modes

Mode I $(0{\sim}t_0)$: it's the initial condition of the whole circuit with V_1, V_2, V_3 on and V_4 off. The circuit is stable and there is a constant reverse current I_{L0} flowing through the resonant inductor L_r.

Mode II $(t_0{\sim}t_1)$: when V_1 is off, L_r and C_r begin their resonance with both the current i_{lr} and voltage u_{cr} decreasing. The energy of L_r feeds back to the battery while that of C_r is transferred to both the winding and the battery. With C_r discharging, the DC bus voltage is declining gradually until zero. The differential equation is as follows:

$$u_{cr} = -L_r \cdot \frac{di_r}{dt} + E \qquad (1)$$

$$C_r \cdot \frac{du_r}{dt} = i_r - I_0 \qquad (2)$$

Then, with the initial value $u_{cr}(t_0) = E$, $i_{cr}(t_0) = -I_{L0}$, we can get

$$u_{cr} = E - \frac{(I_0 + I_{L0})}{\omega_r C_r} \sin[\omega_r(t - t_0)] \qquad (3)$$

$$i_r = I_0 - (I_0 + I_{L0})\cos[\omega_r(t - t_0)] \qquad (4)$$

where, $\omega_r = \frac{1}{\sqrt{L_r C_r}}$, I_0 is the current of the winding load. Therefore, the mode II time,

$$T_2 = \frac{1}{\omega_r} arcsin \frac{\omega_r C_r E}{I_0 + I_{L0}} \qquad (5)$$

Mode III $(t_1{\sim}t_2)$: when u_{cr} reaches zero, diode D_3 conducts which clamps the DC bus voltage at zero and V_2 can be switched off at zero voltage. During this period, L_r

bears a constant voltage of the DC source and i_r decreases linearly until it reaches zero point. So,

$$i_r = \frac{E}{L_r}(t - t_1) - I_1 \qquad (6)$$

Thus, the mode III time,

$$T_3 = \frac{L_r I_1}{E} \qquad (7)$$

Mode IV $(t_2{\sim}t_3)$: IGBT V_4 is turned on to keep i_r positive and the capacitor u_{cx} is conducting. So C_x and L_r start their resonance under the DC bus voltage until it i_r attains its setting value I_2 at the moment t_3. At the end of this period, the switch V_2 is turned on at zero voltage point of the DC bus. Then IGBT V_4 is turned off. We can get the equation,

$$u_{Cx} + L_r \frac{di_r}{dt} = E \qquad (8)$$

$$C_x \frac{du_{Cx}}{dt} = i_r \qquad (9)$$

From (8) and (9), the result is

$$u_{Cx} = E - Ecos[\omega_x(t - t_2)] \qquad (10)$$

$$i_r = \omega_x C_x Esin[\omega_x(t - t_2)] \qquad (11)$$

Therefore, the duration of this mode is

$$T_4 = \frac{1}{\omega_x} arcsin(\frac{I_2}{C_x \omega_x E}) \qquad (12)$$

Mode V $(t_3{\sim}t_4)$: With V_4 switched off, L_r, C_x, C_r begin their resonance together. Because of $C_x \gg C_r$, the series capacitor can be simplified as

$$\frac{C_x C_r}{C_x + C_r} \approx C_r \qquad (13)$$

In this mode, u_{Cx} can be regarded as constant U. During the resonance process, both L_r and C_r are charged, with i_r and u_{cr} increasing till u_{cr} getting to the DC bus voltage. The differential equations are as follows:

$$E - U = L_r \frac{di_r}{dt} + u_{cr} \qquad (14)$$

$$i_r - I_0 = C_r \frac{du_{cr}}{dt} \qquad (15)$$

Then we can calculate,

$$u_{cr} = (E - U)(1 - \cos[\omega_r(t - t_3)]) + \frac{I_2 - I_0}{\omega_r C_r}\sin[\omega_r(t - t_3)] \qquad (16)$$

$$i_r = I_0 + (I_2 - I_0)\cos[\omega_r(t - t_3)] + \omega_r C_r (E - U)\sin[\omega_r(t - t_3)] \qquad (17)$$

From the outcome, when u_{cr} increases to $(E - U)$, i_r reaches its maximum and from then on, u_{cr} augments while i_r decreases. The duration of this mode can be calculated as

978-1-5090-0064-7/15 $31.00 © 2015 IEEE

$$T_5 = \frac{1}{\omega_r}\left(\arcsin\left(\frac{E-U}{M}\right) + \arcsin\left(\frac{U}{M}\right)\right) \tag{18}$$

where, $M = \sqrt{\frac{(I_2 - I_0)^2}{\omega_r^2 C_r^2} + (E - U)^2}$

Mode VI($t_4 \sim t_7$): During this mode, two parts of it need to be analysed. To start with, when i_r is still positive, C_x and L_r start resonance with u_{cx} increasing and i_r declining. Before i_r descends to the winding load current I_0, Diode D1 is conducting during which time the switch V_1 can be turned on under zero voltage condition. The first part ends with u_{cx} arriving at its peak and i_r dropping to zero. The second part begins when i_r is negative, during which, C_x is discharged and i_r increases reversely. It comes to an end when u_{cx} drops to zero voltage and Diode D_2 is on. Later, it comes back to Mode I and start a new circulation. The equation is as follows:

$$u_{cx} + L_r \frac{di_r}{dt} = 0 \tag{19}$$

$$C_x \cdot \frac{du_{cx}}{dt} = i_r \tag{20}$$

Then the value we get is

$$u_{cx} = U\cos[\omega_r(t - t_4)] + \frac{I_3}{C_x \omega_x} \sin[\omega_r(t - t_4)] \tag{21}$$

$$i_r = I_3 \cos[\omega_r(t - t_4)] - C_x \omega_x U \sin[\omega_r(t - t_4)] \tag{22}$$

The length of time is

$$T_6 = \frac{1}{\omega_x}\left(\pi - \arctan\frac{UC_x \omega_x}{I_3}\right) \tag{23}$$

According to the theory analysis above, a simulation of one circulation based on Matlab/Simulink has been done to confirm the theory. The switching signal is applied on the basis of the calculated duration of six modes. The initial current I_{l0} is set as -30A, I_2 is selected as 37.6A and winding load current I_0 is chosen as 12A.

The waveform figure is shown in the following Figure 3:

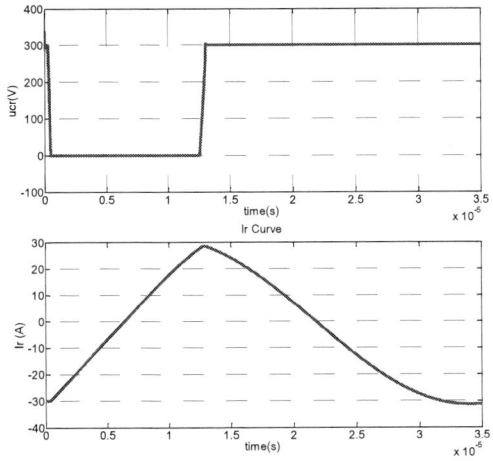

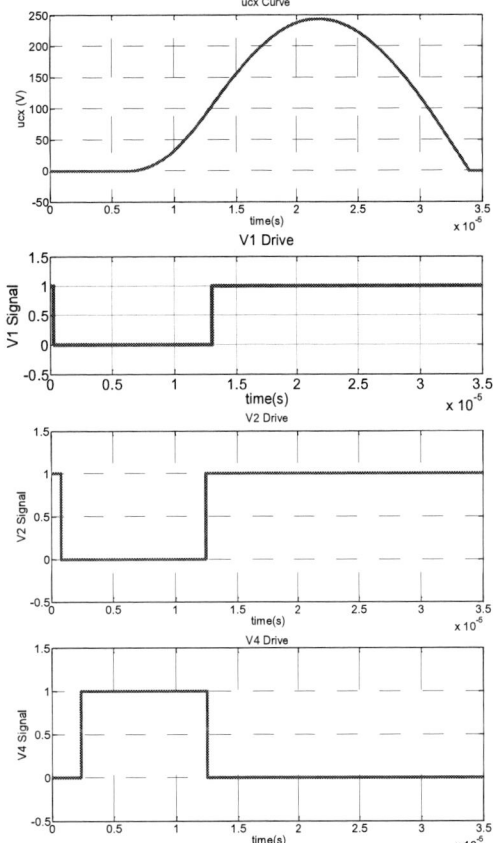

Figure 3. Waveform from Matlab/Simulink

The simulation results agree with the theoretical analysis, confirming the validity of this new topology.

III. AN IMPROVED TOPOLOGY FOR THE SRM

The proposed topology is shown below in Figure 4:

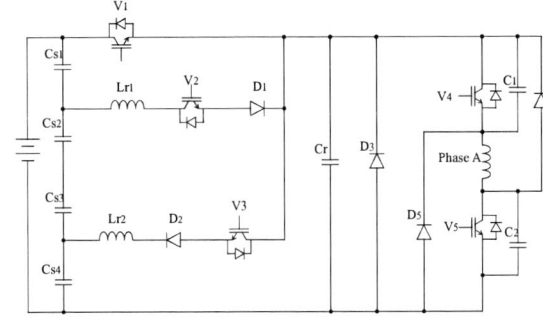

Figure 4. Schematic of the improved soft-switching topology

In this Figure 4, one phase of the SRM is also simplified as a constant inductance during the soft-switching period since it is short enough. And the parameter of the elements in the circuit is in the following Table 2:

Table 2: Circuit Parameters for the Improved Topology

$E(V)$	240	$L_A(mH)$	0.01
$C_r(nF)$	25	$L_r(\mu H)$	10
$C_i(pF)$ (i from 1 to 2)	0.1	$L_{r1}(\mu H)$	10
$C_{si}(\mu F)$ (i from 1 to 4)	25		

V_1, V_2, V_3, V_4, V_5 are all IGBT in the schematic. This improved circuit has the advantage of controlling the zero voltage time by selecting the turning-on time of V_2 [10]. And the theoretical waveforms of the elements in this topology are shown as follows in Figure 5:

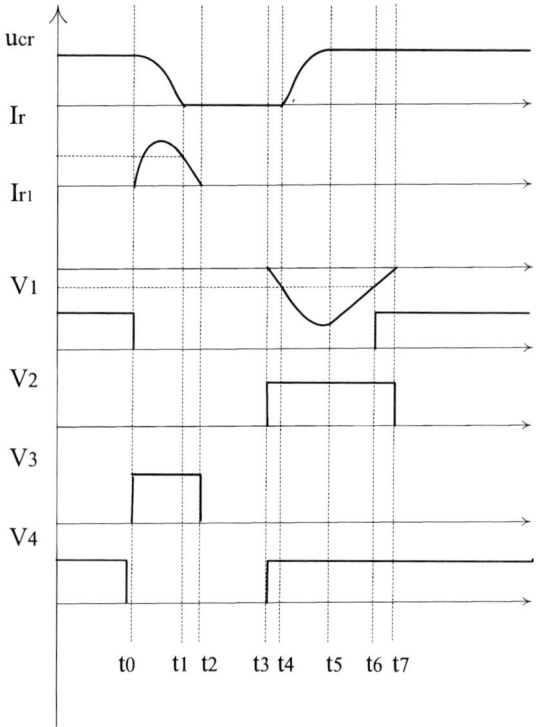

Figure 5. theoretical waveforms for the improved topology

This topology can be separated into eight modes.

Mode I ($0 \sim t_0$): both V_2 and V_3 are off while V_1, V_4, V_5 are on during this period. Phase A is conducting.

Mode II ($t_0 \sim t_1$): At the moment t_0, the switch V_1, V_2 and V_4 are all been turned off with only V_3 and V_5 on. Because of the existence of C_i, V_4 is turned off at zero voltage. Meanwhile, V_1 is also turned on during zero voltage stage. C_r, V_3, L_r, C_4 makes up a loop with C_r and L_r starting their resonance. Because of no sudden change of current in L_r, V_3 is switched on under a zero current condition. At the beginning, C_r is discharged while i_r increases until u_{cr} reaches $\frac{E}{4}$, then i_r starts to decrease ending with u_{cr} getting to the zero voltage point. At the time t_1, C_r is clamped by D_3 to keep a constant zero voltage situation. The equations of it are as follows:

$$u_{cr} = L_r \cdot \frac{di_r}{dt} + \frac{E}{4} \tag{23}$$

$$C_r \frac{du_{cr}}{dt} = -i_r \tag{24}$$

The results are calculated as

$$u_{cr} = \frac{E}{4} + \frac{3}{4}E\cos[\omega_r(t - t_0)] \tag{25}$$

$$i_r = \frac{3}{4}E\omega_r C_r \sin[\omega_r(t - t_0)] \tag{26}$$

Therefore, we can get the duration of this period

$$T_2 = \sqrt{L_r C_r}\arccos(-\frac{1}{3}) \tag{27}$$

Mode III ($t_1 \sim t_2$): since C_r is clamped by D_3, L_r will begin to discharge under the reverse voltage $\frac{E}{4}$ until it arrives at zero point.

$$i_r = I_1 - \frac{E(t - t_1)}{4L_r} \tag{28}$$

The time period is

$$T_3 = \frac{4L_r I_1}{E} \tag{29}$$

Mode IV ($t_2 \sim t_3$): during this mode, no switches are on and the length of period can be controlled by deciding the turning on moment of V_2. V_3 is both zero voltage and zero current turned off. Therefore, the switching on moment of V_4 can be located during the zero voltage stage definitely without thinking about the value of L_r and C_r.

Mode V ($t_3 \sim t_4$): V_2, V_4 and V_5 are on while V_1 and V_3 are off. The phase winding is conducting. Because the DC bus voltage is zero, V_4 is ZVS on and V_2 is ZCS on as a result of the existence of L_{r1}. During this mode, u_{cr} still keeps zero voltage and i_r increases reversely under the voltage of $\frac{3}{4}E$ until it reaches the winding current I_0.

$$i_{r1} = -\frac{3E}{4L_{r1}}(t - t_3) \tag{30}$$

Therefore, the length of this period is

$$T_5 = \frac{4L_{r1}I_0}{3E} \tag{31}$$

Mode VI ($t_4 \sim t_5$): C_r and L_{r1} start their resonance and it lasts until u_{cr} reaches E.

$$\frac{3}{4}E = -L_{r1}\frac{di_{r1}}{dt} + u_{cr} \tag{32}$$

$$C_r \frac{du_{cr}}{dt} = -i_{r1} - I_0 \tag{33}$$

Then we can get the result

$$u_{cr} = \frac{3}{4}E - \frac{3}{4}E\cos[\omega_r(t - t_4)] \tag{34}$$

$$i_{r1} = -I_0 - \frac{3}{4}EC_r\omega_r\sin[\omega_r(t - t_4)] \tag{35}$$

$$T_6 = \frac{1}{\omega_r}\arccos\left(-\frac{1}{3}\right) \tag{36}$$

Mode VII ($t_5 \sim t_6$): Since during this mode, $|i_{r1}| > I_0$, the voltage of $\frac{E}{4}$ is added to L_{r1}. C_{si}, V_1, V_2, L_{r1} form a series loop. V_1 is clamped by the diode, so it's ZVS on during this period.

$$i_{r1} = -I_0 - \frac{\sqrt{2}}{2}EC_r\omega_r + \frac{E}{4L_{r1}}(t - t_5) \tag{37}$$

So the duration is

$$T_7 = 2\sqrt{2}\omega_r C_r L_{r1} \tag{38}$$

Mode VIII ($t_6 \sim t_7$): L_{r1} is still under $\frac{E}{4}$ voltage and the changing rule of i_{r1} is the same as that in Mode VII. When i_{r1} reaches zero, V_2 is ZCS off. Then a new circulation begins.

$$T_8 = \frac{4L_{r1}I_0}{E} \tag{39}$$

The simulation results during one time period of this topology is shown as follows in Figure 6:

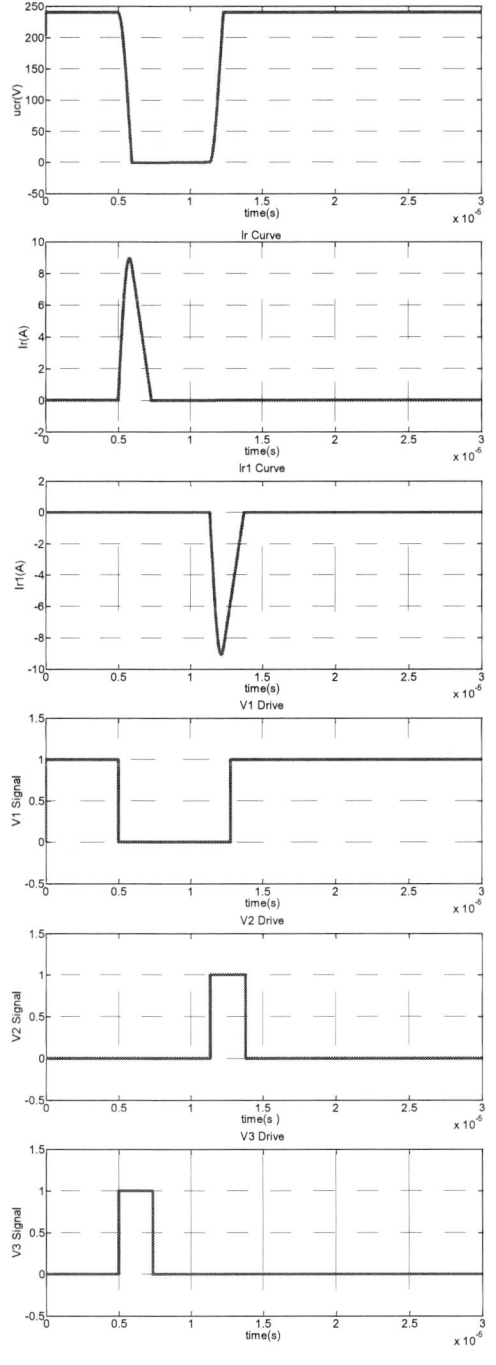

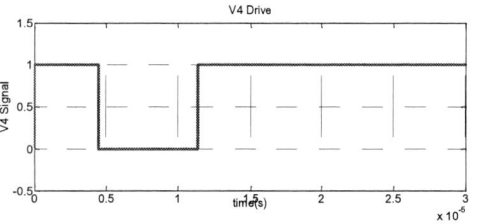

Figure 6. Waveforms from Matlab/Simulink of the improved topology

Then the improved topology is added into a whole current chopping control (CCC) system for a 6/4 SRM. And the DC source voltage is changed to 120V with other parameters unchanged. CCC is a common control method for SRM. When a SRM is running at a low speed, it has a small rotating EMF and a high value of di/dt. Also, because of long period for inductance increasing, we usually utilize CCC to prevent overcurrent in each phase. It regulates the upper and lower limit of the permitted current and keep θ_{on}、θ_{off} constant, in order to limit the value of current within an expected range. Its principle is in the following shown in Figure 7:

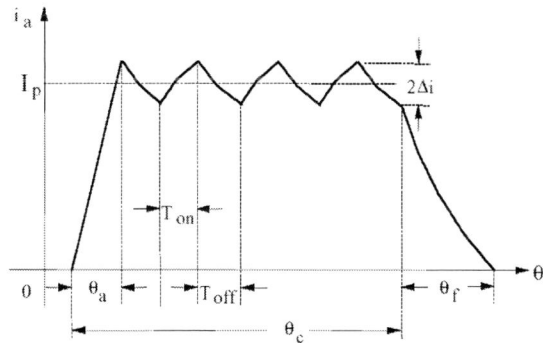

Figure 7. I_p is the chopping current of the winding with the upper limitation $I_p + \Delta i$ and the lower limitation $I_p - \Delta i$

The simulation results of the waveforms on the basis of Matlab/Simulink are as follows in Figure 8:

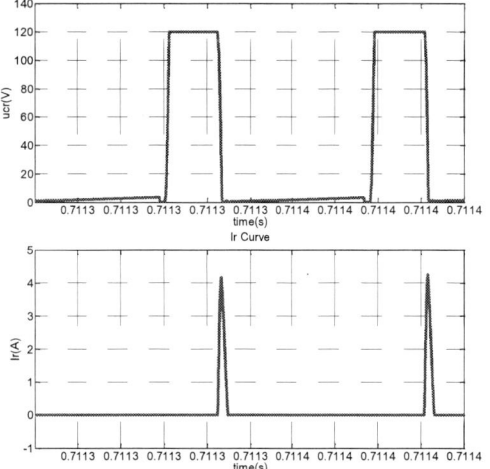

978-1-5090-0064-7/15 $31.00 © 2015 IEEE

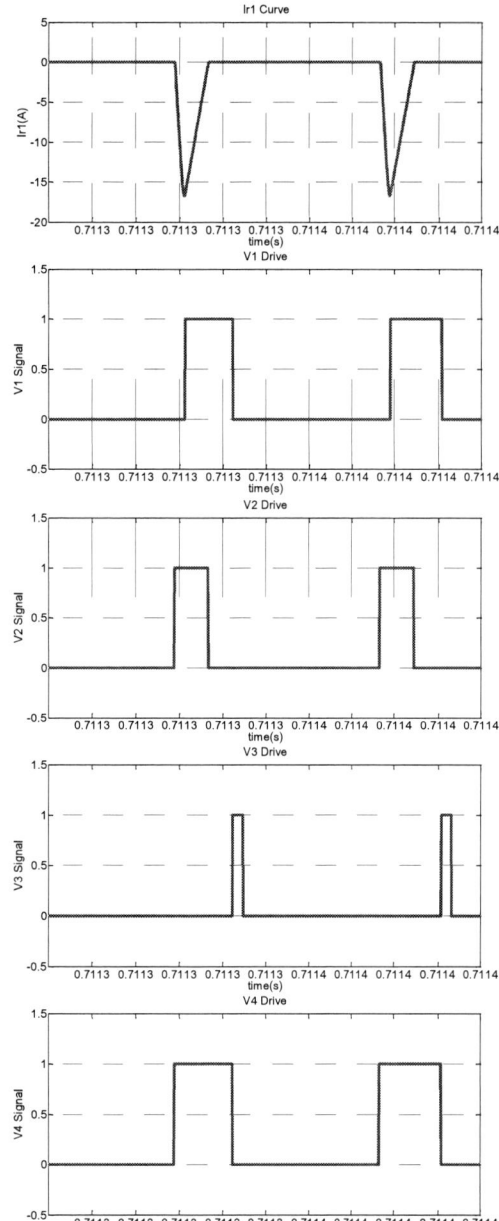

Figure 8. Waveform from Matlab/Simulink of the improved topology when applied to a CCC system for a 6/4 SRM

IV. COMPARISON OF THE IMPROVED SOFT-SWITCHING TOPOLOGY WITH THE CONVENTIONAL ONE IN THE APPLICATIONS OF ELECTRIC VEHICLES

According to the analysis of the mode of the improved topology, all the switching elements achieve the ZVS or ZCS on-off, so the switching loss is zero. However, because of increasing the number of auxiliary switches and diodes, the conducting loss will rise. It is known that the losses include the sorts of switching and conducting. Therefore, a comparison of loss between the proposed improved topology and the traditional one is needed.

The additional conducting losses come from the clamped diode of V_1, diodes D_1, D_2, D_3 and IGBT V_1, V_2, V_3. f_0 in

the following equations is the switching frequency of current chopping control.

$$P_D = f_0 U_{ce} \int_0^{T_7} (-\frac{\sqrt{2}}{2} E C_r \omega_r + \frac{E}{4L_{r1}} t) dt \tag{40}$$

$$P_{D3} = f_0 U_{ce} \int_0^{T_3} (I_1 - \frac{Et}{4L_r}) dt \tag{41}$$

$$P_{D2} = f_0 U_{ce} (\int_0^{T_2} \frac{3}{4} E \omega_r C_r \sin(\omega_r t) dt + \int_0^{T_3} (I_1 - \frac{Et}{4L_r}) dt) \tag{42}$$

$$P_{D1} = f_0 U_{ce} [\int_0^{T_5} \left(-\frac{3E}{4L_{r1}} t\right) dt + \int_0^{T_6} \left(-I_0 - \frac{3}{4} E C_r \omega_r \sin(\omega_r t)\right) dt + \int_0^{T_7} \left(-I_0 - \frac{\sqrt{2}}{2} E C_r \omega_r + \frac{E}{4L_{r1}} t\right) dt + \int_0^{T_8} (-I_0 + \frac{E}{4L_{r1}} t) dt] \tag{43}$$

$$P_{V_1} = I_0^2 R T_1 + \int_0^{T_8} (\frac{E}{4L_{r1}} t)^2 R dt \tag{44}$$

$$P_{V_2} = \int_0^{T_5} (\frac{3E}{4L_{r1}} t)^2 R dt + \int_0^{T_6} [-I_0 - \frac{3}{4} E C_r \omega_r \sin(\omega_r t)]^2 R dt + \int_0^{T_7} \left(-I_0 - \frac{\sqrt{2}}{2} E C_r \omega_r + \frac{E}{4L_{r1}} t\right)^2 R dt + \int_0^{T_8} (-I_0 + \frac{E}{4L_{r1}} t)^2 R dt \tag{45}$$

$$P_{V_3} = \int_0^{T_2} [\frac{3}{4} E \omega_r C_r \sin(\omega_r t)]^2 R dt + \int_0^{T_3} (I_1 - \frac{Et}{4L_r})^2 R dt \tag{46}$$

Where D is the clamped diode of V_1, U_{ce} is the forward voltage drop of a diode.

For the hard switching period in a conventional topology for the SRM, the switching process of the voltage and current can be simplified as the following equation as:

$$u_{ce} = E - \frac{E}{\Delta t} t \tag{47}$$

$$i_{ce} = \frac{I_0}{\Delta t} t \tag{48}$$

$$P_{sw} = \frac{1}{T_0} \int_0^{T_0} u_{ce} \cdot i_{ce} dt$$
$$= \frac{2}{T_0} \int_0^{\Delta t} u_{ce} \cdot i_{ce} dt$$
$$= \frac{1}{3} E I_0 f_0 \Delta t \tag{49}$$

Where, the switching period T_0 is 4×10^{-5}s, voltage rising and current descending time Δt is 2×10^{-7}s.

In Table 3, power loss of both topologies is shown as below:

Table 3: Power Loss Of The Conventional And Improved Topology

n(r/min)	Conventional		Soft-switching		Percentage saved(%)
	I_0(A)	P	I_0(A)	P	
100	9.0	1.80	8.80	1.19	33.9
200	9.8	1.96	9.60	1.30	33.7
300	10.6	2.12	10.4	1.41	33.5
400	11.3	2.26	11.1	1.51	33.2
500	12.0	2.40	11.8	1.60	33.3
600	12.6	2.52	12.3	1.67	33.7
700	13.3	2.66	13.0	1.77	33.5
800	13.8	2.76	13.5	1.84	33.3
900	14.3	2.86	14.0	1.91	33.2

| 1000 | 14.8 | 2.96 | 14.5 | 1.98 | 33.1 |

The comparison curve of both topologies for the power loss is shown in the following Figure 9:

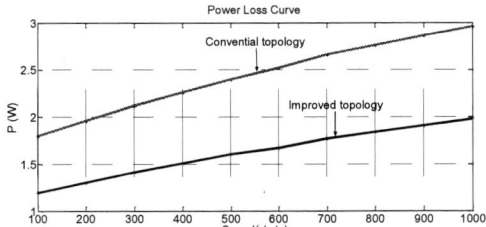

Figure 9. comparison of power loss for different topologies under various speed

From the comparison between these two topologies, less power loss of the proposed topology is seen in the table and figure, about 33 percent of energy saving. Furthermore, with the vehicle speed increasing, the current produced by the DC source ascends, thus increasing the energy or power loss.

VI. CONCLUSION

This paper proposes two topologies of soft-switching for the SRM drive. Each topology is analysed theoretically and calculated by equations to gain the ideal waveforms of the auxiliary components in the soft-switching part. These topologies also have their own deficiencies. The first one has current flowing through the resonant inductance during normal conducting time increasing extra loss and the zero voltage period can't be controlled. Furthermore, the auxiliary switching element is in the main circuit. As for the second improved topology, it solved most of the problems of the former one except one switching device in the main loop by adding another two assistant switching components. The waveforms of the above two topologies are confirmed by simulation based on Matlab/Simulink. Besides, the second improved topology is applied to the CCC system of a 6/4 SRM to prove that power loss of the proposed one is much less than the traditional one under various running speed for EV. Therefore, the proposed topologies has their potential prospects in the application of EV motor drives.

REFERENCE

[1] Mehrdad Ehsani, Khwaja M. Rahman, Maria D. Bellar, Alex J. Severinsky, "Evaluation of Soft Switching for EV and HEV Motor Drives",IEEE Transactions on Industrial Electronics, Vol. 48, No. 1, February 2001

[2] Luo Jianwu, Zhan Qionghua, Deng Qiong, "Study of a Novel Soft-switching Converter for Switched Reluctance Motor [J]". Proceedings of the CSEE, 2005, 25 (17): 142-149

[3] J. Shukla, B.G. Fernandes, "Three-phase soft-switched PWM inverter for motor drive application", Electric Power Applications, IET, 2007: 93-104

[4] Luo Jianwu, Qionghua Zhan, "A Novel Soft-Switching Converter for Switched Reluctance Motor:Analysis, Design and Experimental Results", Electric Machines and Drives, 2005 IEEE International Conference, pages: 1955-1961

[5] Murai Y, Cheng J, Sugimoto S, Yoshida, M, " A Capacitor-boosted Soft-switched Switched Reluctance Motor Drive

[C]", Proceedings of Applied Power Electronics Conference and Exposition. Piscataway: IEEE Inc, 1999: 424-429.

[6] Ming Zhengfeng, Zhong Yanru, Ning Yaobin, "A Novel Transition DC-rail Parallel Resonant Zero Voltage Three Phase PWM Voltage Source Inverter", Transactions of China Electrotechnical Society, 2001, 16(6): 31-35.

[7] Chen Guocheng, Sun Chengbo, Zhang Linglan, "The Analysis of a Novel ZVS Resonant DC-link Inverter Topology", Transactions of China Electrotechnical Society,2001, 16(4): 50-55

[8] Pan Zhiyang, Luo Fanglin, "Novel Soft Switching Inverter for Brushless DC Motor Variable Speed Drive System", IEEE Transactions on Power Electronics, 2004, 19(2): 280-288.

[9] Wang Qiang, Wang Tianshi, " A Parallel Resonant DC Link Inverter Applied to Motor Drives", Electric Machines And Control, 2013, Vol.17, No.1.

[10] Yang Jinling,Zhang Yingjun,Xie Binhong, "A new SRM soft switching power circuit", Journal of China Coal Society,2014,39(1):179-185

BIOGRAPHIES

Jingwei Zhu received his B.S. degree in 2015 in Electrical Engineering & Automation from Zhejiang University, Hangzhou, China.
He joined the department of Electrical Engineering, The Hong Kong Polytechnic University as a postgraduate. His main research interests are in the field of power electronics and motor drives.

K.W.E.Cheng obtained his BSc and PhD degrees both from the University of Bath in 1987 and 1990 respectively. Before he joined the Hong Kong Polytechnic University in 1997, he was with Lucas Aerospace, United Kingdom as a Principal Engineer.
He received the IEE Sebastian Z De Ferranti Premium Award (1995), outstanding consultancy award (2000), Faculty Merit award for best teaching (2003) from the University and Silver award of the 16th National Exhibition of Inventions. Faculty Engineering Industrial and Engineering Services Grant Achievement Award (2006), Brussels Innova Energy Gold medal with Mention (2007), Consumer Product Design Award (2008), Electric vehicle team merit award of the Faculty (2009), Special Prize and Silver Medal of Geneva's Invention Expo (2011) and Eco Star award (2012). He has published over 250 papers and 7 books. He is now the professor and director of Power Electronics Research Centre.

Ideas for Future Electric Aircraft System

S Raghu Raman   K W Eric Cheng

Department of Electrical Engineering, The Hong Kong Polytechnic University, Hong Kong
E-mail: raghu.raman.1990@ieee.org

Abstract – This design paper proposes few improvements on the existing electrical network in the aircraft system. It also suggests improvement in the starter circuitry by introducing super capacitors to reduce the number of batteries onboard to ensure improved safety and reliability. It proposes higher frequency of power distribution owing to several benefits. There are several advantages of high frequency AC power distribution over conventional DC distribution and low frequency AC power distribution. This paper explores the idea of employing switched capacitor and switched inductor converters to design multi-level inverters for high frequency AC power supplies for power distribution.

Keywords – Electric Aircraft, Multi – level inverters, Switched – Capacitor, Switched – Inductor.

I. INTRODUCTION

High frequency AC (HFAC) power distribution systems (PDS) have been popular since the 1980s when NASA proposed a 20 kHz, 440 V_{rms}, for their space station [1] & [2]. Since then HFAC PDS has emerged as a popular research area and has had several applications utilizing it. HFAC PDSs find application in telecommunication, renewable based micro-grid and computer power supply, aerospace and lighting systems. A comprehensive review on HFAC systems has been done in [3].

Basic power distribution architecture of HFAC PDS and DC PDS is clearly shown in Fig. 1 and Fig. 2. HFAC PDSs offer several benefits in comparison to conventional DC distribution systems. They include –
- Cost reduction due to reduction in the number of power conversion stages.
- Overall improved efficiency
- DC PDS target low voltage high current PDSs. Such systems are extremely difficult to design and demand novel control and converter topologies for efficient operation. On the other hand, a high voltage AC, low current system can be easily realized in HFAC system by using a simple HF transformer that easily steps up the voltage. This helps in minimizing the copper loss.
- Improved reliability with the number of power conversion operations decreasing thereby decreasing the semiconductor components
- Galvanic isolation with high frequency transformer
- DCPDS show poor dynamic response in comparison with HFAC PDS
- Higher power density owing to high frequency operation.

In spite of several exceptional advantages, there are a few drawbacks to the system as well, listed out below –

- Higher Electromagnetic Interference (EMI) effect hinders HFAC applications
- At high frequencies, skin and proximity effects increases leading to more loss.
- High frequency power distribution amplifies impedance in the transmission line which makes it difficult to transmit power
- Connecting high frequency inverters in parallel to realize higher power is difficult due to difference in phases of voltage.

This design paper discusses the advantages of employing HFAC PDS systems on aircraft and also explores the idea of switched capacitor and switched inductor based converters' role in designing multi-level voltage source and current source inverters respectively, for more electric aircrafts. Switched inductor converters can be derived using duality principle introduced by Prof. Cheng [4]. Switched inductor based converters are an attractive solution to be used current source inverters. Section II introduces aircraft power system standards and discusses about the existing aircraft power system and suggests improvement on the same. Section III looks into HFAC multilevel inverters employing switched capacitor and possibly switched inductor topologies. Concluding remarks are given in section IV.

II. POWER SYSTEM IN AIRCRAFTS

The power system design and components of an aircraft must be extremely robust and must meet certain stringent standards before they are allowed to be employed onboard. Some of the important standards include:

1. MIL-STD-704F – This standard focuses on the quality of electric power at the input terminals of the utilization equipment. However, this does not include EMI issues [5]
2. MIL-STD-461E – This standard is to control EMI characteristics of electrical equipment of aircraft. [6]
3. MIL-STD-810F – This looks into the stresses that the materials are under during the service period and material system performance requirements [7].
4. MIL-STD-1275D – This standard covers regulations for the 28 V DC power distribution systems in military vehicles including aircraft [8].

All the standards listed above are developed by the team of researchers from the department of defense, USA. The main focus of the standards is to ensure safety and compatibility among different systems onboard. All systems for an aircraft have to pass several such standards before they are employed in the aircraft.

Conventional power distribution system of an aircraft is shown in Fig. 3. With advances in enabling technologies

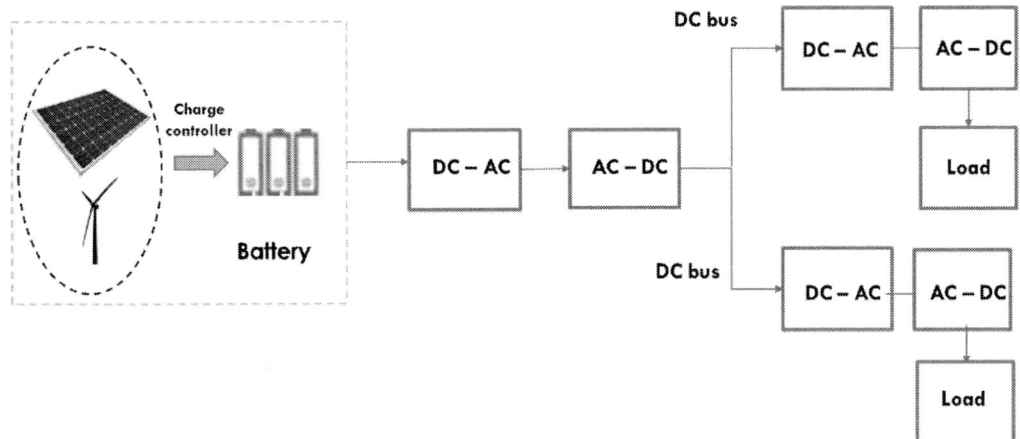

Fig. 1. General DC Power Distribution Architecture

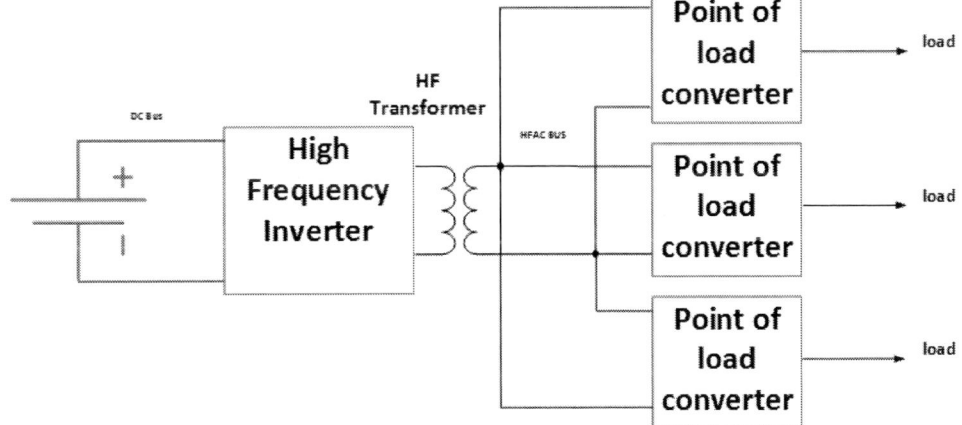

Fig. 2. HFAC Power Distribution Architecture

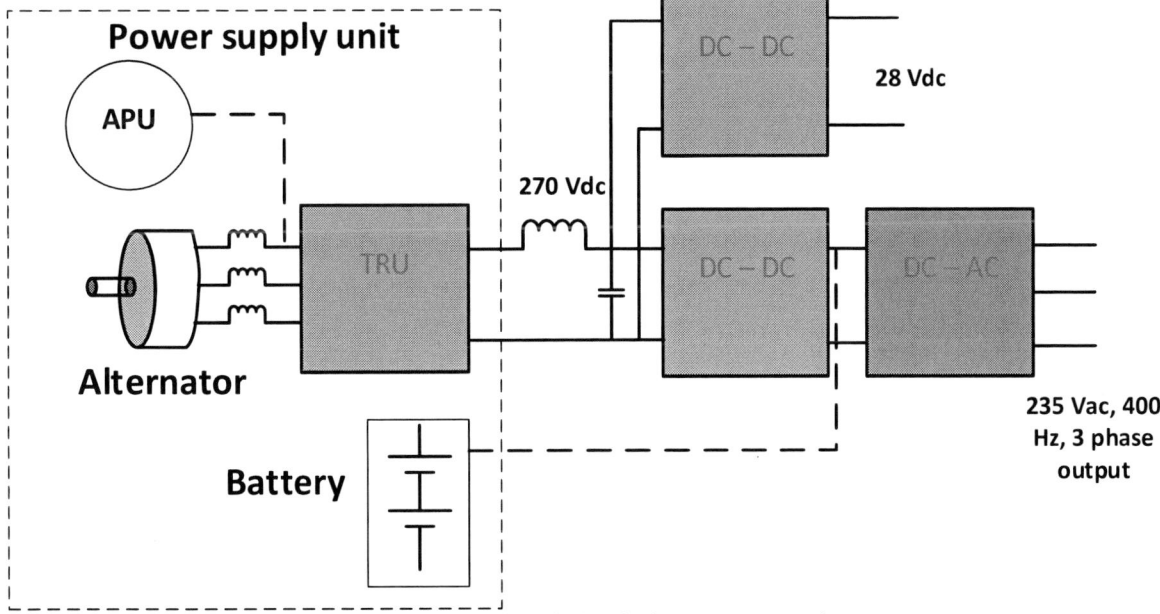

Fig. 3. Present aircraft electrical power system unit

978-1-5090-0064-7/15 $31.00 © 2015 IEEE   343

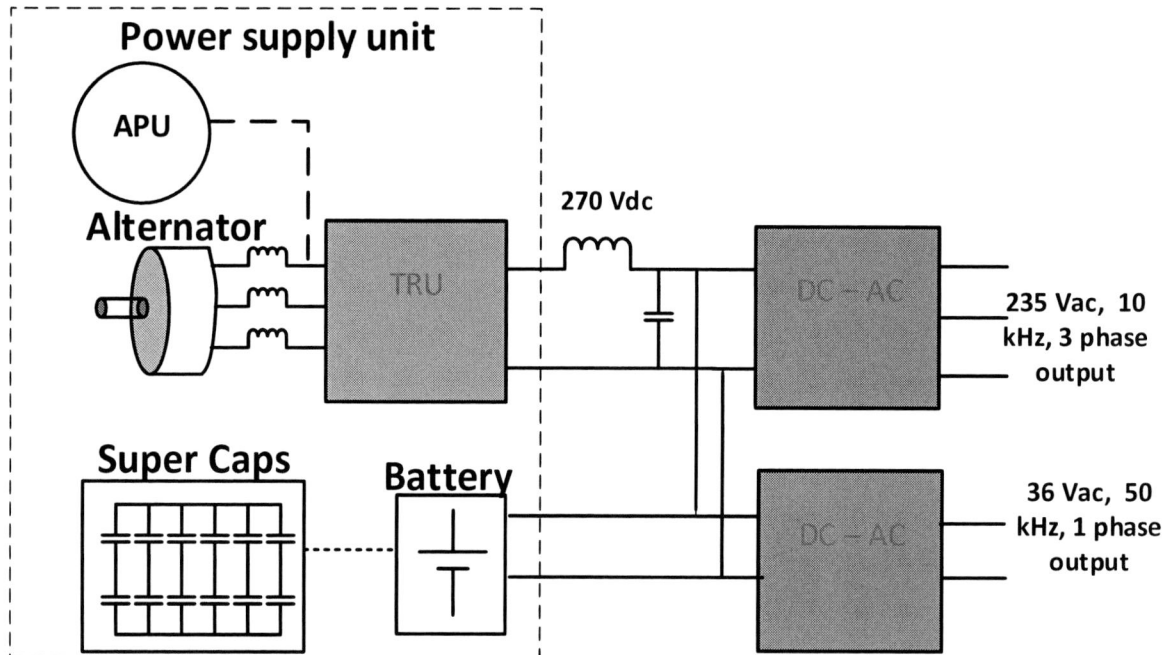

Fig. 4 Proposed electrical system architecture

like power electronics, motor drives, improved thermal management and better materials, all electric aircrafts will soon be a possibility. All electric aircrafts offer improved efficiency, reduction in cost, better reliability, maintenance, improved reliability, batter maneuvering capabilities, enhanced safety and greener systems. A recent example is the electrical system of Boeing's 787 dreamliner [9] [10].

From Fig. 3, it can be observed that the maximum distribution frequency is 400 Hz. There have been several papers on higher frequency range of power distribution for telecommunication, computer power supply, vehicular auxiliary power supply and micro grid applications [3]. It is imperative that we employ such systems inside aircraft for aforementioned advantages. The proposed system, in Fig. 4, replaces 28 Vdc system with a 36 V, 50 kHz power distribution system. The entire aircraft power distribution now is HFAC. This helps in higher power density operation which helps in reducing fuel consumption. HFAC operation, higher than 10 kHz, is safer for human than DC [11].

The new design also incorporates super capacitors into the system. The main purpose of introducing super capacitors is for its high power density. It can be extremely useful during starting of engine where traditionally bulky batteries are used. Additionally, super capacitors are much safer than batteries as they utilize electric field whereas batteries use chemical reaction for energy conversion. Due to the same reason, super capacitors have quicker response time in comparison to batteries. Ideally, one can expect a battery free future electric aircraft. However, this is possible only if research leads us to high energy density super capacitors or reliable fuel cell technology.

Super capacitors also play a crucial role in tapping the regenerative energy when the aircraft lands. Enormous

weight thrusts onto the ground and this energy can be used by using the principle of regeneration of motors. Safe energy harvesting on a large scale is possible by employing large super capacitor banks that can sink in high magnitudes of current. This system design is not easy with only batteries as the power source as batteries have lower charging current rating in comparison with super capacitors and therefore all energy regenerated may not be successfully stored.

III. HF MULTILEVEL INVERTERS

There are plethora of advantages that multilevel inverters offer in comparison to traditional ones [11][12]. In general, they include –

1. Better quality output voltage with lower distortions and *dv/dt*.
2. Input current drawn has low distortion.
3. Lower voltage rating and stress on semiconductor switches

HFAC multilevel inverters are possible by using simple switched capacitor techniques as elaborated in [13][14]. These inverters fully utilize the features of multilevel inverters and apply it to high frequency power distribution systems. These multilevel HFAC inverters can be employed in aircraft systems. A distributed power system consisting of several inverters to cater to different areas of the aircraft loads under a unified controller for the entire aircraft would be a novel and practical design. Using the duality principle [4], future electric aircraft with solar panels embedded onto them can employ multilevel high frequency switched inductor based current source inverters. An example of switched inductor based multilevel CSI derived from switched capacitor based has been shown in Fig.5.

978-1-5090-0064-7/15 $31.00 © 2015 IEEE 344

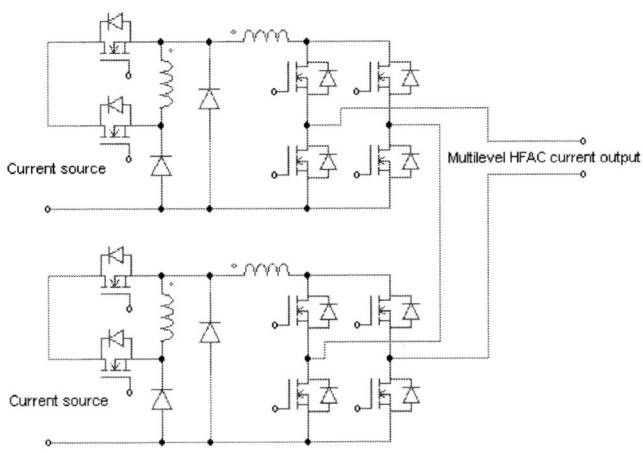

Fig. 5 HFAC 9-level switched inductor based CSI derived from switch cap based from [14] using duality principle

CONCLUSION

This competition paper focuses on simple ideas to improve the existing electrical system of an aircraft and supports the global initiative for more / all electric aircrafts in future. The paper discusses the possibility of HFAC power distribution systems inside the aircraft to utilize several benefits it offers. Incorporating super-capacitors into the system offers several advantages during starting and regenerative braking. Multilevel switched capacitor and switched inductor based HFAC VSI and CSI respectively offer good features that improve the overall design of the power distribution system in an aircraft.

REFERENCES

[1] Status of 20 kHz Space Station Power Distribution Technology. NASA Publication, TM 100781.

[2] Renz, David D., et al. Design considerations for large space electric power systems. No. N-8324552; NASA-TM-83064; E-1535. National Aeronautics and Space Administration, Cleveland, OH (USA). Lewis Research Center, 1983.

[3] Jain, P.; Pahlevaninezhad, M.; Pan, S.; Drobnik, J., "A Review of High-Frequency Power Distribution Systems: For Space, Telecommunication, and Computer Applications," in Power Electronics, IEEE Transactions on , vol.29, no.8, pp.3852-3863, Aug. 2014.

[4] Cheng, K.W.E.; Yuan-mao Ye, "Duality approach to the study of switched-inductor power converters and its higher-order variations," in Power Electronics, IET , vol.8, no.4, pp.489-496, 4 2015

[5] URL - http://everyspec.com/MIL-STD/MIL-STD-0700-0799/MIL_STD_704_1080/ (Last accessed 15th Oct 2015)

[6] URL - http://everyspec.com/MIL-STD/MIL-STD-0300-0499/MIL-STD-461E_8676/ (Last accessed 15th Oct 2015)

[7] URL - http://everyspec.com/MIL-STD/MIL-STD-0800-0899/MIL_STD_810F_949/ (Last accessed 15th Oct 2015)

[8] URL - http://everyspec.com/MIL-STD/MIL-STD-1100-1299/MIL-STD-1275D_5431/ (Last accessed 15th Oct 2015)

[9] URL - (Last accessed 15th Oct 2015) http://www.boeing.com/commercial/aeromagazine/articles/qtr_4_06/article_04_3.html

[10] URL - (Last accessed 15th Oct 2015) http://www.boeing.com/commercial/aeromagazine/articles/qtr_4_07/article_02_3.html

[10] Patel, Mukund R. Spacecraft power systems. CRC press, 2004, ch. 22, sec. 22.7, pp. 539-543.

[11] Kouro, Samir, et al. "Recent advances and industrial applications of multilevel converters." Industrial Electronics, IEEE Transactions on 57.8 (2010): 2553-2580.

[12] Rodriguez, Jose, Jih-Sheng Lai, and Fang Zheng Peng. "Multilevel inverters: a survey of topologies, controls, and applications." Industrial Electronics, IEEE Transactions on 49.4 (2002): 724-738.

[13] Ye, Yuanmao, et al. "A Step-Up Switched-Capacitor Multilevel Inverter With Self-Voltage Balancing." Industrial Electronics, IEEE Transactions on 61.12 (2014): 6672-6680.

[14] Liu, Junfeng, K. W. E. Cheng, and Yuanmao Ye. "A cascaded multilevel inverter based on switched-capacitor for high-frequency AC power distribution system." Power Electronics, IEEE Transactions on 29.8 (2014): 4219-4230.

Application of Cuk converter together with Battery Technologies on the Low Voltage DC supply for Electric Vehicles

Wenzheng Xu[1]   K.W.E.Cheng[2]   K.W.Chan[3]

[1] Department of Electrical Engineering, The Hong Kong Polytechnic University, Hong Kong
E-mail: xuwenzheng2012@gmail.com
[2] Department of Electrical Engineering, The Hong Kong Polytechnic University, Hong Kong
E-mail: eeecheng@polyu.edu.hk
[3] Department of Electrical Engineering, The Hong Kong Polytechnic University, Hong Kong
E-mail: eekwchan@polyu.edu.hk

Abstract–Electric vehicles (EVs) have becoming more and more popular, and although there is no doubt that motor is the key part of a functional EV, it is generally accepted that the battery is the critical component and a main obstacle of the development of EVs. In this paper, some advanced research and technology of batteries is presented and compared one by one, especially those adopted by flagship EV automakers. Some emerging batteries which have great potential for future EVs are also discussed. Based on this, the paper studies the application in EV of Cuk converter, which can be regarded as a combination of boost converter and buck converter, on the low voltage side DC supply electric vehicles together with the batteries, and even super-capacitors which have unique characteristics. Further research and design is given regarding to multi-switching of the DC-DC topology.

Keywords–Electric vehicle, Battery technology, EV power system, Cuk converter, super-capacitors.

I. INTRODUCTION

In recent decades, the implementation of Electric Vehicles (EVs) is a fascinating choice for no pollution, zero roadside emission, higher energy efficiency and active participation in load management in future, compared to traditional Internal Combustion Vehicles (ICVs) [1] because of the worldwide acknowledge of energy saving and renewable energy's importance which has great relationship with environment protection. At this moment, however, the main obstacles of the implementation of EVs are battery technology and construction charging stations. Generally it takes long time to charge the batteries for a normal range of EVs. If we implement the fast charging, not only produces greater burden or impact to the grid, but may also shorten the life cycle of batteries. It is widely accepted that battery is the critical component in the development of EVs. [2]

In this paper, some major basic types of batteries and their performance for EVs are presented and compared. The latest batteries that have been adopted by automakers for their flagship electric vehicles are discussed. Some specific emerging batteries that have higher potential for EVs in the future are also presented and discussed, together with some progress and improvement in the battery technology.

This paper also designs the application of Cuk converter on the low voltage side DC supply electric vehicles together with batteries, especially in starting-up and braking state. A well-designed combination topology of DC-DC converter and batteries is important to the whole EV system. Because of many advantages of Cuk converter, its application is very suitable and flexible in electric vehicle system. This paper further studies the effect of energy variation ratio of the Cuk converter in efficiency perspective and provides guidelines for designing a Cuk converter's application in electric vehicles.

II. BASIC TYPES OF BATTERIES AND COMPARISON

An electric vehicle battery can be either a primary battery or a secondary battery rechargeable battery used for the propulsion of EVs. Electric vehicle batteries consist of many sub-categories, including but is not limited to the lighting (especially emerging LED lighting technology), starting-up, and ignition battery, since they are designed to give power over sustained periods of time. Rechargeable batteries are usually the most expensive component of EVs, being about half the retail cost of the car. On an energy basis, the price of electricity to run an EV is a small fraction of the cost of liquid fuel needed to produce an equivalent amount of energy.

2.1. Types of batteries utilized in EVs at present

In the foreseeable future, batteries are the major energy source for EVs, with their better overall performance than other energy sources like ultra-capacitors, ultrahigh-speed wheels, and fuel cells. Major types of batteries that have been developed for EVs are: valve-regulated lead acid (VRLA), nickel-cadmium (Ni-Cd), nickel-metal hydride (Ni-MH), zinc/air (Zn/air), sodium/sulfur (Na/S), lithium-ion (Li-ion), vanadium redox, and molten salt battery. [3]

Table 1. Batteries comparison for EVs [3]

Type	Energy density (MJ/kg)	Power density (W/kg)	(Dis-)Charge Efficiency (%)	Life Cycles
Lead-acid	0.11-0.14	180	70-92	400-600
Ni-Cd	0.14-0.22	150-350	70-90	600-1200
NiMH	0.11-0.29	250-1000	66	600-1200
Li-ion	0.58	1800	99.9	1200-2000
Molten	N/A	150-220	69	1500
Zn/air	0.36	105	N/A	2000
Na/S	0.32-0.56	200	89-92	800

VRLA is more commonly known as a sealed battery. It is a lead-acid rechargeable battery. Because of their construction, VRLA batteries do not require regular

978-1-5090-0064-7/15 $31.00 © 2015 IEEE

addition of water to the cells, and vent less gas than flooded lead-acid batteries. [4] The reduced venting is an advantage since they can be used in confined or poorly ventilated spaces. It is widely used in low-cost end EVs.. Vehicles used in auto racing may use AGM batteries due to their vibration resistance.

At the moment the Li-Ion technology is the most promising for the usage in EVs. Its high specific energy density surmounts the values of all the other technologies known so far. The higher the energy density the lower is the additional weight of the car, which means a higher range. None of the batteries can offer high energy density and high power density at the same time. Thus, a compromise between the two or a combination of two batteries is often needed. Here is a brief table showing how the various electric car batteries compare with each other. The specific energy density of different technologies is shown in the figure below. Another advantage of Lithium batteries is the high potential for an increasing energy density. [5]

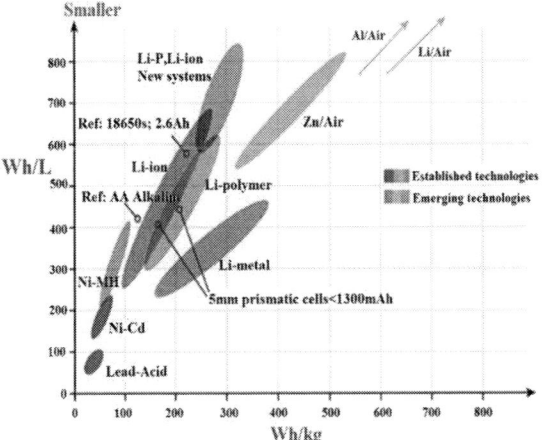

Fig.1: Battery power and energy density [5]

Ni-Cd is a type of rechargeable battery using nickel oxide hydroxide and metallic cadmium as electrodes. Larger flooded cells are used for EVs. Ni-MH is very similar to Ni-Cd. NiMH use positive electrodes of nickel oxyhydroxide (NiOOH), like the NiCd, but the negative electrodes adopts a hydrogen-absorbing alloy rather than cadmium, becoming a practical application of nickel–hydrogen battery chemistry. [6]

Vanadium redox battery is a type of rechargeable flow battery which employs vanadium ions in different oxidation states to store chemical potential energy. [7] The vanadium redox battery exploits the ability of vanadium to exist in solution in four different oxidation states, and uses this property to make a battery that has just one electro active element instead of two.

Li-ion is a member of a family of rechargeable battery types in which lithium ions move from the negative electrode to the positive electrode during discharge, and back when charging. [8] Because of their light weight Li-ion batteries are used for energy storage for many electric vehicles for everything from electric cars to Pedelecs, from hybrid vehicles to advanced electric wheelchairs, from radio-controlled models and model aircraft to the Mars Curiosity rover.

Zn/air is electro-chemical batteries powered by oxidizing zinc with oxygen from the air. These batteries have high energy densities and are relatively inexpensive to produce. Sizes range from very small button cells for hearing aids, larger batteries used in film cameras that previously used mercury batteries, to very large batteries used for electric vehicle propulsion. [10]

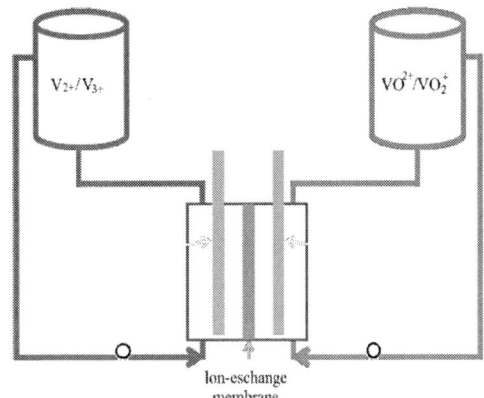

Fig.2: Schematic of a vanadium redox-flow battery. [9]

A sodium–sulfur battery is a type of molten-salt battery constructed from liquid sodium (Na) and sulfur (S). This type of battery has a high energy density, high efficiency of charge/discharge (89–92%) and long cycle life, and is fabricated from inexpensive materials. [11] It needs high temperature to operation.

Molten salt battery is a class of primary cell and secondary cell high-temperature electric battery that uses molten salts as an electrolyte. They offer both a high energy density through the proper selection of reactant pairs as well as a high power density by means of a high-conductivity molten salt electrolyte. [12]

Some researchers have compared many details of various kinds of EV batteries including the size, power and energy density respectively, and life cycle. Which kind of battery is better, however, depends on consumer's criterion and requirement, such as their expected cost and energy density. Some researchers are also considering new type batteries. IBM has declared a research plan to develop various kinds of batteries for EVs [13], including Sodium batteries, Li-O2 batteries, lithium-air batteries, and will adopt nanotechnology.

In total, Lithium-ion batteries are still the most attractive among other ones for EVs. Researchers are trying to develop Li-ion batteries with better quality and less cost, and many governments and companies also pay attention to support the development of Li-ion batteries.

III. DESIGN OF CUK CONVERTER'S APPLICATION IN EV

Nowadays Electric vehicles have becoming more and more popular because of the worldwide acknowledge of energy saving and renewable energy's importance which has great relationship with environment protection. An electric vehicle consists of many systems, while the low voltage dc supply part plays key important role, especially in emergency state.

978-1-5090-0064-7/15 $31.00 © 2015 IEEE

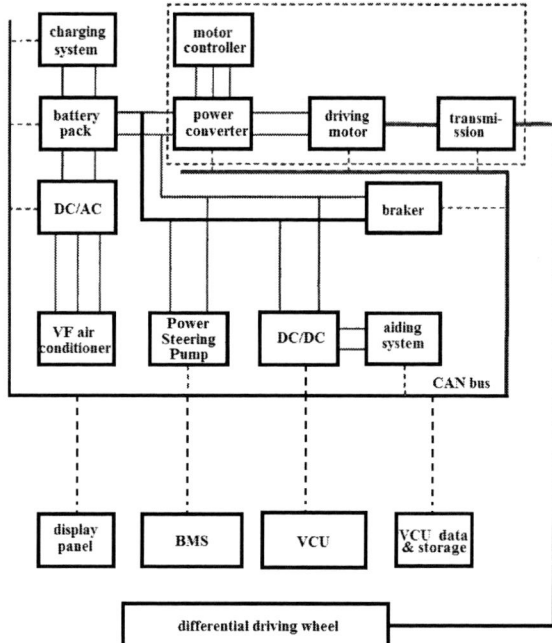

Fig.3: Diagram of EV power flow [14]

Cuk converter is a basic non-isolated DC-DC converter topology which can be regarded as combination of boost converter and buck converter. The output voltage can be adjusted to be larger or smaller than the input voltage, which is very flexible.

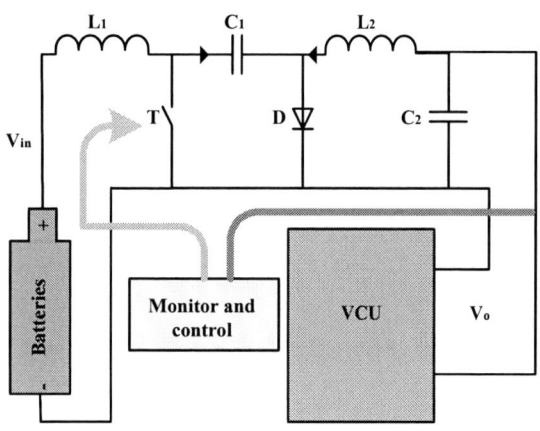

Fig.4: Design of Cuk converter applied on EV

Cuk converter has many advantages among DC-DC topologies. We designed the low voltage side DC power supply by Cuk converter as figure 4 shows. Batteries serve as the input voltage source of the Cuk converter, and the output of the converter is connected to the Vehicle Control Unit (VCU) which normally has a voltage rating of around 20V, and other low-power dc devices. The green line represents feedback and monitoring of the Cuk converter, whereas the yellow line represents the control of the MOSFET of the Cuk converter. No matter what's the value of the battery voltage, Cuk converter can boost or reduce it into an expected voltage level. We select Li-ion

battery as the research target since it is the most popular one in current EVs.

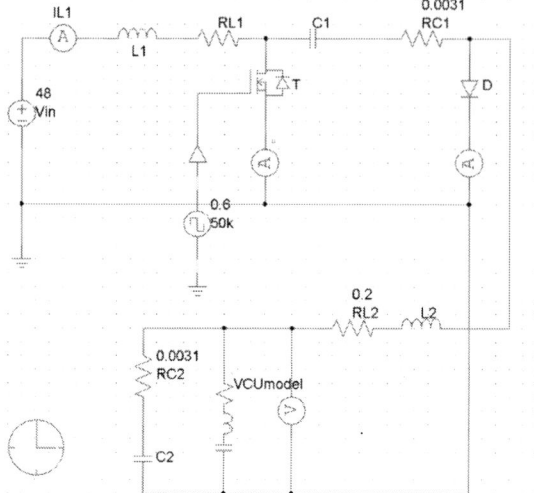

Fig.5: Simulation diagram in PSIM

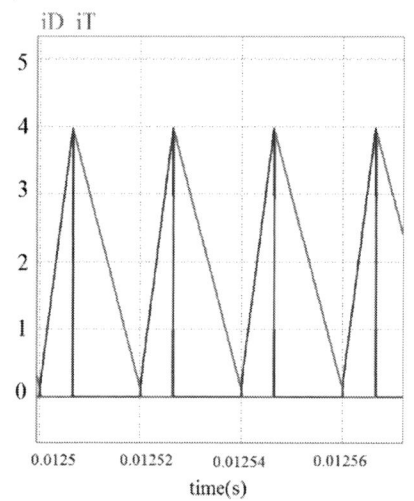

Fig.6: Waveform of diode and transistor current

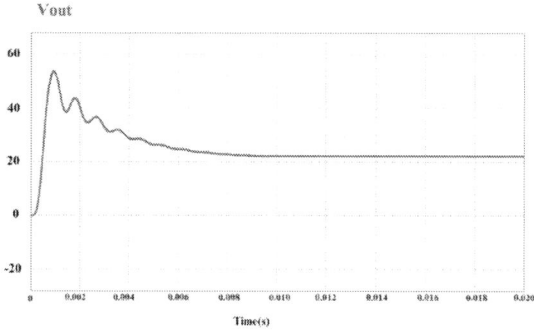

Fig.7: Waveform of the output voltage at the beginning

Simulation of the Cuk converter together with battery and VCU model is conducted in PSIM platform. The diagram is shown in the above figures. The input voltage is set at 48V while the goal of output voltage is 20V. Calculation of the parameters of each device of the Cuk converter is a bit complicated because of the VCU and other devices' model. Combining the simulation result and calculation analysis, we set the inductor at 200uH level, capacitor C1 at 10uF level and C2 at 1mF level, which could bring ideal output. Simulation result is shown as figure XX and XX, and we can see the Cuk converter operates at good condition.

IV. FURTHER RESEARCH OF MULTI-SWITCHING AND SUPERCAPACITORS

The staring-up and braking state during the EV's operation is very important and requires more stable power supply. The instantaneous larger current of the machine would have influence on the battery and the low-voltage DC side's state. In this way, a Cuk converter with multi-switching is proposed to balance this problem.

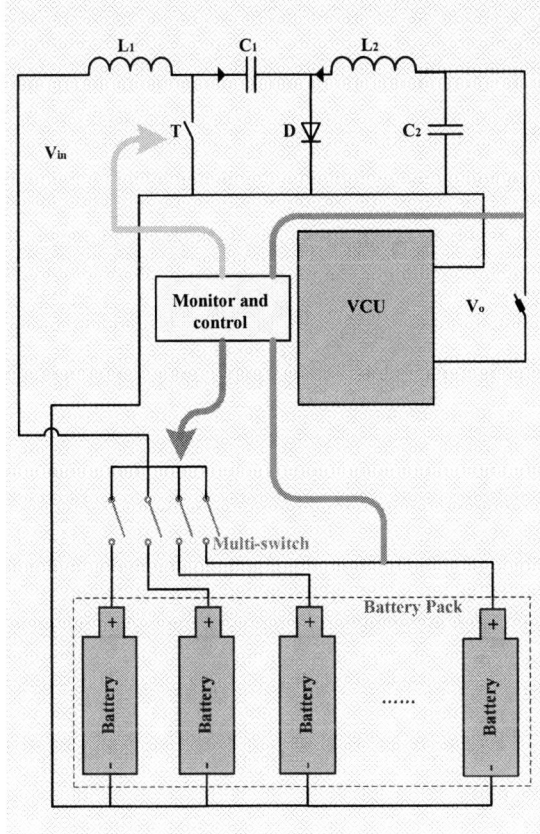

Fig.8: Cuk converter's application with multi-switch

Figure 8 shows the topology diagram of this design. A battery pack is composed of many battery cells. They can be in series with each other and divided into several groups where the voltage of each group is 20-50V. A multi-switch is connected to each group as the figure shows. Also, the green line represents the feedback and monitoring of the condition of output voltage of the Cuk converter and the battery's charging or discharging current and State of Charge (SoC). The red line represents its control over the multi-switch, which plays an important role in the whole design.

During normal driving state of the EV, the multi-switch is connected to one battery at a time. It would switches to another battery after a certain period. During special

condition, the multi-switch would follow rules to handle the connection. For example, during the starting-up and braking state, the multi-switch would connect more than one battery, to avoid unstable power flow. If the SoC of one battery is too low, the multi-switch would not connect it until being charged.

Now researchers are trying to apply super-capacitors to replace part of the EV battery as the power source and storage. Generally one single super-capacitor could only provide 1-5V voltage, thus it must be used in series to improve the output voltage. Cuk converter has the advantage of boosting or reducing DC voltage freely, so it is a good method that combines Cuk converter and super capacitors together. Whereas the disadvantage is the whole system becomes more complicated and the design of multi-switch is a bit harder with higher cost.

V. CONCLUSION

This paper represents some advanced research and technology of batteries and compares them one by one, especially those adopted by flagship EV automakers. Besides, the paper studies the application in EV of Cuk converter, which can be regarded as a combination of boost converter and buck converter, on the low voltage side DC supply electric vehicles together with the batteries, and even super-capacitors which have unique characteristics. Simulation of the Cuk converter is conducted which shows stable operation. Further research of multi-switching together with Cuk converter is given which solve the problem of start-up and braking pulse together with the interrupt of battery state. There is great potential for the application of Cuk converter in EV platform because of a series of unique advantages and feature.

REFERENCES

[1] G. Maggetto, J. Van Mierlo, "Electric and Electric Hybrid Vehicle Technology: A Survey", Electric, Hybrid and Fuel Cell Vehicles, IEE Seminar, Durham, April 2000

[2] Sandeep Dhameja, Electric Vehicle Battery Systems, Newnes, Oxford, October 2001

[3] Electric Vehicle Battery Information, [cited 2015 Apr 5th] Available from http://www.madkatz.com/ev/battery.html

[4] Valve-Regulated Lead Acid (VRLA), [cited 2015 Nov 8th] Available from https://www.gsbattery.com/content/vrla

[5] What is your battery size, [cited 2015 Sep 2nd], Available from http://www.green-and-energy.com/blog/whats-is-your-battery-size/

[6] Electric vehicle batteries – a guide, [cited 2015 Apr 8th] http://www.thegreencarwebsite.co.uk/blog/index.php/electric-vehicle-batteries-a-guide/

[7] Peterson, Scott B., J. F. Whitacre, and Jay Apt. The economics of using plug-in hybrid electric vehicle battery packs for grid storage. Journal of Power Sources 195.8 (2010): 2377-2384.

[8] Kim I S. Nonlinear state of charge estimator for hybrid electric vehicle battery. IEEE Transactions on Power Electronics, 2008, 23(4): 2027-2034.

[9] David Linden, Thomas B. Reddy. Handbook of Batteries 3rd Edition, McGraw-Hill, New York, 2002 ISBN 0-07-135978-8

[10] Wang C S, Stielau O H, Covic G. Design considerations for a contactless electric vehicle battery charger. IEEE

Transactions on Industrial Electronics, 2005, 52(5): 1308-1314.

[11] Sallán, Jesús, et al. Optimal design of ICPT systems applied to electric vehicle battery charge. IEEE Transactions on Industrial Electronics, 56.6 (2009): 2140-2149.

[12] Vanadium Redox-Flow Battery and its advantages & disadvantages, [cited 2015 Sep 8th], Available from http://large.stanford.edu/courses/2011/ph240/xie2/

[13] Kutkut N H, Wiegman H L N, Divan D M, et al. Design considerations for charge equalization of an electric vehicle battery system. IEEE Transactions on Industry Applications, 1999, 35(1): 28-35.

[14] Controlling technology of renewable energy electric vehicles and construction, [cited 2015 Oct 13th], Available from http://gw.greenwheel.com.cn/2015/4933.html

BIOGRAPHY

Wenzheng Xu received his B.E.E. degree from the Department of Electrical Engineering , Beijing Jiaotong University, Beijing, China, in 2012, and received the M.Sc. degree from Department of Electrical and Electronic Engineering, The University of Hong Kong, Hong Kong, in 2013. He is currently working toward the Ph.D. degree in the Department of Electrical Engineering, the Hong Kong Polytechnic University.

From April to September in 2013, he was a part-time research assistant in Department of Electrical and Electronic Engineering in the University of Hong Kong, where he was involved in researches about smart grid development in China. From September 2013 to June 2015, he was a full-time research associate in Department of Electrical Engineering in the Hong Kong Polytechnic University, where he was the team leader for a silicon-carbide power devices based dc-dc converter project. His research interest includes power electronics topologies and control for switch mode converters.

K.W.E.Cheng obtained his BSc and PhD degrees both from the University of Bath in 1987 and 1990 respectively. Before he joined the Hong Kong Polytechnic University in 1997, he was with Lucas Aerospace, United Kingdom as a Principal Engineer.

He received the IEE Sebastian Z De Ferranti Premium Award (1995), outstanding consultancy award (2000), Faculty Merit award for best teaching (2003) from the University, Faculty Engineering Industrial and Engineering Services Grant Achievement Award (2006), Brussels Innova Energy Gold medal with Mention (2007), Consumer Product Design Award (2008), Electric vehicle team merit award of the Faculty (2009). Special Prize and Silver Medal of Geneva's Invention Expo (2011) and Eco Star award (2012) He has published over 250 papers and 7 books. He is now the professor and director of Power Electronics Research Centre of the university. His research interests are all aspects of power electronics, electromagnetics, motor drives, EMI and energy saving.

Dr K.W. Chan received his BSc(Hons) and PhD Degrees in Electronic and Electrical Engineering from the University of Bath (UK) in 1988 and 1992, respectively. His doctoral research study was in the area of real-time power system transient stability simulation using parallel processing techniques.

From 1993 to 1997 Dr Chan was with the Power System and Energy Group at the University of Bath as a research officer. A number of industrial strength real-time power system simulators were developed and applied in the National Grid Control Centre, UK for daily monitoring and protection of the UK National Grid. Dr Chan joined the department of Electrical Engineering of the Hong Kong Polytechnic University in 1998 as a lecturer and subsequently promoted to Assistant Professor in 2005. So far, Dr Chan has secured over 37 research projects, including 6 RGC-ERG, 2 RGC-Large Equipment, and 2 NSF projects with total funding over $15.4M.

978-1-5090-0064-7/15 $31.00 © 2015 IEEE

A New Two-degree of Freedom switched Reluctance Motor for Electric Vessel

S. Y. Li   K. WE. Cheng

[1,2,] Department of Electrical Engineering, The Hong Kong Polytechnic University, Hong Kong

Abstract - This paper introduces a new 2-degree of freedom (2-DOF) switched reluctance motor for an electric vessel. The configuration, operation principle and algorithm of the new motor are described in detail. Moreover, the electromagnetic characteristics have been illustrated by using finite element method (FEM), and the simulation of the electric vessel also has been realized. A modified control method has been suggested as the control scheme to ensure innovation and stability of the electric vessel's operation. Finally, the experimental results have suggested that the innovation and conveniences of the new electric vessel are better than that of conventional electric vessel, then the results of theoretical analysis and experiments has been proved the improvements of the electric vessel.

Index Terms—2-DOF, switched reluctance motor, electric vessel, FEM

INTRODUCTION

Electric vessel, as a new type of electrical propulsion transportation, has been investigated for years. The motors are practically used for the high power transportations with a high degree of reliability, such as electrical vehicles, aircrafts and electric vessel.

A two - dimensional rotating linear motor is usually used for the industrial motion control equipment. The two-degree switched reluctance motor is an alternative to be selected to drive this motion, such as boring mill, drill press and carving machine, etc [1].

The two-degree motions are required in surface motion or multi-axis applications, such as concentrating photovoltaic generation system in capturing the solar direct light to enhance power generation efficiency. The solar tracking is 2-dimension as the solar path varies throughout a year. Conventional 2-degree of freedom motions are realized by combining two rotary or linear (RL) motors in two directions, integrating screw rods and mechanical gears. For example, helical motion induction machines are reported in ref [2] and [3]. By using those mechanical gears, both the position precision and efficiency of the motion system could be reduced because there are backlash, axis-coordination and losses generated by mechanical gears. This 2-degree switched reluctance motor can realize two-degree movements-rotary and linear movements, directly, which is similar to the multilayer SR motor reported in [4]. This proposed direct-drive scheme is able to solve the problem of the low efficiency and position control caused by backlash and extra mechanical losses in conventional 2-degree of freedom motion platform and hence the whole system efficiency is significantly improved.

In addition, due to the simple mechanical structure of the proposed 2-degree of freedom switched reluctance motor characterized by low cost, high robustness, variable speed regulation, etc., the cost of the whole system could be reduced. Meanwhile, the 2-degree of freedom switched reluctance motor can operate well under high temperature as well as deteriorated working environment because this motor has no permanent magnet.

The invention is the 2-degree of freedom switched reluctance actuator that consists of the frame, stator cores, stator coils, mover cores, air gaps between the stator poles and the mover poles, and the mover shaft, that integrates the longitudinal magnetic structure with the transverse magnetic structure, and relates to electromagnetic rotary-linear actuators. Besides, combining pole structure and small tooth structure can further enhance the efficiency of the rotary-linear actuator that can realize linear motion and rotary motion simultaneously.

The traditional form of rotating linear motion adopts two rotating motors or the combination of a linear motor and a rotating motor for implementation [5]-[7]. However, this combination not only increases size and weight of the drive, but also reduces working accuracy of the equipment. Therefore, to explore an integrated rotating linear motor, an optimized rotary-linear motor based on switched reluctance principle is presented. The structure of the motor for the vessel electrical propulsion system is also discussed and the structure of the motor is optimized in this paper. Until now, the integrated rotating linear motor is based on switched reluctance principle, which has been preliminary researched. A rotating linear switched reluctance motor presented in [8], is to integrate rotating motion and linear motion based on the minimum reluctance principle, while to achieve rotating linear motion, and to reduce volume of the driver and increase accurate position control. However, the coupling of this linear motor has occurred when both rotating motion and linear motion, and the motion of rotating and linear are produced by the same coils, so the motor will produce unwanted linear directional force upon the movement of rotating motion, and vice versa. A method of decoupling control for the reduction of coupling effect of the motor's operation has been proposed [9]; to excite reasonable distribution control current of the motor stator coil, and the decoupling of both rotating motion and linear motion have been implemented. Therefore, the control accuracy has been improved. Since the motor is based on the principle of switched reluctance, so the efficiency of switched reluctance is lower than other motors, and the force volume ratio is relatively poor [10]. Even the rotating linear motor has been controlled by the method of the decoupling, but cannot solve the problems of the lower conversion efficiency, the smaller torque and linear direction force [11], [12].

The paper presents an optimized rotating linear motor, and it is applied to the electric vessel. In the paper, firstly, a new structure of the motor is proposed. The mathematical model of the two - dimensional rotating linear is established. Secondly, through the finite element analysis of the motor, internal electromagnetic field distributions of the motor and force characteristics of the motor are obtained. Experiments are carried out to study the accurate control of motor. Thirdly, to control the position of motor and to realize the angle and position control, an optimization is needed to resolve the electrical propulsion performance, and also to improve the force to volume ratio.

CONSTRUCTION OF THE SYSTEM

A. Mechanical structure and basic control method of the electrical propulsion system

The new electrical propulsion system is only based on one two-degree switched reluctance motor to complete all propulsion operation, the two-degree switched reluctance motor have connected with propellers, which is shown in Fig.1. Therefore, in the Fig.2, the direction control is achieved by the linear motion of the motor and the linear acceleration control is achieved by the rotating motion of the motor. To compare with traditional electric vessel propulsion system, not only the operation of electric vessel is more hommization and convenience, but also the operation precision is improved. The electric vessel is readily for industrialized production.

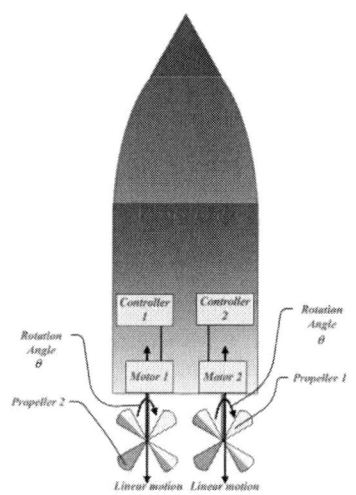

Fig.1 Mechanical structure of the electrical propulsion system

B. Actual structure of the motor

Fig.3 (a) is the mechanical configuration of the two degrees of freedom switched reluctance motor; Fig. 3 (a) is left view of the motor. It mainly consists of a linear guider, the first part and the second part. Both parts have teeth and slots. When each phase is excited, the second part will rotate to the position at which all teeth of the two parts are directly opposite. Fig.3 (b) is front view for the motor. As shown in the figure, the first part has three phases named as phase A, B and C, respectively. The second part has five units fixed on the linear guider. Two bearings are used to guider the movements of the second part both in rotary and linear directions. When phase A is excited, the second part will shift to left. Also, if phase C is excited, it will move to right. Therefore, the second part can realize linear motion to left by exciting phase A, C and B in sequence. The main specifications are listed in Table I.

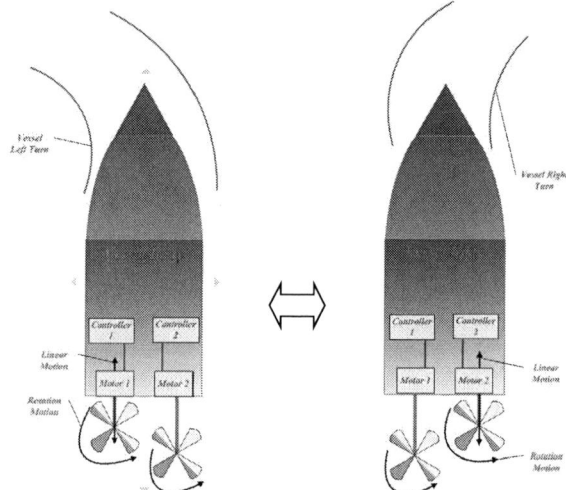

Fig.1 Basic control method of the electrical propulsion system

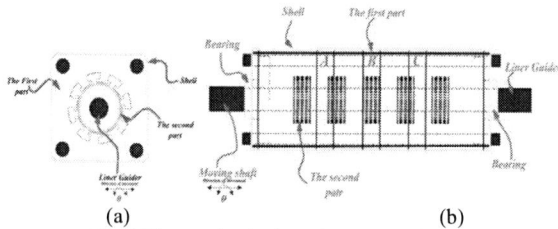

(a) (b)

Fig.3 The mechanical configuration of motor

The magnetic field generated by the concentrated coils passing current, rounding the core in the first part, attracts the second part. Based on the minimum reluctance principle, a unit of the second part will move to the position, which has the minimum reluctance, thus producing electromagnetic propulsion or torque. By controlling currents passing in the coils of the first part, the generated propulsion and torque will drive the second part and then it will realize rotary and linear motions simultaneously. As shown in Fig.3 (b), when the second part shifts linearly, phases A, B and C should be excited in turns. For example, phase B is totally aligned in the figure. At the position the second part would move left if phases A, C and B are excited in sequence and vise versa. For the first part, there are several small teeth in each pole and there are eight poles of each phase of the motor. The working rules probably the same as common switched reluctance drives, namely through controlling the ON/OFF conditions of the first pole. All of A, B and C phrase positions were fixed with one shaft, which would be led by two bearings in order to realize linear and rotational motions.

978-1-5090-0064-7/15 $31.00 © 2015 IEEE

1) Mathematic mode

Equation (1) is the voltage equation for both rotary and linear parts [13].

$$V_\mu = R_\mu I_\mu + \frac{f_\mu dv}{dt} \qquad (1)$$

And V_μ is the input voltage and I_μ is current for the uth coil (u is 1-6). R_μ is winding resistance for the coil, and f_μ is flux-linkage confirmed during excitation.

The rotary-linear machine consists of a coupled RL electromechanical system. From the rotary part with each stator ring, and equation (2) is the machine generalized torque characteristic T as,

$$T = Q\ddot{\theta} + G\dot{\theta} + T_L \qquad (2)$$

where the moment of inertia is Q, rotational friction coefficient is G. Angular position is θ and load torque is T_L. For the linear motion, the generated force F is

$$F = S\ddot{z} + N\dot{z} + F_L \qquad (3)$$

whereas mass of the moving shaft is m, and linear friction factor is L. The linear position and thrust force are x and F_L, respectively. Assume the magnetic circuit is not saturated in the linear motion region. Toward the phase inductance, the influence can be neglected for the phase current [22], also force can be estimated as,

$$\begin{cases} T = \dfrac{1}{2} \cdot \dfrac{\partial L_1}{\partial \theta} \cdot i_1^2 + \dfrac{1}{2} \cdot \dfrac{\partial L_2}{\partial \theta} \cdot i_2^2 \\[2mm] F = \dfrac{1}{2} \cdot \dfrac{\partial L_1}{\partial x} \cdot i_1^2 + \dfrac{1}{2} \cdot \dfrac{\partial L_2}{\partial x} \cdot i_2^2 \end{cases} \qquad (4)$$

And L_1 and L_2 are total inductances, and i_1 and i_2 are the two stator rings current, respectively. It can be found that the torque and force generation are both dependent on phase current of the stators. The stator ring for linear motion in the mean time can generate the torque. Consequently, the magnetic paths are nonlinear and highly coupled.

2) Electromagnetic characteristics

In order to analyze the magnetic field distribution and properties of the motor, FEM is employed. The mesh model of the motor as shown in Fig.4, actually, it is a model with symmetrical structure, which subdivides the computational nodes and reduces calculation time. It can be concluded from the figure, subdivision of the air gap between two parts is intensive, and the more unions of subdivision were set, the higher quality of computation could be obtained.

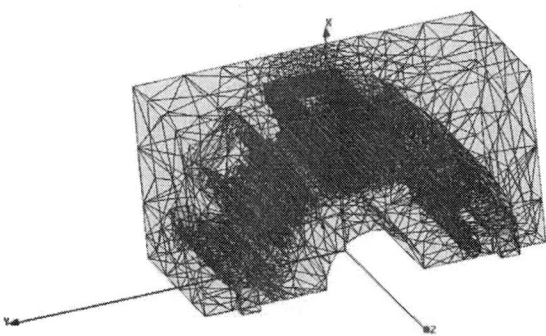

Fig.4 is the distribution of magnetic flux lines in the motor When it works, it is clear to see that the electromagnetic torque or propulsion is produced when the magnetic flux lines are closed from the first part to the second. The flux linkage will be built when the coils are excited. When the motor moves, the magnetic lineation of the field coil is in the same flat with the second part. Fig.5 describes the distribution of magnetic lineation when the second part moves in the position that is not aligned with the first part. When the motor works in the linear direction, magnetic flux line is vertical with the direction of the linear direction, which is different from the principle of producing tension when the motor moves. So the motor can produce not only torque, but also the propulsion in the linear direction, thus decreasing the volume of the two dimensional motions actuators and simplifying the execute components. In addition, there is no need to add any intermediate mechanical converters because both of the torque and linear propulsion of the motor are direct-drive mode, which can realize the precise position control.

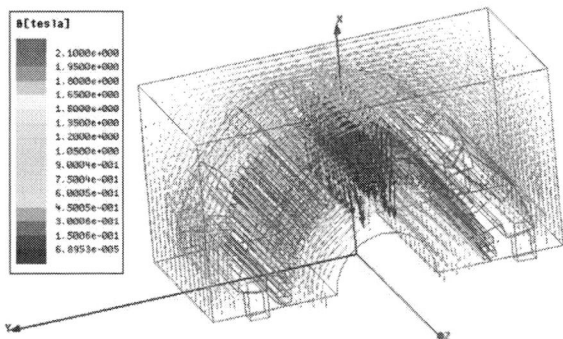

Fig.5 Magnetic field distribution of the motor in rotary direction

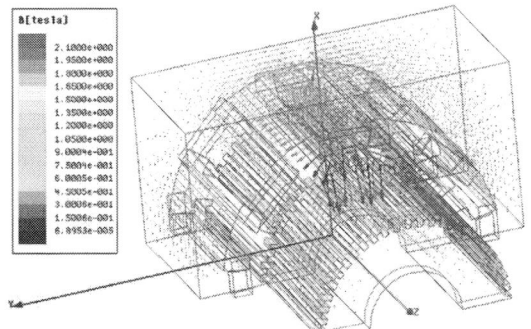

Fig.6 Magnetic field distribution of the motor in linear direction

From the distribution of magnetic field, the rectilinear motion of the motor and the magnetic field both come from the same coil, thus, the motor in rectilinear motion will produce the torque of rotational motion. There is coupling between them, however, it is acceptable to avoid the coupling by a kind of decoupling controlling method, in order to realize the high quality controlling.

Fig.7 and Fig.8 are the waveforms of FEM calculation from the torque and rectilinear directions. Fig.7 is the waveform of measured torque data. Any phrase position excites, the maximal torque is 0.71N.m, and the torque will not increase accordingly under large current, because the motor has saturated, the permeability of magnetic material and the volume of motor limit the increase in magnetic field intensity, which can make the increase in torque inconspicuous.

TheFig.8,that the propulsion of the motor is increased with therising of phase current because it is hard to reach the sa turated point to the motor as it moves from unaligned Posit ionton aligned position, during which the reluctance is big enough. This figure shows the propulsion profiles corresponding to all positions during a stroke. The linear propulsion is relatively low and up to 10N at the rate current.

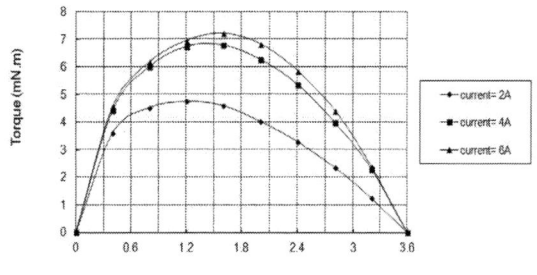

Fig.7 Torque output with different currents

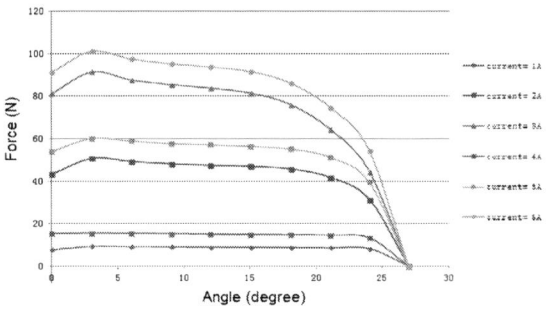

Fig.8 Force output profiles corresponding to position

From the analysis above, this two-degree of freedom stabilized platform motor can directly realize the motions of linear and rotary movements.

POSITION CONTROLLER DESIGN FOR THE TRACKER

Two PID controllers are employed for linear and rotary motions [14] for the tracking system and the whole control scheme is expressed in Fig.9.

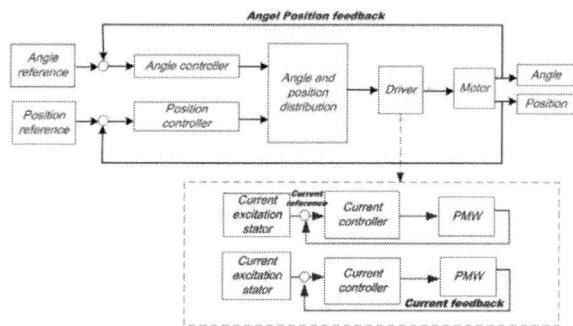

Fig.9 Control block of the motor

The control system is divided into two parts. On one hand, a controller regulates the linear motion whose axis is responsible for the perpendicular motion for the electric vessel. On the other hand, another controller gives the horizontal movements for the system controls in rotary motion. The feedbacks of both the two axis positions of angle and linear displacement are sampled by a sensor which tracks movements of the sun. Controllers output the force and torque reference commands for the force to current and torque to current distribution parts [15]-[17]. Finally, the two distribution parts output current commands for the divers to the motor. The trajectories of the two axes for the motor are obtained in the end as shown in Fig.9. The simple PID controller can be applied in angular and linear position control with control parameters in Table I.

Table I. Parameters of position controllers

Parameters controller	Angular position controller	Linear position
P	0.410	32.5
I	6.006	2.2
D	0.005	0.008

EXPERIMENTAL RESULTS

The achievements of experimental is shown in Fig.10, with A DSPACE DS1104 controller card, the encoders collect the position feedback with two channels of quadrature encoder pulse interface, also for each stator, the controller card generates the current reference of any phase. Lastly, the six current drivers generate the phase current outputs.

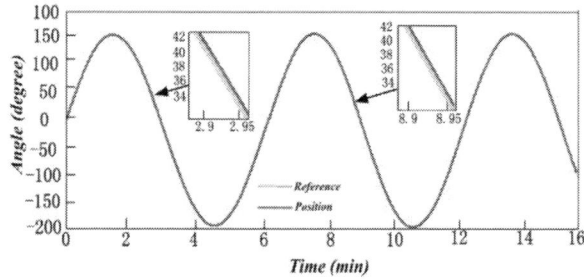

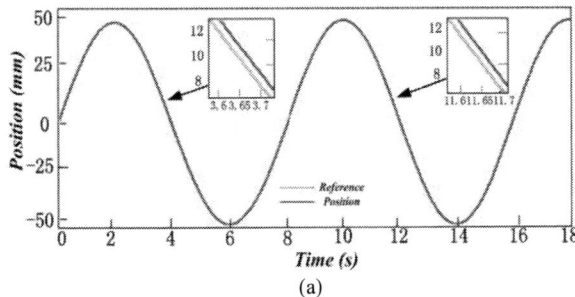

(a)

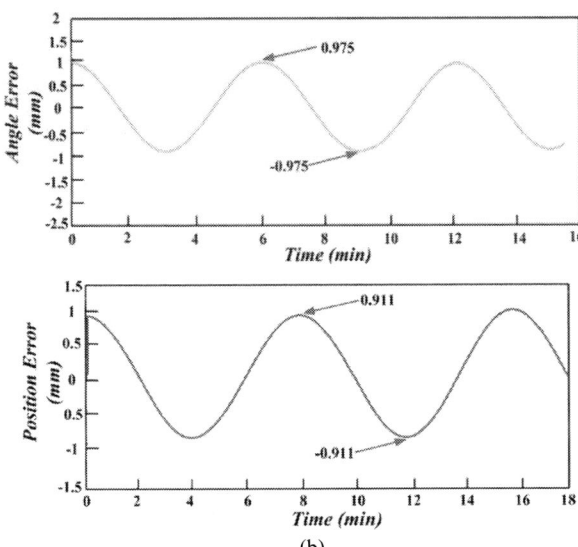

(b)

Fig.10(a) Position control performance for rotary and linear motion (b) Dynamic error response

In Fig.10 (a) and (b) proved the position control performance and the theory of dynamic tracking error response. The results shows that the proposed two-degree freedom of SR motor not only has a high position tracking precision of less than 0.4°, but also the performance of the period of working is stable and reliable. Consequently, the achievements of theory analysis and FEM have been proved by the improvements of the proposed setup

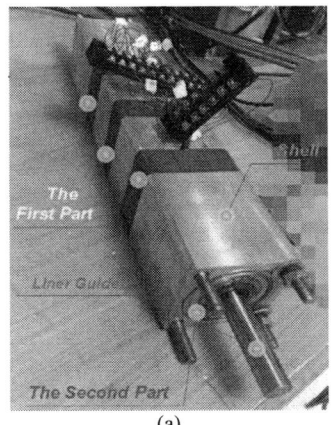

(a)

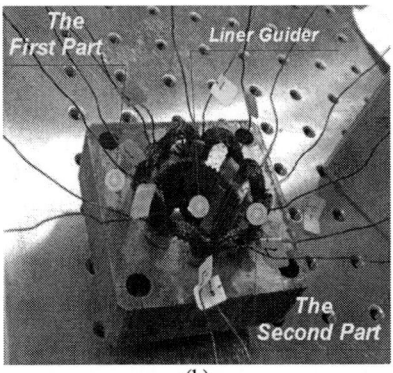

(b)

Fig.10 (a) and (b) are the prototype of the 2-degree freedom of SR motor

The structure of the proposed motor is shown in (b) and (C), its features and operated performance have been discussed in detail above.

CONCLUSION

This paper proposed a modified electric vessel based a new 2-degree of freedom SR motor, also its analysis of theory, configuration and control method are described in detail. Lastly, the achievements of experiment have proved the improvements of the modified electrical propulsion system of the vessel. Using a 2-degree SR motor to replace the two conventional motors is novel for such application. The proposed electric vessel is not only simplifying the operation method, also promoting the direction precision. Consequently, the experiment has proved the improvements of the system, and it will be used for wider industrial applications with its attractive superiority.

REFERENCE

[1] G. Krebs, A. Tounzi, B. Pauwels, D. Willemot, and F. Piriou, "Modeling of a linear and rotary permanent magnet actuator," IEEE Trans. Magn., vol. 44, no. 11, pp. 4357–4360, Nov. 2008.
[2] T. Onuki, et al., "Induction motor with helical motion by phase control," IEEE Transactions on Magnetics, vol. 33, pp. 4218-4220, 1997.
[3] J. Alwash, et al., "Helical motion tubular induction motor," IEEE Transactions on Energy Conversion, vol. 18, pp. 362-369, 2003.
[4] E.S. Afjei and H.A. Toliyat, "A novel multilayer switched reluctance motor," IEEE Transactions on Energy Conversion, vol. 17, pp. 217-221, 2002.
[5] C. T. Liu and T. S. Chiang, "Design and performance evaluation of a microlinear switched-reluctance motor," IEEE Trans. Magn., vol. 40, no. 2, pp. 806–809, Mar. 2004.
[6] Y. Sato, "Development of a 2-degree-of freedom rotational / linear switched reluctance motor," IEEE Trans. Magn., vol.43, no.6, 2007, pp.2564-2566
[7] Pan, Yu Zou, Cheng. N. C, "Performance analysis and decoupling control of an integrated rotary–linear machine with coupled magnetic paths", IEEE Transactions on Magnetics, Vol. 50, Issue: 2, 2014
[8] X.D. Xue, K.W.E. Cheng and S.L. Ho, "Influences of output and Control Parameters on Power Factor of Switched Reluctance Motor Drive Systems", Electric Power Components and Systems, Dec 2004, Vol. 32, No. 12, pp. 1207-1223.
[9] X.D. Xue, K.W.E. Cheng and S.L. Ho, "Improvement of power factor in switched reluctance motor drives through optimizing in switching angles", Electric Power Components and Systems., Dec 2004, Vol. 32, No. 12, pp. 1225-1238.

[10] X.D. Xue , K.W.E. Cheng and S.L. Ho, "Simulation of Switched Reluctance Motor Drives Using Two-dimensional Bicubic Spline", IEEE Tran. Energy Conversion. Dec 2002, Vol. 17, Issue 4, pp. 471-477.

[11] Y. -C. Lai, Y. -L. Lee, and J. -Y. Yen, "Design and servo control of a single-deck planar maglev stage," IEEE Trans. Magn., vol. 43, no. 6, pp. 2600–2602, Jun. 2007

[12] G. Li, J. Ojeda, S. Hlioui, E. Hoang, M. Lecrivain, and M. Gabsi, "Modification in rotor pole geometry of mutually coupled switched reluctance machine for torque ripple mitigating," IEEE Trans. Magn., vol. 48, no. 6, pp. 2025–2034, Jun. 2012.

[13] X.D. Xue, K.W.E. Cheng and S.L. Ho, "Improvement of power factor in switched reluctance motor drives through optimizing in switching angles", Electric Power Components and Systems., Dec 2004, Vol. 32, No. 12, pp. 1225-1238.

[14] M. Bodson, J. Chiasson, R. Novotnak, and R. Ftekowski, "High-per-formance nonlinear feed back control of a permanent magnet stepper motor," IEEE Trans. Control Syst. Technol. , vol. 1, no. 1, pp. 5–14, Mar. 1993.

[15] D. Chen and B. Paden, "Adaptive linearization of hybrid step motors: Stability analysis," IEEE Trans. Autom. Control, vol. 38, no. 6, pp. 874–887, Jun. 1993.

[16] S. A. Stuart, J. E. McInroy, and R. M. Lofthus, "Closed loop low- velocity regulation of hybrid stepping motors amidst torque distur- bances," IEEE Trans. Ind. Electron. , vol. 42, no. 3, pp. 316–324, Jun. 1995.

Design of Electric Vessel Based on Concentrated Photovoltaic for Density Energy Source

S. Y. Li,   K. WE. Cheng,

[1,2,] Department of Electrical Engineering, The Hong Kong Polytechnic University, Hong Kong

Abstract - This paper introduces a new 2-degree of freedom switched reluctance motor to tracking the position of sunlight for a concentrated photovoltaic (CPV) power generation system for the electric vessel. A modified control method has been suggested as the control scheme to ensure stability and accuracy of the tracker's operation. Finally, the experimental results have suggested that the efficiencies and accuracy of the new tracker are better than that of conventional tracking system.

Index Terms—switched reluctance motor, concentrated photovoltaic power generation system, solar tracker, electric vessel

INTRODUCTION

Concentrated photovoltaic (CPV) power generation system, as a new type of solar energy harvesting method, has been investigated for years. One of important parts for the solar power generation is tracking actuator, which is used to regular the position of solar panels for maximum power output by capturing direct sunlight [1]. In order to adjust the position of solar panels in real time, the actuator drives the mechanical support of the photovoltaic power generation system according to the position of the sun, controlling the solar panels always perpendicular with sunlight over a day.

Traditional CPV system is shown in Fig.1 (a) with the advantage of higher photovoltaic conversion efficiency and lower consumption of solar cells [2]. It is a photovoltaic power generation system which converts energy using a small generating device by letting high-density light from the sun that has been concentrated with lenses [3]. For the multi-junction cells, it has been recorded the world's highest efficiency of 44 % at 947 times, as of October 15, 2012. Efficiency greater than 50% can be expected in the near future for these multi-function cells [4-5]. The traditional mechanical tracking systems are usually taking advantage of electrical machines as the executors, including direct current (DC) motors [6] to control the gesture of the solar panel. However, the working environment such as temperature and micro vibration will deteriorate the permanent magnets [7]. The solar panels, presented in CPV are large and heavy and have to been actuated usually through a centralized motor unit. The weight of the whole panel is cumbersome so that a high enough torque output is required to regulate the position of the panel as well as the angle displacement. There are some mechanical devices such as transmission gear boxes, connecting rod etc. [8], are traditionally applied to change the speed of the motor and obtain the high regulating torque simultaneously .

To handle the problems of lower conversion efficiency, and smaller force of linear direction, the paper presents an optimized rotating linear motor, and it is applied to the control parts of concentrated solar photovoltaic power generation system (CPV).

In the paper, firstly, a new structure of the motor is proposed. Linear motor parameters and the mathematical model of linear motor are then established. Secondly, through the finite element analysis of linear motor, internal electromagnetic field distributions of the motor and force characteristics of the motor are obtained. Experiments are carried out to study the accurate control of motor. Thirdly, to control the position of motor and to realize the angle and position control, an optimization is needed to resolve the tracking performance, and also to improve the force to volume ratio.

CONSTRUCTION OF THE SYSTEM

A. Mechanical structure of the CPV system

In Fig.1 (b), the new tracking system is only based on one two-degree switched reluctance motor to complete all tracking operation, the two-degree switched reluctance motor have connected with the CPV module and support group, therefore, the inclination angle tracking is achieved by the linear motion of the motor and the azimuth angle tracking is achieved by the rotating motion of the motor. To compare with inclination-azimuth angle tracking system, not only the size and weight of new tracker is decreased, but also the tracking precision is improved. The system is readily for industrialized production, the cost of CPV power generation system will be significantly reduced.

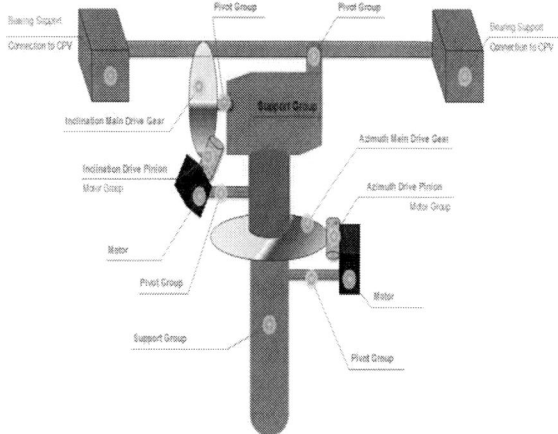

Fig.1 (a) Inclination- azimuth angle tracking system

978-1-5090-0064-7/15 $31.00 © 2015 IEEE

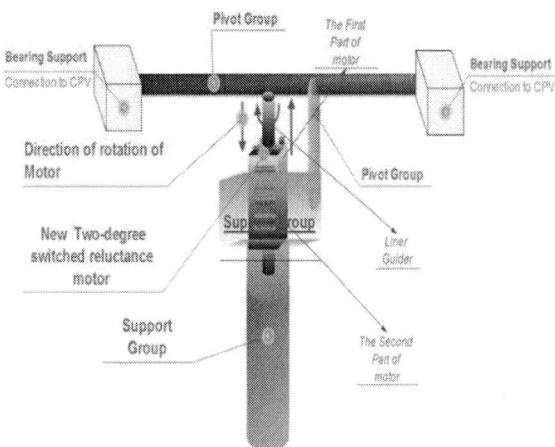

Fig.1 (b) new two-degree switched reluctance motor control tracking system

B. *Mechanical structure of the electric vessel with CPV system*

The Fig.2 shows that the electric vessel with the CPV power generation system. The CPV system is installed to the deck of the electric vessel, which could transfer solar energy into electrical energy by tracking the sun for the electric vessel To compare the traditional PV solar vessel, the proposed vessel not only will save the space, but also the cost is greatly reduced. The efficiency is improved by the new CPV generation system.

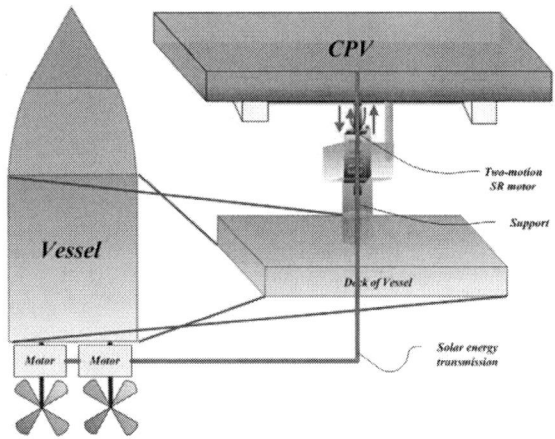

Fig.2 The structure of electric vessel with CPV generation system

CONTROL METHOD DESIGN FOR THE CPV TRACKER

A. *System diagram of the solar tracking*

Auto tracking controller includes photoelectric detector module, micro controller unit (MCU), display modules, memory modules, manual control module, motor drive module and the auxiliary power module. Photoelectric detector module, manual control module and storage module supply the information to the MCU, and then MCU controls the two-motion linear rotary SR motor.

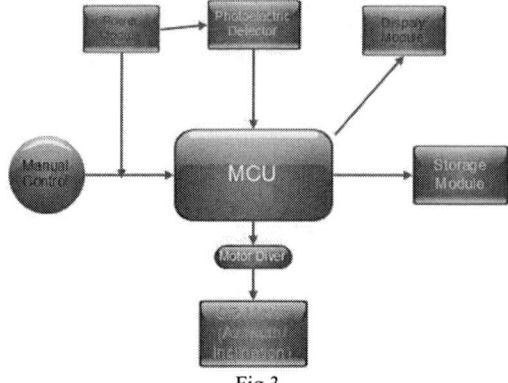

Fig.3

Fig.3 System diagram of the solar tracking

B. *High-precision auto tracking control method*

Solar position detection unit design include: location based on current time and latitude and longitude, astronomical data query and computing, concentrating CPV coarse tracking; accurate solar tracking using sensors and control algorithms.

On the lighting surface of optical concentrator, the angle of incidence and elevating angle θ, the sun declination Angleδ, the sun's Angle of view ω, optical concentrator Inclination angle β, optical concentrator azimuth angle γ, and the local latitude φ, according to the following formula are obtained. The schematic diagram of inclination - azimuth angle solar tracking is shown in Fig.4.

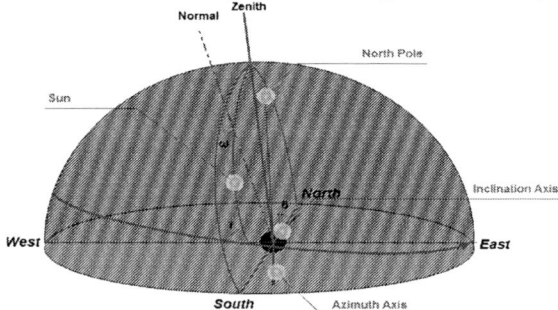

Fig4. Schematic diagram of inclination - azimuth angle solar tracking

Variation in weather is changeable. However, the solar motion track with respect to a point on the earth changes exactly with the time. In other words, from the local geographic position (longitude and latitude), the solar relative position at any time can be determined exactly by using astronomical calculation. The earth rotates15 degrees with the axis per hour, and slows the rotation with the sun [9-11]. Consequently, the solar relative position can be fast and accurately if the weather changes. It is just the control input of the tracing system for CPV solar energy generation. The astronomical calculation is to determine the solar height angle and the azimuth angle. The height angle means the line-plane angle between the rays from the solar center and the local horizontal plane. It changes in the range from 0 degree to 90 degree. It is equal to 90 degree at noon and 0 degree at sunset. The azimuth angle is the angle between the sunlight projections on the horizontal plane and the local meridian. It is zero degree in the direction of right south, gradually becomes large towards west, gradually becomes small towards east, and is 180 degree or -180 degree in the direction of right north. From the local geographic position and time, the solar height angle and azimuth angle can be calculated accurately.

$$\cos q = (\sin d \ \sin j \ \cos b) - (\sin d \ \cos j \ \sin b)$$
$$+ (\cos d \ \cos j \ \cos b \ \cos w) + (\cos d \ \sin b \ \sin g \ \sin b)$$
$$+ (\cos d \ \sin j \ \sin b \ \cos g \ \cos w) \quad (1)$$

In the equations (1), the equation of n days of the sun declination angle of the year is equation (2):

$$d = 23.5° \sin \frac{360(284 + n)}{365} \quad (2)$$

According to the equations (1) and (2), when after the confirmation of d, j and w, the value of b and g is to decide the value of q. So if we can control the inclination and azimuth angle which are in turn controlled by the value of g. The value of q will be ensured be 0, in order to gain the maximum solar energy [12].

The equations of solar altitude and solar azimuth are (3) and (4).

$$h = \sin^{-1}(\sin d \ \sin j \quad + \cos d \ \cos j \quad \cos g) \quad (3)$$

$$A = \cos^{-1}(\sin h \sin j \quad - \sin d \ / (\cos h \cos j \ )) \quad (4)$$

The performance testing method has to solve the trade-off between precision and easiness. The safest method is to conduct one or many years in the field test. The manufacturers have the obligation to demonstrate the real performance of the field at least several years [13-14].
In all types of tracking systems, the accuracy of pure mechanical tracking system is lowest, and the purpose of tracking is to improve the energy density. If the accuracy is low, the tracking efficiency will be low and it deteriorates original meaning in the device of the high efficiency. All system is full-automatically and need not personal interference. The primary control method does the main mission of processing the incoming signals, calculating, and sending out the instruction handle the emergency on time. The secondary control method is just an execution system which can provide the answers back to the primary control method when the instructions are executed. [15-16] Hence, a compromise between high precision of photosensitivity and the cost is needed. With the consideration of tracking accuracy, stability and cost of the different kinds of two-axis tracking system, we can draw a conclusion: polar axis tracking is suitable for application in small-scale CPV system, in other words, for the large-scale CPV system.

POSITION CONTROLLER DESIGN OF MOTOR FOR THE CPV TRACKER

Two PID controllers are employed for linear and rotary motions [17] for the tracking system and the whole control scheme is expressed in Fig.9.

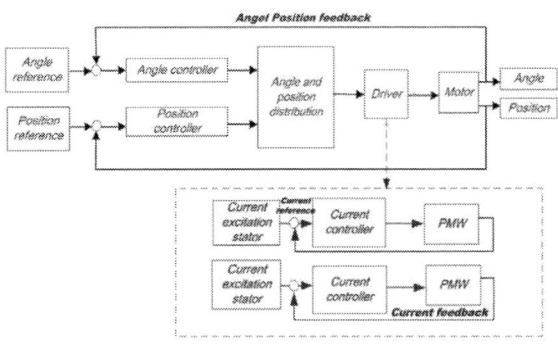

The Motor control system is divided into two parts. On one hand, a controller regulates the linear motion whose axis is responsible for the perpendicular motion for the tracking system. On the other hand, another controller gives the horizontal movements for the system controls in rotary motion. The feedbacks of both the two axis positions of angle and linear displacement are sampled by a sensor which tracks movements of the sun. Controllers output the force and torque reference commands for the force to current and torque to current distribution parts. Finally, the two distribution parts output current commands for the divers to the motor.

EXPERIMENTAL RESULTS

The achievements of experimental is shown in Fig.9, with A DSPACE DS1104 controller card, the encoders collect the position feedback with two channels of quadrature encoder pulse interface, also for each stator, the controller card generates the current reference of any phase. Lastly, the six current drivers generate the phase current outputs.

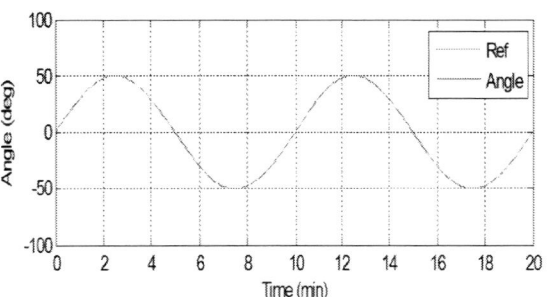

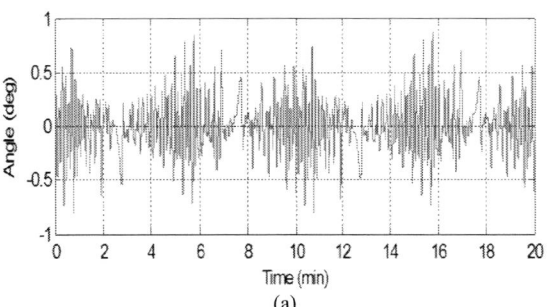

(a)

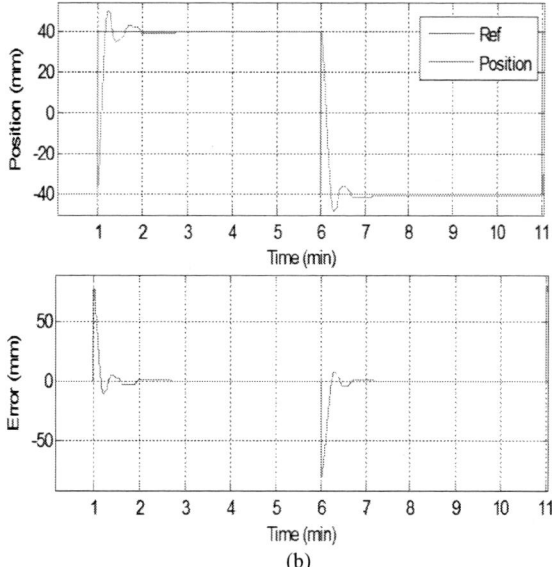

(b)

Fig.9 (a) Position control performance for rotary and linear motion (b) Dynamic error response

In Fig.9 (a), the two-axis position tracking profiles, the controller has improved the performance of tracking upon each axis with the proposed system. Also (b) proves the theory of dynamic tracking error response. The results shows that the proposed a high position tracking precision of less than 0.4°, but also the performance of the period of working is stable and reliable.

Also, the prototype of CPV tracker and two-degree SR motor are proposed in Fig.10, (a) is the configuration of the CPV solar tracking system, which consists of solar sensor for confirmation of the sunlight position, SR motor is the actuator for angle movement to the CPV modules against sunlight. The square block is the CPV modules to give power output.

CONCLUSION

The disadvantages of classical solar tracking system are including complex structure, low tracking precision and efficiency. Those problems must be solved for high utilization rate of the concentrated photovoltaic (CPV) power generation system in industry.

This paper proposed a modified electric vessel based on a new CPV power generation system, also its analysis of theory, configuration and control method are described in detail. Lastly, the achievements of experiment have proved the improvements of the modified tracking system. In two-axle structure of CPV system, in order to improve the accuracy of solar position tracking, many solutions have been explored.

For further study, the improvements of the proposed system efficiency and a proper control algorithm should be explored in later research.

REFERENCE

[1] Pandey, A. ; Nitya Power Technol. Pvt., Ltd., Delhi ; Dasgupta, N. ; Mukerjee, A.K.:" High-Performance Algorithms for Drift Avoidance and Fast Tracking in Solar MPPT System.", IEEE Transactions on Energy Conversion, Vol, 23, No, 2, June 2008

[2] M. Yamaguchi and A. Luque, "High efficiency and high concentration in photovoltaic," IEEE Trans Elec. Dev. 46 (10), (1999), 2139

[3] Masafumi Yamaguchi, "Ultra high efficiency compound solar cells –Materials, Technology trends and challenges," Industrial Materials, Vol.58 No.4, pp49-53, April (2010)

[4] Mingguo Liu, Geoffrey S. Kinsey, William Bagienski, Aditya Nayak, and Vahan Garboushian, "Indooer and Outdoor Comparison of CPV III-V Multi-junction Solar Cells", In: IEEE Journal of Photovoltaics,Vol.3, No.2, April, 2013

[5] K. Sasaki, T. Agui, K. Nakaido, N. Takahashi, R. Onitsuka, and T. Takamoto, in 9th international conference on concentrator photovoltaic system, Miyazaki, Japan (2013)

[6] Alberto Dolara, Francesco Grimaccia, Sonia Leva, Marco Mussetta, Roberto Faranda, and Moris Gualdoni, "Performance Analysis of a Single-Axis Tracking PV System", IEEE Joural of Photovoltaics,VOL. 2, NO. 4, OCTOBER 2012.

[7] X.D.Xue, K.W.E.Cheng, S.L.Ho, "A Position Stepping Method for Predicting Performances of Switched Reluctance Motor Drives", IEEE Trans. Energy Conversion, Vol. 22, No. 4, Dec 2007, pp. 839-847.

[8] Fu-Sheng Pai, Ru-Min Chao, Shin Hong Ko, Tai-Sheng Lee: "Performance Evaluation of Parabolic Prediction to Maximum Power Point Tracking for PV Array", IEEE Trans. Sustainable Energy, vol. 2, no.1, pp. 60-68, Jan.2011

[9] Hanieh, A.A. "Solar Photovoltaic Panels Tracking System":In Proceedings of the 6th WSEAS International Conference on Dynamical Systems and Control, Sousse, Tunisia, 3–6 May 2010; pp. 30–37.

[10] Serhan, M.; El-Chaar, L. "Two Axis Sun Tracking System": Comparison with a Fixed System In Proceedings of International Conference on Renewable Energies and Power Quality, Granada, Spain, 23–25 March 2010.

[11] Yousef, H.A. "Design and Implementation of a Fuzzy Logic Computer-Controlled Sun Tracking System." In Proceedings of the IEEE International Symposium on Industrial Electronics, Bled, Slovenia, 12–16 July 1999; pp. 1030–1034

[12] Chia Seet Chin, Prabhakaran Neelakantan, Hou Pin Yoong, Soo Siang Yang, Kenneth Tze Kin Teo,"Maximum Power Point Tracking for PV Array Under Partially Shaded Conditions," cicsyn, pp.72-77, 2011 Third International Conference on Computational Intelligence, Communication Systems and Networks, 2011

[13] V. Garboushian, "Marketing Concentrating Solar Systems Challenges and Opportunities", SCC2005 (3rd ICSEC), (2005)

[14] K. Araki et al., "500 X to 1000 X - R&D and Market Strategy of Daido Steel", Proc. 4th International Conference on Solar Concentrators for the Generation of Electricity or Hydrogen, (2007)

978-1-5090-0064-7/15 $31.00 © 2015 IEEE

[15] M. X. Qiang, Z. C. Ye, Y. S. Qiang, "Design of Pipeline Monitor Network Based on Zigbee and GPRS," Computer Engineering, China, vol.36, No. 5, pp. 128–130, March 2010

[16] K. K. Chong, C. W. Wong, "Multi-axes sun-tracking system with PLC control for photovoltaic panels in Turkey," Renewable Energy, vol. 34, 2009, pp. 1119-1125

[17] E. Suresh Kumar, Bijan Sarkar, "Impact of Wind and Shading on Energy Contribution by Photovoltaic Panels with Axis Tracking System", International Conference on Microelectronics, Communication and Renewable Energy, ICMiCR, 2013.

Future body design for electric vessel and aircraft

Xiaolin Wang [1] George Sin [2] K.W. Eric Cheng [2]

[1,2,3] Department of Electrical Engineering, The Hong Kong Polytechnic University, Hong Kong
E-mail: xiaolinee.wang@connect.polyu.hk

Abstract–The future electric vehicle design will be substantially different from the fossil fuel based. Multiple motors will be used to replace the single engine as in the conventional vehicle. Even though for the hybrid electric version or the pure electric version the fossil fuel generator is used to charge the battery and in turn the battery powers the motors. Other assistive units for steering will also be used. That will facilitate the 2-D maneuvering and parking for vessel and 3-D primary flight control and taxi for aircraft. This design paper is to examine a number of body designs to meet the new electrified air vehicles. This design paper is to examine a number of body designs to meet the new electrified water and air vehicles.

Keywords–Dream electric aircraft, public transportation, intelligent control, Multiple motors.

I. INTRODUCTION

Greenhouse effect, air pollution and the shortage of fossil fuel become the questions of common concerns. Electric vehicle (EV) can reduce the CO_2 emission and improve the quality of the air. It can also increase the security of the energy and reduce the cost of the fuel [1-3]. The promotion of EV is increasingly necessary and important.

The future design of the electric vehicle is a complex subject that involves the advanced outlook design, the possible motor design and the energy storage system. The key issue of the EV design is about to improve the efficiency of the whole vehicle and the good performance energy storage system [4].

Meanwhile, cities like New York, London, Beijing and Hong Kong all have to deal with thousands of cars running through the streets each day. Traffic congestion is a big problem for everyone within the city. The main reasons why traffic congestion occurs are more cars, poor road management, and poor driving habits. The city's problems do not seem to be improved in the past year - despite repeated measures to limit vehicle numbers. But with the fast development of the cities, the number of the cars is not easy to reduce. In this situation, we should pay more attention on the zero emission EV and the up-ground vehicle.

For the common vehicle there is just one motor for the motive power. But for the multiple motor is nonlinear, multivariable and complex higher-order [5].

This design paper will introduce four types of the vehicle: jet pack, short distance commercial, near space aircraft and vessel.

II. MODEL 1

Due to the limitation of the space on the ground, the space above the ground also can be used in the public transportation.

This design of multi-connective autopilot jet pack electric bus can easily work in the air above our heads. As shown in the Fig.1, it consists of numbers of personal jet pack and a main bus turbine. The jet pack electric bus flies among central districts like shopping mall, school, residential area etc. After landing in the desired area, the personal jet packs can then unplug the bus to fly to their own destinations. The detailed design of the personal jet pack is shown in Figs. 2-4 which is very easy to control. The jet pack electric bus enables wireless recharging, autopilot, and roof-landing. It also avoids formidable CO_2 emission and traffic jams.

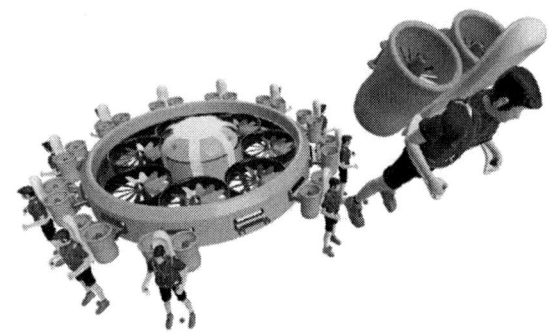

Fig.1 Multi connective autopilot jet pack electric aircraft

Fig.2: Illustration of jet pack with a pilot

Fig.3: Illustration of jet pack

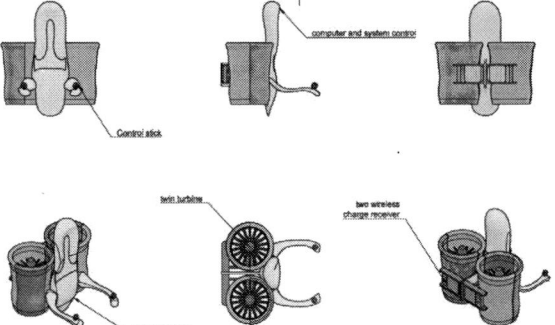

Fig.4: Three-view drawing of jet pack

The main idea is the reduction of the weight of the jet pack. The jet pack in the future, can also be done by high speed electric motor and drive. All transportation devices, in order to enhance energy saving, must to provide high ratio in weight of carrier per unit weight of vehicle. The classical electric vehicle such as SUV is not of high energy efficiency because the weight of the vessel is large as compared with the passenger and therefore nit economic, whereas the present design is of good performance ratio.

III. MODEL 2

Our design of the future near space solar aircraft is shown in Fig. 4. Multi-motor system (Fig. 5) is installed at the middle of the upper surface to increase the propulsion force. The whole surface area of the wings is covered with transparent or low profile thin solar panel (Fig. 6). As the near space aircraft flies high above the clouds, the efficiency of the solar energy absorption can be so high that it can support sustainable flying. Multiple motor can even the weight distribution on the body. It can also reduce the power consumption. Under light low or cursing, portion of motors can be turned off to reduce power usage.

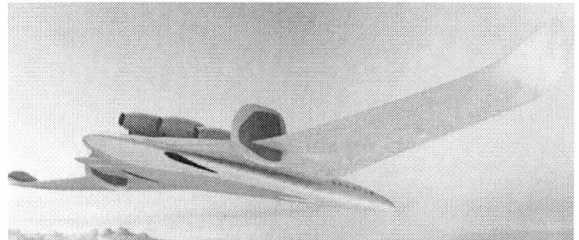

Fig. 4: Appearance of the near space solar aircraft

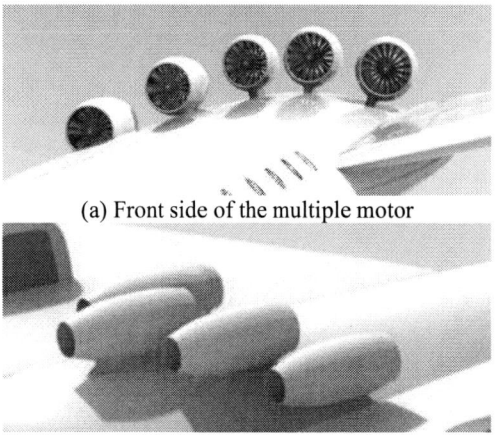

(a) Front side of the multiple motor

(b) Back side of the multiple motor
Fig. 5: Illustration of multiple motor

Fig. 6: Transparent solar panel

IV. MODEL 3

The design of short distance commercial aircraft is shown in Fig. 7. It has two large turbines that make the vertical takeoff and landing possible. The operation angle of the two turbines can be adjusted for various wind conditions. Another multi-motor system (Fig. 8) is used to propel the aircraft in horizontal flying. The advantage of this design is that it can save the space of tracks during takeoff or landing while maintaining large customer capacity.

Fig. 7: Appearance of the short distance commercial with aircraft with vertical takeoff

The motor selected is a switched reluctance motor [6] and the low profile motor topology is used. The rotor does not consist of winding and therefore the low profile motor can be used.

Fig. 8: Illustration of multiple motor

V. MODEL 4

Fig.9 shows the design of a commercial electric vessel with multiple spherical motor. The direction of the spherical motors (Fig. 11) is more flexible than the common motors. Influenced by the fast flow of the ocean current, the vessel does not work in a stable environment. The common vessel always makes passengers feel jolty, but the multiple spherical motor can easily adjust the driving angle of the vessel in accordance to the flow. Fig.10 shows the operation of 6 motors installed at the bottom of vessel to control the different driving angle.

Fig.9: Appearance of the multiple spherical motor vessel

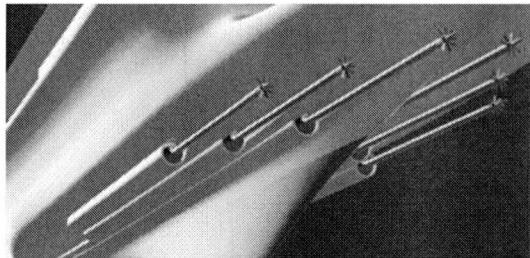

Fig. 10: Illustration of multiple motor

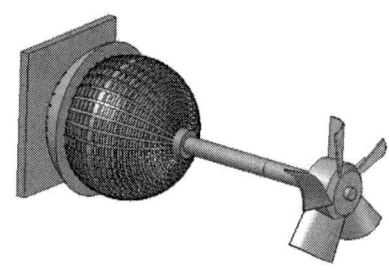

(a)Spherical rotor

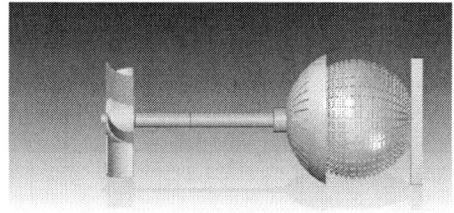

(b)Spherical stator

Fig.11: Smart Propulsion using proposed spherical motor

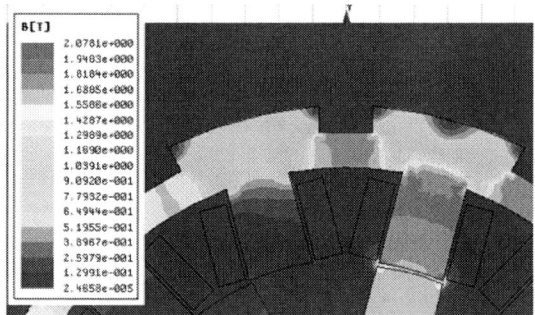

Fig.:12 Magnetic simulation of the spherical rotor

VI. APPLICATION TO ELECTRIC VESSEL

Fig 11 show the multi-dimensional spherical motor can be installed to a propeller of a vessel. Fig 13 shows the design. The motion control allows a better maneuver of the boat for steering. Therefore the tunnel thrusters or a bow thruster can be eliminated [7]. The docking or steering can be done with less motor or position parts. The rotor can be constructed without using winding and the rotation can be realized by switched-reluctance torque.

978-1-5090-0064-7/15 $31.00 © 2015 IEEE

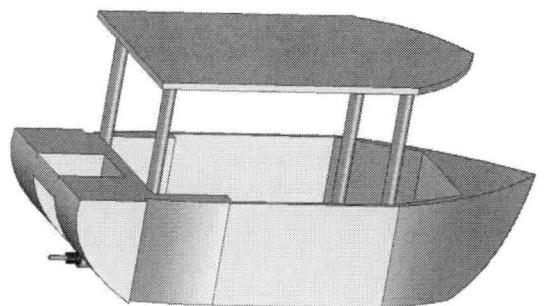

Fig 13: Application of multi-dimensional motor to vessel as the propeller

VII. ENERGY STORAGE SYSTEM

High power energy storage is proposed to be used in the electric vessel. Batteries are not suitable because of the environment condition such as water, sea, or aeronautic environment condition. For aeronautic condition, temperature may vary over a range of 100 degree C and most batteries including Li-ion may not be applicable.

The ideal energy storage system should be high power density, high energy density, long used life, long charging discharging cycle and no self-discharge for vessel and aircraft. But in the common market there is no such energy storage product.

But nowadays supercapacitor (SC) draws the attention of the public. SC is called electrochemical double layer capacitor (EDLC) [8]. It has high power density, long life and charging and discharging cycles. Also it is environmentally friendly. Due to the fact that SC is electrochemical devices, it is necessary to know its unique features during the application. There are two different electrical modelling of supercapacitor.

The super capacitor model shown in Fig.14 consists of three branches, the first branch denotes its fast-respondent behavior, a resistor comparable to the equivalent series resistor in short term, a fixed capacitor and a voltage-variable capacitor, the second branch denotes the long-term characteristics and the third branch including an equivalent parallel resistor simulating self-discharging behaviors. The modeling analysis is a step to assist in overall energy design study. A detailed will give the transient analysis.

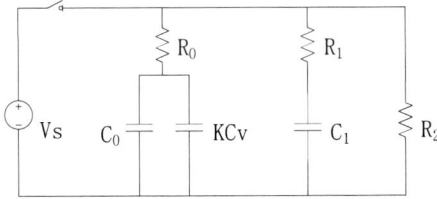

Fig.14: Equivalent circuit of the SC for model 1

As shown in Fig.15, RC model is the simplest model of the SC It only has one branch that consists a resister to simulate the equivalent series resistor (ESR) and a capacitor to model the SC's capacitor during charging and discharging period. For high power and main energy storage, a simple model is usually good enough for the modeling analysis. This reduces the computation time. For electric transportation, the number of capacitor cells are many, therefore a simple model analysis is suitable. This is also a platform for the future design.

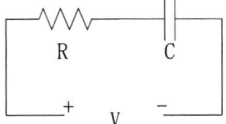

Fig.15: Equivalent electrical circuit for RC model 2

Fig.16: Supercapacitor charging and discharging test

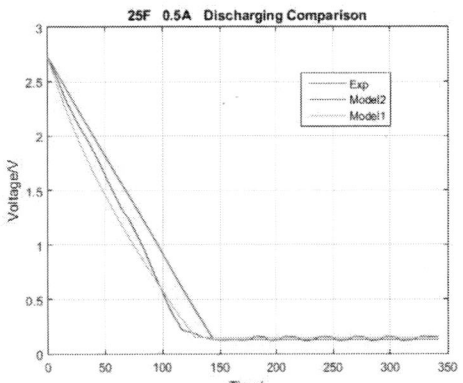

Fig.17: 25F SC, I=0.5A Discharging test

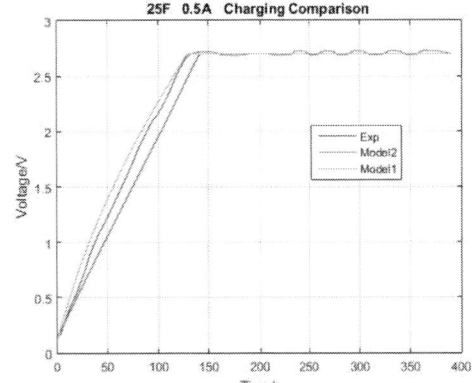

Fig. 18: 25F SC, I=0.5A Charging test

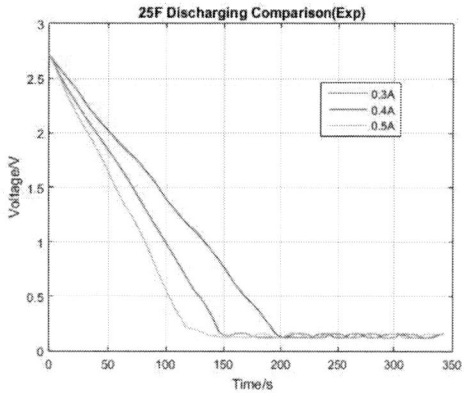

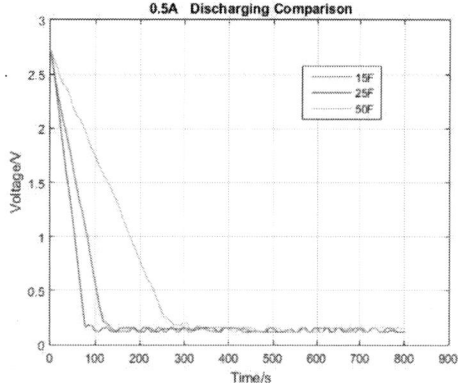

Fig.19: 25F SC discharging test in 0.3A 0.4A 0.5A

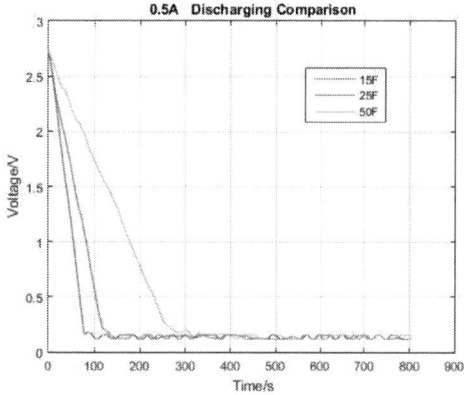

Fig.20: 15F, 25F, 50F SC discharging test in 0.5A

Initial study of the super-capacitor for rapid energy storage has been done. The setup is shown in Fig 16. The measured characteristics are shown in Fig 17-20 for the 25F 0.5A discharge, 25F 0.5A charge, 25F 0.3-0.5A discharge and 15-50F 0,5A discharge.

The discharging characteristics are all linear and is directly co-relate to the basic capacitor equations cdv/dt=i. Using the linear and high power or high current performance feature, the energy storage can be realized in vessel and aircraft. The design is based on shirt distance of travelling and use ultra –charge for super-capacitor charging.

CONTRIBUTION OF DESIGN

The design paper has stressed the following design concept. They are:

- High performance ratio: low vehicle weight
- High outlook design
- Realized by high performance and reliable switched reluctance motor
- Spherical motor gives single motor with multi-dimensional performance.
- Energy storage based on super-capacitor to give high power charging and discharge.
- Renewable solar power panel for body implementation

ACKNOWLEDGE

The authors would like to thank the support of the Hong Kong Polytechnic University with project reference: G-YN27.

REFERENCES

[1] Patrícia Baptista1, Gonçalo Duarte1, Gonçalo Gonçalves1, Tiago Farias1, "Evaluation of low power electric vehicles in demanding urban conditions: an application to Lisbon" EVS27 International Battery, Hybrid and Fuel Cell Electric Vehicle Symposium, Spain, 2013

[2] Baptista, P., Tomás, M., Silva, C., Plug-in hybrid fuel cell vehicles market penetration scenarios, International Journal of Hydrogen Energy, 2010, 35(18), pp. 10024- 10030.

[3] Pina, A., Ioakimidis, C., Ferrão, P., Introduction of electric vehicles in an island as a driver to increase renewable energy penetration, in IEEE International Conference on Sustainable Energy Technologies 2008, 2008, Singapore

[4] IEA, Energy Technology Perspectives, scenarios & strategies to 2050 by International Energy Agency, 2010.

[5] Jinzhao Zhang , Taibin Cao, GouHai Liu, "An Improved Method for Synchronous Control of Complex Multi-motor System" Intelligent Computing and Intelligent Systems,pp.178-182, 2009.

[6] Xue, X. D.; Cheng, K. W. E.; Lin, J. K.; Zhang, Z.; Luk, K. F.; Ng, T. W.; Cheung, N. C.; , "Optimal Control Method of Motoring Operation for SRM Drives in Electric Vehicles", IEEE Transactions on Vehicular Technology, Vol. 59 , Issue: 3, 2010 , pp.1191 – 1204.

[7] Cheng K.W.E., Xue X.D., Chan K.H., "Zero emission electric vessel development", Int. Conf PESA 2015, Hong Kong.

[8] R. Majumder, and B. Chaudhuri, "Improvement of Stability and Load Sharing in an Autonomous Microgrid Using Supplementary Droop Control Loop," *IEEE Trans. on Power Systems*, vol. 25, no. 2, pp. 796-808, Apr. 2010.

978-1-5090-0064-7/15 $31.00 © 2015 IEEE

Hybridization of Energy Storage Systems for Electric Transportation by Means of Bidirectional Power Electronic Converters

Ramy Georgious, Jorge García

Department of Electrical, Electronic, Computers and Systems Engineering,
University of Oviedo, Spain
E-mail: georgiousramy@uniovi.es

Abstract–This paper deals with the design of a Hybrid Energy Storage System (HESSs) for electric transportation such as Electric/Hybrid Vessel and Electric/Hybrid Train. The association of more than one Energy Storage Systems (ESSs) e.g., batteries which have different dynamics permit to take the advantages of the characteristics of both ESSs obtaining simultaneously a high energy density and high power density. This yields to a decrease in terms of the size of the main ESS and the total cost and an increase in terms of life span. The emulation of the batteries and with required control algorithm for the HESS are proposed. The design and the control of the HESSs is validated with the simulation in MATLAB/SIMULINK® environment and also with the real-time emulation of batteries in a laboratory setup of a HESS. The real-time experimental results have been validated against PC simulations showing full consistency. The setup of the hardware of HESS can be used to test any technologies of batteries, being a low cost solution for testing and benchmarking.

Keywords–Energy Storage Systems, Hybrid Energy Storage Systems, Bidirectional Boost Converter, Electrical Transportation.

I. INTRODUCTION

The main purpose of Energy Storage Systems (ESSs) in applications such as electric transportation as well as microgrids or smart grids is to provide a temporary energy buffer between electrical power generators and loads or to provide a permanent energy to the loads in case of islanding mode [1]. These ESS, together with Power Electronic Converters (PECs) and required control algorithms provide the needed power flow versatility in the system. In the case of electrical transportation such as electric vehicles, electric vessel, electric train, etc., the power flow changes very fast depending on the operation mode of motors, generators and ESSs [2].

Hybrid ESS (HESS) provide a solution to obtain a joint enhanced performance of the global ESS with respect to the individual ESS constrained due each individual storage technology. On the other hand in applications of power system related to electric transportation (Electrical Charging Station, Vehicle to Grid (V2G), Power System Operation), the main purpose of ESS is to increase the power quality of the grid in case of line contingencies such as voltage transients, current distortions, phase unbalances, load fluctuations, islanding modes, etc [3]. Unfortunately, as mentioned previously, most current ESS ratings do not allow simultaneously a large energy capability and a fast response. This yields to the need for hybridization of at least two technologies of ESS with different characteristics, one with a high energy density and slow dynamics and other with a high power density and fast dynamics [3-8].

II. DESIGN SPECIFICATIONS

The system designed will be more suitable for Electric/Hybrid vessel and Electric/Hybrid train. It has the following operating conditions: Lithium Ion Battery (LIB) is selected for high power and Fast Dynamics ESS (FDESS) and Vanadium Redox Flow Battery (VRFB) is selected for high energy and Slow Dynamics ESS (SDESS). The control of VRFB is mainly to maintain the DC bus voltage constant and the control of the LIB is to provide or absorb transient power during load variations. This yields to a decrease in the power ratings of the main ESS (VRFB in this design).

The interface of both ESS is carried out by means of bidirectional boost converter as shown in Fig. 1. With this topology, the power flow of each storage device can be controlled independently thanks to the two DC/DC converters offering a high flexibility to manage the HESS [2], [9-16].

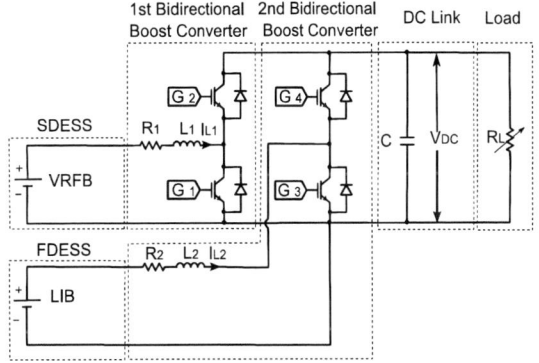

Fig. 1. The power circuit of the designed system.

III. DESIGN METHOD

The selection of the two batteries technologies was according to their advantages over other batteries technologies. The main advantages of VRFB are as following [4], [14-19]:

• It is an energy storage device which intended for energy rather than for power.

• Its storage efficiency is high as it can work for hours.

• It has a high scalability and is suitable for large scale storage applications because of electrolyte tanks.

• Instant recharge by electrolyte exchange.

• Long life cycle up to 10 000 charge/discharge cycles leads to lower through life costs.

978-1-5090-0064-7/15 $31.00 © 2015 IEEE

- Low maintenance requirements because it uses pumps to circulate the electrolytes from the tanks to the cell.

Also, LIB has a lot of advantages as following [20], [21]:

- It is a power storage device compared to the VRFB.
- It has quick response than the VRFB.
- It operates through a wide range of temperatures and it has high efficiency.
- Easy charge controllability and low self-discharge.
- Suitable for short term applications.
- No pollution compared to other battery technologies.

Table 1 shows the parameters of VRFB and LIB used in the designed HESS system.

Table 1: Parameters of VRFB and LIB

VRFB		LIB	
Parameters	Values	Parameters	Values
E	50 KWh	E	13.2 KWh
P	25 KW	P	15 KW
$V_{nominal}$	302.4 V	$V_{nominal}$	311.6 V
V_{min}	150 V	V_{min}	252 V
I_{max}	166.7 A	I_{max}	50 A
Operating region	20-80%	Operating region	20-80%
No. of cells	230	No. of batteries	6

IV. EMULATION OF THE BATTERIES

Before Implementing the HESS with real batteries, which might be expensive, the control can be validated through the emulation of the batteries. Several options can be used to obtain the behavior of batteries.

- The first option is a real time simulation [14], however the equipment associated with this solution is very expensive.
- The second option consists on the emulation of the batteries by hardware construction by their equivalent circuit, nevertheless, if the battery parameters are changed, the hardware components should be modified as well.
- The third option is the emulation of the battery dynamics through real-time software running on a Digital Signal Processing (DSP) to get the virtual battery voltage. This last has the lowest cost and is the one analyzed in this paper.

1. Software part

The dynamic behavior of any battery can be modeled, among other options, by a simple circuit [1], [22], consisting of a series resistor (Rse) standing for the internal resistance of the battery, a capacitor (C_{SOC}) representing the state of charge (SOC) and a parallel capacitor (C_D) with a parallel resistance (R_D) representing the dynamics of the battery as shown in Fig. 2.

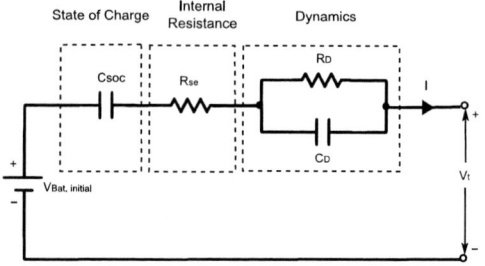

Fig. 2. Dynamic behavior equivalent circuit of any battery.

After some algebraic manipulations, the transfer function of the dynamic equivalent circuit is expressed as following:

$$G(s) = \frac{V_t(s) - V_{Bat.\ init.}}{I(s)}$$
$$= -\frac{R_s s^2 + \left(\frac{1}{C_{SOC}} + \frac{R_{se}}{R_D C_D} + \frac{1}{C_D}\right)s + \frac{1}{R_D C_D C_{SOC}}}{s^2 + \frac{1}{R_D C_D}s} \quad (1)$$

where:

- $_t(s)$: is the terminal voltage of the equivalent circuit,
- (s) : is the current flowing in the equivalent circuit,
- $_{Bat.\ init}$: is the initial battery voltage,
- : is the Laplace transform.

The current is considered positive flowing out of the battery (Discharging mode), thus a negative sign appears in (1). The virtual voltage of the battery is calculated in (2) from the transfer function. Also, the initial conditions of the battery voltage are included in the virtual voltage. Equation (2) is implemented in the DSP to get the virtual battery voltage which presents the same dynamics as the real battery voltage. The parameters of the dynamic equivalents circuit of both batteries are shown in table 2.

$$V_{Bat.\ virtual}(s) = V_t(s)$$
$$= V_{Bat.\ init.} + I(s) * G(s) \quad (2)$$

Table 2: Parameters of the dynamic equivalent circuits of VRFB and LIB

VRFB		LIB	
Parameters	Values	Parameters	Values
C_{SOC}	7594.9 Farads	C_{SOC}	0.0127 Farads
R_{se}	0.06826 Ohms	R_{se}	0.0628 Ohms
C_D	0.025 Farads	C_D	0.0127 Farads
R_D	0.0996 Ohms	R_D	0.1181 Ohms

2. Hardware part

To obtain the initial battery voltage in (2), a three phase uncontrolled rectifier as in Fig. 3. Capacitors are intended to decrease the ripples and smoothing the DC voltage. A charging resistance is used to initially charge the capacitor smoothly and then the bypass switch is closed. A blocking diode is used to prevent power return back to the three phase uncontrolled rectifier. A burning resistance is used to discharge the capacitor and in case of the power flowing from the DC bus through the bidirectional boost converter

978-1-5090-0064-7/15 $31.00 © 2015 IEEE

to the three phase rectifier, the burning resistance will dissipate the power flowing back like in case of regenerative applicants. For this system, the two circuits of the FDESS and SDESS are similar, but with different initial voltage values.

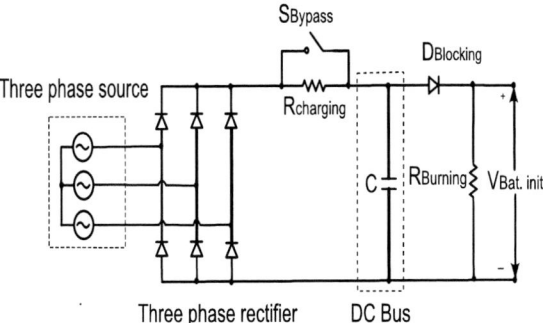

Fig. 3. Emulation of the battery to obtain the initial voltage of the battery and behaves as a battery.

V. CONTROL SCHEME OF THE CONVERTERS

As mentioned, the system consists of two bidirectional DC-DC boost converters. Each converter is connected to one battery and has its own control. One converter controls the DC link voltage to avoid stability problems, while the other converter delivers the peak transient power during load variations.

As discussed in the previous section, the virtual voltage of each battery will be calculated from (2) as in Fig. 4, however, each equivalent circuit has its own parameters. As the control of SDESS bidirectional boost converter is to maintain the DC link voltage constant. A typical cascaded control scheme is used, being the outer loop the DC link voltage control and the inner loop the inductor current control [6], [23]–[27] as shown in Fig. 5.

The control aim of the FDESS bidirectional boost converter is to provide the peak transient power when the load varies. This control will accelerate the recovery of the DC link voltage variations. Also cascaded control scheme is proposed, being the outer loop the power control and the inner loop the inductor current control as shown in Fig. 6.

The power reference is calculated by subtracting to the SDESS measured power its low frequency component by using a Low Pass Filter (LPF) to get the high frequency component.

To assure that the power of LIB is providing transient power during load variations, a PI controller is implemented. To increase the amount of power provided by the LIB, a gain K is multiplied by the power measured of the LIB. This means that more power will be released under sudden load variations, as the power error tends to increase the FDESS current reference. This will be helpful for the fast recovering of the DC link voltage. The value of the gain K goes from 0.1 to 1.0 and it depends on the maximum power of FDESS and the load variations. Table 3 shows the parameters of the control scheme of both converters.

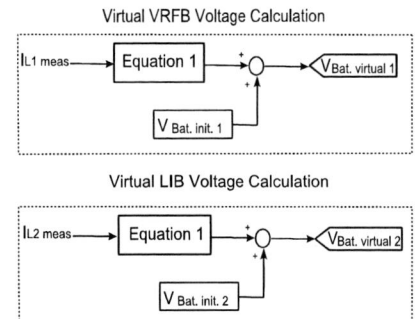

Fig. 4. Virtual voltage calculation for VRFB and LIB to get the battery dynamics.

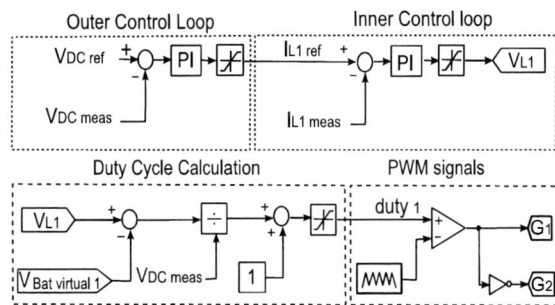

Fig. 5. SDESS converter control scheme to maintain the DC link voltage constant.

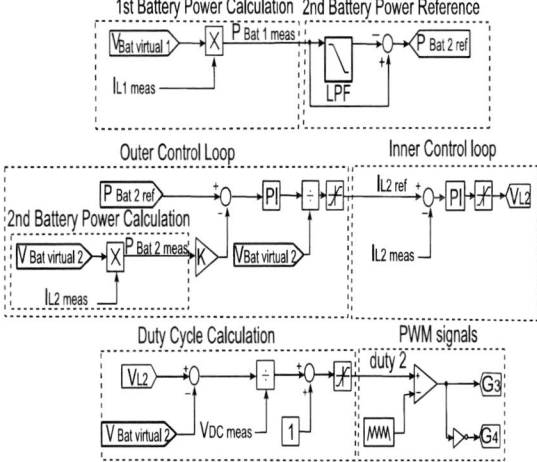

Fig. 6. FDESS converter control scheme to provide transient power during load variations.

Table 3: Parameters of the control scheme

Parameters	Values	Parameters	Values
$V_{VRFB\ init.}$	302.4 V	$V_{LIB\ init}$	311.6 V
SDESS converter control parameters			
$V_{DC\ ref}$	600 V	F_s	10 KHz
K_{pv}	0.19	K_{pi}	79.16
T_{iv}	0.3	T_{ii}	0.042
SDESS converter control parameters			
F_{LPF}	10 Hz	F_s	10 KHz
K_{pp}	1.56	K_{pi}	79.16
T_{ip}	1	T_{ii}	0.042

IV. VERIFICATION OF THE DESIGNED SYSTEM WITH SIMULATIONS

The simulations will consist on applying load variations upon the system once the demanded power has reached the steady state, with different values of gain K, using as a metric the DC link voltage variations.

A voltage reference of 600 V is used to validate the proposed control scheme for the designed HESS. The load is varied from 1.2 KW to 2.4 KW and again to 1.2 KW. The system is tested for two cases to check the effect of connecting LI-IB to VRFB.

- Case 1. Only VRFB is used, being connected to the DC link through the bidirectional boost converter.
- Case 2. Both VRFB and LI-IB are connected to the DC link by means of two bidirectional boost converters.

In Case 1, as shown in Fig. 7 that the DC link voltage drops 18 V when load varies from 1.2 KW to 2.4 KW and increases 19 V when load varies from 2.4 KW to 1.2 KW. These changes in the DC link voltage will increase if the load power is increased as the system designed for 25 KW.

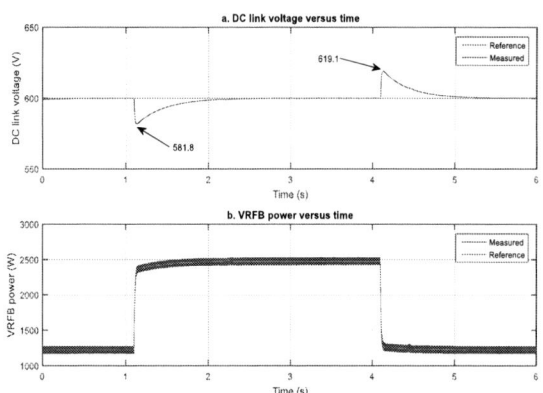

Fig. 7. Simulation Results at VRFB only: a) DC link voltage versus time b) VRFB power versus time.

In case 2, the gain K in Fig. 6 is changed to check the effect on the DC link voltage and the LI-IB current. As starting value, K = 1.0 is selected, the DC link voltage variations already exhibit a decrease, this is due to the power delivered by the LI-IB when load varied. By further decreasing K, the maximum voltage variations also improves, as it is shown in table 4, thus minimizing the effect of the load variations in the DC link voltage. The selected optimal value for K is 0.3 and the results are shown in Fig. 8.

Table 4: Simulation Results

	K	V_{DC_min} (V)	V_{DC_max} (V)	$P_{LIB\ max}$ (W)	$P_{LIB\ min}$ (W)
Case 1		581.9	619.1		
Case 2	1.0	587.9	612.7	465.8	-463.1
	0.7	588.5	612.2	531.3	-529.1
	0.5	588.8	611.2	581.8	-552.3
	0.3	589.9	610.4	655.7	-580.7

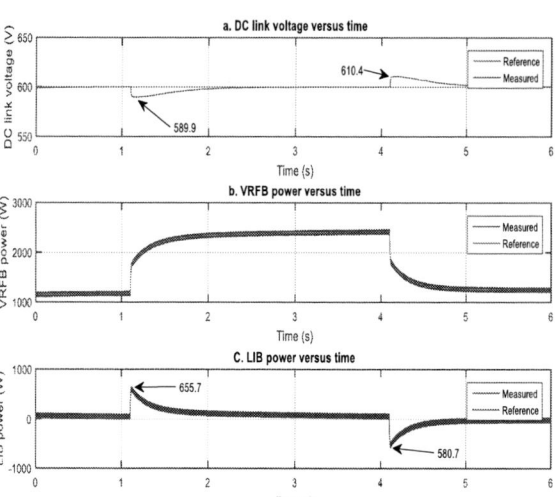

Fig. 8. Simulation Results when VRFB and LIB are connected at K = 0.3: a) DC link voltage versus time. b) VRFB power versus time. c) LIB power versus time.

VI. VERIFICATION OF THE DESIGNED SYSTEM WITH EXPERIMENTAL SETUP

As it has been mentioned, the dynamic behavior of both the LDESS and FDESS will be implemented in real time software with parameters in Table 2, and the power will be absorbed or delivered to a dedicated DC bus. This scheme can be replicated for any ESS technology, provided that the dynamic behavior is known or can be calculated.

The experimental results in Fig. 9 and table 5 are to validate the emulation of the batteries and the proposed control scheme.

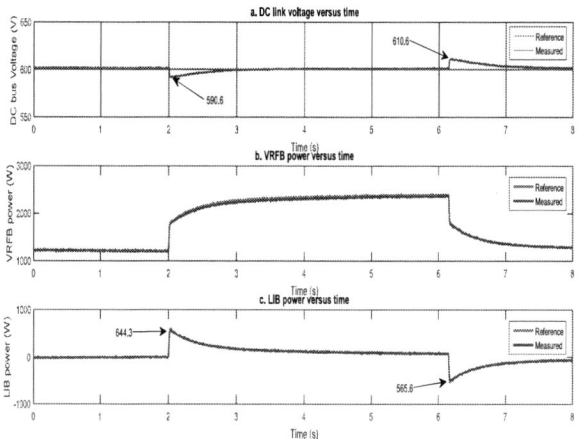

Fig. 9. Experimental Results when VRFB and LIB are connected at K = 0.3: a) DC link voltage versus time b. VRFB power versus time. c) LIB power versus time.

Table 5: Experimental Results

	K	V_{DC_min} (V)	V_{DC_max} (V)	$P_{LIB\ max}$ (W)	$P_{LIB\ min}$ (W)
Case 1		583	619		
Case 2	1.0	587.7	612.7	454.9	-442.5
	0.7	588.3	611.6	502	-503.2
	0.5	589.1	611.1	567.7	-523.8
	0.3	590.6	610.6	644.3	-565.6

V. CONCLUSION AND FUTURE WORK

The designed HESS system for electric transportation decreases the total cost as there is no need to have a main ESS with a high energy density and a high power density simultaneously. Also it increases the life span of the main ESS.

The designed system with the proposed control scheme is validated by simulation and experiments. The simulation results are validated with the experimental results from emulation of the batteries. This emulation of batteries provide a low cost solution to test any technology of batteries and can validate any new control scheme before implementing HESS.

The technique used for emulation can be extended to different battery technologies, only requiring software changes. Further, the proposed control scheme can be used for any ESS technology such as supercapacitors, regular capacitors, etc.

Moreover, future developments of this work would include: increasing the number of ESS in the hybrid system and explore the applicability to other PEC topologies, such as isolated, multilevel, etc.

ACKNOWLEDGMENT

I would like to thank Prof. Pablo Garcia for his guidance and supporting. This work has been partially supported by the Spanish Government, Innovation Development and Research Office (MEC), under research grant ENE2013-44245-R, Project "Microholo", and by the European Union through ERFD Structural Funds (FEDER). Also, this work has been partially supported by the government of Principality of Asturias, Foundation for the Promotion in Asturias of Applied Scientific Research and Technology (FICYT), under Severo Ochoa research grant, PA-13-PF-BP13138.

REFERENCES

[1] A.R. Sparacino, G.F. Reed, R.J. Kerestes, B.M. Grainger, Z.T. Smith, "Survey of battery energy storage systems and modeling techniques," 2012 IEEE in Power and Energy Society General Meeting, 22-26 July 2012, pp.1-8.

[2] Yoo, H., Seung-Ki Sul, Yongho Park, Jongchan Jeong, "System Integration and Power-Flow Management for a Series Hybrid Electric Vehicle Using Supercapacitors and Batteries," IEEE Transactions on Industry Applications, vol.44, no.1, pp.108,114, Jan.-feb. 2008.

[3] J. Cobben, W. Kling, and J. Myrzik, "Power quality aspects of a future micro grid," in International Conference on Future Power Systems. IEEE, Nov. 2005, pp. 1 – 5.

[4] Vechiu, A. Etxeberria, H. Camblong, and J. M. Vinassa, "Three-level neutral point clamped inverter interface for flow battery/supercapacitor energy storage system used for microgrids," in PES International Conference and Exhibition on Innovative Smart Grid Technologies (ISGT Europe), no. 2. IEEE, Dec. 2011, pp. 1 – 6.

[5] B. Wang, B. Zhang, and Z. Hao, "Control of composite energy storage system in wind and pv hybrid microgrid," in International Conference of IEEE Region 10 (TENCON). IEEE, Oct. 2013, pp. 1 – 5.

[6] D. Tran, H. Zhou, and A. M. Khambadkone, "Energy management and dynamic control in composite energy storage system for micro-grid applications," in Annual Conference of IEEE Industrial Electronics Society (IECON), no. 36. IEEE, Nov. 2010, pp. 1818 – 1824.

[7] Etxeberria, I. Vechiu, H. Camblong, A. Etxeberria, J. Vinassa, and H. Camblong, "Hybrid energy storage systems for renewable energy sources integration in microgrids: A review," in IPEC. IEEE, Oct. 2010, pp. 532 – 537.

[8] S. priya.S and Rajakumar.S, "An energy storage system for wind turbine generators- battery and supercapacitor," International Journal of Engineering Research and Applications (IJERA), vol. 3, pp. 1219 – 1223, March - April 2013.

[9] Tummuru, N.R., Mishra, M.K., Srinivas, S., "Dynamic Energy Management of Hybrid Energy Storage System With High-Gain PV Converter," IEEE Transactions on Energy Conversion, vol.30, no.1, pp.150,160, March 2015.

[10] Kollimalla, S.K., Mishra, M.K., Narasamma, N.L., "Design and Analysis of Novel Control Strategy for Battery and Supercapacitor Storage System," IEEE Transactions on Sustainable Energy, vol.5, no.4, pp.1137, 1144, Oct. 2014.

[11] Kollimalla, S.K., Mishra, M.K., Lakshmi Narasamma, N., "Coordinated control and energy management of hybrid energy storage system in PV system," 2014 International Conference on Computation of Power, Energy, Information and Communication (ICCPEIC), pp.363,368, 16-17 April 2014.

[12] Sathishkumar, R., Kollimalla, S.K., Mishra, M.K., "Dynamic energy management of micro grids using battery super capacitor combined storage," 2012 Annual IEEE India Conference (INDICON), pp.1078, 1083, 7-9 Dec. 2012.

[13] Jayasinghe, S.D.G., Vilathgamuwa, D.M., Madawala, U.K., "A direct integration scheme for battery-supercapacitor hybrid energy storage systems with the use of grid side inverter," 2011 Twenty-Sixth Annual IEEE Applied Power Electronics Conference and Exposition (APEC), pp.1388, 1393, 6-11 March 2011.

[14] Wei Li; Joos, G., Belanger, J., "Real-Time Simulation of a Wind Turbine Generator Coupled with a Battery Supercapacitor Energy Storage System," IEEE Transactions on Industrial Electronics, vol.57, no.4, pp.1137, 1145, April 2010.

[15] Wei Li; Joos, G., "A power electronic interface for a battery supercapacitor hybrid energy storage system for wind applications," IEEE Power Electronics Specialists Conference, 2008. PESC 2008, pp.1762, 1768, 15-19 June 2008.

[16] C. A. Smith D. D. Banham-Hall, G. A. Taylor and M. R. Irving, "Frequency control using vanadium redox Flow batteries on wind farms", In IEEE, 2011.

[17] C. Abbey J. Chahwan and G. Joos, "VRB modelling for the study of output terminal voltages, internal losses and performance", In IEEE Canada Electrical Power Conference, 2007.

[18] Wu Bingyin Mao Biao, Zhang Buhan and Xie Guanglong, "Studies on security capacity of wind farms containing vrb energy storage system", In Mao Biao, pages 1704 - 1708.

[19] Daqiang Bi Wenliang Wang, Baoming Ge and Dongsen Sun, "Grid-connected wind farm power control using VRB-based energy storage system", In IEEE, pages 3772 - 3777, 2010.

[20] John A. Chahwan, "Vanadium-Redox Flow and lithium-ion battery modelling and performance in wind energy applications", Master's thesis, McGill University, Montreal, Quebec, May 2007.

[21] Ye Zhang Li Guo and Cheng Shan Wang, "A new battery energy storage system control method based on SOC and variable filter time constant", In IEEE, 2011.

[22] M. Gonzaleza V.M. Garciab C. Blancoa D. Anseana, J.C. Vieraa and J.L. Antunaa, "Meaurement and study of dc internal resistance in lifepo4 batteries", In EEVC European Electric Vehicle Congress, Brussels, Belgium, November 2012.

[23] L. Baoquan, Z. Fang, and B. Xianwen, "Control method of the transient compensation process of a hybrid energy storage system based on battery and ultra-capacitor in micro-grid," in International Symposium on Industrial Electronics (ISIE). IEEE, May 2012, pp. 1325 – 1329.

[24] D. D. Banham-Hall, G. A. Taylor, C. A. Smith, and M. R. Irving, "Flow batteries for enhancing wind power integration," IEEE Transactions on Power Systems, vol. 27, no. 3, pp. 1690 – 1697, Aug. 2012.

[25] H. Kakigano, Y. Miura, T. Ise, and R. Uchida, "Dc micro-grid for super high quality distribution - system configuration and control of distributed generations and energy storage devices -," in Power Electronics Specialists Conference. IEEE, June 2006, pp. 1 – 7.

[26] J. Liang and C. Feng, "Stability improvement of micro-grids with coordinate control of fuel cell and ultracapacitor," in Power Electronics Specialists Conference (PESC). IEEE, June 2007, pp. 2472 – 2477.

[27] L. Barote, C. M. Weissbach, R. Teodorescu, C. Marinescu, and M. Cirstea, "Stand-alone wind system with vanadium redox battery energy storage," in International Conference on Optimization of Electrical and Electronic Equipment (OPTIM), no. 11. IEEE, May 2008, pp. 407 – 412.

Load analysis for the asymmetric bilateral linear switched reluctance generator

Qianlong Li[1], Jun Xia[2], J. F. Pan[1], Bo Zhang[1] and Norbert Cheung[3]

[1] Shenzhen Key Laboratory of Electromagnetic Control, College of Mechatronics and Control Engineering, Shenzhen University, Shenzhen
2 Shenzhen Institute of Electronics
3 Department of Electrical Engineering, The Hong Kong Polytechnic University, Hong Kong

Abstract–In this paper, to study the load characteristics of the asymmetric bilateral linear switched reluctance generator, theoretical analysis is first derived. Then the characteristics of the machine are investigated by the finite element method. Experiments at different loads are carried out for the asymmetric bilateral linear switched reluctance generator. The experiments are based on dSPACE DS1104 controller board and Control Desk software package combined with Matlab/Simulink toolbox. The results demonstrate that each phase is decoupled from the magnetic circuit, and with the increment of load, the rate of discharging current increases

Keywords–Asymmetric bilateral linear switched reluctance generator, load analysis, finite element method.

I. INTRODUCTION

The linear switched reluctance machine, as a new type of direct-drive machine, presents more advantages, such as a simple structure, robustness and low cost, etc. It can be applied in some special areas with severe temperature variations [1-3]. Asymmetric bilateral linear switched reluctance generator (ABLSRG) has a separated winding structure with no coupling effects and more force-to-volume ratio [4-5]. In recent years, there are more attentions paid on the ABLSRG. In this paper, load analysis experiments of ABLSRG at different loads are carried out, and the experiment results at variable load are presented.

II. PRINCIPLE OF ABLSRG

1. Mechanical Structure of ABLSRG

This generator is mainly composed of the mover, stator base, six stators, and a pair of linear guides. The generator adopts an asymmetric structure of the mover [5]. As shown in Fig. 1(a), there are three phases named as phase AA', BB', and CC', respectively, and each phase has two windings connected on two stators. Both the mover and the stators possess a toothed structure with silicon-steel laminations. Six stators are fixed on an aluminum base connected by a pair of linear guides, as shown in Fig. 1(b). The stators and windings are shown in Fig. 1(c) and major specifications are listed in Table 1.

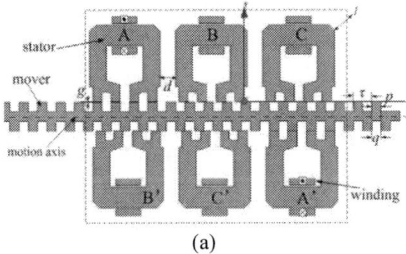

(a)

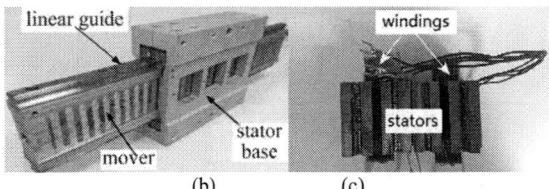

(b)   (c)

Fig. 1: (a) Machine sketch, (b) mover and stator base, and (c) stators

Table 1: Major specification of the ABLSRG

Parameter	Value
Number of phase windings (N)	440 turns
Phase division (d)	10 mm
Air gap (g)	0.3 mm
Stroke length	350 mm
Stack length (l)	100 mm
Stator/mover pole width (p)	6 mm
Stator/mover slot width (q)	20 mm
Pole-pitch (τ)	12 mm
Mover mass	3.8 Kg

2. Theoretical Background of ABLSRG

The ABLSRG can be depicted as a typical energy transformation system as one mechanical input and three electrical power output terminals. From the mechanical perspective [6], we have,

$$F = M\frac{d^2x}{dt^2} + D\frac{dx}{dt} + f \qquad (1)$$

where F is the mechanical force input, M is the mass of the mover track, D is the friction coefficient, f stands for the load force and x is the position of the mover. With respect to the electrical terminal, the ABLSRG model can be described in the form of the voltage balance equation as [7],

$$u_k = R_k i_k + \frac{d\lambda_k}{dt} \quad (k = 1, 2, 3) \qquad (2)$$

with $\lambda_k = \sum_{j=1}^{3} L_k i_j$, where u_k and λ_k stand for voltage drop and flux-linkage of the $\underline{k}$-th winding, respectively. i_k, R_k and L_k are the phase current, resistance and inductance, respectively. Equation (2) can be further depicted in the inductance form as [8],

$$u_k = R_k i_k + \sum \left\{ \left[\frac{\partial L_{kj}}{\partial i_j} i_j + L_{kj} \right] \frac{di_j}{dt} + v_x \cdot \frac{\partial L_{kj}}{\partial x} \cdot i_j \right\} (k = 1, 2, 3) \quad (3)$$

$$v_x = \frac{dx}{dt} \qquad (4)$$

978-1-5090-0064-7/15 $31.00 © 2015 IEEE   373

where v_x is the speed of mover. The electromagnetic force can be denoted as [8],

$$f\left(i_1,i_2,i_3,x\right)=\frac{1}{2}\left(i_1^2\frac{dL_{11}}{dx}+i_2^2\frac{dL_{22}}{dx}+i_3^2\frac{dL_{33}}{dx}\right) \quad (5)$$

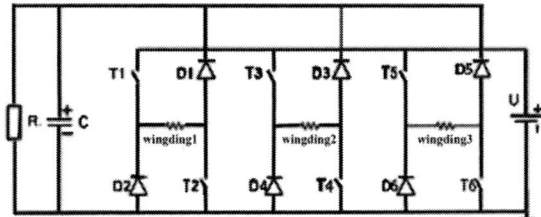

Fig. 2: Topology of the power converter

If the power converter employs the three-phase asymmetrical half bridge as the drive topology as shown in Fig. 2, for any one phase, current can be described as,

$$i(x)=\frac{U(x-x_{on})}{v_x\cdot\left[L_{\min}+k_x(x-x_1)\right]} \quad (6)$$

$$i(x)=\frac{U(x-x_{on})}{v_x\cdot\left[L_{\max}-k_x(x-x_2)\right]} \quad (7)$$

$$i(x)=\frac{U(2x_{off}-x_{on}-x)}{v_x\cdot\left[L_{\max}-k_x(x-x_2)\right]} \quad (8)$$

where k_x is the slope of inductance increase. Equation (6) is the current from the power switches at the turn-on position (x_{on} mm) to the fully aligned position (x_2 mm). Equation (7) is the current from the fully aligned position to the turn-off position (x_{off} mm). Equation (8) is the current from the turn-off position to the beginning position of the next period. The entire phase current changing process is similar to the dotted curve as shown in Fig. 3.

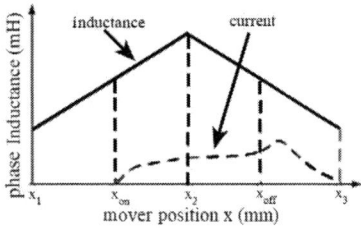

Fig. 3: Turn-on and turn-off position sketch

3. Electromagnetic Characteristic Analysis

To fully predict the generator's performance, two-dimensional (2D) finite element analysis is performed with MAXWELL software package. Test of phase coupling effect is first carried out to guarantee that each phase can be regulated individually so that no additional decoupling methods shall be required [8]. Fig. 4 is the magnetic field distribution diagram with the excitation of 10A. Fig. 5 (a) and (b) show the self-inductance coefficient and mutual-inductance coefficient of phase A respect to phase B with the change of mover's position. As shown in Fig. 5 (b), mutual-inductance coefficient of phase A and phase B is

smaller than one percentage of the self-inductance from fully aligned positions to fully unaligned positions within one pole-pitch of the mover respective to the stators of phase A. In the same way, the mutual-inductance coefficient of phase A respect to phase C and the mutual-inductance coefficient of phase B respect to phase C are also much smaller than the self-inductance coefficients. Therefore, each phase is decoupled in magnetic circuit for ABLSRG.

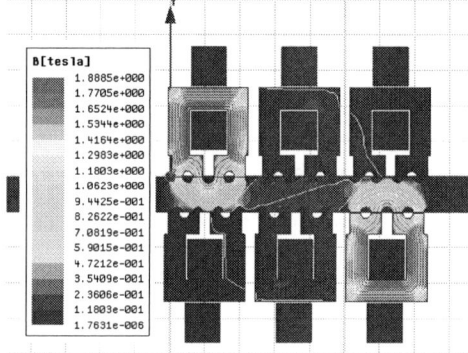

Fig. 4: Distribution contour of magnetic flux

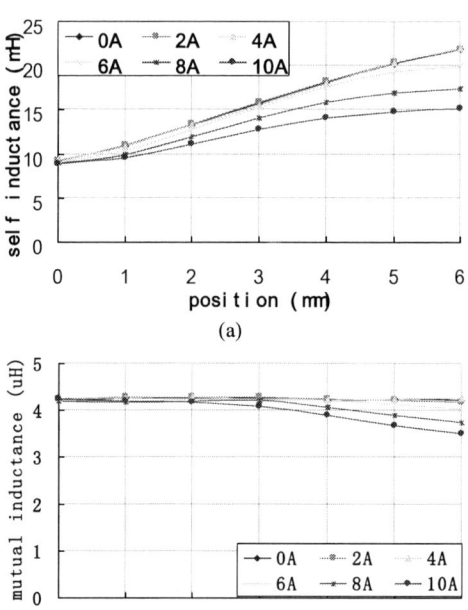

Fig. 5: (a) Self-inductance and (b) mutual inductance profiles.

III. EXPERIMENTAL SETUP AND RESULTS

The experiment for the load analysis of the ABLSRG are carried out base on the dSPACE DS1104 controller card. The experimental setup is shown in Fig. 6. The generation system mainly consists of the ABLSRG, power converters, controller, and electrical energy storage module. The power converter of the generation system employs an asymmetrical half-bridge structure for full isolation. The topology structure of power converter is shown as Fig. 2. The control block diagram is set up in Matlab/Simulink with a sampling frequency of 1 KHz.

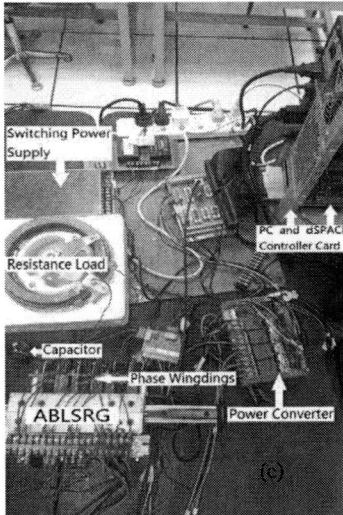

Fig. 6: Experiment setup

The experiment is carried out with the load settings as shown in Table 2 and the results are plotted as shown from Fig.7. (a) to (f). The current waveform of phase A with the excitation switch signal is shown in Fig. 7. The blue signal is the excitation switch signal and the red signal is the current signal of phase A.

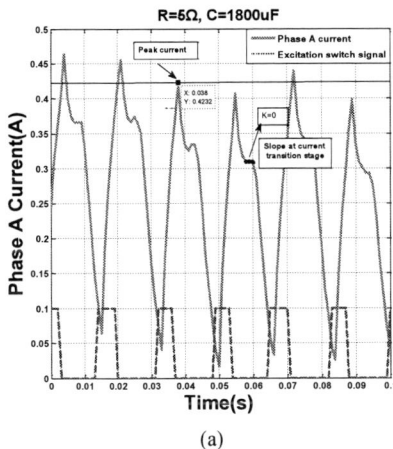

(a)

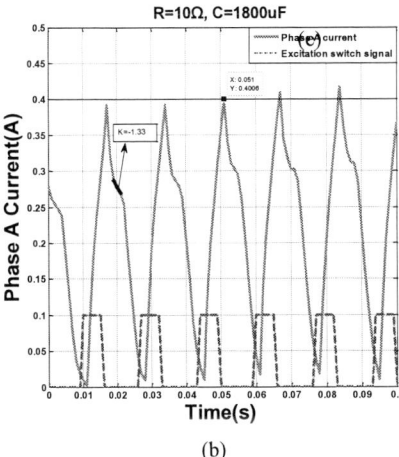

(b)

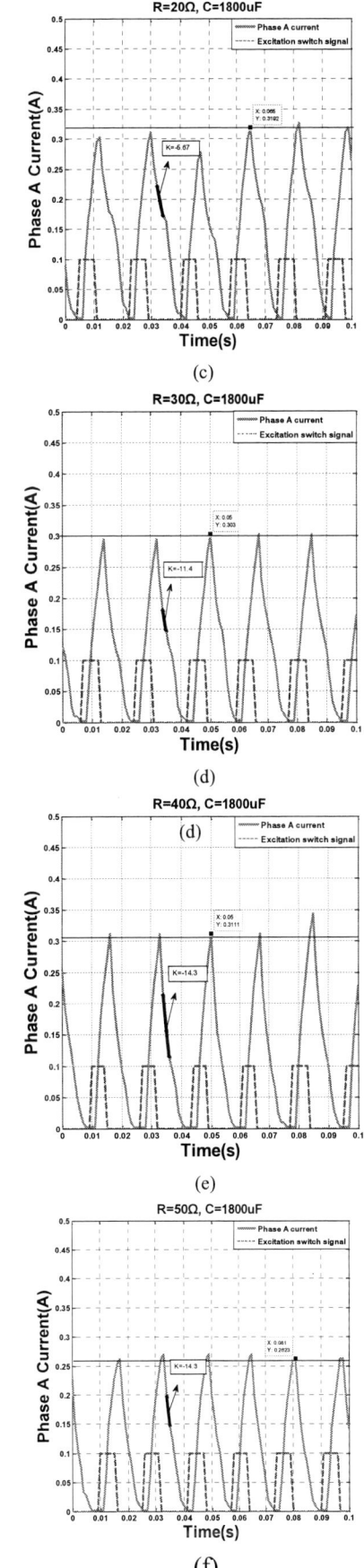

(c)

(d)

(d)

(e)

(f)

Fig. 7: (a)-(f) the current profiles of phase A at different loads

As shown in Fig.7 (a), most of the peak current values reach around 0.42 A, and the slope of current at the transition stage is almost 0 A. The peak current value and the slope value of current at the transition stage are two key indexes to estimate the performance of ABLSRG and they are collected in Table 2.

Table 2: Load settings and two indexes values

Load Settings	Peak Current(A)	Slope at current transition stage
C=1800uF R=5Ω	0.42	0
C=1800uF R=10Ω	0.4	-1.33
C=1800uF R=20Ω	0.32	-5.67
C=1800uF R=30Ω	0.3	-11.4
C=1800uF R=40Ω	0.31	-14.3
C=1800uF R=50Ω	0.26	-14.3

The profile of peak current value versus load and the profile of slope of current at the transition stage versus load are plotted in Fig.8 and Fig.9, respectively.

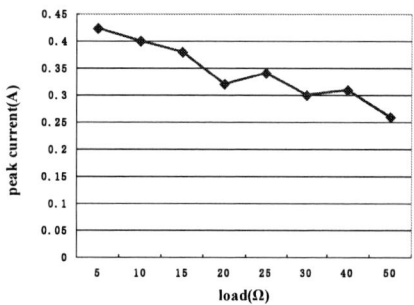

Fig. 8: Profile of peak current at different loads

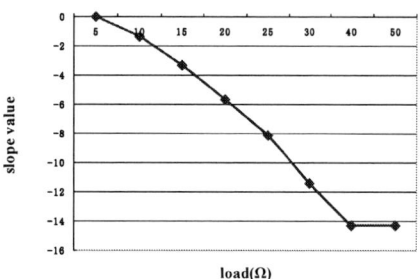

Fig. 9: Profile of slope of current at the transition stage at different loads

Fig.8 and Fig.9 show that the two indexes are decreased as the load rises up. The results prove that with the resistance load increasing, the consumption of power energy is augmented, so the rate of the discharging current increases.

IV. Conclusion

This paper investigates the characteristics including the electromagnetic behaviors for the ABLSRG. The results demonstrate that each phase can be controlled independently so that no decoupling mechanism shall be introduced for the three-phase power generation. This paper also studies the load characteristics for the ABLSRG power generation system. The results prove that with the resistance load increasing, the consumption of power

energy is augmented, and the rate of discharging current increases.

Acknowledgment

This work was supported in part by the National Natural Science Foundation of China under Grant 51477103 and 51577121, and in part by the Guangdong Natural Science Foundation under Grant S2014A030313564 and S2015A010106017.

References

[1] Kwok, Antares San-Chin, Wai-Chuen Gan, and Norbert C. Cheung, "Improvements in the Mechanical Structure of the Linear Switched Reluctance Motor", IEEE International Conference on Power Electronics Systems and Applications, pp. 186-189, 2006.

[2] Jinhua Du, Deliang Liang, Longya Xu, and Qingfu Li, "Modeling of a linear switched reluctance machine and drive for wave energy conversion using matrix and tensor approach", IEEE Trans. Magn., vol. 46, no. 16, pp. 1334-1337, 2010.

[3] Ji-Hoon Park, Seok-Myeong Jang, Jang-Young Choi, SoYoung Sung, and Il-Jung Kim, "Dynamic and experimental performance of linear-switched reluctance machine with inductance variation according to airgap length", IEEE Tran. Magn., vol.46, no. 6, pp. 2334-2337, 2010.

[4] J. F. Pan, Yu Zou, Guangzhong Cao, Norbert C. Cheung, and Bo Zhang, "High-Precision Dual-Loop Position Control of an Asymmetric Bilateral Linear Hybrid Switched Reluctance Motor", IEEE Trans. Magn., vol. 51, no. 11, 8600405, 2015.

[5] J. F. Pan, Y. Zou, and G. Cao, "An asymmetric linear switched reluctance motor", IEEE Trans. Energy Convers., vol. 28, no. 2, pp. 444–451, 2013.

[6] Jinhua, D., Deliang, L., Longya, X., Qingfu, L, "Modeling of a linear switched reluctance machine and drive for wave energy conversion using matrix and tensor approach", IEEE Trans. Magn, vol.46 , no.6, pp. 1334–1337, 2010

[7] Miller, T.J.E, "Switched reluctance motor and their control", Oxford University Press, 1993, 1st edition.

[8] Jianfei. Pan, Yu Zou, Guangzhong Cao, "Investigation of a low-power, double-sided switched reluctance generator for wave energy conversion", IET. Renewable Power Generation, Vol. 7, no. 2, pp. 98 - 109, 2013.

Coordination position control of linear switched reluctance machines

Weiyu Wang[1], Jun Xia[2], J. F. Pan[1], Bo Zhang[1] and Norbert Cheung[3]

[1] Shenzhen Key Laboratory of Electromagnetic Control, College of Mechatronics and Control Engineering, Shenzhen University, Shenzhen
[2] Shenzhen Institute of Electronics
[3] Department of Electrical Engineering, The Hong Kong Polytechnic University, Hong Kong

Abstract-This paper investigates the coordination control system using linear switched reluctance motors (LSRMs). LSRMs have the advantages of fast response, high precision with no backlash. In this paper, a coordination control system using LSRMs based on dSPACE real-time control platform is built. This system uses serial communication to transmit position information among LSRMs. The experiment results show that the dynamic error of this coordination control system can be controlled within 1mm.

Keywords- coordination control, linear switched reluctance motor, serial communication

I. INTRODUCTION

Modern manufacturing system is information-based, networked and distributed. In this background, cooperative control systems have become a hot topic. The traditional standalone control system is informationally encapsulated, difficult to manage and cooperate with others to complete one ultimate the complex tasks [1]. Generally, traditional distributed control system has a clear master-slave relationship. In order to complete collaboration tasks, an integrated system has to be designed. The system is easy to become invalid because of a key node's failure [2]. The structure of the system is difficult to modify for demands of the diversification and individualization of industry.

In this paper, we focus on the realization of the coordination control system based on linear switched reluctance machines, verified by experimental results. The structure of this paper is as follows. The structure and principles are given in the section II. Section III proposes the coordination control system. Section IV provides the experiment results. Section V offers the conclusions.

II. THE STRUCTURE AND PRINCIPLE OF THE LSRM

The principle of the linear switched reluctance motor (LSRM) is: the magnetic lines circulate into a loop according to the magnetic path according to the pricinple of minimum reluctance [1]. The LSRM mainly contains the mover and stators. AA', BB', CC' are the three phases. The structure of the LSRM is shown in the Fig.1.

Table 1 shows the major specifications of the LSRM.

Table 1: Specifications of the designed LSRM

Parameters	Value	Unit
Stator slot radius	3	mm
Stator pole width	6	mm
Mover slot radius	3	mm
Mover pole width	6	mm
Stack length	50	mm
Air gap length	0.3	mm

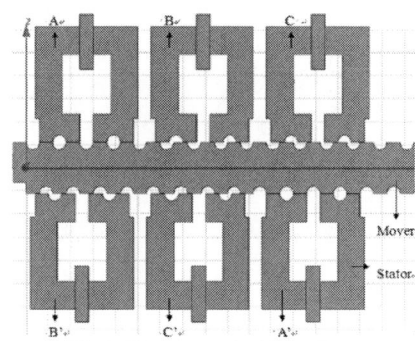

Fig.1: The structure of the LSRM

The mathematical models for the three-phase LSRM include the voltage balancing equation (1), the electromagnetic equation (2) and the mechanical movement equation (3) as follows with $j=1,2,3$ [4].

$$V_j = R_j + \frac{\partial \varphi_j(x, i_j)}{\partial x}\frac{dx}{dt} + \frac{\partial \varphi_j(x, i_j)}{\partial i_j}\frac{di_j}{dt} \tag{1}$$

$$f_e = \sum_{j=1}^{3} \frac{\partial \int_0^j \varphi_j(x, i_j) di_j}{\partial x} \tag{2}$$

$$f_e = M\frac{d^2x}{dt^2} + B\frac{dx}{dt} + f_l \tag{3}$$

where V_j is the supply phase voltage, i_j is the phase current, R_j is the phase resistance, φ_j is the phase flux linkage, x is displacement, f_e is the generated electromechanical force, f_l is the external load force, M is the total mass of the moving platform , B is the friction constant.

The electromagnetic force production equation for any one phase can be denoted as Equation (4).

$$f(x, i) = \frac{\pi i^2 \Delta L}{y} \cdot \sin(\frac{2\pi x}{y}) \tag{4}$$

where y is the polar distance, i is the phase current, and ΔL is the half of the difference between the maximum inductance and the minimum inductance. It is clear that the magnitude of the electromagnetic force is related to the phase current and the relative displacement. [5]

III. COORDINATION CONTROL SYSTEM

Fig.2 (a) shows a single control model of one LSRM. In this system, the reference signal is the sinusoidal wave. PID control algorithm is selected as P=2, I=0.01, D=0.12, respectively, based on a trial and error process. The sampling time is set as 0.001s. Fig.2 (b) and Fig.2 (c) show the results of this experiment. The dynamic error

978-1-5090-0064-7/15 $31.00 © 2015 IEEE   377

between the reference signal and the actual position is within the range of 1mm.

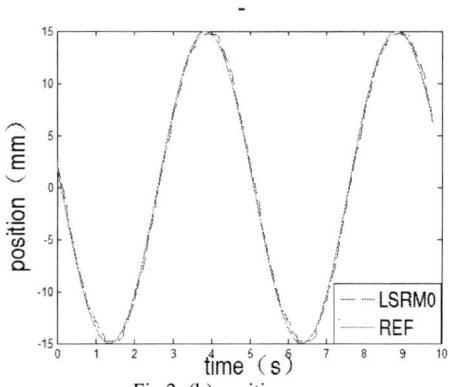

Fig .2: (a) One LSRM control system

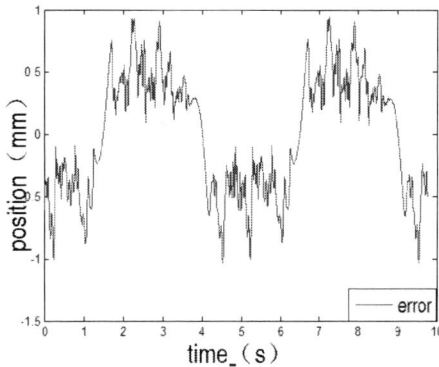

Fig.2: (b) position curve

Fig.2: (c) error curve

On the basis of a better performance for one motor control, the coordination control system is constructed as shown in Fig.3. In this system one LSRM needs to transfer its current position information to other LSRMs, realized by the serial ports. To meet the requirements of transmission speed, the bode rate is set as 115200b/s, the data bit is 8, the parity bit is 0 and the stop bit is 1.

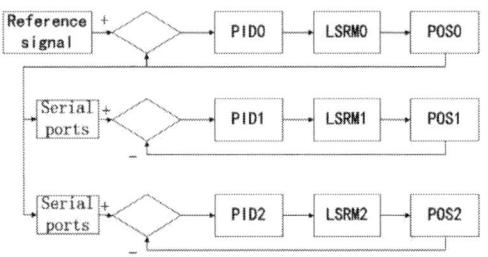

Fig.3: The coordination control system

In this way, LSRM0 takes the sinusoidal wave as the reference signal, LSRM1 and LSRM2 take LSRM0's position signal as the reference signal.

The coordination control system mainly consists of 3 LSRMs, drivers, linear encoders, power converters, three dSPACE platforms. The hardware of this system is shown in Fig.4 and Fig.5, respectively.

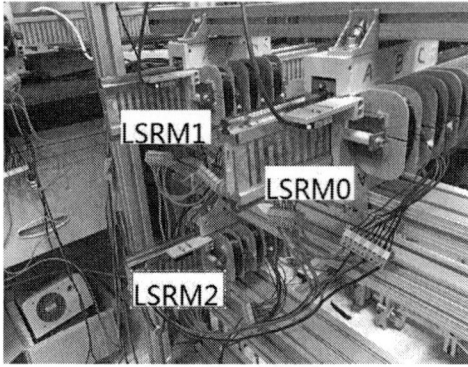

Fig.4: Three LSRMs

Fig.5: The coordination control system

The dSPACE1104 contains a digital signal processer with a high speed data computing power, 8 digital-to-analog converter channels, 20-bit I/O, 2incremental encoder interfaces, and 2 serial communication ports, etc. [6]

The drivers are ADVANCED MOTION CONTROLS drives and current can be controlled through the input voltage of the drivers internally [7]. In this paper, a linear magnetic encoder with 0.001mm-resolution is used. [8].

IV. EXPERIMENTAL RESULTS

In this experiment, we set the frequency of the reference signal as 0.2 Hz, and the amplitude of the signal as 15mm. Fig.6 shows the position response curves, REF is the reference position and LSRM0, LSRM1 and LSRM2 are the actual positions of three LSRMs. Fig.7 shows the dynamic error between any of two LSRMs. error0 is the dynamic error between LSRM1 and LSRM2. The error between LSRM0 and LSRM1 is error1 and the rest is the error between LSRM0 and LSRM1.

978-1-5090-0064-7/15 $31.00 © 2015 IEEE

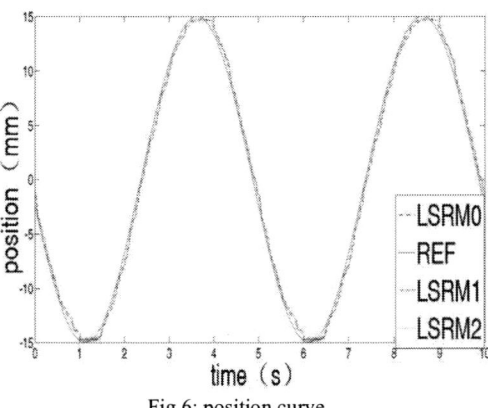

Fig.6: position curve

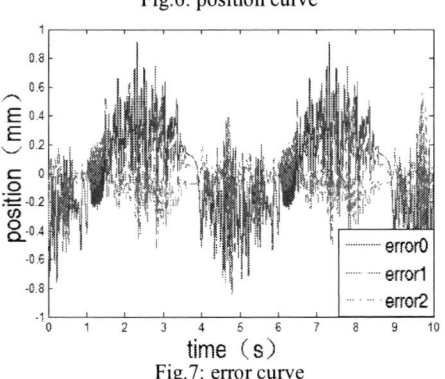

Fig.7: error curve

From these results, multiple machines coordination control system based on dSPACE platform is realized. It can be seen in Fig.7 that the dynamic error between each LSRM is within the range of 1mm.

V. CONCLUSION

This paper proposes a coordination control system using linear switched reluctance motors based on dSPACE real-time control platform. The basic principle of LSRMs and the structure of the control system are presented. The experiments and the results indicate that the coordination control system is effective, and the dynamic error range between each LSRM can be controlled within 1mm.

ACKNOWLEDGEMENT

This work was supported in part by the National Natural Science Foundation of China under Grant 51477103 and 51577121, and in part by the Guangdong Natural Science Foundation under Grant S2014A030313564 and S2015A010106017.

REFERENCES

[1] Miller, T. J. E., Switched reluctance motors and theircontrol. New York, Oxford University Press, 1993.

[2] Ramu, Krishnan, Switched reluctance motor drives:modeling, simulation, analysis, design, and applications. Boca Raton, FL : CRC Press, 2001.

[3] Ye Yun-yue, Principle and Application of LinearMotor .Beijing, China Machine Press, 2000.

[4] Kwok, Antares San-Chin, Wai-Chuen Gan, and NorbertC.Cheung, "Improvements in the Mechanical Structure of the Linear Switched Reluctance Motor", IEEE International Conference on Power Electronics Systems and Applications, ICPESA'06., 2006, pp.

186-189.

[5] Guang-zhong CAO Lv-ming LIN Hong QIU J.F. PAN, "Design and Analysis of a dSPACE-based Position Control System for a Linear Switched Reluctance Motor" 2009

[6] dSPACE User's Manual, Germany, dSPACE Company,2004.

[7] Advanced motion controls website: http://www.a-m-c.com/

[8] Givi website: http://www.givimisure.it/ita/

Cluster Flight Algorithms for Distributed Satellite based on Cyclic Pursuit

Bo Zhang[1,2], Yanhui Yun[1], Jianjun Luo[2], Norbert C. Cheung[3], J. F. Pan[1]

[1] College of Mechatronics and Control Engineering, Shenzhen University, Shenzhen,
[2] National Key Laboratory of Aerospace Flight Dynamics, Northwestern Polytechnical University, Xi'an,
[3] Department of Electrical Engineering, Hong Kong Polytechnic University, Hong Kong

Abstract- This paper investigates the cluster flight algorithm via the cyclic pursuit strategy in order to execute the coordination periodic motion of the distributed satellites. The proposed coordinated control law combining the cyclic pursuit strategy and diffeomorphism mapping is applied to match the trajectory of the cyclic pursuit and the natural periodic relative motion at low earth orbit environment considering the earth oblateness perturbation. The ultimate goal of the proposed method is to implement a cluster flight of multiple modules in the distributed satellite. Finally, a simulation example is presented to demonstrate the effectiveness of the proposed cluster flight algorithms.

Keywords- distributed satellite, cluster flight, cyclic pursuit, relative motion.

I. INTRODUCTION

Traditional satellites are composed of functional modules assembled in a single monolithic structure. When severe faults occur, the satellite usually cannot continue the preplanned mission, and significant losses, both technical and financial, are unavoidable. For instance, if the payload fails, the whole system disorders, and either the substitution or update of the functional modules of the satellite may be required. To traditional satellites, these operations are expensive and time-consuming, especially by re-launching a completely new satellite, since the functional modules are often highly integrated into the body of the satellite.

The above problem can be solved effortlessly by using the emerging concept termed as distributed satellite. The distributed satellite consists of multiple free-flying, physically separated modules interacting through the wireless cross links, and the distributed satellite has one or more specific function modules, such as the modules of navigation execution, attitude control, power generation, and payload, etc. Each module can be equipped with simple propulsion systems for the relative motion maneuver. For extending the entire mission lifetime, the relative motion between any two modules has to match the natural periodic orbit as far as possible, called the cluster flight.

Unlike the formation flying of the traditional satellite applications, the cluster flight of multiple modules of the distributed satellites is usually unnecessary to track the predetermined relative motion orbits. Moreover, the periodic motion property of the cyclic pursuit is appropriate to be applied to achieve a periodic motion. However, the periodic motion produced by the cyclic pursuit algorithm is not a naturally relative periodic orbit for the cluster flight of modules, because the natural periodic relative motion of the cluster flight under the

earth oblateness perturbation (J_2 is only considered in this paper.) is the complex closed trajectory and it is not a standard circle generally. Therefore, the similarity transformation approach is used to match the periodic orbits of the cyclic pursuit algorithm to the natural trajectories for the cluster flight modules of the distributed satellite.

II. DYNAMIC OF RELATIVE MOTION

Supposing the reference module S_0 is running in the circular orbit, in other words, the orbital eccentricity $e_0 = 0$. Under the earth oblateness perturbation, the relative motion between each module $S_i, i = 1, \ldots, N$ and S_0 is modeled as below,

$$
\begin{cases}
\ddot{x} - 2(nc)\dot{y} - (5c^2 - 2)n^2 x = -3n^2 J_2 (R_e^2/r_0) \times \\
\quad \left\{ \dfrac{1}{2} - [3\sin^2 i_0 \sin^2(kt)/2] - [(1 + 3\cos 2i_0)/8] \right\} + u_x \\
\ddot{y} + 2(nc)\dot{x} = -3n^2 J_2 (R_e^2/r_0)\sin^2 i_0 \sin(kt)\cos(kt) \\
\quad + u_y \\
\ddot{z} + q^2 z = 2lq\cos(qt + \phi) + u_z
\end{cases} \tag{1}
$$

where R_e is the earth radius and n , r_0 , and i_0 are the angular velocity, radius and the inclination of the reference orbit, respectively. Moreover, $\boldsymbol{u} = \begin{bmatrix} u_x & u_y & u_z \end{bmatrix}^T$ is the control force. k, c, q, l, ϕ can be calculated step-by-step. Firstly, k, c can be calculated respectively as follows,

$$
\begin{aligned}
k &= nc + \frac{3nJ_2 R_e^2}{2r_0^2}\cos^2 i_0 \\
c &= \sqrt{1 + s} \\
s &= \frac{3J_2 R_e^2}{8r_0^2}(1 + 3\cos^2 i_0)
\end{aligned} \tag{2}
$$

Moreover, q is calculated as,

$$
\begin{aligned}
q &= nc - \delta\dot{\gamma} \\
\delta\dot{\gamma} &= \left(\frac{\partial \gamma}{\partial \Delta\Omega}\right)_0 \left(\dot{\Omega}_1 - \dot{\Omega}_0\right) + \dot{\Omega}_1 \cos i_1 \\
\frac{\partial \gamma}{\partial \Delta\Omega} &= \cos\gamma \sin\gamma \cot \Delta\Omega - \sin^2 \gamma \cos i_1
\end{aligned} \tag{3}
$$

978-1-5090-0064-7/15 $31.00 © 2015 IEEE

where γ, $\Delta\Omega$, $\dot{\Omega}_1$, $\dot{\Omega}_0$, i_1 are obtained respectively as,

$$i_1 = \frac{\dot{z}(0)}{kr_0} + i_0$$

$$\Delta\Omega(t) = \Omega_1 - \Omega_2 = \Delta\Omega(0) + \left(\dot{\Omega}_1 - \dot{\Omega}_2\right)t$$

$$\Delta\Omega(0) = \frac{z(0)}{r_0 \sin i_0} \tag{4}$$

$$\dot{\Omega}_1 = -\frac{3nJ_2R_e^2}{2r_0^2}\cos i_1$$

$$\dot{\Omega}_0 = -\frac{3nJ_2R_e^2}{2r_0^2}\cos i_0$$

$$\gamma(t) = \cot^{-1}\left[\frac{\cot i_0 \sin i_1 - \cos i_1 \cos\Delta\Omega(t)}{\sin\Delta\Omega(t)}\right]$$

Moreover, l is derived as,

$$l = r_0\dot{\Phi} = -r_0\frac{\sin i_0 \sin i_1 \sin\Delta\Omega(0)}{\sin\Phi(0)}\left(\dot{\Omega}_1 - \dot{\Omega}_0\right)$$

$$\Phi(t) = \arccos\left[\cos i_0 \sin i_1 - \sin i_1 \sin i_0 \cos\Delta\Omega(t)\right]$$

The calculation of the coefficient ϕ is,

$$\begin{aligned} lm\sin\phi &= z(0) \\ l\sin\phi + qm\cos\phi &= \dot{z}(0) \end{aligned} \tag{5}$$

where $m \approx r_0\Phi(0)$.

The periodic solution of (1) can thus be derived analytically as below,

$$\begin{cases} x = x(0)\cos\left(nt\sqrt{1-s}\right) + \\ \quad \dfrac{\sqrt{1-s}}{2\sqrt{1+s}}y(0)\sin\left(nt\sqrt{1-s}\right) + p_x \\ y = -x(0)\dfrac{2\sqrt{1+s}}{\sqrt{1-s}}\sin\left(nt\sqrt{1-s}\right) + \\ \quad y(0)\cos\left(nt\sqrt{1-s}\right) + p_y \\ z = \left(lt+m\right)\sin\left(qt+\phi\right) + p_z \end{cases} \tag{6}$$

Here $\boldsymbol{p} = \begin{bmatrix} p_x & p_y & p_z \end{bmatrix}^T$ is defined as the auxiliary reference point, and it is depicted as follows,

$$\boldsymbol{p} = \begin{bmatrix} \alpha\left[\cos(2kt) - \cos\left(nt\sqrt{1-s}\right)\right] \\ \beta\cos(2kt) + \alpha\dfrac{2\sqrt{1+s}}{\sqrt{1-s}}\sin\left(nt\sqrt{1-s}\right) \\ 0 \end{bmatrix} \tag{7}$$

where α and β are derived as,

$$\alpha = -\frac{3nJ_2R_e^2n^2}{4kr_0}\frac{(3k-2n\sqrt{1+s})}{[n^2(1-s)-4k^2]}\sin^2 i_0$$

$$\beta = -\frac{3nJ_2R_e^2n^2}{4kr_0}\frac{[2k(2k-3n\sqrt{1+s})+n^2(3+5s)]}{2k[n^2(1-s)-4k^2]}\sin^2 i_0 \tag{8}$$

III. CYCLIC PURSUIT STRATEGY FOR CLUSTER FLIGHT

1. Cyclic pursuit algorithm

The cyclic pursuit algorithm is a particular consensus algorithm, which updates the state information of all modules through the directed ring communication topology, and all dimension components of the state information of each module are coupled by the rotation matrix. Thereby, the multiple modules can form the specific periodic circularity trajectory by the cyclic pursuit with the different pursuit angles. The cyclic pursuit algorithm has numerous benefits including a simple structure, less information requirement and the convenient implementation.

The typical cyclic pursuit algorithm tracks the dynamic target for the second-order linear system is applied as follows,

$$\boldsymbol{u}_i = -k_d\dot{\boldsymbol{r}}_i + \dot{\boldsymbol{v}}_i^d + k_d\boldsymbol{v}_i^d, i=1,\ldots,N \tag{9}$$

here k_d is a positive gain, $\boldsymbol{u}_i = \begin{bmatrix} u_{x,i} & u_{y,i} & u_{z,i} \end{bmatrix}^T$ is the control input of the second-order linear systems, and $\boldsymbol{r}_i = \begin{bmatrix} x_i & y_i & z_i \end{bmatrix}^T$, $\dot{\boldsymbol{r}}_i = \begin{bmatrix} \dot{x}_i & \dot{y}_i & \dot{z}_i \end{bmatrix}^T$ are the position and velocity of module S_i relative to the reference module S_0, respectively. Moreover, $\boldsymbol{v}_i^d$ holds as below,

$$\boldsymbol{v}_i^d = \begin{cases} \boldsymbol{R}(\alpha)(\boldsymbol{r}_{i+1}-\boldsymbol{r}_i) - k_c(\boldsymbol{r}_i-\boldsymbol{r}_c) + \dot{\boldsymbol{r}}_c, i=1,\ldots,N-1 \\ \boldsymbol{R}(\alpha)(\boldsymbol{r}_1-\boldsymbol{r}_i) - k_c(\boldsymbol{r}_i-\boldsymbol{r}_c) + \dot{\boldsymbol{r}}_c, i=N \end{cases} \tag{10}$$

where $\boldsymbol{R}(\alpha)$ is the rotation matrix and α is the pursuit angle, and k_c is a scalar gain. $\boldsymbol{r}_c = \begin{bmatrix} x_c & y_c & z_c \end{bmatrix}^T$ and $\dot{\boldsymbol{r}}_c = \begin{bmatrix} \dot{x}_c & \dot{y}_c & \dot{z}_c \end{bmatrix}^T$ are the position and velocity of the center of the geometric configuration of the desired relative motion, respectively.

Substituting (10) into (9), we have,

$$\boldsymbol{u}_i = \begin{cases} k_d\boldsymbol{R}(\alpha)(\boldsymbol{r}_{i+1}-\boldsymbol{r}_i) + \boldsymbol{R}(\alpha)(\dot{\boldsymbol{r}}_{i+1}-\dot{\boldsymbol{r}}_i) - k_dk_c(\boldsymbol{r}_i-\boldsymbol{r}_c) \\ \quad -(k_d+k_c)(\dot{\boldsymbol{r}}_i-\dot{\boldsymbol{r}}_c) + \ddot{\boldsymbol{r}}_c, \qquad i=1,\ldots,N-1 \\ k_d\boldsymbol{R}(\alpha)(\boldsymbol{r}_1-\boldsymbol{r}_i) + \boldsymbol{R}(\alpha)(\dot{\boldsymbol{r}}_1-\dot{\boldsymbol{r}}_i) - k_dk_c(\boldsymbol{r}_i-\boldsymbol{r}_c) \\ \quad -(k_d+k_c)(\dot{\boldsymbol{r}}_i-\dot{\boldsymbol{r}}_c) + \ddot{\boldsymbol{r}}_c, \qquad i=N \end{cases} \tag{11}$$

2. Homeomorphism mapping

According to (6) and (7), to achieve a natural relative orbit, a transformation matrix $\boldsymbol{T}_s$ is defined as the following,

978-1-5090-0064-7/15 $31.00 © 2015 IEEE   381

$$T_s = \begin{bmatrix} \dfrac{\sqrt{1-s}}{2\sqrt{1+s}} & 0 & 0 \\ 0 & 1 & 0 \\ z_0\cos(\phi_z) & z_0\sin(\phi_z) & 1 \end{bmatrix} \qquad (12)$$

where z_0,ϕ_z are two adjustable parameters. Utilizing (12), (11) is modified as follows,

$$\boldsymbol{u}_i = \begin{cases} -\boldsymbol{f}_s(\boldsymbol{r}_i)+k_s\Big[k_d\boldsymbol{T}_s\boldsymbol{R}(\alpha)\boldsymbol{T}_s^{-1}(\alpha)(\boldsymbol{r}_{i+1}-\boldsymbol{r}_i)+ \\ \quad \boldsymbol{T}_s\boldsymbol{R}(\alpha)\boldsymbol{T}_s^{-1}(\alpha)(\dot{\boldsymbol{r}}_{i+1}-\dot{\boldsymbol{r}}_i)+\boldsymbol{g}_s(\boldsymbol{r}_i,\boldsymbol{r}_c)\Big],\ i=1,\dots,N-1 \\ -\boldsymbol{f}_s(\boldsymbol{r}_i)+k_s\Big[k_d\boldsymbol{T}_s\boldsymbol{R}(\alpha)\boldsymbol{T}_s^{-1}(\alpha)(\boldsymbol{r}_1-\boldsymbol{r}_i)+ \\ \quad \boldsymbol{T}_s\boldsymbol{R}(\alpha)\boldsymbol{T}_s^{-1}(\alpha)(\dot{\boldsymbol{r}}_1-\dot{\boldsymbol{r}}_i)+\boldsymbol{g}_s(\boldsymbol{r}_i,\boldsymbol{r}_c)\Big],\ i=N \end{cases} \quad (13)$$

where $\boldsymbol{f}_s(\boldsymbol{r}_i)$ represents the dynamic model of the relative motion as (1), and,

$$\boldsymbol{g}_s(\boldsymbol{r}_i,\boldsymbol{r}_c)=-k_dk_c(\boldsymbol{r}_i-\boldsymbol{r}_c)-(k_d+k_c/k_s)(\dot{\boldsymbol{r}}_i-\dot{\boldsymbol{r}}_c)+\ddot{\boldsymbol{r}}_c$$
$$k_s=n\sqrt{1-s}\big/\big[2\sin(\pi/n)\big],\quad n=\sqrt{\mu/r_0^3}$$

where μ is the earth constant.

Therefore, the periodic relative motion of module $S_i, i=1,\dots,N$ applied by the cyclic pursuit algorithm (13) holds on the following geometric configuration as,

$$\boldsymbol{r}_i(t)=\boldsymbol{T}_s\begin{bmatrix} a_E\sin(nt+\alpha_0,i) \\ a_E\cos(nt+\alpha_0,i) \\ 0 \end{bmatrix}+\boldsymbol{p} \qquad (14)$$

This geometric configuration (14) matches fully with the periodic solution in (6) of the relative motion between module S_i and the reference module S_0.

IV. SIMULATION EXAMPLE OF DISTRIBUTED SATELLITE

The following example illustrates the application of (13) to implement the specific periodic motion between multiple modules and the reference module. The example assumes that the orbit of the reference module is a circular orbit, and its orbit altitude is 679.9718 km, the inclination is 75° and the distance of to the reference satellite S_0 is within the range of 100~500 meters

We assume the reference module is placed at the circular orbit with an altitude of 679.972 km and an inclination angle of 75°. The instability initial position and velocity are selected, in other words, the periodic relative orbits are formed from the initial state automatically. The position

and initial velocity of the distributed satellite comprising five modules is tabulated in Table 1.

The pursuit angle is $\alpha_0=1.5\pi/N$, and the parameters are set to $k_d=1$, $k_g=n/(2\sin(\pi/n))$, $z_0=2$, and $\phi_0=4$, respectively.

Table 1. Initial Position and Velocity of 5 Modules

Module	S_1	S_2	S_3	S_4	S_5
x (m)	105.872	162.337	176.453	148.221	190.570
y (m)	0	611.235	611.235	564.655	564.651
z (m)	611.206	304.206	304.916	305.617	306.323
$\dot{x}$ (m/s)	0	0.324	0.325	0.325	0.326
$\dot{y}$ (m/s)	0.233	0.380	0.376	0.376	0.380
$\dot{z}$ (m/s)	0	0.564	0.564	0.564	0.564

The relative motion trajectories of five modules of the distributed satellites with respect to the auxiliary reference point $\boldsymbol{p}$ are illustrated as shown in Fig.1. After spending the time of 2 periods of the reference orbit, the periodic motion of distributed satellite is established. The trajectory owned by all five modules is a natural elliptical relative orbit with the semimajor axis of 1000 m. Then, after maintaining the time of 10 reference orbital periods, the five modules of distributed satellites transfer from $\boldsymbol{p}$ to the location of 2044.5m with respect to $\boldsymbol{p}$. Finally, after 9 reference orbital periods, the five modules rendezvous at the center of cluster flight.

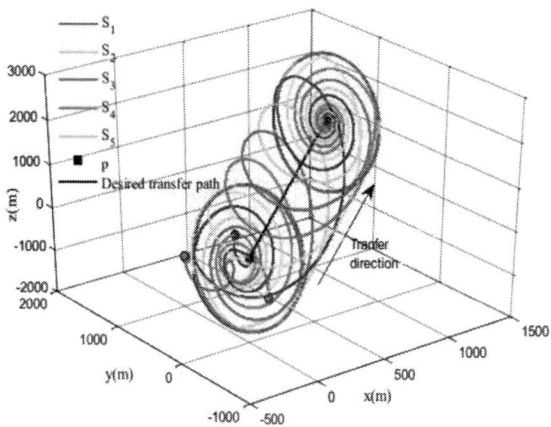

Fig.1 periodic relative motion of 5 modules with respect to the auxiliary reference point $\boldsymbol{p}$.

The position velocity and control acceleration of the five modules relative to $\boldsymbol{p}$ is demonstrated in Fig. 2. It can be seen that the five modules are able to accurately track the desired motion with respect to $\boldsymbol{p}$, and achieve the cluster flight of five modules. The control acceleration is bigger at the task switching moment, whereas the control acceleration sustains at the magnitude of a 10^{-4}-th order.

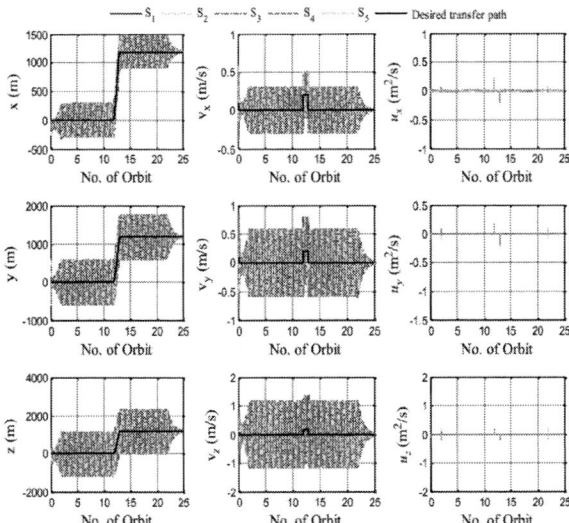

Fig. 2 position, velocity and acceleration of the five modules relative to the reference point P in the loop tracking control

Fig. 3 and Fig. 4 are the periodic motion of p and five modules relative to the reference module S_0, respectively. It is clear that the relative motion of the reference point p is also periodic relative to the reference point S_0.

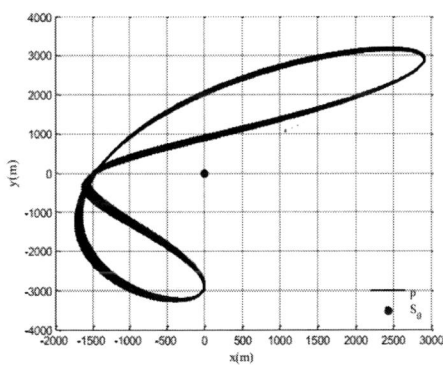

Fig.3 relative motion of the auxiliary reference point P relative to the reference satellite

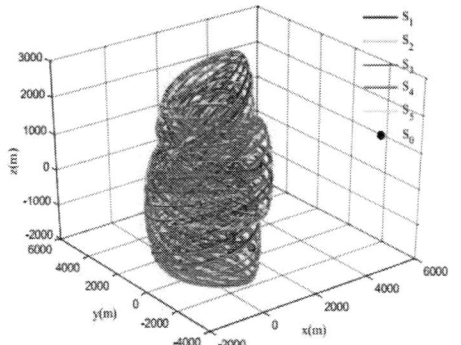

Fig.4 relative motion of five modules relative to the reference satellite S_0.

Figure.5 illustrates the relative motion of five modules and the auxiliary reference point p relative to the reference module S_0 under the control algorithm, respectively. The relative motion of five modules is the composite motion overlapping the motion of five modules relative to p and

the motion of p relative to the reference module S_0. It is clear that the motion configuration of the five modules relative to p is controlled by the control law only, and then the motion of p relative to the reference module S_0 is uncontrolled.

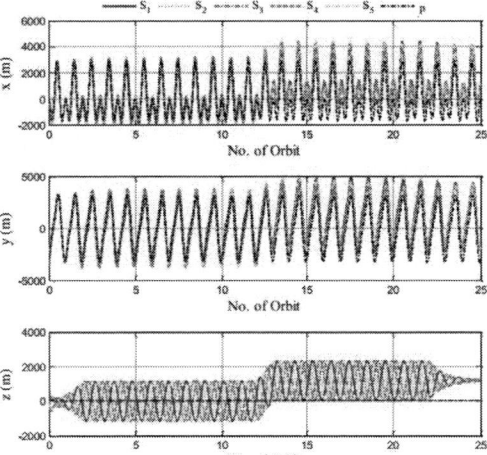

Fig.5 periodic relative motion of five module relative to p

The following formula is used as the index of the fuel consumption of the control,

$$\Delta V_i = \int_0^{25t} \sqrt{u^2_{x,i} + u^2_{y,i} + u^2_{z,i}}\, dt, \ i = 1, \ldots, N \qquad (15)$$

According to (15), the calculated fuel consumption of the five satellites in the whole task is tabulated in table 2. It can be seen that the fuel consumption level is normal and reasonable in the process of the five-satellite control.

Table 2.energy consumption of control

	S1	S2	S3	S4	S5	Cluster
ΔV m/s	18.85	18.82	18.84	18.84	18.82	94.15

The above simulation results demonstrate that the cluster flight can be realized by the proposed cyclic tracking control algorithm.

The configuration of five modules is dependent on the initial state of the distributed satellite, the pursuit angle, and the cyclic topology, instead of defined previously. In addition, the relative motion of five modules is the composite motion overlapping the motion of five modules relative to p and p relative to the reference module S_0, respectively.

V. CONCLUSION

This paper focuses on the cyclic tracking control strategy for the cluster flight of the distributed satellite. First, the relative motion between with the earth oblateness perturbation considered is modeled. Then, the cyclic pursuit algorithm combining diffeomorphism mapping is applied to match the trajectory of the cyclic pursuit and the natural periodic relative motion at low earth orbit environment considering the earth oblateness perturbation.

ACKNOWLEDGEMENT

This work was supported in part by the National Natural Science Foundation of China under Grant 51477103 and 51577121, and in part by the Guangdong Natural Science Foundation under Grant S2014A030313564 and S2015A010106017.

REFERENCES

[1] P. Gurfil, D. Mishne, " Cyclic satellite formations: relative motion control using line-of-sight measurements only. Journal of Guidance, Control, and Dynamics, 30(1), pp:214-226,2007

[2] J. L. Ramirez-Riberos, M. Pavone, E. Frazzoli, D.W. Miller. "Distributed control of satellite formations via cyclic pursuit: Theory and experiments". Journal of Guidance, Control, and Dynamics, 33 (5)PP: 1655-1669, 2010

[3] M. Pavone, E. Frazzoli, "Decentralized policies for geometric pattern formation and path coverage", Journal of Dynamic Systems, Measurement, and Control, 129(2007), pp: 633-643.

[4] L. Ma, N. Hovakimyan, "Vision-based cyclic pursuit for cooperative target tracking", Journal of Guidance, Control and Dynamics, 36 (2), pp: 617-622, 2013.

[5] J. Ramirez, "New Decentralized algorithms for satellite formation control based on a cyclic approach". PhD Dissertation, MIT, Boston, 2010.

[6] G. Xu, D. Wang, "Nonlinear dynamic equations of satellite relative motion around an oblate earth", Journal of Guidance, Control, and Dynamics, 31(5) ,pp: 521-1524, 2008.

[7] S. A.Schweighart and R. J. Sedwick , "A perturbative analysis of geopotential disturbances for satellite formation flying", Proceedings of the IEEE Aerospace Conference, 2(2001), NJ: 1001–1019.

[8] L. S. Breger and J. P. How, Gauss's variational equation-based dynamics and control for formation flying satellite", Journal of Guidance, Control, and Dynamics, 30(2), pp: 437–448, 2007.

[9] D. W. Gim and K. T. Alfriend, "State transition matrix of relative motion for the perturbed noncircular reference orbit", Journal of Guidance, Control, and Dynamics, 26(6), pp: 956-971 , 2003.

[10] M. Paluszek and P. Bhatta. Satellite Attitude and Orbit Control (2nd Edition), Princeton Satellite Systems, Inc. Plainsboro, 2009.

Development of the Equivalent Magnetic Circuit Model for a Surface-Interior Permanent Magnet Synchronous Motor

Meng Si[1], Xiangyu Yang[1], Shiwei Zhao[1], Jikai Si[2]

[1]School of Electric Power, South China University of Technology, Guangzhou 510640, China
Email: simeng@live.com, yangxyu@scut.edu.cn, epswzhao@scut.edu.cn
[2]Key Laboratory of Control Engineering of Henan Province, Henan Polytechnic University, Jiaozuo 454000, China
Email: sijikai527@126.com

Abstract–This paper presents an analytical method for analyzing the magnetic field of a Surface-interior Permanent Magnet Synchronous Motor (SIPMSM). The Equivalent Magnetic Circuit Model (EMCM) of the SIPMSM is established first according to the distribution characteristics of the main magnetic flux and leakage magnetic fluxes of the motor. On this basis, the methods for calculating various leakage magnetic permeances are determined according to the motor structure and thus the leakage magnetic reluctances of the magnetic circuit are obtained. A comparative analysis of the calculation results of the analytical method and the Three-dimensional Finite Element Method (3D FEM) reveals that those results are in good agreement, it shows that the EMCM has certain reasonability and feasibility. Compared with the finite element method, the analytical method has the advantage of easy and fast calculation, it can save a lot of time when analyzing many motor designs with different dimensions, and it can therefore be used as a powerful analytical tool during the design processes of SIPMSMs.

Keywords–Analytical method, equivalent magnetic circuit model (EMCM), surface-interior permanent magnet synchronous motor (SIPMSM), three-dimensional finite element method (3D FEM).

I. INTRODUCTION

SIPMSMs differ from conventional Surface Permanent Magnet Synchronous Motors (SPMSM) and Interior Permanent Magnet Synchronous Motors (IPMSM) in that their main magnetic fields are created jointly by Surface Permanent Magnets (SPM) and Interior Permanent Magnets (IPM) [1], [2]. The flux barriers in the rotor core of a SIPMSM makes the synchronous inductances of the motor different for d-axis and q-axis. Compared with SPMSMs, the motor has a higher overload capacity, a higher power density and better field weakening performance. The flux barriers are arranged at appropriate positions in the rotor core, the SPMs can provide effective guidance to the

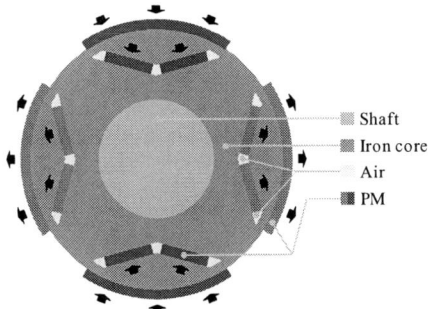

Fig. 1: The rotor structure of a SIPMSM

magnetic flux in the rotor core, hence the leakage flux of the IPMs can be reduced significantly, thin bridges are not

necessary for the rotor core. Compared with IPM rotors, the surface-interior PM rotor has much less leakage flux and a significantly higher mechanical strength, therefore SIPMSMs are more appropriate for high speed operation. The rotor of a SIPMSM is shown in Fig. 1, the magnetization directions of the PMs are denoted by arrows. Many researchers analyze PM motors by using EMCM during design processes, because the equivalent magnetic circuit method has the advantage of easy and fast calculation [3]–[6]. However, magnetic circuits are oversimplified in some of those references, this may lead to considerable calculation errors. This paper develops an EMCM for SIPMSMs, the saturation of the iron cores, the magnetic reluctances of the assembly gaps and the leakage fluxes are fully taken into account. The accuracy of the model is verified by a 3D FEM analysis.

II. MAGNETIC CIRCUIT MODEL

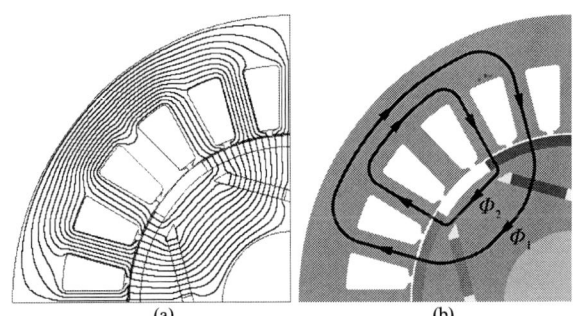

(a)   (b)

Fig. 2: Magnetic flux distribution in a SIPMSM (a) Magnetic flux distribution (b) Two parts of the magnetic flux

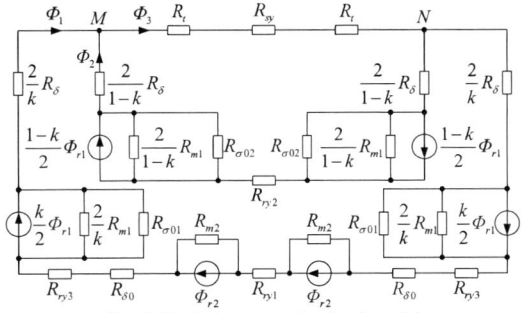

Fig. 4: Equivalent magnetic circuit model

The typical distribution of magnetic flux in a four-pole SIPMSM is shown in Fig. 2(a), the total flux can be divided into two parts which are denoted as Φ_1 and Φ_2 respectively in Fig. 2(b). In order to establish the EMCM, the ratio (denoted as k) between the average arc length of the part of each SPM that is in series with IPMs and the full average arc length of each SPM should be obtained first. As shown in Fig. 3, the flux lines that pass through the ends of the IPMs can be represented by curves (*a-b-c-d* and *e-f-g-h*) that consist of line segments and circular arcs, then the approximation of k can be calculated as the ratio between two central angles: $k = \beta/\alpha$.

Based on Fig. 2, the EMCM can be obtained and it is shown in Fig. 4, the PMs are represented by Norton equivalent circuits consisting of flux sources in parallel with internal reluctances. The leakage fluxes of the IPMs can be nearly eliminated hence they are not taken into account.

As shown in Fig. 4, Φ_{r1} is the remanent flux of each SPM, Φ_{r2} is the remanent flux of each IPM; R_{m1} is the magnetic reluctance of each SPM, R_{m2} is the magnetic reluctance of each IPM; $R_{\delta0}$ is the magnetic reluctance of the assembly gap of each IPM; R_δ is the magnetic reluctance of the air gap of each pole (The slotting can be taken into account by means of Carter's coefficient [7]). Those parameters are constant. R_{ry1} is the magnetic reluctance of the part of the rotor core between two adjacent poles through which Φ_1 passes; R_{ry2} is the magnetic reluctance of the part of the rotor core through which Φ_2 passes; R_{ry3} is the magnetic reluctance of the part of the rotor core between each SPM and an IPM through which Φ_1 passes; R_t is the magnetic reluctance of the stator teeth; R_{sy} is the magnetic reluctance of the stator yoke. Those reluctances are not constant, therefore the relevant calculations should be conducted according to the *B-H* curve of the iron core material. $R_{\sigma01}$ is the leakage magnetic reluctance of one half of the part of each SPM that is in series with an IPM, $R_{\sigma02}$ is the leakage magnetic reluctance of one half of the part of each SPM that is not in series with any IPMs.

III. ANALYSIS OF FLUX LEAKAGES

1. Assumptions
Because the distribution ranges of the leakage magnetic fluxes are always very small, several assumptions can be made to simplify the analysis:

(1) The parts of the iron core through which the leakage magnetic fluxes pass are infinite permeable;
(2) The surface of the stator core is smooth;
(3) Flux lines consist of line segments and circular arcs;
(4) The permeability of the permanent magnets is the same as that of air.

2. Magnetic flux paths
The magnetic fluxes inside the range of one pole pitch (τ) are shown in Fig. 5 and Fig. 6, where δ is the air gap length, h_{m1} is the length of each SPM in the magnetization direction, r_1 is the radius of the rotor core, θ is the angle between the lateral surfaces of two adjacent SPMs. Possible paths of magnetic fluxes that come from a point (A) on a lateral surface of a SPM are represented by three dotted lines: L_1, L_2 and L_3, possible paths of magnetic fluxes that come from a point (B) on an end surface of a SPM are represented by three dotted lines: L_4, L_5 and L_6. The corresponding magnetic fluxes of L_1, L_2, L_4 and L_5 are parts of the total leakage flux, the corresponding magnetic fluxes of L_3 and L_6 are parts of the main magnetic flux. There are five kinds of leakage fluxes around each SPM: the leakage flux between

each lateral surface of each SPM and the rotor core surface ($\Phi_{\sigma1}$), the leakage flux between the lateral surfaces of two adjacent SPMs ($\Phi_{\sigma2}$), the leakage flux between the top surfaces of two adjacent SPMs ($\Phi_{\sigma3}$), the leakage flux between the end surfaces of two adjacent SPMs ($\Phi_{\sigma4}$), one half of the leakage flux between each end surface of each SPM and the end surface of the rotor core ($\Phi_{\sigma5}$).

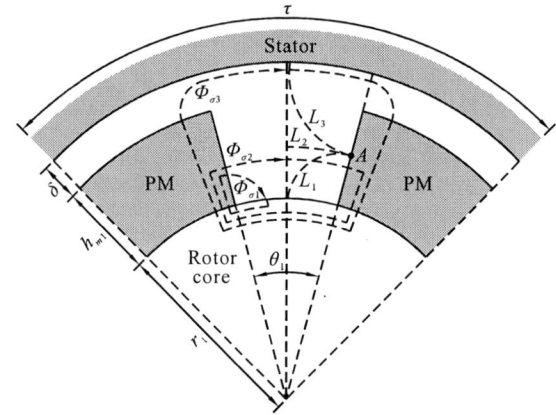

Fig. 5: Possible flux paths inside the axial length range of the iron cores

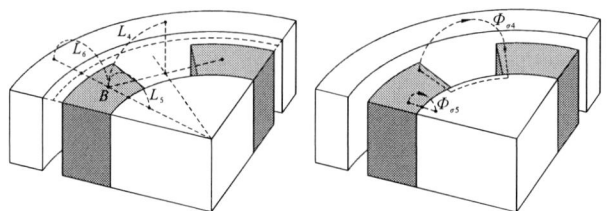

Fig. 6: Possible flux paths outside the axial length range of the iron cores

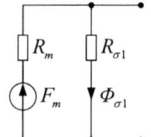

Fig. 7: Ideal condition

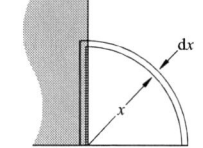

Fig. 8: Differential element of $\Phi_{\sigma1}$

3. Calculations of leakage magnetic reluctances
Take one part of the leakage flux ($\Phi_{\sigma1}$) for example, we assume that an ideal condition shown in Fig. 7 exists, the SPM is represented by a Thévenin equivalent circuit consisting of a Magneto Motive Force (MMF) source (F_m) in series with an internal reluctance (R_m), and the total MMF of the SPM is applied to R_m and the leakage magnetic reluctance ($R_{\sigma1}$) corresponding to $\Phi_{\sigma1}$. The leakage flux around a SPM is generally very little, hence the leakage reluctance of the SPM is much larger than the internal reluctance of the SPM, the total MMF of the SPM can be considered to be applied only to $R_{\sigma1}$. The differential

element of one half of $\Phi_{\sigma1}$ is shown in Fig. 8, its cross section consists of one line segments with length x and one circular arc with radius x subtended by the central angle of $\pi/2$, it can be expressed as

$$\mathrm{d}\Phi_{\sigma1} = F_{\sigma1x}\mathrm{d}\lambda_{\sigma1} \tag{1}$$

where $F_{\sigma1x}$ is the MMF for the flux element, $\mathrm{d}\lambda_{\sigma1}$ is the differential element of the magnetic permeance of the circuit of $\Phi_{\sigma1}$:

$$F_{\sigma1x} = F_m \frac{x}{h_{m1}} \tag{2}$$

$$\mathrm{d}\lambda_{\sigma1} = \frac{\mu_0 l_m}{x(1+\dfrac{\pi}{2})}\mathrm{d}x \tag{3}$$

$\Phi_{\sigma1}$ can be expressed as

$$\Phi_{\sigma1} = \int_{x_1}^{x_2} \frac{F_m \mu_0 l_m}{h_{m1}(1+\dfrac{\pi}{2})}\mathrm{d}x \tag{4}$$

x_1 and x_2 are determined by the dimensions of the motor. The magnetic permeance of the circuit of $\Phi_{\sigma1}$ can be calculated by

$$\lambda_{\sigma1} = \frac{\Phi_{\sigma1}}{F_m} = \frac{2\mu_0 l_m}{h_{m1}(2+\pi)}\int_{x_1}^{x_2}\mathrm{d}x \tag{5}$$

The magnetic permeances corresponding to $\Phi_{\sigma2}$, $\Phi_{\sigma3}$, $\Phi_{\sigma4}$ and $\Phi_{\sigma5}$ can be calculated in similar ways and thus the values of $R_{\sigma01}$ and $R_{\sigma02}$ can be obtained.

IV. MAGNETIC CIRCUIT SOLVING METHOD

The magnetic circuit can be solved by the following method according to Fig. 4.

(1) Assume a value for Φ_1, then calculate the total magneto motive force drop (F_{MN1}) across the branch through which Φ_1 passes.

The magneto motive force drop ($F_{\delta1}$) across $\dfrac{2}{k}R_\delta$ can be calculated by

$$F_{\delta1} = \frac{2}{k}R_\delta\Phi_1 \tag{6}$$

The magneto motive force drop (F_{r11}) across $\dfrac{k}{2}\Phi_{r1}$ can be calculated by

$$F_{r11} = (\frac{k}{2}\Phi_{r1}-\Phi_1)\frac{\dfrac{2}{k}R_{m1}R_{\sigma01}}{\dfrac{2}{k}R_{m1}+R_{\sigma01}} \tag{7}$$

The average magnetic flux density (B_{ry3}) in the part of the rotor core which is denoted by R_{ry3} can be calculated by

$$B_{ry3} = \frac{\Phi_1}{b_{ry3}l_c k_{Fe}} \tag{8}$$

where k_{Fe} is the staking factor of the iron cores, b_{ry3} is the equivalent width of that part, l_c is the axial length of the iron cores.

The magneto motive force drop (F_{ry3}) across R_{ry3} can be calculated by

$$F_{ry3} = H_{ry3}l_{ry3} \tag{9}$$

where l_{ry3} is the equivalent length of that part, H_{ry3} is the magnetic field strength in that part found from the B-H curve of the iron core material according to B_{ry3}.

The magneto motive force drop ($F_{\delta0}$) across $R_{\delta0}$ can be calculated by

$$F_{\delta0} = R_{\delta0}\Phi_1 \tag{10}$$

The magneto motive force drop (F_{r2}) across Φ_{r2} can be calculated by

$$F_{r2} = (\Phi_{r2}-\Phi_1)R_{m2} \tag{11}$$

The average magnetic flux density (B_{ry1}) in the part of the rotor core which is denoted by R_{ry1} can be calculated by

$$B_{ry1} = \frac{\Phi_1}{b_{ry1}l_c k_{Fe}} \tag{12}$$

where b_{ry1} is the equivalent width of that part.

The magneto motive force drop (F_{ry1}) across R_{ry1} can be calculated by

$$F_{ry1} = H_{ry1}l_{ry1} \tag{13}$$

where l_{ry1} is the equivalent length of that part, H_{ry1} is the magnetic field strength in that part found from the B-H curve of the iron core material according to B_{ry1}.

F_{MN1} can be calculated by

$$F_{MN1} = 2F_{r11}+2F_{r2}-2F_{\delta1}-2F_{ry3}-2F_{\delta0}-F_{ry1} \tag{14}$$

(2) Assume a value for Φ_2, then calculate the total magneto motive force drop (F_{MN2}) across the branch through which Φ_2 passes in a way similar to that of F_{MN1}. If the value of $|F_{MN1}-F_{MN2}|$ is not small enough then assume a new value for Φ_2 and calculate the value of $|F_{MN1}-F_{MN2}|$ again, repeat this until the value is small enough and thus the approximation of Φ_2 that matches the assumed value for Φ_1 in step (1) can be obtained.

(3) Φ_3 can be calculated by
$$\Phi_3 = \Phi_1 + \Phi_2 \tag{15}$$

Then calculate the total magneto motive force drop (F_{MN3}) across the branch through which Φ_3 passes in a way similar to that of F_{MN1}. If the value of $|F_{MN1}-F_{MN3}|$ is not small enough then repeat the above steps until the value is small enough and thus the approximations of Φ_1 and Φ_2 can be obtained. The air gap magnetic flux (Φ_δ) of each pole can be calculated by

$$\Phi_\delta = 2\Phi_3 \tag{16}$$

The magnetic flux (Φ_m) leaving the SPM of each pole can be calculated by

$$\Phi_m = \Phi_{r1} - \frac{k(k\Phi_{r1} - 2\Phi_1)R_{\sigma 01}}{2R_{m1} + kR_{\sigma 01}} - \frac{(1-k)[(1-k)\Phi_{r1} - 2\Phi_2]R_{\sigma 02}}{2R_{m2} + (1-k)R_{\sigma 02}}$$

(17)

The flux leakage coefficient can be calculated by

$$\sigma_0 = \frac{\Phi_m}{\Phi_\delta}$$

(18)

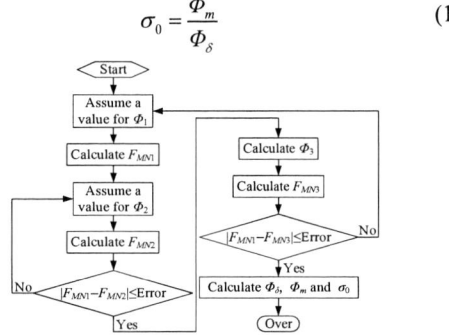

Fig. 9: Flowchart of the solving method

The solving method can be represented by the flowchart shown in Fig. 9, it has been implemented via a computer program.

V. VERIFICATION OF THE MODEL

The equivalent magnetic circuit analysis on open circuit of a SIPMSM was done by using the analytical method, the main design parameters of the SIPMSM are shown in Table I and Fig. 10. The comparison between the analytical results and the results from the 3D FEM analysis is summarized in TABLE II, the comparison between the calculation results of the flux leakage coefficient from the EMCM and the 3D FEM analysis for different pole arc coefficients with other design parameters keeping constant is shown in Fig. 11.

Table 1: Design parameters

Item	Value	Unit
Number of poles	4	
Stator lamination diameter	155	mm
Stator bore	91	mm
Number of stator slots	24	
Stator tooth width	6.6	mm
Stator yoke width	13.5	mm
Air gap length	0.5	mm
Permanent magnet	NdFeB (B_r=1.15T)	

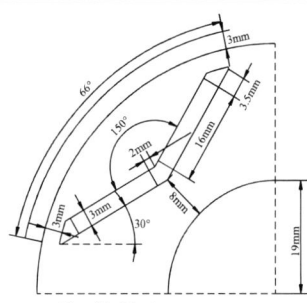

Fig. 10: Design parameters

It can be seen from TABLE II and Fig. 11 that the calculation results of the analytical method are in good agreement with those of 3D FEM, it shows that the EMCM has certain reasonability and feasibility. Fig. 11 also shows that the flux leakage coefficient increases due to the increase of pole arc coefficient, and the rate of change increases

remarkably when pole arc coefficient is getting close to 1, because the distance between the lateral parts of every two adjacent SPMs becomes very small so that the leakage fluxes between SPMs become much more.

Table 2: Comparison between Results from Magnetic Circuit Analysis and 3D FEM Analysis

	Magnetic circuit analysis resluts	3D FEM analysis results	Error (%)*
Φ_m(Wb)	6.0513×10^{-3}	6.2215×10^{-3}	2.74
Φ_δ(Wb)	6.0033×10^{-3}	6.1770×10^{-3}	2.81
σ_0	1.0080	1.0072	0.08
B_t(T)	1.5072	1.5358	1.86
B_{sy}(T)	1.6211	1.6305	0.58
B_{ry1}(T)	0.9966	0.9982	0.16
B_{ry2}(T)	0.8690	0.9009	3.54
B_{ry3}(T)	0.9940	0.9985	0.45

*: Error = |Magnetic circuit analysis result − 3D FEM analysis result| × 100 / 3D FEM analysis result

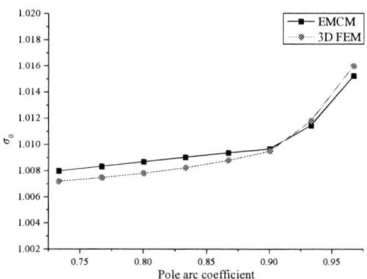

Fig. 11: Leakage flux coefficient versus pole arc coefficient

VI. CONCLUSION

In this paper, a SIPMSM is analyzed by using the proposed model. The calculation results of the model match with those of 3D FEM well. Therefore, the model could be used as a powerful analytical alternative during the design processes of SIPMSMs.

REFERENCES

[1] J. K. Si, M. Si and H. C. Feng, "Research on the Field of New Type Rotor Structure Permanent Magnet Synchronous Motors," Int. Conf. on Electrical Information and Mechatronics, Jiaozuo, China, 2012, pp. 154-158.

[2] J. K. Si, Z. F. Liu, M. Si, X. Z. Xu and X. D. Wang, "Magnetic Field Analysis and Characteristics Research on Permanent Magnet Synchronous Motors with New Structure Rotor," J. China Coal Soc., vol. 38, no. 2, pp. 348-352, Feb. 2013.

[3] V. Ostovic, "Pole-changing Permanent-magnet Machines," IEEE Trans. Ind. Appl., vol. 38, no. 6, pp. 1493-1499, Nov/Dec. 2002.

[4] C. C. Hwang, Y. H. Cho, "Effects of Leakage Flux on Magnetic Fields of Interior Permanent Magnet Synchronous Motors," IEEE Trans. Magn., vol. 37, no. 4, pp. 3021-3024, July. 2001.

[5] C. T. Mi, M. Filippa, W. G. Liu, R. Q. Ma, "Analytical Method for Predicting the Air-gap Flux of Interior-type Permanent-magnet Machines," IEEE Trans. Magn., vol. 40, no. 1, pp. 50-58, Jan. 2004.

[6] X. M. Lu, K. L. V. Iyer, K. Mukherjee, N. C. Kar, "Development of a Novel Magnetic Circuit Model for Design of Premium Efficiency Three-Phase Line Start Permanent Magnet Machines With Improved Starting Performance," IEEE Trans. Magn., vol. 49, no. 7, pp. 3965-3968, July. 2013.

[7] S. K. Chen, Electric Machine Design. Beijing, China: Mechanical Press, 1993.

Implementation of the Derivative-based Technique for Solving a 2x25kV AC Bivoltage Traction System

Mariia Plakhova

Abstract—Modelling and simulation of electrified traction systems are widely used nowadays to provide an efficient design, operation and service of the railways. Various railroad systems models have been developed recently intending to study power flow process. Moreover, most of current software products model the train and the power system separately. The scope of this paper is to analyse the power flow and to perform the power system simulation, applying a commercial algorithm "fsolve" of MATLAB software. For this purpose the complex train model has not been built and trains are represented as non-linear power sources. This paper describes the steps, which have been taken in order to develop the working simulator and builds as following: first, the brief review of the existing algorithms to solve the power flow problem is presented. Second, design specification and solving procedure are explained. Finally, the test procedure validation of results are given.

Index Terms—Power system, power system modelling, railway system, derivative-based technique.

I. INTRODUCTION

NOWADAYS AC electrical railways powered by 2x25kV 50Hz power supplies is one of key mode of urban transportation system. However, new railroads encounter the variety of technical considerations [1]:

- Operational requests (depending on the operation field: urban metro, high-speed passenger or heavy-haul freight).
- Physical route characteristics (gradients, slope, curves, and bridge and tunnel clearances).
- Closeness of power generating utility or railway-owned grids.
- Traction technology (power electronics, traction motors and regenerative capability).

These requests are aimed not only to provide the successful design and implementation of the railways, but also to improve further control and management of the load flow. Solving the power flow problem of the traction systems is not an easy task and requires proper tools and numerical techniques to be applied. There are different approaches to obtain the railway system, which are based on several numerical algorithms, such as: Gauss-Seidel methods, Newton-Raphson methods, decoupled methods, multiple-conductor techniques and forward/backward sweep method. The brief review of each approach will be given further.

Gauss-Seidel (GS) technique has the advantages of clarity, good performance and there is no need to store previous values

The author is with the Department of Electrical Engineering, University of Oviedo, Gijón, Spain (e-mail: mariiaplakhova@gmail.com).

[2]. It is a procedure for solving n equations of the linear system of equations Ax = b one at the time in sequence, and uses previously obtained results as soon as they are available. There are two key aspects of the GS method, that should be noted. Firstly, the estimations are serial, so each component of new iterations depends on all previous results. Secondly, the new iteration x^k depends on the order in which the equations are calculated.

Applying GS technique to the traction system, the value that will be updated with each iteration, is usually voltage. This method has been implemented in probabilistic load flow technique (PLF) [3], [4], [5], [6]. The first mention of this technique has been done in [7]. As it is very common in this field, the first research was done for the DC system, followed by simulation for the AC systems [8]. Solving the power load flow problem with the help of the PLF gives as a result a full range of all possible values of bus voltages, power flows and losses, with their respective probabilities, with respect to load uncertainties and control-action effects [3]. Among with GS this study has also adopted Monte-Carlo simulation technique in order to set probability power distribution function of each train [4].

Newton-Raphson (NR) method is sometimes known as **Newton's iteration** is an algorithm for solving a set of nonlinear equations in an equal number of unknowns [2]. It is a root-finding technique that applies the Taylor series expansion of a function f(x) in proximity of a suspected root. Applying this method to the power flow question in railway systems, the real and reactive power flow equations are approximated by collecting the first two terms and neglecting the other high-order components, though some interaction between them is still presented.

The power flow solution by NR was firstly described in [9], and later implemented in [3], [10], [11]. Compare to GS technique, that operates slowly due to updating each bus voltage separately, from the first to the last, NR quadratic convergence proceeds faster [2]. However, with the current existing railway traction systems it is still not enough.

Decoupled power flow or Fast Decoupled power flow approach. Any electrified traction system that operates in steady state has a strong interconnection between active and reactive powers and bus voltage angles and magnitudes respectively [2]. The coupling between these P-θ and Q-V components is not strong, thus for power flow calculation it has been decided to solve these tasks separately ("decoupling" them).

This method is simple, fast and more reliable than NR or

GS. It is used to solve DC traction systems [12] and also for AC railway systems [13], [14].

Multiple-conductor technique is based on representation of the whole power system as the combination of small separated in frequency domain network components. There are different types of conductors, presented in feeding circuits, such as: trolleys, feeders, rails, soils and protective wires.

This approach has been used to calculate power flow for railway vehicles in Japan [15] and in United Kingdom [16].

Backward/Forward sweep (BFS) method. There are a lot of load flow calculation that are based on this technique [17], [18], [19], [20]. It consists two steps, backward sweep and forward sweep, which are repeated until the calculation is done. Current or power flow summation with possible voltage updates are represents by the backward sweep, while a voltage drop calculation with possible current or power flow updates are introduced by the forward sweep. [21] describes the implementation and prove the effectiveness of this method for Tehran-Golshahr suburban electrical railway.

There are a few of commercial railway system simulators, based on the numerical techniques described above, and they are successfully used today, such as: OpenTrack, TracFeed/Simtrac, Sitrac Sidytrac etc.

II. DESIGN SPECIFICATION

The work has been performed for a 2x25kV bivoltage traction system. Such configuration was described in [21], [22], [23], [24], [25]. Figure 1 illustrates a 2x25kV 50 Hz bivoltage traction system with two cells and one train.

The network represented on the Figure 1 can be divided into two parts: the high voltage feeding network, which is connected to the AC traction network through the primary side of the power transformer, and the AC traction network by itself, that includes the secondary side of the power transformer, ATs and trains. The numeration of the nodes and lines has been done, following the procedure that was described in [26].

Applying the graph theory approach for the dynamic system (by dynamic means systems that consists trains), four node incidence matrices can be built: the first one Γ_{line} represents the real connection between ATs, between the secondary side of the power transformer and ATs, and between trains and ATs and the secondary side of the power transformer, the second and the third matrices (Γ_{w1} and Γ_{w2}) define the connections within winding groups of the ATs and the secondary side of the power transformer, while the fourth matrix Γ_{train} defines the connections of the trains. The number of columns in these matrices are equal to the number of nodes in the network, but the number of rows is different: for the incidence matrix Γ_{line} it is equal to the total number of lines[1], for the transformer matrices Γ_{w1} and Γ_{w2} are equal to the number of transformers (it includes ATs and the secondary windings of the power transformer), and for the train incidence matrix Γ_{train} the number of columns is equal to the number of the trains in the network.

[1]by total means the sum of the static and dynamic lines, where static lines are all lines of the static system and dynamic lines are created by the trains.

As it was mentioned in [25] the mathematical model of the bivoltage traction system can be represented by the set of equations that describe system. This set of equations covers Kirchhoff Voltage and Current Laws (KVL and KCL) for all branches and nodes, power transformer and ATs' equations, and includes the equation of the train model.

III. SOLVING PROCEDURE

The task of the developed script is to calculate the node voltages and line currents for all nodes and all branches. The problem has been solved by applying a derivative method (with help of a function "fsolve" of MATLAB software).

There are several derivative-based techniques, but for this work the Newton-Raphson method has been used. This strategy has been implemented many times by different researches [9], [3], [10], [11]. An algorithm requires to provide the same number of equations and unknowns.

Based the graph theory that has been implemented in order to define incidence matrices Γ, the set of equations, which must be solved, is following:

1. KVL of the feeding system (primary side of the power transformer):

$$V_{src} - I_p * Z_p - V_p = 0 \qquad (1)$$

where V_{src} is the source voltage, V_p, I_p and Z_p are the primary side voltage, current and impedance respectively.

2. The relation between the primary side of the power transformer and the first winding group of the secondary side of the power transformer:

$$V_p - N_{p1} V_1^{PT} = 0 \qquad (2)$$

where N_{p1} is the number of turns of the power transformer secondary winding first group, V_1^{PT} is voltage of the first winding group of the secondary side of the power transformer.

3. The relation between the currents of the primary and secondary sides of the power transformer, which can be derived from the equations (3.1) - (3.3):

$$\frac{1}{N_{p1}} I_1^{PT} + \frac{1}{N_{p2}} I_2^{PT} = 0 \qquad (3)$$

where where N_{p2} is the number of turns of the power transformer first winding first group, I_1^{PT} and I_2^{PT} are currents through the power transformer secondary windings.

4. Relation between the voltages of the secondary side of the power transformer and ATs' winding groups:

$$V_2^{AT} - N * V_1^{AT} = 0 \qquad (4)$$

where V_1^{AT} and V_2^{AT} are the single column vectors that include voltages of the first and second winding groups of the ATs, while N a single column vector that includes turn ratio of the ATs.

5. ATs' current equation:

$$I_1^{AT} + N * I_2^{AT} = 0 \qquad (5)$$

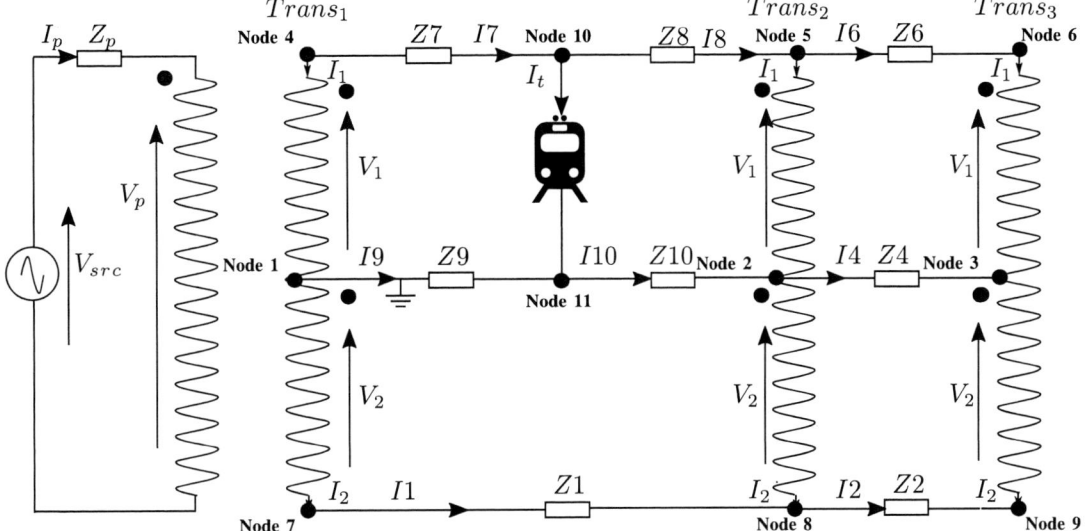

Fig. 1. AC bivoltage system with one cell and one train

where I_1^{AT} and I_2^{AT} are the single column vectors that include currents through the first and second winding groups of the ATs.

6. KVL for all lines of the AC traction system:

$$\Gamma_{line} * V_N - Z * I_{line} = 0 \qquad (6)$$

The vector V_N includes the voltages in the nodes of the traction network and the vector I_{line} contains line currents. Z is the impedance vector that includes the impedances of the lines.

7. KCL for all nodes of the AC traction system:

$$(\Gamma_{line})^T * I_{line} + (\Gamma_{w1})^T * I_1 + (\Gamma_{w2})^T * I_2 \\ + (\Gamma_{train})^T * I_{train} = 0 \qquad (7)$$

where vectors I_1 and I_2 hold the values of the power transformer and ATs currents for the first and the second winding groups respectively, and vector I_{train} contains trains' currents.

It can be noticed that first three equations ((1) - (3)) are related to the power transformer, while next two equations ((4) and (5)) describe AT.

An initial approximation to the voltage and current solutions is assigned. Thus, all nodes of the ground line have an initial voltage 0kV, while all nodes of the positive and negative feeders have an initial voltages +25kV and -25kV respectively. Therefore, voltage across the primary side of the power transformer is initially equal to the source voltage.

Function "fsolve" of MATLAB software solves a problem, defined by F(X) = 0 for X, where F(X) is a function that returns a vector value, and X is a vector or a matrix of unknowns. Solver syntax for the developed scrip looks like this [X,fval,exitflag,output] = fsolve(@myfun, X0, Opt), where @myfun is the set of equation, written in the form, required by the solver, X0 is the initial values for the solution, and Opt is a list of the optimisation options.

IV. TEST AND VERIFICATION OF RESULTS

The developed simulator is based on the application of the Newton-Raphson method using MATLAB software. All tests have been done in a laptop with a processor Intel Core i5 @ 2.5GHz.

The initial conditions are following:

- There are no losses on the primary side of the power transformer, thus the voltage across the primary side of the power transformer is equal to the source voltage.
- The voltage of the positive feeder, ground rail and the negative feeder are equal to 25kV, 0kV and -25kV respectively.
- All branch currents are equal to zero.

To analyse the performance of the simulator, various tests with the different network parameters have been done. The source voltage is 400kV in all cases for all tests.

First test has been run for the small system (2 cells and 1 train), represented on the Figure IV.

Each cell from the Figure IV has a length of 40 km. Resistance and the inductance of each line of the cell are assumed to be to be 93.63 mΩ/km and 0.92 mH [27]. The turn ratio of all ATs is 1 and the turn ratio of the power transformer is 16. The train is located in the cell number 2 at the distance 0.1576 (p.u.) from the beginning of the line and its active and reactive power of the train are 7.6177 MW and 2.5 MW respectively.

It took 11 iterations to obtain the system, taking for each of them only 0.3779 ms. The calculated values are given in the Table I.

In Test 2 the simulated system has been increased. Figure 3 illustrates the network topology for Test 2. In this particular case the network consists 3 cells, each of them has a length of 40 km. Resistance and the inductance of each line of the cell are assumed to be to be 93.63 mΩ/km and 0.92 mH [27]. The turn ratio of all ATs is 1 and the turn ratio of the power transformer is 16. There are 6 trains in the system, which

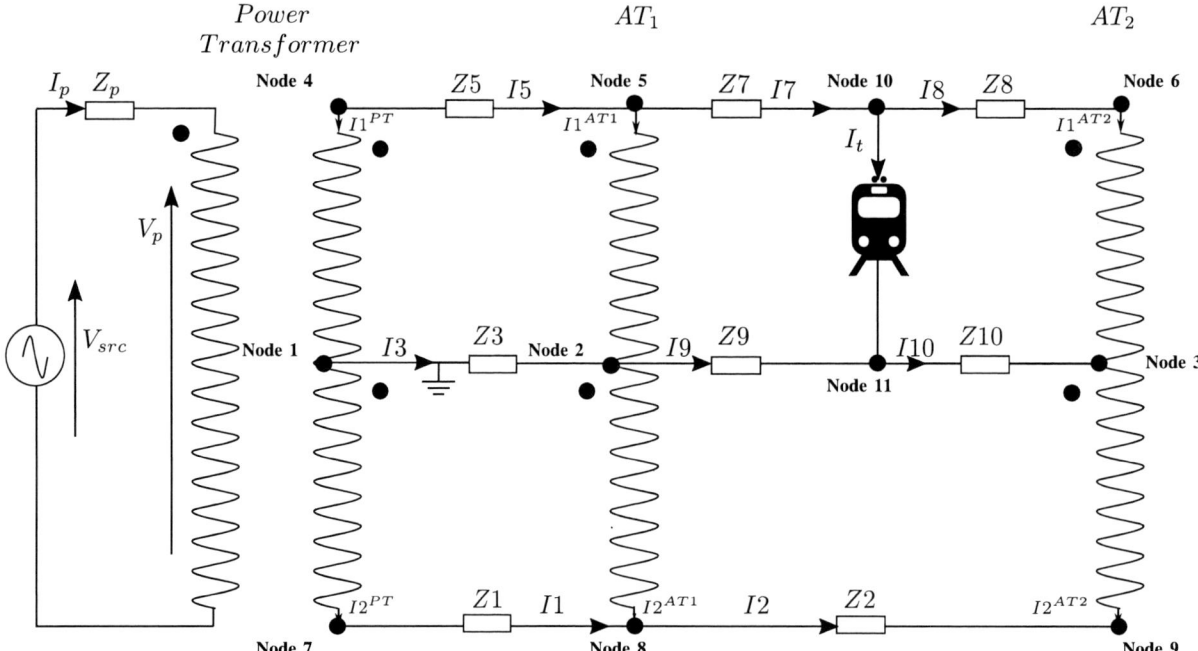

Fig. 2. Test 1 network topology

TABLE I
TEST 1. RESULTS

Unknown	Calculated Value
V_p	$399.64*10^3 - 0.49*10^3i$
I_p	19.07 - 6.32i
$I1^{PT}$	-152.54 + 50.56i
$I1^{AT1}$	-147.67 + 48.95i
$I1^{AT2}$	-4.87 + 1.61i
$I2^{PT}$	-152.54 + 50.56i
$I2^{AT1}$	147.67 - 48.95i
$I2^{AT2}$	4.87 - 1.61i
V_{Node1}	0 + 0i
V_{Node2}	0 + 0i
V_{Node3}	0 + 0i
V_{Node4}	$24.977*10^3 - 0.031*10^3i$
V_{Node5}	$24.949*10^3 - 0.07*10^3i$
V_{Node6}	$24.948*10^3 - 0.071*10^3i$
V_{Node7}	$-24.977*10^3 + 0.031*10^3i$
V_{Node8}	$-24.949*10^3 + 0.07*10^3i$
V_{Node9}	$-24.948*10^3 + 0.071*10^3i$
V_{Node10}	$24.947*10^3 - 0.072*10^3i$
V_{Node11}	$0.001*10^3 + 0.002*10^3i$
I1	-152.54 + 50.56i
I2	- 4.87 + 1.61i
I5	152.54 - 50.56i
I7	300.21 - 99.51i
I8	-4.87 + 1.61i
I3	-0 + 0i
I9	-295.34 + 97.89i
I10	9.73 - 3.23i
I_{train}	305.07 - 101.12i

forth and fifth columns provide an information about active and reactive power respectively.

TABLE II
TRAIN DATA

Train Index	Cell Number	Length (p.u.)	Active Power (MW)	Reactive Power (MW)
1	3	0.2703	4.9985	2.5
2	3	0.1971	0.1583	2.5
3	2	0.8217	5.5107	2.5
4	1	0.4299	4.816	2.5
5	1	0.8878	- 0.0939	2.5
6	2	0.3912	- 2.1918	2.5

Figure 4, given in the Appendix, represent the results of the case, described by Figure 3 and Table II. Figure 4 shows the currents in the system, obtained with help of the derivative method. It takes 0.36 sec and 13 iterations to obtain this system.

Next step to test the simulator is to increase the number of cells and the number of trains in the system and run the test for more different cases. The following tested system has 10 cells, each of them has a length of 40 km. Resistance and inductance of each line of the cell are assumed to be 93.63 mΩ/km and 0.92 mH [27]. The turn ratio of all ATs is 1 and the turn ratio of the power transformer is 16. There are 10 trains in the system, which locations (in which cell and at what distance) and power are generated randomly. The power varies from -5 MW up to 8 MW. In total 10^3 cases with random train data have been generated to check the work of simulator.

The Table III illustrates the summary of results, obtained by the derivative-based method, in terms of time and iterations, spent to solve the cases.

locations (in which cell and at what distance) and power are generated randomly. The active power can vary from -5 MW up to 8 MW, while reactive power remains constant 2.5 MW. Table II summarises all data that has been generated for the trains. The first column shows the index of the train, the second column represents the number of the cells, where trains are located, while the third column shows trains' distance. The

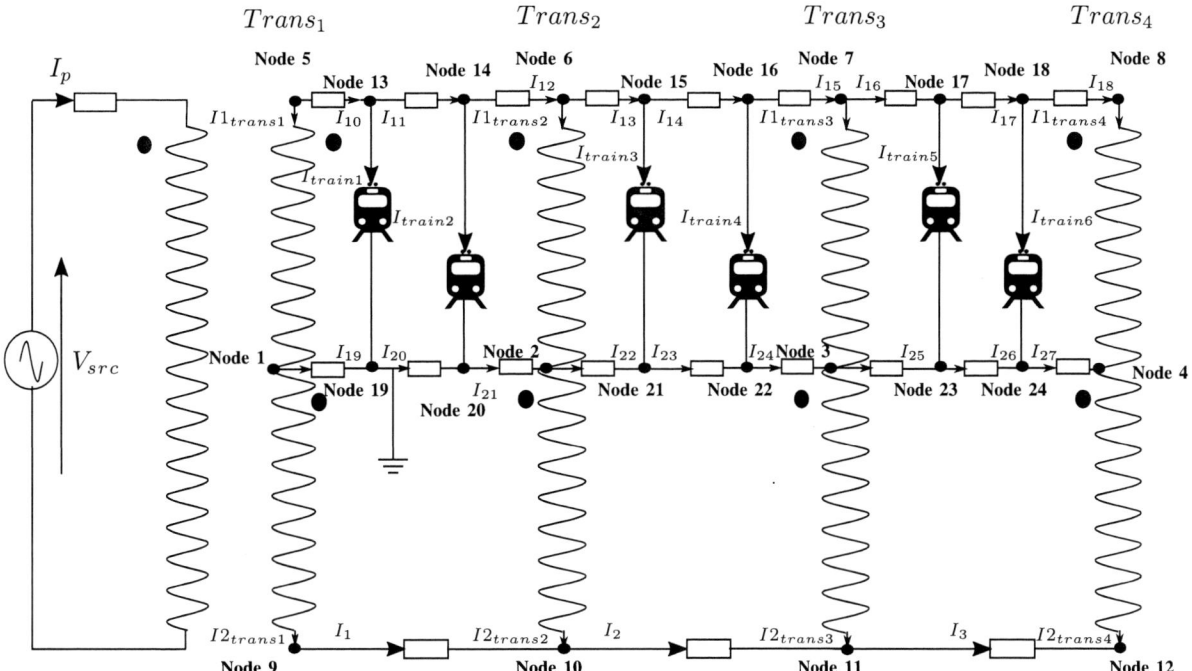

Fig. 3. Test 2 network topology

TABLE III
SUMMARY OF THE DERIVATIVE-BASED METHOD RESULTS

	Derivative-based method
Minimum total time (ms)	500.43
Mean total time (ms)	446.63
Maximum total time (ms)	1153.83
Minimum number of iterations	13
Mean number of iterations	10.57
Maximum number of iterations	15
Minimum time per iteration (ms)	38.5
Mean time per iteration (ms)	42.3
Maximum time per iteration (ms)	76.9

V. CONCLUSIONS

The objective of this work was to to develop a simulator and to find an efficient algorithm to solve a power flow in a 2x25kV bivoltage traction system. The objective has been achieved through the implementation of the Newton-Raphson algorithm. The main advantage of the developed simulator is the compact matrix formulation of the performed method due to the application of graph theory. Developed simulator can be effectively used to solve the load flow for the high speed railway. Aside from the further testing of the simulator and the algorithm, there are several modifications, which can be explored, such as: implementation of another numerical technique for the same purpose, extension of the train and network models, implementation for the industry use. There are also some improvements that can be made. In particular, with the proper data form the railway company, it may be possible to obtain more accurate results that will be closer to the real life scenarios.

REFERENCES

[1] R. Hill, "Electric railway traction. part 3: Traction power supplies," *Power Engineering Journal*, Dec. 1994.

[2] B. Scott, "Review of load-flow calculation methods," *Proc. IEEE*, vol. 62, no. 7, 1974.

[3] T. Ho, Y. Chi, J. Wang, K. Leung, L. Siu, and T. C.T., "Probablistic load flow in ac electrified railways," *IEE Proc. Electr. Power Appl.*, 2005.

[4] T. Ho, Y. Chi, J. Wang, and K. Leung, "Load flow in electrified railway," *2nd IEE International Conference in Power Electronics, Machines and Drives*, 2004.

[5] T. Ho, Y. Chi, L. Siu, and L. Ferreira, "Traction power system simulation in electrified railways," *Journal of Transportation System Engineering and Information Technology*, 2005.

[6] N. Hatziargyriou and T. Karakatsanis, "Probablistic load flow for assessment of voltage instability," *IEE Proc. C*, 1998.

[7] B. Borkowska, "Probablistic load flow," *IEE Trans.*, 1974.

[8] R. Allan and M. Al-Shakarchi, "Probablistic techniques in ac load flow analysis," *IEE Proc. C*, 1998.

[9] W. F. Tinney and C. E. Hart, "Power flow solution by newton's method," *IEEE Transaction on Power Apparatus and Systems*, 1967.

[10] D. Das, D. Kothari, and A. Kalam, "Simple and efficient method for load flow solution of radial distribution networks," *Electrical Power & Energy Systems*, vol. 17, no. 5, 1995.

[11] H. L. Nguen, "Newton-raphson method in complex form," *IEEE Trans. Power Syst.*, 1997.

[12] C. Chang, W. Wang, A. Lieu, F. Wen, and D. Srinivasan, "Genetic algorithm based bicriterion optimisation for tract substation in dc railway system," *IEEE Transaction on Power Apparatus and Systems*, 1974.

[13] B. Scott and O. Alsac, "Fast decoupled load flow," *IEEE International Conference Evolutionary Computation*, 1995.

[14] P. His, S. Chen, and R. Li, "Simulating on-line dynamic voltages of multiple trains under real operating conditions for ac railways," *IEEE Transaction on Power Systems*, vol. 14, no. 2, 1999.

[15] "Integrated simulator for ac traction power supply," http://www.uic.org/cdrom/2008/11_wcrr2008/pdf/PS.2.4.pdf.

[16] M. Chymera, A. Renfrew, M. Barnes, and J. Holden, "Modelling electrified transit system," *IEEE Transaction on Vehicular Technology*, vol. 59, no. 6, 2010.

[17] W. Kersting, "A method to teach the design and operation of distribution system," *IEEE Trans. on Power Appl. Syst.*, 1984.

[18] C. Cheng and D. Shirmohammadi, "A three-phase power flow method for real-time distribution system analysis," *IEEE Trans. on Power Syst.*, 1995.

[19] E. Ramos, A. Exposito, and G. Cordero, "Quasi-coupled three-phase radial load flow," *IEEE Trans. on Power Syst.*, 2004.

[20] Z. Wnag, F. Chen, and J. Li, "Implementing transformer nodal admittance matrices into backward/forward sweep-based power flow ananlysis for unbalanced radial distribution systems," *IEEE Transaction on Power Systems*, 2004.

[21] S. Raygani, A. Tahavorgar, S. Fazel, and B. Moaveny, "Load flow analysis and future development study for an ac electric railway," *IET Electrical System in Transportation*, 2012.

[22] E. Pilo, L. Rouco, and A. Fernandez, "A reduced representation of 2x25kv electrical systems for high-speed railways," *IEEE/ASME Joint Rail Conference*, 2003.

[23] E. Pilo, *Power Supply, Energy Management and Catenary Problems*. WIT Press, 2010.

[24] V. Zakarukin and A. Krukov, "Methods of joint simulation for external power supplies and ac traction systems (written in russian)," Ph.D. dissertation, State Railway University, Irkutsk, 2011.

[25] M. Plakhova, B. Mohamed, and P. Arboleya, "Static model of a 2x25kv ac traction system," in *PESA Conference, 2015 IEEE*, December 2015.

[26] ——, "Graph theory approach for a 2x25kv ac bivoltage traction sytem," in *PESA Conference, 2015 IEEE*, December 2015.

[27] F. Kiessling, *Contact lines for electric railways: Planning - Design - Implementation - Maintenance*. Siemens, 2009.

BIOGRAPHY

Mariia Plakhova received the B.Sc in Industrial Electronics and Automation and the M.Sc in Energy Management from the National Aviation University (Ukraine) in 2012 and 2014. She also received the M.Sc degree in Sustainable Transportation and Electrical Poweer Systems from the University of Oviedo, Gijon, Spain, in 2015. Her master thesis was focused on the development of an AC High speed traction system simulator. Now, she is working on modelling and simulation of AC railway traction networks.

APPENDIX

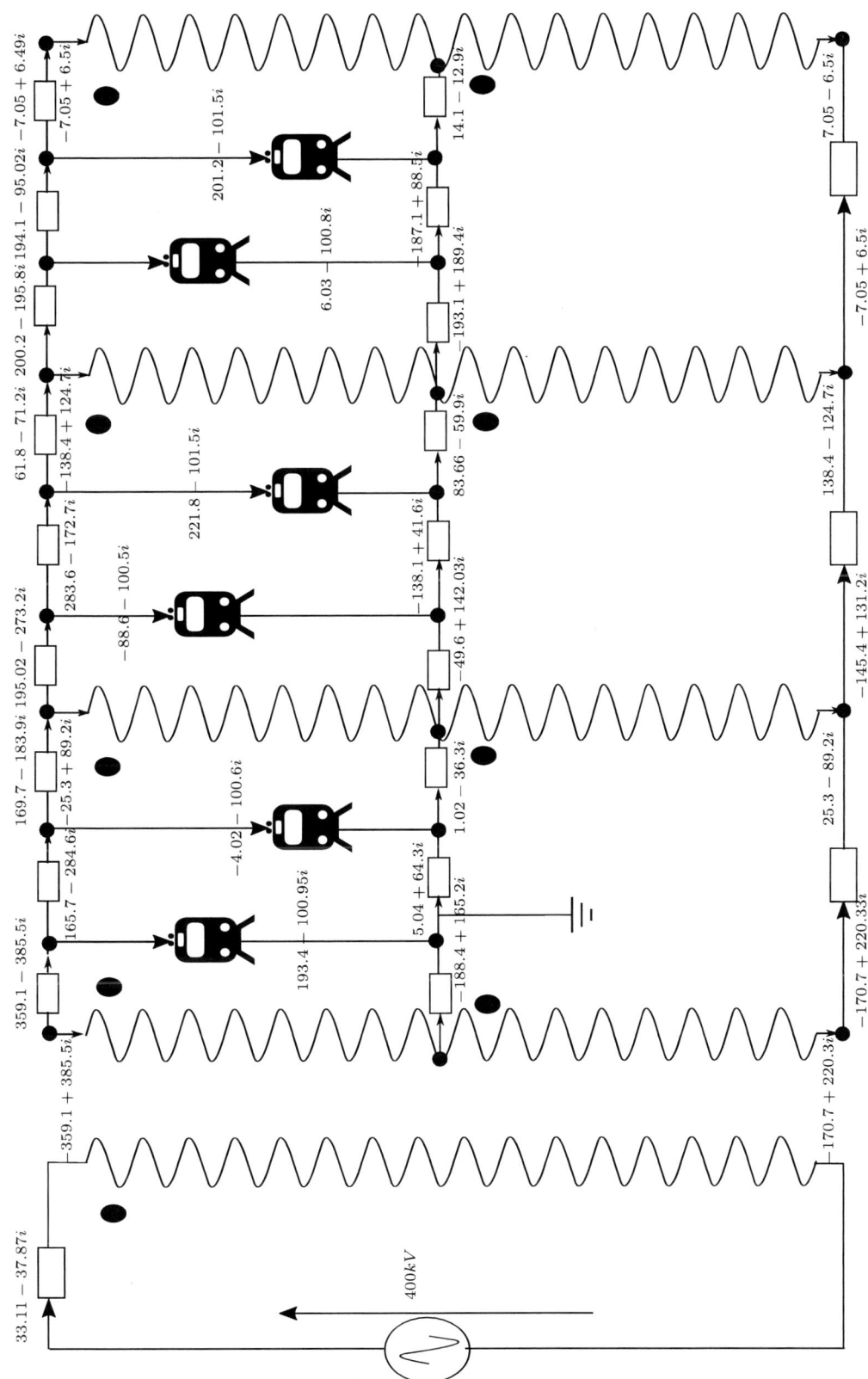

Fig. 4. Current distribution in an AC bivoltage traction system, obtained by derivative method

Design and Construction of a DAB Converter for Integration of Energy Storage Systems in Power Electronic Applications

Sarah Saeed

Abstract–The paper studies the design and construction of a Dual-Active-Bridge isolated DC/DC converter for tying energy storage systems to the DC link of a Power Electronic Converter. Among the keystone applications of such configuration, grid support for power quality systems, on/off board battery charger systems for plugged-in hybrid electric vehicles or Charging Station systems outstand as the most interesting from the technical and economic points of view. All these applications allow for the development of the Microgrid concept and boost the advancement towards electrical transportation.
The converter is analytically characterized, a control structure is developed to ensure a constant battery voltage with a fast stable ride-through during transient events, and an experimental test platform is designed to assess the studied modulation and control strategies under 2kW power level.

Keywords–DAB, Electric Transportation, Energy Storage, and Microgrid.

I. INTRODUCTION

The enhancement in efficiency, reliability and versatility in the transportation of both people and commodities, together with a decrease in its overall costs and polluting emissions can be understood as one of the main enabling keystones for a more sustainable and rational development of modern societies. The boost in electrical transportation depends on the technical development of the electrical vehicles as well as in the final deployment of the electrical Smart Grid [1] [2] [3].

From the point of view of the vehicles themselves, significant technical challenges related with rational use of energy arise in aspects such as size and energy capacity of the on-board stage systems –such as batteries or supercapacitors-, as well as in the performance and design of the dedicated bidirectional Power Electronic Converters (PEC) that aim to control the power flow in the vehicle and to ensure an adequate operation of these storage systems.

However, also infrastructures, as for instance power systems or charging stations, play an important role in the development of electric transportation [4] [5]. Indeed, the modernization of the current grid facility into the Smart Grid concept –including its decentralization into many MicroGrids catering to clusters of loads-provides increased reliability of the system, allowing, among other features, for a swift integration of the renewable sources, such as wind or photovoltaic plants, with the existing power plants. In this scenario, Energy Storage Systems (ESS) enable for an increased power quality in the system, as well as a more reliable and versatile system performance against random variation in these renewable sources and in the existing stochastic loads. For transportation applications, PEC also allow for the integration of the ESS in a distributed manner, in order to allow for the charging of the electric vehicles.

A MicroGrid comprises distributed renewable energy sources, energy storage systems, and load clusters which includes, but not limited to, charging stations for EVs/PHEVs, facilities, lighting systems, home appliances, etc., all connected to the main grid through the Point of Common Coupling (PCC), as clarified by Figure 1.

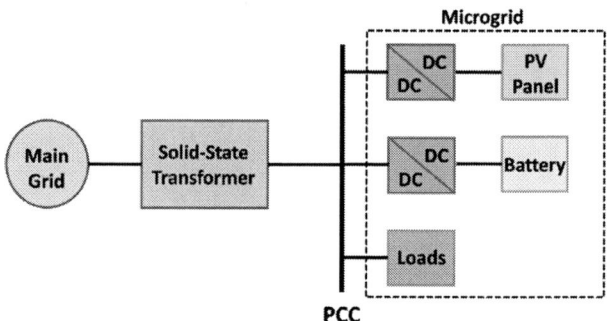

Fig. 1. Microgrid Structure.

All of the aforementioned characteristics and demanded features cannot be fulfilled by a standard Low-Frequency (LF) transformer, 50/60 Hz. The consequent development of the Solid-State-Transformer (SST) concept, first introduced in [6], being a High-Frequency (HF) transformer -tens of kHz and beyond- as the core of an AC/AC PEC system, provides simultaneously galvanic isolation along with full power flow control.

Additionally, higher frequency brings about a major transformer size reduction, and lighter weight. Along with efficiency improvement, the later stated features reply to the need of space-critical applications such as traction systems. Also with the increasing interest towards an effective reduction in fuel consumption, a recent trend in automotive industry towards Hybrid Electric Vehicles (HEVs) and Plugged-in Hybrid Electric Vehicles (PHEVs) has prevailed [7] which raises the need for high power density solutions [8].

Motivated by these issues, this paper addresses the design and construction of an isolated bi-directional SST structure, where a voltage level of 500V, and a power level of 2kW are sought. Through literature review, a single-phase single-stage Dual-Active-Bridge (DAB) converter is the topology of most convenience for the studied application. The topology was first introduced in the early 1990's by *Rik W. A. A. De Doncker et al.* in a paper [9], a few months before which a patent was released [10]. The early publications [11] [12] [13] [14] [15] characterized the topology and tried to develop it. However, the performance limitations of power switches at that time rendered the topology out of attention. With the advances in new power devices and magnetic materials, the

978-1-5090-0064-7/15 $31.00 © 2015 IEEE

DAB has been recalled again as a feasible topology to replace LF transformer [16] [17]. Furthermore, [18], [19], [20] and [21] have discussed employing the DAB converter for automobile applications.

The circuit diagram in Figure 2 depicts this DAB converter, consisting of two H-bridge inverters connected across an intermediate HF transformer.

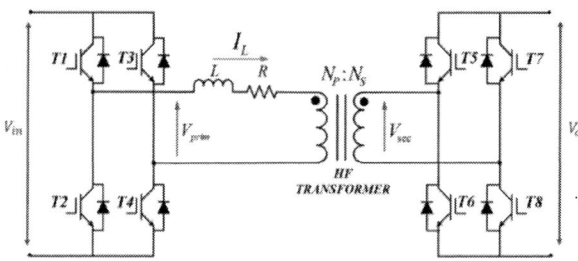

Fig. 2. Dual-Active-Bridge Converter Topology.

The central topic of the work is to characterize the DAB converter and validate its practical operation for the specified application [22]. The objectives guiding the work process can thus be listed to:

• Analysis of the DAB converter topology. The analysis results in the basic expressions that characterize the converter operation, and allows for further implementation of the control structure.
• Develop a Closed-Loop Control structure. The controller design is intended to allow for proper converter operation under the specific studied application.
• Simulations of the DAB converter operation within the studied application. This involves exploration of the studied concepts, modeling of the power and control stages, and development of the control structure as a prior step to practical realization.
• Design and modification of an experimental setup for assessing the studied converter design. The design comprises the construction of the power stage, the implementation in a digital controller of the control system developed, as well as adaptation of the interface between the control and the power stage.
• Verification of the built prototype DAB converter performance. This includes the validation of the converter open-loop and closed-loop behaviors through matching with practical obtained results.

II. OPERATION OF THE DAB CONVERTER

The section analyses the steady-state behavior of the DAB converter, deriving the expressions characterizing the topology.

Recalling the circuit configuration of the converter, Figure 2, the input and output H-bridge converters can be represented by square waveform voltage sources, thus, conceptually, the converter circuit can be viewed as an inductor, the series power transfer inductor, driven at either end by a controlled square-wave voltage source, as shown in Figure 3. The voltage sources are phase shifted from each other by a controlled angle, θ.

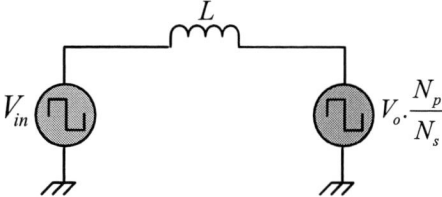

Fig. 3. Exact model of DAB dc/dc Converter

These two phase shifted voltage waveforms, generate a non-zero net voltage, across the inductor. This voltage applied to the power transfer inductor generates a current. The voltage sources can thus generate or receive the respective instantaneous powers.

The power level of the DAB converter is thus typically adjusted using one or more out of four control parameters:

• The phase shift, θ, between the AC voltages

$$-\frac{\pi}{2} \leq \theta \leq +\frac{\pi}{2}$$

• The phase shift between legs of primary bridge, or between legs of secondary bridge.
• The duty cycle of AC voltage at primary bridge, D_1, or of AC voltage at secondary bridge, D_2,

$$0 < D_1, D_2 < \frac{1}{2}$$

• The switching frequency, f_s.

The Phase Shift Modulation varies only the phase shift, θ, between the primary and secondary bridges to control the transferred power, and operates the DAB converter at constant switching frequency, with maximum duty cycles $D_1 = D_2 = 0.5$. Figure 4 thus illustrates the AC square-voltage waveforms are generated at the outputs of each bridge, the voltage generated across the inductor and the current flowing through it.

Studying the inductor current expressions through different intervals, an analytical expression for the power flow can be reached as mentioned below.

$$\therefore P = \frac{N_p}{N_s} \cdot \frac{V_{in} V_o}{X} \cdot \theta \left(1 - \frac{|\theta|}{\pi} \right)$$

where X is the impedance of the inductor, $2\pi f_s L$.

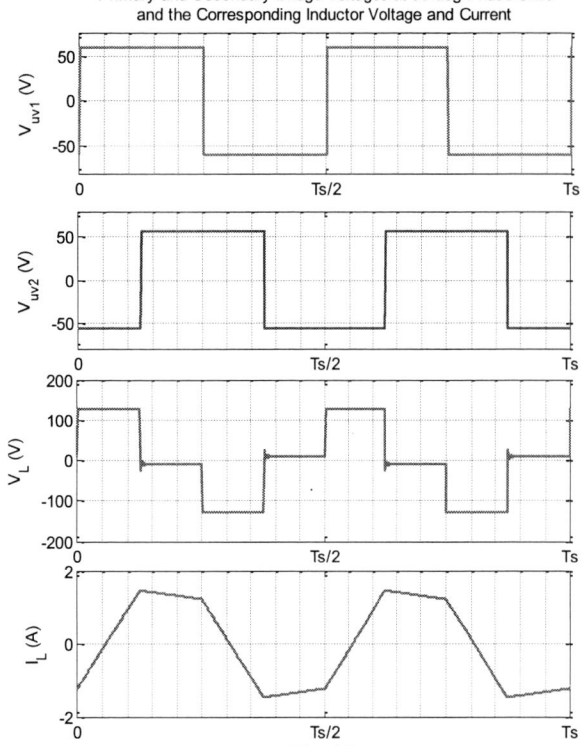

Fig. 4. Dual-Active-Bridge Converter Operation Waveforms – AC Voltages at the output of Primary and Secondary Bridges Respectively, Voltage across the Inductor, and the Current through it.

III. SYSTEM DESIGN

The DAB has been analyzed thoroughly, considering the existing technical literature. After this study, the design of the DAB converter prototype is carried out considering the system equations and verified through the required simulation results. Figure 5 shows a circuit diagram of the converter platform. The platform can be mainly divided into the power

stage, the programmer software, and the programmer hardware. Figure 6 illustrates the final constructed experimental platform.

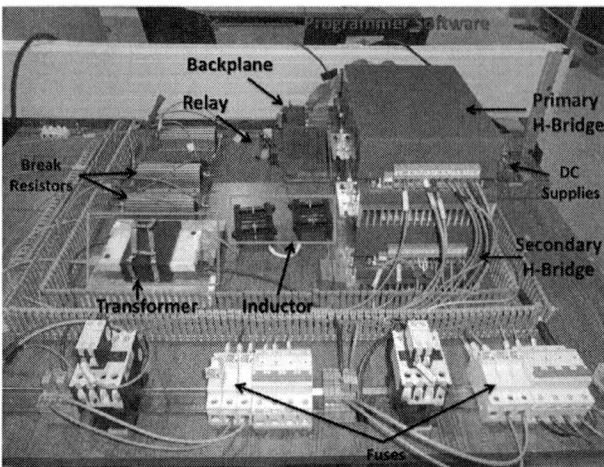

Fig. 6. Constructed Experimental Platform for Practical Test of DAB Converter.

IV. EXPERIMENTAL VALIDATION AT LOW POWER LEVEL

Initially, a series of experiments were carried out at low voltage level values, in order to assess the operation of the full system, prior to scale to higher power levels.

Closed Loop Control Performance
The closed loop performance is checked at an input voltage of 200V. The output commanded voltage reference is set to match the transformer turns ratio, i.e. 160V, in order to minimize circulating currents. The employed load is a 150Ohms-load, thus an output power level of around 200W.

a) Disturbance Rejection
Figure 7 shows the block diagram for this control structure.

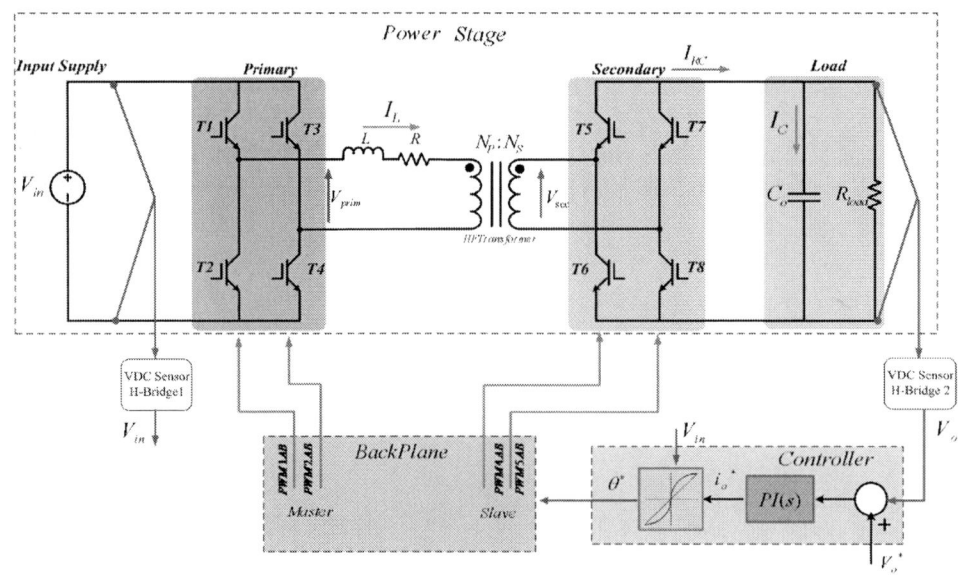

Fig. 5. DAB Converter System Diagram.

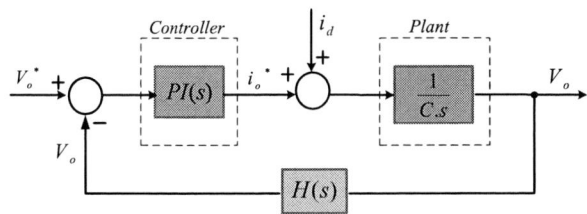

Fig. 7. Closed-Loop-Controller Structure for Studying Controller Response to a Transient Variation of Load Condition.

A disturbance is introduced by load variation from 150Ohms to 100Ohms. The controller is expected to fix the output voltage to the commanded reference of 160V. Thus, a step increase in power of 33% takes place.

Figure 8a illustrates the measured output voltage against the reference one. A voltage drop of 1.3 V is suffered during the transient load variation. Referred to the reference voltage level of 160V, this drop is 0.8%. Compromising this drop with the time the controller takes to recover the voltage reference, this is a good response. The error input to the PI regulator is thus shown in Figure 8b, being zero at steady state. The reference current is observed, in Figure 8c, to increase by a step of nearly 33% and consequently resulting in the expected power step.

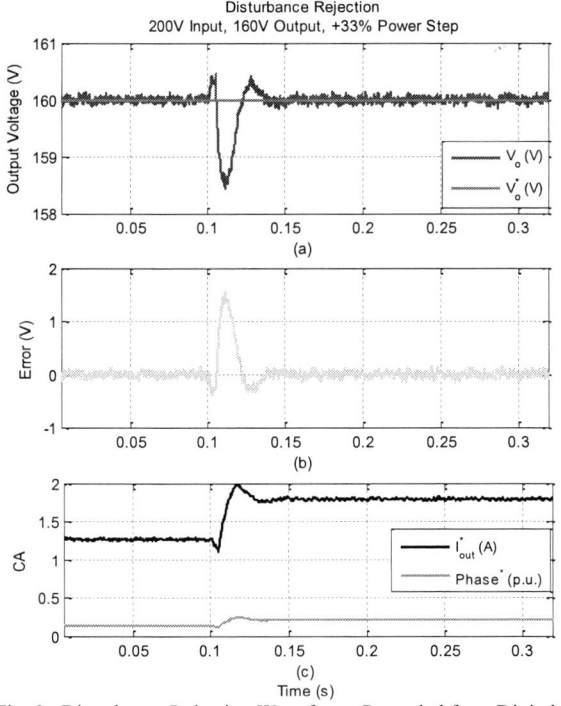

Fig. 8. Disturbance Rejection Waveforms Recorded from Digital Signal Controller (DSC).

b)   Reference Tracking
To study the reference tracking, Figure 9 shows the block diagram for this control structure.

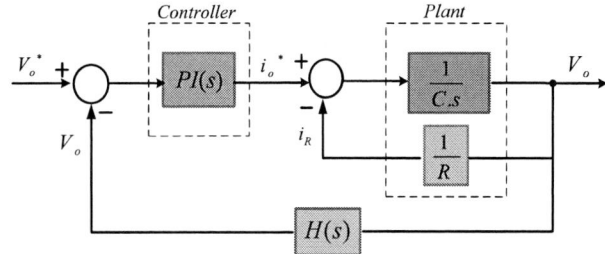

Fig. 9. Closed-Loop-Controller Structure assuming Fixed Load Condition for Studying Controller Response to a Transient Variation of Reference Output Voltage.

A variation in the commanded output voltage reference is introduced from 160V to 170V, thus a 10V step. Figure10a shows the measured output voltage against the reference one. An overshoot of nearly 2.5% is experienced during the transient event. This overshoot is acceptable taking into consideration the compromise carried out in tuning the controller for both types of transient events.

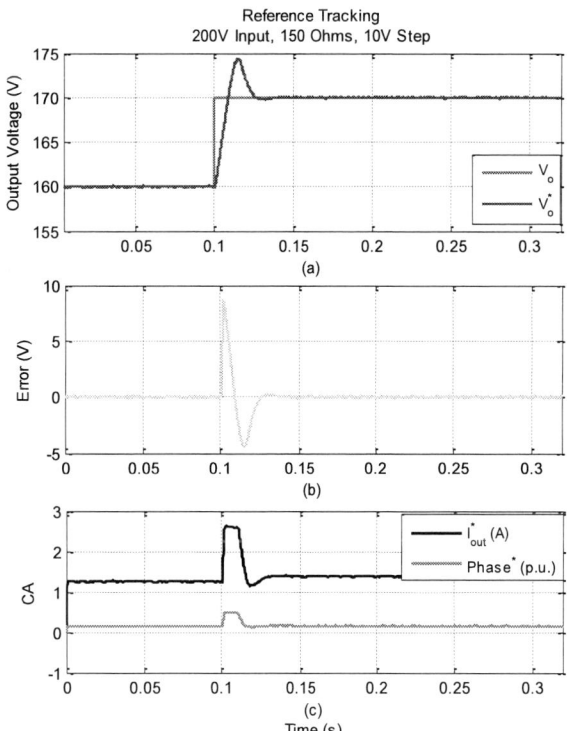

Fig. 10.   Reference Tracking Waveforms Recorded from DSC.

V.   HIGHER POWER TESTS

Some design modifications are taken into consideration for the higher power level tests which are summarized hereafter.

• Input supply is an autotransformer of a maximum AC voltage of 440V which can be supplied through a rectifier stage.

• Inductor design is split into two-core Litz-wired design output on a single layer, which targets minimization of overall

978-1-5090-0064-7/15 $31.00 © 2015 IEEE   399

losses, as clarified by Figure 11.

- The normal 300W resistive loads are replaced by a fan whose rated power is 3kW to withstand the target power level. Also to avoid any asymmetry in voltages applied to the transformer primary and secondary sides, two techniques have been developed and digitally implemented using software code.

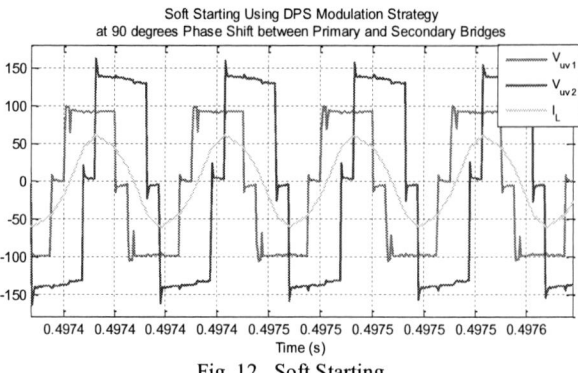

Fig. 12. Soft Starting.

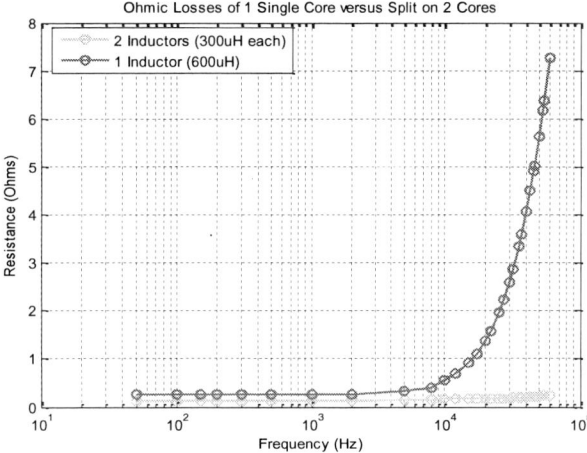

Fig. 11. One versus Two-Core Design.

a) Soft Starting

An internal phase is inserted between the legs of each H-bridge converter, known as Dual-Phase-Shift (DPS) modulation strategy. This modulation is applied to both primary and secondary bridge converters for the first seconds of converter operation to insure smooth variation of applied voltage, and consequently current, Figure 12.

b) Gradual Reference Voltage Command

With the start of the control process, the regulator takes control of the phase shift angle between bridges, adjusting it to reach the commanded output voltage reference. The reference output voltage is also adjusted to increase in steps to the commanded one, Figure 13.

c) Closed Loop Control Test

Disturbance rejection is again checked for validating controller performance at a higher power level. The test is carried out at an input voltage of 500V, output commanded voltage reference of 330V at a 150Ohms-load, thus an output power level of around 1kW.

The measured waveforms are depicted in Figure 14. The top figure illustrates the measured output voltage compared to the reference one. Compromising the drop in voltage, observed to be 1%, during transient ride-through with the time the controller takes to recover the voltage reference, this is a good response.

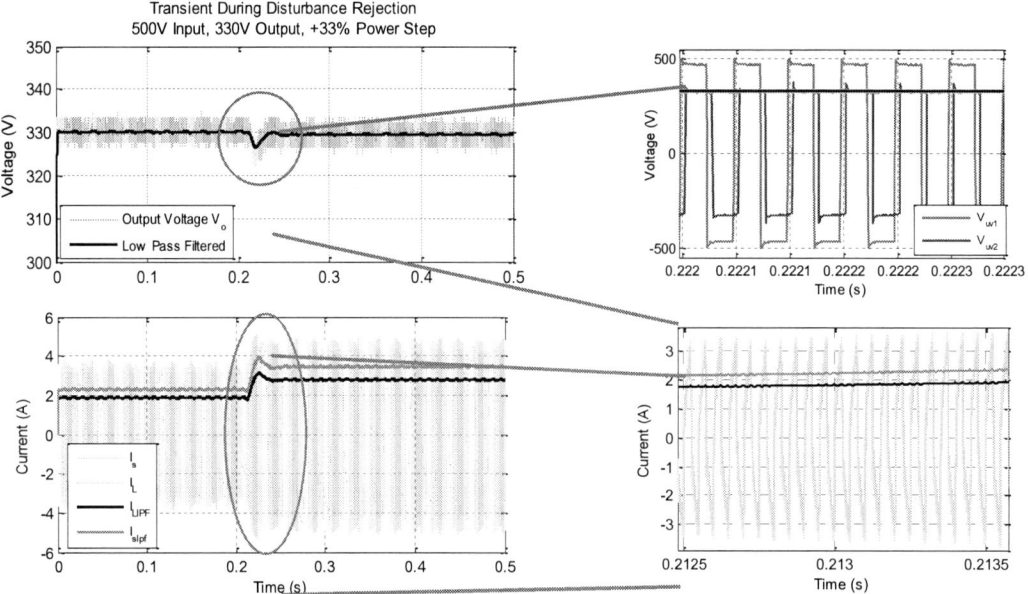

Fig. 14. DAB Converter Response to a 33% Increase in Power due to Load Variation, at a Nominal Power level of 800W.

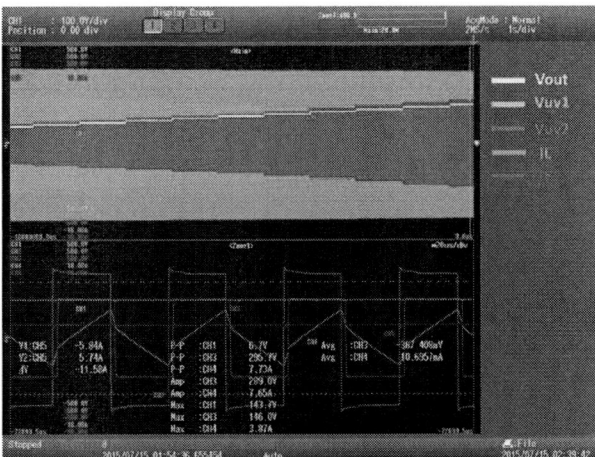

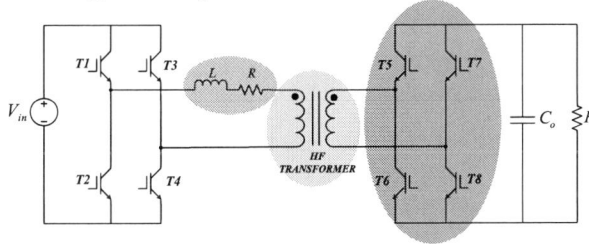

Fig. 13. Gradual Increase of Voltage Reference Command, and Corresponding Reduction in Current Ripple.

The current envelope, bottom figure, is seen to increase and thus an expected power step of nearly 33% is encountered while successfully holding on the system output voltage to the commanded reference.

VI. POWER BREAKOUT AND EFFICIENCY ESTIMATION

An efficiency study is carried out for the DAB converter concerning the main elements marked on Figure 15. The test is carried out at 500V input voltage, and controlling the output voltage to 350V, while employing a 150Ohms-load thus an approximate power level of 800W.

Fig. 15. DAB Converter Main Loss Elements.

a) Power Transfer Inductor

In the case of the power transfer inductor, the power loss has been estimated by considering only the fundamental equivalent resistance at 20kHz switching frequency, recording an average of 0.77W. Compared to an average recorded input power of 823.65W, an estimated efficiency can be calculated to be

$$\eta_{Inductor} = \frac{823.65 - 0.77}{823.65} = 99.99\%$$

b) High Frequency Transformer

The total input power to the transformer and the output power from it are recorded and averaged resulting in an estimated efficiency of

$$\eta_{Transforme} = \frac{849.43}{853.99} = 99.465\%$$

c) Power Converter

Figure 16 depicts the total input power to the secondary H-bridge converter and the output power from it, indicating the average of these powers. The difference is recorded to be

exactly 43.15W. Thus an estimated efficiency of

$$\eta_{Bridge} = \frac{806.22}{849.37} = 94.92\%$$

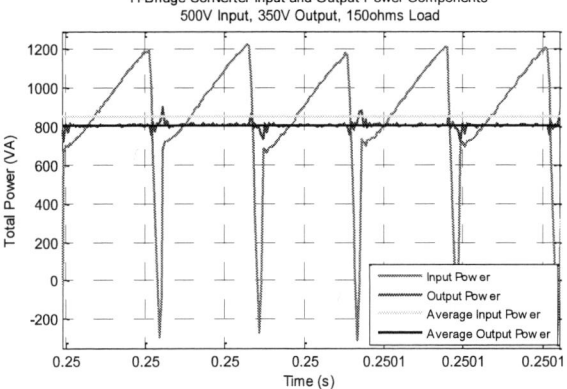

Fig. 16. Total Input and Output Power Components of the Real H-Bridge Converter in the DAB Converter Prototype.

The overall efficiency of the system is thus calculated to be 98.13%. It is concluded that the HF transformer causes 0.5% drop in efficiency, and the highest drop is caused by the bridge converter. To target a more efficient bridge converter, an alternative modulation can be deployed. These modulation schemes, besides reducing the circulating currents, also allow and extend the Zero-Voltage-Switching (ZVS) range of operation, cutting down the switching losses associated with IGBT devices, which becomes more important as the switching frequency increases.

VII. CONCLUSIONS

The DAB converter topology has been studied with an approach of employing the converter in a DC/DC conversion application. The conclusions reached are stated hereafter.

• The DAB converter topology has been analytically studied, deriving the characteristic equations that define its operation.
• The control of the DAB has been structured and designed for suiting the application under test. In this aspect, an output voltage control with a single-loop structure was adopted, aiming to a compromise between the controller performance to reference tracking and disturbance rejection.
• A simulation platform has been implemented for the DAB converter prior to constructing the experimental one. The models include most loss elements of the real system, to assure as close match as possible with practical results.
• A DAB converter experimental test platform has been constructed. The design also discussed the main loss elements in the system, pushing forward a study for power limitations under the designed platform, which was concluded to be 2kW.
• As a final step, tests have been carried out to validate the converter practical operation under the design and control constraints. After characterizing the steady state converter operation waveforms, and matching those with the simulation ones, an approach to increase the power has been taken. The converter was verified for operation at open loop under 1.7kW power level. Also, the controller performance has been

tested under load disturbance and found to ride-through the disturbance with a good transient response while successfully holding the system output voltage to the commanded reference one.

In general, the DAB converter is concluded to be an efficient power electronic converter with many optimization aspects available for boosting the use of ESS in high demanding performance applications.

REFERENCES

[1] L. Heinemann and G. Mauthe, "The Universal Power Electronics Based Distribution Transformer, an Unified Approach," in IEEE Power Electronics Specialists Conference (PESC), 2001, pp. 504–509.

[2] UNIFLEX, "UNIFLEX Project," 2013. [Online]. Available: http://www.eee.nott.ac.uk/uniflex/Project.htm

[3] M. Simoes, R. Roche, E. Kyriakides, A. Miraoui, B. Blunier, K. McBee, S. Suryanarayanan, P. Nguyen, and P. Ribeiro, "Smart-Grid Technologies and Progress in Europe and the USA," in IEEE Energy Conversion Congress and Exposition (ECCE), Sep. 2011, pp. 383–390.

[4] Y. Chen, A. Oudalov, and J. S. Wang, "Integration of electric vehicle charging system into distribution network," in Power Electronics and ECCE Asia (ICPE & ECCE), 2011 IEEE 8th International Conference on, 2011, pp. 593-598.

[5] J. A. P. Lopes, F. J. Soares, and P. M. R. Almeida, "Integration of Electric Vehicles in the Electric Power System," Proceedings of the IEEE, vol. 99, pp. 168-183, 2011.

[6] Power converter circuits having a high frequency link, William McMurray http://www.google.com/patents/US3517300U. S.

[7] Patent 3517300, June 23, 1970.

[8] M. Ehsani, A. Gao, S. E. Gay, and A. Emadi, Modern Electric, Hybrid Electric, and Fuel Cell Vehicles: Fundamentals, Theory, and Design. Boca Raton: CRC Press, 2005.

[9] Lingxiao Xue; Diaz, D.; Zhiyu Shen; Fang Luo; Mattavelli, P.; Boroyevich, D., "Dual active bridge based battery charger for plug-in hybrid electric vehicle with charging current containing low frequency ripple," in Applied Power Electronics Conference and Exposition (APEC), 2013 Twenty-Eighth Annual IEEE , vol., no., pp.1920-1925, 17-21 March 2013

[10] doi: 10.1109/APEC.2013.6520557

[11] R. W. A. A. De Doncker, D. M. Divan, and M. H. Kheraluwala, "A three-phase soft-switched high-power-density dc-dc converter for high-power applications," IEEE Trans. Ind. Appl., vol. 27, no. 1, pp. 63--73, 1991, 0093-9994.

[12] Power conversion apparatus for DC/DC conversion using dual active bridges, Rik W. De Doncker, Mustansir H. Kheraluwala, Deepakraj M. Divan http://www.google.com/patents/US5027264

[13] U.S. Patent 5027264, June 25, 1991.

[14] M. H. Kheraluwala, R. W. Gascoigne, D. M. Divan, and E. D. Baumann, "Performance characterization of a high-power dual active bridge dc-to-dc converter," IEEE Trans. Ind. Appl., vol. 28, no. 6, pp. 1294–1301,Nov./Dec. 1992.

[15] M. H. Kheraluwala and R. W. De Doncker, "Single phase unity power factor control for dual active bridge converter," Conf. Rec. of the IEEE industry Applications Society Annual Meeting (IAS 1993), Toronto, Canada, 2–8 Oct. 1993, pp. 909–916.

[16] H. L. Chan, K. W. E. Cheng, and D. Sutanto, "A novel square-wave converter with bidirectional power flow," in Proc.

IEEE Int. Conf. Power Electron. Drive Syst., 1999, pp. 966–971.

[17] H. L. Chan, K. W. E. Cheng, and D. Sutanto, "Phase-shift controlled dc-dc converter with bidirectional power flow," IEE Proc. Electric Power Appl., vol. 148, no. 2, pp. 193–201, Mar. 2001.

[18] H. L. Chan, K. W. E. Cheng, and D. Sutanto, "ZCS-ZVS bidirectional phase-shifted dc-dc converter with extended load range," IEE Proc. Electric Power Appl., vol. 150, no. 3, pp. 269–277, May 2003.

[19] J. Biela, M. Schweizer, S. Waffler, and J. W. Kolar, "SiC versus Si-Evaluation of potentials for performance improvement of inverter and dc-dc converter systems by SiC power semiconductors," IEEE Trans. Ind. Electron., vol. 58, no. 7, pp. 2872–2882, Jul. 2011.

[20] M. C. Lee, C. Y. Lin, S. H. Wang, and T. S. Chin, "Soft-magnetic Fe based nano-crystalline thick ribbons," IEEE Trans. Magn., vol. 44, no. 11,pp. 3836–3838, Nov. 2008.

[21] G. J. Su and L. Tang, "A three-phase bidirectional dc-dc converter for automotive applications," in Proc. IEEE Ind. Appl. Soc. Annu. Meet., 2008, pp. 1–7.

[22] J. H. Jung, C. K. Kwon, J. P. Hong, E. C. Nho, H. G. Kim, and T. W.C Hun, "Power control and transformer design method of bidirectional dc-dc converter for a hybrid generation system," in Proc. IEEE Vehicle Power Propulsion Conf., 2012, pp. 1512–1515.

[23] Yen-Ching Wang; Yen-Chun Wu; Tzung-Lin Lee, "Design and implementation of a bidirectional isolated dual-active-bridge-based DC/DC converter with dual-phase-shift control for electric vehicle battery," in Energy Conversion Congress and Exposition (ECCE), 2013 IEEE , vol., no., pp.5468-5475, 15-19 Sept. 2013

[24] doi: 10.1109/ECCE.2013.6647443

[25] Krismer, F.; Kolar, J.W., "Efficiency-Optimized High-Current Dual Active Bridge Converter for Automotive Applications," in Industrial Electronics, IEEE Transactions on , vol.59, no.7, pp.2745-2760, July 2012

[26] Sarah Saeed, "Design and Construction of an Isolated DC to DC Switching Converter for Integration of Energy Storage Systems in Power Electronic Applications," M.S. thesis, Elect. Electron. Comp. and Sys. Eng., Univ. of Oviedo, Spain, 2015.

[27] http://hdl.handle.net/10651/33257

Half-Bridge Current Source Bidirectional Resonant Converter for Supercapacitor Storage in Traction Systems

Mert Karadeniz
Email: eexmk16@nottingham.ac.uk

Abstract—The project consists on the analysis, design, simulation and implementation of a 3kW half bridge current source resonant converter, for energy storage applications in electric/hybrid vehicles. There is an existing prototype of a 3kW Half-Bridge Current-Source Hard-Switching converter in the electrical machines laboratory of the University of Rome. However this circuit is not working under resonant operation. The project consists on modify and adapt the full circuit to attain resonant operation of the system. In addition, to adapt the power stage, and in order to implement the control action, a filter and measurement circuit will also be designed and implemented. Furthermore, the existing system and the control algorithm will be combined with a fuzzy logic controller which is designed by another student from the University of Rome (Stefano Leonori), so an energy management system for an electric/hybrid car is going to be designed.

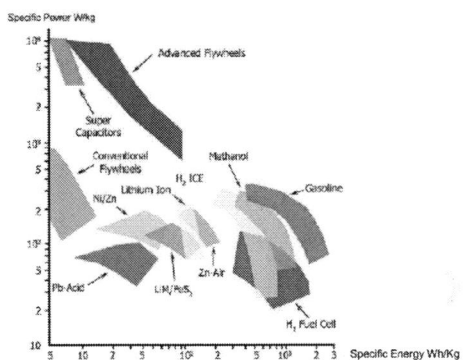

Fig. 1. Specific Power(W/kg) vs. Specific Energy(Wh/kg) [9]

I. INTRODUCTION

During the last decades due to the environmental effects and climate change, sustainable solutions are increasing their usage in every area of life. From industry to energy generation, from urban planning to transportation, sustainable aproaches are evident. As a part of environmental revolution, electric vehicles are also getting more popular day by day. Most manufacturers produces at least one full electric model.

Electric vehicles are superior to internal combustion engine (ICE) cars by some aspects; efficiency, environmental effects, noise, torque speed performance. On the other hand their shorter range, energy storage problems are the biggest disadvantages of the electric vehicles compare to ICE cars.

When the energy storage structure of an electric vehicle is considered, there are four different commonly used storage devices; batteries, fuel cells, super capacitors (SC), flywheels.

In Figure 1. specific energy versus specific power comparison of the devices can be seen [9].

As it is shown in the figure 1, while the specific power of flywheel and SC is much higher than battery's and fuel cell's, energy density is significantly smaller. Therefore flywheels and SC are mostly used as a secondary storage devices which mostly activated during transients of the vehicle like, accelerating or braking. On the other hand, since the specific energy of the battery and flywheel is much higher than the rest, they are used as primary storage device which is mostly in use at steady state.

In this paper, a 3kW charging system for a SC charging application is investigated and this circuit application is

converted into a resonant converter. The results related to resonant operation are given and comparison between resonant operation and existing operation is shown.

II. MOTIVATIONS FOR RESONANT SUPER CAPACITOR CHARGER

The main motivation after moving to a resonant switching scheme is to increase the efficiency of the original converter. As it is known, desired specifications of a power converter are higher power density and higher efficiency. These quantities are highly related with switching frequency. However when the switching frequency is considered, there is an inverse ratio between power density and efficiency. While the switching frequency is increasing, the size of the output filter of the converter is getting smaller which leads a higher power density. On the other hand, the switching losses increase due to the frequency, so the efficiency is getting smaller, furthermore since the losses are increased a larger and bulky heat sink or cooler must be added to circuit which also decreases the power density.

In resonant converter switching losses are reduced by two mechanisms called zero voltage switching (ZVS) and zero current switching (ZCS) [1].

A. Half Bridge Bi-Directional Current Source Converter (HBCS)

Before presenting the designed resonant converter, the existing circuit is shown. In order to control the power flow and the energy of the SC, a DC/DC converter which is half bridge

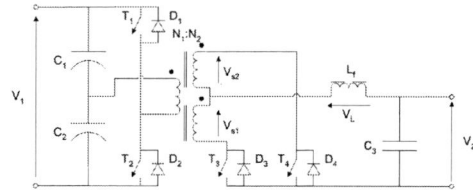

Fig. 2. The circuit schematics of HBCS [4]

bi-directional converter for this application, should be placed in between the battery and SC. For this purpose, three different connection schemes are proposed in the literature; series connection, cascaded connection and parallel connection [4,5]. Since the power sharing control and SC voltage utilization performance is much better, parallel configuration is suggested [4,5].

However, when the peak power of the SC is considered, voltage value of the SC is 6 to 10 times less thane the DC bus which means that in order to make the DC bus voltage constant gain of the converter should be around 6-10 [4]. Furthermore this high value of voltage gain causes an increment at the size of components and decrement at the energy extracted from SC [4,6].

In order to achieve these disadvantages mentioned above, a particular structure has been suggested in the literature, called Half bridge current source converter (HBCS) [4,5,6]. The circuit schematic is given in figure 2.

In order to understand the operation of HBCS two different modes are considered; charging and discharging.

When the SCs are charging the HBCS behaves as a buck converter where it reduces the battery voltage at the SC side. The power flow is from battery to SC side. In figure 2 the power flow is shown from V_1 to V_2. In this mode only the primary side switches (T_1 and T_2) are triggered and secondary side switches (T_3 and T_4) are held open [7]. The duty ratio of the primary side switches named d_1 and varies from 0 to 0.5 [7]. However in the application in order to prevent the cross conduction the d_1 is always less than 0.5. In this situation T_1 and T_2 operates 180° phase shift.

When T_1 is conducting, the voltage at the secondary side windings is:

$$V_{s1} = V_{s2} = \frac{N_1}{N_2} \frac{V_1}{2} \qquad (1)$$

When the T_1 conducts, D_3 is forward biased and the D_4 is reverse biased so the voltage applied across the inductor is:

$$V_L = \frac{N_1}{N_2} \frac{V_1}{2} - V_2 \qquad (2)$$

Since the voltage induced at the secondary side of the transformer is higher than V_2 the voltage across the inductor is positive which causes an increment on the inductor current [7].

When the T_1 is off, the transformer current goes to zero which also changes the polarity of the secondary side voltage

V_{s2} and both of the diodes at the secondary side starts conducting [7]. Therefore voltage induced secondary side of the transformer goes to zero and the inductor voltage is:

$$V_L = -V_2 \qquad (3)$$

Because of that inductor current decreases.

At the second half of switching cycle T_2 turns on and voltage across the secondary side turns D_4 on and the D_3 off [7]. Therefore the voltage across the inductance reaches same value as in the equation 2. After this step same procedure is repeated.

When the second mode -discharging- is considered, the switches at the secondary side are triggered with a gate signal called d_2 which varies between 0.5 to 1 and primary side switches are held open [7]. Therefore the switches T_3 and T_4 never turns off at the same time by shifting the phase of T_4 with respect to T_3.

In this case converter acts as a step-up converter since the power flow is from V_2 to V_1 where voltage value at V_2 is lower than V_1. When the both of the switches start conducting at the secondary side, the inductor stores energy. On the other hand both of the switches are conducting and all of the inductor current is flowing over the switches, secondary side of the transformer is shorted. Therefor there is no voltage across the primary side windings of the transformer [7].

During the time interval when the T_3 is on and T_4 is off current flows from the inductor to the transformer. Since there is change in the current, the voltage is induced over the secondary windings and so the primary windings hence D_2 is forced to be turn on [7]. Therefore the power is sent to the battery. The voltage across the inductor same as in the equation 2. However since the current direction is opposite (from capacitor to the transformer) inductor voltage is not increasing but decreasing [7].

When the T_4 is conducting but T_3, same conditions are repeated, however this time instead of D_2, D_3 starts conducting at the primary side [7].

Voltage transfer functions related to operation mode are given in equation 4 and 5.

$$G_{charge} = \frac{V_2}{V_1} = \frac{V_{SC}}{V_{BAT}} = \frac{N_2}{N_1} d_1 \qquad (4)$$

$$G_{discharge} = \frac{V_1}{V_2} = \frac{V_{BAT}}{V_{SC}} = \frac{N_1}{N_2} \frac{1}{1-d_2} \qquad (5)$$

The waveforms depending on the operation mode is given in figure 3.

B. Resonant Converter

The resonant operation is a technique which increases the converter efficiency by allowing the switching devices to perform ZVS and/or ZCS. Mainly there are two issues which causes switching loss in MOSFET and IGBT. Diode recovery charge and the output capacitance is the main issues which causes the switching losses in a MOSFET [1]. In addition in an IGBT, current tailing problem and stray inductances can

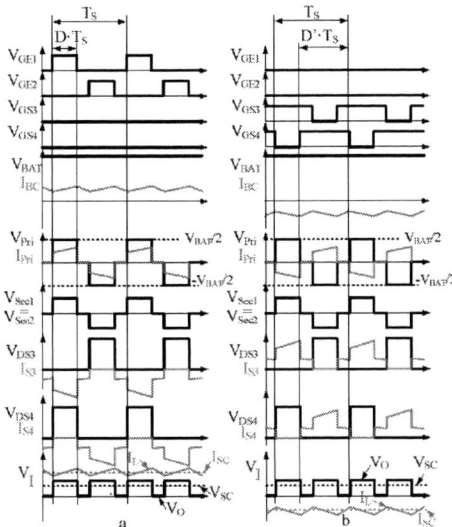

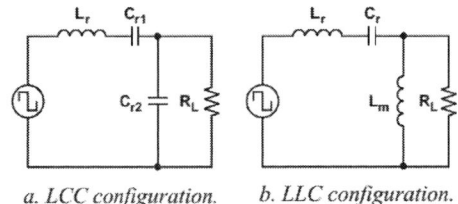

a. LCC configuration.   *b. LLC configuration.*

Fig. 5. The Basic Schematics of LCC (a) and LLC (b) [2]

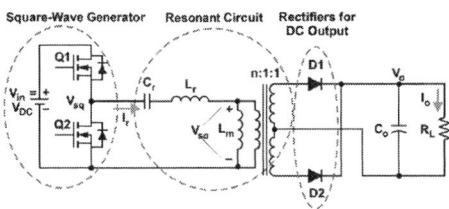

Fig. 6. The Schematics of the typical LLC resonant Half-Bridge Converter [2]

Fig. 3. Voltage (black) and Current (red) Waveforms Depending on the Circuit Operation a-) Charging b-) Discharging [4,5]

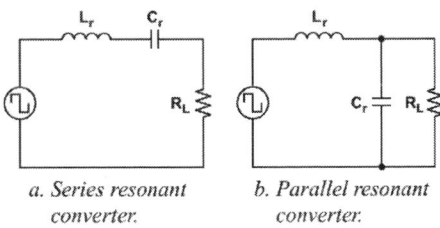

a. Series resonant converter.   *b. Parallel resonant converter.*

Fig. 4. Schematics of Series and Parallel Resonant Converter [2]

cause the switch action losses [1]. The ZVS and/or the ZCS (depends on operating point) can mitigate these losses [1].

In a resonant converter, the switch (or switch network) produces square wave voltages which contains the harmonics at switching frequency (fundamental) and odd harmonics as in the traditional PWM converters. When a resonant tank is implemented to circuit, it acts like a filter and a sinusoidal current which has the same frequency with the switching frequency starts flowing over the switches [1]. By changing the switching frequency the phase of the current with respect to switch voltage and the magnitude of the load current and voltage can be changed [1, 2].

There are two main type of resonant converter; series and resonant related to the connection of the resonant tank inductor and capacitor [2]. The equivalent circuit schematics are shown.

As seen in figure 4 resonant tank acts as a voltage divider with load. Since the impedance of the resonant tank varies with the frequency, the voltage drop over the tank and so the voltage drop on the load changes. Therefore, the voltage gain can be controlled by changing the frequency.

However these basic topologies have some drawbacks. First of all for SRC since the load series to tank the voltage gain never reaches unity [2]. Secondly, when the load is too low, output resistance value dominates the resonant tank

impedance. Therefore in order to change the output gain, the frequency converges to infinity [2]. In addition, even with nominal loads, the frequency range should be very large in order to control the output voltage [2]. When the PRC is considered, it is hard to apply this circuit for large load variations since the circulating current is too high because of the parallel connection of the resonant capacitor [2].

III. HALF BRIDGE LLC RESONANT CONVERTER

In order to overcome the problems of the SRC and PRC another structure of the resonant converter is proposed; series-parallel resonant converter (SRPL) [2]. It has an another capacitor or inductor compare to SRC and PRC. Basically, the series structure is same with SRC; an inductor and a capacitor is connected in series. Additionally, a capacitor or an inductor is connected in series with the load. Depending on the parallel component these structures called LLC (when inductor is in parallel) or LCC (when a capacitor is in parallel). The basic schematics are given in figure 5.

Where C_{r2} is the parallel capacitor, and the L_r is the magnetizing inductance of the high frequency transformer.

In this paper LLC resonant converter is discussed, since the thesis topic is related to LLC resonant converter. In addition LLC resonant converter has some advantages compare to LCC resonant converter.

First of all as seen in figure 5 the LCC has a capacitor parallel to load, which causes high AC currents circulating in the tank [2]. Therefore big and bulky capacitor should be placed in order to operate under high AC currents [2]. Secondly, L_r and L_m can be placed in one physical inductor when LLC is considered [2].

The schematic of the typical LLC resonant half bridge converter is shown in figure 6.

The operation principle of LLC resonant is more or less same with the SRC resonant converter. The resonant tank has a very low impedance value at resonant frequency, and

978-1-5090-0064-7/15 $31.00 © 2015 IEEE

the magnitude of the current and voltage values at load side depends on the switching frequency. The resonant frequency equals to;

$$f_0 = \frac{1}{2\pi\sqrt{L_r C_r}} \qquad (6)$$

However in LLC resonant since the L_m is added a new frequency value will appear. This frequency is called pole frequency and it is;

$$f_p = \frac{1}{2\pi\sqrt{(L_m + L_r)C_r}} \qquad (7)$$

When there is no load the peak frequency equals to pole frequency and while the load increases peak frequency goes towards f_0 [2]. Hence the gain control is much easier since the frequency range is reduced [2].

A. Design Specifications

The design specifications are given in table 1.

TABLE I
INITIAL PARAMETERS OF HALF BRODGE LLC RESONANT CONVERTER

Parameter	Value
Input Voltage	300 VDC
Rated Output Power	3000 W
Output Voltage	45V to 30 V
Rated Output Current	66.7 A
Efficiency	>90%
Output Capacitor	4700 uF
Output Inductor	27 uH
Transformer Ratio	3.5
Magnetizing Inductance	2.3 mH
Switching Frequency	10 to 30 kHz
Resonant Frequency	20 kHz

As seen in the table 1 there are some parameters related to output filter and high frequency transformer. Since the aim of the project is converting the existing HBCS converter into resonant converter, output filter and high frequency transformer is not changed. On the other hand these design constraints can cause the final values or results be significantly away from the recommended margins of the design procedures found in the literature. For instance, in [2] the maximum inductor ratio is 20, while the calculated one is 86.46.

IV. MEASUREMENT AND FILTER CIRCUIT DESIGN

In order to implement the control action, measurements related to circuit should be taken in all applications. For HBCS converter, SC voltage, SC current, battery voltage and current are measured. However since there is switch action in the circuit, the noise at around switching frequency exists at all the measured currents and voltages. In addition, there can be other noises due to different effects in the measurements. Therefore this noise must be filtered in order to measure true values. In addition the signal coming from the measurement device

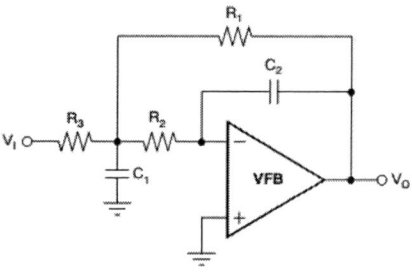

Fig. 7. MFB Filter Topology [3]

can be in the range of milivolts and the polarity of the output of the measurement device can be plus or negative related to measurement technique. For instance, when the current is measured, since the direction of current can change depending on charging or discharging mode, the polarity of the current transducer's output changes. Therefore a scaling circuit must be placed in between the measurement device and filter circuit.

As mentioned above for the control action, measured signals must be filtered. Since the switching frequency is 20 kHz in order to send the true value of the measurement to the DSP a low pass filter should be applied to the signal. A filter structure called Multiple FeedBack (MFB) is suggested for the analog to digital (ADC) conversion applications [3]. MFB filter topology is shown in figure 7.

A. Measurement Circuit Calculations

Six different measurement nodes are selected; battery voltage (VBAT), battery current (ICAP), SC voltage (VCAP), SC current (ICAP) and a series inductor voltage measurement (VIND), load current (ILOAD) for fuzzy logic controller. In order to measure the VBAT and VCAP LV 25-P voltage transducers are used. This transducer is basically a current transformer. A shunt resistor is connected in between the voltage node and the input of the voltage transducer. Therefore a current starts flowing from the measurement node to the transducer. Fundamentally this transducer changes the current gain which is basically 2.5:1. At the secondary side of the transducer another resistor is connected in parallel so a voltage drop is created proportional to the measured voltage. The voltage measurement and scaling circuit schematics is given in figure 8.

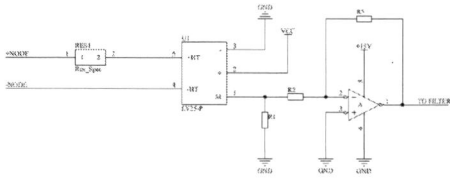

Fig. 8. The Voltage Measurement and Scaling Circuit Schematics

The measurement circuits shown above is for voltage measurements. However current measurement and scaling circuit is a little bit different than the voltage measurement circuit. First

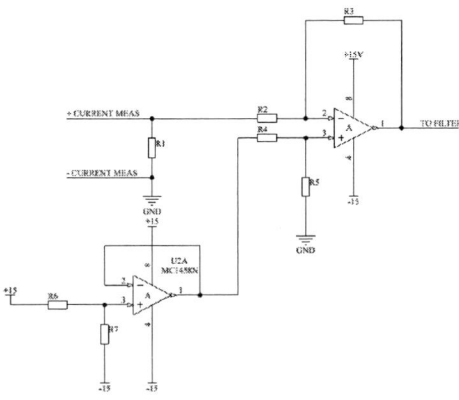

Fig. 9. The Current Measurement and Scaling Circuit

of all measurement transducer is different. LA125-P current transducer is used in order to measure the IBAT, ILOAD and ICAP so there is no need to use a shunt resistor between the measured node and transducer input. Secondly, the current direction changes related to the operation mode, charge or discharge. Therefore the output voltage of the scaling circuit varies from positive to negative. Since it is not possible to apply negative voltages to DSP, negative current measurement signal should be shifted. The current measurement and scaling circuit is shown in figure 9.

Finally the Inductor voltage measurement circuit is shown in figure 10. This circuit is designed to calculate the di/dt ratio for the fuzzy logic controller. However this controller is not the subject of this thesis. Since any of the inductor's terminal is not grounded, a voltage transducer can not measure the voltage induced over the inductor. Therefore an instrumentation amplifier circuit is designed. The topology shown in figure 10 is called classic 3 op-amp instrumentation amplifier circuit [8].

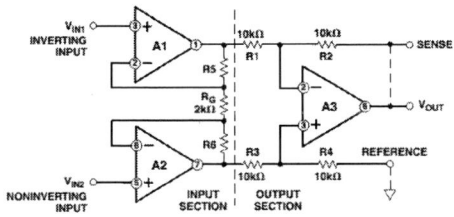

Fig. 10. Inductor Voltage Measurement Circuit [8]

V. SIMULATION AND TEST RESULTS

Before the bench tests are performed, a simulation is run in PSIM. The circuit built in the program is given in figure 11.

The values used in the simulation are shown in table 2.

A. Operation at Resonant Frequency

The first simulation is performed at resonant point (17.8 kHz). Most important waveform to observe the ZVS is the voltage between collector and emitter (V_{CE}) of the IGBT versus tank current (I_r). The simulation waveform is given in figure 12.

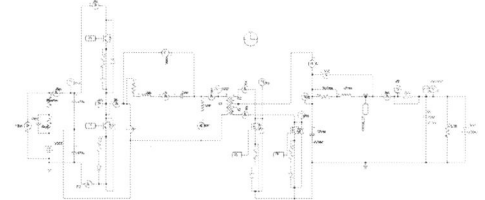

Fig. 11. Schematics of the Simulated Circuit

TABLE II
VALUES USED IN THE SIMULATION

Parameter	Value
Input Voltage	300 VDC
R_{dyn}	0.1 Ω
C_{dyn}	1000 mF
R_{batser}	10 Ω
Input Capacitor	470 uF
IGBT R_{on}	0.032 Ω
IGBT Diode Forward Voltage	1.7 V
Resonant Inductor	34.5 uH
ESR of the Resonant Inductor	0.028 Ω
Resonant Capacitor	2.344 uH
Parallel Inductor	2.3 mH
Primary Winding Resistor	8.06 mΩ
Secondary Winding Resistor	1.1 mΩ
Primary Leakage Inductance	1 uH
Secondary Leakage Inductance	0.025 uH
MOSFET R_{on}	0.02 mΩ
MOSFET Diode Forward Voltage	1.4 V
Output Capacitor	4700 uF
Output Inductor	27 uH
Series Resistor of the Output Inductor	0.002 Ω
Load Resistance	0.78

The test waveform is given in figure 13.

The waveforms are consistent except the oscillations at the I_r during turn-on and off. However since in the simulation behavior of the all components are ideal this effect can be neglected. The RMS value of I_r is 15.61 A, while it is 15.99 A for the simulation. The error is 2.37%.

B. Operation below Resonant Frequency

At this operation mode the frequency is set to 13.16 kHz. V_{CE} versus I_r waveforms are shown in figure 14 and 15.

In this mode the waveforms are also consistent. For simulation I_r is 18.18 A, while the test result is 18.66 A. The error value of the current is 2.64%.

C. Operation above Resonant Frequency

The frequency is 29.41 kHz for this mode. V_{CE} versus I_r waveforms are shown in figure 16 and 17.

The simulation I_r is 12.89 A and the test result is 12.58 A. The error is 2.4%.

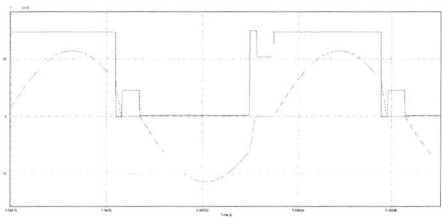

Fig. 12. The Simulation Waveform of V_{CE} (blue) and I_r (red) at resonant frequency (Duty=0.462)

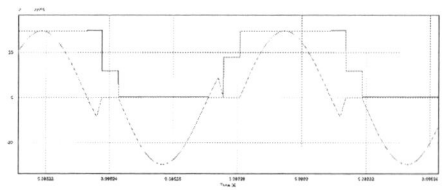

Fig. 14. The Simulation Waveform of V_{CE} (blue) and I_r (red) below resonant (Duty=0.462)

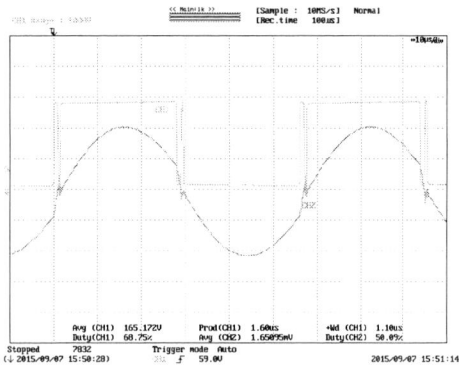

Fig. 13. The Test Bench Waveform of V_{CE} (yellow, 100 V/div) and I_r (green, 10 A/div) at resonant

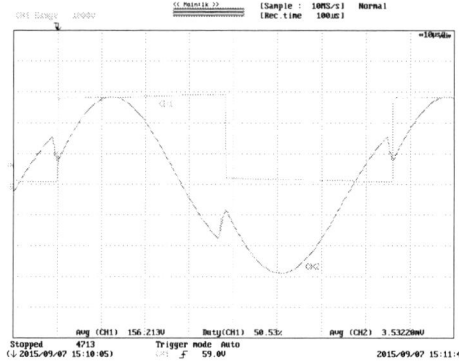

Fig. 15. he Test Bench Waveform of V_{CE} (yellow, 100 V/div) and I_r (green, 10 A/div) below resonant

As a result except the magnitudes of the current and voltage values are consistent, while there are such differences between the waveforms. The main reason is the MOSFET and the IGBT existing in the PSIM library have ideal behaviors. Therefore, turn-on, turn-off, rise and conduction delays will not appear in the simulations. Because of this, such differences are observed between the simulation and the test bench results.

D. Gain Comparison of the Simulation and Test Bench

In order to the calculate the gain values equation 8 is used.

$$M_g = \frac{nV_o}{V_{in}/2} \qquad (8)$$

The gain values both calculated from simulation and test bench are shown in table 3.

The error value is always less than 5 % in between 14.29 kHz and 23.81 kHz. After that point the error in the gain calculation exceeds 10 %. Therefore the equations related to LLC converter given in chapter 2 may not be valid after these two boundary. 23.81 kHz is assumed the maximum frequency limit.

At this point , it should be stated that LLC resonant converter must be operated in the vicinity of resonant frequency [2]. If the LLC converter is designed to operate exactly at f_0, then the results will be accurate. For the values far from the f_0 the higher harmonics begin to affect the whole system. Therefore the calculations will not be valid anymore. On the other hand there will be error in percentage between the actual values and calculated and/or simulated values.

E. Efficiency Comparison of the LLC Resonant Converter and HBCS

The highest efficiency reported in [4] is 92.3% at 25 V and 1250 W (load resistance=0.5 Ω) for HBCS converter. The efficiency curves of HBCS converter for 25 V, 30 V and 34 V are shown in figure 18. For the resonant converter the efficiency values with respect to frequency is given in figure 19. During the test input voltage is fixed to 300 V, the load resistor is 0.78 Ω, the duty is 0.462 and the switching frequency is changed between 12.82 kHz and 29.41 kHz in order to see the efficiency results. The minimum frequency is chosen randomly, however the maximum frequency is limited to 30 kHz due to the suggestion in [1] for IGBTs.

The highest efficiency value 93% when the output power is 1940 W and the output voltage is 38.68 V. Furthermore when the efficiency vs. load curve of the HBCS PWM converter is considered, the efficiency performance of resonant converter for higher output voltage values (when the gain is in between 0.87 and 0.92) is much better than HBCS converter. However for lower output voltages efficiency starts decreasing.

On the other hand, when the frequency is 15.15 kHz the ZVS is lost and the current becomes capacitive. Therefore for the application the lowest frequency value is set to 15.87 kHz where the ZVS can still be achieved. For the upper limit the frequency value 22.73 kHz is chosen since the efficiency is decreasing (less than 90%) and the gain error between the simulation and the test bench results is more than 5% above this frequency value.

In addition for different load values the efficiency characteristic of the resonant converter varies. While the load is

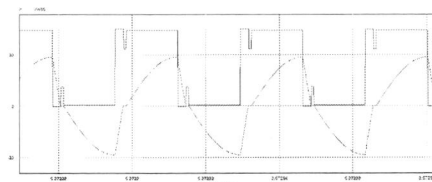

Fig. 16. The Simulation Waveform of V_{CE} (blue) and I_r (red) above resonant (Duty=0.462)

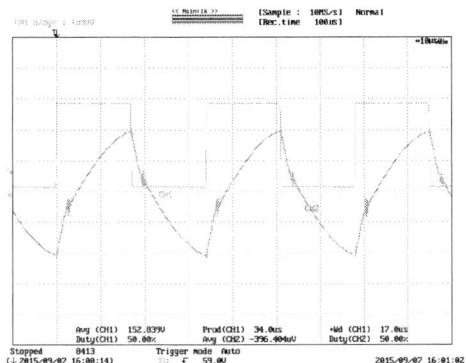

Fig. 17. The Test Bench Waveform of V_{CE} (yellow, 100 V/div) and I_r (green, 10 A/div) above resonant (D=0.462)

decreasing the efficiency of the converter is getting higher. Another loss mechanism of the converter is the conduction losses and in LLC converter high current stress can reduce the efficiency benefit of ZVS due to the conduction losses. The efficiency behavior of the LLC converter for different loads and frequencies is shown in figure 20.

As it is shown in figure 20 the highest efficiency value is 98.6% when the load resistor is 2.11 Ω and the frequency is 20 kHz. On the other hand the lowest efficiency is achieved (85.02%) when the load resistor is 0.645 Ω at 23.8 kHz as expected.

1) Subsubsection Heading Here: Subsubsection text here.

VI. CONCLUSION

In conclusion 3 kW Half-bridge LLC resonant converter is designed and tested. The main outcomes are;

- The efficiency is improved around 2 points for nominal load (load resistance=0.78 Ω) at 35 V output voltage. In the case of higher loads efficiency is less than HBCS converter. However for the load resistance higher than 0.7 Ω LLC resonant converter has higher efficiency values than HBCS for the 15.87 kHz to 22.73 kHz switching region. In general HBCS converter has a better efficiency performance than LLC converter for higher loads than the nominal load while LLC converter is better at nominal load and lower load values.
- The maximum efficiency is achieved by the LLC converter is 98.06% when the load is 2.11 Ω.
- A measurement circuit is designed for the LLC and HBCS converter which includes the measurement of

TABLE III
GAIN AND ERROR VALUES FOR SIMULATION AND TEST BENCH

Frequency (kHz)	Simulation Gain	Test Bench Gain	Error (%)
12.82	0.88	0.75	15.65
13.16	0.90	0.8	11.81
13.89	0.933	0.88	5.75
14.29	0.939	0.9	4.22
15.15	0.939	0.92	2.11
15.87	0.934	0.93	0.44
16.67	0.924	0.924	0.08
17.24	0.916	0.92	0.37
17.86	0.906	0.915	0.89
19.23	0.89	0.908	1.92
20	0.87	0.9	3.37
20.8	0.856	0.87	1.57
21.74	0.841	0.84	0.17
22.73	0.825	0.8	3.06
23.81	0.808	0.76	6
25	0.791	0.72	9
27.78	0.753	0.65	13.78
29.41	0.734	0.62	15.63

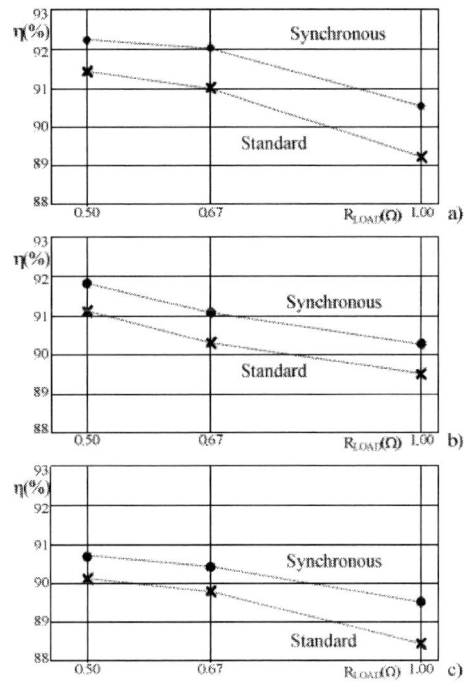

Fig. 18. Efficiency Curves of HBCS converter. a-) 25 V, b-) 30 V, c-) 34 V [4]

VBAT, VCAP, IBAT, ICAP, VIND and ILOAD measurement.

REFERENCES

[1] Robert W. Ercikson, Dragan Maksimovic. *Fundamentals of Power Electronics*. Kluwer Academic Publishers. Second Edition, 2001.

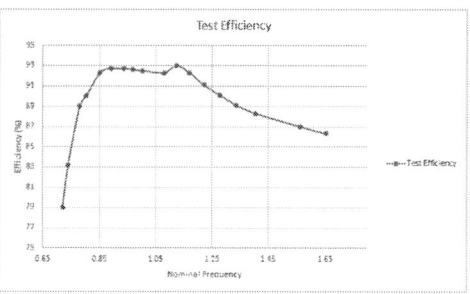

Fig. 19. Efficiency vs. Frequency Curve for Resonant Converter

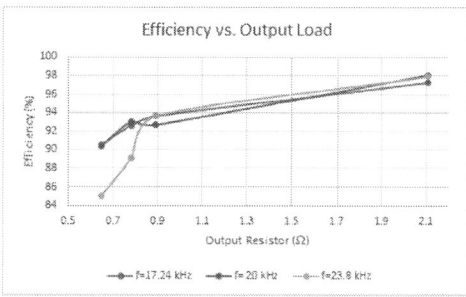

Fig. 20. The Efficiency Behavior of LLC Converter for Different Loads and Frequencies

[2] Hong Huang. Designing an LLC Resonant Half-Bridge Power Converter. Texas Instruments Power Supply Design Seminar. 2010.

[3] Michael Steffes. Design Methodology for MFB Filters in ADC Interface Applications. Texas Instruments Application Report, 2006.

[4] J. Garcia, P. Garcia, F. G. Capponi, G, Borocci, G. D. Donato. Analysis, Modeling and Control of Half-Bridge Current-Source Converter for Supercapacitor Applications. 2014.

[5] J. Garcia, F. G. Capponi, G. Borocci, P. Garcia. Control Strategy for Bidirectional HBCS Converter for Supercapacitor Applications. 2013.

[6] F. G. Capponi, P. Santoro, E. Crescenzi. HBCS Converter: A Bidirectional DC/DC Converter For Optimal Power Flow Regulation in Supercapacitor Applications.

[7] F. G. Capponi, L. D. Ferraro. DC/DC Converter For Coupling Supercapacitors With Fuel Cells For Distruted Generation.

[8] C. Kitchin, L. Counts. A Designer's Guide to Instrumentation Amplifiers. Analog Devices, Third Edition, 2006.

[9] Alberto Martin Pernia. ESHEV. ClassNotes distributed by Alberto Martin Pernia.

Design Optimizations of Outer-Rotor Permanent Magnet Synchronous Machines with Fractional-Slot and Concentrated-Winding Configurations in Lightweight Electric Vehicles

Demin Wu, Weizhong Fei, and Patrick Chi-Kwong Luk

Electric Power and Drive Group, Power Engineering Centre, Cranfield University, Cranfield, MK43 0AL, U.K.
Email: p.c.k.luk@cranfield.ac.uk

Abstract–This paper concerns the design optimization of an outer-rotor permanent magnet synchronous machine (PMSM) with fractional-slot and concentrated-winding (FSCW) configuration for a lightweight direct-drive electric vehicle. Based on the rigid space envelope limits and performance specifications required by the vehicle, four candidate versions of a FSCW PMSM have been determined by initial machine sizing analysis. All versions share the same rotor outer diameter, rotor back iron thickness, permanent magnet thickness, air gap length, and magnet pole arc width ratio. A window-zoom-in response surface method based on two-dimensional parametric finite element analysis is proposed for comprehensive electromagnetic optimization for maximum torque density. Additionally, the machine losses at rated torque output including copper joule loss, rotor eddy current loss, and stator core loss maintain the same in all the candidates. It is found that significant correlation exists between stator slot depth and active axial length in the optimal outer-rotor PMSMs. Finally, the performance characteristics of the four optimal machines are comprehensively evaluated and compared to confirm the best candidate, and its utility for the vehicle's mission.

Keywords–Permanent magnet synchronous machine, fractional-slot and concentrated-winding, direct drive, in-wheel motor, lightweight electric vehicles, design optimization.

I. INTRODUCTION

All major cities worldwide are facing the challenges caused by transport and traffic. In the EU, urban mobility accounts for 40 % of all CO_2 emissions of road transport and up to 70 % of other pollutants from transport, whilst congestion costs nearly EUR 100 billion, or 1 % of the EU's annual GDP [1]. Recent statistics show that vehicle speeds in City of London at busy hours are - passenger cars at 3 mph, buses and conventional taxis at 5 mph and dispatch riders at 15 mph [2]. The dispatch riders, with their narrow footprint and agility, are particular suitable to combat congestion. The benefits of front/rear 2-seater lightweight electric vehicles (EVs), such as the electric quadricycle shown in Fig.1(a), become increasingly evident in enhancing urban mobility and achieving zero emission. One well-known commercial example is Renault's Twizy. However, current EVs including lightweight ones such as Twizy, invariably use conventional centralized drivetrains that take no advantages electric drives have to offer. When configured in direct-drive in-wheel form as shown in Fig.1(b), electric motors can reduce vehicle part count, complexity and cost, incorporate integrated power electronics, give complete vehicle design freedom and enhance regenerative braking [3-4]. With independent control, direct-drive in-wheel motor drivetrain has superior driving characteristics

especially in the way torque is delivered to each wheel, with highest torque available from standstill and excellent dynamic response to torque demand [5]. To capitalize these advantages, it is vital to design an electric motor with maximum torque density and high efficiency.

(a) Narrow car concept (1m wide, 2.5 long, 1.9m tall) Courtesy of Hugh Kemp, Naro Car Company

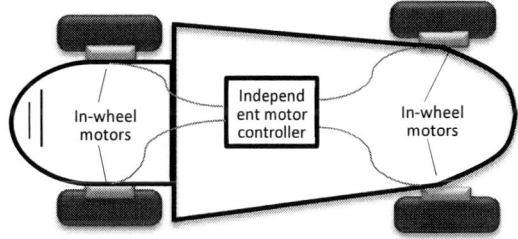

(b) Independent direct drive configuration
Fig.1 Electric quadricycle under study

This paper concerns a comprehensive design study of an in-wheel permanent magnet synchronous machine (PMSM) used for the lightweight vehicle shown in Fig.1(a). It has a narrow footprint (1m x 2.5m) designed for urban mobility, speed limit of 45km/h and weight limit of 250kg to be qualified as a quadricycle that is exempt from crash tests for normal cars. There is also a power output limit of 4kW for the electric drivetrain. Since the outer-rotor radial flux motor configuration has superior torque density over all other common configurations [6], and is particularly conducive to in-wheel integration of other mechanical parts, the study will focus on the optimization and comparison of the four versions of a PMSM motor with outer rotor. Besides, fractional-slot and concentrated-winding (FSCW) configuration normally results in compact end windings with minimal overhang, coupled with superior electromagnetic and thermal performance over conventional integral-slot counterpart [7]. Hence, it is not surprising that the outer-rotor surface mounted PMSM with FSCW configurations has become the preferred choice for direct-drive applications such as in-wheel traction drives [8] and unmanned aerial vehicle propulsion

[9].

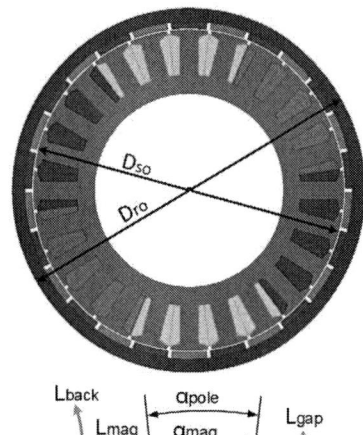

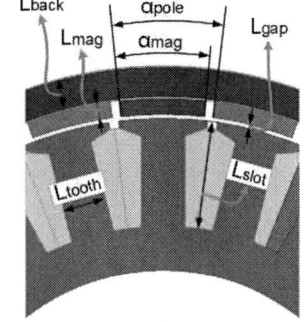

Fig.2 Geometric parameters of the outer-rotor PMSM with
FSCW configuration under study

The paper is organized as follows. Section II provides a literature review of optimization methods in PMSMs. Section III describes the outer-rotor FSCW PMSM configuration and the key design parameters. Section IV introduces the window moving zoom-in method proposed for accurate and effective optimization. Section V provides comprehensive comparisons of the four optimal designs, and shows how well the final choice is benchmarked against commercial options. Section VI offers concluding remarks.

II. REVIEW OF MACHINE OPTIMIZATION

Normally, the design optimization of PMSM is a nonlinear multi-parametric and multi-objective problem. Recent advances in computer hardware and software enable finite element analysis (FEA) to be so routinely used as a machine design and optimization tool that the need of traditional closed-form analytical equations or magnetic equivalent circuit methods becomes trivial. The PMSM design optimization problems can be divided into direct and stochastic search algorithms and computationally efficient surrogate models [10]. The direct search algorithms are most popular design optimization methods for PMSM, especially those concerning magnet shapes [11]–[14]. Moreover, particle swarm opti¬mization and genetic algorithm are very common stochastic search algorithms in PMSM optimization to deal with multi-parametric and multi-objective issues [15]–[18]. On the other hand, computationally efficient surrogate models can drastically reduce the repetitive computational efforts by estimating the correlation between the design input parameters and output characteristics. The response surface methodology (RSM) is one of the most efficient surrogate models for parametric modeling and design optimization [19], [20]. The RSM coefficients can be effectively obtained through regression methods based on relatively few samples of the set of input parameters and their corresponding response values. The sample selection process is normally guided by the design of experiment theory. The RSM based design optimization research has been widely reported for PMSM with various design parameters and objects [21]–[25]. However, design optimization based on FEA can be computationally very intensive and normally involves manual supervision due to the large number of machine parameters involved. Therefore, it would be of great interest to develop an efficient method to optimize the machine parameters with minimum manual involvement. The literature shows a lack of study in this.

Table I: the 1kw-motor parameters of FSCW PMSM after initial sizing analysis

Symbol	Machine Parameter	Values	Unit
-	Phase number	3	-
N_s	Stator tooth number	24	-
N_p	Rotor pole number	20, 22, 26, 28	-
D_{so}	Stator outer diameter	160	mm
D_{ro}	Rotor outer diameter	180	mm
L_{back}	Rotor back iron thickness	5.7	mm
L_{mag}	Magnet thickness	3.5	mm
L_{gap}	Air gap length	0.8	mm
α_p	Magnet pole arc width	0.9	-
-	Permanent magnet material	NdFe35	-
-	Lamination material	35WW270	-
-	Rotor backiron material	Carbon steel 1010	-
T_r	Rated torque	10	N·m
n_r	Rated rotational speed	1000	rpm
V_{dc}	Direct current link voltage	48	V

III. OUTER-ROTOR FSCW PMSM CONFIGURATION

The vehicle has four independently controlled 1kW (4kW total output) traction motors based on the FSCW PMSM type. The geometric parameters of the motor under study are shown in Fig.2. Based on the given performance requirements and space envelope limits, initial sizing analysis has been carried out to establish the scope of the optimization, and four candidate machine versions are chosen that are characterized by having the same rotor outer diameter, rotor back iron thickness, permanent magnet thickness, air gap length, and magnet pole arc width ratio; and under the same rated operational speed, torque, and efficiency. The stator slot number is twenty four while rotor pole number ranges from twenty to twenty eight for four different FSCW configurations. Table I shows the full list of the key motor design parameters. Outer-rotor topology lends itself to a direct drive machine since it provides the maximum torque density among other stator and rotor topologies. FSCW configurations with similar stator slot and rotor pole numbers offer distinct merits of high efficiency, compactness, and good fault tolerance [26]. Thus, the outer-rotor PMSM with FSCW configurations, which have the same stator tooth number of 24 and different rotor pole numbers of 20, 22, 26, and 28 with double-layer concentrated windings, are studied. The rotor outer diameters of all four configurations are limited to 180mm, and the rated torque and speed of the machine are 10Nm and 1000rpm respectively. High strength parallel magnetized rare earth magnets (NdFe35) with 3.5mm thickness are mounted on the inner surface of the back iron to achieve high magnetic load. Solid back iron ring of 5.7mm, made of common carbon steel 1010, is used to simplify rotor structure, ease of manufacture and assembly.

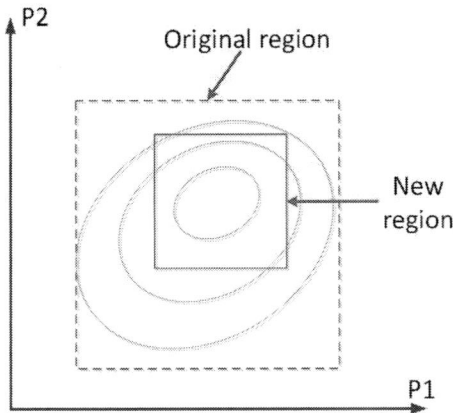

Fg.3 Window moving and zoom-in

Moreover, lamination material 35WW270 of 0.35mm thickness is adopted for stator core to mitigate core loss and achieve reasonable stack factor. The main geometric parameters of the machine are incorporated into the RSM to optimize the electromagnetic design with same electromagnetic loss including copper resistive loss, rotor eddy current loss, and stator core loss under the rated torque output. The objective function is based on the minimum active material weight and the final optimal designs for all four configurations are derived, compared and appraised to determine the best candidate.

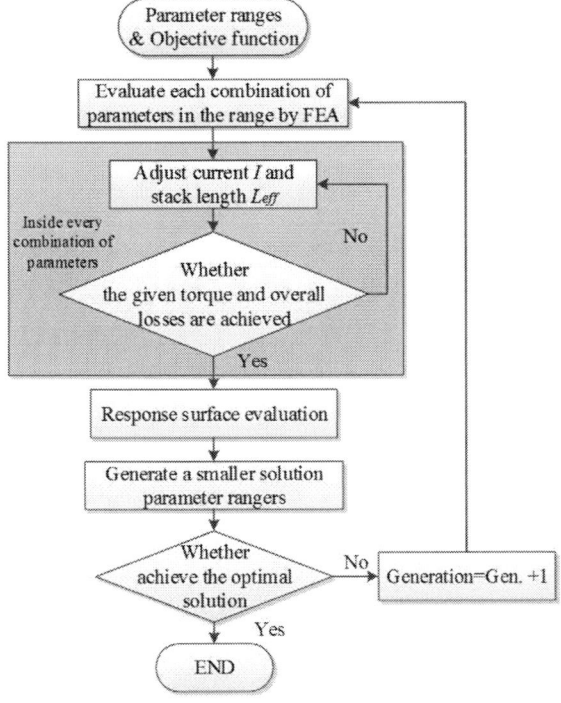

Fig.4 Flow chart of window-zoom-in in RSM.

In order to minimize the design space for optimization, the reference value of stator tooth width can be roughly calculated by analytical method based open-circuit condition as

$$L_{toothref} = \frac{16 B_r D_{ro} L_{mag}}{1.8 N_p \pi (L_{mag} + L_{gap})} \sin(\frac{\alpha_p \pi}{2}) \sin(\frac{N_p \pi}{2 N_s}) \quad (1)$$

where Br is the residual flux density of the permanent magnet, Dro is the rotor outer diameter, Nd and Np are the stator tooth and rotor pole numbers, Lback, Lmag, and Lgap are the rotor back iron thickness, permanent magnet thickness, and air gap length, respectively. αp is the magnet pole arc width ratio, which can be expressed as

$$\alpha_p = \frac{\alpha_{mag}}{\alpha_{pole}} \quad (2)$$

where α_{mag} and α_{pole} are the permanent magnet pole arc width and rotor pole arc width, as shown in Fig. 1, respectively. The actual optimization space of stator tooth width would be confined between 0.7 and 1.6 times the reference value derived from (1). The stator yoke thickness is kept the same as the tooth width in order to further confine the design space. As a result of stator dimension limitation and geometric constraints, the maximum slot depth can be easily derived as

$$L_{slotmax} = \frac{D_{ro}}{2} - \frac{N_s}{2\pi} L_{tooth} \quad (3)$$

The actual optimization range of slot depth is between 0.2 and 0.8 times the maximum one from (3). The overall electromagnetic loss is set as 80W, which can be used to determine the active actual length of the machine.

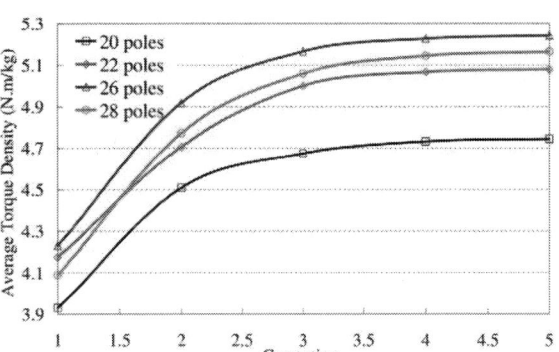

Fig.5 The average torque density of each generation during the optimization process.

IV. OPTIMIZATION OF OUTER-ROTOR PMSM WITH FSCW CONFIGURATIONS

In RSM optimization, a response surface is the relationship establishment between and objective function and design parameters. The combina¬tion of the design parameters to maximize or minimize the objective functions can be achieved by a sensitivity analysis with approximate gradient through a regression analysis, as detailed in [27]. In order to improve the accuracy of the fitted regression model and obtain near-optimal performance, a window-zoom¬-in approach is incorporated with RSM [20]. Basically, it is assumed that

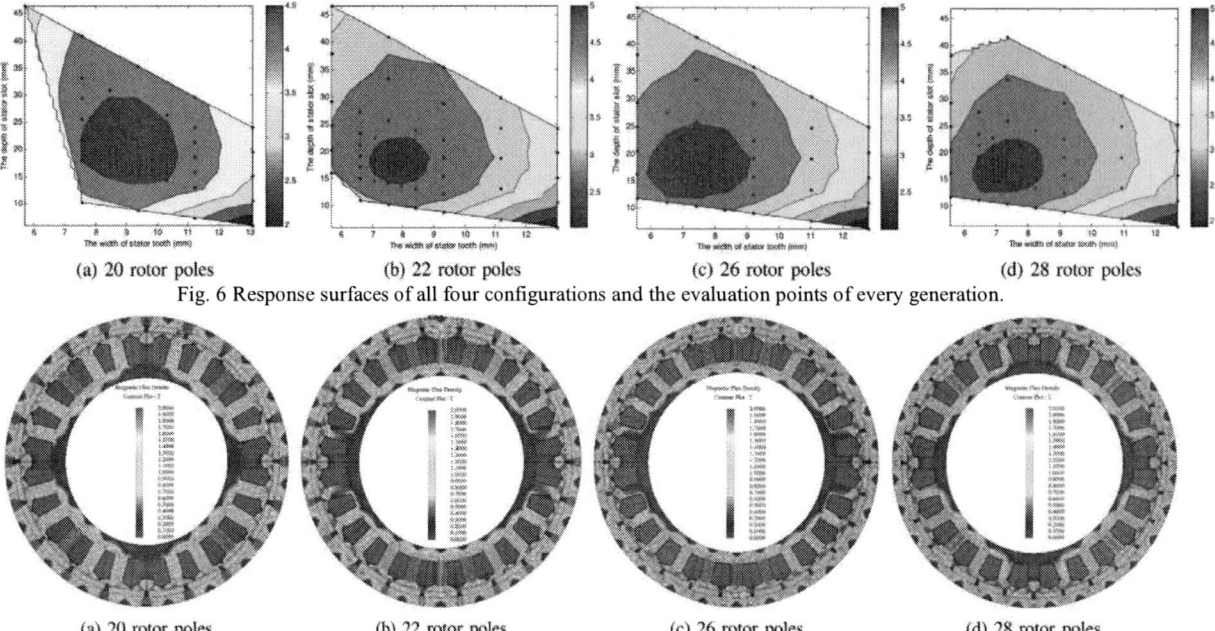

(a) 20 rotor poles   (b) 22 rotor poles   (c) 26 rotor poles   (d) 28 rotor poles

Fig. 6 Response surfaces of all four configurations and the evaluation points of every generation.

(a) 20 rotor poles   (b) 22 rotor poles   (c) 26 rotor poles   (d) 28 rotor poles

Fig.7 Open-circuit flux density distributions of the four optimal designs from 2-D FEA results.

the regions of variables can be scanned by a window. The movement of the window to the optimal point is tightly associated with the response surface. As depicted in Fig.3, new range will therefore be closer to the optimal area where the response surface of the objective function is the highest or lowest. This particular window-zoom-in technique can effectively reduce the region of the factors so that more accurate and near-optimal response value can be located. It is envisaged that the solutions obtained from the initial generations are normally not the real optimal value, but can provide the possible directions towards the optimal value by subsequent trial runs.

Two parameters, stator tooth width and stator slot depth, are optimized for all four configurations by 2-D FEA using window-zoom-in RSM. The detailed optimization process is illustrated as Fig.4. The objective function of the opti¬mization is the torque density of each configuration. During the optimization, the applied phase currents and active axial length are adjusted accordingly with the respective stator tooth width and stator slot depth in order to ensure the same torque output and overall electromagnetic loss. The edge of each window is evenly divided by four intervals with five points so that the window is divided into 16 subsections with 25 junction points. Therefore, there are 25 models required to be evaluated in each window or generation. The parameter ranges of first generation cover all the effective space, while the ones of the following generations gradually shrunk. The optimizations for the four configurations of stator slot and rotor pole combination are carried out respectively. After the first optimization generation, 9 junction points out of the 25 ones in the new window are directly inherited from the previous window so that only 16 new generated junction points are evaluated by 2-D FEA for each new generation. Hence the computational time could be further reduced.

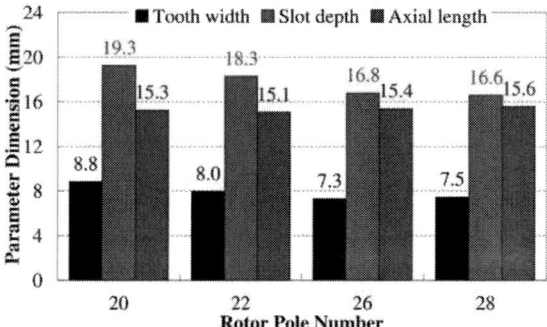

Fig.8 Parameter dimensions of the four optimal designs

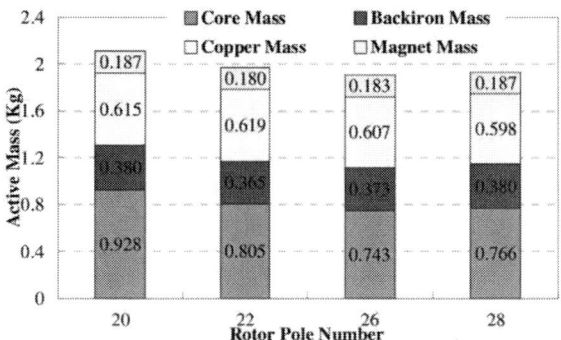

Fig.9 Active material weights of the four optimal designs

After five generations of window-zoom-in optimization, the objective function of torque density does not exhibit any obvious further improvement as shown in Fig.5. Therefore, the design parameters with the highest torque density in the final generation are considered as the optimal ones. Moreover, it can be easily observed that the value of the objective function has been effectively improved through the optimal process for all four configurations. The machine with 26 magnet poles has the highest torque density of 5.25N.m/Kg while the one with 20 magnet poles delivers the lowest one of 4.74N.m/kg.

The response surfaces and evaluation points for all four configurations are compiled and depicted in Fig.6, which demonstrates that the parameter ranges of the generation are gradually shrunk and dwell on the optimal area.

V. COMPARISONS OF THE FOUR OPTIMAL DESIGNS

Based on the results of the optimizations from last section, the optimal designs for all four configurations are obtained accordingly and the corresponding flux density distributions of the machines under open-circuit condition from 2-D FEA results are shown in Fig.7. It can be seen that some moderate saturations occur at very small parts of the back iron and tooth tip areas, especially in the 20- and 22-pole machines. Since double-layer concentrated coils are wound around the stator teeth to achieve high fill factor and flux linage in the proposed four machines, each phase winding would have eight coils connected in series. Moreover, the fact that the direct current (DC) link voltage for the machine drive is 48V would determine the resultant turn number of 13, 13, 12, and 12 in each coil for the four optimal machines with 20, 22, 26, and 28 rotor poles, respectively. The flux density in majority of the stator and rotor is very close to saturation level so that the torque density of the machines is maximized. The stator tooth width, stator slot depth, and active axial length of the optimal machines are obtained and compared in Fig.8, which shows the four optimal machines have almost the same active actual length but the stator slot depth declines as the rotor pole number rises. The core loss density in the stator increases as the electric frequency rises along with the rotor pole number. Hence the machine with high rotor pole number tends to have shallow slot in order to balance the copper joule loss and stator core loss in the stator. Moreover, the resultant active material weights for the four optimal PMSMs are derived and given in Fig.9. Generally, the machine with low magnet pole number would have high stator core weight and winding copper weight due to its deep slot. Based on the premises of achieving the same performance, it is preferred to use less material to improve torque density and reduce the cost. Since the rare earth magnet material is the most expensive component, more attention should be paid to magnet reduction during the design.

The electromagnetic loss components including stator core loss, rotor eddy current loss, and winding copper joule loss, of the final four optimal machines are derived from the 2-D FEA results and compared in Fig.10. Although high magnet pole number results in low stator core weight, the stator core loss generally increases with the rotor magnet pole number due to the rising electrical frequency. The winding copper resistive loss, which is proportional to the product of current density and copper volume, is the main loss component accounting for nearly two thirds. Since the current density is nearly the same in all four machines, the copper resistive loss more or less follows the same trend as the copper weight. Overall, the 28-pole machine has the lowest copper joule loss, while the 22-pole one has highest. The eddy current loss in permanent magnets and rotor back iron is the least component, at about 10% of the overall loss. The 22-pole machine has the lowest rotor eddy current loss, while the 28-pole one has the highest.

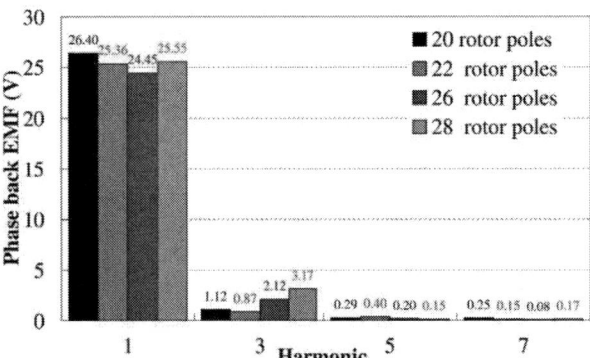
Fig. 12 The main harmonic components of the phase back EMF of the four optimal machines.

Table II
The main parameters of the optimal outer-rotor PMSM with FSCW configurations

Parameters	Rotor pole number			
	20	22	26	28
PM flux linkage (mWb)	24.86	22.40	18.32	17.70
Phase resistance (mΩ)	25.91	24.37	21.12	21.90
d-axis inductance (μH)	303.8	304.3	264.5	264.3
q-axis inductance (μH)	303.8	304.3	264.5	264.3
Rated current (A)	27.26	27.65	28.71	27.48
Rated Efficiency (%)	92.80	92.78	92.93	92.72
Rated Power factor	0.949	0.936	0.924	0.925

Furthermore, torque pulsation, which is one of the most important parasitic effects in direct-drive PMSM, has to be mitigated during the design process. The peak-to-peak (P-P) cogging torque from open-circuit condition and P-P torque ripple from rated load condition are evaluated from 2-D FEA results and compared in Fig. 11, which confirms all the four configurations have very low level of torque pulsation and meet the design requirement. The torque ripple of the machine normally should be minimized during the design stage. Generally, the magnitude of cogging torque will decrease as the least common multiple of the numbers of stator slots and rotor poles rises. Consequently, the machine with 28 poles has the largest cogging torque, followed by the one with 20 poles, while the ones with 26 and 22 poles have the lowest cogging torque. Besides the cogging torque, the torque pulsations generated by the interactions between the harmonics of the back electromotive force (EMF) and excitation current, are another main contributor to the overall torque ripple. The main harmonic components of the phase back EMF in the optimal machines are obtained and demonstrated in Fig. 12, which reveals that all four machines have very small fifth and seventh harmonic components and implies that the associated torque ripple components would be quite trivial. However, the machine with 28 poles has relatively large cogging torque and the machine with 20 poles have relatively high fifth and seventh back EMF harmonic components, the two machines exhibit relatively large torque ripple. And the machines with 22 and 26 poles possess lower torque ripple are due to the small cogging torque and low harmonic contents. The power factor under rated load condition would normally decrease as the rotor pole number increases. Despite the highest speed and efficiency, the machine with 26 rotor poles has the lowest power factor.

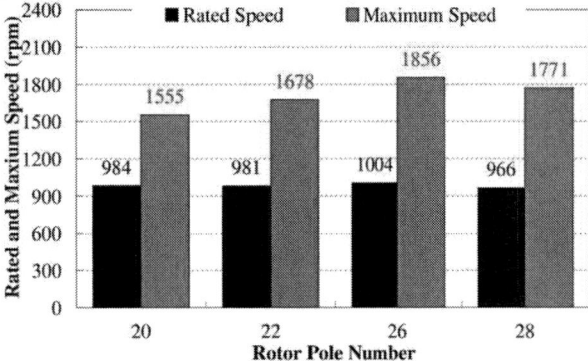

Fig. 13 Maximum rotational speed under rated torque and current loads the four optimal machines

The main parameters of the final four optimal machines are evaluated from the 2-D FEA results and given in Table.II. As the PMs are mounted on the inner surface of the solid back iron, the d- and q-axis inductances in the machines are the same so that no reluctance toque exists. Therefore, the electromagnetic torque in the machines is proportional to the product of PM flux linkage and q-axis current, as well as rotor pole number. The PM flux linkage in the machine is almost reverse-proportional to the rotor pole number, and the rated current keeps more or less constant in order to maintain the same torque output. The FSCW configuration results in short end winding and hence relatively small resistance. The phase resistance in the machine would generally decline as the rotor pole number rises. However, the machine with 26 rotor poles possesses the smallest phase resistance. Moreover, the inductances follow the same trend as the resistance. The maximum speeds of the four machines under same rated torque output are obtained and shown together with the corresponding maximum speeds under the rated current excitations in Fig. 13, which shows the achievable speeds with the rated torque, are all very close to the targeted speed of 1000rpm. However, small deviations occur as the turn number of the coil has to be an integer. With flux weakening operation, the maximum speeds under the rated current excitation can be further extended and generally the extension rate increases along with the rotor pole number. This can be explained by the corresponding ratio of PM flux linkage and d-axis inductance. Overall, the machine with 26 rotor poles owns the highest rated speed and maximum speed and the flux weakening capabilities of the all four machines are quite limited. As a result of the minor differences on the rated speed, the rated efficiency slightly deviates.

Last but not the least, the torque and speed characteristics of all four optimal machines are derived from 2-D FEA results under the conditions of given DC link voltage and corresponding rated currents. Furthermore, comprehensive 2-D FEA models are carried out to evaluate the performance of the machine over the whole operational range. The efficiency maps of the four optimal machines are computed as shown in Fig.14, which shows that the high efficiency regions of the four machines all dwell on the speed of near 1000rpm and torque of around 6Nm. The highest efficiencies are 93.917%, 93.783%, 93.843%, and

93.885%, in the machines with 20, 22, 26, and 28 poles, respectively. Moreover, the efficiencies of any of the four machines are higher than 80% even at start up speed. This means the overall efficiency for any drive cycle of the vehicle will be very high, which is particularly important in reducing battery weight.

Considering collectively the key performance indicators of torque density, efficiency, torque quality and power factor, the 26-pole machine is found to be the most optimal among the four versions in meeting the requirements for the quadricycle's mission. Finally, the 26-pole machine is compared against some commercially available 1kW traction motors. These motors, shown in Fig.15, are chosen based on their relevancy in terms of power output, speed and voltage. For fair comparison, the proposed motor's active mass of 1.91kg is scaled up by 30% to 2.48kg to reflect a realistic packaged motor as shown in Fig.15 (d). Table III shows the performance comparison between the motors. It is evident that the proposed motor's torque density of 4.03 Nm/kg is much higher than the rest of field, and it also tops the efficiency at 93.8%.

(a)UU Motor　　　　(b)Future Energy

(c)Compact Power Motion GmbH　(d)Cranfield's proposed motor

Fig.15 Selected 1kW commercial motors and proposed motor

Table III:Performance comparison with commercial 1kw traction motors

	Torque (Nm)	Weight (kg)	Torque density (Nm/kg)	Efficiency
UU Motor	17Nm@600rpm @48V	10	1.7	>80%
FutureEnergy	20Nm@500rpm @48V	7	2.8	?
Compact Power Motion GmbH	7Nm@3500rpm @24V	3.3	2.12	90%
Proposed 26-pole Motor	10Nm@1000rpm @48V	(1.91)* 2.48	4.03	93.8%

VI. CONCLUSION

The design optimizations of a 1kW outer-rotor FSCW PMSM are comprehensively carried out by the proposed window-zoom-in RSM based on 2-D FEA. From the

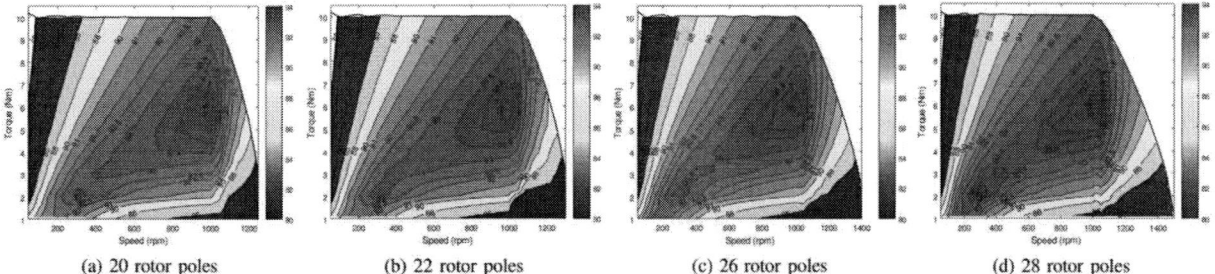

| (a) 20 rotor poles | (b) 22 rotor poles | (c) 26 rotor poles | (d) 28 rotor poles |

Fig. 14 Efficiency maps of the four optimal designs from 2-D FEA results

specific requirements due to the vehicle's specifications, four optimal PMSM versions with 20, 22, 26, and 28 rotor poles are pre-determined by sizing analysis for further vigorous performance optimization and evaluation. Results show that the stator slot depth and active axial length which have critical influence on loss distributions inside the machine, and hence the subsequent machine performance. Comprehensive investigations on the characteristics of the four optimal machines, including torque ripple, PM flux linkage, phase back EMF, inductance, phase resistance, speed, power factor, and efficiency map, are performed based on 2-D FEA results. Detailed comparisons and discussions are given to reveal the impacts of rotor pole number on the machine performance. The optimized 26-pole machine is selected to be the most suitable and is compared with commercially available ones, and its distinct advantages of high torque density and efficiency are evident. The proposed motor thus meets and exceeds the requirements of the vehicle's mission. Future work will involve the manufacture and experimental validation of the proposed motor.

ACKNOWLEDGMENT

This work is supported by EPRSC, U.K., under Grant Ref.EP/I038543/1.

REFERENCES

[1] http://ec.europa.eu/transport/themes/urban/urban_mobility [accessed Oct 2015]

[2] https://www.heacademy.ac.uk [accessed Oct 2015]

[3] I. A. Smadi, H. Omori, and Y. Fujimoto, "Development, analysis, and experimental realization of a direct-drive helical motor," IEEE Trans. Ind. Electron., vol. 59, no. 5, pp. 2208–2216, May 2012.

[4] G. Patterson, T. Koseki, Y. Aoyama, and K. Sako, "Simple modeling and prototype experiments for a new high-thrust low-speed permanent-magnet disk motor," IEEE Trans. Ind. Appl., vol. 47, no. 1, pp. 65–7 1, Jan. 2011.

[5] S Sakai, H Sado, Y Hori, "Motion control in an electric vehicle with four independently driven in-wheel motors," IEEE Tran Mechatronics, vol. 4, no. 1, pp. 9–16, Mar. 199

[6] W.Fei, P.C.K.Luk, "Torque ripple reduction of a direct-drive permanent-magnet synchronous machine by material-efficient axial pole pairing," IEEE Trans. Ind. Electron., vol. 59, no. 6, pp. 2601 - 2611, Jun. 2012

[7] J. Nerg, M. Rilla, V. Ruuskanen, J. Pyrhonen, and S. Ruotsalainen, "Direct-driven interior magnet permanent-magnet synchronous motors for a full electric sports car,"

IEEE Trans. Ind. Electron., vol. 61, no. 8, pp. 4286–4294, Aug. 2014.

[8] J. Cros and P. Viarouge, "Synthesis of high performance pm motors with concentrated windings," IEEE Trans. Energy Convers., vol. 17, no. 2, pp. 248–253, Jun. 2002.

[9] D. Wu, W. Fei, P. C. K. Luk, and B. Xia, "Design considerations of outer-rotor permanent magnet synchronous machines for in-wheel electric drivetrain using particle swarm optimization," in Proc. IET Int. Conf. Power Electron. Machines and Drives, Apr. 2014, pp. 1–6.

[10] B. C. Mecrow, J. W. Bennett, A. G. Jack, D. J. Atkinson, and A. J. Freeman, "Drive topologies for solar-powered aircraft," IEEE Trans. Ind. Electron., vol. 57, no. 1, pp. 457–464, Jan. 2010.

[11] Y. Duan and D. M. Ionel, "A review of recent developments in elec¬trical machine design optimization methods with a permanent-magnet synchronous motor benchmark study," IEEE Trans. Ind. Appl., vol. 49, no. 3, pp. 1268–1275, May 2013.

[12] [11] F. Scuiller, "Magnet shape optimization to reduce pulsating torque for a five-phase permanent-magnet low-speed machine," IEEE Trans. Magn., vol. 50, no. 4, pp. 1–9, Apr. 2014.

[13] K. I. Laskaris and A. G. Kladas, "Permanent-magnet shape optimization effects on synchronous motor performance," IEEE Trans. Ind. Electron., vol. 58, no. 9, pp. 3776–3783, Sep. 2011.

[14] H. Hong and J. Yoo, "Shape design of the surface mounted permanent magnet in a synchronous machine," IEEE Trans. Magn., vol. 47, no. 8, pp. 2109–2117, Aug. 2011.

[15] M. Ashabani and Y. A.-R. I. Mohamed, "Multiobjective shape optimiza¬tion of segmented pole permanent-magnet synchronous machines with improved torque characteristics," IEEE Trans. Magn., vol. 47, no. 4, pp. 795–804, Apr. 2011.

[16] R. Wrobel and P. H. Mellor, "Particle swarm optimisation for the design of brushless permanent magnet machines," in Proc. IEEE Ind. Appl. Soc. Annu. Meeting, vol. 4, Oct. 2006, pp. 1891–1897.

[17] Y. Duan, R. G. Harley, and T. G. Habetler, "Multi-objective design optimization of surface mount permanent magnet machine with particle swarm intelligence," in Proc. IEEE Swarm Intell. Symp., Sep. 2008, pp. 1–5.

[18] "Method for multi-objective optimized designs of surface mount permanent magnet motors with concentrated or distributed stator wind¬ings," in Proc. IEEE Int. Electr. Mach. Drives Conf., May 2009, pp. 323–328.

[19] "A useful multi-objective optimization design method for pm motors considering nonlinear material properties," in

Proc. IEEE Energy Convers. Congr. Expo., Sep. 2009, pp. 187–193.

[20] X. K. Gao, T. S. Low, S. X. Chen, and Z. J. Liu, "Structural robust design for torque optimization of bldc spindle motor using response surface methodology," IEEE Trans. Magn., vol. 37, no. 4, pp. 2814– 2817, Jul 2001.

[21] X. K. Gao, T. S. Low, Z. J. Liu, and S. X. Chen, "Robust design for torque optimization using response surface methodology," IEEE Trans. Magn., vol. 38, no. 2, pp. 1141–1144, Mar. 2002.

[22] D.-K. Hong, B.-C. Woo, D.-H. Koo, and D.-H. Kang, "Optimum design of transverse flux linear motor for weight reduction and improvement thrust force using response surface methodology," IEEE Trans. Magn., vol. 44, no. 11, pp. 4317–4320, Nov. 2008.

[23] L. Fang, J.-W. Jung, J.-P. Hong, and J.-H. Lee, "Study on high-efficiency performance in interior permanent-magnet synchronous motor with double-layer pm design," IEEE Trans. Magn., vol. 44, no. 11, pp. 4393–4396, Nov. 2008.

[24] L. Jolly, M. A. Jabbar, and Q. Liu, "Optimization of the constant power speed range of a saturated permanent-magnet synchronous motor," IEEE Trans. Ind. Appl., vol. 42, no. 4, pp. 1024–1030, Jul. 2006.

[25] S. Vivier, F. Gillon, and P. Brochet, "Optimization techniques derived from experimental design method and their application to the design of a brushless direct current motor," IEEE Trans. Magn., vol. 37, no. 5, pp. 3622–3626, Sep. 2001.

[26] J.-M. Park, S.-I. Kim, J.-P. Hong, and J.-H. Lee, "Rotor design on torque ripple reduction for a synchronous reluctance motor with concentrated winding using response surface methodology," IEEE Trans. Magn., vol. 42, no. 10, pp. 3479–3481, Oct. 2006.

[27] C. C. Hwang, S. P. Cheng, and C. M. Chang, "Design of high-performance spindle motors with concentrated windings," IEEE Trans. Magn., vol. 41, no. 2, pp. 971–973, Feb. 2005.

[28] H.-H. Kim, D.-K. Kim, Y.-J. Lee, and B.-T. Kim, "Efficiency optimiza¬tion design of a bldc motor driving fans using a design of experiment method," in Proc. Int. Electr. Mach. Syst. Conf., Oct. 2010, pp. 1147– 115.

Analysis and Design of V-Spoke Ferrite Interior Permanent Magnet Machine for Traction Applications

Bing Xia, Weizhong Fei, Patrick Luk

Electric Power and Drive Group, Power Engineering Centre, Cranfield University, Cranfield, MK43 0AL, U.K.
Email: p.c.k.luk@cranfield.ac.uk

Abstract–Due to the low residual flux density of ferrite magnets, flux-focusing and multi-layer configurations have been deployed to harness both the permanent magnet torque and reluctance torque. Here, a new configuration featuring a V-spoke rotor structure embracing a combined configuration of the conventional V- and spoke-shaped interior permanent magnet (IPM) machines is proposed for its superior torque density over equivalent counterparts. The magnetic open-circuit characteristic of the V-spoke machine is first analysed and compared with that of the equivalent V- and spoke-shaped machines. Then finite element analysis (FEA) simulations are performed for on-load conditions to confirm the superior performance of the proposed design. The frozen permeability method is carried out to provide further insights into the torque production mechanism. Operating efficiency, torque profiles and power-speed performance are also be investigated and evaluated to confirm the utility of the proposed ferrite machine design.

Keywords–Ferrite, V-spoke, interior permanent magnet, traction motor.

I. INTRODUCTION

With ever increasing concerns over carbon emission and global warming, the electrification of road transportation will continue to stay at the top of the agenda for environmental policies worldwide. The electric propulsion system, which is at the heart of the electrification of transport, has been receiving increasing attention. The permanent magnet synchronous machine (PMSM) is one of the most promising solutions for traction applications, with the remarkable features of high torque density, excellent controllability and good efficiency [1]. Most high performance PMSMs have been using high energy rare-earth permanent magnets (PMs) to deliver high power density and efficiency. However, the high price and uncertain supply chain of the rare-earth materials have been a major concern for researchers and the car industry. As a result, there have been a surge of interests in seeking alternative solutions to rare-earth PM based high performance electric machines. As a non-strategic material of low price and abundant supply worldwide, ferrite PM has been considered a promising potential alternative to replace rare-earth PM for traction applications [2,3]. Compared to rare-earth PM, ferrite has PM a very low residual flux density, which will reduce the power density of ferrite-based machines. To achieve comparable torque ability with rare-earth PM machines, new machine configurations need to be explored.

The total torque output in an IPM machine consists of two components: PM torque and reluctance torque. There are two ways to overcome the challenges that low flux ferrite PM motors face. On the one hand, larger amount of ferrite materials can be placed in the rotor to boost flux. On the

other hand, reluctance torque can be enhanced by certain rotor geometries that result in increasing the difference between d- and q-axis inductances. By exploiting reluctance torque, not only the torque density can be increased, but also the operating speed range can be extended. Thus, regardless of the PM materials to be used, one can conclude IPM topologies are inherently suitable for traction applications [4,5].

V-shaped configurations are commonly used in industry and widely reported in the literature [6-8]. As an improved design of radially magnetized IPM, V-shaped IPMs have excellent torque capabilities and wide speed range with good rotor saliency ratio [8,9]. But most of the existing work with V-shaped designs harnesses rare-earth PMs, and few has discussed the possibility of using ferrite V-shaped structures. As for spoke-type configurations, they are eminently appropriate for ferrite PM machines and have been investigated by many researchers [4,10], since high airgap flux density can be achieved with flux squeeze technique. However, there are some shortcomings of spoke-type IPMs, such as limited flux-weakening abilities and large torque ripples [11,12]. Based on the investigations on V-shaped and spoke-shaped machines, a novel V-spoke rotor configuration is proposed in this paper so as to combine the advantages of both V- and spoke-shaped rotor structures for traction applications.

The paper is organized as follows. Section II discusses and analyses the models of the V- and spoke types. r structures, a novel configuration of V-spoke shaped IPM structure is proposed by combining both the advantages of spoke-type and V-shaped machines. To further understand the influence of rotor geometry on torque characteristics of the three promising ferrite IPM configurations, comprehensive comparisons of the performance of the V-shaped, spoke-shaped, and proposed V-spoke machines are carried out under various working conditions, and results indicate that the novel rotor configuration can deliver sufficient torque and power for traction application with a rare-earth-free solution.

Table I. Properties common dimensions and specifications

Parameters	Value	Unit
Stator Outer Diameter (D_{so})	264	mm
Rotor Outer Diameter (D_{so})	160.4	mm
Airgap Length (g)	0.75	mm
Stack Length (l)	50	mm
Number of Poles (p)	8	
Number of Slots (q)	48	
DC Link Voltage	300	V
Maximum Current (I_{pk})	250	A_{pk}
Amount of Ferrite (V_m)	280	cm^3
Base Rotating Speed	3000	rpm
Lamination Material	50CS470	
Ferrite Material	Y40	

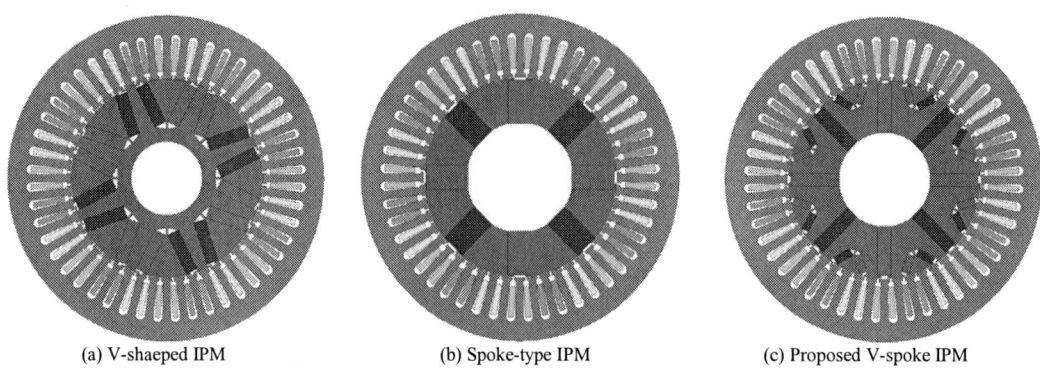

(a) V-shaeped IPM   (b) Spoke-type IPM   (c) Proposed V-spoke IPM

Fig. 1. Geometry of different IPM models under study.

II. MACHINE MODELS AND SPECIFICATIONS

Fig. 1 demonstrates the two rotor configurations chosen as potential solution for high performance ferrite motor designs as well as the proposed new rotor structure V-spoke structure.

In order to concentrate on the influence of different rotor structure on torque production, all the designs under study here are with the same stator of 48 slots, short-pitched single-layer distributed windings, airgap length. And without the loss of generality, the pole numbers, total amount of ferrite material and driving conditions are kept the same. Some key parameters of the designs are shown in Table 1. The shape and locations of different models are optimized at rated current conditions individually to achieve the highest torque capability.

Since torque characteristics are the major criteria to evaluate the performance of a traction motor, the main objective to improve the design for ferrite IPMs to delivered higher torque density and wider constant power speed range.

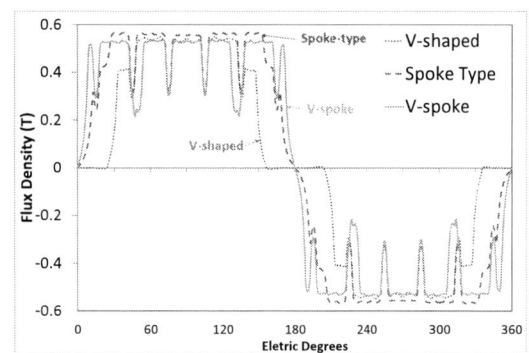

Fig. 2. Comparison of open-circuit airgap flux density.

For PMSMs, the total torque production can be derived by

$$
\begin{aligned}
T_{em} &= \frac{m}{2} p \left[\Psi_m i_q + (L_d - L_q) i_d i_q \right] \\
&= \frac{mp}{2} \Psi_m i_q + \frac{mp}{2} (L_d - L_q) i_d i_q \\
&= T_{pm} + T_r
\end{aligned}
\tag{1}
$$

where m the is number of phases, p is the number of poles, Ψm is the armature flux linkage by PM, i_d, i_q are d, q-axis current,

L_d, L_q are d, q-axis inductance, and T_pm, T_r represent PM torque and reluctance torque respectively. Equation (1) reveals the two torque components in PMSMs. The first part indicates PM torque, which results from the interaction of PM and armature fields; the second represents the reluctance torque owing to the rotor saliency.

Though it is easy to distinguish different torque components from (1), in actual to accomplish the torque decomposition is extremely difficult with analytical methods. Because Ψ_m, L_d and L_q rely heavily on armature excitation in a non-linear system due to the saturation under load conditions, the torque components are highly coupled together. However, with the frozen permeability technique, it is much easier to separate the torque components individually using finite element analysis (FEA) methods. By doing this, this paper provides a revealing insight into the generation of PM and reluctance torque and the impact on total electromagnetic torque.

III. OPEN-CIRCUIT CHARACTERISTICS

Although the loading has great influences on the performance of these three machines, preliminary prediction can be made based on the open-circuit results. Thus, no-load simulations are first carried out to derive the initial results.

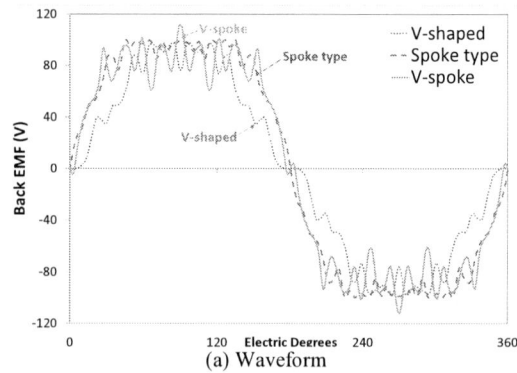

(a) Waveform

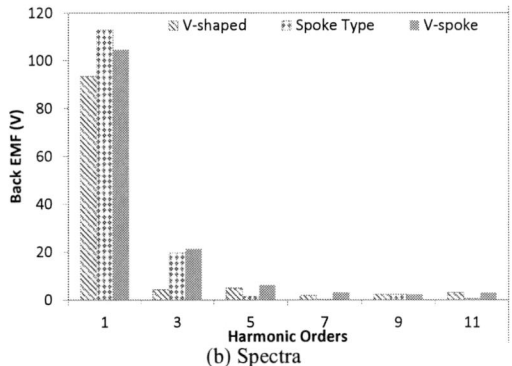

(b) Spectra

Fig. 3. Comparison of back EMF.

A. Airgap flux density and back EMF

The magnitude of airgap flux density and back EMF imply the PM torque, and the harmonics in them would contribute the torque ripple. The corresponding airgap flux density waveforms under open-circuit conditions are depicted in Fig. 2. Though the total amount of ferrite PM is the same for the three configurations, spoke-type motor shows highest flux density. That is because the spoke-type structure has the advantage of strong flux-focusing ability. As a result, the magnitude of the fundamental back EMF of spoke type is 21% higher than that of V-shaped one, and 12% higher than V-spoke type, as illustrated in Fig. 3. That means potentially spoke-type is able to achieve highest PM torque. But from harmonic point of view, the V-shaped design has most sinusoidal waveform with the lowest harmonic content, while spoke type has a noticeable third harmonic content. Since the proposed V-spoke type applied the spoke structure, it inherits the merit of high flux density and defect of high harmonics, with similar harmonics as spoke type machine. By using star-connection, the third harmonic can be suppressed, but the third order flux linkage may cause extra iron loss during operation.

B. Cogging torque

The comparison of cogging torque is demonstrated in Fig. 4. The peak-to-peak (P-P) cogging torque of V-shaped is the highest among all, which is 1.35Nm. The spoke type is slight smaller with a cogging torque of 0.9Nm. As for V-spoke machine, the PM poles are divided into spoke poles and v-shaped poles, the cogging torque can be reduced. Thus, the *V-spoke* configuration exhibits the lowest P-P value of 0.48 Nm. However, the cogging torque of all three designs is well under 2% of the rated torque.

C. No-load inductance

The inductance of armature windings indicates the motor saliency for PMSMs. Fig. 5(a) depicted the phase self and mutual inductances of the machines at open-circuit condition. The spoke type machine has the lowest phase self and mutual inductance, while the V-shaped one has highest absolute average values. According to (1), the reluctance torque is dependent on the difference between L_d and L_q. To have a straightforward view of the impact of these difference, L_d and L_q are calculated by Park Transformation based on self and mutual inductances, and the results are shown in Fig. 5(b). As expected, the V-shaped motor has the largest L_q and relatively small L_d, and achieves a highest dq-axis difference. On the other hand, the spoke type exhibits lowest L_d, L_q and difference, which implied a poor reluctance torque capability. The V-spoke has a slightly lower d- and q-axis inductance difference compared with V-shaped, but still much higher than the spoke type one.

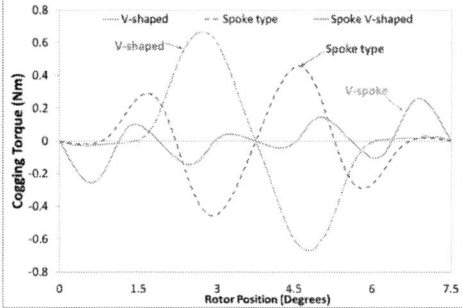

Fig. 4. Comparison of cogging torque.

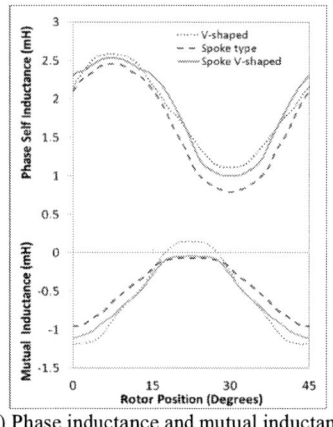

(a) Phase inductance and mutual inductance

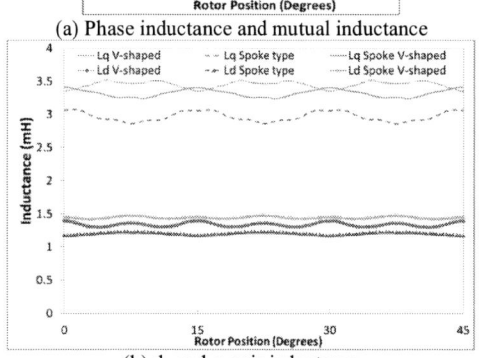

(b) d- and q-axis inductance

Fig. 5. Comparison of inductance.

IV. ON-LOAD TORQUE CHARACTERISTICS

Under loaded conditions, especially with heavy loading, the induced voltage, inductance even cogging torque could change dramatically due to the armature reaction and saturation, the three machine configurations are compared with various loadings in this section.

A. Overall average torque output

The overall average electromagnetic torque characteristics of the machines under different current and advantaged angle driving conditions are illustrated in Fig. 6. From the maximum torque capability point of view, the combination of spoke and V structure has a great improvement in the torque production, and the proposed V-spoke model is able to deliver the highest torque of 160 Nm, which is

978-1-5090-0064-7/15 $31.00 © 2015 IEEE

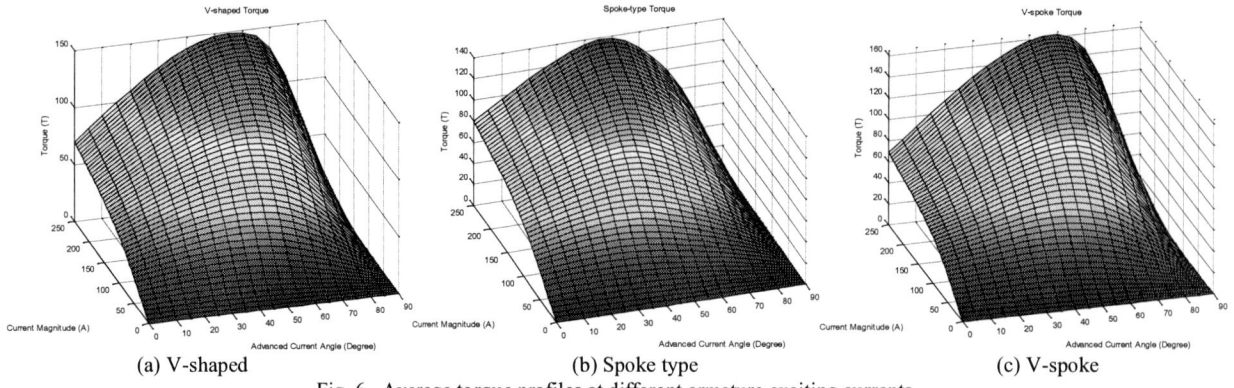

(a) V-shaped   (b) Spoke type   (c) V-spoke

Fig. 6. Average torque profiles at different armature exciting currents.

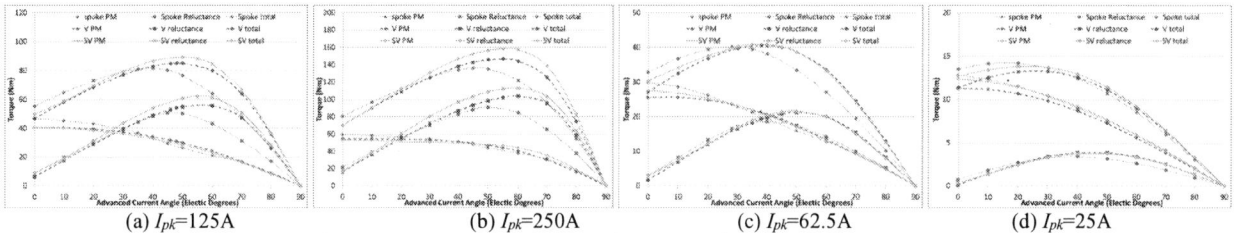

(a) I_{pk}=125A   (b) I_{pk}=250A   (c) I_{pk}=62.5A   (d) I_{pk}=25A

Fig. 7. Torque components comparison among the three configurations.

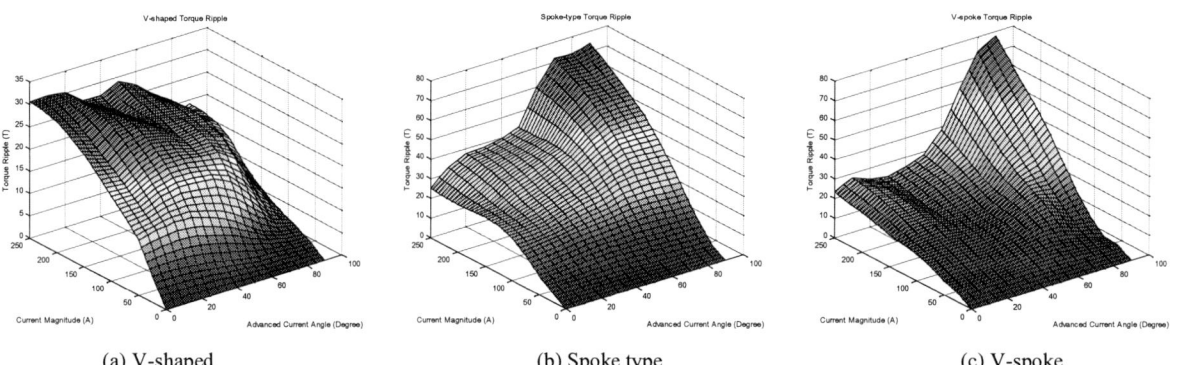

(a) V-shaped   (b) Spoke type   (c) V-spoke

Fig. 8. Comparison of P-P torque ripple.

about 16% higher than spoke type and 9% than V-shaped one. For spoke type configuration, maximum torque is achieved at a lower advanced current angle than the others, which indicates poorer reluctance torque ability compared with the other two. However, lower current angle also means a larger portion of PM torque and higher power factor and efficiency.

According to equation (1), the PM torque is proportional to current, and reluctance torque is proportional to current square. So with the increase of armature current, the reluctance torque grows quicker than PM torque. Consequently, the performance could be different under different loading conditions.

B. Torque segregation

To further investigate the torque production of the three models, PM torque and reluctance torque are segregated using frozen permeability techniques. The torque components of the machines under four typical loading conditions are derived and demonstrated in Fig. 7.

At the rated current excitation as shown in Fig. 7(a), reluctance torque contributes larger part for the overall torque output. Though the spoke type machine shows better PM torque ability, the reluctance component becomes much smaller as the current angle is over 45 electric degrees. On the other hand, with the better salient ratio, the V-spoke configuration exhibits the highest reluctance torque and good PM torque. As a result, the best overall torque capability can be achieved, as depicted in Fig 7(b). With the increase of armature current, the increase of reluctance torque is more significant than PM component. Under overload operation with double rated current, reluctance torque contributes major part of the overall torque output, and V-spoke shows even better torque performance due to its high rotor saliency even at highly saturated situation.

When it comes to low load conditions, the results are a little different. With half of the rated driving current, the PM torques account 50% of total torques. And thus the maximum torque is almost the same for the three machines. Under very light low conditions, the torque output is basically from PM torque to achieve maximum torque per ampere operation, as illustrated in Fig 7(c). As a result, spoke type exhibits the highest torque of the three.

978-1-5090-0064-7/15 $31.00 © 2015 IEEE

C. Torque ripple

Torque P-P torque ripple under various loading conditions are depicted in Fig. 6. V-shaped structures exhibits lowest overall torque ripple, and the P-P ripple decreases with current angle. On the contrary, the ripple for spoke type is much larger than V-shaped, and increases dramatically with the current angle. As for V-spoke machine, the ripple keeps at a low value when the advanced current angle is below 55 degrees, which is similarly to V-shaped one. But the torque ripple can increase dramatically at large armature advanced current angle.

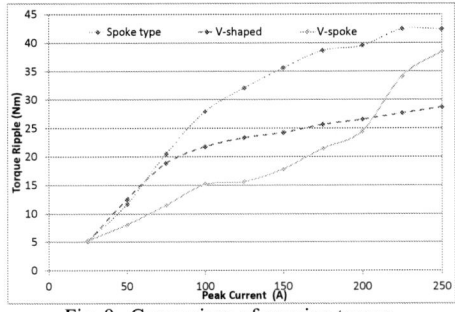

Fig. 9. Comparison of cogging torque.

It is noteworthy that the largest optimal current angle to achieve the maximum torque is 55 electric degrees, and thus higher advanced current angle situation will be avoided by control algorithm to accomplish low torque-ripple operation. Fig. 9 depicts the torque ripple at optimal current phase angle. V-spoke has the lowest torque ripple within rated operation range, though the torque ripple will increase dramatically under heavy overload condition. As for spoke type configurations, the large torque ripple over the whole operation range is a disadvantage for high performance applications.

D. Efficiency map

The efficiency maps for the machines under maximum power control algorithm are demonstrated in Fig. 10. V-shaped machine can achieve constant torque at 3200 rpm due to the low back EMF, while the torque begins to drop at 2500 rpm for spoke type, which means the maximum power is much lower than the other two. However, spoke type has a higher overall efficiency, with 11.7% and 52.4% of the operation range over the efficiency of 95% and 90%, respectively. V-shaped machine has the lowest efficiency with only 8.2% of the operation range exceeding the efficiency of 95%. As for the V-spoke design, 11.4% of the operation range can attain the efficiency over 95%, which is a little lower than spoke type. But the V-spoke has a larger operational range with over 90% efficiency. It is noteworthy that under low and light load conditions, the efficiency is higher for spoke type machine due to its higher PM torque component and lower. But under medium and higher loadings, higher efficiency can be achieved for V-shaped and V-spoke type machines. For EV applications, the motor operates at medium to rated loading conditions. Thus, the V-spoke machine would be more preferred for the driving cycle point of view.

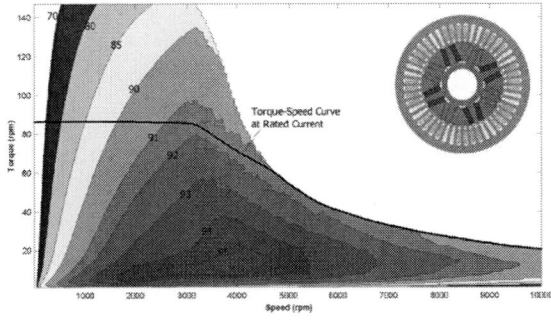

(a) V-shaped

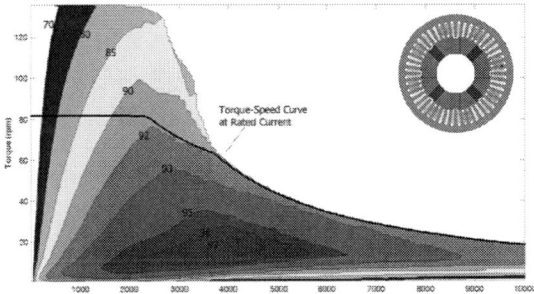

(b) Spoke type

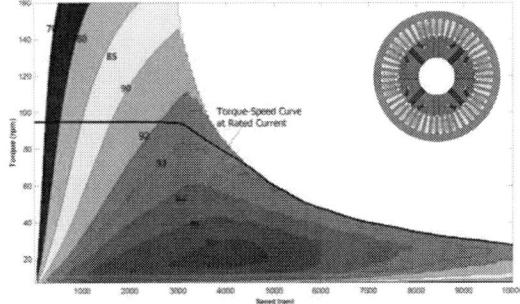

(c) V-spoke

Fig. 10. Comparison of efficiency maps.

E. Constant-power speed range

Flux-weakening capability is an important indicator for traction motors. For a PMSM machine, the magnetic field excited by the rotor is fixed. Flux-weakening control need be applied to extend the operating speed range.

The power-speed envelope of the machines are derived and shown in Fig. 11. Due to low torque ability and high back EMF, spoke type can barely reach the rated power requirement of 25 kW under the inverter DC link voltage of 300V. And then the power output drops with speed. V-shaped design has quite a standard power-speed feature. The output power is almost constant after rated power obtained, and rated power can still be achieved at 7200 rpm. As for V-spoke machine, the power output increases over base speed and achieves 30 kW, then begins to drop slowly after 4500 rpm. When the speed exceeds 7200 rpm, the power V-spoke machine can deliver is below the rated. Though the maximum speed achieved for the two V-shaped0 machines under rated current condition is the same, V-spoke type has higher torque density and overload capability.

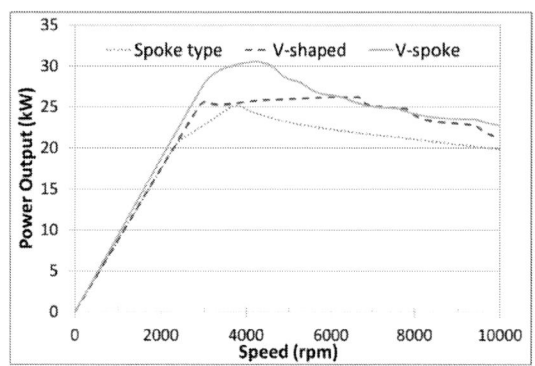

Fig. 11. Comparison of power-speed curve.

Based on the investigations of performances in this section, it can be concluded that compared to the other two designs, the proposed V-spoke machine has higher torque density, relatively low torque ripple, good efficiency and better overload capability, and is a better solution for high performance ferrite PM traction machines.

V. CONCLUSION

To develop a non-rare-earth solution of high performance PMSM for traction applications, two potential candidates are chosen from existing literatures, namely V-shaped and spoke-type machines. Based on these two designs, a novel V-spoke configuration is proposed in this paper to combine the advantage of the two structures and achieve a higher torque density. The electromagnetic performances of the three machines are investigated and compared thoroughly. Frozen permeability technique is applied with 2D FEA to decompose the torque components and reveal the difference in torque production for these three designs. Comprehensive comparisons of torque ripple, operating efficiency and constant-power speed range are also carried out. Results indicate that the proposed V-spoke configuration can deliver higher torque with low torque ripple, excellent efficiency and wide operation range, and is among the most potential alternatives for non-rare-earth EV applications.

ACKNOWLEDGMENT

This work is supported by EPRSC, U.K., under Grant Ref.EP/I038543/1.

REFERENCES

[1]   C.C. Chan, "An overview of electric vehicle technology," Proc. IEEE, vol. 81, no. 9, pp. 1202–13, Sep. 1993.

[2]   S.I. Kim, J. Cho, S. Park, T. Park, and S. Lim, "Characteristics Comparison of a Conventional and Modified Spoke-Type Ferrite Magnet Motor for Traction Drives of Low-Speed Electric Vehicles," IEEE Trans. Ind. Appl., vol. 49, no. 6, pp. 2516–2523, Nov./Dec. 2013.

[3]   I. Petrov and J. Pyrhonen, "Performance of Low-Cost Permanent Magnet Material in PM Synchronous Machines,"

IEEE Trans. Ind. Electron., vol. 60, no. 6, pp. 2131–2138, Jun. 2013.

[4]   K.T. Chau, C.C. Chan, and C. Liu, "Overview of Permanent-Magnet Brushless Drives for Electric and Hybrid Electric Vehicles," IEEE Trans. Ind. Electron, vol. 55, no. 6, pp. 2246–2257, Jun. 2008.

[5]   Z.Q. Zhu and D. Howe, "Electrical Machines and Drives for Electric, Hybrid, and Fuel Cell Vehicles," Proc. IEEE, vol. 95, no. 4, pp. 746–765, Apr. 2007.

[6]   L. Chong, R. Dutta, M.F. Rahman, H. Lovatt, "Experimental verification of core and magnet losses in a concentrated wound IPM machine with V-shaped magnets used in field weakening applications," Proc. IEEE Electric Machines & Drives Conf., pp.977-982, 15-18 May 2011

[7]   S.-I. Kim, J. Cho, S. Park, T. Park, and S. Lim, "Characteristics Comparison of a Conventional and Modified Spoke-Type Ferrite Magnet Motor for Traction Drives of Low-Speed Electric Vehicles," IEEE Tran. Ind. Appl., vol. 49, pp. 2516–2523, 2013.

[8]   A. Wang, Y. Jia, and W.L. Soong, "Comparison of Five Topologies for an Interior Permanent-Magnet Machine for a Hybrid Electric Vehicle," IEEE Trans. Magn., vol. 47, no. , pp. 3606–3609, Oct. 2011.

[9]   S. Wu, L. Tian, and S. Cui, "A comparative study of the interior permanent magnet electrical machine's rotor configurations for a single shaft hybrid electric bus," in Proc. IEEE Vehicle Power and Propulsion Conf., pp. 1-4, Sep. 2008.

[10]   Lee, G.-H. Kang, J. Hur, and D.-W. You, "Design of spoke type BLDC motors with high power density for traction applications," in Rec. IEEE Ind. Appl. Conf., vol. 2, pp. 1068–1074, Oct. 2004.

[11]   E.E. Montalvo-Ortiz, S.N. Foster, J.G. Cintron-Rivera, and E.G. Strangas, "Comparison between a spoke-type PMSM and a PMASynRM using ferrite magnets," Proc. IEEE Electric Machines & Drives Conf., pp. 1080–1087, 2013.

[12]   B. Lee, G.-H. Kang, J. Hur, and D.-W. You, "Design of spoke type BLDC motors with high power density for traction applications," in Rec. IEEE Ind. Appl. Conf., vol. 2, pp. 1068–1074, 2004.

The Design and Application of an Unmanned Surface Vehicle Powered by Solar and Wind Energy

Zhou X.Q.   Ling L.L.   Ma J.M.   Tian H.L.   Yan Q.S.   Bai G.F.   Liu S.Y.   Dong L.

Department of Engineering and Automation, Beijing Institute of Technology, Beijing
E-mail: Correspondent_dong@163.com

Abstract–The development of unmanned surface vehicles (USVs) for reconnaissance or other littoral missions is attracting increasing attention. In order to increase the sustainability of USVs, a new hybrid driving system powered by solar energy and wind energy is designed for the proposed USV. The developed USV is based on a catamarans prototype boat, and is equipped with a wing sail. In addition, the superficial of the USV is covered with PV panels for providing sustainable energy generation. The batteries based storage system is used for storing extra electricity as well as providing electricity for the control system and electrical system. To achieve tasks independently and intelligently, the USV is integrated with an automatic control system. The USV is discussed in detail, and some testing results are presented to validate its feasibility.

Keywords–USV, solar energy, indirect wind power.

I. INTRODUCTION

In the 21st century, the unmanned operation of transportation tools such as airplanes, automobile and ship has been the research hotspot for a long time. Among the three transportation tools, the research of unmanned aerial vehicles (UAVs) is the first to be started, and the associated technologies are the most mature. In recent years, several breakthrough techniques have been obtained in the field of driverless car. Comparatively, the recent achievements in unmanned surface vehicles (USVs) are less remarkable because of the complexity of surface vehicles and environmental conditions. The potential of the use of USVs for tasks such as shallow-water surveying, weapon delivery, environmental data gathering and surveillance is attracting increasing attention.

Many countries especially those with vast waters are vigorously developing USVs [1]. Countries such as America lead the world in the research and application of USVs and are developing USVs in the direction of intelligence, systematization and standardization. The most representative USV is the Spartan Scout USV which was created as an Advanced Concept Technology Demonstration (ACTD) in 2002 by the US Naval Undersea Warfare Center in Newport, Rhode Island, in conjunction with Radix Marine, Northrop Grumman, and Raytheon. The Spartan Scout USV is based on commercial off the shelf 7m rigid hull inflatable boat, which weighs two tons [2]. A prototype Spartan Scout was successfully launched and remotely operated in the Persian Gulf from the USS Gettysburg in December 2003. There are some other successful USVs developed in the United States. For example, the Ghost Guardians USV, developed by U.S. Robotics Ship company, can accomplish the missions of force protection, delivery of goods, collection of information, marine monitoring and so on [3].

Outside the U.S., there are also some outstanding USVs developed for different purposes. In 2005, Israel Elbit Systems Company came out with the Stingray USV. Stingray attracted the world wide attention for the maximum speed at 40 knots, the total laden weight of 150kg, and self-sustainability for more than 8 hours [3]. Stingray can perform autonomously or be remotely controlled by a single operator located at the shore station or onboard the ship. In 2010, Singapore Technologies Electronics launched Venus, a 9-meter USV at the Singapore Airshow 2010. In Japan, Yamaha developed two USVs, the Unmanned Marine Vehicle High-Speed UMV-H and the Unmanned Marine Vehicle Ocean type UMV-O [4]. The UMV-H is a reworked design based on a high-speed powerboat hull that enables it to move on the water and make mobile observations. The UMV-O is an ocean-going USV with displacement hull. It is used primarily in applications involving monitoring of bio-geo-chemical, physical parameters of the oceans and atmosphere.

Although lots of breakthrough techniques for developing USVs have been obtained, the existing USVs are suffering from some disadvantages. Firstly, the internal-combustion engine is adopted as the main driving force in traditional USVs, therefore, a large amount of fuel should be carried which reduces the effective load of USVs. Secondly, because of the limited capacity of USVs, the carried fuel is limited so that the sustainability of USVs cannot be long. In other words, its endurance mileage is short, which makes traditional USVs incompetent for applications in large-scale sea area. Moreover, the resulting tail gas and oil contamination due to fuel consumption may degrade the accuracy of environmental measurement associated with air and/or water. In addition, the fuel consumption is not cost-effective. It is worth mentioning that the intellectualization of USVs is always attracting the most attention. To be specific, USVs should have the capability of autonomous path planning and autonomous navigation, and can independently complete tasks such as environmental perception and target detection. In terms of the intellectualization level, the research of USVs still has a long way to go in the future.

To increase the endurance mileage of USVs, the Eco Marine Power (EMP) Company adopted a hybrid power supply scheme for the established Aquarius USV. The batteries based storage system of Aquarius could be charged with the shore grid or solar energy while sailing. To achieve this goal, the surface of Aquarius is covered with solar cells to provide sustainable electricity supply. However, the solar power generation cannot provide enough electricity because the superficial area of the USV

978-1-5090-0064-7/15 $31.00 © 2015 IEEE

is limited. Furthermore, the effective time for solar power generation is about 5 hours in each sunny day, and zero solar energy could be produced in rainy days. As a result, the application of Aquarius in long-term and long-distance sailings is still challenging. To solve this power insufficient problem, the wind energy is used as the main driving force in the USV SD1 built by Saildrone. SD1 is 19 feet in length and 7 feet in width, and is driven by the 20 feet high wing sail. Although the endurance mileage of SD1 is relatively long, it is still finite because the whole control system of SD1 relies on the batteries whose capacity is limited. Considering the aforementioned factors, the USV with a hybrid driving system powered by solar and wind is designed in this paper. The designed USV is named Adak, and has the following features:

1) By adopting both solar energy and wind energy as driving force, the driving system reliability can be enhanced and the endurance mileage can be considerably increased.

2) By adopting solar power generation and indirect wind power generation technologies, the storage system based lead-acid batteries can obtain continual electricity charging, which can ensure long-term electricity supply for the control system.

3) Two side thrusters are installed to provide ancillary driving force. The resulting benefits are that the veer of the USV becomes easy and the pivot steering is possible.

4) Multilevel optimization strategy according to wind speed, wind direction, solar irradiation, current position and posture information of the USV and the upper command is designed for improving the system performance as well as maximizing the energy using efficiency.

II. STRUCTURAL SYSTEM DESCRIPTION

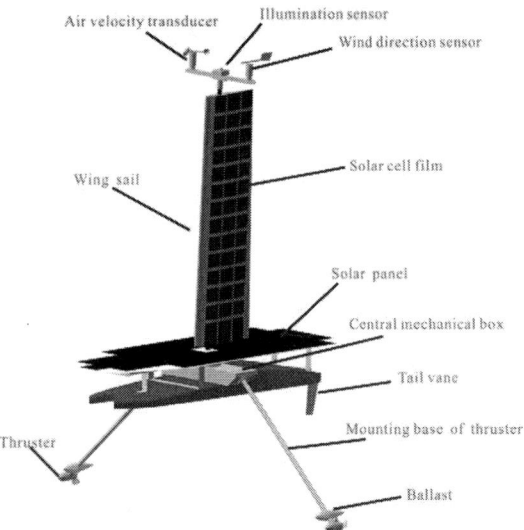

Fig. 1: The proposed USV powered by solar and wind energy

Fig. 2: The RCMK catamarans prototype

Table 1: Parameters of the proposed USV

Parameter	Value
Displacement	0.0293(m^3)
Designed speed	0.8(m/s)
Length of the hull	1900(mm)
Width of the hull	400(mm)
Height of the hull	200(mm)
Length of the wing sail	1450(mm)
Width of the wing sail	500(mm)
Height of the wing sail	1450(mm)
Waterline length	1800(mm)
Waterline breadth	380(mm)
Water plane area	0.641(m^2)
Wetted surface	1.26(m^2)
Prismatic coefficient	0.8
Block coefficient	0.5
Waterplane coefficient	0.84
Froude number	0.18
Sail area	0.725(m^2)
Rudder area	0.018(m^2)
Propeller diameter	190(mm)
Length of the fixed bracket	1000(mm)
Weight	34.3(kg)
Ballast	5.3(kg)

The proposed USV is shown in Fig. 1, where it can be found that the hull and the wing sail constitute the main structure of the USV. The parameters of the proposed USA are shown in table 1. The USV is fabricated based on the RCMK catamarans prototype, which is 1900 mm in length, 400 mm in width, and 200 mm in height as depicted in Fig. 2. The wing sail is self-designed according to some basic requirements such as:

1) The aspect ratio λ of the wing sail should be in the range of [3, 6].
2) The wing sail should have a camber between 7% and 13%.
3) The mast should locate in the aerodynamic center.

Considering that the stalling angle cannot be too small, we adopt the NACA4 wing sail with relatively large stalling angle. As well known, it needs to change the ship's rail right and left during Z glyph navigation, which implies that the wing sail should be symmetric. After detailed comparison between different kinds of wing sails, the NACA0012 wing sail is finally selected in this project. The designed wing sail is 1450 mm in length and 500 mm in width. The carbon fiber tube with external diameter 25 mm is used for the mast mounted at 1/4 of the wing chord. It should be mentioned that a servo driving system is used to rotate the wing sail for obtaining the maximum propulsion for the boat during the sailing. Unlike traditional sailing boats, two lead blocks are installed separately close to the bilateral propulsion devices. It can simplify the design, while effectively preventing the boat from overturn.

It can be observed from Fig.1 that both the wing sail and deck are covered with PV panels, which could provide continual electricity for the whole control and monitoring system as well as the electric driving system. In this way, the endurance mileage of the proposed USV can be infinite theoretically under ideal conditions. The parameters of the PV panels installed in different positions are shown in table 2, table 3 and table 4 respectively. In addition, the meteorological data and position information are measured by using according sensors for the self-sailing purpose.

Table 2: Parameters of the solar panel in the bow

Parameter	Value
Maximum power	104(W)
Open-circuit voltage	18(V)
Open-circuit current	6.166(A)

Table 3: Parameters of the solar panel in the stern

Parameter	Value
Maximum power	81.5(W)
Open-circuit voltage	14(V)
Open-circuit current	5.82(A)

Table 4: Parameters of the solar panel on the sail

Parameter	Value
Maximum power	31(W)
Open-circuit voltage	10.5(V)
Open-circuit current	5(A)

III. CONTROL SYSTEM DESCRIPTION

In order to implement the self-sailing function, an intelligent control system is designed, as depicted in Fig. 3. In what follows, the whole system will be divided into four sub-systems according to different functions for introduction.

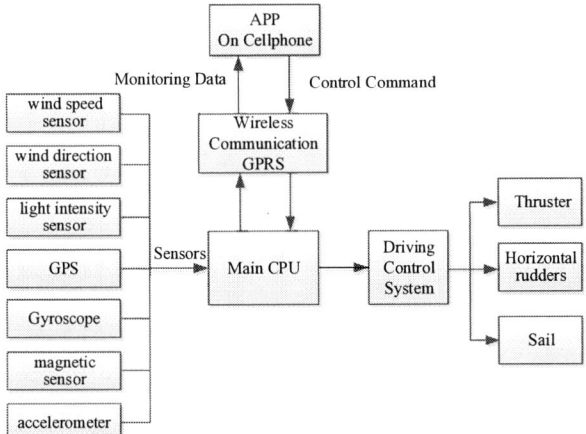

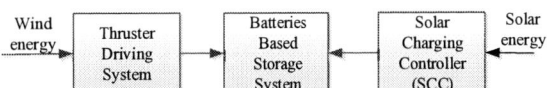

Fig. 3: The block diagram of the whole control system and storage system

3.1. Sensor system

Since the USV is designed to sail with the propulsion generated from the wind energy, the wind speed and wind direction are of our interest, and are measured with the wind speed sensor and wind direction sensor, respectively. Because the proposed USV employs PV panels for providing long-term electricity service, the light intensity sensor is adopted for measuring the solar irradiation.

As a USV, it should always know its own position and posture for sailing as required routes. Consequently, another task of the sensor system is providing the required position and posture information of the USV. As shown in Fig. 3, the Global Position System (GPS) is responsible for detecting the current position information such as present longitude and latitude information of the USV. While the other three kinds of sensors, gyroscope, magnetic sensor and accelerator, are employed to obtain its posture information.

All the data collected by the sensor system is transmitted to the center processing unit (CPU), where these data can be processed and analyzed to generate corresponding command for other sub-systems.

3.2. Driving control system

The driving system is the key component for providing required propulsion for the USV. It can be observed from Fig. 3 that the discussed USV can obtain driving power from three devices. The first one is the sail, by adjusting its angle with servo motor, the USV can be driven by the wind. The used servo motor is GW4468 type worm geared motor, the motor's outline dimensional drawing is shown in figure. 4 and its parameters are shown in table 5. Thus, the sail driving system is designed to rotate the sail to the required angle command from CPU. As depicted in Fig. 1, there are two thrusters at both sides of the USV, the outline dimensional drawing of the thruster is shown in figure. 5. The thruster employs a kind of DC brushless motor, whose parameters are shown in table 6. When the wind is very small and cannot provide enough propulsion for the USV, the thrusters could operate for compensating the propulsion deficit. Therefore, the thruster driving system is used to let both thrusters work reasonably for offering ancillary propulsion. In addition, in order to make the veering of the boat easy, the horizontal rudder and corresponding driving system are equipped. All the above mentioned driving systems are operating coordinately for ensuring the reliable sailing of the USV under the instructions from CPU.

978-1-5090-0064-7/15 $31.00 © 2015 IEEE

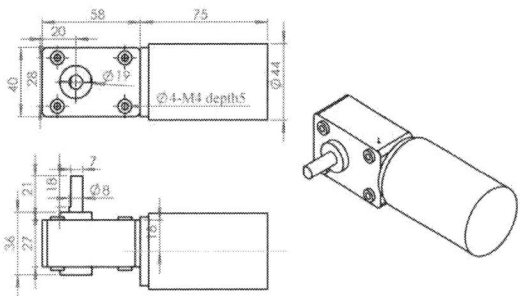

Fig. 4: GW4468 type worm geared motor's outline dimensional drawing

Table 5: GW4468 type worm geared motor's parameters

Parameter	Value
Rated voltage	12(V)
Rated current	1.5(A)
No-load speed	9(rpm)
Rated speed	6.8(rpm)
Rated torque	40(kg.cm)
Weight	0.6(kg)

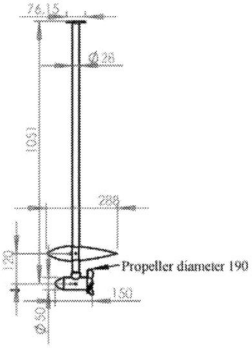

Fig. 5: Thruster's outline dimensional drawing

Table 6: DC brushless motor's parameters

Parameter	Value
Rated voltage	24(V)
Rated current	11(A)
Rated power	210(W)
Rated speed	3000(rpm)

3.3. Remote communication system

The USV is an autonomous system, which can implement specific sailing route command received from the console without external help. In this design, the APP program installed in cellphones is used to send the mission information to the CPU and receive the weather and sailing data for monitoring purpose via the GPRS communication system. On the one hand, with the APP, we can control the start and stop of the USV, set the starting point and ending point of sailing, schedule task plans, etc. On the other hand, we can monitor the working states of the USV and critical system parameters including the current position and posture information, meteorological information, and the state of charge of batteries.

3.4. Center processing unit

The center processing unit (CPU) is devoted to coordinating each sub-system for completing received command from the APP with the minimum energy consumption as well as sending the system information to the upper monitoring platform. During the self-sailing process, the first thing of CPU is to implement the route planning according to the upper command, present wind speed and direction, and current position and posture information of the boat. Subsequently, an optimization algorithm for maximizing energy using efficiency determines the rotating angle of the sailing, rotating speed of the thrusters, as well as the action command for the horizontal rudders. All the commands are sent to each driving system via the CAN bus, which is of high transmission rate and reliability.

3.5. Batteries based storage system

All the electric devices of the boat are powered by the lead-acid batteries, which are charged mainly by the mounted PV panels. To achieve this goal, a solar charging controller based on buck converter is adopted to extract the solar energy from PV panels and charge the batteries. A close observation of Fig. 3 reveals that the batteries can also be charged with the wind energy. This function is realized by letting the DC brushless motor of the thrusters work in generator mode when the wind speed is high enough. It is because the DC brushless motor can work as a generator under the force of the water flow, when the USV is experiencing fast navigation with the strong propulsion from the sail.

IV. WORKING PRINCIPLE DESCRIPTION

4.1. Propulsion analysis of the sailing boat

As the main propulsion device, the wing sail has an important influence on the system performance such as the sailing speed and stability of the boat. In order to provide required propulsion for the USV, the CPU calculates the optimal angle of attack of the wing sail according to the wind speed, wind direction, sailing direction and the posture of the boat, and the corresponding rotating angle of the sail. Subsequently, the angle command is sent to the sailing driving system for letting the sail rotate to the required position.

The force analysis of the wing sail is illustrated in Fig. 6. Assuming that the USV is navigating at speed V_c along with the X axis, V_z is the real wind speed, and V_x is the relative wind speed. For illustration purpose, the angle of the real wind is denoted with θ_z, the angle of the relative wind is denoted with θ_x, the rotating angle of the wing is β and its angle of attack is α, and ω is the drift angle. It follows from the Bernoulli principle that the lift force F_s produced by the wing sail is orthogonal to the wind direction, and the generated resistance F_z is contrary to the wind direction. The force of the wing sail to the sailing boat is the resultant force F_r of F_s and F_z, and the projection of F_r on the sailing direction is the effective driving propulsion T_c as

$$T_C = F_s \sin\theta_x - F_Z \cos\theta_x \qquad (1)$$

and the projection of F_r on the direction orthogonal to the sailing direction is known as the heeling force T_H and it can be given as

$$T_H = F_s \cos\theta_x - F_Z \sin\theta_x \qquad (2)$$

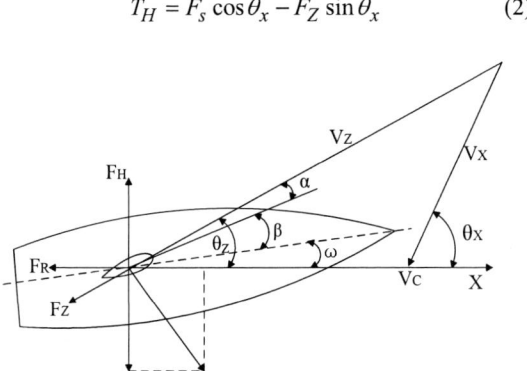

Fig. 6: Force analysis of the wing sail

As a sailing boat, the propulsion generated by the wing sail is expected to be the main driving force for the boat. However, no effective propulsion could be produced when the wind speed is too small, or when the wind direction is contrary to the sailing direction. In this work, the installed two thrusters could be started for providing ancillary sailing force for the boat. Another benefit is that the veer of the boat becomes easy and the pivot steering is possible with the assistance of thrusters. From the view point of maximizing the wind energy utilization, the thrusters only operate in necessary scenarios, which are determined by the CPU for saving the electricity consumption.

4.2. Operation modes of thrusters

In this study, the employed DC brushless motor is driven by using the following three-phase electronic circuit as shown in Fig. 7. The dc-bus of the driving circuit is connected to the 24V storage system, and its output is connected to the thruster.

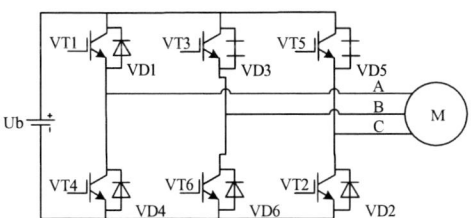

Fig. 7: The three-phase driving circuit for thrusters

As mentioned above, there are two operating modes for the equipped thrusters in this study. To be specific, the thruster should operate in motor mode for providing ancillary propulsion for the USV when the driving force generated by the wind is not enough, and work in generator mode when the USV is experiencing fast navigation with enough propulsion from the wind. Since the motor mode operation of DC brushless motors has been discussed in previous papers, it therefore will be skipped in this study. In what follows, the generator mode operation of thrusters will be introduced in detail.

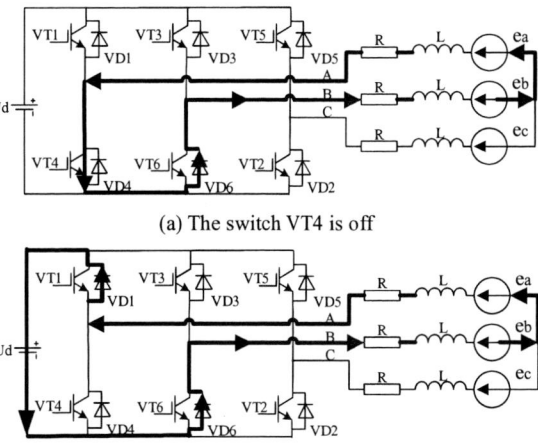

(a) The switch VT4 is off

(b) The switch VT4 is on

Fig. 8: The current flow in generator mode when the electrical degree is between 0 and 60 degrees

In generator mode, the bottom switches of Fig. 7 operate at PWM mode while the upper switches are off and only the anti-parallel diodes are used for freewheeling. Fig. 8 (a) and (b) describe the current flow in generator mode when the electrical angle of thrusters is between 0 and 60 degrees. From Fig. 8 (a), it can be observed that the A- and B-phase current of the thruster is flowing through the switch VT4 and the anti-parallel diode VD6. In this mode, the energy produced by the water flow is stored in the winding inductors of thrusters, and the A- and B-phase current increases. Then, the stored energy of the winding inductors is charged into the dc-bus batteries when the VT4 is off as shown in Fig. 8 (b), and the A- and B-phase current decreases. The duty cycle of VT4 determines the charging current. A close observation of Fig. 8 reveals that the three-phase driving circuit is equivalent to a boost converter in generator mode. It should be pointed out that the rotator position and electrical angle are measured via the internal hall sensors in the thruster. Therefore, the energy could be transferred to the batteries based storage system from the thrusters in generator mode by producing reasonable switching signals according to the collected hall signals. The three-phase signals are depicted in Fig. 9, where it can be concluded that there are six sectors in one period. During each sector, only one bottom switch is designed to work at PWM mode and all other switches are off when the thruster operates in generator mode. The relationship between the hall signals and activated power switch in generator mode are listed in Table 7.

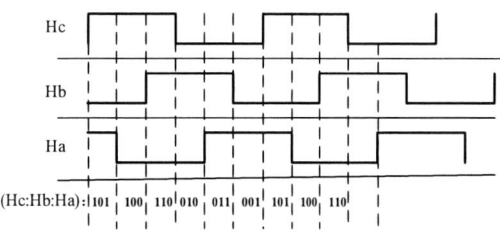

Fig. 9: Three-phase hall signals of the thruster

978-1-5090-0064-7/15 $31.00 © 2015 IEEE

Table 7: The relationship between the hall signals and activated power switch

Hall signals (Hc : Hb : Ha)	Activated power switch
Sector : 010,110	V4
Sector : 100, 101	V6
Sector : 001, 011	V2

4.3. Solar charging controller (SCC)

As depicted in Fig. 3, the PV panels, solar charging controller (SCC), and 24V storage system constitute the whole power system for the USV. The storage system is two 12V batteries connected in series. The available solar energy is transferred to the batteries based storage system by using SCC, while the battery voltage is always monitored for safety. Before introducing the working principle of SCC, it is necessary to explore the power characteristic of PV panels. Fig. 10 (a) and (b) depict the power characteristics of PV panels during varying solar irradiation and temperature, respectively.

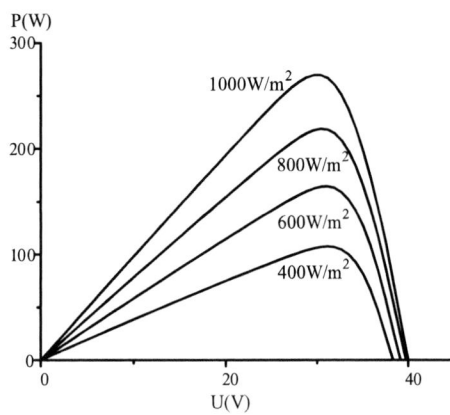

(a) P-V characteristics under different solar irradiations

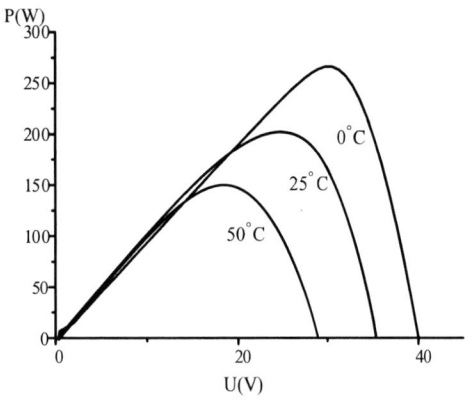

(b) P-V characteristics under different temperature
Fig. 10: P-V characteristics

It can be observed from Fig. 10 (a) that each plot has a maximum power point (MPP), and the maximum solar power increases with the solar irradiation. The similar P-V characteristics can be found in Fig. 10 (b), while the maximum solar power increases with the decrease of temperature. It is worth mentioning that the open-circuit voltage of PV panels approximately remains constant under different solar irradiations as shown in Fig. 10 (a), while it significantly increases as the temperature reduces as shown in Fig. 10 (b). When the operating point deviates from the MPP, the obtained solar energy decreases below the available maximum power, i.e., the undesirable solar energy loss is produced. In order to achieve the maximum solar energy harvesting, a MPP tracking (MPPT) strategy should be developed for maintaining the operating point at the MPP under varying conditions. Owing to its simplicity and excellent performance, perturbation and observation (P&O) method is adopted in the designed SCC to track the MPP by perturbing the terminal voltage of PV panels [5]. The flow chart of the P&O method is depicted in Fig. 11. If a given perturbation ΔV leads to the increase (decrease) of PV power, the next perturbation is made in the same (opposite) direction. This way, the MPPT controller continuously seeks the maximum power point (MPP). Once the MPP is achieved, the output voltage of solar cells remains close to the MPP voltage.

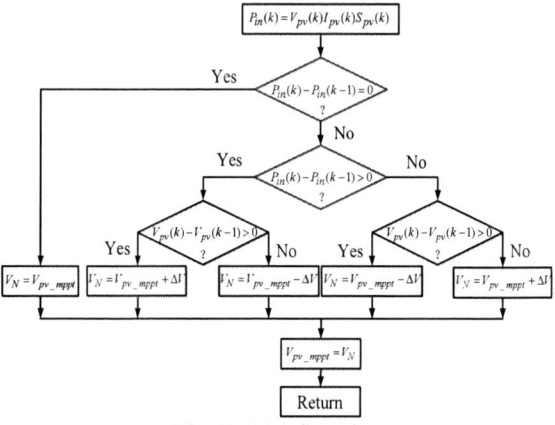

Fig. 11: P&O flow chart

However, the PV panels of the proposed USV are constituted by some serial-parallel connected cells. Each of them has different output characteristics, which lead to multiple peaks in the P-V curve of the synthesized PV panel. When adopting P&O method, it is easy to trap into a local maximum power point. To deal with this problem, the SCC will perform a global search to seek the global maximum power point in advance, and then, P&O method is executed by perturbing the terminal voltage of the solar panel around the global maximum power point.

Generally, the SCC is preferred to operate in MPPT mode for enhancing the solar energy using efficiency, while it is not always valid in practice. For example, when the batteries are full, no or very little solar energy should be transferred to the storage system regardless of the available maximum solar energy. That is to say, the MPPT function of SCC should be disabled for the security of batteries in this scenario. Considering the practical characteristics of the adopted lead-acid batteries, three charging modes are integrated into SCC for the storage system, i.e., constant current mode, constant voltage mode and tiny current mode. During constant current mode, SCC operates in MPPT mode and all the available solar energy is collected to charge the batteries. While SCC works in Non-MPPT mode in both constant voltage mode and tiny current mode. The input solar power is regulated by SCC for maintaining the battery voltage at a constant value in the constant voltage mode, while a tiny charging current is provided for the batteries during the tiny current mode.

4.4. Center processing unit (CPU)

The CPU is in a central position linking all the sub-systems closely together by unscrambling the sensor information and then controlling the driving control system. According to the project requirement, the CPU should have the following functions:

- Receive remote command and execute.
- Path planning and navigation control.
- Engine control and condition monitoring.
- Electrical control and condition monitoring.
- Emergency safety protection.
- Send data to the upper monitoring platform.

Due to the paper length, only the path planning algorithm will be discussed here and other functions would be skipped. In order to implement the path planning, the grid method is used to perform the environment modeling. Based on the established grid model, both the blindness search and heuristic search algorithms can be used for local route searching. According to the wind speed, wind direction, position and posture of the USV at present, the path planning for the USV in a local area can be implemented. Compared with blindness search algorithm, the heuristic search technique is more competent because it has high searching efficiency and can obtain the globally optimal solutions online. A* algorithm is a custom heuristic search method and is widely used in technological industries such as robotics and video game designing [6-8]. The evaluation of the F value in A* is depicted in the following equation:

$$f(p) = g(p) + h(p) \qquad (3)$$

where $g(p)$ is the time spent from the start to the current node p, $h(p)$ is the estimated time which would be spent from p to the target node, and $f(p)$ denotes the total time spent during the search. As shown in Fig. 12, the triangle represents the USV, and its vertex is the sailing direction. The USV locates at the node A at present, and the nodes B, C, D, E and F locate at possible searching directions. G is the target node, and circle denotes the obstacle. This is the polar curve of the real wind for the USV. In what follows, the calculation of $f(B)$ is used for illustration. The $g(B)$ can be given by

$$g(B) = \frac{S_{AB}}{f(\xi_{AB}, V_T)} \qquad (4)$$

in which, ξ_{AB} is the angle between the sailing direction and the wind direction during the journey from A to B, and S_{AB} is the Euclidean distance from A to B. Similarly, $h(B)$ can be calculated as

$$h(B) = \frac{S_{BG}}{f(\xi_{BG}, V_T)} \qquad (5)$$

in which, ξ_{BG} is the angle between the sailing direction and the wind direction during the journey from B to G, and S_{BG} is the Euclidean distance from B to G. Then, the totally spent time $f(B)$ can be achieved. In the same way, the evaluation of the F value in A* for other nodes can be obtained. Subsequently, the optimal node can be selected by minimizing f and considering some practical constraints as well as the optimal course angle. With the attained course angle, corresponding command is generated in the CPU and sent to the driving control system for letting the USV move to the target position.

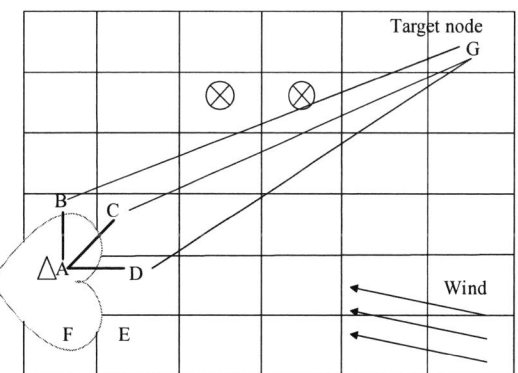

Fig. 12: The route searching map for the USV

V. SIMULATION AND EXPERIMENTAL RESULTS

Simulations and experiments are performed to verify the proposed MPPT method. The maximum power output of the PV panel used in simulation is 50W and the transient responses of the solar charging system are shown in figure. 13. The transient responses of the solar charging system in experiment are shown in figure. 14, where the displayed power values are 100 times the actual values and the displayed current values are 10 times the actual values. The results show the PV panel can be controlled to work around the maximum power point with acceptable errors, indicating the validity of the proposed MPPT method.

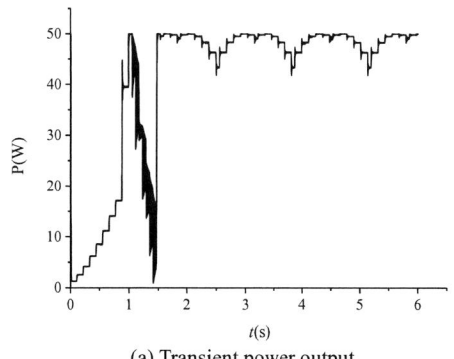

(a) Transient power output

978-1-5090-0064-7/15 $31.00 © 2015 IEEE

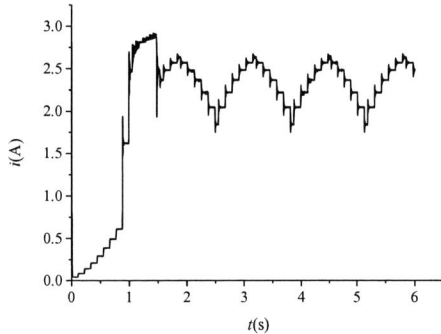

(b) Transient current of the PV panel
Fig. 13 Simulation results of the MPPT

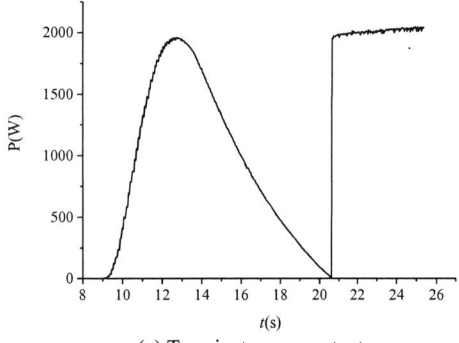

(a) Transient power output

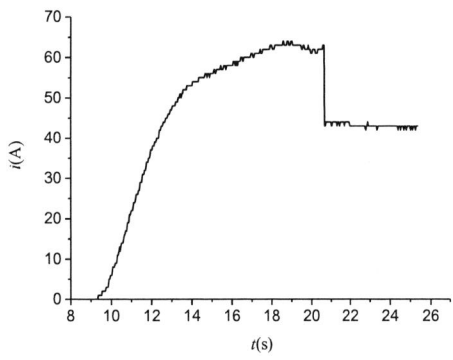

(b) Transient current of the PV panel
Fig. 14 Experiment results of the MPPT

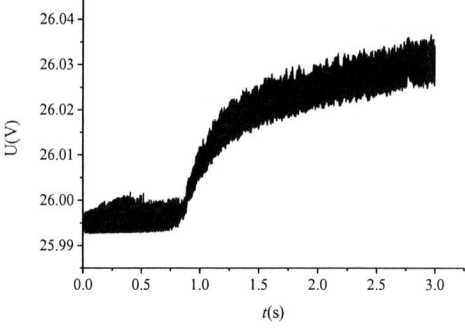

(a) Transient response of the battery voltage

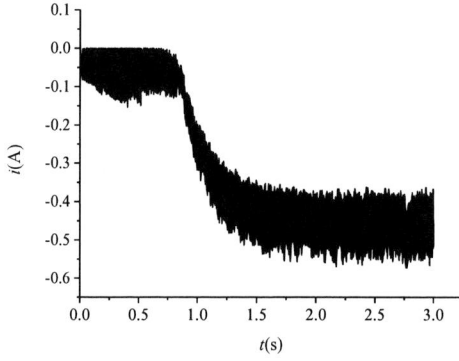

(b) Transient response of the current charging into the battery

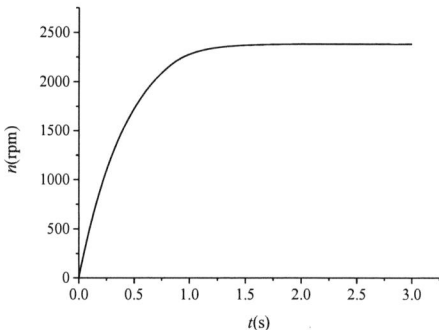

(c) Transient response of the motor speed
Fig. 15 Transient responses of the system when the motor works
in motor mode

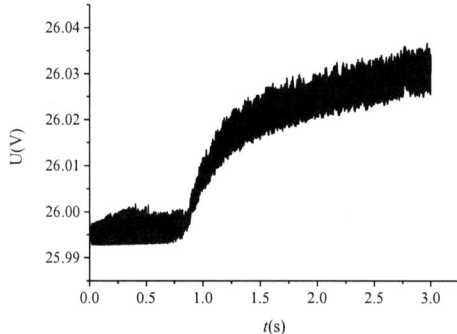

(a) Transient response of the battery voltage

As mentioned above, batteries can also be charged with the wind energy by letting the DC brushless motor work in generator mode when the wind speed is high enough. Simulations are performed to verify this charging method. Figure. 15 shows the transient responses of the system when the motor works in motor mode. Figure. 15 shows if power switches operate as an uncontrolled rectifier, the battery will not be charged until the motor speed exceeds 2000rpm. Because when the motor speed is low, the counter electromotive force produced by the motor is lower than the voltage of the battery. When the motor works in generator mode, the power switches operate in the boost mode, the counter electromotive force will be increased to be higher than the battery voltage. In this mode, the motor can feedback energy to the battery even if the motor speed is low. As Figure. 16 shows, the feedback current is also very low.

978-1-5090-0064-7/15 $31.00 © 2015 IEEE

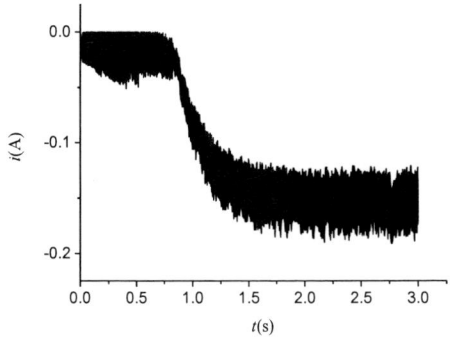

(b) Transient response of the current charging into the battery

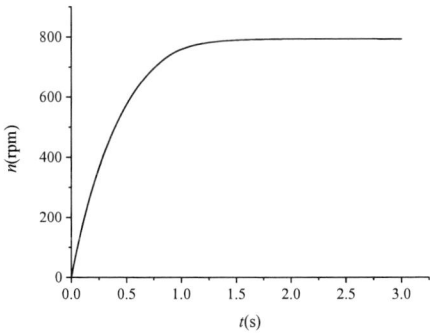

(c) Transient response of the motor speed
Fig. 16 Transient responses of the system when the motor works in generator mode

The prototyped USV is shown in figure. 17. Figure. 18 is the image when it is sailing. In the water trials, based on A* algorithm, its route is adjusted according to the direction and speed of the wind in real-time. The boat has strong anti-overturning stability in strong wind and can sail to the destination smoothly.

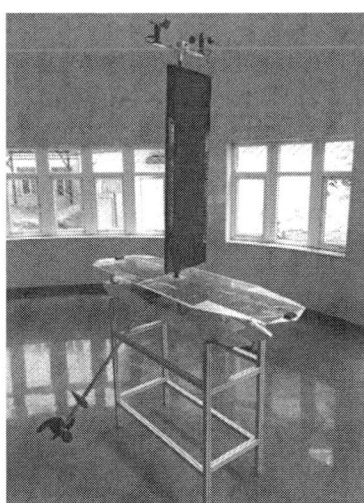

Fig. 17 The prototyped USV

Fig. 18 The proposed USV in water

VI. CONCLUSION

In this paper, an USV based on a catamarans prototype boat was designed. It is powered by solar energy and indirect wind energy, with batteries to store excess energy, which ensures a relatively long-term power supply for its control system. The charging method, three driven methods and path planning algorithm have been introduced. The adopted solar charging method and indirect wind charging method were analyzed in theory. They were then verified with simulations and experiments. The proposed USV has been prototyped and tested. The experimental results agreed very well with the theoretical prediction.

ACKNOWLEDGMENT

The authors appreciate the contribution of PhD candidates Xiao F.R and Ju X.L toward this design. The authors' sincere thank also go to laboratory of new energy and energy conversion in Beijing Institute of Technology who provides the solar panels and RCMK catamarans prototype to complete this work.

REFERENCES

[1] J.E. Manley, "Unmanned Surface Vehicles, 15 Years of Development", Proc. Oceans 2008 In: MTS/IEEE Quebec Conference and Exhibition (Ocean' 08), Sep 2008, pp. 1-4.

[2] A. Maguer, D. Gourmelon, M. Adatte, F. Dabe, "Flash and/or flash-s dipping sonars on Spartan unmanned surface vehicle (USV): A new asset for littoral waters", Turkish International Conference Acoustics, Istanbul, Turky, 2005.

[3] R.J. Yan, S. Pang, H.B. Sun, Y.J. Pang, "Development and Missions of Unmanned Surface Vehicle", Journal and Marine Science and Application, Vol. 4, No. 4, 2010, pp. 451-457.

[4] B. Enderle, T. Yanagihara, M. Suemori, H. Imai, A. Sato, "Recent developments in a total unmanned integration system", AUVSI Unmanned Systems Conference, Anaheim, CA, USA, 2004.

[5] C. Hua, J. Lin, and C. Shen, "Implementation of a DSP-controlled photovoltaic system with peak power tracking", IEEE Trans. Ind. Electron., Vol. 45, No. 1, 1998, pp. 99-107.

[6] H. Choset, "Principles of robot motion: theory, algorithms and implementations", Massachusetts: MIT Press, 2005.

[7] R. Murphy, "Introduction to AI robotics", Massachusetts: MIT Press, 2000.

[8] Amit's Game Programming Information. 2000. http://theory.stanford.edu/?amitp/Game Programming/MapRepresentations.html

Integrated Piezoelectric Energy Harvesting and Structural Health Monitoring for Transportation Infrastructure

Kaur N.[1]   Balguvhar S.[2]

[1] Department of Civil and Environmental Engineering, The Hong Kong Polytechnic University, Hong Kong
E-mail: n.kaur@polyu.edu.hk
[2] Department of Civil Engineering, Indian Institute of Technology IIT Delhi, India
E-mail: ersumitest@gmail.com

Abstract– This paper, based on lead author's doctoral research at IIT Delhi, presents the feasibility of integrated structural health monitoring (SHM) and energy harvesting using a same thin Piezoelectric patch. The structure is assumed to be operating in two states, idle state and SHM state. During the idle state (when SHM is not being performed), the PZT patches will harvest the energy and store it in an appropriate storage device, such as a battery or a capacitor. In the SHM state, the same stored energy will be utilized for the SHM of the host structure by the same PZT patch, either in the global mode (standard vibration techniques) or the local mode (EMI technique) or both. It is assumed that the total duration of the SHM state will be very small as compared to the idle state.

Keywords–Piezoelectricity, energy harvesting, d_{31} mode, structural health monitoring

I. INTRODUCTION

Structural health monitoring (SHM) is currently attracting huge research funding across the world. Experts have identified SHM as one of the top ten technologies having the potential of driving the global economy [1]. SHM, especially for large civil structures, needs a mega network of sensors distributed throughout the structure. Powering these sensors is a critical issue. It either requires a distributed network of wiring or warrants installation of individual batteries, both of which call for huge investment. Fortunately, the power requirement of the sensors generally is low, ranging from micro to milli watts, and reducing further day by day. This has catalysed research efforts to explore the possibility of employing renewable energy sources to power the sensors. The use of piezoelectric materials to act as generators for converting the vibrational energy of structures into electrical energy is one such possibility being explored rigorously over the last three decades. Piezoelectric ceramic (PZT) patches, operating in d_{31}-mode, are considered best for SHM. However, for energy harvesting, built up configurations such as stack actuators are somewhat more preferred. The main objective of this paper is to explore the potential of employing the same piezo sensor, operating in d_{31}-mode, for SHM as well as energy harvesting on real-life transportation infrastructure including bridges, flyovers etc.

II. DESIGN METHOD AND SPECIFICATION

The idea of integrated SHM and energy harvesting using a thin piezoelectric patch is explained in Figure 1 [2]. Consider any structure with PZT patches either surface bonded or embedded in it. The structure is assumed to be operating in two states, *idle* state and *SHM* state. During the idle state (when SHM is not being performed), the structure will experience the ambient vibrations which will be transferred to the PZT patches attached to it. By the virtue of piezoelectric effect, the PZT patch will convert the mechanical energy stored in the structure to the electric energy. The harvested electric energy can be stored in any appropriate storage device, such as a battery or a capacitor. In the *SHM* state, the same stored energy will be utilized for the SHM of the host structure by the same PZT patch, either in the global mode (standard vibration techniques) or the local mode (EMI technique) or both. It is assumed that the total duration of the SHM state will be very small as compared to the idle state.

Fig. 2 represents the above idea extended to a real life scenario where PZT patches are attached to a typical bridge under moving vehicle. So during the idle stage, the PZT patches will keep harvesting the energy from the vibrations experienced by the bridge due to the vehicle movement. Now, consider damage has occurred in the bridge, and then the same stored energy can be used for the SHM of the bridge. The location of the damage in the bridge can be identified and a warning signal can be issued by illuminating the lights across the particular region of the bridge span where damage has been identified using the same stored energy.

III. VERIFICATION

Proof-of-concept demonstration of energy harvesting for SHM using embedded concrete vibration sensor (CVS), carried out in the Smart Structures and Dynamics Laboratory (SSDL), IIT Delhi, has been presented by the lead author in her Doctoral research thesis [2, 3, 4]. The RC beam was subjected to a sinusoidal loading at a frequency of 16.5 Hz using the inertial-type shaker placed at the centre of the beam. It was operated at a frequency equal to the natural frequency of the beam to attain maximum deflection in the beam at the centre. The corresponding acceleration to attain maximum deflection, hence the maximum strain and voltage in the PZT patch, at the beam centre was measured to be 3.14 ms⁻². An energy harvesting circuit [shown in Fig. 3(a)], consisting of a full wave bridge rectifier built of Zener diodes and a 1000 μF capacitor, was employed for harvesting and storing the energy generated by the CVS embedded at location 11 in the RC beam. The variation of the voltage across the capacitor during its charging and discharging is shown in Fig. 3(b). It can be observed that the capacitor was charged to a maximum voltage of 97 mV in 187 seconds.

Using the relation, $E_c = 1/2\,CV^2$, the energy stored in the capacitor (E_c) was computed as 4.753 µJ, from which it can be derived that a continuous harvesting for 15 days is sufficient for one time operation of AD5933 (Table 1), which requires an energy of 33 mJ for one time operation. Hence, SHM of the real-life structure can be performed twice a month using the same CVS for SHM and energy harvesting. Other low power consuming circuits available in market for various applications, which can also be powered using the embedded CVS, are summarized in Table 1. Hence, combined energy harvesting and SHM are feasible using embedded CVS. It can carry out energy harvesting during the idle time and can use this energy to carry out SHM after regular intervals.

Table 1. Charging cycles time for different circuits for various applications.

Circuit / IC	Energy Required	Charging Cycles	Application
Typical A/D Convertor, TMP 112 [5]	25.2 µJ	6	Industrial Application
RE46C800 [6]	3 mJ	631	CO/heat Detector
AD5933 [7]	33 mJ	6943	SHM via EMI Technique

IV. Technical Developments to Enhance Piezoelectric Energy Harvesting

Piezoelectric energy harvesting technology has significant advantages over other renewable energy sources such as solar energy as ambient vibrations are a widespread source of energy in the environment. Especially, where solar or thermal energy may not be consistently available, this can be of high importance. Although solar cells have come up as sustainable power source, their limitation however, especially in the Indian context, is the exorbitant investment and a very long period of 8 to 10 years for recovery. On the other hand, PE harvesting element can generate sufficient power to recover cost in 6 to 12 months.

It is clear from the literature to date that much more attention has been paid to the transducer itself than the power conditioning. One of the limitations of the existing PE harvesters is in their interface circuitry. Commonly used full-bridge rectifiers severely limit the electrical power extractable from a PE harvesting element. Further, the power consumed in the control circuits of these harvesters reduces the amount of usable electrical power. Secondly, the ambient vibrations normally encountered in bridges/ flyovers are typically low-frequency and low-acceleration vibrations (Fig. 4), often characterized by non-periodic nature.

In all of the applications, the use of a diode to provide the rectification has the serious drawback of having to overcome the threshold voltage of the diodes. This problem prevents the rectification of signals below a voltage of about 0.6 V. The Precision rectifier can improve the power extraction capability of existing full-bridge rectifiers (Fig. 5). The technique offers several advantages which support the IC implementation. All these elements will lead to reduced costs of the suggested solution. The circuit provides linear variation of the DC output voltage with the input voltage, with the output voltage amplitude being almost the same as the peak input voltage.

V. Conclusions

This paper has presented the proof-of-concept demonstration of combined SHM and energy harvesting using specially designed embedded PZT patch (CVS) operating in the axial mode. Experimental demonstration for harvesting and storage of energy has been described. With the ongoing developments in electronics, as lesser power consuming circuits are emerging, it is believed that the necessary minimum energy scavenging time will drastically come down. Hence, using PZT patch in the form of CVS both for SHM and energy harvesting in real-life structures is expected to emerge as a new and useful contribution. For effective dissemination of the idea, a simulation based experiment has been developed for the Virtual Smart Structures and Dynamics Laboratory, IIT Delhi [8]. Also, the current developments to improve the piezoelectric energy harvesting from the ambient vibrations have been proposed here, which involves the use of precision rectifier in place of conventional full-bridge rectifier. The technique offers several advantages which support the IC implementation. All these elements will lead to reduced costs of the suggested solution.

References

[1] Majcher, K. (2011), "10 technologies to watch", Aviation Week.

[2] N. Kaur, "Integrated structural health monitoring and energy harvesting potential of adhesively bonded thin piezo patches operating in d_{31} mode", PhD Thesis, Indian Institute of Technology Delhi, India.

[3] N. Kaur and S. Bhalla, "Feasibility of energy harvesting from thin piezo patches via axial strain actuation mode", *Journal of Civil Structural Health Monitoring*, Vol. 4, 2014a, pp. 1-15.

[4] N. Kaur and S. Bhalla, "Combined energy harvesting and structural health monitoring potential of embedded piezo concrete vibration sensors", *Journal of Energy Engineering, American Society of Civil Engineers* ASCE, 2014b, (accepted, in press).

[5] Texas Instrument (2015), www.ti.com. (Date of access: 04 July, 2015)

[6] Microstrain Inc., http://www.microstrain.com. (Date of access: 13 July, 2015)

[7] An Analog Devices (2014), AD5933 Tech note. www.analog.com. (Date of access: 07 Oct, 2014)

[8] Virtual Smart Structures and Dynamics Laboratory (2015), http://ssdl.iitd.ac.in

APPENDIX

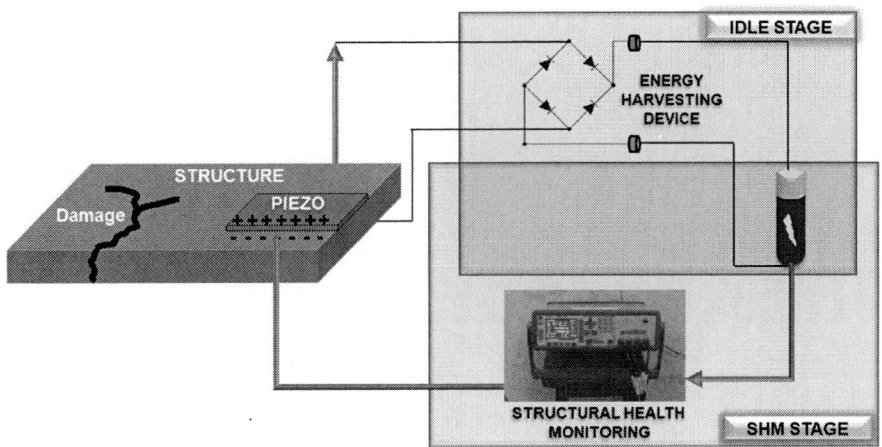

Fig. 1: Principle of integrated SHM and energy harvesting.

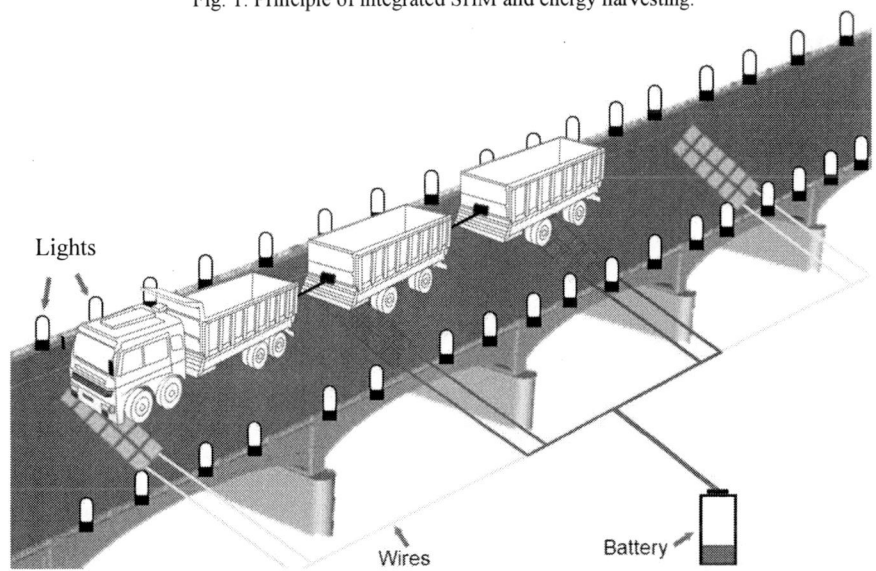

Fig. 2: Typical Bridge with moving vehicle during energy harvesting stage.

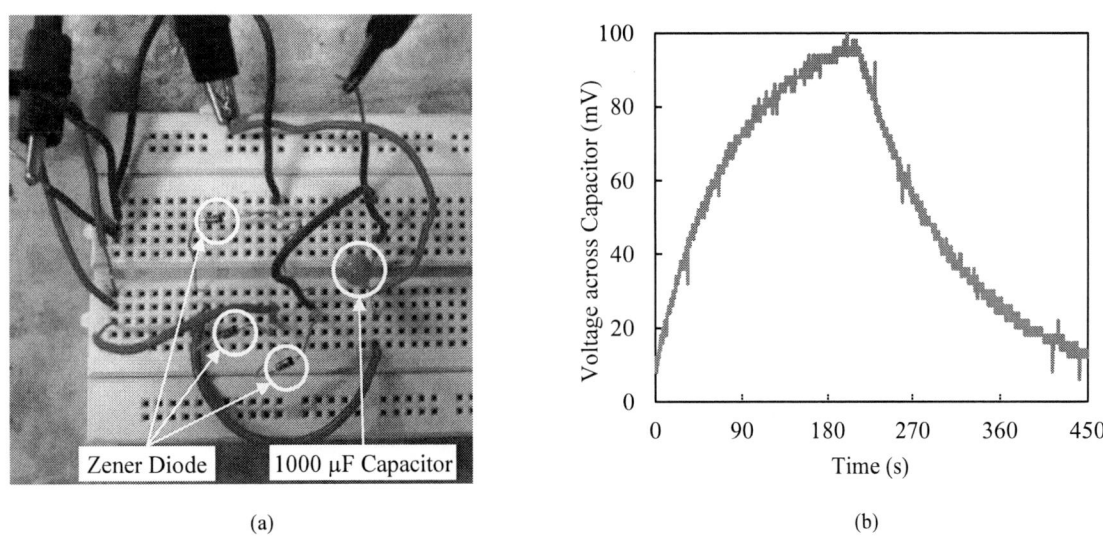

(a)   (b)

Fig. 3: (a) Full-wave bridge rectifier circuit used for storing energy in capacitor.
(b) Charging and discharging voltage across capacitor.

978-1-5090-0064-7/15 $31.00 © 2015 IEEE   437

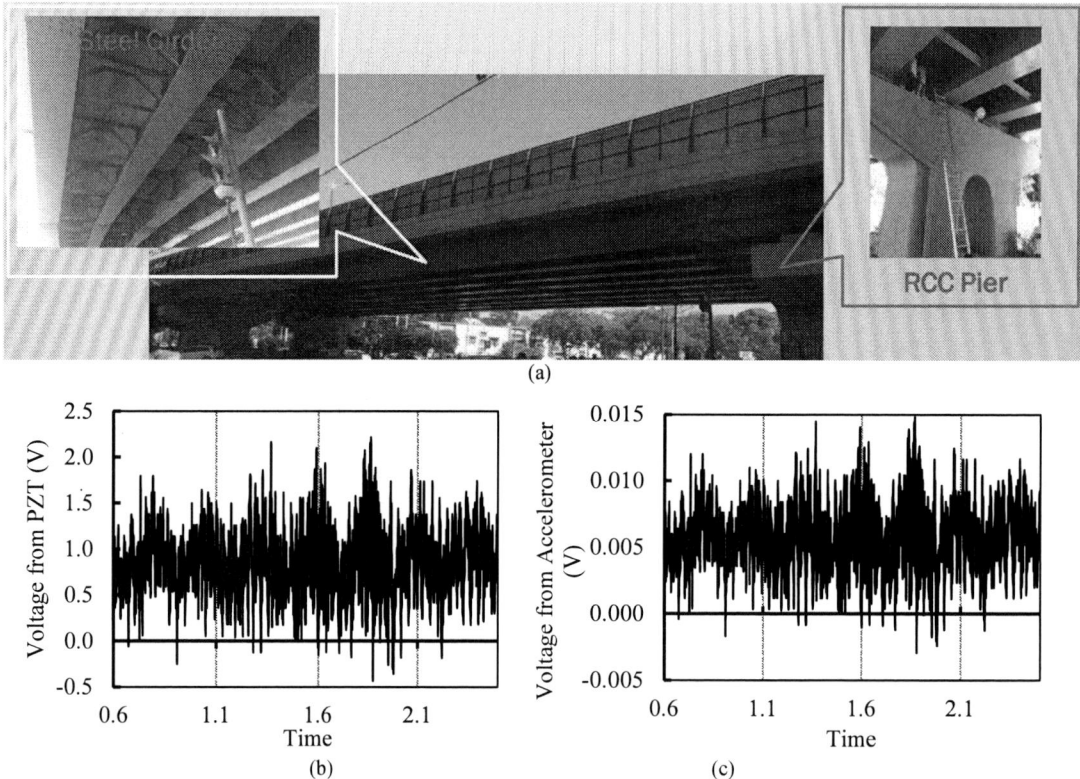

Fig. 4: (a) Flyover in front of IIT Delhi, India
(b) Vibration data acquired from the bridge girder using accelerometer
(c) Harvestable voltage from PZT if attached to the bridge girder

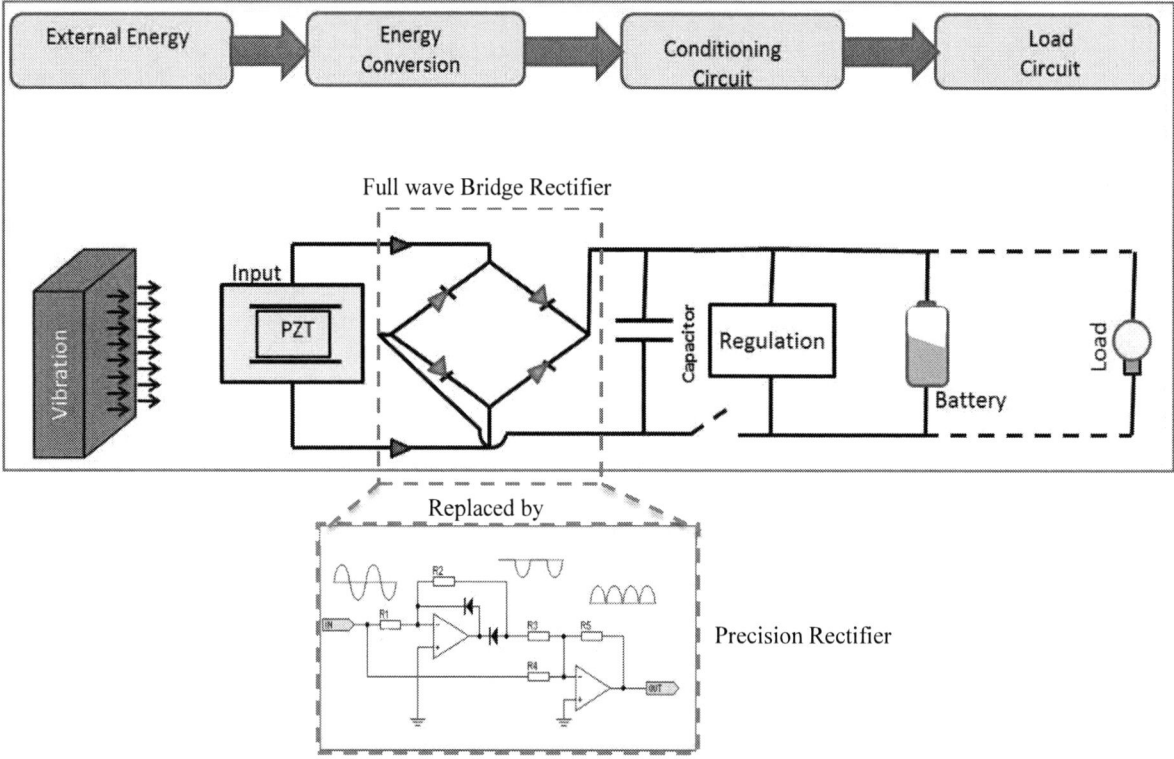

Fig. 5: Electrical components involved in improved piezoelectric energy harvesting where conventional full wave bridge rectifier will be replaced by precision rectifier

Advances in the Application of Power Electronics to Railway Traction

Kang-kuen LEE

Department of Electrical Engineering,
The Hong Kong Polytechnic University, Hong Kong

Abstract-Building on the advancement of power electronics technology there has been an on- going evolution of railway traction equipment, both for the traction drives on board the trains or the traction power supply systems on the track side. This paper describes how systems based on power electronics have emerged and then become a standard provision in the railway industry. New systems and more powerful power electronic devices which have emerged in the past few years and applied to railway traction have also been presented. Key players and their development efforts on the application of power electronics to railway traction have also been introduced.

I. INTRODUCTION

The past 20 years have witnessed phenomenal growth in railway transportation, in both areas of urban rail transits and high speed passenger railways. It is also noted that electric traction has become the dominant mode of railway traction and there has been on going and significant evolution in railway traction during the past 20 years, mainly driven by the advancement in power electronic technology. This paper describes some of the more significant emerging applications and some of the key players in the railway industry.

II. THE PAST

Up to the mid-1980s the application of power electronics to railway traction had been very limited. Typical areas of application include:

(a)For railways with ac traction power supply system power diodes were used to rectify the ac voltage which was stepped down through the heavy trainborne transformers for driving the dc traction motors. The use of thyristors to replace the tap changer of the traction transformer for varying the values of the rectified dc voltage to control motor speeds was emerging towards the mid-1980s,

(b)For railways with dc traction power supply system the use of thyristors as choppers to replace the armature resistors for controlling the speed of the dc traction motors was emerging since the early 1980s.

(c)For railways with dc traction power supply system power diodes were used as rectifiers to convert the ac supply voltage to dc traction power supply for the trains. As power diodes were fairly expensive in those days 6-pulse rectifier was the mainstream configuration.

III. THE PRESENT

Before the mid-1980s railway traction technology had been fairly stagnant and improvements were piecemeal and confined to those achieved through electrical and mechanical components. The rapid advance in power electronic technology since the beginning of the 1980s has drastically changed railway traction technology and the following applications of power electronic technology have now become standard provisions for railway traction:

(a)The use of ac traction drive based on IGBTs has replaced dc traction drive so as to reap the many advantages of 3-phase induction motors or synchronous motors over the conventional dc series motors,

(b)The use of 4-quadrant convertors to improve the control and power factor performance of the traction equipment,

(c)Rotary convertors have been replaced by solid state ones, 12 or 24-pulse rectifiers for the dc traction supply system have now been used to improve the ripple performance so as to avoid the use of the bulky filters. Regenerative braking in railways with either ac or dc traction power supply system

IV. EMERGING APPLICATIONS

Both tertiary institutions and the railway industry have invested substantially in the deployment of advances in power electronics in recent years to railway traction. The more prominent areas of R&D effort include:

(a) The use of medium frequency transformers to reduce the weight and size of the bulky and heavy trainborne traction transformers for railways with ac traction power supply systems,

(b)The use of Silicon Carbide based devices, feedback of energy generated through regenerative braking in dc traction power system back to the grid when the line is not receptive,

(c)Catenary free traction power supply for light rail and trams, Power supply quality improvement devices at feeder stations for railways with ac traction power supply system, and co-phase power supply for railways with ac traction power supply system.

V. MEDIUM FREQUENCY TRACTION TRANSFORMERS

For railways with ac traction power supply system the voltage collected from the overhead catenary is single phase with RMS values of 15 kV or 25 kV and frequencies of 60 Hz, 50 Hz or 16.7 Hz. A trainborne traction transformer is used to step down the RMS value of the voltage to a range between 600V and 1000V for rectification to dc voltage to drive the traction motors.

978-1-5090-0064-7/15 $31.00 © 2015 IEEE

This traction transformer is heavy (4 to 6 tons per unit) as the frequency is fairly low. The traction transformer is also hazardous because it is usually oil cooled. As the volume of the transformer core materials is inversely proportional to the frequency the size of the traction transformer can be reduced by increasing the value of the frequency. With the advance in converter technology it has become practicable to convert the frequency of the supply voltage to a few kHz before the voltage transformation takes place so that medium frequency transformers can be used, resulting in substantial reduction of transformer size. Due to limitations on the voltage rating of the converter units several levels of converters have to be employed as the primary voltage can range from 15 kV to 25 kV.

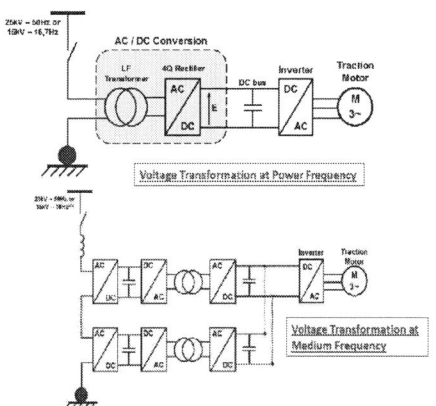

Fig.1: Traction Drives Schematics

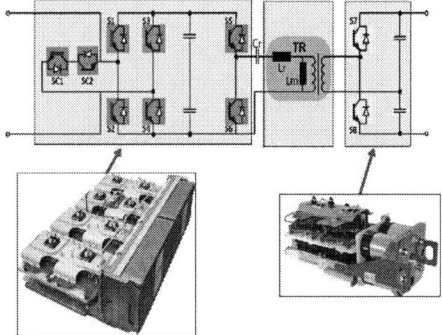

Fig. 2: One level of converters for frequency conversion

VI. SILICON CARBIDE BASED DEVICES

Silicon technology is reaching an asymptote and prospects for substantial advancement on breakdown voltage and switching performance are limited. For the past ten years there have been significant investment of R&D efforts on Silicon Carbide (SiC) technology as the prospects for advancement on blocking voltage, operating temperature and switching frequency are very attractive. In recent years convertors with hybrid modules of SiC diodes and IGBTs have been launched up to the 3.3 kV level. Recent trials conducted by Hitachi in a Japanese line have obtained the following results when comparing with the conventional IGBT modules:

a)   Inverter losses have been reduced by 35% by cutting diode switching loss to one-sixth and IGBT turn-on loss to less than one-half.

b)   Smaller size and lighter weight: because of the low losses described above, the SiC hybrid module is only two-thirds the size of previous devices with the same current capacity. Features such as the reduction in heat generation and the smaller size of the cooling system, which was achieved by using thermo fluid analysis to optimize the heat dissipating fin and heat pipe layout, which have succeeded in reducing the volume and weight of the inverter by 40% compared to the previous model.

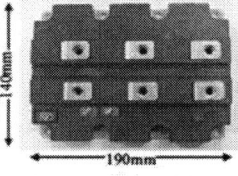

Fig. 3: IGBT Module

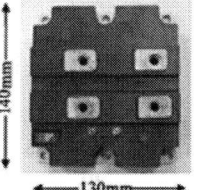

Fig. 4: Hybrid SiC Diode/IGBT Module

VI. DIRECT FEEDBACK OF ENERGY FROM DC TRACTION SUBSTATION TO THE GRID

For railways with dc traction power supply system when the line is not receptive the energy generated through regenerative braking has to be dissipated either through trackside or trainborne resistors, or stored in either trackside or trainborne energy storage devices. With the advance in converter technology another option of direct feedback of this redundant energy to the grid has now been developed. As shown in the following diagram this is achieved through the addition of an inverter unit to convert the redundant dc energy into 3-phase ac and synchronizing with the supply side voltage. A set of selector switch will connect the inverter unit to the supply side when the overhead catenary voltage is above a preset threshold which signifies that the line is not receptive and there is redundant energy arising from regenerative braking of trains.

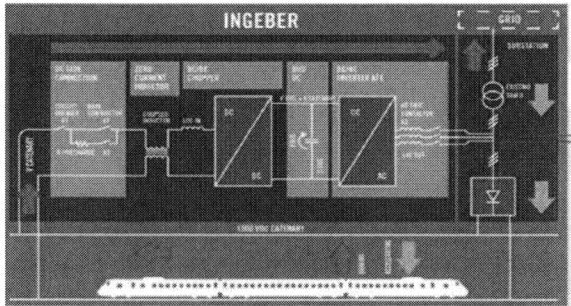

Fig. 5: Feedback of redundant energy to the supply side

VII. CATENARY FREE POWER SUPPLY FOR LIGHT RAIL AND TRAMS

A significant part of light rail or tram routes could be through densely built up areas or historical/scenic areas where the presence of overhead catenary for traction power collection would be most undesirable. With the advance in solid state switching technology safe system to enable the light rail vehicles or trams to collect traction current at ground level has now been developed and has also been deployed in some lines. The following figure shows the Alstom APS ground power collection system for light rail vehicles and trams:

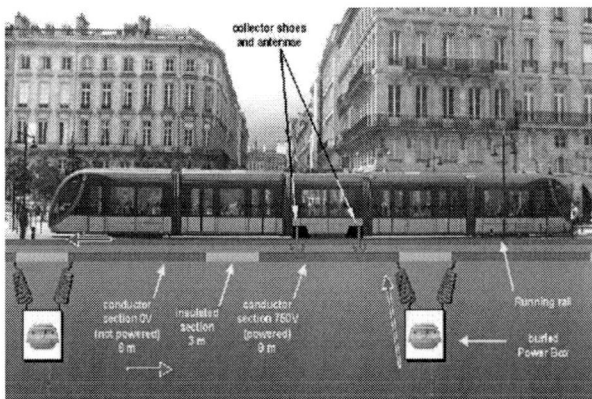

Fig. 6: Alstom APS Ground Power Collection System

The buried third rail system is divided into sections. Each of these sections would be normally in the switched off mode and would only be switched on upon the arrival of the collector shoes of the light rail vehicle or tram. This is made possible through the solid state power switches accommodated in buried boxes.

VIII. POWER SUPPLY QUALITY FOR AC TRACTION POWER SUPPLY SYSTEMS

For railways with ac traction power supply system impacts of the railway traction loadings upon the grid have always been a major concern, in particular in the areas of harmonics, phase unbalance and power factor. In the past various electrical methods have been employed to reduce such impacts, such as the Scott transformer. With the advancement in converter technology rail power conditioner (RPC) have now been employed to improve the quality of railway traction loading. As shown in the following diagram the basic principle of a rail power conditioner is to enhance the functionality of the load balancing transformer arrangement through the use of converters connected between the two legs on the LV side of the transformer. The Scott transformer balances loads presented to the grid provided that the loads taken by each of the two legs on the traction side are equal. In most cases the traction loads on each of the two legs are in fact not the same and the difference could be quite substantial. Under such circumstances the converters could compensate the reactive load drawn by the trains, such that only active load is seen by the Scott transformer. The rail power conditioner then equalizes the two resultant active loads drawn by each leg through transferring half of the load difference from the higher loaded leg to the other then the load presented to the grid on the HV side will be purely active and balanced.

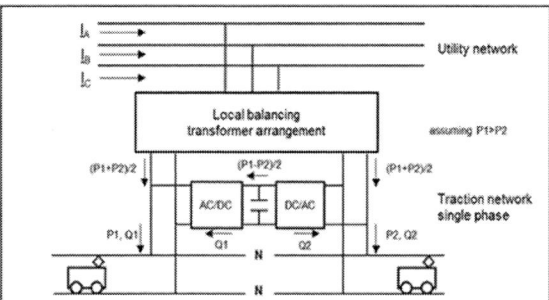

Fig. 7: Rail Power Conditioner Employing Two Converters

IX. CO-PHASE POWER SUPPLY

Due to the phase difference between the two adjacent feeding arms of ac traction power supply systems a neutral section has to be inserted in the overheard catenary to electrically segregate the two adjacent sections. As the neutral section is earthed the trainborne VCB has to be opened prior to entering the neutral section and then reclosed after passing through the neutral section. There is therefore a temporary loss of traction power supply to the train when it is passing through the neutral section and the train has to rely on the momentum of the train and traction drive of the other motive units of the train to maintain its speed. Severe arcing would occur if the trainborne VCB fails to open prior to entry to the neutral section, often leading to a dewirement incident.

As shown in the diagram below instead of the conventional approach of an open delta configuration on the LV side of the supply transformer feeding two adjacent legs of overhead catenary, a converter is used to receive the three phase balance output from the supply or step down transformer and then convert it into a single phase output, stepped up and then feed the overhead catenary. A step up transformer has to be used before the output of the converter can feed the overhead catenary because the converter's voltage rating is far below the rated voltage of 15 kV or 25 kV for the overhead catenary. In this way the need for neutral section could be avoided. In January this year the first set of co-phase power supply for commercial operation was successfully commissioned for a heavy haul line in Mainland China.

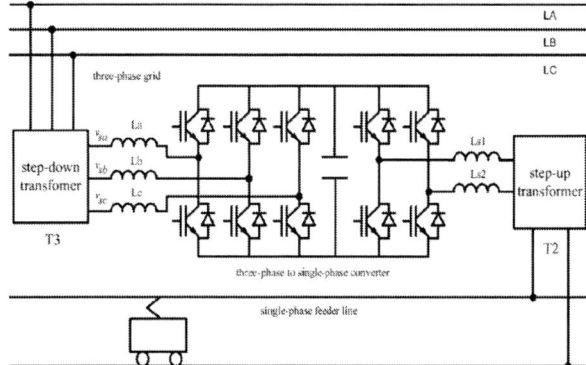

Fig. 8: Co-phase traction power supply based on converters

X. KEY PLAYERS IN THE RAILWAY INDUSTRY

Key players in the railway industry including ABB, Hitachi, Alstom, Siemens and Bombardier have invested substantially in the R & D works of the following areas:

a) Medium frequency traction transformers,
b) Co-phase power supply and rail power conditioners
c) Catenary free power supply for light rail and trams
d) Upgrading of the traction drive control and electronics
e) Semiconductor devices with higher voltage and current ratings and improved methods of heat dissipation

XI. Conclusions

The key to the ongoing evolution of railway traction is more power electronics. New converter concepts developed for transmission, distribution, and motor drives and the availability of power electronic devices with ever increasing voltage and current ratings and reduction in size are quickly finding their way into railway traction.

References

[1] Chuanhong Zhao, Drazen Dujic, Akos Mester, Juergen K. Steinke, Michael Weiss, Silvia Lewdeni-Schmid, Toufann Chaudhuri, and Philippe Stefanutti 'Power Electronic Traction Transformer—Medium Voltage Prototype ', IEEE TRANSACTIONS ON INDUSTRIAL ELECTRONICS, VOL. 61, NO. 7, JULY 2014

[2] Katsumi Ishikawa, Kazutoshi Ogawa, Seigo Yukutake, Norifumi Kameshiro, Yasuhiko Kono, 'Traction Inverter that Applies Compact 3.3 kV / 1200 A SiC Hybrid Module', 2014 International Power Electronics Conference

[3] [3] Priya Sharma, 'Reversible Substation in DC Traction', Journal of the International Association of Advanced Technology and Science Vol. 16, March 2015

[4] Margarita Novales, 'Overhead wires free light rail systems', Transportation Research Board 90th Annual Meeting

[5] Igor Perin, Peter F Nussey, Dr Umberto M Cella, Truc V Tran, 'Application of Power Electronics in Improving Power Quality and Supply Efficiency of AC Traction Networks', IEEE PEDS 2015, Sydney, Australia

[6] Xiaoqiong He, Zeliang Shu, Xu Peng, Qi Zhou, Yingying Zhou, Qijun Zhou, Shibin Gao, 'Advanced Co-phase Traction Power Supply System Based on Three-Phase to Single-Phase Converter', IEEE TRANSACTIONS ON POWER ELECTRONICS, VOL. 29, NO. 10, Oct 2014

978-1-5090-0064-7/15 $31.00 © 2015 IEEE

Achieving Mobility on Demand using Autonomous Vehicles

Marcelo H Ang Jr

Ag Director, Advanced Robotics Centre,
Associate Professor, Department of Mechanical Engineering National University of Singapore
mpeangh@nus.edu.sg

Abstract-Easy mobility is an important capability to enhance the quality of all our lives. Self-driving cars provide mobility-on-demand anytime and anywhere. Besides the convenience, autonomous driving provides a safe and productive environment, and an efficient use of resources. This talk shares our research group's current activities in achieving mobility-on-demand using autonomous vehicles, in both pedestrian (malls, airports, parks, etc) and road environments. The current state-of-the-art will be presented together with current challenges and our approach to solving these challenges.

I. INTRODUCTION

Transportation is a key service that has direct impact to the quality of our lives. It affects everybody from all walks of life: young or old, physically challenged or healthy, rich or poor. Being able to go places freely anytime without relying on a driver translates to independence, which in turn improves the quality of our lives. This is especially so for those who are not able to drive. These include the young and aged members of our society, and those recovering from accidents or illnesses. This motivates the work towards realizing autonomous vehicles or driverless cars that can self-drive.

There are many benefits of autonomous vehicles. On top of the list is safety, which is contrary to what some may believe. Autonomous vehicles are "driven" by computers. Computers never get tired, is always alter, consistent in the decision making, and always consistent. They key challenge is developing robust algorithms for perception of the environment and the deciding what actions to do depending on the situation and context of the environment. Productivity would be increased with autonomous vehicles. Imagine the extra hour that could be gained checking e-mails, doing work, etc while inside the car. This will also change the design of the car. A driver's seat is no longer needed. Perhaps the interior of car should be designed like a small 3 person office with a conference table and chairs around it; or a small living room with sofas. Fuel efficiency will also be improved since the computer is able to better driving the car at maximum efficiency, without unnecessarily accelerating and braking the vehicle.

Another issue is number of vehicles on the road is rapidly increasing and we cannot just keep expanding our road network. As of 2010, the number of vehicles in use in the world is estimated to be 1.015 billion [1], while the world population is estimated to be 6.916 billion [2]. This translates to one vehicle for every seven persons. Major cities around the world that experience rapid population growth are finding it difficult for their infrastructure to keep up. Let's look at the usage of our cars. How many hours a day do we use our cars. Maybe it's 2 hours and the car sits idle for the rest of the 22 hours. This poor equipment utilization is a waste of resource.

Mobility-on-Demand (MoD) transportations systems [3], such as car sharing or taxi services, results in maximizing the usage of cars. They can also be used to address the "first and last mile" problem by complementing and encouraging the use of public transport. This will lead to reduced private vehicle ownership and greater transportation network connectivity, which in turn will result in reduced traffic congestion as well as reduced overall commuting time. One of the main challenges of a MoD system is in the rebalancing of the vehicles to ensure minimal waiting time for the customers at a sustainable cost. An optimal and real-time rebalancing policy that can operate under stochastic customer demand is presented in [4]. One means of rebalancing the vehicles in a MoD system is to utilise autonomous vehicles [5], [6]. Autonomous vehicles offer additional safety, increased productivity, greater accessibility, better road efficiency, and have a positive impact to the environment. While the merits to using autonomous vehicles are aplenty, allowing this transportation paradigm shift to materialise will require the concurrence of (a) technology maturity, (b) government support, and (c) public acceptance of autonomous vehicles.

For the purpose of raising public awareness of autonomous vehicles and gaining user acceptance of the technology, a public trial involving two autonomous golf cars was conducted at the Chinese and Japanese Gardens in Singapore over the course of two weeks [7] and [8]. Members of the public were invited to experience the autonomous vehicles firsthand and could select any of ten destinations within the gardens. Visitors to the gardens could call for a vehicle via a website as well as monitor the status and positions of all vehicles in the gardens in real-time. As the golf cars had to continuously operate for at least six continuous hours with members of the public, which include children and the elderly, riding in them, we had to make the golf cars to be robust, reliable and safe.

II. MULTI-CLASS MOBILITY

While most current MoD systems furthermore are limited in the range of vehicle classes available to customers (perhaps small car to large sport-utility-vehicle sizes range), driverless vehicle classes can range from personal-transport-vehicle size, to full-size busses. Furthermore, there also exists autonomous aquatic and aerial vehicles which are often considered for separate usage cases. Each of these vehicle classes has strengths and weaknesses that offer advantages in certain environments. In this work we investigate the utility of a multi-class AV fleet for a MoD

978-1-5090-0064-7/15 $31.00 © 2015 IEEE

system, such that driverless technology would be available across various vehicle classes, where each class of vehicle would be chosen based on intended environment. By cooperatively allowing interchanges within such a multi-class AV fleet, a wider variety of environments could be serviced, yielding better area coverage and offering more flexible, convenient transportation options to the public.

A simple usage case is demonstrated through a flexible MoD system involving one Mitsubishi iMIEV car and one Yamaha G22E golf car on the National University of Singapore campus. Both vehicles are designed to utilize the same software architecture (with only low-level controls differing) and general sensor configuration, which are chosen for ease of fleet expansion. It is shown that while the car can operate at higher speeds on the road, the golf car has the flexibility of operating in pedestrian areas where cars are not allowed, thereby expanding the area coverage of the MoD service.

III. OUR AUTONOMOUS VEHICLES

The system architecture, which is common to both golf cars and our Mitsubishi iMIEV, comprises of four main modules: (1) perception, (2) planning, (3) control, and (4) external communication, as shown in Fig. 1. Note that only some portion of the internal sensors and vehicle controls may vary between vehicles, thus those portions of the software would need to be updated. The external sensors and bulk of the software would be maintained; this includes all high-level algorithms, such sensor data fusion and localization [4], mapping [5], and motion planning with RRT* [6] in our case. The Robot Operating System (ROS) is employed to standardize communication across modules [7].

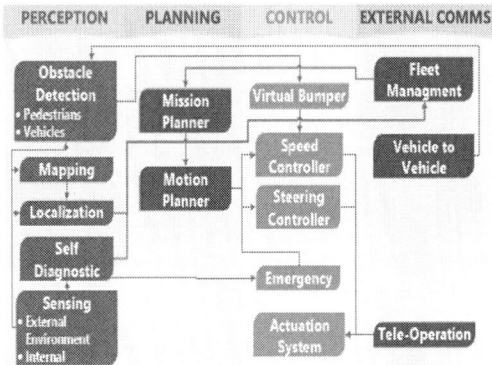

Fig. 1: System Architecture of our Autonomous Vehicle

The sensor configuration used was initially tested on a converted Yamaha golf car, with details provided in (1) and (2). The Mitsubishi iMIEV was then retrofitted with the same camera and lidar configuration, with odometry obtained by the same inertial measurement unit (IMU) now fused with CAN bus data, and all actuation performed via a Kairos Autonomy Pronto 4 kit. Both vehicle classes are shown in Fig. 2.

Fig. 3 shows the Yamaha YDREX3 electric golf car, which we used the vehicle base platform for pedestrian environments, and further retrofitted by our team to incorporate necessary actuation, sensing, computing, and

power systems along with various additional features to enhance passengers' comfort and safety.

Our vehicles do not use GPS. GPS is not reliable, has poor accuracies and are therefore not suitable for autonomous vehicles. Furthermore, we would like to operate our vehicles in GPS-denied environments, such as inside shopping malls and passenger terminal buildings. In a new environment, we have the manually drive our vehicles to all possible routes and locations to map building. We do simultaneous localization and map building [5].

Fig. 2: Our Fleet of Autonomous Vehicles

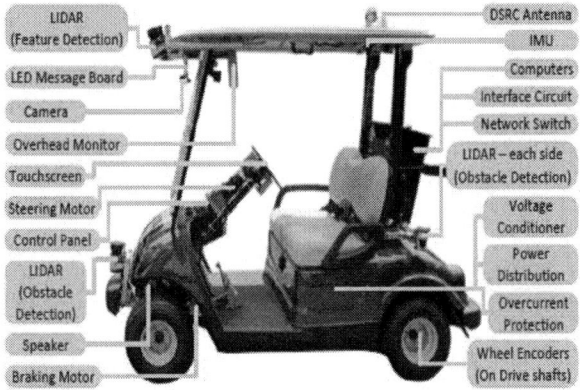

Fig. 3: Hardware overview, highlighting primary retrofit additions to a Yamaha YDREX3 golf car in order to enable autonomous operation. Similar modifications are done on a Mitsubishi Electric Vehicle (iMieV)

IV. SOME RESULTS

For the purpose of raising public awareness of autonomous vehicles and gaining user acceptance of the technology, a public trial involving two autonomous golf cars was conducted at the Chinese and Japanese Gardens in Singapore over the course of two weeks [8, 9]. Members of the public were invited to experience the autonomous vehicles firsthand and could select any of ten destinations within the gardens. Visitors to the gardens could call for a vehicle via a website as well as monitor the status and positions of all vehicles in the gardens in real-time. During the whole duration of the trial, the golf cars navigated reliably through all parts of the gardens. Fig. 4 shows the standard deviation plots of the localization system relative to the vehicle's orientation. The plot is

978-1-5090-0064-7/15 $31.00 © 2015 IEEE

drawn with a 0:5 m grid resolution. In each grid, the false colour represents the average standard deviation value reported by the localization system. Localization variance remains well under 0:4 m with the exception of a section of the path near the center of the map. This is where the worst standard deviation values are reported. This section of the path consists of a 13-Arch bridge (~45 m) crossing over a lake connecting the Chinese and Japanese Gardens. It is a particularly challenging section due to the lack of features and contains only repetitive geometrical shapes similar to that of a long corridor. Longitudinally, the largest deviation is found to be 1:1 m and 0:8 m lateral deviation is reported. Overall, the average deviation is 0:24 m_0:12 longitudinally and 0:16 m_0:10 laterally. It can be seen that the deviation value is larger longitudinally. This is also true when comparing two plots visually where the lateral plot stays on the colder spectrum than the longitudinal plot. During the course of operations at the Chinese and Japanese gardens we estimate that the SMART vehicles consumed approximately 100 kWh of power. The estimate was calculated based on the assumption that vehicles were charged from an empty state to 100% capacity for each of the six days of operations. The total combined distance traveled by the two golf cars was 351.6 km, as recorded by their odometry systems. The total number of trips was 220. More details can be found in. [9]

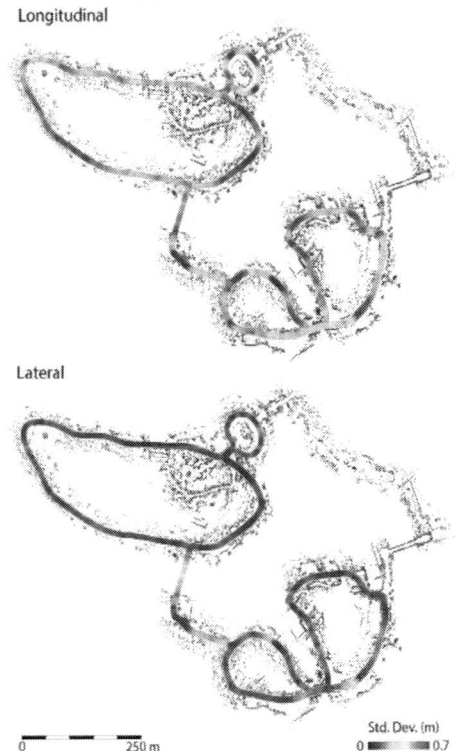

Fig. 4: Localization standard deviation results relative to vehicle's orientation
(Top: longitudinal standard deviation, bottom: lateral standard deviation)

Multi-Class Mobility on Demand was also tested in the National University of Singapore campus, highlighted in Fig. 5. Here, two vehicles are used, one Mitsubishi car, and one Yamaha golf car. As golf cars are used by security personnel in this area, they are allowed on the plaza and

even inside some buildings alongside pedestrian traffic. The car however has higher top speed, thus it is suited better for on-road use. With this system, a passenger travelled from the door of our office building to a coffee shop via a driverless golf car passing through the Stephen Riady Centre, then interchanged to the driverless car to be carried to our workshop. Images of the vehicles in operation along the route are shown in Fig. 6. By using two classes of vehicles in this case, the service area of the MoD system was greatly expanded over either single-class case. Both vehicles safely avoided real pedestrian and vehicular traffic in all runs [10].

Fig. 5: Experimental routes are highlighted on the university map, with golf car route in yellow, car route in green, and pick-up/drop-off locations marked as stars (Top). The interchange station is marked as a half yellow, half green star, with the bottom image showing both test vehicles at the interchange.

Fig. 6: Golf car is shown in Stephen Riady Centre (upper left) with visualization of lidar readout superimposed over the front camera image (upper right). Car is shown on road at roundabout (lower left), with the road surface shown by the lidar readout (lower right).

V. CONCLUSIONS

We have developed a multi-class driverless vehicle fleet for MoD and demonstrated this through successful experimental operation within a university campus. In our scenario, a driverless golf car was able to provide transportation in an unstructured pedestrian environment and stop at a transfer station for the passenger to board the driverless car, which travelled over roads alongside vehicular traffic.

We have also developed and deployed two autonomous golf buggies for mobility in demand for the public in Singapore's Chinese and Japanese gardens in Oct 2014. Since this is for public deployment and unlike typical experimental scenarios, the golf cars had to be designed to be safe, robust, and reliable, under prolonged operations while being subjected to the elements. The different implemented algorithms (e.g., mapping and localization, obstacle avoidance, Dynamic Virtual Bumper, booking system) have proven reliable over the prolonged operations. The purposes to raise public awareness about autonomous vehicle technology and to gain user acceptance were achieved as seen from the survey results.

We are currently developing a more efficient fleet management system. We are also testing our Mitsubishi Vehicle in One North, Singapore. One North is a designated area by Singapore's Land Transport Authority for Automatic Vehicle testing. As of Oct 2015, two vehicles have passed the safety test. One of the vehicles is SCOT (Shared Computer Operated Transport), our group's autonomous car for road environments, which is a modified Mitsubishi Vehicle (the iMiev) to make it autonomous. SCOT can now negotiate intersections and pedestrian crossings and recognize and obey traffic lights. We are now working on better algorithms for prediction of other vehicles' and pedestrians' motions, and improved motion planning algorithms.

ACKNOWLEDGEMENT

This research was supported by the Future Urban Mobility project of the Singapore-MIT Alliance for Research and Technology (SMART) Center, with funding from Singapore's National Research Foundation (NRF).

This is a project of the SMART-NUS Autonomous Vehicle Group: Scott Pendleton, Tawit Uthaicharoenpong, Zhuang Jie Chong, Guo Ming James Fu, Baoxing Qin, Wei Liu, Xiaotong Shen, Zhiyong Weng, Cody Kamin, Mark Adam Ang, Katarzyna Anna Marczuk, Hans Andersen, Mengdan Feng, , Zhuang Zhi Chong, Emilio Frazzoli, and Daniela Rus, and Marcelo H Ang Jr.

REFERENCES

[1] John Sousanis. World vehicle population tops 1 billion units. http://wardsauto.com/ar/world_vehicle_population_110815. Accessed: 2015-02-25.

[2] Population Division of the Department of Economic and Social Affairs of the United Nations Secretariat. World population prospects: The 2012 revision. http://esa.un.org/unpd/wpp/index.htm. Accessed: 2015-02-25.

[3] William J Mitchell. Reinventing the automobile: Personal urban mobility for the 21st century. MIT press, 2010.

[4] Z. J. Chong, B. Qin, T. Bandyopadhyay, M. H. Ang, E. Frazzoli, and D. Rus, D. "Synthetic 2D LIDAR for precise

vehicle localization in 3D urban environment." 2013 IEEE International Conference on Robotics and Automation (ICRA), IEEE, 2013.

[5] Z. J. Chong, B. Qin, T. Bandyopadhyay, M. H. Ang, E. Frazzoli, and D. Rus, "Mapping with synthetic 2D LIDAR in 3D urban environment." 2013 IEEE/RSJ International Conference on Intelligent Robots and Systems (IROS), IEEE, 2013.

[6] S. Karaman, and E. Frazzoli, "Sampling-based Algorithms for Optimal Motion Planning", International Journal of Robotics Research, Vol. 30, No. 7, 2011, pp 846-894.

[7] M. Quigley, K. Conley, B. P. Gerkey, J. Faust, T. Foote, J. Leibs, R. Wheeler, and A. Y. Ng, "ROS: an open-source Robot Operating System," ICRA Workshop on Open Source Software, 2009.

[8] Singapore. First driverless vehicles for public launched. http://smart.mit.edu/news-a-events/pressroom/article/51-first-driverless-vehicles-forpublic-launched.html. Accessed: 2015-02-25.

[9] Scott Pendleton , Tawit Uthaicharoenpong, Zhuang Jie Chong, Guo Ming James Fu, Baoxing Qin, Wei Liu, Xiaotong Shen, Zhiyong Weng, Cody Kamin, Mark Adam Ang, Lucas Tetsuya Kuwae, Katarzyna Anna Marczuk, Hans Andersen, Mengdan Feng, Gregory Butron, Zhuang Zhi Chong, Marcelo H. Ang Jr., Emilio Frazzoli, Daniela Rus, "Autonomous Golf Cars for Public Trial of Mobility-on-Demand Service", 2015 IEEE/RSJ International Conference on Intelligent Robots and Systems (IROS), Hamburg, Germany, 28 Sept to 2 Oct 2015.

[10] Scott Pendleton, Zhuang Jie Chong, Baoxing Qin, Wei Lin, Tawit Uthaicharoenpong, Xiaotong Shen, Guo Ming James Fu, Marcello Scarnecchia, Seong-Woo Kim, Marcelo H Ang Jr, Emilio Frazzoli, "Multi-Class Driverless Vehicle Cooperation for Mobility on Demand", the 21st World Congress on Intelligent Transport Systems, Detroit, USA, Oct 2014

BIOGRAPHY

Marcelo H Ang Jr received his BSc and MSc degrees in Mechanical Engineering from the De La Salle University in the Philippines and University of Hawaii, USA in 1981 and 1985, respectively, and his PhD in Electrical Engineering from the University of Rochester, New York in 1988 where he was an Assistant Professor of Electrical Engineering. In 1989, he joined the Department of Mechanical Engineering of the National University of Singapore where he is currently an Associate Professor and Acting Director of the Advanced Robotics Center. His research interests span the areas of robotics, mechatronics, autonomous systems, and applications of intelligent systems. He teaches robotics; creativity and innovation; applied electronics and instrumentation; computing; design and related areas. In addition to academic and research activities, he is also actively involved in the Singapore Robotic Games as its founding chairman, and the World Robot Olympiad as member of its Advisory Council.

Recent Progress in Developments of On-line Electric Vehicles

Su Y. Choi[1] and Chun T. Rim[2]

[1] Department of Nuclear and Quantum Engineering, KAIST, Korea
E-mail: suchoi@kaist.ac.kr
[2] Department of Nuclear and Quantum Engineering, KAIST, Korea
E-mail: ctrim@kaist.ac.kr

Abstract—Wireless electric vehicles (WEVs), classified into roadway powered electric vehicles (RPEVs) and stationary charging electric vehicles (SCEVs), are in the spotlight as future mainstream transportations. RPEVs are free from serious battery problems such as large, heavy, and expensive battery packs and long charging time because they get power directly from the road while moving. The power transfer capacity, efficiency, lateral tolerance, EMF, air-gap, size, weight, and cost of the WPTSs have been improved by virtues of innovative semiconductor switches, better coil designs, roadway construction techniques, and higher operating frequency. In this paper, fundamentals of WPTSs for RPEVs as well as recent progress in developments of on-line electric vehicles (OLEVs) that have been commercialized firstly in 2013 and made by KAIST, Korea are summarized. The fifth- (5G) and sixth-generation (6G) OLEVs, which can reduce construction cost and time for its commercialization, and the interoperability issue between RPEVs and SCEVs are addressed in detail.

Keywords—Wireless electric vehicle (WEV), roadway powered electric vehicle (RPEV), stationary charging electric vehicle (SCEV), on-line electric vehicle (OLEV)

I. INTRODUCTION

Wireless electric vehicles (WEVs) using wireless power transfer systems (WPTSs), which are classified into roadway powered electric vehicles (RPEVs), also referred to as dynamic charging electric vehicles, and stationary charging electric vehicles (SCEVs), have recently been in the spotlight as attractive alternatives to internal combustion vehicles (ICVs). Pure battery EVs (PEVs) [1]-[4], hybrid EVs (HEVs) [5]-[8], plug-in hybrid EVs (PHEVs) [9]-[11], and battery replace EVs (BREVs) [12]-[15] are also proposed. Among them, the charging of PEVs and PHEVs has gradually changed from wired type to wireless type, which is the background of actively studied SCEVs. Generally, WPTSs for WEVs can be divided into inductive power transfer systems (IPTSs) [16]-[94], coupled magnetic resonance systems (CMRSs) [95]-[97], and capacitive power transfer systems (CPTSs) [98]. It seems that these three WPTSs are totally different from each other; however, CMRSs are found to be just a special form of IPTSs having an extremely high quality factor (Q) [97]. CMRSs not only have difficulty in maintaining resonance conditions due to their extremely high Q but are also too bulky to be installed at the bottom of a vehicle. For these reasons, CMRSs are far from the right candidate for WEVs [99]. Meanwhile, even though CPTSs have their own strong advantages such as large tolerances and simple structures, they are only available for small-size WEVs, which require only several hundred W output power. Therefore, IPTSs have been considered as the most appropriate WPTS and are commonly used for RPEVs because they are not constrained from the aforementioned problems of CPTS and CMRS [99]. Among WEVs, RPEVs

are becoming the most promising candidate for future transportation because they are ideally free from large, heavy, and expensive batteries, and get power directly while moving on a road. Despite the fact that RPEVs are free from battery problems, RPEVs have not been widely used so far due to high initial investment cost for commercialization. To cope with this problem, in addition to new RPEV technology to effectively reduce the cost, strong motivation to build national infrastructures for RPEVs with public consent is needed. So the best deployment scenario would be that SCEVs are completely compatible with the IPTSs built for RPEVs. SCEVs will be widely deployed in the near future to replace the wired EV chargers of PHEVs and PEVs due to the convenience and safe operation of SCEVs. As a result of the growing interest in SCEVs, in 2010, the society of automotive engineers (SAE) established the SAE J2954 wireless charging task force to set the comprehensive standards for SCEVs such as transmitter and receiver coils, a feasible operating frequency band, air-gaps, power transfer levels and efficiencies, control strategies, foreign metal and living object detections, communication protocols, and magnetic and electric field regulations [100]. The standards are planned to be published in 2016, and it would be, therefore, worthwhile for us to follow up the SAE J2954 standard and design the IPTSs for RPEVs in accordance with the SAE J2954 for the interoperability between RPEVs and SCEVs.

As a follow-up to the previous review paper that dealt with a full history of WPTSs for RPEVs from their advent in the 1890s and important design considerations [99], recent studies of the past couple of years are newly summarized in this review paper. Above all, important technical issues in the new developments of the fifth- (5G) and sixth-generation (6G) on-line electric vehicles (OLEVs), which highly focus on the reduction of initial investment costs for commercialization as well as the interoperability between RPEVs and SCEVs, are addressed, while major milestones of the developments of other RPEVs are newly summarized. In the rest of this paper, some important technical issues, such as coil structures, power supply schemes, and segmentation switching techniques of a lumped inductive power transfer system (IPTS) for RPEVs, have been addressed.

II. FUNDAMENTALS OF RPEVS

1. Overall Configuration of RPEVs

As mentioned in the introduction section, most of the recent research teams for RPEVs adopt IPTSs for wireless power transfer purposes due to their highly efficient and robust power transfer characteristics to lateral displacements as well as relatively small transmitter and receiver coils compared to CMRSs and CPTSs. In general, the IPTS consists two subsystems [99]: one is the roadway subsystem to transfer power, which includes an HF inverter, a primary capacitor bank, and a power supply rail (or transmitter). The other one is the on-board subsystem to receive power from the roadway subsystem, and this includes a pick-up coil (or receiver), a secondary capacitor bank, a rectifier, and a regulator for a battery pack, as shown in Fig. 1.

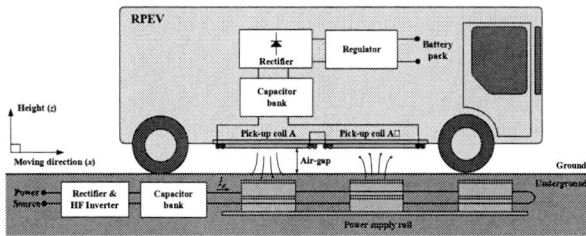

Fig. 1: The configuration of the IPTS for an RPEV.

2. Fundamental Principles of the IPTS

The IPTSs are basically governed by Ampere's and Faraday's laws among four Maxwell equations, and its basic operation can be briefly explained as follows:

1) If a high frequency AC current flows into a power supply rail, a time-varying magnetic flux is generated by Ampere's law;
2) In accordance with Faraday's law, voltage is induced across a pick-up coil inductively coupled with the power supply rail;
3) Through the inductive coupling, wireless power transfer is achieved while capacitors are used to nullify inductive reactance.

As explained above, the time-varying magnetic flux is generated from the AC current and unwanted leakage magnetic flux exposed to pedestrians and other electronic systems should be effectively mitigated to meet the ICNIRP guidelines [101]-[102] and EMC tests for commercialization. The EMF cancellation techniques are divided into two methods: one is passive EMF cancel methods using 1) high-conductivity materials to cancel the unwanted magnetic flux by using induced eddy-current on the material surface or using 2) high-relative-permeability materials to guide the unwanted magnetic flux to the intended directions by providing a low magnetic reluctance path. The other one is active EMF cancellation methods [103]-[115], which include several complex systems such as additional coils, power sources, phasor detectors, and controllers to generate the opposite magnetic flux to cancel the unwanted magnetic flux. In addition, compensation circuits are necessary for an IPTS of RPEVs to maximize its output power. It is critical to select an appropriate compensation circuit at the beginning of IPTS designs, considering critical electrical characteristics such as maximum efficiency conditions, maximum load power transfer conditions, load-independent output power conditions, coupling coefficient independent compensation conditions, and allowance for the absence of RPEVs. Among many compensation schemes, the current source series-series compensation (I-SS) scheme, as shown in Fig. 2, as well as the current source series-parallel compensation (I-SP) scheme, are viable solutions to meet all the mentioned electrical characteristics [116]. Moreover, the switching frequency of the inverter f_s for IPTSs does not match the resonant frequency of LC circuits f_r because the zero voltage switching (ZVS) of inverters [55]-[56] can be guaranteed as f_s becomes a little bit higher than f_r. For those who are interested in the detailed requirements and design issues of the IPTS for RPEVs, it is recommended to refer to a previous review paper [99].

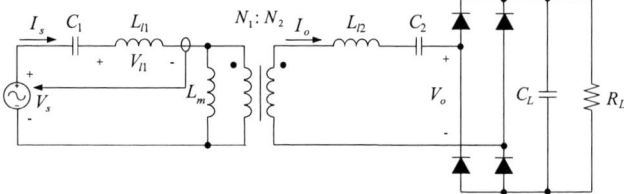

Fig. 2: An equivalent circuit of the current source series-series compensation (I-SS) scheme for an IPTS for RPEVs.

3. The Early History of RPEV

The origin of the RPEV is found in the US patent of the transformer system for electric railways by M. Hustin and M. Leblanc in France in 1894 [16], and the basic configuration of this transformer system is almost the same as modern IPTSs. Moreover, several important design features of IPTSs for RPEVs, such as the deployment of power supply rails, compensation schemes, high-power transfer to pick-up coils, and reduction of conduction and eddy current losses, have been claimed in the patent. About 100 year later, as a result of the growing interest in RPEVs due to the oil crisis in the 1970s, three projects were undertaken in the US to investigate RPEVs in order to minimize the petroleum use of vehicles [17]-[39]. The first project was started in 1976 by the Lawrence Berkeley National Laboratory [19]-[20], [22] while the Santa Barbara Electric Bus Project was started in 1979. During these two projects, several prototype RPEVs were developed but not appropriately operated [25]-[30]. After the previous two projects of RPEVs, the University of California, Berkeley undertook the Partners for Advanced Transit and Highways (PATH) program in 1992 to validate the technical viability of RPEVs in [37]-[38]. Through the PATH program, broad investigation and field tests on RPEVs were performed and the PATH team developed the first RPEV buses, which achieved an efficiency of 60% at an output power of 60 kW with a 7.6 cm air-gap. The RPEV buses, however, had not been commercialized because they used a low operating frequency of 400 Hz due to the absences of fast power switches, good ferrite cores, and litz wires. As a result, a high-power rail construction cost of around 1 M$/km, a large power rail current of thousands amperes, a heavy power supply and pick-up coils, acoustic noises, and low system efficiencies were obtained. In addition, the small air-gap, lateral tolerance, and large EMF level are not acceptable for its commercialization.

III. DEVELOPMENTS OF ON-LINE ELECTRIC VEHICLES

Since 2009, the OLEV project has been conducted by a research team led by KAIST [40]-[73], and this project has solved most of the remaining problems of the PATH project for commercialization. Throughout the OLEV project, innovative coil designs and roadway construction techniques as well as a reasonably high operating frequency of 20 kHz realized the highest power efficiency of 83% at an

978-1-5090-0064-7/15 $31.00 © 2015 IEEE

output power of 60 kW with a large air-gap of 20 cm and a fairly good lateral tolerance of 24 cm [56]. Moreover, the power rail construction cost of the OLEV, which accounts for more than 80% of the total commercialization cost for RPEVs [99], has been dramatically reduced to at least a third of that of the PATH project. By virtue of the high operating frequency, the power supply current has also been reasonably mitigated by 200 A, and the battery size has been significantly reduced to 20 kWh, which can be further reduced by increasing the length of power supply rails.

Throughout the OLEV project, the first- (1G), second- (2G), third- (3G), upgraded third- (3⁺G), and fourth-generation (4G) OLEVs have been developed by KAIST [99] and extensively tested at the test sites at KAIST since 2009, as shown in Fig. 3.

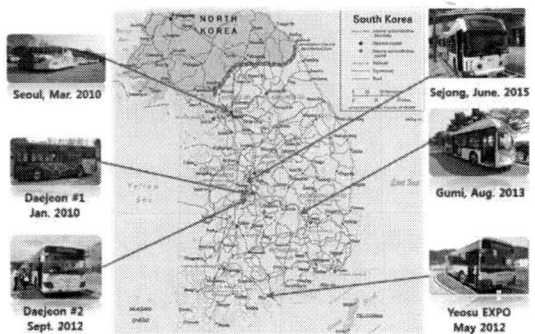

Fig. 3: The deployment status of OLEVs in Korea.

Among them, the 3⁺G OLEV has been widely deployed in Korea and firstly commercialized at a 48km route in Gumi, Korea. In addition, the 3⁺G OLEV bus has been newly commercialized in two bus routes, 12 km in length, respectively, in Sejong, Korea since June 2015. In the rest of this section, the new developments of the fifth- (5G) and sixth-generation (6G) OLEVs, which are highly focused on the reduction of the initial investment cost for its commercialization as well as the interoperability between RPEVs and SCEVs to strongly promote the commercialization of RPEVs, will be primarily addressed since the full development history of the 1G, 2G, 3G, 3⁺G, and 4G OLEVs were already addressed in a previous review paper [99]. Now, the development of the 5G OLEV, which adopted an ultra slim S-type power supply module of 4 cm width to further reduce the power supply rail construction cost and time, is in the final stage of development [71], while that of the 6G OLEV using a new coreless power supply rail for the interoperability between RPEVs and SCEVs is in the early stages of development [117].

1. The Fifth-generation (5G) OLEV
Through the developments of the previous OLEVs, a significant improvement in lateral tolerance as well as a large air-gap, high power efficiency, lower construction time and cost have been achieved. The construction cost and time of the power supply rail, however, should be further reduced for better commercialization because the construction cost of the power supply rail is critical for deploying the RPEVs and long construction time results in more traffic jams and extra deployment costs.

In order to mitigate these problems, the 5G OLEV adopted an ultra slim S-type power supply rail for RPEVs with a maximum output power of 22 kW for a flat pick-up, having an air-gap of 20 cm and a lateral tolerance of 30 cm [71]. The S-type power supply rail, where the name "S-type" stems from the front shape of the power supply rail, as shown in Fig. 4(c), has an ultra slim width of only 4 cm by virtue of the S-type configuration, which has been decreased more than two times compared to that of the I-type power supply module for the 4G OLEV.

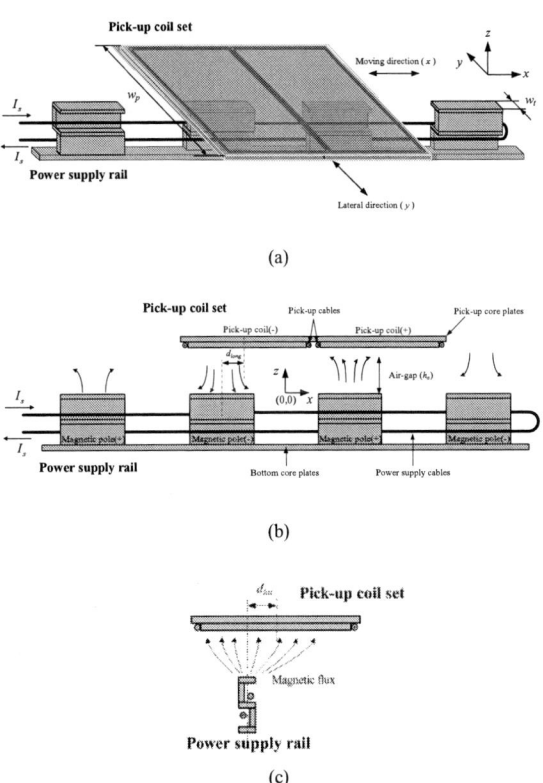

Fig. 4: Proposed ultra-slim S-type power rail and the flat pick-up coil of the IPTS [71]. (a) Bird's eye view. (b) Side view. (c) Front view.

Each magnetic pole of the S-type power rail consists of ferrite core plates and power cables, and the adjacent magnetic poles are connected by bottom core plates, as shown in Fig. 4(a). The EMF for pedestrians around the power supply rail can be significantly reduced due to the opposite magnetic polarity of adjacent poles, as shown in Fig. 4(b). The EMF generated from the I-type power supply rail and a flat type pick-up coil set, which is basically the same structure of the proposed S-type one, was as low as 1.5 µT at a distance of 1 m from the power supply rail [71]. Therefore, the S-type one has a similar EMF to the I-type one, which is well below the ICNIRP guideline of 27 µT. Moreover, a large lateral displacement d_{lat} is obtained [71] due to the small width w_t of the S-type power supply rail for a given pick-up width w_p, as in (1).

$$d_{lat} \cong \frac{w_p}{2} - \frac{w_t}{2} \qquad (1)$$

From the experimental verifications, a large lateral tolerance of 30 cm at an air gap of 20 cm was experimentally obtained, which is 6 cm larger than that of the I-type power supply rail

978-1-5090-0064-7/15 $31.00 © 2015 IEEE

[71]. The proposed S-type power supply rail adopts a module concept, as shown in Fig. 5(a), and makes it easier to fold itself by virtue of flexible thin power cables. Therefore, no power cable connection is required after being deployed, as shown in Fig. 5(b).

In summary, the S-type power supply rail leads to lower construction cost and deployment time because the impact of deploying the S-type power supply modules on the road surface is minimal and the change of existing road operation conditions is also greatly reduced. As a result, the S-type power supply rail has the following three merits:

1) Substantial reduction in construction cost and time for commercialization of RPEVs,
2) Larger lateral tolerance of an IPTS,
3) Further reduction in the EMF generated from a power supply rail for pedestrians.

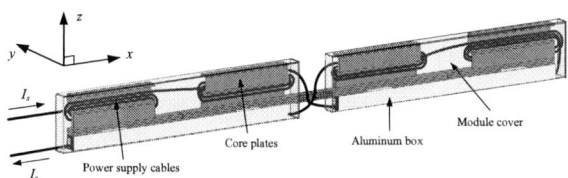

(a)

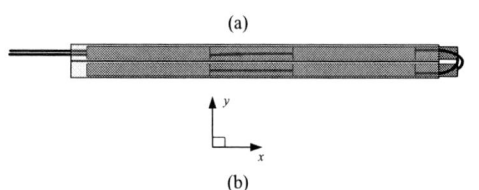

(b)

Fig. 5: Configuration of the ultra slim S-type power supply modules including two magnetic poles [71]. (a) Bird's eye view for two unfolded modules. (b) Top view of a folded module.

The S-type power supply module was fabricated, as shown in Fig. 6. The S-type power supply module includes S-type power supply rails, a transparent module cover, and an aluminum box for capacitor banks for better heat transfer to the ground. As shown in Fig. 7, the capacitor bank should be smaller than the aluminum box to be inserted into it and electrically isolated from the aluminum box because aluminum is a conductive material. In addition to this, the aluminum box also plays a role in canceling the leakage magnetic flux generated by power supply rails. Moreover, each module, which has two magnetic poles, is serially connected to the capacitor bank installed in an adjacent module in order to mitigate high voltage stress for a capacitor bank due to the large self-inductance of the power supply rail. In general, thinner power supply cables for the S-type power supply rail are more suitable choice with respect to the construction cost of a power supply rail. This is because the space usability for cable windings can be maximized by thinner power cables for a given size and ampere-turn of a power supply rail. However, it should be noted that there are several demerits of using many turns for the power supply rail such as higher conduction loss due to the proximity effect and worsened heat transfer ability to ambient.

With the flexible thin power cables having a diameter of 0.9 cm, it is possible to connect the cables for power supply

modules at the factory instead of at the construction site, thus avoiding a long construction time, which leads to traffic jams and additional construction costs.

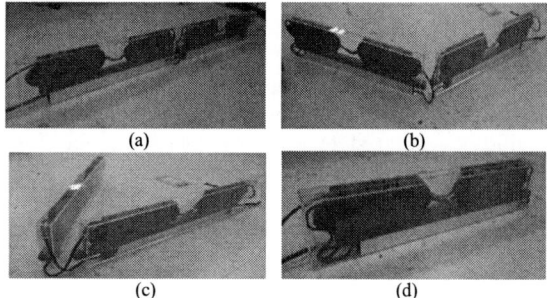

Fig. 6: Fabricated ultra slim S-type power supply modules [71]. (a) Fully deployed case. (b) One-third folded case. (c) Two-thirds folded case. (d) Completely folded case.

For commercialization, power supply modules should be robust to high humidity as well as repetitive external mechanical impacts for at least 10 years [99]. As a remedy, the S-type power supply module can be filled with epoxy to reinforce the S-type power supply rail and to protect the module from high humidity while the module cover endures external mechanical impacts.

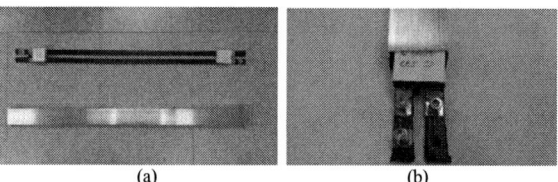

Fig. 7: Aluminum box and capacitor banks for the S-type power supply modules [71]. (a) Separated case. (b) Inserted case.

In conclusion, the characteristics of the I-type and S-type power supply rails are summarized, as shown in the Table I.

TABLE I:
CHARACTERISTICS OF I-TYPE AND S-TYPE POWER SUPPLY RAILS [71]

	I-type (4G OLEV)	S-type (5G OLEV)
Rail width	10 cm	4 cm
Lateral tolerance	24 cm	30 cm
Air-gap	20 cm	20 cm
Output power	27 kW/pick-up	22 kW/pick-up
Efficiency	74 % at 27kW	71 % at 22kW

Except for the rail width reduction and lateral tolerance increase, the output power and efficiency of the S-type one are slightly inferior to those of the I-type one, assuming that the pick-up size and ampere-turn of the power supply rail are the same. For readers who are interested in the detailed design issues and experimental verifications for the 5G OLEV of the S-type power supply rail, it is recommended to refer to the reference [71].

2. The Sixth-generation (6G) OLEV
Although RPEVs are free from the battery problems and are ready for commercialization, RPEVs have not been widely used so far due to the huge initial investment cost. To cope with this problem, we may need strong motivation to build the national infrastructure for RPEVs with public consent. So one of the best deployment scenarios would be that SCEVs are designed to be completely compatible with IPTSs for RPEVs to be wirelessly charged while moving on a road, as shown in

978-1-5090-0064-7/15 $31.00 © 2015 IEEE 450

Fig. 8, because there is no doubt that SCEVs will be widely deployed all over the world in the near future to replace wired EV chargers due to SCEVs' convenience and safe operation. However, the design considerations of an IPTS for RPEVs are quite different from those of SCEVs because the IPTS for RPEVs should meet additional system requirements such as low construction cost and time, low voltage stress, large lateral tolerance, high power delivery capability, and continuous power delivery while moving on the road.

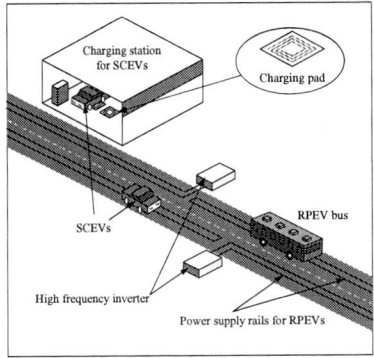

Fig. 8: Ideal concepts for SCEVs compatible with power supply rails for RPEVs [117].

In order to manage the interoperability issue between RPEVs and SCEVs to the satisfaction of the design goals for RPEVs, the 6G OLEVs, using a new coreless power supply rail, were recently proposed [117]. As shown in Fig. 9, the shape of the proposed coreless power supply rail is basically the same as the U- and W-type power supply rails, which were used for the 3G and 3G+ OLEVs, but there is no core plate. Therefore, the proposed coreless power supply rail can generate a uniform magnetic field along a road, and a rectangular pick-up coil determined by the SAE J2954 for SCEVs, as shown in Fig. 9(a), is completely compatible with the power supply rail for RPEVs and continuously get uniform output power when moving on the road. Moreover, the construction cost and time of the proposed coreless power supply rails can be further reduced compared to conventional with-core power supply rails because core plates are totally eliminated in the proposed coreless power supply rails.

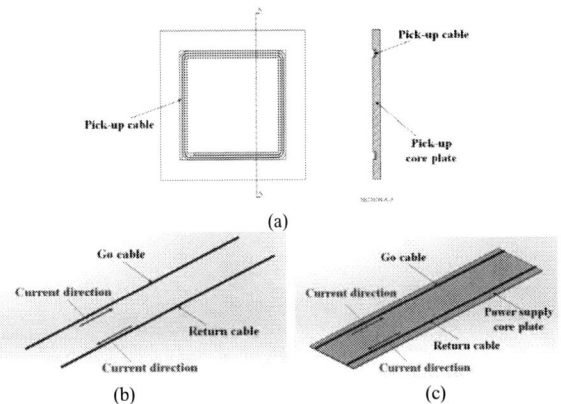

Fig. 9: Conceptual scheme of the proposed coreless power supply rail for both RPEVs and SCEVs [117]. (a) A rectangular pick-up coil for SCEVs in accordance with the SAE J2954. (b) Proposed coreless power supply rail. (c) Conventional power rail used for the 3G and 3G+ OLEVs.

Meanwhile, the large voltage stresses on both the compensation capacitor bank and distributed power supply rail V_{l1} are a unique feature of RPEVs, and this is one of the main reasons that the operating frequency is limited by around 20 kHz [99] because the voltage stress is directly proportional to its operating frequency f_s as well as its power supply current I_s as follows:

$$V_{l1} = j\omega_s L_{l1} I_s \quad \because \omega_s = 2\pi f_s \;, \qquad (2a)$$

$$\frac{\partial V_{l1}}{\partial x} = j\omega_s I_s \frac{\partial L_{l1}}{\partial x} \;. \qquad (2b)$$

In accordance with the magnetic mirror model [66], however, it is well known that the self-inductance of a power supply rail with core plates can become two times that without core plates when core plates are infinitely long and its relative permeability is infinite; hence, it is expected that the proposed coreless power supply rail will have about a half of its previous voltage stress due to its reduced inductance so that the operating frequency can be increased from 20 kHz to 85 kHz to meet the SAE J2954 standard for SCEVs with only about two times the voltage stress compared to that of 20 kHz for a given rail current. Moreover, it is also expected that the proposed coreless power supply rail can guarantee a large lateral tolerance compared to the conventional with-core power supply rails because the self-inductance variation of the pick-up coil along to lateral displacements is small enough to be negligible when a coreless power supply rail is used.

In order to validate those unique characteristics of the proposed coreless power supply rail, two simulation models, which are with-core and coreless power supply rails with a rectangular pick-up for SCEVs, are proposed, as shown in Fig. 10.

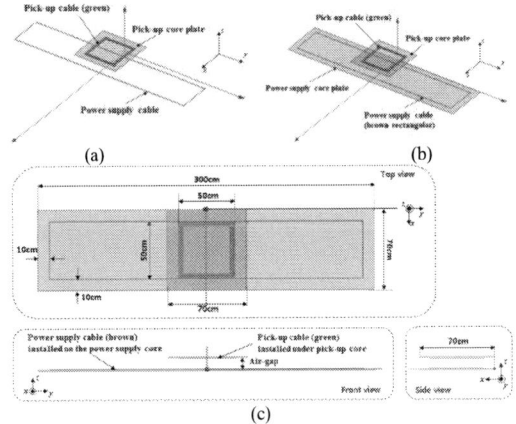

Fig. 10: Maxwell simulation models for the proposed coreless power supply rail with a rectangular pick-up [117]. (a) Bird's eye view of the proposed coreless power supply rail. (b) Bird's eye view of the conventional with-core power supply rail. (c) Dimensions for the simulation models.

From the simulation results, it is found that the self-inductance of the proposed coreless rail is about half that of the conventional with-core power supply rail, as shown in Fig. 11(a). At the same time, the mutual inductance of the proposed coreless power rail also becomes half that of the conventional with-core power rail, as shown in Fig. 11(b).

978-1-5090-0064-7/15 $31.00 © 2015 IEEE

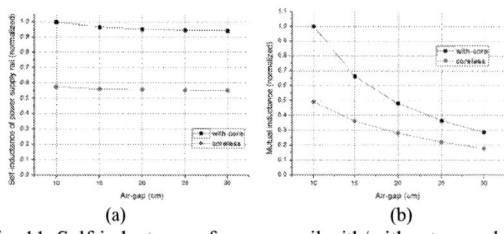

(a)   (b)

Fig. 11: Self-inductance of a power rail with/without core plates along to an air-gap [117]. (a) Self-inductance and (b) mutual inductance variations.

Moreover, the self-inductance variation of a pick-up coil with the proposed coreless power supply rail is negligible along lateral displacements, as shown in Fig. 12, which means that the proposed coreless power supply rail can guarantee a larger lateral tolerance of RPEVs and SCEVs compared to the conventional with-core power supply rail [117].

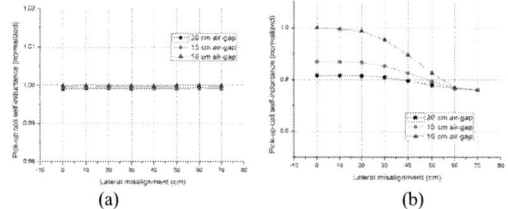

(a)   (b)

Fig. 12: Self-inductance of the pick-up coil along lateral displacements [117]. (a) Proposed coreless rail. (b) Conventional with-core rail.

IV. CONCLUSION

As a follow-up to the previous review paper that dealt with a full history of WPTSs for RPEVs from its advent in the 1890s to its modern status [99], recent studies over the past couple of years have been newly summarized in this review paper. For the more than 100-year history of RPEVs, the power transfer capacity, efficiency, lateral tolerance, EMF, air-gap, size, weight, and cost of the WPTSs have been significantly improved by virtues of innovative semiconductor switches, better coil designs, enhanced roadway construction techniques, and higher operating frequency. Thus, there is no doubt that RPEVs are becoming viable solutions for future transportation and that the 6G OLEV, reducing infrastructure cost for commercialization and increasing the interoperability between RPEVs and SCEVs, will be an especially strong candidate for the near-future widespread use of RPEVs in public transportation.

References

[1] J. T. Salihi, P. D. Agarwal, and G. J. Spix, "Induction motor control scheme for battery-powered electric car (GM-Electrovair I)," *IEEE Trans. on Industry and General Applications*, Vol. IGA-3, No. 5, pp. 463-469, Sept. 1967.

[2] J. R. Bish, and G. P. Tietmeyer, "Electric vehicle field test experience," *IEEE Trans. on Vehicular Technology*, Vol. 32, No. 1, pp. 81-89, Feb. 1983.

[3] J. Dixon, I. Nakashima, E. F. Arcos, and M. Ortuzar, "Electric vehicle using a combination of ultra capacitors and ZEBRA battery," *IEEE Trans. on Ind. Electron.*, Vol. 57, No. 3, pp. 943-949, Mar. 2010.

[4] C. H. Kim, M. Y. Kim, and G. W. Moon, "A modularized charge equalizer using a battery monitoring IC for series-connected Li-Ion battery string in electric vehicles," *IEEE Trans. Power Electron.*, Vol. 28, No. 8, pp. 3779-3787, Nov. 2012.

[5] J. Malan and M. J. Kamper, "Performance of hybrid electric vehicle using reluctance synchronous machine technology," *IEEE Industrial Applications Conference*, 2000, pp. 1881-1887.

[6] N. A. Rahim, H. W. Ping, and M. Tadjuddin, "Design of axial flux permanent magnet brushless DC motor for direct drive of electric vehicle," *IEEE Power Electronics Society General meeting*, 2007, pp. 1-6.

[7] M. Ceraolo, A. Donato, and G. Franceschi, "A general approach to energy optimization of hybrid electric vehicles," *IEEE Trans. on Vehicular Technology*, Vol. 57, No. 3, pp. 1433-1441, May 2008.

[8] Y. Cheng, R. Trigui, C. Espanet, A. Bouscayrol, and S. Cui, "Specifications and designs of a PM electric variable transmission for Toyota Prius II," *IEEE Trans. on Vehicular Technology*, Vol. 60, No. 9, pp. 4106-4114, Nov. 2011.

[9] F. L. Mapelli, D. Tarsitano, and Macro Mauri, "Plug-in hybrid electric vehicle: modeling, prototype, realization, and inverter losses reduction analysis," *IEEE Tran. on Ind. Electron.*, Vol. 57, No. 2, pp. 598-607, Feb. 2010.

[10] E. Tara, S. Shahidinejad, and Eric Bibeau, "Battery storage sizing in a retrofitted plug in hybrid electric vehicle," *IEEE Trans. on Vehicular Technology*, Vol. 59, No. 6, pp. 2786-2794, July 2010.

[11] S. G. Li, F. C. Walsh, and C. N. Zhang, "Energy and battery management of a plug-in series hybrid electric vehicle using fuzzy logic," *IEEE Tran. on Vehicular Technology*, Vol. 60, No. 8, pp. 3571-3585, Oct. 2011.

[12] P. Lombardi, M. Heuer, and Z. Styczynski, "Battery switch station as storage system in an autonomous power system: optimization issue," *IEEE Power and Energy Society General Meeting*, 2010, pp. 1-6.

[13] M. Takagi. Y. Iwafune, K. Yamaji, H. Yamato, K. Okano, R. Hiwatari, and T. Ikeya, "Economic value of PV energy storage using batteries of battery-switch stations," *IEEE Trans. on Sustainable Energy*, Vol. 4, No. 1, pp. 164-173, Jan. 2013.

[14] J. J. Jamian, M. W. Mustafa, Z. Muda, and M. M. Aman, "Effect of load models on battery-switching station allocation in distribution network," *IEEE International Conference on Power and Energy*, 2012, pp. 189-193.

[15] M. Takagi, Y. Iwafune, H. Yamamoto, K. Yamaji, K. Okano, R. Hiwatari, and T. Ikeya, "Energy storage of PV using batteries of battery-switch stations," *IEEE International Symposium on Industrial Electronics* (*ISIE*), 2010, pp. 3413-3419.

[16] M. Hutin and M. Leblanc, "Transformer system for electric railways," Patent US 527857, 1894.

[17] J. G. Bolger, "Supplying power to vehicles," Patent US 3914562, 1975.

[18] J. G. Bolger, "Roadway for supplying power to vehicle and method of using the same," Patent US 4007817, 1977.

[19] J. G. Bolger and F. A. Kirsten, Investigation of the feasibility of a dual mode electric transportation system, Lawrence Berkeley National Laboratory Report, LBL6301, May 1977.

[20] J. G. Bolger, M. I. Green, L. S. Ng, and R. I. Wallace, Test of the performance and characteristics of a prototype inductive power coupling for electric highway systems, Lawrence Berkeley National Laboratory Report, LBL7522, July 1978.

[21] J. G. Bolger, F. A. Kirsten, and L. S. Ng, "Inductive power coupling for an electric highway system," *in Proc. IEEE 28th Vehicular Technology Conference*, 1978, pp. 137-144.

[22] J. G. Bolger, L. S. Ng, D. B. Turner, and R. I. Wallace, "Testing a prototype inductive power coupling for an electric vehicle highway system," *in Proc. IEEE 29th Vehicular Technology Conference*, 1979, pp. 48-56.

[23] C. E. Zell and J. G. Bolger, "Development of an engineering prototype of a roadway powered electric transit vehicle system," *in Proc. 32nd IEEE Vehicular Technology Conference*, 1982, pp. 435-438.

[24] J. G. Bolger, "Power control system for electrically driven vehicle," Patent US 4331225, 1982.

[25] Santa Barbara Electric Bus Project, Phase 3A-final report, Santa Barbara Research Paper, Sept. 1983.

[26] Santa Barbara Electric Bus Project, Prototype development and testing program phase 3B-final report, Santa Barbara Research Paper, Sept. 1984.

[27] Santa Barbara Electric Bus Project, Test facilities development and testing program-static test report, Santa Barbara Research Paper, June 1985.

[28] Santa Barbara Electric Bus Project, Prototype development and testing program phase 3C-final report, Santa Barbara Research Paper, May 1986.

[29] K. Lashkari, S. E. Schladover, and E. H. Lechner, "Inductive power transfer to an electric vehicle," *in Proc. of 8th International Electric Vehicle Symposium*, 1986.

[30] E. H. Lechner and S. E. Schladover, "The roadway powered electric vehicle-an all-electric hybrid system," *in Proc. of 8th International Electric Vehicle Symposium*, 1986.

978-1-5090-0064-7/15 $31.00 © 2015 IEEE

[31] J. G. Bolger and L. S. Ng, "Inductive power coupling with constant voltage output," Patent US 0253345, 1988.

[32] S. E. Schladover, "Systems engineering of the roadway powered electric vehicle technology," *in Proc. of 9ᵗʰ International Electric Vehicle Symposium*, 1988.

[33] J. G. Bolger, "Roadway power and control system for inductively coupled transportation system," Patent US 4836344, 1989.

[34] M. Eghtesadi, "Inductive power transfer to an electric vehicle-an analytical model," *in Proc. 40ᵗʰ IEEE Vehicular Technology Conference*, 1990, pp. 100-104.

[35] K. W. Klontz, D. M. Divan, D. W. Novotny, and R. D. Lorenz, "Contactless battery charging system," Patent US 5157319, 1992.

[36] K. W. Klontz, D. M. Divan, D. W. Novotny, and R. D. Lorenz, "Contactless coaxial winding transformer power transfer system," Patent US 5341280, 1994.

[37] J. G. Bolger, "Urban electric transportation systems: the role of magnetic power transfer," *IEEE WESCON94 Conference*, 1994, pp. 41-45.

[38] California PATH Program, *Roadway powered electric vehicle project track construction and testing program phase 3D*, California PATH Research Paper, Mar. 1994.

[39] K. W. Klontz, D. M. Divan, D. W. Novotny, and R. D. Lorenz, "Contactless power delivery system for mining applications," *IEEE Trans. on Ind. Appli.*, Vol. 31, No. 1, pp. 27-35, Jan. 1995.

[40] KAIST OLEV team, *Feasibility studies of On-Line Electric Vehicle (OLEV) Project*, KAIST Internal Report, Aug. 2009.

[41] N. P. Suh, D. H. Cho, and Chun T. Rim, "Design of on-line electric vehicle (OLEV)," *Plenary lecture at the 2010 CIRP Design Conference*, 2010, pp. 3-8.

[42] S. W. Lee, J. Huh, C. B. Park, N. S. Choi, G. H. Cho, and Chun T. Rim, "On-line electric vehicle (OLEV) using inductive power transfer system," *IEEE Energy Conversion Congress and Exposition* (*ECCE*), 2010, pp. 1598-1601.

[43] J. Huh, S. W. Lee, C. B. Park, G. H. Cho, and Chun T. Rim, "High performance inductive power transfer system with narrow rail width for on-line electric vehicles," *IEEE Energy Conversion Congress and Exposition* (*ECCE*), 2010, pp. 647-651.

[44] Chun T. Rim, "The difficult technologies in wireless power transfer," *Trans. of the Korean Institute of Power Electronics* (*KIPE*), Vol. 15, No. 6, pp. 32-39, Dec. 2010.

[45] S. W. Lee, C. B. Park, J. G. Cho, G. H. Cho, and Chun T. Rim, "Ultra slim U & W power supply and pick-up coil design for OLEV," *Korean Institute of Power Electronics* (*KIPE*) *Annual Summer Conference*, 2010, pp. 353-354.

[46] G. H. Jung, K. H. Lee, H. G. Kim, Y. J. Cho, B. Y. Song, Y. D. Son, E. H. Park, J. Y. Park, J. Y. Choi, B. O. Kong, J. Huh, H. S. Son, J. G. Cho, Chun T. Rim, and S. J. Jeon, "Non-touch inductive power transfer system for OLEV," *Korean Institute of Electrical Engineering* (*KIEE*) *Annual Summer Conference*, 2010, pp. 1054-1055.

[47] C. B. Park, S. W. Lee, and Chun T. Lim, "Dynamic phasor transformation using complex Laplace transformation," *Korean Institute of Power Electronics* (*KIPE*) *Annual Fall Conference*, 2010, pp. 46-47.

[48] G. H. Jung, K. H. Lee, H. G. Kim, Y. J. Cho, B. Y. Song, Y. D. Son, E. H. Park, J. Y. Park, J. Y. Choi, B. O. Kong, J. Huh, H. S. Son, J. G. Cho, Chun T. Rim, and S. J. Jeon, "Power supply and pick-up system for OLEV," *Korean Institute of Power Electronics* (*KIPE*) *Annual Summer Conference*, 2010, pp. 218-219.

[49] J. Huh, W. Y. Lee, J. G. Cho, G. H. Cho, and Chun T. Rim, "A study on current source-transformer resonance inductive power transfer system," *Korean Institute of Power Electronics Annual Summer Conference*, 2010, pp. 355-356.

[50] Chun T. Rim, "Electric vehicle system," Patent WO 2010 076976, 2010.

[51] Chun T. Rim, "Unified general phasor transformation for AC converters," *IEEE Trans. on Power Electron.*, Vol. 26, No. 9, pp. 2465–2745, Sep. 2011.

[52] S. W. Lee, C. B. Park, J. G. Cho, G. H. Cho, and Chun T. Rim, "Ultra slim supply and pick-up coils for on-line electric vehicles (OLEV)," *Trans. of the Korean Institute of Power Electronics* (*KIPE*), Vol. 16, No. 3, pp. 274-282, Aug. 2011.

[53] J. Huh and Chun T. Rim, "KAIST wireless electric vehicles - OLEV," *JSAE Annual Congress*, 2011.

[54] S. W. Lee, W. Y. Lee, J. Huh, H. J. Kim, C. B. Park, G. H. Cho, and Chun T. Rim, "Active EMF cancellation method for I-type pick-up of On-Line Electric Vehicles (OLEV)," *IEEE Applied Power Electronics Conference & Exposition* (*APEC*), 2011, pp. 1980-1983.

[55] J. Huh, W. Y. Lee, G. H. Cho, B. H. Lee, and Chun T. Rim, "Characterization of novel inductive power transfer systems for on-line electric vehicles (OLEV)," *IEEE Applied Power Electronics Conference & Exposition* (*APEC*), 2011, pp. 1975-1979.

[56] J. Huh, S. W. Lee, W. Y. Lee, G. H. Cho, and Chun T. Rim, "Narrow-width inductive power transfer system for on-line electrical vehicles (OLEV)," *IEEE Trans. on Power Electron.*, Vol. 26, No. 12, pp. 3666-3679, Dec. 2011.

[57] S. W. Lee, C. B. Park, and Chun T. Rim, "An analysis of DQ inverter for wireless power transfer by complex Laplace-phasor transformation," *Korean Institute of Power Electronics Annual Summer Conference*, 2011, pp. 192-193.

[58] N. P. Suh, D. H. Cho, G. H. Cho, J. G. Cho, Chun T. Rim, and S. H. Jang, "Ultra slim power supply and collector device for electric vehicle," Patent KR 1010406620000, 2011.

[59] S. Z. Jeon, D. H. Cho, Chun T. Rim, and G. H. Jeong, "Load-segmentation-based full bridge inverter and method for controlling same," Patent WO 2011 078424, 2011.

[60] S. Z. Jeon, D. H. Cho, Chun T. Rim, and G. H. Jeong, "Load-segmentation-based 3-level inverter and method for controlling the same," Patent WO 2011 078425, 2011.

[61] N. P. Suh, D. H. Cho, J. G. Cho, Chun T. Rim, J. Huh, J. H Kim, C. S. Choi, K. H. Lee, B. Y. Song, Y. J. Cho, and C. H. Rim, "Monorail type power supply device for electric vehicle including EMF cancellation apparatus," Patent WO 2011 046374, 2011.

[62] N. P. Suh, S. H. Jang, D. H. Cho, G. H. Cho, Chun T. Rim, S. W. Lee, C. B. Park, J. Huh, and H. J. Kim, "Space division multiplexed power supply and collector device," Patent WO 2011 152678, 2011.

[63] N. P. Suh, S. H. Jang, D. H. Cho, G. H. Cho, Chun T. Rim, J. Huh, B. H. Lee, and Y. H. Kim, "Cross-type segment power supply," Patent WO 2011 152677, 2011.

[64] J. Huh, and Chun T. Rim, "A new coil set with core for magnetic resonant systems," *Korean Institute of Power Electronics Annual Summer Conference*, 2012, pp. 625-626.

[65] N. P. Suh, S. H. Jang, D. H. Cho, G. H. Cho, Chun T. Rim, S. W. Lee, and C. B. Park, "EMI cancellation device in power supply and collector device for magnetic induction power transmission," Patent WO 2011 149263, 2012.

[66] W. Y. Lee, J. Huh, S. Y. Choi, X. V. Thai, J. H. Kim, E. A. Al-Ammar, M. A. El-Kady, and Chun T. Rim, "Finite-width magnetic mirror models of mono and dual coils for wireless electric vehicles," *IEEE Trans. on Power Electron.*, Vol. 28, No. 3, pp. 1413-1428, Mar. 2013.

[67] Chun T. Rim, "Trend of roadway powered electric vehicle technology," *Magazine of the Korean Institute of Power Electronics* (*KIPE*), Vol. 18, No. 4, pp. 45-51, Aug. 2013.

[68] Chun T. Rim, "The development and deployment of on-line electric vehicles (OLEV)," *IEEE Energy Conversion Congress and Exposition* (*ECCE*), 2013.

[69] Su Y. Choi, J. Huh, W. Y. Lee, S. W. Lee, and Chun T. Rim, "New cross-segmented power supply rails for roadway powered electric vehicles," *IEEE Trans. on Power Electron.*, Vol. 28, no. 12, pp. 5832-5841, Dec. 2013.

[70] S. Lee, B. Choi, and Chun T. Rim, "Dynamics characterization of the inductive power transfer system for On-Line Electric Vehicles (OLEV) by Laplace phasor transform," *IEEE Trans. on Power Electron.*, Vol. 28, No. 12, pp. 5902-5909, Dec. 2013.

[71] Su Y. Choi, Beom W. Gu, Seog Y. Jeong, and Chun T. Rim, "Ultra slim S-type power supply rails for roadway-powered electric vehicles," *IEEE Trans. on Power Electron.*, vol. 30, no. 11, pp. 6456–6468, June 2015.

[72] Su Y. Choi, J. Huh, Woo Y. Lee, Jung G. Cho, and Chun T. Rim, "Asymmetric coil sets for wireless stationary EV chargers with large lateral tolerance by dominant field analysis," *IEEE Trans. on Power Electron.*, vol. 29, no. 12, pp. 6406–6420, Feb. 2014.

[73] Su Y. Choi, Beom W. Gu, Sung W. Lee, Woo Y. Lee, J. Huh, and Chun T. Rim, "Generalized active EMF cancel methods for wireless electric vehicles," *IEEE Trans. on Power Electron.*, vol. 29, no. 11, pp. 5770-5783, Nov. 2014.

[74] M. Budhia, G. A. Covic, and J. T. Boys, "Design and optimization of magnetic structures for lumped inductive power transfer systems," *IEEE Trans. on Power Electron.*, vol. 26, no. 11, pp. 3096–3108, Nov. 2011.

[75] M. Budhia, G. A. Covic, and J. T. Boys, "A new magnetic coupler for inductive power transfer electric vehicle charging systems," *in Proc. 36ᵗʰ Annu. Conf. IEEE Ind. Electron.*, Nov. 2010, pp. 2487–2492.

[76] M. Budhia, J. T. Boys, G. A. Covic, and C.-Y. Huang, "Development of a single-sided flux magnetic coupler for electric vehicle IPT charging systems," *IEEE Trans. on Ind. Electron.*, vol. 60, no. 1, pp. 318–328, Jan. 2013.

[77] G. A. Covic, L. G. Kissin, D. Kacprzak, N. Clausen, and H. Hao,"A bipolar primary pad topology for EV stationary charging and

highway power by inductive coupling," *IEEE Energy Conversion Congress and Exposition (ECCE)*, Sep. 2011, pp. 1832–1838.

[78] M. Budhia, G. A. Covic, J. T. Boys, and C.-Y. Huang, "Development and evaluation of single sided flux couplers for contactless electric vehicle charging," *IEEE Energy Conversion Congress and Exposition (ECCE)*, Sep. 2011, pp. 614–621.

[79] A. Zaheer, D. Kacprzak, and G. A. Covic, "A bipolar receiver pad in a lumped IPT system for electric vehicle charging applications," *IEEE Energy Conversion Congress and Exposition (ECCE)*, Sep. 2012, pp. 283–290.

[80] G. A. Covic, J. T. Boys, M. Kissin, and H. Lu, "A three-phase inductive power transfer system for roadway power vehicles," *IEEE Trans. on Ind. Electron.*, vol. 54, no. 6, pp. 3370–3378, Dec. 2007.

[81] C. S. Wang, O. H. Stielau, and G. A. Covic, "Design considerations for a contactless electric vehicle battery charger," *IEEE Trans. on Ind. Electron.*, vol. 52, no. 5, pp. 1308–1314, Oct. 2005.

[82] Boys J.T., Covic G.A. and. Green A.W. "Stability and Control of inductively coupled power transfer systems," *IEE Proc. EPA*, 147. pp 37-43.

[83] Wang, C.S, Covic G.A. and Stielau, O. H. "Power transfer capability and bifurcation phenomena of loosely coupled inductive power transfer systems," *IEEE Trans. on Ind. Electron.*, 51 no. 1, pp. 148-157, 2004.

[84] G. A. Covic and J. T. Boys, "Modern trends in inductive power transfer for transportation applications," *IEEE Journal of Emerging and Selected Topics in Power Electron.*, vol. 1, no. 1, pp. 28-41, Mar. 2013.

[85] J. Meins and S. Carsten, "Transferring energy to a vehicle," Patent WO 2010 000494, 2010.

[86] J. Meins and K. Vollenwyder, "System and method for transferring electrical energy to a vehicle," Patent WO 2010 000495, 2010.

[87] K. Vollenwyder, J. Meins, and C. Struve, "Inductively receiving electric energy for a vehicle," Patent US 0055751, 2012.

[88] M. Zengerle, "Transferring electric energy to a vehicle using a system which comprises consecutive segments for energy transfer," Patent US 0217112, 2012.

[89] K. Vollenwyder and J. Meins, "Producing electromagnetic fields for transferring electric energy to a vehicle," Patent US 8544622, 2013.

[90] R. Czainski, J. Meins, and J. Whaley, "Transferring electric energy to a vehicle by induction," Patent US 0248311, 2013.

[91] J. Meins, "German activities on contactless inductive power transfer," *IEEE Energy Conversion Congress and Exposition (ECCE), 2013.*

[92] Bombardier Website, "http://primove.bombardier.com/media/news/".

[93] Omer C. Onar, John M. Miller, Steven L. Campbell, Chester Coomer, Cliff. P. White, and Larry E. Seiber, "A novel wireless power transfer for in-motion EV/PHEV charging," *IEEE Applied Power Electronics Conference & Exposition (APEC)*, 2013, pp. 3073-3080.

[94] John M. Miller, Omer C. Onar, and P. T. Jones, "ORNL developments in stationary and dynamic wireless charging," *IEEE Energy Conversion Congress and Exposition (ECCE), 2013.*

[95] Jin Huh, Woo Y. Lee, Su Y. Choi, Gyu H. Cho, and Chun T. Rim, "Explicit static circuit model of coupled magnetic resonance system," *IEEE Energy Conversion Congress and Exposition (ECCE)-Asia*, May 2011, pp. 2233-2240.

[96] Jin Huh, Woo Y. Lee, Su Y. Choi, Gyu H. Cho, and Chun T. Rim, "Frequency-domain circuit model and analysis of coupled magnetic resonance systems," *Journal of Power Electronics*, Vol. 13, No. 2, pp. 275-286, March, 2013.

[97] Eun S. Lee, J. Huh, Xuan V. Thai, Su Y. Choi, and Chun T. Rim, "Impedance transformers for compact and robust coupled magnetic resonance systems," *IEEE Energy Conversion Congress and Exposition (ECCE)*, Sep. 2013, pp. 2239-2244.

[98] M. Hanazawa, N. Sakai, and T. Ohira, "SUPRA: Supply underground power to running automobiles," *IEEE International Electric Vehicle Conference, (IEVC2012)*, Greenville, Mar. 2012.

[99] Su Y. Choi, Beom W. Gu, Seog Y. Jeong, and Chun T. Rim, "Advances in wireless power transfer systems for roadway-powered electric vehicles," *IEEE Journal of Emerging and Selected Topics in Power Electron.*, vol. 3, no. 1, pp. 18-36, March 2015.

[100] J. Schneider, "SAE J2954 overview and path forward," *Society of Automotive Engineers (SAE) International, 2012.*

[101] Guidelines for limiting exposure to time-varying electric and magnetic fields (Up to 300 *GHz*), ICNIRP Guidelines, 1998.

[102] Guidelines for limiting exposure to time-varying electric and magnetic fields (Up to 100 k*Hz*), ICNIRP Guidelines, 2010.

[103] P. R. Bannister, "New theoretical expressions for predicting shielding effectiveness for the plane shield case," *IEEE Trans. on Electro. Comp.*, Vol. EMC-10, No. 1, pp. 2-7, Mar. 1968.

[104] P. Moreno and R. G. Olsen, "A simple theory for optimizing finite width ELF magnetic field shields for minimum dependence on source orientation," *IEEE Trans. on Electro. Comp.*, Vol. 39, No. 4, pp. 340-348, Nov. 1997.

[105] Yaping Du, T. C. Cheng, and A. S. Farag, "Principles of power-frequency magnetic field shielding with flat sheets in a source of long conductors," *IEEE Trans. on Electro. Comp.*, Vol. 38, No. 3, pp. 450-459, Aug. 1996.

[106] Seung Y. Ahn, Jun S. Park, and Joung H. Kim, "Low frequency electromagnetic field reduction techniques for the on-line electric vehicles (OLEV)," *IEEE International Symposium on Electromagnetic Compatibility,* 2010, pp. 625-630.

[107] Jong H. Kim and Joung H. Kim, "Analysis of EMF noise from the receiving coil topologies for wireless power transfer," *IEEE Asia-Pacific Symposium on Electromagnetic Compatibility*, 2012, pp. 645-648.

[108] H. S. Kim and Joung H. Kim, "Shielded coil structure suppressing leakage magnetic field from 100W-class wireless power transfer system with higher efficiency," *IEEE Microwave Workshop Series on Innovative Wireless Power Transmission Technologies, Systems and Applications*, 2012, pp. 83-86.

[109] Seung Y. Ahn, Huyn H. Park, and Joung H. Kim, "Reduction of electromagnetic field (EMF) of wireless power transfer system using quadruple coil for laptop applications," *IEEE Microwave Workshop Series on Innovative Wireless Power Transmission Technologies, Systems and Applications*, 2012, pp. 65-68.

[110] S. C. Tang, S. Y. R. Hui, and H. S. Chung, "Evaluation of the shielding effects on printed circuit board transformers using ferrite plates and copper sheets," *IEEE Trans. on Power Electron.*, Vol. 17, No. 6, pp. 1080-1088, Nov. 2002.

[111] X. Liu and S. Y. R. Hui, "An analysis of a double-layer electromagnetic shield for a universal contactless battery charging platform," *IEEE Power Electronics Specialists Conference (PESC)*, June 2005, pp. 1767 – 1772.

[112] P. Wu, F. Bai, Q. Xue, and S. Y. R. Hui, "Use of frequency selective surface for suppressing radio-frequency interference from wireless charging pads," *IEEE Trans. on Ind. Electron.*, vol. 61, no. 8, pp. 3969-3977, Aug. 2014.

[113] M. L. Hiles and K. L. Griffing, "Power frequency magnetic field management using a combination of active and passive shielding technology," *IEEE Trans. on Power De.*, pp. 171-179, 1998.

[114] C. Buccella and V. Fuina, "ELF magnetic field mitigation by active shielding," *IEEE International Symposium on Industrial Electronics*, 2002, pp. 994-998.

[115] J. Kim, Jong H. Kim, and Seung Y. Ahn, "Coil design and shielding methods for a magnetic resonant wireless power transfer system," *Proceeding of the IEEE*, Vol. 101, No. 6, pp. 1332-1342, June 2013.

[116] Sohn Y. Hoon, Bo H. Choi, Eun S. Lee, and Chun T. Rim, "General unified analyses of two-capacitor inductive power transfer systems: equivalence of current-source SS and SP compensations," *IEEE Trans. on Power Electron.*, vol. 30, no. 11, pp. 6030-6045, Nov. 2015.

[117] Van X. Thai, Su Y. Choi, Seog Y. Jeong, and Chun T. Rim, "Coreless power supply rails compatible with both stationary and dynamic charging of electric vehicles," *2015 IEEE International Future Energy Electronics Conference (IEEE IFEEC 2015)*, 978-1-4799-7657-7.

ZERO EMISISON ELECTRC VESSEL DEVELOPMENT

CHENG K.W.E.[1]   XUE X.D.[2]   CHAN K.H.[3]

[1,2,3] Department of Electrical Engineering, The Hong Kong Polytechnic University, Hong Kong
[1] E-mail: eeecheng@polyu.edu.hk

Abstract–Emission from the vessel has not been taken serious in the past. Recently the concern has been intensified due to the fluctuation of oil price, mature in technology and commercial successful in other similar electric transportation such as train and road vehicle. Based on the modern power electronics, the technology is now ready to be used in vessel. Various type of vessel including boat, ship, tank and submarine are now ready for this revolution. The paper provide a design and technology discussion of the electric vessel development.

Keywords–Power converter, electric vessel, electric boat, motor, battery, ground power.

I. INTRODUCTION

The emission due to transportation accounts for a large portion of the emission. In the past commercial and research development, concentration has been made on the electric vehicle for road and nearly each of the automotive company has one model in electric vehicle (EV). The development has also been extended to electric bus and vans. The rapid growth in electric vehicle has been recorded. In the last 5 years, most of the cities have recorded a growth of 10 times in EV. It seems that the awareness of the public in EV is certain.

On the other hand, the emission for vessel or watercraft has not been taken care seriously in the past. If we take a walk to the coast, pier, river or lake front, the emission from the various types of vessel including boat, ship of ferries are astonishing. The emission to both air and the water surface are serious. Most of the emissions are of highly pollution and the leakage of the oil from the vessel may form a layer of oil on the water surface. The engines that most of the vessel using are not effective. They may not use any ultra-low sulfur diesel (ULSD).

The emission from vessels are Sulphur oxide, Nitrogen oxide, volatile organic compound (VoC), Carbon oxide and other suspended particles including respirable and fine (RSP and FSP). In compact city such as Hong Kong, The sulphur contented in diesel was initially switched from 0.5% in 2011 to become 0.1% in 2015. The ultimate goal is the diesel fuel with sulphur content of 0.05% in 2020 or later [1].

The analysis from [2] shows that the emissions among road, aviation and vessels in Hong Kong that has a large variations in emission. Table 1 shows the summary. The portion from vessels accounts for a large portion. Therefore the emissions due to navigation cannot be ignored.

The technology developed for electric vehicle can well be used for electric vessel.

II. ADVANTAGE OF ELECTRIC VESSEL

The obvious advantage of Electric vessel is the zero emission. The fuel saving in electric vessel is much more effectively than EV that will be discussed later. Other advantages are listed as follows:

1. Vessel maneuver

The water current, waves and wind all natural elements govern the power level needed for a vessel. The design also depends on the vessel structure and weight. A vessel has to provide station keeping and steering. An electric vessel power driving is to overcome all these issues. Using variable speed drive (VSD), detailed torque and speed control to the propeller can be made.

The parking or maneuver of a vessel is not simple because most of the vessels consist of a single or couple of engines only. During docking, the rudder has to provide the main duty to move a vessel to the pier and it is not a simple task and therefore large ship, typically over 10m, is installed with a tunnel thrusters or a bow [3] thruster to assist the side movement during the docking. This can also reduce the help from tugboat. This thruster is usually driven by electric motor of hydraulic system. For large ship, a number of thruster may be installed such that a sign of X with a red circle to identify the thruster above the waterline. Using VSD, the thruster can be control much easier. With the concept of electric vessel (E-vessel), the power source and the control of the thruster motor are provided by the electrical system of the E-vessel.

Table 1: Pollution in Hong Kong

Pollutant Sources	SO$_2$	CO	NO$_x$	RSP	FSP	VOC
Public Electricity Generation	47%	6%	31%	16%	9%	2%
Road Transport	< 1%	59%	23%	18%	21%	23%
Aviation	2%	5%	6%	< 1%	1%	2%
Vessels	**50%**	**1%**	**31%**	**36%**	**42%**	**11%**
Others	< 1%	10%	10%	30%	26%	62%
Total Emissions (KiloT)	31	61	113	6	5	29

2. Regenerative braking

There is no braking for vessel as the electric vehicle does on road. But the vessel can operate on reverse to reduce the speed with the expense of high power needed. This is known as the plugging mode. When a vessel operates in deceleration, the dynamic energy can be returned to energy storage. Also when a vessel is in cruising or stationary, the water stream can also provide regeneration

978-1-5090-0064-7/15 $31.00 © 2015 IEEE

for additional power charging to energy storage unit.

The electric bow thruster can also be a water stream power generator for energy storage unit charging.

3. Regenerative energy

Unlike electric vehicle, an electric vessel has much more opportunity for installation or makes use of renewable energy. In general, a vessel has a large surface area for installation of solar energy panels. The water or sea environments provide a better exposure to sunlight whereas in city, the sunlight is usually blocked or restricted by buildings. Even water wave power generator using linear generator installed in a vessel is a good source of renewable energy. Other alternatives are wind power. When a vessel is in stationary, the energy captured by wind is a good power source.

4. Energy storage

In most of the vessels, the compartment is more flexible for installation of energy storage units such as batteries. The heavy batteries also provide a good mechanism for center of gravity balancing. The weight balancing is easier to be managed in vessel. In the counterpart, electric vehicle is usually very tight in weight and space. The weight balancing and packaging are usually difficult tasks.

5. Thermal management

The loss generated by batteries, motor and power electronics is enormous. In electric vehicle, the thermal management is usually done by water cooling and only small vehicle is done by natural cooling. For electric vessel, the water cooling is easily available and the performance is usually very good.

6. Air conditioning

One of the difficult issues in electric vehicle is the power requirement for air conditioning especially for a vehicle operated in summer or near topical area. The global temperature rise has triggered high power needed for air conditioning. Air conditioning reduces the driving range and consumes energy during traffic congestion. For the case of electric vessel, the general temperature environment is better and the demand for air conditioning is less. The cool water can also be used as a water cooling based green air-conditioning rather than using electric compressor.

When a vessel is in stationary, the diesel or petrol engine is no longer needed to be turned on. Energy storage system provides energy to the hotel load and air-conditioning. A good design of the power system can manage with the hotel load and to provide a quiet and smell-free environment.

7. Simple design

Electric vessel provides an overall simple design for the system. The conventional engine, oil tank, transmission, gear box, and many mechanical parts can be eliminated. The weight and the lubrication system can also be removed as well. It is expected the maintenance cost and the material consumed can be reduced. It is also well known that the life time or motor and the maintenance cost of electric motor and drive system is low.

8. Safety

With the low maintenance, the safety of electric vessel will be increased. Mechanical fatigue or latent damage is difficult to be detected in the past whereas it is established technology for power electronics drive and battery detection. The reliable switched-reluctance motor is one of the candidates for propeller. Direct drive also reduces the mechanical loss.

III. ENERGY SAVING

This efficiency of diesel or petrol engine for vessel is low and of around 15%. There are losses in the mechanical subsystem such as transmission and its efficiency is of 60%. Therefore the actual efficiency from fuel to propeller is 9%. For electric vessel, the electricity is derived mainly from power station, the efficiency is 45%. Electric motor efficiency is 90% and therefore the fuel to propeller efficiency is 40%.

There are other saving using electric motors for vessels. They are the regenerative deceleration. This will provide additional 35% of extra energy advantage. Therefore the actual efficiency is 54%.

Table 2: Performance analysis

Parameter	Value	Unit
Diesel Vessel		
Efficiency of diesel engine	0.15	
Efficiency of mechanical system	0.60	
Diesel energy density	9.87	kWh/L
Actual energy per Litre	0.89	KWh/L
Fuel Cost	1.22	$/L
Actual Cost per kWh	1.38	$/kWh
Electric Vessel		
Efficiency of motor	0.9	
Regenerative power saving	0.35	
Electricity cost	0.13	$/kWh
Actual cost per kWh	0.11	$/kWh
Performance ratio	13	

A performance analysis between electric vessel to diesel vessel is estimated in Table 2. The performance ratio in term of US$ is 13 when electric vessel is used instead of diesel. The saving for petrol engine will be even more.

The analysis has not taken account of the torque control performance of electric motor. Using variable speed drive and constant torque control, the motor power can be further reduced. If renewable energy such as photovoltaic system and wave power generator is installed, the performance or saving can further be increased. It is estimated that the ratio can be increased to 15-20. The performance for electric vehicles is in general 5-7.

V. Research Opportunity

Using electric vessel is not new. It has been done in partly in many vessel and power distribution of vessel. Military vessel has been using extensively. This research in electric vessel includes the motor, energy storage, control, packaging, renewable energy, power distribution and electric hotel load.

1. Motor and drive

The electric motor is well developed for transportation. Even for electric vessel, there are numerous reports addressed for defense and other large ship application. The typical motors selected are permanent magnet motor, induction motor. Switched-reluctance motor [4] is a good choice as it provides fault tolerance and potential low cost because of only simple materials needed. Direct drive reduces the mechanical system. Magnetic gear to replace the mechanical gear is also important research potential for electric vessel. The motor should be designed under humid environment. The multiple electric motor to propeller is an alternative design concept.

Using active suspension, the motion sickness can be minimized. The electric active suspension using linear motor has been used extensively in vehicle [5] and it is a new research for vessel. Torque control [6] for vessel is different from electric vehicles. Dynamic control against the water stream, wind and wave are needed.

2. Energy storage

The energy storage includes super-capacitor and batteries are commonly used devices. The future fuel should be a combination of a number of fuel storages in order to optimize the energy storage requirement. Recently fuel cells are being re-visited by users. The energy storage devices should be metal-oxide fuel cell with combination of super-capacitors and Li-ion batteries. Cells should be implemented with balancing to avoid the cell unbalancing [7] that may cause overvoltage and under voltage issue.

3. DC distribution

The energy storage are all DC based. Even the alternative energy sources such as photovoltaic cells are DC therefore it is more straight forward to use DC distribution on vessel. The motor drives are inverter driven that is also DC input. Hotel loads such as air conditioning are inverter based DC system, lighting including LED driver and electronic ballast are DC input, audio and video entertainment equipment and communication system are DC driven electronics. Therefore DC power distribution [8] should be a good candidate for electric vessel and the associated research work on the power management, smart distribution and energy management are now being examined by electric vessel designers.

Other important research and development are dynamic control of the vessel operation. The acceleration, deceleration and other motion control are now researched by many machines and drive specialists.

VI. Design of Electric Boat

1. Structure of Propulsion System in Boat

The typical propulsion system in a boat consists of the energy storage, the motor controller, the motor, the transmission, and the propeller. The typical structure and the power flow of the propulsion system in the boat is illustrated in Fig. 1. The battery or other energy storage components are used to store the electric energy for supplying the total electric power (P_b) required by all the electric equipment in boat. The motor controller is used to change the output power of the battery to the electric input power of the motor (P_{m_in}) according to the requirement. The motor is used to convert the electric input power to the mechanical output power (P_{m_out}). The transmission is used to regulate the rotary speed and transfer the mechanical output power of the motor to the propeller. The propeller is used to convert the input power of the propeller to the effective power, which is employed to propel the linear motion of the boat.

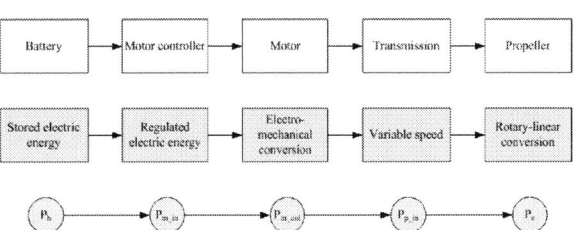

Fig. 1 Structure and power flow of propulsion system in boat

2. Key Design Equations

The relationship between P_b and P_{m_in} can be expressed as

$$P_{m\_in} = \eta_c P_b \tag{1}$$

where η_c represents the efficiency of the motor controller. Assuming that the efficiency of the motor is η_m, P_{m_out} is calculated as

$$P_{m\_out} = \eta_m P_{m\_in} \tag{2}$$

The input power of the propeller is given as

$$P_{p\_in} = \eta_t P_{m\_out} \tag{3}$$

where η_t is the efficiency of the transmission?
If the propulsion efficiency of the propeller is η_p, the output power of the propeller, i.e., the effective power of the propeller is expressed as

$$P_e = \eta_p P_{p\_in} \tag{4}$$

Taking into account (3) and (4), the propulsion coefficient (C_p) is defined as

$$C_p = \frac{P_e}{P_{m\_out}} = \frac{P_e}{P_{p\_in}} \frac{P_{m\_in}}{P_{m\_out}} = \eta_p \eta_t \tag{5}$$

The propulsion coefficient indicates the propulsion performance of a boat. The higher propulsion coefficient results in the better propulsion performance.
Assuming that the resistance force of a boat is F_r if a boat moves linearly with the constant velocity V, the effective propulsion force developed by the effective power of the propeller F_e is calculated as

$$F_e = \frac{P_e}{V} \tag{6}$$

Furthermore, the relationship between F_r and F_e is given as

$$F_e = F_r \tag{7}$$

If the boat moves during the acceleration, the dynamic

978-1-5090-0064-7/15 $31.00 © 2015 IEEE

motion equation is expressed as

$$F_e = F_r + m\frac{dv}{dt} \tag{8}$$

where v is the instant velocity of the boat and m is the total mass of the boat.

C. Case Study

In this case, a conventional boat propelled by a diesel engine is changed into the electric boat. The main parameters of the boat with the diesel engine are given as follows.

Power of the diesel engine = 45 kW
Boat velocity = 20 knots (36.5 km/h)
The basic requirement on the electric boat is as follows.
Boat velocity = 20 knots (36.5 km/h)
Maximum travel duration = 1 h

The designed components of the electrical system for electric boats should include the motor, the motor controller, the battery, and the charger for the battery.

(1) Motor

Based on the power of the diesel engine, the power of an induction motor is selected as 45 kW. The IP degree of the motor is IP65 and the cooling mode is cooled-water mode. The nominal speed of the motor is 2970 rpm.

(2) Motor controller

Using (2), the input power of the motor is calculated as 50 kW if the efficiency is assumed as 0.9. It is also the output power of the motor controller. The input DC voltage is selected as 515 V, to match the AC voltage of the motor. The IP degree of the motor controller is IP20.

(3) Battery

The Li-ion battery is preferred here. Using (1), the power of the battery is required as 52.6 kW if the efficiency of the motor controller is assumed as 0.95. Due to the maximum travel duration of 1 h, thus, the energy of the battery is 52.6 kWh. The DC bus voltage of the battery is selected as 515 V according to the input DC voltage of the motor controller. Consequently, the capacity of the battery is calculated as 102 Ah. Thus, the nominal capacity of the battery can be selected as 120 Ah. The IP degree of the battery pack is IP20.

(4) Charger

For the above battery, the maximum DC voltage is selected as 585 V and the charging current is selected as 0.33C, i.e., 40 A. Thus, the nominal power of the charger is selected as 23 kW.

VII. ENERGY SAVING ANALYSIS OF AN ELECTRIC FERRY

The harbour, channel or river crossing using a diesel ferry usually operate under slow speed and the travelling is around 10 min 1.25km of a trip with speed around 4 knots. The ferries are designed to be charged for short duration using speed charging for every round trip and the energy storage can use 4C or above Li-ion battery or super-capacitor. The designs are estimated I shown in Table 3.

The electric ferry design for short range vessel is acceptable because the charging current and charging time

are realistic. Therefore using fast charging for each round trip scheme can reduce the battery weight and cost for an electric ferry.

Table 3 : Electric Ferry design analysis

Parameter	Value	Unit
Motor		
Motor power	200	kW
Estimated operating hours per round trip	0.33	hr
Estimated Energy Needed	74	kWh
Bus voltage	600	V
Battery Pack	123	Ah
Docking time	15	min
Charging requirement		
Charging rate	4	C
Charging power	296	kW
Electricity cost	0.13	$/kWh
Actual cost per kWh	0.11	$/kWh
Charging AC current	450	A
Fuel		
Fuel cost	130	$
Saving compared with electric version	93%	

VIII. CONCLUSION

A study of the emission for the vessels has been conducted. It is obviously that the energy saving is very high and better than electric vehicle on road. An estimation of 13 in saving performance can be achieved. It is expected that by installation of renewable energy source in the vessel, further saving can be made. The new research areas of electric vessels are control, motor and energy storage. It is expected that electric vessel will be replacing a large number of short range vessels in next 10 years.

ACKNOWLEDGMENT

Thanks to China Dynamics Ltd and the Sunlight Ltd contributed to the paper.

REFERENCES

[1] Transport & Housing Bureau, Food & Health Bureau and Development Bureau and Environment Bureau, "A Clean air plan for Hong Kong",, Mar 2013.

[2] Air Science group, Environment Protection Department, HKSAR, Hong Kong Emission Inventory Report, June 2015.

[3] Amitava Chakrabarty, How Bow Thruster is used for Maneuvering a Ship? Marine Insight, 13 Sep 2012, http://www.marineinsight.com/marine/marine-news/headline/how-bow-thruster-is-used-for-maneuvering-a-ship/

[4] X.D.Xue, K.W.E.Cheng, S.L.Ho, "A Position Stepping Method for Predicting Performances of Switched Reluctance Motor Drives", IEEE Trans. Energy Conversion, Vol. 22, No. 4, Dec 2007, pp. 839-847.

[5] J.K.Lin, K.W.E.Cheng, "Active Suspension System Based on Linear Switched Reluctance Actuator and Control Schemes", IEEE Transactions on Vehicular Technology, Vol. 62, Issue: 2, 2013, pp. 562 – 572.

[6] X.D.Xue, K.W.E.Cheng, S.L.Ho, "Online and Offline Rotary Regressive Analysis of Torque Estimator for SRM

Drives", IEEE Trans Energy conversion, Vol. 22, No. 4, Dec 2007, pp. 810-818.

[7] Ye Yuanmao and K.W.E.Cheng, "Zero-Current Switching Switched-Capacitor Zero-Voltage-Gap Automatic Equalization System for Series Battery String", IEEE Transactions on Power Electronics, Jul 2012, Vol 27, No. 7, Jul 2012, pp. 3234-3242.

[8] Cheng K.W.E, "Overview of the DC Power Conversion and Distribution", Asian Power Electronics Journal, Vol 2, No. 2, pp. 75-82.

AUTHOR INDEX

Akhtar, M. J. 51, 109
Ang, M. H. 443
Arboleya, P. 203, 208
Arora, V. 197
Bai, G. F. 425
Balguvhar, S. 435
Banerjee, A. 19, 122
Behera, R. K. 51, 104, 109
Biswas, P. K. 19, 122
Bodlak, E. 310
Cao, Guang-Zhong....................152, 158, 163
Chan, K. H. 455
Chan, K. W. 346
Che, Yanbo 81, 286, 301
Chen, Sizhe 218, 223
Chen, W. F. 276, 331
Chen, Zhen 98
Cheng, E. K. W. 261
Cheng, K. We 351, 357
Cheng, K. W. Eric 185, 236, 242, 250, 255, 266,
    270, 276, 280, 320, 326, 331, 335, 342, 346,
    362, 455
Cheung, N. 373, 377, 380
Chin, C. S. 173
Choi, Hyeon-Gyu 56
Choi, S. Y. 447
Da Toh, Wei..173
Deng, Banglin 24
Deng, Hanbo 168
Deng, Hui 152
Dong, L. 425
Dong, P. 291, 295
Fang, Kaijie 92
Fang, Y. T. 139
Fei, W. 411, 419
Feng, Y. 291, 295
Fong, Y. C. 280, 320
Gao, Congzhe 98
Gao, Dawei..128
Gao, Zuchang 173
Garcia, J. 367
Georgious, R. 367
Gour, R. 179
Guangqiang, Kan 77, 193, 230
Guo, X. 31, 39, 45, 60
Guo, Xiao-Qin 163
Ha, Jung-Ik..56, 66, 72
Hasan, M. A. 51
He, Ting 39
He, Wei 81, 87, 92
He, Zhi-Ming 152
Hu, Hengzai 98
Hu, Xiaodong..193, 230
Huang, C. F. 233
Huang, Shenghua 193, 230

Huang, Su-Dan 152, 158, 163
Huang, X. Y. 139
Huihui, Wang.. 11
Jia, Junbo 173
Jiang, Z. Y. 261
Karadeniz, M. 403
Kaur, N...435
Kumar, A. 51
Lee, Jun 66, 72
Lee, K.-K. 439
Lee, Kyung-Hwan 66
Lei, J. 31, 39, 45, 60
Li, J. 139
Li, Qianlong 373
Li, S. 134
Li, S. Y. 351, 357
Li, Wen-Bo 163
Liang, Yu-Xin 158
Lin, Y. 291, 295
Ling, L. L. 425
Liu, Guojian 301
Liu, J. F. 242, 266
Liu, Jiexun 128
Liu, Junfeng 185, 255
Liu, S. Y. 425
Liu, Xiangdong 98
Liu, Xiaokun 286, 301
Luan, Shuaijie 218, 223
Luk, P. 411, 419
Luo, Jianjun 380
Luo, Wenguang 168
Ma, J. M. 425
Ma, X. 60
Mei, J. 242, 261
Mi, C. 306
Misal, S. 114
Mohamed, B..203, 208
Nie, Yongquan 250
Ojo, O. 104
Pan, J. F. 373, 377, 380
Parida, S. K. 51, 109
Peng, Jiajun 60
Ping, Wang... 11
Plakhova, M. 203, 208, 389
Qi, G. 134
Qin, C. 291, 295
Qiu, Hong 158
Raghu, R. S. 242, 255, 266
Raman, S. R. 342
Rim, C. T. 447
Saeed, S. 396
Sain, C. 19, 122
Shenghua, Huang 77
Shi, Lai-Juan 152
Si, J. 385

AUTHOR INDEX

Si, M. 385
Sin, G. 362
Swetha, T. 114
Tai, Chang Yan 114
Tang, Xiongmin 218, 223
Tian, B. 39
Tian, H. L. 425
Verma, V. 179, 197
Wandalkar, M. 114
Wang, Jidong 87, 92
Wang, K. 1
Wang, Le 218, 223
Wang, Wei 1, 168
Wang, Weiyu 377
Wang, Xiaolin 250, 362
Wang, Yue 128
Wang, Zhihao 147
Wei, W. 31
Weng, Yu-Wei 233
Woo, Wai Lok 173
Wu, D. 411
Xia, Bing 419
Xia, Jun 373, 377
Xie, Guangming 168
Xu, C. D. 185, 266
Xu, Gang 24
Xu, Wenzheng 236, 346
Xu, Ying 24, 213
Xu, Z. 31, 39, 45, 60
Xue, X. D. 242, 255, 261, 455
Xunhao, Zou 77, 193, 230
Yan, Q. S. 425
Yang, D. 134, 147
Yang, P. 31, 39, 45, 60
Yang, Xiangyu 385
Yang, Yuhao 87
Yang, Zhangang 286, 301
Ye, Yuanmao 326
Yin, Zhaojing 81
Yu, L. 45
Yuan, H. 39
Yun, Yanhui 380
Zeng, Jun 185
Zeng, Z. 45
Zhang, Bo 373, 377, 380
Zhang, Hancheng 98
Zhang, Tie 213
Zhang, Y. 31
Zhao, Jing 98
Zhao, S. F. 139
Zhao, Shiwei 385
Zheng, Q. 45
Zhou, S. 31, 39, 60
Zhou, Shou-Qin 158
Zhou, X. Q. 425

Zhu, Jingwei 270, 335
Zhuqian, Qin 77, 193, 230

IEEE
445 Hoes Lane
Piscataway, NJ 08854-4141

ISBN 978-1-5090-0064-7